FUNDAMENTALS OF AERODYNAMICS

McGraw-Hill Series in Aeronautical and Aerospace Engineering

Consulting Editor

John D. Anderson, Jr., *University of Maryland*

Anderson: *Fundamentals of Aerodynamics*
Anderson: *Hypersonic and High Temperature Gas Dynamics*
Anderson: *Introduction to Flight*
Anderson: *Modern Compressible Flow With Historical Perspective*
D'Azzo and Houpis: *Linear Control System Analysis and Design*
Kane, Likins and Levinson: *Spacecraft Dynamics*
Katz and Plotkin: *Low-Speed Aerodynamics: From Wing Theory to Panel Methods*
Nelson: *Flight Stability and Automatic Control*
Peery and Azar: *Aircraft Structures*
Rivello: *Theory and Analysis of Flight Structures*
Schlichting: *Boundary Layer Theory*
White: *Viscous Fluid Flow*
Wiesel: *Spaceflight Dynamics*

FUNDAMENTALS OF AERODYNAMICS

Second Edition

John D. Anderson, Jr.

Professor of Aerospace Engineering
University of Maryland

McGraw-Hill, Inc.
New York St. Louis San Francisco Auckland Bogotá
Caracas Lisbon London Madrid Mexico Milan
Montreal New Delhi Paris San Juan Singapore
Sydney Tokyo Toronto

This book was set in Times Roman.
The editors were Lyn Beamesderfer and Eleanor Castellano;
the production supervisor was Janelle S. Travers.
The cover designer was Joan O'Connor.
New drawings were done by Oxford Illustrators Limited.
Arcata Graphics/Martinsburg was printer and binder.

FUNDAMENTALS OF AERODYNAMICS

    6 7 8 9 0 AGM AGM 9 5

ISBN 0-07-001679-8

Library of Congress Cataloging-in-Publication Data

Anderson, John David.
    Fundamentals of aerodynamics/John D. Anderson, Jr.—2nd ed.
        p.   cm.—(McGraw-Hill series in aeronautical and aerospace engineering)
        Includes bibliographical references and index.
        ISBN 0-07-001679-8
    1. Aerodynamics.   I. Title.   II. Series.
    TL570.A677   1991
    629.132'3—dc20   90-42291

ABOUT THE AUTHOR

John D. Anderson, Jr., was born in Lancaster, Pennsylvania, on October 1, 1937. He attended the University of Florida, graduating in 1959 with high honors and a bachelor of aeronautical engineering degree. From 1959 to 1962, he was a lieutenant and task scientist at the Aerospace Research Laboratory at Wright-Patterson Air Force Base. From 1962 to 1966, he attended the Ohio State University under the National Science Foundation and NASA Fellowships, graduating with a Ph.D. in aeronautical and astronautical engineering. In 1966, he joined the U.S. Naval Ordnance Laboratory as Chief of the Hypersonic Group. In 1973, he became Chairman of the Department of Aerospace Engineering at the University of Maryland, and since 1980 has been professor of Aerospace Engineering at Maryland. In 1982, he was designated a Distinguished Scholar/Teacher by the University. During 1986–1987, while on sabbatical from the University, Dr. Anderson occupied the Charles Lindbergh chair at the National Air and Space Museum of the Smithsonian Institution.

Dr. Anderson has published five books: *Gasdynamic Lasers: An Introduction,* Academic Press (1976), and under McGraw-Hill: *Introduction to Flight* (1978, 1985, and 1989), *Modern Compressible Flow* (1982 and 1990), *Fundamentals of Aerodynamics* (1984 and 1990), and *Hypersonic and High Temperature Gas Dynamics* (1989). He is the author of over 90 papers on radiative gas dynamics, reentry aerothermodynamics, gas dynamic and chemical lasers, computational fluid dynamics, applied aerodynamics, and hypersonic flow. Dr. Anderson is in *Who's Who in America,* and is a Fellow of the American Institute of Aeronautics and Astronautics. He is also a fellow of the Washington Academy of Sciences, and a member of Tau Beta Pi, Sigma Tau, Phi Kappa Phi, Phi Eta Sigma, the American Society for Engineering Education, and the American Physical Society. In 1989, he was awarded the John Leland Atwood Medal for excellence in Aerospace Engineering Education, given jointly by the American Institute of Aeronautics and Astronautics and the American Society for Engineering Education.

DEDICATED TO MY FAMILY
Sarah-Allen, Katherine, and Elizabeth

CONTENTS

Part II Inviscid, Incompressible Flow

Part III Inviscid, Compressible Flow

Part IV Viscous Flow

PREFACE TO THE SECOND EDITION

The purpose of this second edition is the same as the first—to be read, understood, and enjoyed. Due to the extremely favorable comments from readers and users of the first edition, virtually all the first edition has been carried over intact to the second edition. Therefore, all the basic philosophy, approach, and content discussed and itemized by the author in the Preface to the First Edition is equally applicable now. Since that preface is repeated on the following pages, no further elaboration will be given here.

Question: What distinguishes the second edition from the first? *Answer:* Much *new* material has been added in order to enhance and expand that covered in the first edition. In particular, the second edition has:

1. A much larger number of worked examples. These new worked examples provide the reader with a better understanding of how the fundamental principles of aerodynamics are used for the solution of practical problems. In the second edition, we continue the emphasis of the first edition on the fundamentals, but at the same time, add a stronger *applied* flavor.

2. New sections subtitled *applied aerodynamics* at the end of many of the chapters. In concert with the enhanced number of worked examples, these new applied aerodynamics sections are designed to provide the reader with some feeling for "real-world" aerodynamics.

3. Some additional historical content. This continues the author's strong belief that students of aerodynamics should have an appreciation for some of the historical origins of the tools of the trade.

4. An expanded treatment of hypersonic aerodynamics. Specifically, Chap. 14 has been lengthened to cover additional aspects of this subject in response to the increased importance of hypersonic flight in recent years.

5. An expanded treatment of viscous flow. Part of this expansion is a new chapter on parallel viscous flows (such as Couette flow), because such flows illustrate many of the basic features of a general viscous flow without many of the complexities of a general case.

6. New homework problems added to those carried over from the first edition.
7. A large number of new graphic and photographic illustrations in order to help the reader visualize the content as given by the written word.

McGraw-Hill and I would like to thank the following reviewers for their many helpful comments and suggestions: Eugene E. Covert, Massachusetts Institute of Technology; Russell M. Cummings, NASA Ames Research Center; Jack E. Fairchild, University of Texas–Arlington; James McDaniel, University of Virginia; Pasquale M. Sforza, Polytechnic University; and C. P. Van Dam, University of California–Davis.

As a final comment, aerodynamics is a subject of intellectual beauty, composed and drawn by many great minds over the centuries. *Fundamentals of Aerodynamics* is intended to portray and convey this beauty. Do you feel challenged and interested by these thoughts? If so, then read on, and enjoy!

John D. Anderson, Jr.

PREFACE TO THE FIRST EDITION

This book is for students—to be read, understood, and enjoyed. It is consciously written in a clear, informal, and direct style designed to *talk* to the reader and to gain his or her immediate interest in the challenging and yet beautiful discipline of aerodynamics. The explanation of each topic is carefully constructed to make sense to the reader. Moreover, the structure of each chapter is highly organized in order to keep the reader aware of where we are, where we were, and where we are going. Too frequently the student of aerodynamics loses sight of what is trying to be accomplished; to avoid this, we attempt to keep the reader informed of our intent at all times. For example, virtually each chapter contains a road map—a block diagram designed to keep the reader well aware of the proper flow of ideas and concepts. The use of such chapter road maps is one of the unique features of this book. Also, to help organize the reader's thoughts, there are special summary sections at the end of most chapters.

The material in this book is at the level of college juniors and seniors in aerospace or mechanical engineering. It assumes no prior knowledge of fluid dynamics in general, or aerodynamics in particular. It does assume a familiarity with differential and integral calculus, as well as the usual physics background common to most students of science and engineering. Also, the language of vector analysis is used liberally; a compact review of the necessary elements of vector algebra and vector calculus is given in Chap. 2 in such a fashion that it can either educate or refresh the reader, whichever may be the case for each individual.

This book is designed for a 1-year course in aerodynamics. Chapters 1 to 6 constitute a solid semester emphasizing inviscid, incompressible flow. Chapters 7 to 14 occupy a second semester dealing with inviscid, compressible flow. Finally, Chaps. 15 to 18 introduce some basic elements of viscous flow, mainly to serve as a contrast to and comparison with the inviscid flows treated throughout the bulk of the text.

This book contains several unique features:

1. The use of chapter road maps to help organize the material in the mind of the reader, as discussed earlier.

2. An introduction to computational fluid dynamics as an integral part of the beginning study of aerodynamics. Computational fluid dynamics (CFD) has recently become a third dimension in aerodynamics, complimenting the previously existing dimensions of pure experiment and pure theory. It is absolutely necessary that the modern student of aerodynamics be introduced to some of the basic ideas of CFD—he or she will most certainly come face to face with either its "machinery" or its results after entering the professional ranks of practicing aerodynamicists. Hence, such subjects as the source and vortex panel techniques, the method of characteristics, and explicit finite-difference solutions are introduced and discussed as they naturally arise during the course of our discussions. In particular, Chap. 13 is devoted exclusively to numerical techniques, couched at a level suitable to an introductory aerodynamics text.

3. A short chapter is devoted entirely to hypersonic flow. Although hypersonics is at one extreme end of the flight spectrum, it has current important applications to the design of the space shuttle, hypervelocity missiles, and planetary entry vehicles. Therefore, hypersonic flow deserves some attention in any modern presentation of aerodynamics. This is the purpose of Chap. 14.

4. Historical notes are placed at the end of many of the chapters. This follows in the tradition of the author's previous books, *Introduction to Flight: Its Engineering and History* (McGraw-Hill, 1978), and *Modern Compressible Flow: With Historical Perspective* (McGraw-Hill, 1982). Although aerodynamics is a rapidly evolving subject, its foundations are deeply rooted in the history of science and technology. It is important for the modern student of aerodynamics to have an appreciation for the historical origin of the tools of the trade. Therefore, this book addresses such questions as who were Bernoulli, Euler, d'Alembert, Kutta, Joukowski, and Prandtl; how was the circulation theory of lift developed; and what excitement surrounded the early development of high-speed aerodynamics? The author wishes to thank various members of the staff of the National Air and Space Museum of the Smithsonian Institution for opening their extensive files for some of the historical research behind these history sections. Also, a constant biographical reference was the *Dictionary of Scientific Biography*, edited by C. C. Gillespie, Charles Schribner's Sons, New York, 1980. This is a 16-volume set of books which is a valuable source of biographic information on the leading scientists in history.

This book has developed from the author's experience in teaching both incompressible and compressible flow to undergraduate students at the University of Maryland. Such courses require careful attention to the structure and sequence of the presentation of basic material, and to the manner in which sophisticated subjects are described to the uninitiated reader. This book meets the author's needs at Maryland; it is hoped that it will also meet the needs of others, both in the formal atmosphere of the classroom and in the informal pleasure of self-study.

Readers who are already familiar with the author's *Introduction to Flight* will find the present book to be a logical sequel. Many of the aerodynamic concepts first introduced in the most elementary sense in *Introduction to Flight* are revisited and greatly expanded in the present book. For example, at Maryland, *Introduction to Flight* is used in a sophomore-level introductory course, followed by the material of the present book in junior- and senior-level courses in incompressible and compressible flow. On the other hand, the present book is entirely self-contained; no prior familiarity with aerodynamics on the part of the reader is assumed. All basic principles and concepts are introduced and developed from their beginnings.

The author wishes to thank his students for many stimulating discussions on the subject of aerodynamics—discussions which ultimately resulted in the present book. Special thanks go to two of the author's graduate students, Tae-Hwan Cho and Kevin Bowcutt, who provided illustrative results from the source and vortex panel techniques. Of course, all of the author's efforts would have gone for nought if it had not been for the excellent preparation of the typed manuscript by Ms. Sue Osborn.

Finally, special thanks go to two institutions: (1) the University of Maryland for providing a challenging intellectual atmosphere in which the author has basked for the past 9 years and (2) the Anderson household—Sarah-Allen, Katherine, and Elizabeth—who have been patient and understanding while their husband and father was in his ivory tower.

John D. Anderson, Jr.

PART
I

FUNDAMENTAL PRINCIPLES

In Part I, we cover some of the basic principles that apply to aerodynamics in general. These are the pillars on which all of aerodynamics is based.

CHAPTER
1

AERODYNAMICS: SOME INTRODUCTORY THOUGHTS

The term "aerodynamics" is generally used for problems arising from flight and other topics involving the flow of air.

Ludwig Prandtl, 1949

Aerodynamics: The dynamics of gases, especially of atmospheric interactions with moving objects.

The American Heritage
Dictionary of the English
Language, 1969

1.1 IMPORTANCE OF AERODYNAMICS: HISTORICAL EXAMPLES

On August 8, 1588, the waters of the English Channel churned with the gyrations of hundreds of warships. The great Spanish Armada had arrived to carry out an invasion of Elizabethan England and was met head-on by the English fleet under the command of Sir Francis Drake. The Spanish ships were large and heavy; they were packed with soldiers and carried formidable cannons that fired 50-lb round shot that could devastate any ship of that era. In contrast, the English ships were smaller and lighter; they carried no soldiers and were armed with lighter, shorter-range cannons. The balance of power in Europe hinged on the outcome of this naval encounter. King Philip II of Catholic Spain was attempting to squash Protestant England's rising influence in the political and religious affairs of Europe; in turn, Queen Elizabeth I was attempting to defend the very existence of England as a sovereign state. In fact, on that crucial day in 1588, when the English floated six fire ships into the Spanish formation and then drove headlong into the ensuing confusion, the future history of Europe was in the balance. In the final outcome, the heavier, sluggish, Spanish ships were no match for the faster, more maneuverable, English craft, and by that evening the Spanish Armada lay in disarray, no longer a threat to England. This naval battle is of particular importance because it was the first in history to be fought by ships on both sides powered completely by sail (in contrast to earlier combinations of oars and sail) and it taught the world that political power was going to be synonymous with naval power. In turn, naval power was going to depend greatly on the speed and maneuverability of ships. To increase the speed of a ship, it is important to reduce the resistance created by the water flow around the ship's hull. Suddenly, the drag on ship hulls became an engineering problem of great interest, thus giving impetus to the study of fluid mechanics.

This impetus hit its stride almost a century later, when, in 1687, Isaac Newton (1642–1727) published his famous *Principia*, in which the entire second book was devoted to fluid mechanics. Newton encountered the same difficulty as others before him, namely, that the analysis of fluid flow is conceptually more difficult than the dynamics of solid bodies. A solid body is usually geometrically well defined, and its motion is therefore relatively easy to describe. On the other hand, a fluid is a "squishy" substance, and in Newton's time it was difficult to decide even how to qualitatively model its motion, let alone obtain quantitative relationships. Newton considered a fluid flow as a uniform, rectilinear stream of particles, much like a cloud of pellets from a shotgun blast. As sketched in Fig. 1.1, Newton assumed that upon striking a surface inclined at an angle θ to the stream, the particles would transfer their normal momentum to the surface but their tangential momentum would be preserved. Hence, after collision with the surface, the particles would then move along the surface. This led to an expression for the hydrodynamic force on the surface which varies as $\sin^2 \theta$. This is Newton's famous sine-squared law (described in detail in Chap. 14). Although its accuracy left much to be desired, its simplicity led to wide application in naval architecture.

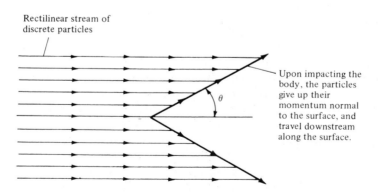

Rectilinear stream of discrete particles

θ

Upon impacting the body, the particles give up their momentum normal to the surface, and travel downstream along the surface.

FIGURE 1.1
Isaac Newton's model of fluid flow in the year 1687. This model was widely adopted in the seventeenth and eighteenth centuries but was later found to be conceptually inaccurate for most fluid flows.

Later, in 1777, a series of experiments was carried out by Jean Le Rond d'Alembert (1717–1783), under the support of the French government, in order to measure the resistance of ships in canals. The results showed that "the rule that for oblique planes resistance varies with the sine square of the angle of incidence holds good only for angles between 50 and 90° and must be abandoned for lesser angles." Also, in 1781, Leonhard Euler (1707–1783) pointed out the physical inconsistency of Newton's model (Fig. 1.1) consisting of a rectilinear stream of particles impacting without warning on a surface. In contrast to this model, Euler noted that the fluid moving toward a body "*before* reaching the latter, bends its direction and its velocity so that when it reaches the body it flows past it along the surface, and exercises no other force on the body except the pressure corresponding to the single points of contact." Euler went on to present a formula for resistance which attempted to take into account the shear stress distribution along the surface, as well as the pressure distribution. This expression became proportional to $\sin^2 \theta$ for large incidence angles, whereas it was proportional to $\sin \theta$ at small incidence angles. Euler noted that such a variation was in reasonable agreement with the ship-hull experiments carried out by d'Alembert.

This early work in fluid dynamics has now been superseded by modern concepts and techniques. (However, amazingly enough, Newton's sine-squared law has found new application in very high speed aerodynamics, to be discussed in Chap. 14.) The major point here is that the rapid rise in the importance of naval architecture after the sixteenth century made fluid dynamics an important science, occupying the minds of Newton, d'Alembert, and Euler, among many others. Today, the modern ideas of fluid dynamics, presented in this book, are still driven in part by the importance of reducing hull drag on ships.

Consider a second historical example. The scene shifts to Kill Devil Hills, 4 mi south of Kitty Hawk, North Carolina. It is summer of 1901, and Wilbur and Orville Wright are struggling with their second major glider design, the first being a stunning failure the previous year. The airfoil shape and wing design of their glider are based on aerodynamic data published in the 1890s by the great German aviation pioneer Otto Lilienthal (1848–1896) and by Samuel Pierpont Langley (1834–1906), secretary of the Smithsonian Institution—the most prestigious scientific position in the United States at that time. Because their first glider in 1900 produced no meaningful lift, the Wright brothers have increased the wing area from 165 to 290 ft^2 and have increased the wing camber (a measure of the airfoil curvature—the larger the camber, the more "arched" is the thin airfoil shape) by almost a factor of 2. But something is still wrong. In Wilbur's words, the glider's "lifting capacity seemed scarcely one-third of the calculated amount." Frustration sets in. The glider is not performing even close to their expectations, although it is designed on the basis of the best available aerodynamic data. On August 20, the Wright brothers despairingly pack themselves aboard a train going back to Dayton, Ohio. On the ride back, Wilbur mutters that "nobody will fly for a thousand years." However, one of the hallmarks of the Wrights is perseverance, and within weeks of returning to Dayton, they decide on a complete departure from their previous approach. Wilbur later wrote that "having set out with absolute faith in the existing scientific data, we were driven to doubt one thing after another, until finally after two years of experiment, we cast it all aside, and decided to rely entirely upon our own investigations." Since their 1901 glider was of poor aerodynamic design, the Wrights set about determining what constitutes good aerodynamic design. In the fall of 1901, they design and build a 6-ft-long, 16-in square wind tunnel powered by a two-bladed fan connected to a gasoline engine. An original photograph of the Wrights' tunnel in their Dayton bicycle shop is shown in Fig. 1.2a. In this wind tunnel they test over 200 different wing and airfoil shapes, including flat plates, curved plates, rounded leading edges, rectangular and curved planforms, and various monoplane and multiplane configurations. A sample of their test models is shown in Fig. 1.2b. The aerodynamic data is taken logically and carefully. It shows a major departure from the existing "state-of-the art" data. Armed with their new aerodynamic information, the Wrights design a new glider in the spring of 1902. The airfoil is much more efficient; the camber is reduced considerably, and the location of the maximum rise of the airfoil is moved closer to the front of the wing. The most obvious change, however, is that the ratio of the length of the wing (wingspan) to the distance from the front to the rear of the airfoil (chord length) is increased from 3 to 6. The success of this glider during the summer and fall of 1902 is astounding; Orville and Wilbur accumulate over a thousand flights during this period. In contrast to the previous year, the Wrights return to Dayton flushed with success and devote all their subsequent efforts to powered flight. The rest is history.

The major point here is that good aerodynamics was vital to the ultimate success of the Wright brothers and, of course, to all subsequent successful airplane designs up to the present day. The importance of aerodynamics to successful manned flight goes without saying, and a major thrust of this book is to present the aerodynamic fundamentals that govern such flight.

Consider a third historical example of the importance of aerodynamics, this time as it relates to rockets and space flight. High-speed, supersonic flight had become a dominant feature of aerodynamics by the end of World War II. By this time, aerodynamicists appreciated the advantages of using slender, pointed body shapes to reduce the drag of supersonic vehicles. The more pointed and slender the body, the weaker the shock wave attached to the nose, and hence the smaller the wave drag. Consequently, the German V-2 rocket used during the last stages of World War II had a pointed nose, and all short-range rocket vehicles flown during the next decade followed suit. Then, in 1953, the first hydrogen bomb was exploded by the United States. This immediately spurred the development of long-range intercontinental ballistic missiles (ICBMs) to deliver such bombs. These vehicles were designed to fly outside the region of the earth's atmosphere for distances of 5000 mi or more and to reenter the atmosphere at suborbital speeds of from 20,000 to 22,000 ft/s. At such high velocities, the aerodynamic heating of the reentry vehicle becomes severe, and this heating problem dominated the minds of high-speed aerodynamicists. Their first thinking was conventional—a sharp-pointed, slender reentry body. Efforts to minimize aerodynamic heating centered on the maintenance of laminar boundary layer

(a)

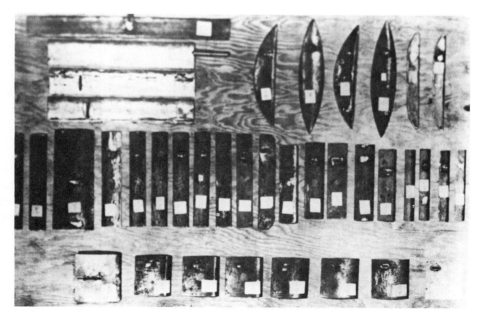

(b)

FIGURE 1.2
(a) Wind tunnel designed, built, and used by the Wright brothers in Dayton, Ohio, during 1901–1902.
(b) Wing models tested by the Wright brothers in their wind tunnel during 1901–1902.

flow on the vehicle's surface; such laminar flow produces far less heating than turbulent flow (discussed in Chaps. 15 and 17). However, nature much prefers turbulent flow, and reentry vehicles are no exception. Therefore, the pointed-nose reentry body was doomed to failure because it would burn up in the atmosphere before reaching the earth's surface.

However, in 1951, one of those major breakthroughs that come very infrequently in engineering was created by H. Julian Allen at the NACA Ames Aeronautical Laboratory—he introduced the concept of the *blunt* reentry body. His thinking was paced by the following concepts. At the beginning of reentry, near the outer edge of the atmosphere, the vehicle has a large amount of kinetic energy due to its high velocity and a large amount of potential energy due to its high altitude. However, by the time the vehicle reaches the surface of the earth, its velocity is relatively small and its altitude is zero; hence, it has virtually no kinetic or potential energy. Where has all the energy gone? The answer is that it has gone into (1) heating the body and (2) heating the airflow around the body. This is illustrated in Fig. 1.3. Here, the shock wave from the nose of the vehicle heats the airflow around the vehicle; at the same time, the vehicle is heated by the intense frictional dissipation within the boundary layer on the surface. Allen reasoned that if more of the total reentry energy could be dumped into the airflow, then less would be available to be transferred to the vehicle itself in the form of heating. In turn, the way to increase the heating of the airflow is to create a stronger shock wave at the nose, i.e., to use a blunt-nosed body. The contrast between slender and blunt reentry bodies is illustrated in Fig. 1.4. This was a stunning conclusion—to minimize aerodynamic heating, you actually want a blunt rather than a slender body. The result was so important that it was bottled up in a secret government document. Moreover, because it was so foreign to contemporary intuition, the blunt-reentry-body concept was accepted only gradually by the technical community. Over the next few years, additional

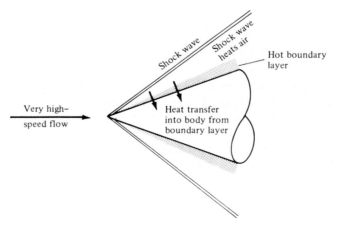

FIGURE 1.3
Energy of reentry goes into heating both the body and the air around the body.

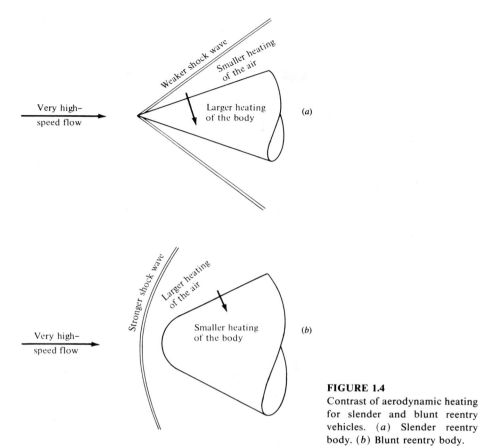

FIGURE 1.4
Contrast of aerodynamic heating for slender and blunt reentry vehicles. (*a*) Slender reentry body. (*b*) Blunt reentry body.

aerodynamic analyses and experiments confirmed the validity of blunt reentry bodies. By 1955, Allen was publicly recognized for his work, receiving the Sylvanus Albert Reed Award of the Institute of the Aeronautical Sciences (now the American Institute of Aeronautics and Astronautics). Finally, in 1958, his work was made available to the public in the pioneering document NACA Report 1381 entitled "A Study of the Motion and Aerodynamic Heating of Ballistic Missiles Entering the Earth's Atmosphere at High Supersonic Speeds." Since Harvey Allen's early work, all successful reentry bodies, from the first Atlas ICBM to the manned Apollo lunar capsule, have been blunt. Incidentally, Allen went on to distinguish himself in many other areas, becoming the director of the NASA Ames Research Center in 1965, and retiring in 1970. His work on the blunt reentry body is an excellent example of the importance of aerodynamics to space vehicle design.

In summary, the purpose of this section has been to underscore the importance of aerodynamics in historical context. The goal of this book is to introduce the fundamentals of aerodynamics and to give the reader a much deeper insight to many technical applications in addition to the few described above. Aerodynamics is also a subject of intellectual beauty, composed and drawn by many great minds over the centuries. If you are challenged and interested by these thoughts, or even the least bit curious, then read on.

1.2 AERODYNAMICS: CLASSIFICATION AND PRACTICAL OBJECTIVES

A distinction between solids, liquids, and gases can be made in a simplistic sense as follows. Put a solid object inside a larger, closed container. The solid object will not change; its shape and boundaries will remain the same. Now put a liquid inside the container. The liquid will change its shape to conform to that of the container and will take on the same boundaries as the container up to the maximum depth of the liquid. Now put a gas inside the container. The gas will completely fill the container, taking on the same boundaries as the container.

The word "fluid" is used to denote either a liquid or a gas. A more technical distinction between a solid and a fluid can be made as follows. When a force is applied tangentially to the surface of a solid, the solid will experience a *finite* deformation, and the tangential force per unit area—the shear stress—will usually be proportional to the amount of deformation. In contrast, when a tangential shear stress is applied to the surface of a fluid, the fluid will experience a *continuously increasing* deformation, and the shear stress usually will be proportional to the rate of change of the deformation.

The most fundamental distinction between solids, liquids, and gases is at the atomic and molecular level. In a solid, the molecules are packed so closely together that their nuclei and electrons form a rigid geometric structure, "glued" together by powerful intermolecular forces. In a liquid, the spacing between molecules is larger, and although intermolecular forces are still strong they allow enough movement of the molecules to give the liquid its "fluidity." In a gas, the spacing between molecules is much larger (for air at standard conditions, the spacing between molecules is, on the average, about 10 times the molecular diameter). Hence, the influence of intermolecular forces is much weaker, and the motion of the molecules occurs rather freely throughout the gas. This movement of molecules in both gases and liquids leads to similar physical characteristics, the characteristics of a fluid—quite different from those of a solid. Therefore, it makes sense to classify the study of the dynamics of both liquids and gases under the same general heading, called *fluid dynamics*. On the other hand, certain differences exist between the flow of liquids and the flow of gases; also, different species of gases (say, N_2, He, etc.) have different properties. Therefore, fluid dynamics is subdivided into three areas as follows:

Hydrodynamics—flow of liquids

Gas dynamics—flow of gases

Aerodynamics—flow of air

These areas are by no means mutually exclusive; there are many similarities and identical phenomena between them. Also, the word "aerodynamics" has taken on a popular usage that sometimes covers the other two areas. As a result, this author tends to interpret the word "aerodynamics" very liberally, and its use throughout this book does *not* always limit our discussions just to air.

Aerodynamics is an applied science with many practical applications in engineering. No matter how elegant an aerodynamic theory may be, or how mathematically complex a numerical solution may be, or how sophisticated an aerodynamic experiment may be, all such efforts are usually aimed at one or more of the following practical objectives:

1. The prediction of forces and moments on, and heat transfer to, bodies moving through a fluid (usually air). For example, we are concerned with the generation of lift, drag, and moments on airfoils, wings, fuselages, engine nacelles, and, most importantly, whole airplane configurations. We want to estimate the wind force on buildings, ships, and other surface vehicles. We are concerned with the hydrodynamic forces on surface ships, submarines, and torpedoes. We need to be able to calculate the aerodynamic heating of flight vehicles ranging from the supersonic transport to a planetary probe entering the atmosphere of Jupiter. These are but a few examples.

2. Determination of flows moving internally through ducts. We wish to calculate and measure the flow properties inside rocket and air-breathing jet engines and to calculate the engine thrust. We need to know the flow conditions in the test section of a wind tunnel. We must know how much fluid can flow through pipes under various conditions. A recent, very interesting application of aerodynamics is high-energy chemical and gas-dynamic lasers (see Ref. 1), which are nothing more than specialized wind tunnels that can produce extremely powerful laser beams. Figure 1.5 is a photograph of an early gas-dynamic laser designed in the late 1960s.

The applications in item 1 come under the heading of *external aerodynamics* since they deal with external flows over a body. In contrast, the applications in item 2 involve *internal aerodynamics* because they deal with flows internally within ducts. In external aerodynamics, in addition to forces, moments, and aerodynamic heating associated with a body, we are frequently interested in the details of the flow field away from the body. For example, the communication blackout experienced by the space shuttle during a portion of its reentry trajectory is due to a concentration of free electrons in the hot shock layer around the body. We need to calculate the variation of electron density throughout such flow fields. Another example is the propagation of shock waves in a supersonic flow; for instance, does the shock wave from the wing of a supersonic airplane impinge upon and interfere with the tail surfaces? Yet another example is the flow associated with the strong vortices trailing downstream from the wing tips of large subsonic airplanes such as the Boeing 747. What are the properties of these vortices, and how do they affect smaller aircraft which happen to fly through them?

FIGURE 1.5
A CO_2-N_2 gas-dynamic laser, circa 1969. (*Courtesy of the Avco-Everett Research Laboratory.*)

The above is just a sample of the myriad applications of aerodynamics. One purpose of this book is to provide the reader with the technical background necessary to fully understand the nature of such practical aerodynamic applications.

1.3 ROAD MAP FOR THIS CHAPTER

When learning a new subject, it is important for you to know where you are, where you are going, and how you can get there. Therefore, at the beginning of each chapter in this book, a road map will be given to help guide you through the material of that chapter and to help you obtain a perspective as to how the material fits within the general framework of aerodynamics. For example, a road map for Chap. 1 is given in Fig. 1.6. You will want to frequently refer back to these road maps as you progress through the individual chapters. When you reach the end of each chapter, look back over the road map to see where you started, where you are now, and what you learned in between.

1.4 SOME FUNDAMENTAL AERODYNAMIC VARIABLES

A prerequisite to understanding physical science and engineering is simply learning the vocabulary used to describe concepts and phenomena. Aerodynamics

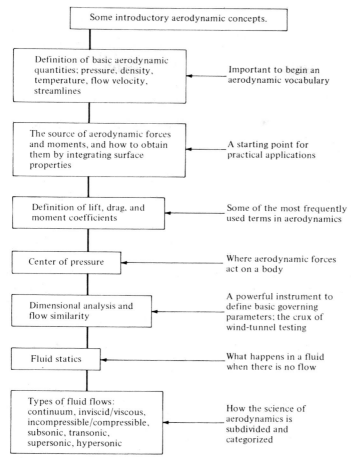

FIGURE 1.6
Road map for Chap. 1.

is no exception. Throughout this book, and throughout your working career, you will be adding to your technical vocabulary list. Let us start by defining four of the most frequently used words in aerodynamics: "pressure," "density," "temperature," and "flow velocity."†

Consider a surface immersed in a fluid. The surface can be a real, solid surface such as the wall of a duct or the surface of a body; it can also be a free surface which we simply imagine drawn somewhere in the middle of the fluid. Also, keep in mind that the molecules of the fluid are constantly in motion. *Pressure* is the normal force per unit area exerted on a surface due to the time rate of change of momentum of the gas molecules impacting on (or crossing)

---

† A basic introduction to these quantities is given on pages 51–55 of Ref. 2.

that surface. It is important to note that even though pressure is defined as force "per unit area," you do not need a surface that is exactly 1 ft^2 or 1 m^2 to talk about pressure. In fact, pressure is usually defined as a *point* in the fluid or a *point* on a solid surface and can vary from one point to another. To see this more clearly, consider a point B in a volume of fluid. Let

$$dA = \text{elemental area at } B$$

$$dF = \text{force on one side of } dA \text{ due to pressure}$$

Then, the pressure at point B in the fluid is defined as

$$p = \lim \left(\frac{dF}{dA} \right) \qquad dA \to 0$$

The pressure p is the limiting form of the force per unit area, where the area of interest has shrunk to nearly zero at the point B.† Clearly, you can see that pressure is a *point property* and can have a different value from one point to another in the fluid.

Another important aerodynamic variable is *density*, defined as the mass per unit volume. Analogous to our discussion on pressure, the definition of density does not require an actual volume of 1 ft^3 or 1 m^3. Rather, it is a *point property* that can vary from point to point in the fluid. Again, consider a point B in the fluid. Let

$$dv = \text{elemental volume around } B$$

$$dm = \text{mass of fluid inside } dv$$

Then, the density at point B is

$$\rho = \lim \frac{dm}{dv} \qquad dv \to 0$$

Therefore, the density ρ is the limiting form of the mass per unit volume, where the volume of interest has shrunk to nearly zero around point B. (Note that dv cannot achieve the value of zero for the reason discussed in the footnote concerning dA in the definition of pressure.)

Temperature takes on an important role in high-speed aerodynamics (introduced in Chap. 7). The temperature T of a gas is directly proportional to the average kinetic energy of the molecules of the fluid. In fact, if KE is the mean molecular kinetic energy, then temperature is given by $\text{KE} = \frac{3}{2} kT$, where k is the Boltzmann constant. Hence, we can qualitatively visualize a high-temperature gas as one in which the molecules and atoms are randomly rattling about at high

---

† Strictly speaking, dA can never achieve the limit of zero, because there would be no molecules at point B in that case. The above limit should be interpreted as dA approaching a very small value, near zero in terms of our macroscopic thinking, but sufficiently larger than the average spacing between molecules on a microscopic basis.

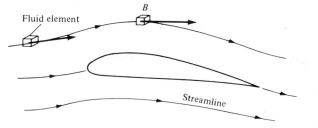

FIGURE 1.7
Illustration of flow velocity and streamlines.

speeds, whereas in a low-temperature gas, the random motion of the molecules is relatively slow. Temperature is also a point property, which can vary from point to point in the gas.

The principal focus of aerodynamics is fluids in motion. Hence, flow velocity is an extremely important consideration. The concept of the velocity of a fluid is slightly more subtle than that of a solid body in motion. Consider a solid object in translational motion, say, moving at 30 m/s. Then all parts of the solid are simultaneously translating at the same 30 m/s velocity. In contrast, a fluid is a "squishy" substance, and for a fluid in motion, one part of the fluid may be traveling at a different velocity from another part. Hence, we have to adopt a certain perspective, as follows. Consider the flow of air over an airfoil, as shown in Fig. 1.7. Lock your eyes on a specific, infinitesimally small element of mass in the gas, called a *fluid element*, and watch this element move with time. Both the speed and direction of this fluid element can vary as it moves from point to point in the gas. Now fix your eyes on a specific fixed point in space, say, point B in Fig. 1.7. *Flow velocity* can now be defined as follows: The velocity of a flowing gas at any fixed point B in space is the velocity of an infinitesimally small fluid element as it sweeps through B. The flow velocity **V** has both magnitude and direction; hence, it is a vector quantity. This is in contrast to p, ρ, and T, which are scalar variables. The scalar magnitude of **V** is frequently used and is denoted by V. Again, we emphasize that velocity is a point property and can vary from point to point in the flow.

Referring again to Fig. 1.7, a moving fluid element traces out a fixed path in space. As long as the flow is steady, i.e., as long as it does not fluctuate with time, this path is called a *streamline* of the flow. Drawing the streamlines of the flow field is an important way of visualizing the motion of the gas; we will frequently be sketching the streamlines of the flow about various objects. A more rigorous discussion of streamlines is given in Chap. 2.

1.5 AERODYNAMIC FORCES AND MOMENTS

At first glance, the generation of the aerodynamic force on a giant Boeing 747 may seem complex, especially in light of the complicated three-dimensional flow field over the wings, fuselage, engine nacelles, tail, etc. Similarly, the aerodynamic resistance on an automobile traveling at 55 mi/h on the highway involves a

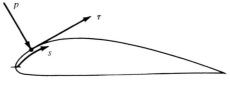

$p = p(s)$ = surface pressure distribution
$\tau = \tau(s)$ = surface shear stress distribution

FIGURE 1.8
Illustration of pressure and shear stress on an aerodynamic surface.

complex interaction of the body, the air, and the ground. However, in these and all other cases, the aerodynamic forces and moments on the body are due to only two basic sources:

1. *Pressure distribution* over the body surface
2. *Shear stress distribution* over the body surface

No matter how complex the body shape may be, the aerodynamic forces and moments on the body are due entirely to the above two basic sources. The *only* mechanisms nature has for communicating a force to a body moving through a fluid are pressure and shear stress distributions on the body surface. Both pressure p and shear stress τ have dimensions of force per unit area (pounds per square foot or newtons per square meter). As sketched in Fig. 1.8, p acts *normal* to the surface, and τ acts *tangential* to the surface. Shear stress is due to the "tugging action" on the surface, which is caused by friction between the body and the air (and is studied in great detail in Chaps. 15 to 17).

The net effect of the p and τ distributions integrated over the complete body surface is a resultant aerodynamic force R and moment M on the body, as sketched in Fig. 1.9. In turn, the resultant R can be split into components, two sets of which are shown in Fig. 1.10. In Fig. 1.10, V_∞ is the *relative wind*, defined as the flow velocity far ahead of the body. The flow far away from the body is called the *freestream*, and hence V_∞ is also called the freestream velocity. In Fig. 1.10, by definition,

$$L \equiv \text{lift} \equiv \text{component of } R \text{ perpendicular to } V_\infty$$
$$D \equiv \text{drag} \equiv \text{component of } R \text{ parallel to } V_\infty$$

The chord c is the linear distance from the leading edge to the trailing edge of the body. Sometimes, R is split into components perpendicular and parallel to

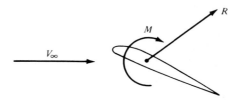

FIGURE 1.9
Resultant aerodynamic force and moment on the body.

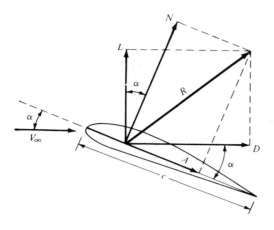

FIGURE 1.10
Resultant aerodynamic force and the components into which it splits.

the chord, as also shown in Fig. 1.10. By definition,

$N \equiv$ normal force $\equiv$ component of R perpendicular to c

$A \equiv$ axial force $\equiv$ component of R parallel to c

The angle of attack α is defined as the angle between c and V_∞. Hence, α is also the angle between L and N and between D and A. The geometrical relation between these two sets of components is, from Fig. 1.10,

$$L = N \cos \alpha - A \sin \alpha \qquad (1.1)$$

$$D = N \sin \alpha + A \cos \alpha \qquad (1.2)$$

Let us examine in more detail the integration of the pressure and shear stress distributions to obtain the aerodynamic forces and moments. Consider the two-dimensional body sketched in Fig. 1.11. The chord line is drawn horizontally, and hence the relative wind is inclined relative to the horizontal by the angle of attack α. An xy coordinate system is oriented parallel and perpendicular, respectively, to the chord. The distance from the leading edge measured along the body surface to an arbitrary point A on the upper surface is s_u; similarly, the distance to an arbitrary point B on the lower surface is s_l. The pressure and shear stress on the upper surface are denoted by p_u and τ_u, respectively; both p_u and τ_u are functions of s_u. Similarly, p_l and τ_l are the corresponding quantities on the lower surface and are functions of s_l. At a given point, the pressure is normal to the surface and is oriented at an angle θ relative to the perpendicular; shear stress is tangential to the surface and is oriented at the same angle θ relative to the horizontal. In Fig. 1.11, the sign convention for θ is positive when measured *clockwise* from the vertical line to the direction of p and from the horizontal line to the direction of τ. In Fig. 1.11, all thetas are shown in their positive direction. Now consider the two-dimensional shape in Fig. 1.11 as a cross section of an infinitely long cylinder of uniform section. A unit span of such a cylinder is shown in Fig. 1.12. Consider an elemental surface area dS of this cylinder, where

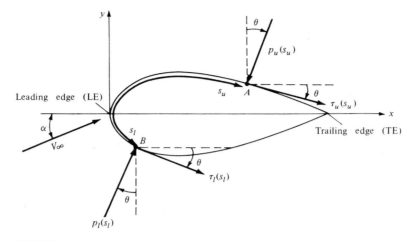

FIGURE 1.11
Nomenclature for the integration of pressure and shear stress distributions over a two-dimensional body surface.

$dS = (ds)(1)$ as shown by the shaded area in Fig. 1.12. We are interested in the contribution to the total normal force N' and the total axial force A' due to the pressure and shear stress on the elemental area dS. The primes on N' and A' denote force per unit span. Examining both Figs. 1.11 and 1.12, we see that the elemental normal and axial forces acting on the elemental surface dS on the *upper* body surface are

$$dN'_u = -p_u \, ds_u \cos\theta - \tau_u \, ds_u \sin\theta \tag{1.3}$$

$$dA'_u = -p_u \, ds_u \sin\theta + \tau_u \, ds_u \cos\theta \tag{1.4}$$

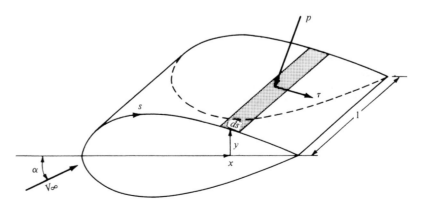

FIGURE 1.12
Aerodynamic force on an element of the body surface.

On the *lower* body surface, we have

$$dN'_l = p_l\,ds_l\cos\theta - \tau_l\,ds_l\sin\theta \tag{1.5}$$

$$dA'_l = p_l\,ds_l\sin\theta + \tau_l\,ds_l\cos\theta \tag{1.6}$$

In Eqs. (1.3) to (1.6), the positive directions of N' and A' are those shown in Fig. 1.10. In these equations, the positive clockwise convention for θ must be followed. For example, consider again Fig. 1.11. Near the leading edge of the body, where the slope of the upper body surface is positive, τ is inclined upward, and hence it gives a positive contribution to N'. For an upward inclined τ, θ would be counterclockwise, hence negative. Therefore, in Eq. (1.3), $\sin\theta$ would be negative, making the shear stress term (the last term) a positive value, as it should be in this instance. Hence, Eqs. (1.3) to (1.6) hold in general (for both the forward and rearward portions of the body) as long as the above sign convention for θ is consistently applied.

The total normal and axial forces *per unit span* are obtained by integrating Eqs. (1.3) to (1.6) from the leading edge (LE) to the trailing edge (TE):

$$N' = -\int_{LE}^{TE} (p_u\cos\theta + \tau_u\sin\theta)\,ds_u + \int_{LE}^{TE} (p_l\cos\theta - \tau_l\sin\theta)\,ds_l \tag{1.7}$$

$$A' = \int_{LE}^{TE} (-p_u\sin\theta + \tau_u\cos\theta)\,ds_u + \int_{LE}^{TE} (p_l\sin\theta + \tau_l\cos\theta)\,ds_l \tag{1.8}$$

In turn, the total lift and drag per unit span can be obtained by inserting Eqs. (1.7) and (1.8) into (1.1) and (1.2); note that Eqs. (1.1) and (1.2) hold for forces on an arbitrarily shaped body (unprimed) and for the forces per unit span (primed).

The aerodynamic moment exerted on the body depends on the point about which moments are taken. Consider moments taken about the leading edge. By convention, moments which tend to increase α (pitch up) are positive, and moments which tend to decrease α (pitch down) are negative. This convention is illustrated in Fig. 1.13. Returning again to Figs. 1.11 and 1.12, the moment per unit span about the leading edge due to p and τ on the elemental area dS on the upper surface is

$$dM'_u = (p_u\cos\theta + \tau_u\sin\theta)x\,ds_u + (-p_u\sin\theta + \tau_u\cos\theta)y\,ds_u \tag{1.9}$$

On the bottom surface,

$$dM'_l = (-p_l\cos\theta + \tau_l\sin\theta)x\,ds_l + (p_l\sin\theta + \tau_l\cos\theta)y\,ds_l \tag{1.10}$$

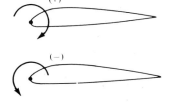

FIGURE 1.13
Sign convention for aerodynamic moments.

In Eqs. (1.9) and (1.10), note that the same sign convention for θ applies as before and that y is a positive number above the chord and a negative number below the chord. Integrating Eqs. (1.9) and (1.10) from the leading to the trailing edges, we obtain for the moment about the leading edge per unit span

$$M'_{LE} = \int_{LE}^{TE} [(p_u \cos \theta + \tau_u \sin \theta)x - (p_u \sin \theta - \tau_u \cos \theta)y] \, ds_u$$

$$+ \int_{LE}^{TE} [(-p_l \cos \theta + \tau_l \sin \theta)x + (p_l \sin \theta + \tau_l \cos \theta)y] \, ds_l \qquad (1.11)$$

In Eqs. (1.7), (1.8), and (1.11), θ, x, and y are known functions of s for a given body shape. Hence, if p_u, p_l, τ_u, and τ_l are known as functions of s (from theory or experiment), the integrals in these equations can be evaluated. Clearly, Eqs. (1.7), (1.8), and (1.11) demonstrate the principle stated earlier, namely, *the sources of the aerodynamic lift, drag, and moments on a body are the pressure and shear stress distributions integrated over the body.* A major goal of theoretical aerodynamics is to calculate $p(s)$ and $\tau(s)$ for a given body shape and freestream conditions, thus yielding the aerodynamic forces and moments via Eqs. (1.7), (1.8), and (1.11).

As our discussions of aerodynamics progress, it will become clear that there are quantities of an even more fundamental nature than the aerodynamic forces and moments themselves. These are *dimensionless force and moment coefficients,* defined as follows. Let ρ_∞ and V_∞ be the density and velocity, respectively, in the freestream, far ahead of the body. We define a dimensional quantity called the freestream *dynamic pressure* as

Dynamic pressure: $\qquad\qquad\qquad q_\infty \equiv \tfrac{1}{2}\rho_\infty V_\infty^2$

The dynamic pressure has the units of pressure (i.e., pounds per square foot or newtons per square meter). In addition, let S be a reference area and l be a reference length. The dimensionless force and moment coefficients are defined as follows:

Lift coefficient: $\qquad\qquad\qquad\qquad C_L \equiv \dfrac{L}{q_\infty S}$

Drag coefficient: $\qquad\qquad\qquad\qquad C_D \equiv \dfrac{D}{q_\infty S}$

Normal force coefficient: $\qquad\qquad C_N \equiv \dfrac{N}{q_\infty S}$

Axial force coefficient: $\qquad\qquad\quad C_A \equiv \dfrac{A}{q_\infty S}$

Moment coefficient: $\qquad\qquad\qquad C_M \equiv \dfrac{M}{q_\infty S l}$

In the above coefficients, the reference area S and reference length l are chosen to pertain to the given geometric body shape; for different shapes, S and l may be different things. For example, for an airplane wing, S is the planform area, and l is the mean chord length, as illustrated in Fig. 1.14a. However, for a sphere, S is the cross-sectional area, and l is the diameter, as shown in Fig. 1.14b. The particular choice of reference area and length is not critical; however, when using force and moment coefficient data, you must always know what reference quantities the particular data are based upon.

The symbols in capital letters listed above, i.e., C_L, C_D, C_M, C_N, and C_A, denote the force and moment coefficients for a complete three-dimensional body such as an airplane or a finite wing. In contrast, for a two-dimensional body, such as given in Figs. 1.11 and 1.12, the forces and moments are per unit span. For these two-dimensional bodies, it is conventional to denote the aerodynamic coefficients by lowercase letters; e.g.,

$$c_l \equiv \frac{L'}{q_\infty c} \qquad c_d \equiv \frac{D'}{q_\infty c} \qquad c_m \equiv \frac{M'}{q_\infty c^2}$$

where the reference area $S = c(1) = c$.

Two additional dimensionless quantities of immediate use are

Pressure coefficient: $$C_p \equiv \frac{p - p_\infty}{q_\infty}$$

Skin friction coefficient: $$c_f \equiv \frac{\tau}{q_\infty}$$

where p_∞ is the freestream pressure.

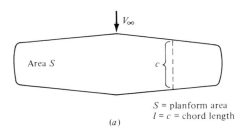

S = planform area
$l = c$ = chord length

(a)

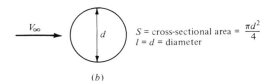

S = cross-sectional area = $\frac{\pi d^2}{4}$
$l = d$ = diameter

(b)

FIGURE 1.14
Some reference areas and lengths.

The most useful forms of Eqs. (1.7), (1.8), and (1.11) are in terms of the dimensionless coefficients introduced above. From the geometry shown in Fig. 1.15,

$$dx = ds \cos \theta \tag{1.12}$$

$$dy = -(ds \sin \theta) \tag{1.13}$$

$$S = c(1) \tag{1.14}$$

Substituting Eqs. (1.12) and (1.13) into Eqs. (1.7), (1.8), and (1.11), dividing by q_∞, and further dividing by S in the form of Eq. (1.14), we obtain the following integral forms for the force and moment coefficients:

$$c_n = \frac{1}{c} \left[\int_0^c (C_{p,l} - C_{p,u}) \, dx + \int_0^c \left(c_{f,u} \frac{dy_u}{dx} + c_{f,l} \frac{dy_l}{dx} \right) dx \right] \tag{1.15}$$

$$c_a = \frac{1}{c} \left[\int_0^c \left(C_{p,u} \frac{dy_u}{dx} - C_{p,l} \frac{dy_l}{dx} \right) dx + \int_0^c (c_{f,u} + c_{f,l}) \, dx \right] \tag{1.16}$$

$$c_{m_{LE}} = \frac{1}{c^2} \left[\int_0^c (C_{p,u} - C_{p,l}) x \, dx - \int_0^c \left(c_{f,u} \frac{dy_u}{dx} + c_{f,l} \frac{dy_l}{dx} \right) x \, dx \right.$$
$$\left. + \int_0^c \left(C_{p,u} \frac{dy_u}{dx} + c_{f,u} \right) y_u \, dx + \int_0^c \left(-C_{p,l} \frac{dy_l}{dx} + c_{f,l} \right) y_l \, dx \right] \tag{1.17}$$

The simple algebraic steps are left as an exercise for the reader. When evaluating these integrals, keep in mind that y_u is directed above the x axis, and hence is positive, whereas y_l is directed below the x axis, and hence is negative. Also, dy/dx on both the upper and lower surfaces follow the usual rule from calculus, i.e., positive for those portions of the body with a positive slope and negative for those portions with a negative slope.

The lift and drag coefficients can be obtained from Eqs. (1.1) and (1.2) cast in coefficient form:

$$c_l = c_n \cos \alpha - c_a \sin \alpha \tag{1.18}$$

$$c_d = c_n \sin \alpha + c_a \cos \alpha \tag{1.19}$$

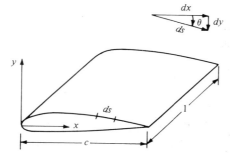

FIGURE 1.15
Geometrical relationship of differential lengths.

Integral forms for c_l and c_d are obtained by substituting Eqs. (1.15) and (1.16) into (1.18) and (1.19).

It is important to note from Eqs. (1.15) through (1.19) that the aerodynamic force and moment coefficients can be obtained by integrating the pressure and skin friction coefficients over the body. This is a common procedure in both theoretical and experimental aerodynamics. In addition, although our derivations have used a two-dimensional body, an analogous development can be presented for three-dimensional bodies—the geometry and equations only get more complex and involved—the principle is the same.

Example 1.1. Consider the supersonic flow over a 5° half-angle wedge at zero angle of attack, as sketched in Fig. 1.16a. The freestream Mach number ahead of the wedge is 2.0, and the freestream pressure and density are 1.01×10^5 N/m^2 and 1.23 kg/m^3, respectively (this corresponds to standard sea level conditions). The pressures on the upper and lower surfaces of the wedge are constant with distance s and equal to each other, namely, $p_u = p_l = 1.31 \times 10^5$ N/m^2, as shown in Fig. 1.16b. The pressure exerted on the base of the wedge is equal to p_∞. As seen in Fig. 1.16c, the shear stress varies over both the upper and lower surfaces as $\tau_w = 431 \ s^{-0.2}$. The chord length, c, of the wedge is 2 m. Calculate the drag coefficient for the wedge.

Solution. We will carry out this calculation in two equivalent ways. First, we calculate the drag from Eq. (1.8), and then obtain the drag coefficient. In turn, as an illustration of an alternate approach, we convert the pressure and shear stress to pressure coefficient and skin friction coefficient, and then use Eq. (1.16) to obtain the drag coefficient.

Since the wedge in Fig. 1.16 is at zero angle of attack, then $D' = A'$. Thus, the drag can be obtained from Eq. (1.8) as

$$D' = \int_{LE}^{TE} (-p_u \sin \theta + \tau_u \cos \theta) \, ds_u + \int_{LE}^{TE} (p_l \sin \theta + \tau_l \cos \theta) \, ds_l$$

Referring to Fig. 1.16c, recalling the sign convention for θ, and noting that integration over the upper surface goes from s_1 to s_2 on the inclined surface and from s_2 to s_3 on the base, whereas integration over the bottom surface goes from s_1 to s_4 on the inclined surface and from s_4 to s_3 on the base, we find that the above integrals become

$$\int_{LE}^{TE} -p_u \sin \theta \, ds_u = \int_{s_1}^{s_2} -(1.31 \times 10^5) \sin(-5°) \, ds_u$$

$$+ \int_{s_2}^{s_3} -(1.01 \times 10^5) \sin 90° \, ds_u$$

$$= 1.142 \times 10^4 (s_2 - s_1) - 1.01 \times 10^5 (s_2 - s_3)$$

$$= 1.142 \times 10^4 \left(\frac{c}{\cos 5°} \right) - 1.01 \times 10^5 (c)(\tan 5°)$$

$$= 1.142 \times 10^4 (2.008) - 1.01 \times 10^5 (0.175) = 5260 \text{ N}$$

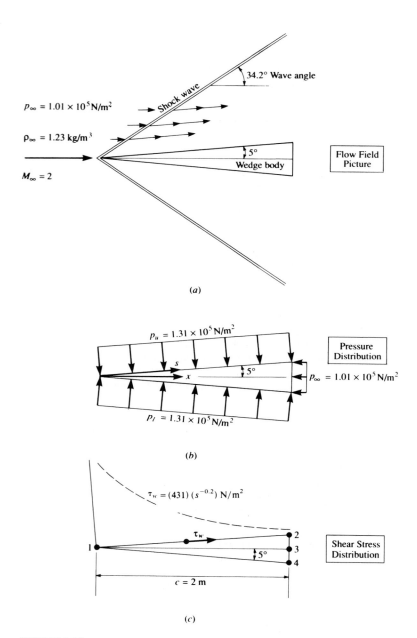

$p_\infty = 1.01 \times 10^5\,\text{N/m}^2$

$\rho_\infty = 1.23\,\text{kg/m}^3$

$M_\infty = 2$

Shock wave

34.2° Wave angle

5°

Wedge body

Flow Field Picture

(a)

$p_u = 1.31 \times 10^5\,\text{N/m}^2$

s

x

5°

$p_\infty = 1.01 \times 10^5\,\text{N/m}^2$

$p_l = 1.31 \times 10^5\,\text{N/m}^2$

Pressure Distribution

(b)

$\tau_w = (431)\,(s^{-0.2})\,\text{N/m}^2$

τ_w

1

2
3
4

5°

$c = 2\,\text{m}$

Shear Stress Distribution

(c)

FIGURE 1.16
Illustration for Example 1.1.

$$\int_{LE}^{TE} p_l \sin\theta \, ds_l = \int_{s_1}^{s_4} (1.31 \times 10^5) \sin(5°) \, ds_l + \int_{s_4}^{s_3} (1.01 \times 10^5) \sin(-90°) \, ds_l$$

$$= 1.142 \times 10^4 (s_4 - s_1) + 1.01 \times 10^5 (-1)(s_2 - s_3)$$

$$= 1.142 \times 10^4 \left(\frac{c}{\cos 5°}\right) - 1.01 \times 10^5 (c)(\tan 5°)$$

$$= 2.293 \times 10^4 - 1.767 \times 10^4 = 5260 \text{ N}$$

Note that the integrals of the pressure over the top and bottom surfaces, respectively, yield the same contribution to the drag—a result to be expected from the symmetry of the configuration in Fig. 1.16:

$$\int_{LE}^{TE} \tau_u \cos\theta \, ds_u = \int_{s_1}^{s_2} 431 s^{-0.2} \cos(-5°) \, ds_u$$

$$= 429 \left(\frac{s_2^{0.8} - s_1^{0.8}}{0.8}\right)$$

$$= 429 \left(\frac{c}{\cos 5°}\right)^{0.8} \frac{1}{0.8} = 936.5 \text{ N}$$

$$\int_{LE}^{TE} \tau_l \cos\theta \, ds_l = \int_{s_1}^{s_4} 431 s^{-0.2} \cos(5°) \, ds_l$$

$$= 429 \left(\frac{s_4^{0.8} - s_1^{0.8}}{0.8}\right)$$

$$= 429 \left(\frac{c}{\cos 5°}\right)^{0.8} \frac{1}{0.8} = 936.5 \text{ N}$$

Again, it is no surprise that the shear stress acting over the upper and lower surfaces, respectively, give equal contributions to the drag; this is to be expected due to the symmetry of the wedge shown in Fig. 1.16. Adding the pressure integrals, and then adding the shear stress integrals, we have for total drag

$$D' = \underbrace{1.052 \times 10^4}_{\substack{\text{pressure} \\ \text{drag}}} + \underbrace{0.1873 \times 10^4}_{\substack{\text{skin friction} \\ \text{drag}}} = \boxed{1.24 \times 10^4 \text{ N}}$$

Note that, for this rather slender body, but at a supersonic speed, most of the drag is pressure drag. Referring to Fig. 1.16a, we see that this is due to the presence of an oblique shock wave from the nose of the body, which acts to create pressure drag (sometimes called "wave drag"). In this example, only 15 percent of the drag is skin friction drag; the other 85 percent is the pressure drag (wave drag). This is typical of the drag of slender supersonic bodies. In contrast, as we will see later, the drag of a slender body at subsonic speed, where there is no shock wave, is mainly skin friction drag.

The drag coefficient is obtained as follows. The velocity of the freestream is twice the sonic speed, which is given by

$$a_\infty = \sqrt{\gamma R T_\infty} = \sqrt{(1.4)(287)(288)} = 340.2 \text{ m/s}$$

(See Chap. 8 for a derivation of this expression for the speed of sound.) Note that, in the above, the standard sea level temperature of 288 K is used. Hence, $V_\infty = 2(340.2) = 680.4$ m/s. Thus,

$$q_\infty = \tfrac{1}{2}\rho_\infty V_\infty^2 = (0.5)(1.23)(680.4)^2 = 2.847 \times 10^5 \text{ N/m}^2$$

Also,

$$S = c(1) = 2.0 \text{ m}^2$$

Hence,

$$c_d = \frac{D'}{q_\infty S} = \frac{1.24 \times 10^4}{(2.847 \times 10^5)(2)} = \boxed{0.022}$$

An alternate solution to this problem is to use Eq. (1.16), integrating the pressure coefficients and skin friction coefficients to obtain directly the drag coefficient. We proceed as follows:

$$C_{p,u} = \frac{p_u - p_\infty}{q_\infty} = \frac{1.31 \times 10^5 - 1.01 \times 10^5}{2.847 \times 10^5} = 0.1054$$

On the lower surface, we have the same value for C_p, i.e.,

$$C_{p,l} = C_{p,u} = 0.1054$$

Also,

$$c_{f,u} = \frac{\tau_w}{q_\infty} = \frac{431 s^{-0.2}}{q_\infty} = \frac{431}{2.847 \times 10^5}\left(\frac{x}{\cos 5°}\right)^{-0.2} = 1.513 \times 10^{-3} x^{-0.2}$$

On the lower surface, we have the same value for c_f, i.e.,

$$c_{f,l} = 1.513 \times 10^{-3} x^{-0.2}$$

Also,

$$\frac{dy_u}{dx} = \tan 5° = 0.0875$$

and

$$\frac{dy_l}{dx} = -\tan 5° = -0.0875$$

Inserting the above information into Eq. (1.16), we have

$$c_d = c_a = \frac{1}{c}\int_0^c \left(C_{p,u}\frac{dy_u}{dx} - C_{p,l}\frac{dy_l}{dx}\right) dx + \frac{1}{c}\int_0^c (C_{f,u} + C_{f,l}) \, dx$$

$$= \frac{1}{2}\int_0^2 (0.1054)(0.0875) - (0.1054)(-0.0875) \, dx$$

$$+ \frac{1}{2}\int_0^2 2(1.513 \times 10^{-3})x^{-0.2} \, dx$$

$$= 0.009223 x\big|_0^2 + 0.00189 x^{0.8}\big|_0^2$$

$$= 0.01845 + 0.00329 = \boxed{0.022}$$

This is the same result as obtained earlier.

Example 1.2. Consider a cone at zero angle of attack in a hypersonic flow. (Hypersonic flow is very high-speed flow, generally defined as any flow above a Mach number of 5; hypersonic flow is further defined in Sec. 1.10.) The half-angle of the cone is θ_c, as shown in Fig. 1.17. An approximate expression for the pressure coefficient on the surface of a hypersonic body is given by the newtonian sine-squared law (to be derived in Chap. 14):

$$C_p = 2 \sin^2 \theta_c$$

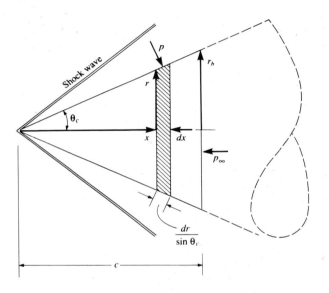

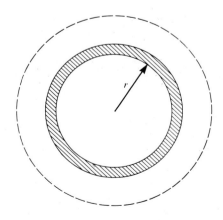

FIGURE 1.17
Illustration for Example 1.2.

Note that C_p, hence, p, is constant along the inclined surface of the cone. Along the base of the body, we assume that $p = p_\infty$. Neglecting the effect of friction, obtain an expression for the drag coefficient of the cone, where C_D is based on the area of the base, S_b.

Solution. We cannot use Eqs. (1.15) to (1.17) here. These equations are expressed for a two-dimensional body, such as the airfoil shown in Fig. 1.15, whereas the cone in Fig. 1.17 is a shape in three-dimensional space. Hence, we must treat this three-dimensional body as follows. From Fig. 1.17, the drag force on the shaded strip of surface area is

$$(p \sin \theta_c)(2\pi r) \frac{dr}{\sin \theta_c} = 2\pi r p \, dr$$

The total drag due to the pressure acting over the total surface area of the cone is

$$D = \int_0^{r_b} 2\pi r p \, dr - \int_0^{r_b} 2\pi r p_\infty \, dr$$

The first integral is the horizontal force on the inclined surface of the cone, and the second integral is the force on the base of the cone. Combining the integrals, we have

$$D = \int_0^{r_b} 2\pi r (p - p_\infty) \, dr = \pi (p - p_\infty) r_b^2$$

Referenced to the base area, πr_b^2, the drag coefficient is

$$C_D = \frac{D}{q_\infty \pi r_b^2} = \frac{\pi r_b^2 (p - p_\infty)}{\pi r_b^2 q_\infty} = C_p$$

(*Note*: The drag coefficient for a cone is equal to its surface pressure coefficient.) Hence, using the newtonian sine-squared law, we obtain

$$\boxed{C_D = 2 \sin^2 \theta_c}$$

1.6   CENTER OF PRESSURE

From Eqs. (1.7) and (1.8), we see that the normal and axial forces on the body are due to the *distributed* loads imposed by the pressure and shear stress distributions. Moreover, these distributed loads generate a moment about the leading edge, as given by Eq. (1.11). *Question*: If the aerodynamic force on a body is specified in terms of a resultant single force, R, or its components such as N and A, *where* on the body should this resultant be placed? The answer is that the resultant force should be located on the body such that it produces the same effect as the distributed loads. For example, the distributed load on a two-dimensional body such as an airfoil produces a moment about the leading edge given by Eq. (1.11); therefore, N' and A' must be placed on the airfoil at such a location to generate the same moment about the leading edge. If A' is placed on the chord line as shown in Fig. 1.18, then N' must be located a distance x_{cp} downstream of the leading edge such that

$$M'_{LE} = -(x_{cp})N'$$

$$\boxed{x_{cp} = -\frac{M'_{LE}}{N'}} \tag{1.20}$$

In Fig. 1.18, the direction of the curled arrow illustrating M'_{LE} is drawn in the positive (pitch-up) sense. (From Sec. 1.5, recall the standard convention that aerodynamic moments are positive if they tend to increase the angle of attack.) Examining Fig. 1.18, we see that a positive N' creates a negative (pitch-down) moment about the leading edge. This is consistent with the negative sign in Eq. (1.20). Therefore, in Fig. 1.18, the actual moment about the leading edge is negative, and hence is in a direction opposite to the curled arrow shown.

In Fig. 1.18 and Eq. (1.20), x_{cp} is defined as the *center of pressure*. It is the location where the resultant of a distributed load effectively acts on the body. If moments were taken about the center of pressure, the integrated effect of the distributed loads would be zero. Hence, an alternate definition of the center of pressure is that point on the body about which the aerodynamic moment is zero.

In cases where the angle of attack of the body is small, $\sin \alpha \approx 0$ and $\cos \alpha \approx 1$; hence, from Eq. (1.1), $L' \approx N'$. Thus, Eq. (1.20) becomes

$$x_{cp} \approx -\frac{M'_{LE}}{L'} \tag{1.21}$$

Examine Eqs. (1.20) and (1.21). As N' and L' decrease, x_{cp} increases. As the forces approach zero, the center of pressure moves to infinity. For this reason, the center of pressure is not always a convenient concept in aerodynamics. However, this is no problem. To define the force-and-moment system due to a distributed load on a body, the resultant force can be placed at *any* point on the body, as long as the value of the moment about that point is also given. For example, Fig. 1.19 illustrates three equivalent ways of specifying the force-and-moment system on an airfoil. In the left figure, the resultant is placed at the leading edge, with a finite value of M'_{LE}. In the middle figure, the resultant is placed at the quarter-chord point, with a finite value of $M'_{c/4}$. In the right figure,

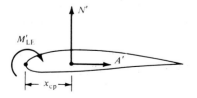

FIGURE 1.18
Center of pressure for an airfoil.

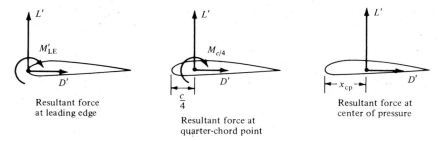

FIGURE 1.19
Equivalent ways of specifying the force-and-moment system on an airfoil.

the resultant is placed at the center of pressure, with a zero moment about that point. By inspection of Fig. 1.19, the quantitative relation between these cases is

$$M'_{LE} = -\frac{c}{4} L' + M'_{c/4} = -x_{cp} L' \tag{1.22}$$

Example 1.3. In low-speed, incompressible flow, the following experimental data are obtained for an NACA 4412 airfoil section at an angle of attack of 4°: $c_l = 0.85$ and $c_{m,c/4} = -0.09$. Calculate the location of the center of pressure.

Solution. From Eq. (1.22),

$$x_{cp} = \frac{c}{4} - \frac{M'_{c/4}}{L'}$$

$$\frac{x_{cp}}{c} = \frac{1}{4} - \frac{(M_{c/4}/q_\infty c^2)}{(L'/q_\infty c)} = \frac{1}{4} - \frac{c_{m,c/4}}{c_l}$$

$$= \frac{1}{4} - \frac{(-0.09)}{0.85} = \boxed{0.356}$$

(*Note:* In Chap. 4, we will learn that, for a thin, symmetric airfoil, the center of pressure is at the quarter-chord location. However, for the NACA 4412 airfoil, which is not symmetric, the center-of-pressure location is behind the quarter-chord point.)

1.7 DIMENSIONAL ANALYSIS: THE BUCKINGHAM PI THEOREM

The aerodynamic forces and moments on a body, and the corresponding force and moment coefficients, have been defined and discussed in Sec. 1.5. *Question:* What physical quantities determine the variation of these forces and moments? The answer can be found from the powerful method of *dimensional analysis*, which is introduced in this section.†

---

† For a more elementary treatment of dimensional analysis, see Chap. 5 of Ref. 2.

Consider a body of given shape at a given angle of attack, e.g., the airfoil sketched in Fig. 1.10. The resultant aerodynamic force is R. On a physical, intuitive basis, we expect R to depend on:

1. Freestream velocity V_∞.
2. Freestream density ρ_∞.
3. Viscosity of the fluid. We have seen that shear stress τ contributes to the aerodynamic forces and moments. In turn, in Chap. 15, we will see that τ is proportional to the velocity gradients in the flow. For example, if the velocity gradient is given by $\partial u / \partial y$, then $\tau = \mu \, \partial u / \partial y$. The constant of proportionality is the viscosity coefficient μ. Hence, let us represent the influence of viscosity on aerodynamic forces and moments by the freestream viscosity coefficient μ_∞.
4. The size of the body, represented by some chosen reference length. In Fig. 1.10, the convenient reference length is the chord length c.
5. The compressibility of the fluid. The technical definition of compressibility is given in Chap. 7. For our present purposes, let us just say that compressibility is related to the *variation* of density throughout the flow field, and certainly the aerodynamic forces and moments should be sensitive to any such variation. In turn, compressibility is related to the speed of sound, a, in the fluid, as shown in Chap. 8.† Therefore, let us represent the influence of compressibility on aerodynamic forces and moments by the freestream speed of sound, a_∞.

In light of the above, and without any a priori knowledge about the variation of R, we can use common sense to write

$$R = f(\rho_\infty, V_\infty, c, \mu_\infty, a_\infty) \tag{1.23}$$

Equation (1.23) is a general functional relation, and as such is not very practical for the direct calculation of R. In principle, we could mount the given body in a wind tunnel, incline it at the given angle of attack, and then systematically measure the variation of R due to variations of ρ_∞, V_∞, c, μ_∞, and a_∞, taken one at a time. By cross-plotting the vast bulk of data thus obtained, we might be able to extract a precise functional relation for Eq. (1.23). But it would be hard work, and would certainly be costly in terms of a huge amount of required wind-tunnel time. Fortunately, we can simplify the problem and considerably reduce our time and effort by first employing the method of dimensional analysis. This method will define a set of dimensionless parameters which governs the aerodynamic forces and moments; this set will considerably reduce the number of independent variables as presently occurs in Eq. (1.23).

---

† Common experience tells us that sound waves propagate through air at some finite velocity, much slower than the speed of light; you see a flash of lightning in the distance, and hear the thunder moments later. The speed of sound is an important physical quantity in aerodynamics and is discussed in detail in Sec. 8.3.

Dimensional analysis is based on the obvious fact that in an equation dealing with the real physical world, each term must have the same dimensions. For example, if

$$\psi + \eta + \zeta = \phi$$

is a physical relation, then ψ, η, ζ, and ϕ must have the same dimensions. Otherwise we would be adding apples and oranges. The above equation can be made dimensionless by dividing by any one of the terms, say, ϕ:

$$\frac{\psi}{\phi} + \frac{\eta}{\phi} + \frac{\zeta}{\phi} = 1$$

These ideas are formally embodied in the Buckingham pi theorem, stated below without derivation. (See Ref. 3, pages 21–28, for such a derivation.)

Buckingham pi theorem. Let K equal the number of fundamental dimensions required to describe the physical variables. (In mechanics, all physical variables can be expressed in terms of the dimensions of *mass*, *length*, and *time*; hence, $K = 3$.) Let $P_1, P_2, \ldots, P_N$ represent N physical variables in the physical relation

$$f_1(P_1, P_2, \ldots, P_N) = 0 \qquad (1.24)$$

Then, the physical relation Eq. (1.24) may be reexpressed as a relation of $(N - K)$ dimensionless products (called Π products),

$$f_2(\Pi_1, \Pi_2, \ldots, \Pi_{N-K}) = 0 \qquad (1.25)$$

where each Π product is a dimensionless product of a set of K physical variables plus one other physical variable. Let $P_1, P_2, \ldots, P_K$ be the selected set of K physical variables. Then

$$\Pi_1 = f_3(P_1, P_2, \ldots, P_K, P_{K+1})$$

$$\Pi_2 = f_4(P_1, P_2, \ldots, P_K, P_{K+2}) \qquad (1.26)$$

$$\cdots\cdots\cdots\cdots\cdots\cdots\cdots\cdots\cdots\cdots$$

$$\Pi_{N-K} = f_5(P_1, P_2, \ldots, P_K, P_N)$$

The choice of the repeating variables, $P_1, P_2, \ldots, P_K$ should be such that they include *all* the K dimensions used in the problem. Also, the dependent variable [such as R in Eq. (1.23)] should appear in only one of the Π products.

Returning to our consideration of the aerodynamic force on a given body at a given angle of attack, Eq. (1.23) can be written in the form of Eq. (1.24):

$$g(R, \rho_\infty, V_\infty, c, \mu_\infty, a_\infty) = 0 \qquad (1.27)$$

Following the Buckingham pi theorem, the fundamental dimensions are

$$m = \text{dimensions of mass}$$

$$l = \text{dimension of length}$$

$$t = \text{dimension of time}$$

Hence, $K = 3$. The physical variables and their dimensions are

$$[R] = mlt^{-2}$$
$$[\rho_\infty] = ml^{-3}$$
$$[V_\infty] = lt^{-1}$$
$$[c] = l$$
$$[\mu_\infty] = ml^{-1}t^{-1}$$
$$[a_\infty] = lt^{-1}$$

Hence, $N = 6$. In the above, the dimensions of the force R are obtained from Newton's second law, force = mass × acceleration; hence, $[R] = mlt^{-2}$. The dimensions of μ_∞ are obtained from its definition, e.g., $\mu = \tau/(\partial u/\partial y)$, and from Newton's second law. (Show for yourself that $[\mu_\infty] = ml^{-1}t^{-1}$.) Choose ρ_∞, V_∞, and c as the arbitrarily selected set of K physical variables. Then Eq. (1.27) can be reexpressed in terms of $N - K = 6 - 3 = 3$ dimensionless Π products in the form of Eq. (1.25):

$$f_2(\Pi_1, \Pi_2, \Pi_3) = 0 \tag{1.28}$$

From Eq. (1.26), these Π products are

$$\Pi_1 = f_3(\rho_\infty, V_\infty, c, R) \tag{1.29a}$$
$$\Pi_2 = f_4(\rho_\infty, V_\infty, c, \mu_\infty) \tag{1.29b}$$
$$\Pi_3 = f_5(\rho_\infty, V_\infty, c, a_\infty) \tag{1.29c}$$

For the time being, concentrate on Π_1 from Eq. (1.29a). Assume that

$$\Pi_1 = \rho_\infty^d V_\infty^b c^e R \tag{1.30}$$

where d, b, and e are exponents to be found. In dimensional terms, Eq. (1.30) is

$$[\Pi_1] = (ml^{-3})^d (lt^{-1})^b (l)^e (mlt^{-2}) \tag{1.31}$$

Because Π_1 is dimensionless, the right side of Eq. (1.31) must also be dimensionless. This means that the exponents of m must add to zero, and similarly for the exponents of l and t. Hence,

For m: $\qquad\qquad\qquad d + 1 = 0$

For l: $\qquad\qquad\qquad -3d + b + e + 1 = 0$

For t: $\qquad\qquad\qquad -b - 2 = 0$

Solving the above equations, we find that $d = -1$, $b = -2$, and $e = -2$. Substituting these values into Eq. (1.30), we have

$$\Pi_1 = R\rho_\infty^{-1} V_\infty^{-2} c^{-2}$$

$$= \frac{R}{\rho_\infty V_\infty^2 c^2} \tag{1.32}$$

The quantity $R/\rho_\infty V_\infty^2 c^2$ is a dimensionless parameter in which c^2 has the dimensions of an area. We can replace c^2 with any reference area we wish (such as the planform area of a wing, S), and Π_1 will still be dimensionless. Moreover, we can multiply Π_1 by a pure number, and it will still be dimensionless. Thus, from Eq. (1.32), Π_1 can be redefined as

$$\Pi_1 = \frac{R}{\frac{1}{2}\rho_\infty V_\infty^2 S} = \frac{R}{q_\infty S} \tag{1.33}$$

Hence, Π_1 is a force coefficient, C_R, as defined in Sec. 1.5. In Eq. (1.33), S is a reference area germane to the given body shape.

The remaining Π products can be found as follows. From Eq. (1.29b), assume

$$\Pi_2 = \rho_\infty V_\infty^h c^i \mu^j \tag{1.34}$$

Paralleling the above analysis, we obtain

$$[\Pi_2] = (ml^{-3})(lt^{-1})^h (l)^i (ml^{-1}t^{-1})^j$$

Hence,

For m: $\qquad\qquad\qquad\qquad 1 + j = 0$

For l: $\qquad\qquad\qquad\qquad -3 + h + i - j = 0$

For t: $\qquad\qquad\qquad\qquad -h - j = 0$

Thus, $j = -1$, $h = 1$, and $i = 1$. Substitution into Eq. (1.34) gives

$$\Pi_2 = \frac{\rho_\infty V_\infty c}{\mu_\infty} \tag{1.35}$$

The dimensionless combination in Eq. (1.35) is defined as the freestream *Reynolds number* $\mathrm{Re} = \rho_\infty V_\infty c / \mu_\infty$. The Reynolds number is physically a measure of the ratio of inertia forces to viscous forces in a flow and is one of the most powerful parameters in fluid dynamics. Its importance is emphasized in Chaps. 15 and 16.

Returning to Eq. (1.29c), assume

$$\Pi_3 = V_\infty \rho_\infty^k c^r a_\infty^s \tag{1.36}$$

$$[\Pi_3] = (lt^{-1})(ml^{-3})^k (l)^r (lt^{-1})^s$$

For m: $\qquad\qquad\qquad\qquad k = 0$

For l: $\qquad\qquad\qquad\qquad 1 - 3k + r + s = 0$

For t: $\qquad\qquad\qquad\qquad -1 - s = 0$

Hence, $k = 0$, $s = -1$, and $r = 0$. Substituting into Eq. (1.36), we have

$$\Pi_3 = \frac{V_\infty}{a_\infty} \tag{1.37}$$

The dimensionless combination in Eq. (1.37) is defined as the freestream *Mach number* $M = V_\infty/a_\infty$. The Mach number is the ratio of the flow velocity to the speed of sound; it is a powerful parameter in the study of gas dynamics. Its importance is emphasized in subsequent chapters.

The results of our dimensional analysis may be organized as follows. Inserting Eqs. (1.33), (1.35), and (1.37) into (1.28), we have

$$f_2\left(\frac{R}{\frac{1}{2}\rho_\infty V_\infty^2 S}, \frac{\rho_\infty V_\infty c}{\mu_\infty}, \frac{V_\infty}{a_\infty}\right) = 0$$

or

$$f_2(C_R, \text{Re}, M_\infty) = 0$$

or

$$\boxed{C_R = f_6(\text{Re}, M_\infty)} \qquad (1.38)$$

This is an important result! Compare Eqs. (1.23) and (1.38). In Eq. (1.23), R is expressed as a general function of five independent variables. However, our dimensional analysis has shown that:

1. R can be expressed in terms of a dimensionless force coefficient, $C_R = R/\frac{1}{2}\rho_\infty V_\infty^2 S$.
2. C_R is a function of only Re and M_∞, from Eq. (1.38).

Therefore, by using the Buckingham pi theorem, we have reduced the number of independent variables from five in Eq. (1.23) to two in Eq. (1.38). Now, if we wish to run a series of wind-tunnel tests for a given body at a given angle of attack, we need only to vary the Reynolds and Mach numbers in order to obtain data for the direct formulation of R through Eq. (1.38). With a small amount of analysis, we have saved a huge amount of effort and wind-tunnel time. More importantly, we have defined two dimensionless parameters, Re and M_∞, which govern the flow. They are called *similarity parameters*, for reasons to be discussed in the following section. Other similarity parameters are introduced as our aerodynamic discussions progress.

Since the lift and drag are components of the resultant force, corollaries to Eq. (1.38) are

$$C_L = f_7(\text{Re}, M_\infty) \qquad (1.39)$$

$$C_D = f_8(\text{Re}, M_\infty) \qquad (1.40)$$

Moreover, a relation similar to Eq. (1.23) holds for the aerodynamic moments, and dimensional analysis yields

$$C_M = f_9(\text{Re}, M_\infty) \qquad (1.41)$$

Keep in mind that the above analysis was for a given body shape at a given angle of attack, α. If α is allowed to vary, then C_L, C_D, and C_M will in general depend on the value of α. Hence, Eqs. (1.39) to (1.41) can be generalized to

$$C_L = f_{10}(\text{Re}, M_\infty, \alpha) \tag{1.42}$$

$$C_D = f_{11}(\text{Re}, M_\infty, \alpha) \tag{1.43}$$

$$C_M = f_{12}(\text{Re}, M_\infty, \alpha) \tag{1.44}$$

Equations (1.42) to (1.44) assume a given body shape. Much of theoretical and experimental aerodynamics is focused on obtaining explicit expressions for Eqs. (1.42) to (1.44) for specific body shapes. This is one of the practical applications of aerodynamics mentioned in Sec. 1.2, and it is one of the major thrusts of this book.

For mechanical problems that also involve thermodynamics and heat transfer, the temperature, specific heat, and thermal conductivity of the fluid, as well as the temperature of the body surface (wall temperature), must be added to the list of physical variables, and the unit of temperature (say, kelvin or degree Rankine) must be added to the list of fundamental dimensions. For such cases, dimensional analysis yields additional dimensionless products such as heat transfer coefficients, and additional similarity parameters such as the ratio of specific heat at constant pressure to that at constant volume c_p/c_v, the ratio of wall temperature to freestream temperature T_w/T_∞, and the Prandtl number $\text{Pr} = \mu_\infty c_p/k_\infty$, where k_∞ is the thermal conductivity of the freestream.† Thermodynamics is essential to the study of compressible flow (Chaps. 7 to 14), and heat transfer is part of the study of viscous flow (Chaps. 15 to 17). Hence, these additional similarity parameters will be emphasized when they appear logically in our subsequent discussions. For the time being, however, the Mach and Reynolds numbers will suffice as the dominant similarity parameters for our present considerations.

1.8 FLOW SIMILARITY

Consider two different flow fields over two different bodies. By definition, different flows are *dynamically similar* if:

---

† The *specific heat* of a fluid is defined as the amount of heat added to a system, δq, per unit increase in temperature; $c_v = \delta q/dT$ if δq is added at constant volume, and similarly, for c_p if δq is added at constant pressure. Specific heats are discussed in detail in Sec. 7.2. The thermal conductivity relates heat flux to temperature gradients in the fluid. For example, if $\dot{q}_x$ is the heat transferred in the x direction per second per unit area and $\partial T/\partial x$ is the temperature gradient in the x direction, then thermal conductivity k is defined by $\dot{q}_x = -k(\partial T/\partial x)$. Thermal conductivity is discussed in detail in Sec. 15.3.

1. The streamline patterns are geometrically similar.
2. The distributions of V/V_∞, p/p_∞, T/T_∞, etc., throughout the flow field are the same when plotted against common nondimensional coordinates.
3. The force coefficients are the same.

Actually, item 3 is a consequence of item 2; if the nondimensional pressure and shear stress distributions over different bodies are the same, then the non-dimensional force coefficients will be the same.

The definition of dynamic similarity was given above. *Question*: What are the *criteria* to ensure that two flows are dynamically similar? The answer comes from the results of the dimensional analysis in Sec. 1.7. Two flows will be dynamically similar if:

1. The bodies and any other solid boundaries are geometrically similar for both flows.
2. The similarity parameters are the same for both flows.

So far, we have emphasized two parameters, Re and M_∞. For many aerodynamic applications, these are by far the dominant similarity parameters. Therefore, in a limited sense, but applicable to many problems, we can say that flows over geometrically similar bodies at the same Mach and Reynolds numbers are dynamically similar, and hence the lift, drag, and moment coefficients will be identical for the bodies. This is a key point in the validity of wind-tunnel testing. If a scale model of a flight vehicle is tested in a wind tunnel, the measured lift, drag, and moment coefficients will be the same as for free flight as long as the Mach and Reynolds numbers of the wind-tunnel test-section flow are the same as for the free-flight case. As we will see in subsequent chapters, this statement is not quite precise because there are other similarity parameters that influence the flow. In addition, differences in freestream turbulence between the wind tunnel and free flight can have an important effect on C_D and the maximum value of C_L. However, direct simulation of the free-flight Re and M_∞ is the primary goal of many wind-tunnel tests.

Example 1.4. Consider the flow over two circular cylinders, one having four times the diameter of the other, as shown in Fig. 1.20. The flow over the smaller cylinder has a freestream density, velocity, and temperature given by ρ_1, V_1, and T_1, respectively. The flow over the larger cylinder has a freestream density, velocity, and temperature given by ρ_2, V_2, and T_2, respectively, where $\rho_2 = \rho_1/4$, $V_2 = 2V_1$, and $T_2 = 4T_1$. Assume that both μ and a are proportional to $T^{1/2}$. Show that the two flows are dynamically similar.

Solution. Since $\mu \propto \sqrt{T}$ and $a \propto \sqrt{T}$, then

$$\frac{\mu_2}{\mu_1} = \sqrt{\frac{T_2}{T_1}} = \sqrt{\frac{4T_1}{T_1}} = 2$$

and

$$\frac{a_2}{a_1} = \sqrt{\frac{T_2}{T_1}} = 2$$

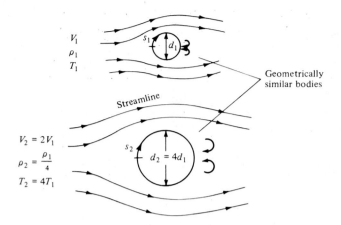

FIGURE 1.20
Example of dynamic flow similarity. Note that as part of the definition of dynamic similarity, the streamlines (lines along which the flow velocity is tangent at each point) are geometrically similar between the two flows.

By definition,

$$M_1 = \frac{V_1}{a_1}$$

and

$$M_2 = \frac{V_2}{a_2} = \frac{2V_1}{2a_1} = \frac{V_1}{a_1} = M_1$$

Hence, the Mach numbers are the same. Basing the Reynolds number on the diameter, d, of the cylinder, we have by definition,

$$\mathrm{Re}_1 = \frac{\rho_1 V_1 d_1}{\mu_1}$$

and

$$\mathrm{Re}_2 = \frac{\rho_2 V_2 d_2}{\mu_2} = \frac{(\rho_1/4)(2V_1)(4d_1)}{2\mu_1} = \frac{\rho_1 V_1 d_1}{\mu_1} = \mathrm{Re}_1$$

Hence, the Reynolds numbers are the same. Since the two bodies are geometrically similar and M_∞ and Re are the same, we have satisfied all the criteria; the two flows are dynamically similar. In turn, as a consequence of being similar flows, we know from the definition that:

1. The streamline patterns around the two cylinders are geometrically similar.
2. The nondimensional pressure, temperature, density, velocity, etc., distributions are the same around the two cylinders. This is shown schematically in Fig. 1.21, where the nondimensional pressure distribution p/p_∞ is shown as a function of the nondimensional surface distance, s/d. It is the same curve for both bodies.
3. The drag coefficients for the two bodies are the same. Here, $C_D = D/q_\infty S$, where $S = \pi d^2/4$. As a result of the flow similarity, $C_{D1} = C_{D2}$. (*Note*: Examining Fig. 1.20, we see that the lift on the cylinders is zero because the flow is symmetrical

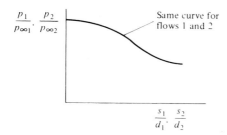

FIGURE 1.21
One aspect of the definition of dynamically similar flows. The nondimensional flow variable distributions are the same.

about the horizontal axis through the center of the cylinder. The pressure distribution over the top is the same as over the bottom, and they cancel each other in the vertical direction. Therefore, drag is the only aerodynamic force on the body.)

Example 1.5. Consider a Boeing 747 airliner cruising at a velocity of 550 mi/h at a standard altitude of 38,000 ft, where the freestream pressure and temperature are 432.6 lb/ft^2 and 390°R, respectively. A one-fiftieth scale model of the 747 is tested in a wind tunnel where the temperature is 430°R. Calculate the required velocity and pressure of the test airstream in the wind tunnel such that the lift and drag coefficients measured for the wind-tunnel model are the same as for free flight. Assume that both μ and a are proportional to $T^{1/2}$.

Solution. Let subscripts 1 and 2 denote the free-flight and wind-tunnel conditions, respectively. For $C_{L,1} = C_{L,2}$ and $C_{D,1} = C_{D,2}$, the wind-tunnel flow must be dynamically similar to free flight. For this to hold, $M_1 = M_2$ and $\text{Re}_1 = \text{Re}_2$:

$$M_1 = \frac{V_1}{a_1} \propto \frac{V_1}{\sqrt{T_1}}$$

and

$$M_2 = \frac{V_2}{a_2} \propto \frac{V_2}{\sqrt{T_2}}$$

Hence,

$$\frac{V_2}{\sqrt{T_2}} = \frac{V_1}{\sqrt{T_1}}$$

or

$$V_2 = V_1 \sqrt{\frac{T_2}{T_1}} = 550 \sqrt{\frac{430}{390}} = \boxed{577.5 \text{ mi/h}}$$

$$\text{Re}_1 = \frac{\rho_1 V_1 c_1}{\mu_1} \propto \frac{\rho_1 V_1 c_1}{\sqrt{T_1}}$$

and

$$\text{Re}_2 = \frac{\rho_2 V_2 c_2}{\mu_2} \propto \frac{\rho_2 V_2 c_2}{\sqrt{T_2}}$$

Hence,

$$\frac{\rho_1 V_1 c_1}{\sqrt{T_1}} = \frac{\rho_2 V_2 c_2}{\sqrt{T_2}}$$

or

$$\frac{\rho_2}{\rho_1} = \left(\frac{V_1}{V_2}\right)\left(\frac{c_1}{c_2}\right)\sqrt{\frac{T_2}{T_1}}$$

However, since $M_1 = M_2$, then

$$\frac{V_1}{V_2} = \sqrt{\frac{T_1}{T_2}}$$

Thus,

$$\frac{\rho_2}{\rho_1} = \frac{c_1}{c_2} = 50$$

The equation of state for a perfect gas is $p = \rho R T$, where R is the specific gas constant. Thus,

$$\frac{p_2}{p_1} = \frac{\rho_2}{\rho_1}\frac{T_2}{T_1} = (50)\left(\frac{430}{390}\right) = 55.1$$

Hence,

$$p_2 = 55.1 p_1 = (55.1)(432.6) = \boxed{23{,}836 \text{ lb/ft}^2}$$

Since 1 atm $= 2116$ lb/ft^2, then $p_2 = 23{,}836/2116 = \boxed{11.26 \text{ atm}}$.

In Example 1.5, the wind-tunnel test stream must be pressurized far above atmospheric pressure in order to simulate the proper free-flight Reynolds number. However, most standard subsonic wind tunnels are not pressurized as such, because of the large extra financial cost involved. This illustrates a common difficulty in wind-tunnel testing, namely, the difficulty of simulating both Mach number and Reynolds number simultaneously in the same tunnel. It is interesting to note that the NACA (National Advisory Committee for Aeronautics, the predecessor of NASA) in 1922 began operating a pressurized wind tunnel at the NACA Langley Memorial Laboratory in Hampton, Virginia. This was a subsonic wind tunnel contained entirely inside a large tank pressurized to as high as 20 atm. Called the variable density tunnel (VDT), this facility was used in the 1920s and 1930s to provide essential data on the NACA family of airfoil sections at the high Reynolds numbers associated with free flight. A photograph of the NACA variable density tunnel is shown in Fig. 1.22; notice the heavy pressurized shell in which the wind tunnel is enclosed. A cross section of the VDT inside the pressure cell is shown in Fig. 1.23. These figures demonstrate the extreme measures sometimes taken in order to simulate simultaneously the free-flight values of the important similarity parameters in a wind tunnel. Today, for the most part, we do not attempt to simulate all the parameters simultaneously; rather, Mach number simulation is achieved in one wind tunnel, and Reynolds number simulation in another tunnel. The results from both tunnels are then analyzed and correlated to obtain reasonable values for C_L and C_D appropriate for free flight. In any event, this example serves to illustrate the difficulty of full free-flight simulation in a given wind tunnel and underscores the importance given to dynamically similar flows in experimental aerodynamics.

FIGURE 1.22
The NACA variable density tunnel (VDT). Authorized in March of 1921, the VDT was operational in October 1922 at the NACA Langley Memorial Laboratory at Hampton, Virginia. It is essentially a large, subsonic wind tunnel entirely contained within an 85-ton pressure shell, capable of 20 atm. This tunnel was instrumental in the development of the various families of NACA airfoil shapes in the 1920s and 1930s. In the early 1940s, it was decommissioned as a wind tunnel and used as a high-pressure air storage tank. In 1983, due to its age and outdated riveted construction, its use was discontinued altogether. Today, the VDT remains at the NASA Langley Research Center; it has been officially designated as a National Historic Landmark. (*Courtesy of NASA.*)

1.9 FLUID STATICS: BUOYANCY FORCE

In aerodynamics, we are concerned about fluids in motion, and the resulting forces and moments on bodies due to such motion. However, in this section, we consider the special case of *no* fluid motion, i.e., *fluid statics.* A body immersed in a fluid will still experience a force even if there is no relative motion between the body and the fluid. Let us see why.

To begin, we must first consider the force on an element of fluid itself. Consider a stagnant fluid above the xz plane, as shown in Fig. 1.24. The vertical direction is given by y. Consider an infinitesimally small fluid element with sides of length dx, dy, and dz. There are two types of forces acting on this fluid element: pressure forces from the surrounding fluid exerted on the surface of the element, and the gravity force due to the weight of the fluid inside the element. Consider forces in the y direction. The pressure on the bottom surface of the element is

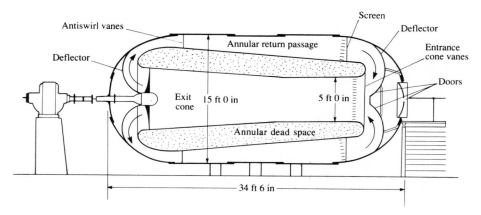

FIGURE 1.23
Schematic of the variable density tunnel. (*From Baals, D. D. and Corliss, W. R.*, Wind Tunnels of NASA, *NASA SP-440, 1981.*)

p, and hence the force on the bottom face is $p(dx\,dz)$ in the upward direction, as shown in Fig. 1.24. The pressure on the top surface of the element will be slightly different from the pressure on the bottom because the top surface is at a different location in the fluid. Let dp/dy denote the rate of change of p with respect to y. Then the pressure exerted on the top surface will be $p+(dp/dy)dy$, and the pressure force on the top of the element will be $[p+(dp/dy)\,dy](dx\,dz)$ in the downward direction, as shown in Fig. 1.24. Hence, letting upward force be positive, we have

$$\text{Net pressure force} = p(dx\,dz) - \left(p + \frac{dp}{dy}\,dy\right)(dx\,dz)$$

$$= -\frac{dp}{dy}(dx\,dy\,dz)$$

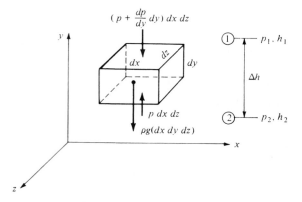

FIGURE 1.24
Forces on a fluid element in a stagnant fluid.

Let ρ be the mean density of the fluid element. The total mass of the element is $\rho(dx\,dy\,dz)$. Therefore,

$$\text{Gravity force} = -\rho(dx\,dy\,dz)g$$

where g is the acceleration of gravity. Since the fluid element is stationary (in equilibrium), the sum of the forces exerted on it must be zero:

$$-\frac{dp}{dy}(dx\,dy\,dz) - g\rho(dx\,dy\,dz) = 0$$

or

$$\boxed{dp = -g\rho\,dy} \qquad (1.45)$$

Equation (1.45) is called the *hydrostatic equation*; it is a differential equation which relates the change in pressure, dp, in a fluid with a change in vertical height, dy.

The net force on the element acts only in the vertical direction. The pressure forces on the front and back faces are equal and opposite and hence cancel; the same is true for the left and right faces. Also, the pressure forces shown in Fig. 1.24 act at the center of the top and bottom faces, and the center of gravity is at the center of the elemental volume (assuming the fluid is homogeneous); hence, the forces in Fig. 1.24 are colinear, and as a result, there is no moment on the element.

Equation (1.45) governs the variation of atmospheric properties as a function of altitude in the air above us. It is also used to estimate the properties of other planetary atmospheres such as for Venus, Mars, and Jupiter. The use of Eq. (1.45) in the analysis and calculation of the "standard atmosphere" is given in detail in Ref. 2; hence, it will not be repeated here.

Let the fluid be a liquid, for which we can assume ρ is constant. Consider points 1 and 2 separated by the vertical distance Δh as sketched on the right side of Fig. 1.24. The pressure and y locations at these points are p_1, h_1, and p_2, h_2, respectively. Integrating Eq. (1.45) between points 1 and 2, we have

$$\int_{p_1}^{p_2} dp = -\rho g \int_{h_1}^{h_2} dy$$

or

$$p_2 - p_1 = -\rho g(h_2 - h_1) = \rho g\,\Delta h \qquad (1.46)$$

where $\Delta h = h_1 - h_2$. Equation (1.46) can be more conveniently expressed as

$$p_2 + \rho g h_2 = p_1 + \rho g h_1$$

or

$$\boxed{p + \rho g h = \text{constant}} \qquad (1.47)$$

Note that in Eqs. (1.46) and (1.47), increasing values of h are in the positive (upward) y direction.

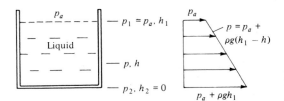

FIGURE 1.25
Hydrostatic pressure distribution on the walls of a container.

A simple application of Eq. (1.47) is the calculation of the pressure distribution on the walls of a container holding a liquid, and open to the atmosphere at the top. This is illustrated in Fig. 1.25, where the top of the liquid is at a height h_1. The atmospheric pressure p_a is impressed on the top of the liquid; hence, the pressure at h_1 is simply p_a. Applying Eq. (1.47) between the top (where $h = h_1$) and an arbitrary height h, we have

$$p + \rho g h = p_1 + \rho g h_1 = p_a + \rho g h_1$$

or
$$p = p_a + \rho g (h_1 - h) \tag{1.48}$$

Equation (1.48) gives the pressure distribution on the vertical sidewall of the container as a function of h. Note that the pressure is a *linear* function of h, as sketched on the right of Fig. 1.25, and that p increases with depth below the surface.

Another simple and very common application of Eq. (1.47) is the liquid-filled U-tube manometer used for measuring pressure differences, as sketched in Fig. 1.26. The manometer is usually made from hollow glass tubing bent in the shape of the letter U. Imagine that we have an aerodynamic body immersed in an

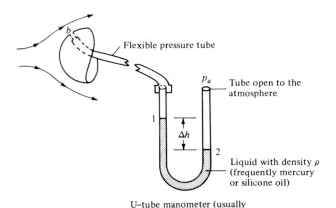

Flexible pressure tube

p_a Tube open to the atmosphere

Δh

Liquid with density ρ (frequently mercury or silicone oil)

U-tube manometer (usually made from glass tubing)

FIGURE 1.26
The use of a U-tube manometer.

airflow (such as in a wind tunnel), and we wish to use a manometer to measure the surface pressure at point b on the body. A small pressure orifice (hole) at point b is connected to one side of the manometer via a long (usually flexible) pressure tube. The other side of the manometer is open to the atmosphere, where the pressure p_a is a known value. The U tube is partially filled with a liquid of known density ρ. The tops of the liquid on the left and right sides of the U tube are at points 1 and 2, with heights h_1 and h_2, respectively. The body surface pressure p_b is transmitted through the pressure tube and impressed on the top of the liquid at point 1. The atmospheric pressure p_a is impressed on the top of the liquid at point 2. Because in general $p_b \neq p_a$, the tops of the liquid will be at different heights; i.e., the two sides of the manometer will show a displacement $\Delta h = h_1 - h_2$ of the fluid. We wish to obtain the value of the surface pressure at point b on the body by reading the value of Δh from the manometer. From Eq. (1.47) applied between points 1 and 2,

$$p_b + \rho g h_1 = p_a + \rho g h_2$$

or

$$p_b = p_a - \rho g (h_1 - h_2)$$

or

$$p_b = p_a - \rho g \, \Delta h \qquad (1.49)$$

In Eq. (1.49), p_a, ρ, and g are known, and Δh is read from the U tube, thus allowing p_b to be measured.

At the beginning of this section, we stated that a solid body immersed in a fluid will experience a force even if there is no relative motion between the body and the fluid. We are now in a position to derive an expression for this force, henceforth called the *buoyancy force*. We will consider a body immersed in either a stagnant gas or liquid, hence ρ can be a variable. For simplicity, consider a rectangular body of unit width, length l, and height $(h_1 - h_2)$, as shown in Fig. 1.27. Examining Fig. 1.27, we see that the vertical force F on the body due to the pressure distribution over the surface is

$$F = (p_2 - p_1)l(1) \qquad (1.50)$$

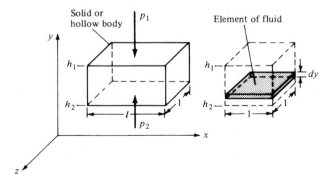

FIGURE 1.27
Source of the buoyancy force on a body immersed in a fluid.

There is no horizontal force because the pressure distributions over the vertical faces of the rectangular body lead to equal and opposite forces which cancel each other. In Eq. (1.50), an expression for $p_2 - p_1$ can be obtained by integrating the hydrostatic equation, Eq. (1.45), between the top and bottom faces:

$$p_2 - p_1 = \int_{p_1}^{p_2} dp = - \int_{h_1}^{h_2} \rho g \, dy = \int_{h_2}^{h_1} \rho g \, dy$$

Substituting this result into Eq. (1.50), we obtain for the buoyancy force

$$F = l(1) \int_{h_2}^{h_1} \rho g \, dy \tag{1.51}$$

Consider the physical meaning of the integral in Eq. (1.51). The weight of a small element of fluid of height dy and width and length of unity as shown at the right of Fig. 1.27 is $\rho g \, dy \, (1)(1)$. In turn, the weight of a column of fluid with a base of unit area and a height $(h_1 - h_2)$ is

$$\int_{h_2}^{h_1} \rho g \, dy$$

which is precisely the integral in Eq. (1.51). Moreover, if we place l of these fluid columns side by side, we would have a volume of fluid equal to the volume of the body on the left of Fig. 1.27, and the *weight* of this total volume of *fluid* would be

$$l \int_{h_2}^{h_1} \rho g \, dy$$

which is precisely the right-hand side of Eq. (1.51). Therefore, Eq. (1.51) states in words that

$$\frac{\text{Buoyancy force}}{\text{on body}} = \frac{\text{weight of fluid}}{\text{displaced by body}}$$

We have just proved the well-known *Archimedes principle,* first advanced by the Greek scientist, Archimedes of Syracuse (287–212 B.C.). Although we have used a rectangular body to simplify our derivation, the Archimedes principle holds for bodies of any general shape. (See Prob. 1.14 at the end of this chapter.) Also, note from our derivation that the Archimedes principle holds for both gases and liquids and does not require that the density be constant.

The density of liquids is usually several orders of magnitude larger than the density of gases; e.g., for water $\rho = 10^3$ kg/m^3, whereas for air $\rho = 1.23$ kg/m^3. Therefore, a given body will experience a buoyancy force a thousand times greater in water than in air. Obviously, for naval vehicles buoyancy force is all important, whereas for airplanes it is negligible. On the other hand, lighter-than-air vehicles, such as blimps and hot-air balloons, rely on buoyancy force for sustentation;

they obtain sufficient buoyancy force simply by displacing huge volumes of air. For most problems in aerodynamics, however, buoyancy force is so small that it can be readily neglected.

Example 1.6. A hot-air balloon with an inflated diameter of 30 ft is carrying a weight of 800 lb, which includes the weight of the hot air inside the balloon. Calculate (a) its upward acceleration at sea level the instant the restraining ropes are released and (b) the maximum altitude it can achieve. Assume that the variation of density in the standard atmosphere is given by $\rho = 0.002377(1 - 7 \times 10^{-6}h)^{4.21}$, where h is the altitude in feet and ρ is in slug/ft^3.

Solution

(a) At sea level, where $h = 0$, $\rho = 0.002377$ slug/ft^3. The volume of the inflated balloon is $\frac{4}{3}\pi(15)^3 = 14{,}137$ ft^3. Hence,

$$\text{Buoyancy force} = \text{weight of displaced air}$$

$$= g\rho \mathcal{V}$$

where g is the acceleration of gravity and $\mathcal{V}$ is the volume.

$$\text{Buoyancy force} \equiv B = (32.2)(0.002377)(14{,}137) = 1082 \text{ lb}$$

The net upward force at sea level is $F = B - W$, where W is the weight. From Newton's second law,

$$F = B - W = ma$$

where m is the mass, $m = \frac{800}{32.2} = 24.8$ slug. Hence,

$$a = \frac{B - W}{m} = \frac{1082 - 800}{24.8} = \boxed{11.4 \text{ ft/s}^2}$$

(b) The maximum altitude occurs when $B = W = 800$ lb. Since $B = g\rho\mathcal{V}$, and assuming the balloon volume does not change,

$$\rho = \frac{B}{g\mathcal{V}} = \frac{800}{(32.2)(14{,}137)} = 0.00176 \text{ slug/ft}^3$$

From the given variation of ρ with altitude, h,

$$\rho = 0.002377(1 - 7 \times 10^{-6}h)^{4.21} = 0.00176$$

Solving for h, we obtain

$$h = \frac{1}{7 \times 10^{-6}}\left[1 - \left(\frac{0.00176}{0.002377}\right)^{1/4.21}\right] = \boxed{9842 \text{ ft}}$$

1.10 TYPES OF FLOW

An understanding of aerodynamics, like that of any other physical science, is obtained through a "building-block" approach—we dissect the discipline, form the parts into nice polished blocks of knowledge, and then later attempt to reassemble the blocks to form an understanding of the whole. An example of this process is the way that different types of aerodynamic flows are categorized and visualized. Although nature has no trouble setting up the most detailed and complex flow with a whole spectrum of interacting physical phenomena, we must attempt to understand such flows by modeling them with less detail, and neglecting some of the (hopefully) less significant phenomena. As a result, a study of aerodynamics has evolved into a study of numerous and distinct types of flow. The purpose of this section is to itemize and contrast these types of flow, and to briefly describe their most important physical phenomena.

1.10.1 Continuum Versus Free Molecule Flow

Consider the flow over a body, say, e.g., a circular cylinder of diameter d. Also, consider the fluid to consist of individual molecules, which are moving about in random motion. The mean distance that a molecule travels between collisions with neighboring molecules is defined as the *mean-free path* λ. If λ is orders of magnitude smaller than the scale of the body measured by d, then the flow appears to the body as a continuous substance. The molecules impact the body surface so frequently that the body cannot distinguish the individual molecular collisions, and the surface feels the fluid as a continuous medium. Such flow is called *continuum flow*. The other extreme is where λ is on the same order as the body scale; here the gas molecules are spaced so far apart (relative to d) that collisions with the body surface occur only infrequently, and the body surface can feel distinctly each molecular impact. Such flow is called *free molecular flow*. For manned flight, vehicles such as the space shuttle encounter free molecular flow at the extreme outer edge of the atmosphere, where the air density is so low that λ becomes on the order of the shuttle size. There are intermediate cases, where flows can exhibit some characteristics of both continuum and free molecule flows; such flows are generally labeled "low-density flows" in contrast to continuum flow. By far, the vast majority of practical aerodynamic applications involve continuum flows. Low-density and free molecule flows are just a small part of the total spectrum of aerodynamics. Therefore, in this book we will always deal with continuum flow; i.e., we will always treat the fluid as a continuous medium.

1.10.2 Inviscid Versus Viscous Flow

A major facet of a gas or liquid is the ability of the molecules to move rather freely, as explained in Sec. 1.2. When the molecules move, even in a very random

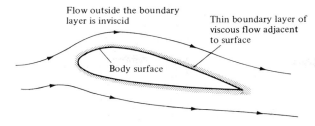

Flow outside the boundary
layer is inviscid

Thin boundary layer of
viscous flow adjacent
to surface

Body surface

FIGURE 1.28
The division of a flow into two regions: (1) the thin viscous boundary layer adjacent to the body surface and (2) the inviscid flow outside the boundary layer.

fashion, they obviously transport their mass, momentum, and energy from one location to another in the fluid. This transport on a molecular scale gives rise to the phenomena of mass diffusion, viscosity (friction), and thermal conduction. Such "transport phenomena" will be discussed in detail in Chap. 15. For our purposes here, we need only to recognize that all real flows exhibit the effects of these transport phenomena; such flows are called *viscous flows*. In contrast, a flow that is assumed to involve no friction, thermal conduction, or diffusion is called an *inviscid flow*. Inviscid flows do not truly exist in nature; however, there are many practical aerodynamic flows (more than you would think) where the influence of transport phenomena is small, and we can *model* the flow as being inviscid. For this reason, more than 70 percent of this book (Chaps. 3 to 14) deals with inviscid flows.

Theoretically, inviscid flow is approached in the limit as the Reynolds number goes to infinity (to be proved in Chap. 15). However, for practical problems, many flows with high but finite Re can be assumed to be inviscid. For such flows, the influence of friction, thermal conduction, and diffusion is limited

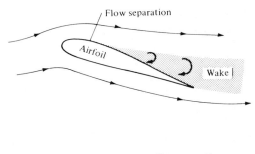

Flow separation

Airfoil

Wake

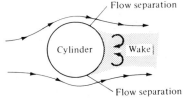

Flow separation

Cylinder

Wake

Flow separation

FIGURE 1.29
Examples of viscous-dominated flow.

to a very thin region adjacent to the body surface (the boundary layer, to be defined in Chap. 17), and the remainder of the flow outside this thin region is essentially inviscid. This division of the flow into two regions is illustrated in Fig. 1.28. Hence, the material discussed in Chaps. 3 to 14 applies to the flow outside the boundary layer. For flows over slender bodies, such as the airfoil sketched in Fig. 1.28, inviscid theory adequately predicts the pressure distribution and lift on the body and gives a valid representation of the streamlines and flow field away from the body. However, because friction (shear stress) is a major source of aerodynamic drag, inviscid theories by themselves cannot adequately predict total drag.

In contrast, there are some flows that are dominated by viscous effects. For example, if the airfoil in Fig. 1.28 is inclined to a high incidence angle to the flow (high angle of attack), then the boundary layer will tend to separate from the top surface, and a large wake is formed downstream. The separated flow is sketched at the top of Fig. 1.29; it is characteristic of the flow field over a "stalled" airfoil. Separated flow also dominates the aerodynamics of blunt bodies, such as the cylinder at the bottom of Fig. 1.29. Here, the flow expands around the front face of the cylinder, but separates from the surface on the rear face, forming a rather fat wake downstream. The types of flow illustrated in Fig. 1.29 are dominated by viscous effects; no inviscid theory can independently predict the aerodynamics of such flows. They require the inclusion of viscous effects, to be presented in Part IV.

1.10.3   Incompressible Versus Compressible Flows

A flow in which the density ρ is constant is called *incompressible*. In contrast, a flow where the density is variable is called *compressible*. A more precise definition of compressibility will be given in Chap. 7. For our purposes here, we will simply note that all flows, to a greater or lesser extent, are compressible; truly incompressible flow, where the density is precisely constant, does not occur in nature. However, analogous to our discussion of inviscid flow, there are a number of aerodynamic problems that can be modeled as being incompressible without any detrimental loss of accuracy. For example, the flow of homogeneous liquids is treated as incompressible, and hence most problems involving hydrodynamics assume ρ = constant. Also, the flow of gases at a low Mach number is essentially incompressible; for $M < 0.3$, it is always safe to assume ρ = constant. (We will prove this in Chap. 8.) This was the flight regime of all airplanes from the Wright brothers' first flight in 1903 to just prior to World War II. It is still the flight regime of most small, general aviation aircraft of today. Hence, there exists a large bulk of aerodynamic experimental and theoretical data for incompressible flows. Such flows are the subject of Chaps. 3 to 6. On the other hand, high-speed flow (near Mach 1 and above) must be treated as compressible; for such flows ρ can vary over wide latitudes. Compressible flow is the subject of Chaps. 7 to 14.

1.10.4 Mach Number Regimes

Of all the ways of subdividing and describing different aerodynamic flows, the distinction based on the Mach number is probably the most prevalent. If M is the local Mach number at an arbitrary point in a flow field, then by definition the flow is locally:

Subsonic if $M < 1$

Sonic if $M = 1$

Supersonic if $M > 1$

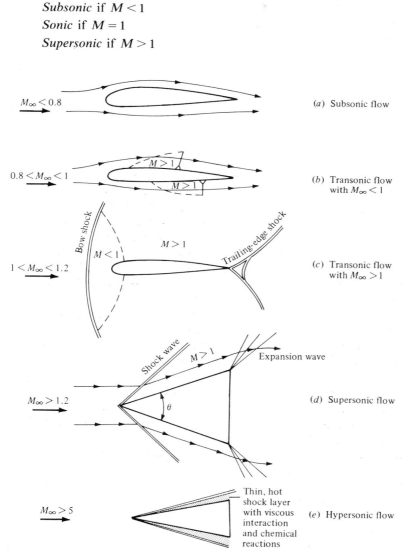

FIGURE 1.30
Different regimes of flow.

Looking at the whole field simultaneously, four different speed regimes can be identified using Mach number as the criterion:

1. *Subsonic flow* ($M < 1$ everywhere). A flow field is defined as *subsonic* if the Mach number is less than 1 at every point. Subsonic flows are characterized by smooth streamlines (no discontinuity in slope), as sketched in Fig. 1.30a. Moreover, since the flow velocity is everywhere less than the speed of sound, disturbances in the flow (say, the sudden deflection of the trailing edge of the airfoil in Fig. 1.30a) propagate both upstream and downstream, and are felt throughout the entire flow field. Note that a freestream Mach number, M_∞, less than 1 does not guarantee a totally subsonic flow over the body. In expanding over an aerodynamic shape, the flow velocity increases above the freestream value, and if M_∞ is close enough to 1, the local Mach number may become supersonic in certain regions of the flow. This gives rise to a rule of thumb that $M_\infty < 0.8$ for subsonic flow over slender bodies. For blunt bodies, M_∞ must be even lower to ensure totally subsonic flow. (Again, emphasis is made that the above is just a loose rule of thumb, and should not be taken as a precise quantitative definition.) Also, we will show later that incompressible flow is a special limiting case of subsonic flow where $M \to 0$.

2. *Transonic flow* (mixed regions where $M < 1$ and $M > 1$). As stated above, if M_∞ is subsonic but is near unity, the flow can become locally supersonic ($M > 1$). This is sketched in Fig. 1.30b, which shows pockets of supersonic flow over both the top and bottom surfaces of the airfoil, terminated by weak shock waves behind which the flow becomes subsonic again. Moreover, if M_∞ is increased slightly above unity, a bow shock wave is formed in front of the body; behind this shock wave the flow is locally subsonic, as shown in Fig. 1.30c. This subsonic flow subsequently expands to a low supersonic value over the airfoil. Weak shock waves are usually generated at the trailing edge, sometimes in a "fishtail" pattern as shown in Fig. 1.30c. The flow fields shown in Fig. 1.30b and c are characterized by mixed subsonic-supersonic flows, and are dominated by the physics of both types of flows. Hence, such flow fields are called *transonic flows*. Again, as a rule of thumb for slender bodies, transonic flows occur for freestream Mach numbers in the range $0.8 < M_\infty < 1.2$.

3. *Supersonic flow* ($M > 1$ everywhere). A flow field is defined as *supersonic* if the Mach number is greater than 1 at every point. Supersonic flows are frequently characterized by the presence of shock waves across which the flow properties and streamlines change discontinuously (in contrast to the smooth, continuous variations in subsonic flows). This is illustrated in Fig. 1.30d for supersonic flow over a sharp-nosed wedge; the flow remains supersonic behind the oblique shock wave from the tip. Also shown are distinct expansion waves, which are common in supersonic flow. (Again, the listing of $M_\infty > 1.2$ is strictly a rule of thumb. For example, in Fig. 1.30d, if θ is made large enough, the oblique shock wave will detach from the tip of the wedge, and will form a strong, curved bow shock ahead of the wedge with a substantial region of

subsonic flow behind the wave. Hence, the totally supersonic flow sketched in Fig. 1.30d is destroyed if θ is too large for a given M_∞. This shock detachment phenomenon can occur at any value of $M_\infty > 1$, but the value of θ at which it occurs increases as M_∞ increases. In turn, if θ is made infinitesimally small, the flow field in Fig. 1.30d holds for $M_\infty \geq 1.0$. These matters will be considered in detail in Chap. 9. However, the above discussion clearly shows that the listing of $M_\infty > 1.2$ in Fig. 1.30d is a very tenuous rule of thumb, and should not be taken literally.) In a supersonic flow, because the local flow velocity is greater than the speed of sound, disturbances created at some point in the flow *cannot* work their way upstream (in contrast to subsonic flow). This property is one of the most significant physical differences between subsonic and supersonic flows. It is the basic reason why shock waves occur in supersonic flows, but do not occur in steady subsonic flow. We will come to appreciate this difference more fully in Chaps. 7 to 14.

4. *Hypersonic flow* (very high supersonic speeds). Refer again to the wedge in Fig. 1.30d. Assume θ is a given, fixed value. As M_∞ increases above 1, the shock wave moves closer to the body surface. Also, the strength of the shock wave increases, leading to higher temperatures in the region between the shock and the body (the shock layer). If M_∞ is sufficiently large, the shock layer becomes very thin, and interactions between the shock wave and the viscous boundary layer on the surface occur. Also, the shock layer temperature becomes high enough that chemical reactions occur in the air. The O_2 and N_2 molecules are torn apart; i.e., the gas molecules dissociate. When M_∞ becomes large enough such that viscous interaction and/or chemically reacting effects begin to dominate the flow (Fig. 1.30e), the flow field is called *hypersonic*. (Again, a somewhat arbitrary but frequently used rule of thumb for hypersonic flow is $M_\infty > 5$.) Hypersonic aerodynamics received a great deal of attention during the period 1955–1970 because atmospheric entry vehicles encounter the atmosphere at Mach numbers between 25 (ICBMs) and 36 (the Apollo lunar return vehicle). Today, hypersonic aerodynamics is just part of the whole spectrum of realistic flight speeds. Some basic elements of hypersonic flow are treated in Chap. 14.

In summary, we attempt to organize our study of aerodynamic flows according to one or more of the various categories discussed in this section. The block diagram in Fig. 1.31 is presented to help emphasize these categories and to show how they are related. Indeed, Fig. 1.31 serves as a road map for this entire book. All the material to be covered in subsequent chapters fits into these blocks, which are lettered for easy reference. For example, Chap. 2 contains discussions of some fundamental aerodynamic principles and equations which fit into both blocks C and D. Chapters 3 to 6 fit into blocks D and E, Chap. 7 fits into blocks D and F, etc. As we proceed with our development of aerodynamics, we will frequently refer to Fig. 1.31 in order to help put specific, detailed material in proper perspective relative to the whole of aerodynamics.

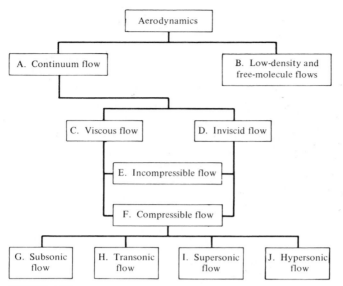

FIGURE 1.31
Block diagram categorizing the types of aerodynamic flows.

1.11 APPLIED AERODYNAMICS: THE AERODYNAMIC COEFFICIENTS—THEIR MAGNITUDES AND VARIATIONS

With the present section, we begin a series of special sections in this book under the general heading of "applied aerodynamics." The main thrust of this book is to present the *fundamentals* of aerodynamics, as is reflected in the book's title. However, *applications* of these fundamentals are liberally sprinkled throughout the book, in the text material, in the worked examples, and in the homework problems. The term *applied aerodynamics* normally implies the application of aerodynamics to the practical evaluation of the aerodynamic characteristics of real configurations such as airplanes, missiles, and space vehicles moving through an atmosphere (the earth's, or that of another planet). Therefore, to enhance the reader's appreciation of such applications, sections on applied aerodynamics will appear near the end of many of the chapters. To be specific, in this section, we address the matter of the aerodynamic coefficients defined in Sec. 1.5; in particular, we focus on lift, drag, and moment coefficients. These nondimensional coefficients are the primary language of applications in external aerodynamics (the distinction between external and internal aerodynamics was made in Sec. 1.2). It is important for you to obtain a feeling for typical values of the aerodynamic coefficients. (For example, do you expect a drag coefficient to be as low as 10^{-5}, or maybe as high as 1000—does this make sense?) The purpose of this section is to begin to provide you with such a feeling, at least for some common aerodynamic body shapes. As you progress through the remainder of this book, make every effort to note the typical magnitudes of the aerodynamic coefficients that are discussed in various sections. Having a realistic feel for these magnitudes is part of your technical maturity.

Question: What are some typical drag coefficients for various aerodynamic configurations? Some basic values are shown in Fig. 1.32. The dimensional analysis described in Sec. 1.7 proved that $C_D = f(M, \text{Re})$. In Fig. 1.32, the drag-coefficient values are for low speeds, essentially incompressible flow; therefore, the Mach number does not come into the picture. (For all practical purposes, for an incompressible flow, the Mach number is theoretically zero, not because the velocity goes to zero, but rather because the speed of sound is infinitely large. This will be made clear in Sec. 8.3.) Thus, for a low-speed flow, the aerodynamic coefficients for a fixed shape at a fixed orientation to the flow are functions of just the Reynolds number. In Fig. 1.32, the Reynolds numbers are listed at the left and the drag-coefficient values at the right. In Fig. 1.32a, a flat plate is oriented

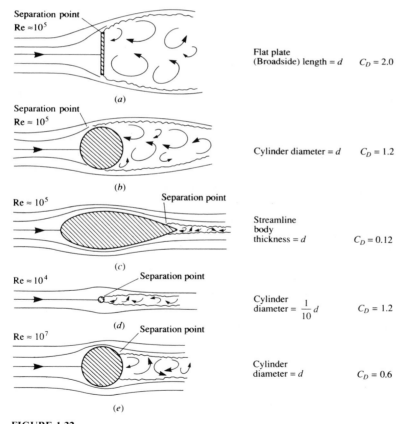

FIGURE 1.32
Drag coefficients for various aerodynamic shapes. (*From Talay, T. A.,* Introduction to the Aerodynamics of Flight, *NASA SP-367, 1975.*)

perpendicular to the flow; this configuration produces the largest possible drag coefficient of any conventional configuration, namely, $C_D = D'/q_\infty S = 2.0$, where S is the frontal area per unit span, i.e., $S = (d)(1)$, where d is the height of the plate. The Reynolds number is based on the height, d; i.e., $\text{Re} = \rho_\infty V_\infty d/\mu_\infty = 10^5$. Figure 1.32$b$ illustrates flow over a circular cylinder of diameter d; here, $C_D = 1.2$, considerably smaller than the vertical plate value in Fig. 1.32a. The drag coefficient can be reduced dramatically by streamlining the body, as shown in Fig. 1.32c. Here, $C_D = 0.12$; this is an order of magnitude smaller than the circular cylinder in Fig. 1.32b. The Reynolds numbers for Fig. 1.32a, b, and c are all the same value, based on d (diameter). The drag coefficients are all defined the same, based on a reference area per unit span of $(d)(1)$. Note that the flow fields over the configurations in Fig. 1.32a, b, and c show a wake downstream of the body; the wake is caused by the flow separating from the body surface, with a low-energy, recirculating flow inside the wake. The phenomenon of flow separation will be discussed in detail in Part IV of this book, dealing with viscous flows. However, it is clear that the wakes diminish in size as we progressively go from Fig. 1.32a, b, and c. The fact that C_D also diminishes progressively from Fig. 1.32a, b, and c is no accident—it is a direct result of the regions of separated flow becoming progressively smaller. Why is this so? Simply consider this as one of the many interesting questions in aerodynamics—a question that will be answered in due time in this book. Note, in particular, that the physical effect of the streamlining in Fig. 1.32c results in a very small wake, hence a small value for the drag coefficient.

Consider Fig. 1.32d, where once again a circular cylinder is shown, but of much smaller diameter. Since the diameter here is 0.1d, the Reynolds number is now 10^4 (based on the same freestream V_∞, ρ_∞, and μ_∞ as Fig. 1.32a, b, and c). It will be shown in Chap. 3 that C_D for a circular cylinder is relatively independent of Re between $\text{Re} = 10^4$ and 10^5. Since the body *shape* is the same between Fig. 1.32d and b, namely, a circular cylinder, then C_D is the same value of 1.2 as shown in the figure. However, since the drag is given by $D' = q_\infty S C_D$, and S is one-tenth smaller in Fig. 1.32d, then the *drag force* on the small cylinder in Fig. 1.32d is one-tenth smaller than that in Fig. 1.32b.

Another comparison is illustrated in Fig. 1.32c and d. Here we are comparing a large streamlined body of thickness d with a small circular cylinder of diameter 0.1d. For the large streamlined body in Fig. 1.32c,

$$D' = q_\infty S C_D = 0.12 q_\infty d$$

For the small circular cylinder in Fig. 1.32d,

$$D' = q_\infty S C_D = q_\infty(0.1d)(1.2) = 0.12 q_\infty d$$

The drag values are the same! Thus, Fig. 1.32c and d illustrate that the drag on a circular cylinder is the same as that on the streamlined body which is ten times thicker—another way of stating the aerodynamic value of streamlining.

As a final note in regard to Fig. 1.32, the flow over a circular cylinder is again shown in Fig. 1.32e. However, now the Reynolds number has been increased to 10^7, and the cylinder drag coefficient has decreased to 0.6—a dramatic factor of two less than in Fig. 1.32b and d. Why has C_D decreased so precipitously at the higher Reynolds number? The answer must somehow be connected with the smaller wake behind the cylinder in Fig. 1.32e compared to Fig. 1.32b. What is going on here? This is one of the fascinating questions we will answer as we progress through our discussions of aerodynamics in this book—an answer that will begin with Sec. 3.18 and culminate in Part IV dealing with viscous flow.

At this stage, pause for a moment and note the *values* of C_D for the aerodynamic shapes in Fig. 1.32. With C_D based on the *frontal projected area* ($S = d(1)$ per unit span), the values of C_D range from a maximum of 2 to numbers as low as 0.12. These are typical values of C_D for aerodynamic bodies.

Also, note the values of Reynolds number given in Fig. 1.32. Consider a circular cylinder of diameter 1 m in a flow at standard sea level conditions ($\rho_\infty = 1.23$ kg/m^3 and $\mu_\infty = 1.789 \times 10^{-5}$ kg/m $\cdot$ s) with a velocity of 45 m/s (close to 100 mi/h). For this case,

$$\mathrm{Re} = \frac{\rho_\infty V_\infty d}{\mu_\infty} = \frac{(1.23)(45)(1)}{1.789 \times 10^{-5}} = 3.09 \times 10^6$$

Note that the Reynolds number is over 3 million; values of Re in the millions are typical of practical applications in aerodynamics. Therefore, the large numbers given for Re in Fig. 1.32 are appropriate.

Let us examine more closely the nature of the drag exerted on the various bodies in Fig. 1.32. Since these bodies are at zero angle of attack, the drag is equal to the axial force. Hence, from Eq. (1.8) the drag per unit span can be written as

$$D' = \underbrace{\int_{\mathrm{LE}}^{\mathrm{TE}} -p_u \sin\theta \, ds_u + \int_{\mathrm{LE}}^{\mathrm{TE}} p_l \sin\theta \, ds_l}_{\text{pressure drag}}$$

$$\underbrace{+ \int_{\mathrm{LE}}^{\mathrm{TE}} \tau_u \cos\theta \, ds_u + \int_{\mathrm{LE}}^{\mathrm{TE}} \tau_l \cos\theta \, ds_l}_{\text{skin friction drag}} \tag{1.52}$$

That is, the drag on any aerodynamic body is composed of pressure drag and skin friction drag; this is totally consistent with our discussion in Sec. 1.5, where it is emphasized that the only two basic sources of aerodynamic force on a body are the pressure and shear stress distributions exerted on the body surface. The division of total drag onto its components of pressure and skin friction drag is frequently useful in analyzing aerodynamic phenomena. For example, Fig. 1.33 illustrates the comparison of skin friction drag and pressure drag for the cases shown in Fig. 1.32. In Fig. 1.33, the bar charts at the right of the figure give the relative drag force on each body; the cross-hatched region denotes the amount of skin friction drag, and the blank region is the amount of pressure drag. The freestream density and viscosity are the same for Fig. 1.33a to e; however, the

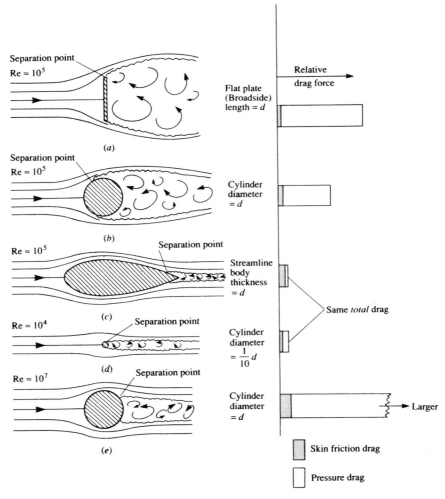

FIGURE 1.33
The relative comparison between skin friction drag and pressure drag for various aerodynamic shapes.
(*From Talay, T. A.*, Introduction to the Aerodynamics of Flight, *NASA SP-367, 1975.*)

freestream velocity, V_∞, is varied by the necessary amount to achieve the Reynolds numbers shown. That is, comparing Fig. 1.33b and e, the value of V_∞ is much larger for Fig. 1.33e. Since the drag force is given by

$$D' = \tfrac{1}{2}\rho_\infty V_\infty^2 S C_D$$

then the drag for Fig. 1.33e is much larger than for Fig. 1.33b. Also shown in the bar chart is the equal drag between the streamlined body of thickness d and the circular cylinder of diameter $0.1d$—a comparison discussed earlier in conjunction with Fig. 1.32. Of most importance in Fig. 1.33, however, is the relative amounts of skin friction and pressure drag for each body. Note that the drag of the vertical flat plate and the circular cylinders is dominated by pressure drag, whereas, in contrast, most of the drag of the streamlined body is due to skin friction. Indeed, this type of comparison leads to the definition of two generic body shapes in aerodynamics, as follows:

Blunt body = a body where most of the drag is pressure drag

Streamlined body = a body where most of the drag is skin friction drag

In Figs. 1.32 and 1.33, the vertical flat plate and the circular cylinder are clearly *blunt bodies.*

The large pressure drag of blunt bodies is due to the massive regions of flow separation which can be seen in Figs. 1.32 and 1.33. The reason why flow separation causes drag will become clear as we progress through our subsequent discussions. Hence, the pressure drag shown in Fig. 1.33 is more precisely denoted as "pressure drag due to flow separation"; this drag is frequently called *form drag.* (For an elementary discussion of form drag and its physical nature, see Ref. 2.)

Let us now examine the drag on a flat plate at zero angle of attack, as sketched in Fig. 1.34. Here, the drag is completely due to shear stress; there is no pressure force in the drag direction. The skin friction drag coefficient is defined as

$$C_f = \frac{D'}{q_\infty S} = \frac{D'}{q_\infty c(1)}$$

where the reference area is the *planform* area per unit span, i.e., the surface area as seen by looking down on the plate from above. C_f will be discussed further in Chap. 16. However, the purpose of Fig. 1.34 is to demonstrate that:

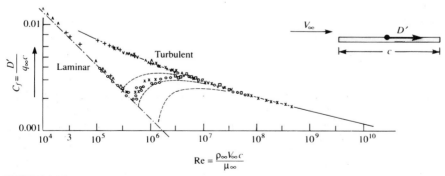

FIGURE 1.34

Variation of laminar and turbulent skin friction coefficient for a flat plate as a function of Reynolds number based on the chord length of the plate. The intermediate dashed curves are associated with various transition paths from laminar flow to turbulent flow.

1. C_f is a strong function of Re, where Re is based on the chord length of the plate, $Re = \rho_\infty V_\infty c / \mu_\infty$. Note that C_f decreases as Re increases.

2. The value of C_f depends on whether the flow over the plate surface is laminar or turbulent, with the turbulent C_f being higher than the laminar C_f at the same Re. What is going on here? What is laminar flow? What is turbulent flow? Why does it affect C_f? The answers to these questions will be addressed in Chaps. 15 and 17.

3. The magnitudes of C_f range typically from 0.001 to 0.01 over a large range of Re. These numbers are considerably smaller than the drag coefficients listed in Fig. 1.32. This is mainly due to the different reference areas used. In Fig. 1.32, the reference area is a cross-sectional area normal to the flow; in Fig. 1.34, the reference area is the *planform* area.

A flat plate is not a very practical aerodynamic body—it simply has no volume. Let us now consider a body with thickness, namely, an airfoil section. An NACA 63-210 airfoil section is one such example. The variation of the drag coefficient, c_d, with angle of attack is shown in Fig. 1.35. Here, as usual, c_d is defined as

$$c_d = \frac{D'}{q_\infty c}$$

where D' is the drag per unit span. Note that the lowest value of c_d is about 0.0045. The NACA 63-210 airfoil is classified as a "laminar-flow airfoil" because it is designed to promote such a flow at small α. This is the reason for the "bucketlike" appearance of the c_d curve at low α; at higher α, transition to turbulent flow occurs over the airfoil surface, causing a sharp increase in c_d. Hence, the value of $c_d = 0.0045$ occurs in a laminar flow. Note that the Reynolds

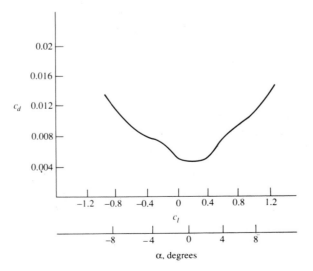

FIGURE 1.35
Variation of section drag coefficient for an NACA 63-210 airfoil. Re = 3×10^6.

number is 3 million. Once again, a reminder is given that the various aspects of laminar and turbulent flows will be discussed in Part IV. The main point here is to demonstrate that typical airfoil drag-coefficient values are on the order of 0.004 to 0.006. As in the case of the streamlined body in Figs. 1.32 and 1.33, most of this drag is due to skin friction. However, at higher values of α, flow separation over the top surface of the airfoil begins to appear and pressure drag due to flow separation (form drag) begins to increase. This is why c_d increases with increasing α in Fig. 1.35.

Let us now consider a complete airplane. In Chap. 3, Fig. 3.2 is a photograph of the Seversky P-35, a typical fighter aircraft of the late 1930s. Figure 1.36 is a detailed drag breakdown for this type of aircraft. Configuration 1 in Fig. 1.36 is the stripped-down, aerodynamically cleanest version of this aircraft; its drag coefficient (measured at an angle of attack corresponding to a lift coefficient of $C_L = 0.15$) is $C_D = 0.0166$. Here, C_D is defined as

$$C_D = \frac{D}{q_\infty S}$$

where D is the airplane drag and S is the planform area of the wing. For configurations 2 through 18, various changes are progressively made in order to bring the aircraft to its conventional, operational configuration. The incremental drag increases due to each one of these additions are tabulated in Fig. 1.36. Note that the drag coefficient is increased by more than 65 percent by these additions; the value of C_D for the aircraft in full operational condition is 0.0275. This is a typical airplane drag-coefficient value. The data shown in Fig. 1.36 were obtained in the full-scale wind tunnel at the NACA Langley Memorial Laboratory just prior to World War II. (The full-scale wind tunnel has test-section dimensions of 30 by 60 ft, which can accommodate a whole airplane—hence the name "full-scale.")

The values of drag coefficients discussed so far in this section have applied to low-speed flows. In some cases, their variation with the Reynolds number has been illustrated. Recall from the discussion of dimensional analysis in Sec. 1.7 that drag coefficient also varies with the Mach number. *Question*: What is the effect of increasing the Mach number on the drag coefficient of an airplane? Consider the answer to this question for a Northrop T-38A jet trainer, shown in Fig. 1.37. The drag coefficient for this airplane is given in Fig. 1.38 as a function of the Mach number ranging from low subsonic to supersonic. The aircraft is at a small negative angle of attack such that the lift is zero, hence the C_D in Fig. 1.38 is called the *zero-lift drag coefficient*. Note that the value of C_D is relatively constant from $M = 0.1$ to about 0.86. Why? At Mach numbers of about 0.86, the C_D rapidly increases. This large increase in C_D near Mach one is typical of all flight vehicles. Why? Stay tuned; the answers to these questions will become clear in Part III dealing with compressible flow. Also, note in Fig. 1.38 that at low subsonic speeds, C_D is about 0.015. This is considerably lower than the 1930s-type airplane illustrated in Fig. 1.36; of course, the T-38 is a more modern, sleek, streamlined airplane, and its drag coefficient should be smaller.

Airplane condition

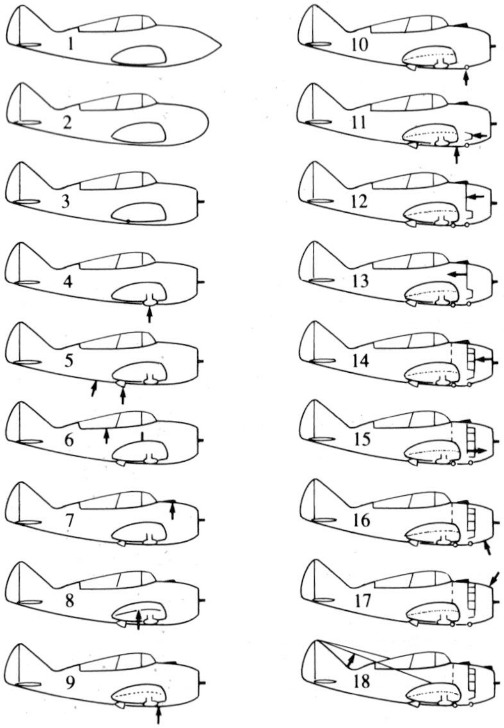

Condition number	Description	C_D ($C_L = 0.15$)	ΔC_D	ΔC_D, %[a]
1	Completely faired condition, long nose fairing	0.0166		
2	Completely faired condition, blunt nose fairing	0.0169		
3	Original cowling added, no airflow through cowling	0.0186	0.0020	12.0
4	Landing-gear seals and fairing removed	0.0188	0.0002	1.2
5	Oil cooler installed	0.0205	0.0017	10.2
6	Canopy fairing removed	0.0203	-0.0002	-1.2
7	Carburetor air scoop added	0.0209	0.0006	3.6
8	Sanded walkway added	0.0216	0.0007	4.2
9	Ejector chute added	0.0219	0.0003	1.8
10	Exhaust stacks added	0.0225	0.0006	3.6
11	Intercooler added	0.0236	0.0011	6.6
12	Cowling exit opened	0.0247	0.0011	6.6
13	Accessory exit opened	0.0252	0.0005	3.0
14	Cowling fairing and seals removed	0.0261	0.0009	5.4
15	Cockpit ventilator opened	0.0262	0.0001	0.6
16	Cowling venturi installed	0.0264	0.0002	1.2
17	Blast tubes added	0.0267	0.0003	1.8
18	Antenna installed	0.0275	0.0008	4.8
	Total		0.0109	

[a] Percentages based on completely faired condition with long nose fairing.

FIGURE 1.36

The breakdown of various sources of drag on a late 1930s airplane, the Seversky XP-41 (derived from the Seversky P-35 shown in Fig. 3.2). [Experimental data from Coe, Paul J., "Review of Drag Cleanup Tests in Langley Full-Scale Tunnel (From 1935 to 1945) Applicable to Current General Aviation Airplanes," NASA TN-D-8206, 1976.]

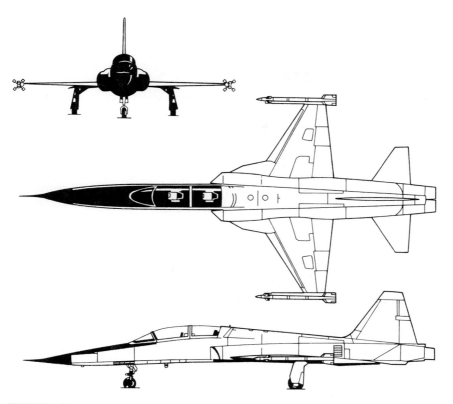

FIGURE 1.37
Three-view of the Northrop T-38 jet trainer. (*Courtesy of the U.S. Air Force.*)

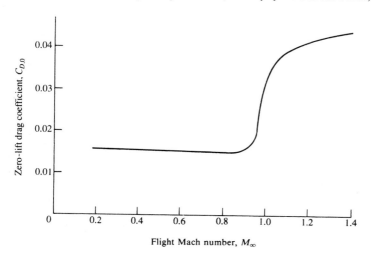

FIGURE 1.38
Zero-lift drag coefficient variation with Mach number for the T-38. (*Courtesy of the U.S. Air Force.*)

We now turn our attention to lift coefficient and examine some typical values. As a complement to the drag data shown in Fig. 1.35 for an NACA 63-210 airfoil, the variation of lift coefficient versus angle of attack for the same airfoil is shown in Fig. 1.39. Here, we see c_l increasing linearly with α until a maximum value is obtained near $\alpha = 14°$, beyond which there is a precipitous drop in lift. Why does c_l vary with α in such a fashion—in particular, what causes the sudden drop in c_l beyond $\alpha = 14°$? An answer to this question will evolve over the ensuing chapters. For our purpose in the present section, observe the *values* of c_l; they vary from about -1.0 to a maximum of 1.5, covering a range of α from -12 to $14°$. Conclusion: For an airfoil, the magnitude of c_l is about a factor of 100 larger than c_d. A particularly important figure of merit in aerodynamics is the *ratio* of lift to drag, the so-called L/D ratio; many aspects of the flight performance of a vehicle are directly related to the L/D ratio (see, e.g., Ref. 2). Other things being equal, a higher L/D means better flight performance. For an airfoil—a configuration whose primary function is to produce lift with as little drag as possible—values of L/D are large. For example, from Figs. 1.35 and 1.39, at $\alpha = 4°$, $c_l = 0.6$ and $c_d = 0.0046$, yielding $L/D = \frac{0.6}{0.0046} = 130$. This value is much larger than those for a complete airplane, as we will soon see.

To illustrate the lift coefficient for a complete airplane, Fig. 1.40 shows the variation of C_L with α for the T-38 in Fig. 1.37. Three curves are shown, each for a different flap deflection angle. (Flaps are sections of the wing at the trailing

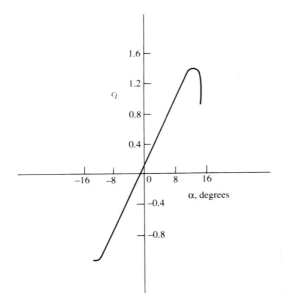

FIGURE 1.39
Variation of section lift coefficient for an NACA 63-210 airfoil. $Re = 3 \times 10^6$. No flap deflection.

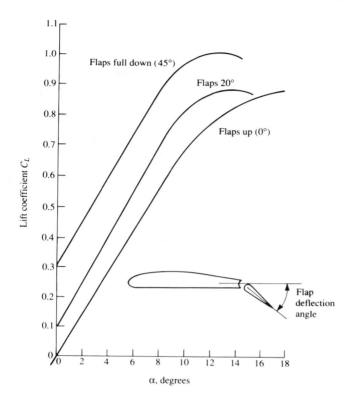

FIGURE 1.40
Variation of lift coefficient with angle of attack for the T-38. Three curves are shown corresponding to three different flap deflections. Freestream Mach number is 0.4. (*Courtesy of the U.S. Air Force.*)

edge which, when deflected downward, increase the lift of the wing. See Sec. 5.17 of Ref. 2 for a discussion of the aerodynamic properties of flaps.) Note that at a given α, the deflection of the flaps increases C_L. The values of C_L shown in Fig. 1.40 are about the same as that for an airfoil—on the order of 1. On the other hand, the maximum L/D ratio of the T-38 is about 10—considerably smaller than that for an airfoil alone. Of course, an airplane has a fuselage, engine nacelles, etc., which are elements with other functions than just producing lift, and indeed produce only small amounts of lift while at the same time adding a lot of drag to the vehicle. Thus, the L/D ratio for an airplane is expected to be much less than that for an airfoil alone. Moreover, the wing on an airplane experiences a much higher pressure drag than an airfoil due to the adverse aerodynamic effects of the wing tips (a topic for Chap. 5). This additional pressure drag is called induced drag, and for short, stubby wings, such as on the T-38, the induced drag can be large. (We must wait until Chap. 5 to find out about the nature of induced drag.) As a result, the L/D ratio of the T-38 is fairly small as most airplanes go. For example, the maximum L/D ratio for the Boeing B-52

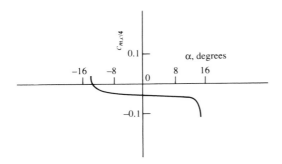

FIGURE 1.41
Variation of section moment coefficient about the quarter chord for an NACA 63-210 airfoil. $Re = 3 \times 10^6$.

strategic bomber is 21.5 (see Ref. 48). However, this value is still considerably smaller than that for an airfoil alone.

Finally, we turn our attention to the values of moment coefficients. Figure 1.41 illustrates the variation of $c_{m,c/4}$ for the NACA 63-210 airfoil. Note that this is a *negative* quantity; all conventional airfoils produce negative, or "pitch-down," moments. (Recall the sign convention for moments given in Sec. 1.5.) Also, notice that its value is on the order of -0.035. This value is typical of a moment coefficient—on the order of hundredths.

With this, we end our discussion of typical values of the aerodynamic coefficients defined in Sec. 1.5. At this stage, you should reread this section, now from the overall perspective provided by a first reading, and make certain to fix in your mind the typical values discussed—it will provide a useful "calibration" for our subsequent discussions.

1.12 HISTORICAL NOTE: THE ILLUSIVE CENTER OF PRESSURE

The center of pressure of an airfoil was an important matter during the development of aeronautics. It was recognized in the nineteenth century that, for a heavier-than-air machine to fly at stable, equilibrium conditions (e.g., straight-and-level flight), the moment about the vehicle's center of gravity must be zero (see Chap. 7 of Ref. 2). The wing lift acting at the center of pressure, which is generally a distance away from the center of gravity, contributes substantially to this moment. Hence, the understanding and prediction of the center of pressure was felt to be absolutely necessary in order to design a vehicle with proper equilibrium. On the other hand, the early experimenters had difficulty measuring the center of pressure, and much confusion reigned. Let us examine this matter further.

The first experiments to investigate the center of pressure of a lifting surface were conducted by the Englishman George Cayley (1773–1857) in 1808. Cayley was the inventor of the modern concept of the airplane, namely, a vehicle with fixed wings, a fuselage, and a tail. He was the first to separate conceptually the functions of lift and propulsion; prior to Cayley, much thought had gone into ornithopters—machines that flapped their wings for both lift and thrust. Cayley rejected this idea, and in 1799, on a silver disk now in the collection of the Science Museum in London, he inscribed a sketch of a rudimentary airplane with all the basic elements we recognize today. Cayley was an active, inventive, and long-lived man, who conducted numerous pioneering aerodynamic experiments and fervently believed that powered, heavier-than-air, manned flight was inevitable. (See chap. 1 of Ref. 2 for an extensive discussion of Cayley's contributions to aeronautics.)

In 1808, Cayley reported on experiments of a winged model which he tested as a glider and as a kite. His comments on the center of pressure are as follows:

> By an experiment made with a large kite formed of an hexagon with wings extended from it, all so constructed as to present a hollow curve to the current, I found that when loaded nearly to 1 lb to a foot and $\frac{1}{2}$, it required the center of gravity to be suspended so as to leave the anterior and posterior portions of the surface in the ratio of 3 to 7. But as this included the tail operating with a double leverage behind, I think such hollow surfaces relieve about an equal pressure on each part, when they are divided in the ratio of 5 to 12, 5 being the anterior portion. It is really surprising to find so great a difference, and it obliges the center of gravity of flying machines to be much forwarder of the center of bulk (the centroid) than could be supposed a priori.

Here, Cayley is saying that the center of pressure is 5 units from the leading edge and 12 units from the trailing edge; i.e., $x_{cp} = \frac{5}{17}c$. Later, he states in addition: "I tried a small square sail in one plane, with the weight nearly the same, and I could not perceive that the center-of-resistance differed from the center of bulk." That is, Cayley is stating that the center of pressure in this case is $\frac{1}{2}c$.

There is no indication from Cayley's notes that he recognized that center of pressure moves when the lift, or angle of attack, is changed. However, there is no doubt that he was clearly concerned with the location of the center of pressure and its effect on aircraft stability.

The center of pressure on a flat surface inclined at a small angle to the flow was studied by Samuel P. Langley during the period 1887–1896. Langley was the secretary of the Smithsonian at that time, and devoted virtually all his time and much of the Smithsonian's resources to the advancement of powered flight. Langley was a highly respected physicist and astronomer, and he approached the problem of powered flight with the systematic and structured mind of a scientist. Using a whirling arm apparatus as well as scores of rubber-band powered models, he collected a large bulk of aerodynamic information with which he subsequently designed a full-scale aircraft. The efforts of Langley to build and fly a successful airplane resulted in two dismal failures in which his machine fell into the Potomac River—the last attempt being just 9 days before the Wright brothers' historic first flight on December 17, 1903. In spite of these failures, the work of Langley helped in many ways to advance powered flight. (See chap. 1 of Ref. 2 for more details.)

Langley's observations on the center of pressure for a flat surface inclined to the flow are found in the *Langley Memoir on Mechanical Flight, Part I, 1887 to 1896*, by Samuel P. Langley, and published by the Smithsonian Institution in 1911—5 years after Langley's death. In this paper, Langley states:

> The center-of-pressure in an advancing plane in soaring flight is always in advance of the center of figure, and moves forward as the angle-of-inclination of the sustaining surfaces diminishes, and, to a less extent, as horizontal flight increases in velocity. These facts furnish the elementary ideas necessary in discussing the problem of equilibrium, whose solution is of the most vital importance to successful flight.
>
> The solution would be comparatively simple if the position of the center-of-pressure could be accurately known beforehand, but how difficult the solution is may be realized from a consideration of one of the facts just stated, namely, that the position of the center-of-pressure in horizontal flight shifts with velocity of the flight itself.

Here, we see that Langley is fully aware that the center of pressure moves over a lifting surface, but that its location is hard to pin down. Also, he notes the correct variation for a flat plate, namely, x_{cp} moves forward as the angle of attack decreases. However, he is puzzled by the behavior of x_{cp} for a curved (cambered) airfoil. In his own words:

> Later experiments conducted under my direction indicate that upon the curved surfaces I employed, the center-of-pressure moves forward with an increase in the angle of elevation, and backward with a decrease, so that it may lie even behind the center of the surface. Since for some surfaces the center-of-pressure moves backward, and for others forward, it would seem that there might be some other surface for which it will be fixed.

Here, Langley is noting the totally opposite behavior of the travel of the center of pressure on a cambered airfoil in comparison to a flat surface, and is indicating ever so slightly some of his frustration in not being able to explain his results in a rational scientific way.

Three-hundred-fifty miles to the west of Langley, in Dayton, Ohio, Orville and Wilbur Wright were also experimenting with airfoils. As described in Sec. 1.1, the Wrights had constructed a small wind tunnel in their bicycle shop with which they conducted aerodynamic tests on hundreds of different airfoil and wing shapes during the fall, winter, and spring of 1901–1902. Clearly, the Wrights had an appreciation of the center of pressure, and their successful airfoil design used on the 1903 Wright Flyer is a testimonial to their mastery of the problem. Interestingly enough, in the written correspondence of the Wright brothers, only one set of results for the center of pressure can be found. This appears in Wilbur's notebook, dated July 25, 1905, in the form of a table and a graph. The graph is shown in Fig. 1.42—the original form as plotted by Wilbur. Here, the center of pressure, given in terms of the percentage of distance from the leading edge, is plotted versus angle of attack. The data for two airfoils are given, one with large curvature (maximum height to chord ratio $= \frac{1}{12}$) and one with more moderate curvature (maximum height to chord ratio $= \frac{1}{20}$). These results show the now

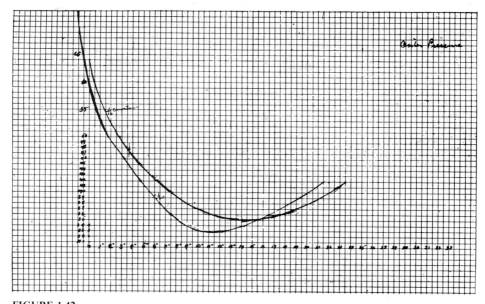

FIGURE 1.42
Wright brothers' measurements of the center of pressure as a function of angle of attack for a curved (cambered) airfoil. Center of pressure is plotted on the ordinate in terms of percentage distance along the chord from the leading edge. This figure shows the actual data as hand plotted by Wilbur Wright, which appears in Wilbur's notebook dated July 25, 1905.

familiar travel of the center of pressure for a curved airfoil, namely, x_{cp} moves forward as the angle of attack is increased, at least for small to moderate values of α. However, the most forward excursion of x_{cp} in Fig. 1.42 is 33 percent behind the leading edge—the center of pressure is always behind the quarter-chord point.

The first practical airfoil theory, valid for thin airfoils, was developed by Ludwig Prandtl and his colleagues at Göttingen, Germany, during the period just prior to and during World War I. This thin airfoil theory is described in detail in Chap. 4. The result for the center of pressure for a curved (cambered) airfoil is given by Eq. (4.66), and shows that x_{cp} moves forward as the angle of attack (hence c_l) increases, and that it is always behind the quarter-chord point for finite, positive values of c_l. This theory, in concert with more sophisticated wind-tunnel measurements that were being made during the period 1915–1925, finally brought the understanding and prediction of the location of the center of pressure for a cambered airfoil well into focus.

Because x_{cp} makes such a large excursion over the airfoil as the angle of attack is varied, its importance as a basic and practical airfoil property has diminished. Beginning in the early 1930s, the National Advisory Committee for Aeronautics (NACA), at its Langley Memorial Aeronautical Laboratory in Virginia, measured the properties of several systematically designed families of airfoils—airfoils which became a standard in aeronautical engineering. These NACA airfoils are discussed in Secs. 4.2 and 4.3. Instead of giving the airfoil data in terms of lift, drag, and center of pressure, the NACA chose the alternate systems of reporting lift, drag, and moments about either the quarter-chord point or the aerodynamic center. These are totally appropriate alternative methods of defining the force-and-moment system on an airfoil, as discussed in Sec. 1.6 and illustrated in Fig. 1.19. As a result, the center of pressure is rarely given as part of modern airfoil data. On the other hand, for three-dimensional bodies, such as slender projectiles and missiles, the location of the center of pressure still remains an important quantity, and modern missile data frequently include x_{cp}. Therefore, a consideration of center of pressure still retains its importance when viewed over the whole spectrum of flight vehicles.

1.13 SUMMARY

Refer again to the road map for Chap. 1 given in Fig. 1.6. Read again each block in this diagram as a reminder of the material we have covered. If you feel uncomfortable about some of the concepts, or if your memory is slightly "foggy" on certain points, go back and reread the pertinent sections until you have mastered the material.

This chapter has been primarily qualitative, emphasizing definitions and basic concepts. However, some of the more important quantitative relations are summarized below:

The normal, axial, lift, drag, and moment coefficients for an aerodynamic body can be obtained by integrating the pressure and skin friction coefficients over the body surface from the leading to the trailing edge. For a two-dimensional body,

$$c_n = \frac{1}{c} \left[\int_0^c (C_{p,l} - C_{p,u})\, dx + \int_0^c \left(c_{f,u} \frac{dy_u}{dx} + c_{f,l} \frac{dy_l}{dx} \right) dx \right] \tag{1.15}$$

$$c_a = \frac{1}{c} \left[\int_0^c \left(C_{p,u} \frac{dy_u}{dx} - C_{p,l} \frac{dy_l}{dx} \right) dx + \int_0^c (c_{f,u} + c_{f,l})\, dx \right] \tag{1.16}$$

$$c_{m_{LE}} = \frac{1}{c^2} \left[\int_0^c (C_{p,u} - C_{p,l}) x\, dx - \int_0^c \left(c_{f,u} \frac{dy_u}{dx} + c_{f,l} \frac{dy_l}{dx} \right) x\, dx \right.$$
$$\left. + \int_0^c \left(C_{p,u} \frac{dy_u}{dx} + c_{f,u} \right) y_u\, dx + \int_0^c \left(-C_{p,l} \frac{dy_l}{dx} + c_{f,l} \right) y_l\, dx \right] \tag{1.17}$$

$$c_l = c_n \cos \alpha - c_a \sin \alpha \tag{1.18}$$

$$c_d = c_n \sin \alpha + c_a \cos \alpha \tag{1.19}$$

The center of pressure is obtained from

$$x_{cp} = -\frac{M_{LE}}{N} \approx -\frac{M_{LE}}{L} \qquad \text{(1.20) and (1.21)}$$

The criteria for two or more flows to be dynamically similar are:

1. The bodies and any other solid boundaries must be geometrically similar.
2. The similarity parameters must be the same. Two important similarity parameters are Mach number $M = V/a$ and Reynolds number $\text{Re} = \rho Vc/\mu$.

If two or more flows are dynamically similar, then the force coefficients C_L, C_D, etc., are the same.

In fluid statics, the governing equation is the hydrostatic equation:

$$dp = -g\rho \, dy \qquad (1.45)$$

For a constant density medium, this integrates to

$$p + \rho g h = \text{constant} \qquad (1.47)$$

or $\qquad\qquad p_1 + \rho g h = p_2 + \rho g h_2$

Such equations govern, among other things, the operation of a manometer, and also lead to Archimedes' principle that the buoyancy force on a body immersed in a fluid is equal to the weight of the fluid displaced by the body.

PROBLEMS

1.1. For most gases at standard or near standard conditions, the relationship among pressure, density, and temperature is given by the perfect gas equation of state: $p = \rho RT$, where R is the specific gas constant. For air at near standard conditions, $R = 287 \text{ J/(kg} \cdot \text{K)}$ in the International System of Units and $R = 1716 \text{ ft} \cdot \text{lb/(slug} \cdot °\text{R)}$ in the English Engineering System of Units. (More details on the perfect gas equation of state are given in Chap. 7.) Using the above information, consider the following two cases:
(*a*) At a given point on the wing of a Boeing 727, the pressure and temperature of the air are $1.9 \times 10^4 \text{ N/m}^2$ and 203 K, respectively. Calculate the density at this point.
(*b*) At a point in the test section of a supersonic wind tunnel, the pressure and density of the air are 1058 lb/ft^2 and $1.23 \times 10^{-3} \text{ slug/ft}^3$, respectively. Calculate the temperature at this point.

1.2. Starting with Eqs. (1.7), (1.8), and (1.11), derive in detail Eqs. (1.15), (1.16), and (1.17).

1.3. Consider an infinitely thin flat plate of chord c at an angle of attack α in a supersonic flow. The pressures on the upper and lower surfaces are different but constant over each surface; i.e., $p_u(s) = c_1$ and $p_l(s) = c_2$, where c_1 and c_2 are constants and $c_2 > c_1$. Ignoring the shear stress, calculate the location of the center of pressure.

1.4. Consider an infinitely thin flat plate with a 1-m chord at an angle of attack of $10°$ in a supersonic flow. The pressure and shear stress distributions on the upper surfaces are given by $p_u = 4 \times 10^4 (x-1)^2 + 5.4 \times 10^4$, $p_l = 2 \times 10^4 (x-1)^2 + 1.73 \times 10^5$, $\tau_u = 288x^{-0.2}$, and $\tau_l = 731x^{-0.2}$, respectively, where x is the distance from the leading edge in meters and p and τ are in newtons per square meter. Calculate the normal and axial forces, the lift and drag, moments about the leading edge, and moments about the quarter chord, all per unit span. Also, calculate the location of the center of pressure.

1.5. Consider an airfoil at $12°$ angle of attack. The normal and axial force coefficients are 1.2 and 0.03, respectively. Calculate the lift and drag coefficients.

1.6. Consider an NACA 2412 airfoil (the meaning of the number designations for standard NACA airfoil shapes is discussed in Chap. 4). The following is a tabulation of the lift, drag, and moment coefficients about the quarter chord for this airfoil, as a function of angle of attack.

α(degrees)	c_l	c_d	$c_{m,c/4}$
−2.0	0.05	0.006	−0.042
0	0.25	0.006	−0.040
2.0	0.44	0.006	−0.038
4.0	0.64	0.007	−0.036
6.0	0.85	0.0075	−0.036
8.0	1.08	0.0092	−0.036
10.0	1.26	0.0115	−0.034
12.0	1.43	0.0150	−0.030
14.0	1.56	0.0186	−0.025

From this table, plot on graph paper the variation of x_{cp}/c as a function of α.

1.7. The drag on the hull of a ship depends in part on the height of the water waves produced by the hull. The potential energy associated with these waves therefore depends on the acceleration of gravity, g. Hence, we can state that the wave drag on the hull is $D = f(\rho_\infty, V_\infty, c, g)$ where c is a length scale associated with the hull, say, the maximum width of the hull. Define the drag coefficient as $C_D \equiv D/q_\infty c^2$. Also, define a similarity parameter called the *Froude number*, $\mathrm{Fr} = V/\sqrt{gc}$. Using Buckingham's pi theorem, prove that $C_D = f(\mathrm{Fr})$.

1.8. The shock waves on a vehicle in supersonic flight cause a component of drag called supersonic wave drag, D_w. Define the wave-drag coefficient as $C_{D,w} = D_w/q_\infty S$, where S is a suitable reference area for the body. In supersonic flight, the flow is governed in part by its thermodynamic properties, given by the specific heats at constant pressure, c_p, and at constant volume, c_v. Define the ratio $c_p/c_v \equiv \gamma$. Using Buckingham's pi theorem, show that $C_{D,w} = f(M_\infty, \gamma)$. Neglect the influence of friction.

1.9. Consider two different flows over geometrically similar airfoil shapes, one airfoil being twice the size of the other. The flow over the smaller airfoil has freestream properties given by $T_\infty = 200$ K, $\rho_\infty = 1.23$ kg/m^3, and $V_\infty = 100$ m/s. The flow over the larger airfoil is described by $T_\infty = 800$ K, $\rho_\infty = 1.739$ kg/m^3, and $V_\infty = 200$ m/s. Assume that both μ and a are proportional to $T^{1/2}$. Are the two flows dynamically similar?

1.10. Consider a Lear jet flying at a velocity of 250 m/s at an altitude of 10 km, where the density and temperature are 0.414 kg/m^3 and 223 K, respectively. Consider also a one-fifth scale model of the Lear jet being tested in a wind tunnel in the laboratory. The pressure in the test section of the wind tunnel is 1 atm $= 1.01 \times 10^5$ N/m^2. Calculate the necessary velocity, temperature, and density of the airflow in the wind-tunnel test section such that the lift and drag coefficients are the same for the wind-tunnel model and the actual airplane in flight. *Note:* The relation among pressure, density, and temperature is given by the equation of state described in Prob. 1.1.

1.11. A U-tube mercury manometer is used to measure the pressure at a point on the wing of a wind-tunnel model. One side of the manometer is connected to the model, and the other side is open to the atmosphere. Atmospheric pressure and the density of liquid mercury are 1.01×10^5 N/m^2 and 1.36×10^4 kg/m^3, respectively. When the displacement of the two columns of mercury is 20 cm, with the high column on the model side, what is the pressure on the wing?

1.12. The German Zeppelins of World War I were dirigibles with the following typical characteristics: volume $= 15,000$ m^3 and maximum diameter $= 14.0$ m. Consider a Zeppelin flying at a velocity of 30 m/s at a standard altitude of 1000 m (look up the corresponding density in any standard altitude table). The Zeppelin is at a small angle of attack such that its lift coefficient is 0.05 (based on the maximum cross-sectional area). The Zeppelin is flying in straight-and-level flight with no acceleration. Calculate the total weight of the Zeppelin.

1.13. Consider a circular cylinder in a hypersonic flow, with its axis perpendicular to the flow. Let ϕ be the angle measured between radii drawn to the leading edge (the stagnation point) and to any arbitrary point on the cylinder. The pressure coefficient distribution along the cylindrical surface is given by $C_p = 2 \cos^2 \phi$ for $0 \le \phi \le \pi/2$ and $3\pi/2 \le \phi \le 2\pi$ and $C_p = 0$ for $\pi/2 \le \phi \le 3\pi/2$. Calculate the drag coefficient for the cylinder, based on projected frontal area of the cylinder.

1.14. Derive Archimedes' principle using a body of general shape.

CHAPTER
2

AERODYNAMICS: SOME FUNDAMENTAL PRINCIPLES AND EQUATIONS

There is so great a difference between a fluid and a collection of solid particles that the laws of pressure and of equilibrium of fluids are very different from the laws of the pressure and equilibrium of solids.

Jean Le Rond d'Alembert, 1768

The principle is most important, not the detail.

Theodore von Karman, 1954

2.1 INTRODUCTION AND ROAD MAP

To be a good craftsperson, one must have good tools and know how to use them effectively. Similarly, a good aerodynamicist must have good aerodynamic tools and must know how to use them for a variety of applications. The purpose of this chapter is "tool-building"; we develop some of the concepts and equations which are vital to the study of aerodynamic flows. However, please be cautioned: A craftsperson usually derives his or her pleasure from the works of art created with the *use* of the tools; the actual building of the tools themselves is sometimes considered a mundane chore. You may derive a similar feeling here. As we proceed to build our aerodynamic tools, you may wonder from time to time why such tools are necessary and what possible value they may have in the solution of practical problems. But rest assured that every aerodynamic tool we develop in this and subsequent chapters is important for the analysis and understanding of practical problems to be discussed later. So, as we move through this chapter,

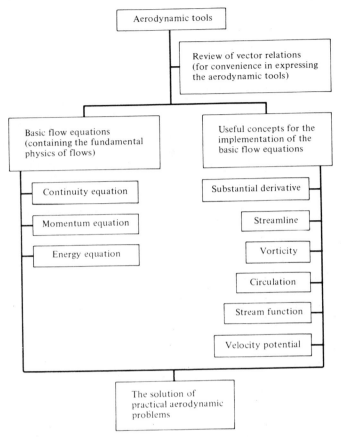

FIGURE 2.1
Road map for Chap. 2.

do not get lost or disoriented; rather, as we develop each tool, simply put it away in the store box of your mind for future use.

To help you keep track of our tool building, and to give you some orientation, the road map in Fig. 2.1 is provided for your reference. As we progress through each section of this chapter, use Fig. 2.1 to help you maintain a perspective of our work. You will note that Fig. 2.1 is full of strange-sounding terms, such as "substantial derivative," "circulation," and "velocity potential." However, when you finish this chapter, and look back at Fig. 2.1, all these terms will be second nature to you.

2.2 REVIEW OF VECTOR RELATIONS

Aerodynamics is full of quantities that have both magnitude and direction, such as force and velocity. These are *vector quantities*, and as such, the mathematics of aerodynamics is most conveniently expressed in vector notation. The purpose of this section is to set forth the basic relations we need from vector algebra and vector calculus. If you are familiar with vector analysis, this section will serve as a concise review. If you are not conversant with vector analysis, this section will help you establish some vector notation, and will serve as a skeleton from which you can fill in more information from the many existing texts on vector analysis (see, e.g., Refs. 4 to 6).

2.2.1 Some Vector Algebra

Consider a vector quantity $\mathbf{A}$; both its magnitude and direction are given by the arrow labeled $\mathbf{A}$ in Fig. 2.2. The absolute magnitude of $\mathbf{A}$ is $|\mathbf{A}|$, and is a scalar quantity. The *unit vector* $\mathbf{n}$, defined by $\mathbf{n} = \mathbf{A}/|\mathbf{A}|$, has a magnitude of unity and a direction equal to that of $\mathbf{A}$. Let $\mathbf{B}$ represent another vector. The *vector addition* of $\mathbf{A}$ and $\mathbf{B}$ yields a third vector, $\mathbf{C}$,

$$\mathbf{A} + \mathbf{B} = \mathbf{C} \tag{2.1}$$

which is formed by connecting the tail of $\mathbf{A}$ with the head of $\mathbf{B}$, as shown in Fig. 2.2. Now consider $-\mathbf{B}$, which is equal in magnitude to $\mathbf{B}$, but opposite in direction. The vector subtraction of $\mathbf{B}$ and $\mathbf{A}$ yields vector $\mathbf{D}$,

$$\mathbf{A} - \mathbf{B} = \mathbf{D} \tag{2.2}$$

which is formed by connecting the tail of $\mathbf{A}$ with the head of $-\mathbf{B}$, as shown in Fig. 2.2.

There are two forms of vector multiplication. Consider two vectors $\mathbf{A}$ and $\mathbf{B}$ at an angle θ to each other, as shown in Fig. 2.2. The *scalar product* (dot

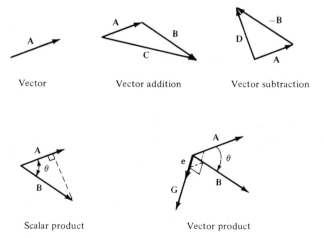

FIGURE 2.2
Vector diagrams.

product) of the two vectors **A** and **B** is defined as

$$\mathbf{A} \cdot \mathbf{B} \equiv |\mathbf{A}||\mathbf{B}| \cos \theta$$

$$= \text{magnitude of } \mathbf{A} \times \text{magnitude of the}$$
$$\text{component of } \mathbf{B} \text{ along the direction of } \mathbf{A} \quad (2.3)$$

Note that the scalar product of two vectors is a scalar. In contrast, the *vector product* (cross product) of the two vectors **A** and **B** is defined as

$$\mathbf{A} \times \mathbf{B} \equiv (|\mathbf{A}||\mathbf{B}| \sin \theta)\mathbf{e} = \mathbf{G} \quad (2.4)$$

where **G** is perpendicular to the plane of **A** and **B** and in a direction which obeys the "right-hand rule." (Rotate **A** into **B**, as shown in Fig. 2.2. Now curl the fingers of your right hand in the direction of the rotation. Your right thumb will be pointing in the direction of **G**.) In Eq. (2.4), **e** is a unit vector in the direction of **G**, as also shown in Fig. 2.2. Note that the vector product of two vectors is a vector.

2.2.2 Typical Orthogonal Coordinate Systems

To describe mathematically the flow of fluid through three-dimensional space, we have to prescribe a three-dimensional coordinate system. The geometry of some aerodynamic problems best fits a rectangular space, whereas others are mainly cylindrical in nature, and yet others may have spherical properties. Therefore, we have interest in the three most common orthogonal coordinate systems: cartesian, cylindrical, and spherical. These systems are described below. (An orthogonal coordinate system is one where all three coordinate directions are mutually perpendicular. It is interesting to note that some modern numerical solutions of fluid flows utilize nonorthogonal coordinate spaces; moreover, for some numerical problems the coordinate system is allowed to evolve and change during the course of the solution. These so-called adaptive grid techniques are beyond the scope of this book. See Ref. 7 for details.)

A *cartesian coordinate system* is shown in Fig. 2.3a. The x, y, and z axes are mutually perpendicular, and **i**, **j**, and **k** are unit vectors in the x, y, and z directions, respectively. An arbitrary point P in space is located by specifying the three coordinates (x, y, z). The point can also be located by the *position vector* **r**, where

$$\mathbf{r} = x\mathbf{i} + y\mathbf{j} + z\mathbf{k}$$

If **A** is a given vector in cartesian space, it can be expressed as

$$\mathbf{A} = A_x\mathbf{i} + A_y\mathbf{j} + A_z\mathbf{k}$$

where A_x, A_y, and A_z are the scalar components of **A** along the x, y, and z directions, respectively, as shown in Fig. 2.3b.

A *cylindrical coordinate system* is shown in Fig. 2.4a. A "phantom" cartesian system is also shown with dashed lines to help visualize the figure. The location of point P in space is given by three coordinates (r, θ, z), where r and θ are measured in the xy plane shown in Fig. 2.4a. The r coordinate direction is the

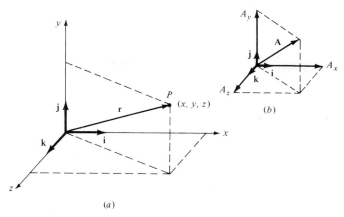

FIGURE 2.3
Cartesian coordinates.

direction of increasing r, holding θ and z constant; $\mathbf{e}_r$ is the unit vector in the r direction. The θ coordinate direction is the direction of increasing θ, holding r and z constant; $\mathbf{e}_\theta$ is the unit vector in the θ direction. The z coordinate direction is the direction of increasing z, holding r and θ constant; $\mathbf{e}_z$ is the unit vector in the z direction. If $\mathbf{A}$ is a given vector in cylindrical space, then

$$\mathbf{A} = A_r\mathbf{e}_r + A_\theta\mathbf{e}_\theta + A_z\mathbf{e}_z$$

where A_r, A_θ, and A_z are the scalar components of $\mathbf{A}$ along the r, θ, and z directions, respectively, as shown in Fig. 2.4b. The relationship, or *transformation*, between cartesian and cylindrical coordinates can be obtained from inspection

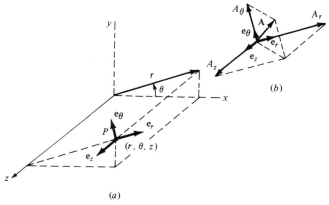

FIGURE 2.4
Cylindrical coordinates.

of Fig. 2.4a, namely,

$$x = r \cos \theta$$

$$y = r \sin \theta \qquad (2.5)$$

$$z = z$$

or inversely,

$$r = \sqrt{x^2 + y^2}$$

$$\theta = \arctan \frac{y}{x} \qquad (2.6)$$

$$z = z$$

A *spherical coordinate system* is shown in Fig. 2.5a. Once again, a phantom cartesian system is shown with dashed lines. (However, for clarity in the picture, the z axis is drawn vertically, in contrast to Figs. 2.3 and 2.4.) The location of point P in space is given by the three coordinates (r, θ, Φ), where r is the distance of P from the origin, θ is the angle measured from the z axis and is in the rz plane, and Φ is the angle measured from the x axis and is in the xy plane. The r coordinate direction is the direction of increasing r, holding θ and Φ constant; $\mathbf{e}_r$ is the unit vector in the r direction. The θ coordinate direction is the direction of increasing θ, holding r and Φ constant; $\mathbf{e}_\theta$ is the unit vector in the θ direction. The Φ coordinate direction is the direction of increasing Φ, holding r and θ constant; $\mathbf{e}_\Phi$ is the unit vector in the Φ direction. The unit vectors $\mathbf{e}_r$, $\mathbf{e}_\theta$, and $\mathbf{e}_\Phi$ are mutually perpendicular. If $\mathbf{A}$ is a given vector in spherical space, then

$$\mathbf{A} = A_r \mathbf{e}_r + A_\theta \mathbf{e}_\theta + A_\Phi \mathbf{e}_\Phi$$

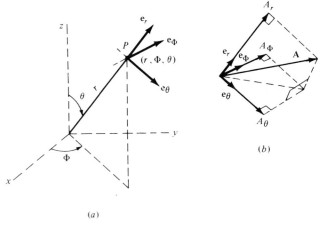

(a)

(b)

FIGURE 2.5
Spherical coordinates.

where A_r, A_θ, and A_Φ are the scalar components of **A** along the r, θ, and Φ directions, respectively, as shown in Fig. 2.5b. The transformation between cartesian and spherical coordinates is obtained from inspection of Fig. 2.5a, namely,

$$x = r \sin \theta \cos \Phi$$
$$y = r \sin \theta \sin \Phi \tag{2.7}$$
$$z = r \cos \theta$$

or inversely,

$$r = \sqrt{x^2 + y^2 + z^2}$$
$$\theta = \arccos \frac{z}{r} = \arccos \frac{z}{\sqrt{x^2 + y^2 + z^2}} \tag{2.8}$$
$$\Phi = \arccos \frac{x}{\sqrt{x^2 + y^2}}$$

2.2.3 Scalar and Vector Fields

A scalar quantity given as a function of coordinate space and time t is called a *scalar field*. For example, pressure, density, and temperature are scalar quantities, and

$$p = p_1(x, y, z, t) = p_2(r, \theta, z, t) = p_3(r, \theta, \Phi, t)$$
$$\rho = \rho_1(x, y, z, t) = \rho_2(r, \theta, z, t) = \rho_3(r, \theta, \Phi, t)$$
$$T = T_1(x, y, z, t) = T_2(r, \theta, z, t) = T_3(r, \theta, \Phi, t)$$

are scalar fields for pressure, density, and temperature, respectively. Similarly, a vector quantity given as a function of coordinate space and time is called a *vector field*. For example, velocity is a vector quantity, and

$$\mathbf{V} = V_x \mathbf{i} + V_y \mathbf{j} + V_z \mathbf{k}$$

where

$$V_x = V_x(x, y, z, t)$$
$$V_y = V_y(x, y, z, t)$$
$$V_z = V_z(x, y, z, t)$$

is the vector field for **V** in cartesian space. Analogous expressions can be written for vector fields in cylindrical and spherical space. In many theoretical aerodynamic problems, the above scalar and vector fields are the unknowns to be obtained in a solution for a flow with prescribed initial and boundary conditions.

2.2.4 Scalar and Vector Products

The scalar and vector products defined by Eqs. (2.3) and (2.4), respectively, can be written in terms of the components of each vector as follows.

CARTESIAN COORDINATES. Let

$$\mathbf{A} = A_x \mathbf{i} + A_y \mathbf{j} + A_z \mathbf{k}$$

and

$$\mathbf{B} = B_x \mathbf{i} + B_y \mathbf{j} + B_z \mathbf{k}$$

Then
$$\mathbf{A} \cdot \mathbf{B} = A_x B_x + A_y B_y + A_z B_z \tag{2.9}$$
and
$$\mathbf{A} \times \mathbf{B} = \begin{bmatrix} \mathbf{i} & \mathbf{j} & \mathbf{k} \\ A_x & A_y & A_z \\ B_x & B_y & B_z \end{bmatrix} = \mathbf{i}(A_y B_z - A_z B_y) + \mathbf{j}(A_z B_x - A_x B_z) + \mathbf{k}(A_x B_y - A_y B_x) \tag{2.10}$$

CYLINDRICAL COORDINATES. Let
$$\mathbf{A} = A_r \mathbf{e}_r + A_\theta \mathbf{e}_\theta + A_z \mathbf{e}_z$$
and
$$\mathbf{B} = B_r \mathbf{e}_r + B_\theta \mathbf{e}_\theta + B_z \mathbf{e}_z$$
Then
$$\mathbf{A} \cdot \mathbf{B} = A_r B_r + A_\theta B_\theta + A_z B_z \tag{2.11}$$
and
$$\mathbf{A} \times \mathbf{B} = \begin{vmatrix} \mathbf{e}_r & \mathbf{e}_\theta & \mathbf{e}_z \\ A_r & A_\theta & A_z \\ B_r & B_\theta & B_z \end{vmatrix} \tag{2.12}$$

SPHERICAL COORDINATES. Let
$$\mathbf{A} = A_r \mathbf{e}_r + A_\theta \mathbf{e}_\theta + A_\Phi \mathbf{e}_\Phi$$
and
$$\mathbf{B} = B_r \mathbf{e}_r + B_\theta \mathbf{e}_\theta + B_\Phi \mathbf{e}_\Phi$$
Then
$$\mathbf{A} \cdot \mathbf{B} = A_r B_r + A_\theta B_\theta + A_\Phi B_\Phi \tag{2.13}$$
and
$$\mathbf{A} \times \mathbf{B} = \begin{vmatrix} \mathbf{e}_r & \mathbf{e}_\theta & \mathbf{e}_\Phi \\ A_r & A_\theta & A_\Phi \\ B_r & B_\theta & B_\Phi \end{vmatrix} \tag{2.14}$$

2.2.5 Gradient of a Scalar Field

We now begin a review of some elements of vector calculus. Consider a scalar field
$$p = p_1(x, y, z) = p_2(r, \theta, z) = p_3(r, \theta, \Phi)$$
The *gradient* of p, ∇p, at a given point in space is defined as a vector such that:

1. Its magnitude is the maximum rate of change of p per unit length of the coordinate space at the given point.

2. Its direction is that of the maximum rate of change of p at the given point.

For example, consider a two-dimensional pressure field in cartesian space as sketched in Fig. 2.6. The solid curves are lines of constant pressure; i.e., they connect points in the pressure field which have the same value of p. Such lines are called *isolines*. Consider an arbitrary point (x, y) in Fig. 2.6. If we move away from this point in an arbitrary direction, p will, in general, change because we are moving to another location in space. Moreover, there will be some direction from this point along which p changes *the most* over a unit length in that direction.

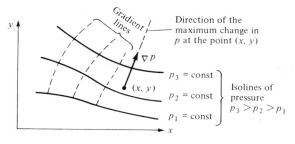

FIGURE 2.6

Illustration of the gradient of a scalar field.

This defines the *direction of the gradient* of p and is identified in Fig. 2.6. The magnitude of ∇p is the rate of change of p per unit length in that direction. Both the magnitude and direction of ∇p will change from one point to another in the coordinate space. A line drawn in this space along which ∇p is tangent at every point is defined as a *gradient line*, as sketched in Fig. 2.6. The gradient line and isoline through any given point in the coordinate space are perpendicular.

Consider ∇p at a given point (x, y) as shown in Fig. 2.7. Choose some arbitrary direction s away from the point, as also shown in Fig. 2.7. Let $\mathbf{n}$ be a unit vector in the s direction. The rate of change of p per unit length in the s direction is

$$\frac{dp}{ds} = \nabla p \cdot \mathbf{n} \tag{2.15}$$

In Eq. (2.15), dp/ds is called the *directional derivative* in the s direction. Note from Eq. (2.15) that the rate of change of p in any arbitrary direction is simply the component of ∇p in that direction.

Expressions for ∇p in the different coordinate systems are given below:

Cartesian:
$$p = p(x, y, z)$$

$$\boxed{\nabla p = \frac{\partial p}{\partial x}\mathbf{i} + \frac{\partial p}{\partial y}\mathbf{j} + \frac{\partial p}{\partial z}\mathbf{k}} \tag{·2.16}$$

Cylindrical:
$$p = p(r, \theta, z)$$

$$\boxed{\nabla p = \frac{\partial p}{\partial r}\mathbf{e}_r + \frac{1}{r}\frac{\partial p}{\partial \theta}\mathbf{e}_\theta + \frac{\partial p}{\partial z}\mathbf{e}_z} \tag{2.17}$$

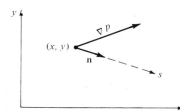

FIGURE 2.7

Sketch for the directional derivative.

Spherical: $$p = p(r, \theta, \Phi)$$

$$\nabla p = \frac{\partial p}{\partial r} \mathbf{e}_r + \frac{1}{r} \frac{\partial p}{\partial \theta} \mathbf{e}_\theta + \frac{1}{r \sin \theta} \frac{\partial p}{\partial \Phi} \mathbf{e}_\Phi \qquad (2.18)$$

2.2.6 Divergence of a Vector Field

Consider a vector field

$$\mathbf{V} = \mathbf{V}(x, y, z) = \mathbf{V}(r, \theta, z) = \mathbf{V}(r, \theta, \Phi)$$

In the above, $\mathbf{V}$ can represent any vector quantity. However, for practical purposes, and to aid in physical interpretation, consider $\mathbf{V}$ to be the flow velocity. Also, visualize a small fluid element of fixed mass moving along a streamline with velocity $\mathbf{V}$. As the fluid element moves through space, its volume will, in general, change. In Sec. 2.3, we prove that the time rate of change of the volume of a moving fluid element of fixed mass, per unit volume of that element, is equal to the *divergence* of $\mathbf{V}$, denoted by $\nabla \cdot \mathbf{V}$. The divergence of a vector is a scalar quantity; it is one of two ways that the derivative of a vector field can be defined. In different coordinate systems, we have

Cartesian: $$\mathbf{V} = \mathbf{V}(x, y, z) = V_x \mathbf{i} + V_y \mathbf{j} + V_z \mathbf{k}$$

$$\nabla \cdot \mathbf{V} = \frac{\partial V_x}{\partial x} + \frac{\partial V_y}{\partial y} + \frac{\partial V_z}{\partial z} \qquad (2.19)$$

Cylindrical: $$\mathbf{V} = \mathbf{V}(r, \theta, z) = V_r \mathbf{e}_r + V_\theta \mathbf{e}_\theta + V_z \mathbf{e}_z$$

$$\nabla \cdot \mathbf{V} = \frac{1}{r} \frac{\partial}{\partial r} (r V_r) + \frac{1}{r} \frac{\partial V_\theta}{\partial \theta} + \frac{\partial V_z}{\partial z} \qquad (2.20)$$

Spherical: $$\mathbf{V} = \mathbf{V}(r, \theta, \Phi) = V_r \mathbf{e}_r + V_\theta \mathbf{e}_\theta + V_\Phi \mathbf{e}_\Phi$$

$$\nabla \cdot \mathbf{V} = \frac{1}{r^2} \frac{\partial}{\partial r} (r^2 V_r) + \frac{1}{r \sin \theta} \frac{\partial}{\partial \theta} (V_\theta \sin \theta) + \frac{1}{r \sin \theta} \frac{\partial V_\Phi}{\partial \Phi} \qquad (2.21)$$

2.2.7 Curl of a Vector Field

Consider a vector field

$$\mathbf{V} = \mathbf{V}(x, y, z) = \mathbf{V}(r, \theta, z) = \mathbf{V}(r, \theta, \Phi)$$

Although $\mathbf{V}$ can be any vector quantity, again consider $\mathbf{V}$ to be the flow velocity. Once again visualize a fluid element moving along a streamline. It is possible for this fluid element to be rotating with an angular velocity $\boldsymbol{\omega}$ as it translates along the streamline. In Sec. 2.9, we prove that $\boldsymbol{\omega}$ is equal to one-half of the *curl of* $\mathbf{V}$, where the curl of $\mathbf{V}$ is denoted by $\nabla \times \mathbf{V}$. The curl of $\mathbf{V}$ is a vector quantity; it is the alternate way that the derivative of a vector field can be defined, the first being $\nabla \cdot \mathbf{V}$ (see Sec. 2.2.6, Divergence of a Vector Field). In different coordinate systems, we have

Cartesian:
$$\mathbf{V} = V_x\mathbf{i} + V_y\mathbf{j} + V_z\mathbf{k}$$

$$\nabla \times \mathbf{V} = \begin{vmatrix} \mathbf{i} & \mathbf{j} & \mathbf{k} \\ \dfrac{\partial}{\partial x} & \dfrac{\partial}{\partial y} & \dfrac{\partial}{\partial z} \\ V_x & V_y & V_z \end{vmatrix} = \mathbf{i}\left(\frac{\partial V_z}{\partial y} - \frac{\partial V_y}{\partial z}\right) + \mathbf{j}\left(\frac{\partial V_x}{\partial z} - \frac{\partial V_z}{\partial x}\right) + \mathbf{k}\left(\frac{\partial V_y}{\partial x} - \frac{\partial V_x}{\partial y}\right)$$

(2.22)

Cylindrical:
$$\mathbf{V} = V_r\mathbf{e}_r + V_\theta\mathbf{e}_\theta + V_z\mathbf{e}_z$$

$$\nabla \times \mathbf{V} = \frac{1}{r}\begin{vmatrix} \mathbf{e}_r & r\mathbf{e}_\theta & \mathbf{e}_z \\ \dfrac{\partial}{\partial r} & \dfrac{\partial}{\partial \theta} & \dfrac{\partial}{\partial z} \\ V_r & rV_\theta & V_z \end{vmatrix}$$

(2.23)

Spherical:
$$\mathbf{V} = V_r\mathbf{e}_r + V_\theta\mathbf{e}_\theta + V_\Phi\mathbf{e}_\Phi$$

$$\nabla \times \mathbf{V} = \frac{1}{r^2 \sin\theta}\begin{vmatrix} \mathbf{e}_r & r\mathbf{e}_\theta & (r\sin\theta)\mathbf{e}_\Phi \\ \dfrac{\partial}{\partial r} & \dfrac{\partial}{\partial \theta} & \dfrac{\partial}{\partial \Phi} \\ V_r & rV_\theta & (r\sin\theta)V_\Phi \end{vmatrix}$$

(2.24)

2.2.8 Line Integrals

Consider a vector field

$$\mathbf{A} = \mathbf{A}(x, y, z) = \mathbf{A}(r, \theta, z) = \mathbf{A}(r, \theta, \Phi)$$

Also, consider a curve C in space connecting two points a and b as shown on the left side of Fig. 2.8. Let ds be an elemental length of the curve, and $\mathbf{n}$ be a

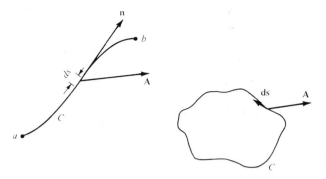

FIGURE 2.8
Sketch for line integrals.

unit vector tangent to the curve. Define the vector $\mathbf{ds} = \mathbf{n}\, ds$. Then, the *line integral* of A along curve C from point a to point b is

$$\oint_a^b \mathbf{A} \cdot \mathbf{ds}$$

If the curve C is closed, as shown at the right of Fig. 2.8, then the line integral is given by

$$\oint_C \mathbf{A} \cdot \mathbf{ds}$$

where the *counterclockwise* direction around C is considered positive. (The positive direction around a closed curve is, by convention, that direction you would move such that the area enclosed by C is always on your left.)

2.2.9 Surface Integrals

Consider an open surface S bounded by the closed curve C, as shown in Fig. 2.9. At point P on the surface, let dS be an elemental area of the surface and $\mathbf{n}$ be a unit vector normal to the surface. The orientation of $\mathbf{n}$ is in the direction according to the right-hand rule for movement along C. (Curl the fingers of your right hand in the direction of movement around C; your thumb will then point in the general direction of $\mathbf{n}$.) Define a vector elemental area as $\mathbf{dS} = \mathbf{n}\, dS$. In

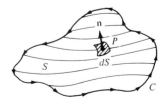

FIGURE 2.9
Sketch for surface integrals. The three-dimensional surface area S is bounded by the closed curve C.

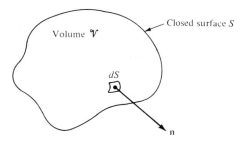

FIGURE 2.10
Volume $\mathscr{V}$ enclosed by the closed surface S.

terms of **dS**, the *surface integral* over the surface S can be defined in three ways:

$$\iint\limits_S p \, \mathbf{dS} = \text{surface integral of a scalar } p \text{ over the open surface } S \text{ (the result is a vector)}$$

$$\iint\limits_S \mathbf{A} \cdot \mathbf{dS} = \text{surface integral of a vector } \mathbf{A} \text{ over the open surface } S \text{ (the result is a scalar)}$$

$$\iint\limits_S \mathbf{A} \times \mathbf{dS} = \text{surface integral of a vector } \mathbf{A} \text{ over the open surface } S \text{ (the result is a vector)}$$

If the surface S is *closed* (e.g., the surface of a sphere or a cube), **n** points out of the surface, away from the enclosed volume, as shown in Fig. 2.10. The surface integrals over the closed surface are

$$\oiint\limits_S p \, \mathbf{dS} \qquad \oiint\limits_S \mathbf{A} \cdot \mathbf{dS} \qquad \oiint\limits_S \mathbf{A} \times \mathbf{dS}$$

2.2.10 Volume Integrals

Consider a volume $\mathscr{V}$ in space. Let ρ be a scalar field in this space. The *volume integral* over the volume $\mathscr{V}$ of the quantity ρ is written as

$$\oiiint\limits_{\mathscr{V}} \rho \, d\mathscr{V} = \text{volume integral of a scalar } \rho \text{ over the volume } \mathscr{V} \text{ (the result is a scalar)}$$

Let **A** be a vector field in space. The volume integral over the volume $\mathscr{V}$ of the quantity **A** is written as

$$\oiiint\limits_{\mathscr{V}} \mathbf{A} \, d\mathscr{V} = \text{volume integral of a vector } \mathbf{A} \text{ over the volume } \mathscr{V} \text{ (the result is a vector)}$$

2.2.11 Relations Between Line, Surface, and Volume Integrals

Consider again the open area S bounded by the closed curve C, as shown in Fig. 2.9. Let $\mathbf{A}$ be a vector field. The line integral of $\mathbf{A}$ over C is related to the surface integral of $\mathbf{A}$ over S by *Stokes' theorem*:

$$\oint_C \mathbf{A} \cdot \mathbf{ds} \equiv \iint_S (\nabla \times \mathbf{A}) \cdot \mathbf{dS} \tag{2.25}$$

Consider again the volume $\mathscr{V}$ enclosed by the closed surface S, as shown in Fig. 2.10. The surface and volume integrals of the vector field $\mathbf{A}$ are related through the *divergence theorem*:

$$\oiint_S \mathbf{A} \cdot \mathbf{dS} \equiv \oiiint_\mathscr{V} (\nabla \cdot \mathbf{A}) \, d\mathscr{V} \tag{2.26}$$

If p represents a scalar field, a vector relationship analogous to Eq. (2.26) is given by the *gradient theorem*:

$$\oiint_S p \, \mathbf{dS} \equiv \oiiint_\mathscr{V} \nabla p \, d\mathscr{V} \tag{2.27}$$

2.2.12 Summary

This section has provided a concise review of those elements of vector analysis which we will use as tools in our subsequent discussions. Make certain to review these tools until you feel comfortable with them, especially the relations in boxes.

2.3 MODELS OF THE FLUID: CONTROL VOLUMES AND FLUID ELEMENTS

Aerodynamics is a fundamental science, steeped in physical observation. As you proceed through this book, make every effort to gradually develop a "physical feel" for the material. An important virtue of all successful aerodynamicists (indeed, of all successful engineers and scientists) is that they have good "physical intuition," based on thought and experience, which allows them to make reasonable judgments on difficult problems. Although this chapter is full of equations and (seemingly) esoteric concepts, now is the time for you to start developing this physical feel.

With this section, we begin to build the basic equations of aerodynamics. There is a certain philosophical procedure involved with the development of these equations, as follows:

1. Invoke three fundamental physical principles which are deeply entrenched in our macroscopic observations of nature, namely,
   a. Mass is conserved, i.e., mass can be neither created nor destroyed.
   b. Newton's second law: force = mass × acceleration.
   c. Energy is conserved; it can only change from one form to another.
2. Determine a suitable *model* of the fluid. Remember that a fluid is a squishy substance, and therefore it is usually more difficult to describe than a well-defined solid body. Hence, we have to adopt a reasonable model of the fluid to which we can apply the fundamental principles stated in item 1.
3. Apply the fundamental physical principles listed in item 1 to the model of the fluid determined in item 2 in order to obtain mathematical equations which properly describe the physics of the flow. In turn, use these fundamental equations to analyze any particular aerodynamic flow problem of interest.

In this section, we concentrate on item 2; namely, we ask the question: What is a suitable model of the fluid? How do we visualize this squishy substance in order to apply the three fundamental physical principles to it? There is no single answer to this question; rather, three different models have been used successfully throughout the modern evolution of aerodynamics. They are (1) finite control volume, (2) infinitesimal fluid element, and (3) molecular. Let us examine what these models involve and how they are applied.

2.3.1 Finite Control Volume Approach

Consider a general flow field as represented by the streamlines in Fig. 2.11. Let us imagine a closed volume drawn within a *finite* region of the flow. This volume

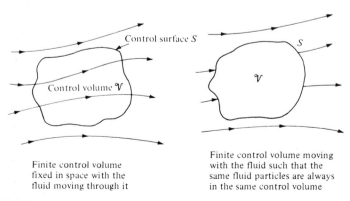

Finite control volume fixed in space with the fluid moving through it

Finite control volume moving with the fluid such that the same fluid particles are always in the same control volume

FIGURE 2.11
Finite control volume approach.

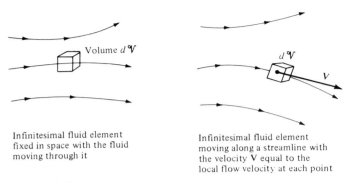

Infinitesimal fluid element
fixed in space with the fluid
moving through it

Infinitesimal fluid element
moving along a streamline with
the velocity **V** equal to the
local flow velocity at each point

FIGURE 2.12
Infinitesimal fluid element approach.

defines a *control volume*, $\mathcal{V}$, and a *control surface*, S, is defined as the closed surface which bounds the control volume. The control volume may be fixed in space with the fluid moving through it, as shown at the left of Fig. 2.11. Alternatively, the control volume may be moving with the fluid such that the same fluid particles are always inside it, as shown at the right of Fig. 2.11. In either case, the control volume is a reasonably large, finite region of the flow. The fundamental physical principles are applied to the fluid inside the control volume, and to the fluid crossing the control surface (if the control volume is fixed in space). Therefore, instead of looking at the whole flow field at once, with the control volume model we limit our attention to just the fluid in the finite region of the volume itself.

2.3.2 Infinitesimal Fluid Element Approach

Consider a general flow field as represented by the streamlines in Fig. 2.12. Let us imagine an infinitesimally small fluid element in the flow, with a differential volume $d\mathcal{V}$. The fluid element is infinitesimal in the same sense as differential calculus; however, it is large enough to contain a huge number of molecules so that it can be viewed as a continuous medium. The fluid element may be fixed in space with the fluid moving through it, as shown at the left of Fig. 2.12. Alternatively, it may be moving along a streamline with velocity **V** equal to the flow velocity at each point. Again, instead of looking at the whole flow field at once, the fundamental physical principles are applied to just the fluid element itself.

2.3.3 Molecular Approach

In actuality, of course, the motion of a fluid is a ramification of the mean motion of its atoms and molecules. Therefore, a third model of the flow can be a microscopic approach wherein the fundamental laws of nature are applied directly to the atoms and molecules, using suitable statistical averaging to define the resulting fluid properties. This approach is in the purview of *kinetic theory*, which is a very elegant method with many advantages in the long run. However, it is beyond the scope of the present book.

In summary, although many variations on the theme can be found in different texts for the derivation of the general equations of fluid flow, the flow model can usually be categorized under one of the approaches described above.

2.3.4 Physical Meaning of the Divergence of Velocity

In the equations to follow, the divergence of velocity, $\nabla \cdot \mathbf{V}$, occurs frequently. Before leaving this section, let us prove the statement made earlier (Sec. 2.2) that $\nabla \cdot \mathbf{V}$ is physically the time rate of change of the volume of a moving fluid element of fixed mass per unit volume of that element. Consider a control volume moving with the fluid (the case shown on the right of Fig. 2.11). This control volume is always made up of the same fluid particles as it moves with the flow; hence, its mass is fixed, invariant with time. However, its volume $\mathcal{V}$ and control surface S are changing with time as it moves to different regions of the flow where different values of ρ exist. That is, this moving control volume of fixed mass is constantly increasing or decreasing its volume and is changing its shape, depending on the characteristics of the flow. This control volume is shown in Fig. 2.13 at some instant in time. Consider an infinitesimal element of the surface dS moving at the local velocity $\mathbf{V}$, as shown in Fig. 2.13. The change in the volume of the control volume $\Delta \mathcal{V}$, due to just the movement of dS over a time increment Δt, is, from Fig. 2.13, equal to the volume of the long, thin cylinder with base area dS and altitude $(\mathbf{V} \, \Delta t) \cdot \mathbf{n}$; i.e.,

$$\Delta \mathcal{V} = [(\mathbf{V}\Delta t) \cdot \mathbf{n}] \, dS = (\mathbf{V}\Delta t) \cdot \mathbf{dS} \tag{2.28}$$

Over the time increment Δt, the total change in volume of the whole control volume is equal to the summation of Eq. (2.28) over the total control surface. In the limit as $dS \rightarrow 0$, the sum becomes the surface integral

$$\oiint_S (\mathbf{V} \, \Delta t) \cdot \mathbf{dS}$$

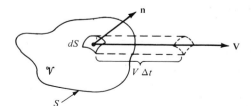

FIGURE 2.13
Moving control volume used for the physical interpretation of the divergence of velocity.

If this integral is divided by Δt, the result is physically the time rate of change of the control volume, denoted by $D\mathcal{V}/Dt$; i.e.,

$$\frac{D\mathcal{V}}{Dt} = \frac{1}{\Delta t} \oiint_S (\mathbf{V}\Delta t) \cdot \mathbf{dS} = \oiint_S \mathbf{V} \cdot \mathbf{dS} \tag{2.29}$$

(The significance of the notation D/Dt is revealed in Sec. 2.9.) Applying the divergence theorem, Eq. (2.26), to the right side of Eq. (2.29), we have

$$\frac{D\mathcal{V}}{Dt} = \oiiint_{\mathcal{V}} (\nabla \cdot \mathbf{V}) \, d\mathcal{V} \tag{2.30}$$

Now let us imagine that the moving control volume in Fig. 2.13 is shrunk to a very small volume, $\delta\mathcal{V}$, essentially becoming an infinitesimal moving fluid element as sketched on the right of Fig. 2.12. Then Eq. (2.30) can be written as

$$\frac{D(\delta\mathcal{V})}{Dt} = \oiiint_{\delta\mathcal{V}} (\nabla \cdot \mathbf{V}) \, d\mathcal{V} \tag{2.31}$$

Assume that $\delta\mathcal{V}$ is small enough such that $\nabla \cdot \mathbf{V}$ is essentially the same value throughout $\delta\mathcal{V}$. Then the integral in Eq. (2.31) can be approximated as $(\nabla \cdot \mathbf{V})\delta\mathcal{V}$. From Eq. (2.31), we have

$$\frac{D(\delta\mathcal{V})}{Dt} = (\nabla \cdot \mathbf{V})\delta\mathcal{V}$$

or

$$\boxed{\nabla \cdot \mathbf{V} = \frac{1}{\delta\mathcal{V}} \frac{D(\delta\mathcal{V})}{Dt}} \tag{2.32}$$

Examine Eq. (2.32). It states that $\nabla \cdot \mathbf{V}$ is physically the *time rate of change of the volume of a moving fluid element, per unit volume*. Hence, the interpretation of $\nabla \cdot \mathbf{V}$, first given in Sec. 2.2.6, Divergence of a Vector Field, is now proved.

2.4 CONTINUITY EQUATION

In Sec. 2.3, we discussed several models which can be used to study the motion of a fluid. Following the philosophy set forth at the beginning of Sec. 2.3, we now apply the fundamental physical principles to such models. Unlike the above derivation of the physical significance of $\nabla \cdot \mathbf{V}$ wherein we used the model of a moving finite control volume, we now employ the model of a *fixed* finite control volume as sketched on the left side of Fig. 2.11. Here, the control volume is fixed in space, with the flow moving through it. Unlike our previous derivation, the volume $\mathcal{V}$ and control surface $\mathbf{S}$ are now constant with time, and the mass of fluid contained within the control volume can change as a function of time (due to unsteady fluctuations of the flow field).

Before starting the derivation of the fundamental equations of aerodynamics, we must examine a concept vital to those equations, namely, the concept of *mass flow*. Consider a given area A arbitrarily oriented in a flow field as shown in Fig. 2.14. In Fig. 2.14, we are looking at an edge view of area A. Let A be small enough such that the flow velocity V is uniform across A. Consider the fluid elements with velocity V that pass through A. In time dt after crossing A, they have moved a distance $V \, dt$ and have swept out the shaded volume shown in Fig. 2.14. This volume is equal to the base area A times the height of the cylinder, $V_n \, dt$, where V_n is the component of velocity normal to A; i.e.,

$$\text{Volume} = (V_n \, dt)A$$

The mass inside the shaded volume is therefore

$$\text{Mass} = \rho(V_n \, dt)A \tag{2.33}$$

This is the mass that has swept past A in time dt. By definition, the *mass flow* through A is the mass crossing A per second (e.g., kilograms per second, slugs per second). Let $\dot{m}$ denote mass flow. From Eq. (2.33),

$$\dot{m} = \frac{\rho(V_n \, dt)A}{dt}$$

or

$$\boxed{\dot{m} = \rho V_n A} \tag{2.34}$$

Equation (2.34) demonstrates that mass flow through A is given by the product

> Area × density × component of flow velocity *normal* to the area

A related concept is that of *mass flux*, defined as the mass flow *per unit area*. From Eq. (2.34),

$$\boxed{\text{Mass flux} = \frac{\dot{m}}{A} = \rho V_n} \tag{2.35}$$

Typical units of mass flux are $\text{kg}/(\text{s} \cdot \text{m}^2)$ and $\text{slug}/(\text{s} \cdot \text{ft}^2)$.

The concepts of mass flow and mass flux are important. Note from Eq. (2.35) that mass flux across a surface is equal to the product of density times the

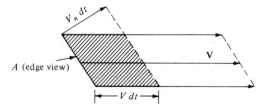

FIGURE 2.14
Sketch for discussion of mass flow through area A in a flow field.

component of velocity perpendicular to the surface. Many of the equations of aerodynamics involve products of density and velocity. For example, in cartesian coordinates, $\mathbf{V} = V_x\mathbf{i} + V_y\mathbf{j} + V_z\mathbf{k} = u\mathbf{i} + v\mathbf{j} + w\mathbf{k}$, where u, v, and w denote the x, y, and z components of velocity, respectively. (The use of u, v, and w rather than V_x, V_y, and V_z to symbolize the x, y, and z components of velocity is quite common in aerodynamic literature; we henceforth adopt the u, v, and w notation.) In many of the equations of aerodynamics, you will find the products ρu, ρv, and ρw; always remember that these products are the mass fluxes in the x, y, and z directions, respectively. In a more general sense, if V is the magnitude of velocity in an arbitrary direction, the product ρV is physically the mass flux (mass flow per unit area) across an area oriented perpendicular to the direction of V.

We are now ready to apply our first physical principle to a finite control volume fixed in space.

Physical principle   Mass can be neither created nor destroyed.

Consider a flow field wherein all properties vary with spatial location and time, e.g., $\rho = \rho(x, y, z, t)$. In this flow field, consider the fixed finite control volume shown in Fig. 2.15. At a point on the control surface, the flow velocity is $\mathbf{V}$ and the vector elemental surface area is $\mathbf{dS}$. Also, $d\mathcal{V}$ is an elemental volume inside the control volume. Applied to this control volume, the above physical principle means

$$
\begin{array}{ccc}
\text{Net mass flow } \textit{out} \text{ of control} & & \text{time rate of decrease of} \\
\text{volume through surface } S & = & \text{mass inside control volume } \mathcal{V}
\end{array} \qquad (2.36a)
$$

or
$$ B = C \qquad (2.36b) $$

where B and C are just convenient symbols for the left and right sides, respectively, of Eq. (2.36a). First, let us obtain an expression for B in terms of the quantities shown in Fig. 2.15. From Eq. (2.34), the elemental mass flow across the area dS is

$$ \rho V_n\, dS = \rho \mathbf{V} \cdot \mathbf{dS} $$

Examining Fig. 2.15, note that by convention, $\mathbf{dS}$ always points in a direction *out* of the control volume. Hence, when $\mathbf{V}$ also points out of the control volume

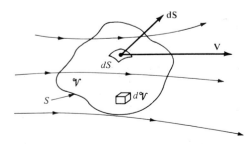

FIGURE 2.15
Finite control volume fixed in space.

(as shown in Fig. 2.15), the product $\rho\mathbf{V} \cdot \mathbf{dS}$ is *positive*. Moreover, when $\mathbf{V}$ points out of the control volume, the mass flow is physically leaving the control volume; i.e., it is an *outflow*. Hence, a positive $\rho\mathbf{V} \cdot \mathbf{dS}$ denotes an outflow. In turn, when $\mathbf{V}$ points into the control volume, $\rho\mathbf{V} \cdot \mathbf{dS}$ is *negative*. Moreover, when $\mathbf{V}$ points inward, the mass flow is physically entering the control volume; i.e., it is an *inflow*. Hence, a negative $\rho\mathbf{V} \cdot \mathbf{dS}$ denotes an *inflow*. The *net* mass flow *out* of the entire control surface S is the summation over S of the elemental mass flows. In the limit, this becomes a surface integral, which is physically the left side of Eqs. (2.36*a* and *b*); i.e.,

$$B = \oiint_S \rho\mathbf{V} \cdot \mathbf{dS} \tag{2.37}$$

Now consider the right side of Eqs. (2.36*a* and *b*). The mass contained within the elemental volume $d\mathcal{V}$ is

$$\rho \, d\mathcal{V}$$

Hence, the total mass inside the control volume is

$$\oiiint_{\mathcal{V}} \rho \, d\mathcal{V}$$

The time rate of *increase* of mass inside $\mathcal{V}$ is then

$$\frac{\partial}{\partial t} \oiiint_{\mathcal{V}} \rho \, d\mathcal{V}$$

In turn, the time rate of *decrease* of mass inside $\mathcal{V}$ is the negative of the above; i.e.,

$$-\frac{\partial}{\partial t} \oiiint_{\mathcal{V}} \rho \, d\mathcal{V} = C \tag{2.38}$$

Thus, substituting Eqs. (2.37) and (2.38) into (2.36*b*), we have

$$\oiint_S \rho\mathbf{V} \cdot \mathbf{dS} = -\frac{\partial}{\partial t} \oiiint_{\mathcal{V}} \rho \, d\mathcal{V}$$

or

$$\boxed{\frac{\partial}{\partial t} \oiiint_{\mathcal{V}} \rho \, d\mathcal{V} + \oiint_S \rho\mathbf{V} \cdot \mathbf{dS} = 0} \tag{2.39}$$

Equation (2.39) is the final result of applying the physical principle of the conservation of mass to a finite control volume fixed in space. Equation (2.39) is called the *continuity equation*. It is one of the most fundamental equations of fluid dynamics.

Note that Eq. (2.39) expresses the continuity equation in integral form. We will have numerous opportunities to use this form; it has the advantage of relating aerodynamic phenomena over a finite region of space without being concerned about the details of precisely what is happening at a given distinct point in the flow. On the other hand, there are many times when we are concerned with the details of a flow and we want to have equations which relate flow properties at a *given point*. In such a case, the integral form as expressed in Eq. (2.39) is not particularly useful. However, Eq. (2.39) can be reduced to another form which does relate flow properties at a given point, as follows. To begin with, since the control volume used to obtain Eq. (2.39) is fixed in space, the limits of integration are also fixed. Hence, the time derivative can be placed inside the volume integral, and Eq. (2.39) can be written as

$$\iiint_{\mathcal{V}} \frac{\partial \rho}{\partial t}\,d\mathcal{V} + \iint_{S} \rho\mathbf{V}\cdot d\mathbf{S} = 0 \tag{2.40}$$

Applying the divergence theorem, Eq. (2.26), we can express the right-hand term of Eq. (2.40) as

$$\iint_{S} (\rho\mathbf{V})\cdot d\mathbf{S} = \iiint_{\mathcal{V}} \nabla\cdot(\rho\mathbf{V})\,d\mathcal{V} \tag{2.41}$$

Substituting Eq. (2.41) into (2.40), we obtain

$$\iiint_{\mathcal{V}} \frac{\partial \rho}{\partial t}\,d\mathcal{V} + \iiint_{\mathcal{V}} \nabla\cdot(\rho\mathbf{V})\,d\mathcal{V} = 0$$

or

$$\iiint_{\mathcal{V}} \left[\frac{\partial \rho}{\partial t} + \nabla\cdot(\rho\mathbf{V})\right]d\mathcal{V} = 0 \tag{2.42}$$

Examine the integrand of Eq. (2.42). If the integrand were a finite number, then Eq. (2.42) would require that the integral over part of the control volume be equal and opposite in sign to the integral over the remainder of the control volume, such that the net integration would be zero. However, the finite control volume is *arbitrarily* drawn in space; there is no reason to expect cancellation of one region by the other. Hence, the only way for the integral in Eq. (2.42) to be zero for an arbitrary control volume is for the integrand to be zero at *all* points within the control volume. Thus, from Eq. (2.42), we have

$$\boxed{\frac{\partial \rho}{\partial t} + \nabla\cdot(\rho\mathbf{V}) = 0} \tag{2.43}$$

Equation (2.43) is the continuity equation in the form of a partial differential equation. This equation relates the flow field variables at a *point in the flow*, as opposed to Eq. (2.39), which deals with a finite space.

It is important to keep in mind that Eqs. (2.39) and (2.43) are equally valid statements of the physical principle of conservation of mass. They are mathematical representations, but always remember that they speak words—they say that mass can be neither created nor destroyed.

Note that in the derivation of the above equations, the only assumption about the nature of the fluid is that it is a continuum. Therefore, Eqs. (2.39) and (2.43) hold in general for the three-dimensional, unsteady flow of any type of fluid, inviscid or viscous, compressible or incompressible. (*Note:* It is important to keep track of all assumptions which are used in the derivation of any equation because they tell you the limitations on the final result, and therefore prevent you from using an equation for a situation in which it is not valid. In all our future derivations, develop the habit of noting all assumptions that go with the resulting equations.)

It is important to emphasize the difference between unsteady and steady flows. In an *unsteady* flow, the flow-field variables are a function of both spatial location and time, e.g.,

$$\rho = \rho(x, y, z, t)$$

This means that if you lock your eyes on one fixed point in space, the density at that point will change with time. Such unsteady fluctuations can be caused by time-varying boundaries (e.g., an airfoil pitching up and down with time or the supply valves of a wind tunnel being turned off and on). Equations (2.39) and (2.43) hold for such unsteady flows. On the other hand, the vast majority of practical aerodynamic problems involve *steady* flow. Here, the flow-field variables are a function of spatial location only, e.g.,

$$\rho = \rho(x, y, z)$$

This means that if you lock your eyes on a fixed point in space, the density at that point will be a fixed value, invariant with time. For steady flow, $\partial/\partial t = 0$, and hence Eqs. (2.39) and (2.43) reduce to

$$\oiint_S \rho \mathbf{V} \cdot d\mathbf{S} = 0 \tag{2.44}$$

and

$$\nabla \cdot (\rho \mathbf{V}) = 0 \tag{2.45}$$

2.5 MOMENTUM EQUATION

Newton's second law is frequently written as

$$\mathbf{F} = m\mathbf{a} \tag{2.46}$$

where **F** is the force exerted on a body of mass m and **a** is the acceleration. However, a more general form of Eq. (2.46) is

$$\mathbf{F} = \frac{d}{dt}(m\mathbf{V}) \qquad (2.47)$$

which reduces to Eq. (2.46) for a body of constant mass. In Eq. (2.47), $m\mathbf{V}$ is the momentum of a body of mass m. Equation (2.47) represents the second fundamental principle upon which theoretical fluid dynamics is based.

Physical principle   Force = time rate of change of momentum.

We will apply this principle [in the form of Eq. (2.47)] to the model of a finite control volume fixed in space as sketched in Fig. 2.15. Our objective is to obtain expressions for both the left and right sides of Eq. (2.47) in terms of the familiar flow-field variables p, ρ, **V**, etc. First, let us concentrate on the left side of Eq. (2.47), i.e., obtain an expression for **F**, which is the force exerted on the fluid as it flows through the control volume. This force comes from two sources:

1. *Body forces*: gravity, electromagnetic forces, or any other forces which "act at a distance" on the fluid inside $\mathscr{V}$
2. *Surface forces*: pressure and shear stress acting on the control surface S

Let **f** represent the net body force per unit mass exerted on the fluid inside $\mathscr{V}$. The body force on the elemental volume $d\mathscr{V}$ in Fig. 2.15 is therefore

$$\rho\mathbf{f} \, d\mathscr{V}$$

and the total body force exerted on the fluid in the control volume is the summation of the above over the volume $\mathscr{V}$:

$$\text{Body force} = \iiint\limits_{\mathscr{V}} \rho\mathbf{f} \, d\mathscr{V} \qquad (2.48)$$

The elemental surface force due to pressure acting on the element of area dS is

$$-p \, \mathbf{dS}$$

where the negative sign indicates that the force is in the direction opposite of **dS**. That is, the control surface is experiencing a pressure force which is directed into the control volume and which is due to the pressure from the surroundings, and examination of Fig. 2.15 shows that such an inward-directed force is in the direction opposite of **dS**. The complete pressure force is the summation of the elemental forces over the entire control surface:

$$\text{Pressure force} = -\iint\limits_{s} p \, \mathbf{dS} \qquad (2.49)$$

In a viscous flow, the shear and normal viscous stresses also exert a surface force. A detailed evaluation of these viscous stresses is not warranted at this stage of our discussion. Let us simply recognize this effect by letting $\mathbf{F}_{viscous}$ denote the total viscous force exerted on the control surface. We are now ready to write an expression for the left-hand side of Eq. (2.47). The total force experienced by the fluid as it is sweeping through the fixed control volume is given by the sum of Eqs. (2.48) and (2.49) and $\mathbf{F}_{viscous}$:

$$\mathbf{F} = \iiint_{\mathcal{V}} \rho \mathbf{f} \, d\mathcal{V} - \oiint_{S} p \, d\mathbf{S} + \mathbf{F}_{viscous} \tag{2.50}$$

Now consider the right side of Eq. (2.47). The time rate of change of momentum of the fluid as it sweeps through the fixed control volume is the sum of two terms:

$$\begin{array}{c} \text{Net flow of momentum } out \\ \text{of control volume across surface } S \end{array} \equiv \mathbf{G} \tag{2.51a}$$

and

$$\begin{array}{c} \text{Time rate of change of momentum due to} \\ \text{unsteady fluctuations of flow properties inside } \mathcal{V} \end{array} \equiv \mathbf{H} \tag{2.51b}$$

Consider the term denoted by $\mathbf{G}$ in Eq. (2.51a). The flow has a certain momentum as it enters the control volume in Fig. 2.15, and, in general, it has a different momentum as it leaves the control volume (due in part to the force $\mathbf{F}$ that is exerted on the fluid as it is sweeping through $\mathcal{V}$). The *net* flow of momentum *out* of the control volume across the surface S is simply this outflow minus the inflow of momentum across the control surface. This change in momentum is denoted by $\mathbf{G}$, as noted above. To obtain an expression for $\mathbf{G}$, recall that the mass flow across the elemental area dS is $(\rho \mathbf{V} \cdot d\mathbf{S})$; hence, the flow of momentum per second across dS is

$$(\rho \mathbf{V} \cdot d\mathbf{S})\mathbf{V}$$

The net flow of momentum out of the control volume through S is the summation of the above elemental contributions, namely,

$$\mathbf{G} = \oiint_{S} (\rho \mathbf{V} \cdot d\mathbf{S})\mathbf{V} \tag{2.52}$$

In Eq. (2.52), recall that positive values of $(\rho \mathbf{V} \cdot d\mathbf{S})$ represent mass flow out of the control volume, and negative values represent mass flow into the control volume. Hence, in Eq. (2.52) the integral over the whole control surface is a combination of positive contributions (outflow of momentum) and negative contributions (inflow of momentum), with the resulting value of the integral representing the net outflow of momentum. If $\mathbf{G}$ has a positive value, there is more momentum flowing out of the control volume per second than flowing in; conversely, if $\mathbf{G}$ has a negative value, there is more momentum flowing into the control volume per second than flowing out.

Now consider **H** from Eq. (2.51*b*). The momentum of the fluid in the elemental volume $d\mathcal{V}$ shown in Fig. 2.15 is

$$(\rho \, d\mathcal{V})\mathbf{V}$$

The momentum contained at any instant inside the control volume is therefore

$$\oiiint_{\mathcal{V}} \rho \mathbf{V} \, d\mathcal{V}$$

and its time rate of change due to unsteady flow fluctuations is

$$\mathbf{H} = \frac{\partial}{\partial t} \oiiint_{\mathcal{V}} \rho \mathbf{V} \, d\mathcal{V} \tag{2.53}$$

Combining Eqs. (2.52) and (2.53), we obtain an expression for the total time rate of change of momentum of the fluid as it sweeps through the fixed control volume, which in turn represents the right-hand side of Eq. (2.47):

$$\frac{d}{dt}(m\mathbf{V}) = \mathbf{G} + \mathbf{H} = \oiint_{S} (\rho \mathbf{V} \cdot d\mathbf{S})\mathbf{V} + \frac{\partial}{\partial t} \oiiint_{\mathcal{V}} \rho \mathbf{V} \, d\mathcal{V} \tag{2.54}$$

Hence, from Eqs. (2.50) and (2.54), Newton's second law,

$$\frac{d}{dt}(m\mathbf{V}) = \mathbf{F}$$

applied to a fluid flow is

$$\boxed{\frac{\partial}{\partial t} \oiiint_{\mathcal{V}} \rho \mathbf{V} \, d\mathcal{V} + \oiint_{S} (\rho \mathbf{V} \cdot d\mathbf{S})\mathbf{V} = -\oiint_{S} p \, d\mathbf{S} + \oiiint_{\mathcal{V}} \rho \mathbf{f} \, d\mathcal{V} + \mathbf{F}_{\text{viscous}}} \tag{2.55}$$

Equation (2.55) is the momentum equation in integral form. Note that it is a vector equation. Just as in the case of the integral form of the continuity equation, Eq. (2.55) has the advantage of relating aerodynamic phenomena over a finite region of space without being concerned with the details of precisely what is happening at a given distinct point in the flow. This advantage is illustrated in Sec. 2.6.

From Eq. (2.55), we now proceed to a partial differential equation which relates flow-field properties at a point in space. Such an equation is a counterpart to the differential form of the continuity equation given in Eq. (2.43). Apply the gradient theorem, Eq. (2.27), to the first term on the right side of Eq. (2.55):

$$-\oiint_{S} p \, d\mathbf{S} = -\oiiint_{\mathcal{V}} \nabla p \, d\mathcal{V} \tag{2.56}$$

Also, because the control volume is fixed, the time derivative in Eq. (2.55) can be placed inside the integral. Hence, Eq. (2.55) can be written as

$$\oiiint_{\mathcal{V}} \frac{\partial(\rho\mathbf{V})}{\partial t} d\mathcal{V} + \oiint_{S} (\rho\mathbf{V}\cdot\mathbf{dS})\mathbf{V} = -\oiiint_{\mathcal{V}} \nabla p \, d\mathcal{V} + \oiiint_{\mathcal{V}} \rho\mathbf{f} \, d\mathcal{V} + \mathbf{F}_{\text{viscous}} \quad (2.57)$$

Recall that Eq. (2.57) is a vector equation. It is convenient to write this equation as three scalar equations. Using cartesian coordinates, where

$$\mathbf{V} = u\mathbf{i} + v\mathbf{j} + w\mathbf{k}$$

the x component of Eq. (2.57) is

$$\oiiint_{\mathcal{V}} \frac{\partial(\rho u)}{\partial t} d\mathcal{V} + \oiint_{S} (\rho\mathbf{V}\cdot\mathbf{dS})u = -\oiiint_{\mathcal{V}} \frac{\partial p}{\partial x} d\mathcal{V} + \oiiint_{\mathcal{V}} \rho f_x \, d\mathcal{V} + (F_x)_{\text{viscous}} \quad (2.58)$$

[*Note*: In Eq. (2.58), the product $(\rho\mathbf{V}\cdot\mathbf{dS})$ is a scalar, and therefore has no components.] Apply the divergence theorem, Eq. (2.26), to the surface integral on the left side of Eq. (2.58):

$$\oiint_{S} (\rho\mathbf{V}\cdot\mathbf{dS})u = \oiint_{S} (\rho u\mathbf{V})\cdot\mathbf{dS} = \oiiint_{\mathcal{V}} \nabla\cdot(\rho u\mathbf{V}) \, d\mathcal{V} \quad (2.59)$$

Substituting Eq. (2.59) into Eq. (2.58), we have

$$\oiiint_{\mathcal{V}} \left[\frac{\partial(\rho u)}{\partial t} + \nabla\cdot(\rho u\mathbf{V}) + \frac{\partial p}{\partial x} - \rho f_x - (\mathcal{F}_x)_{\text{viscous}} \right] d\mathcal{V} = 0 \quad (2.60)$$

where $(\mathcal{F}_x)_{\text{viscous}}$ denotes the proper form of the x component of the viscous shear stresses when placed inside the volume integral (this form will be obtained explicitly in Chap. 15). For the same reasons as stated in Sec. 2.4, the integrand in Eq. (2.60) is identically zero at all points in the flow; hence,

$$\boxed{\frac{\partial(\rho u)}{\partial t} + \nabla\cdot(\rho u\mathbf{V}) = -\frac{\partial p}{\partial x} + \rho f_x + (\mathcal{F}_x)_{\text{viscous}}} \quad (2.61a)$$

Equation (2.61a) is the x component of the momentum equation in differential form. Returning to Eq. (2.57), and writing the y and z components, we obtain in a similar fashion

$$\boxed{\frac{\partial(\rho v)}{\partial t} + \nabla\cdot(\rho v\mathbf{V}) = -\frac{\partial p}{\partial y} + \rho f_y + (\mathcal{F}_y)_{\text{viscous}}} \quad (2.61b)$$

and

$$\boxed{\frac{\partial(\rho w)}{\partial t} + \nabla\cdot(\rho w\mathbf{V}) = -\frac{\partial p}{\partial z} + \rho f_z + (\mathcal{F}_z)_{\text{viscous}}} \quad (2.61c)$$

where the subscripts y and z on f and $\mathcal{F}$ denote the y and z components of the body and viscous forces, respectively. Equations (2.61a to c) are the scalar x, y, and z components of the momentum equation, respectively; they are partial differential equations which relate flow-field properties at any point in the flow.

Note that Eqs. (2.55) and (2.61a to c) apply to the unsteady, three-dimensional flow of any fluid, compressible or incompressible, viscous or inviscid. Specialized to a steady ($\partial/\partial t \equiv 0$), inviscid ($\mathbf{F}_{viscous} = 0$) flow with no body forces ($\mathbf{f} = 0$), these equations become

$$\oiint_S (\rho\mathbf{V} \cdot \mathbf{dS})\mathbf{V} = -\oiint_S p\,\mathbf{dS} \qquad (2.62)$$

and

$$\nabla \cdot (\rho u \mathbf{V}) = -\frac{\partial p}{\partial x} \qquad (2.63a)$$

$$\nabla \cdot (\rho v \mathbf{V}) = -\frac{\partial p}{\partial y} \qquad (2.63b)$$

$$\nabla \cdot (\rho w \mathbf{V}) = -\frac{\partial p}{\partial z} \qquad (2.63c)$$

Since most of the material in Chaps. 3 through 14 assumes steady, inviscid flow with no body forces, we will have frequent occasion to use the momentum equation in the forms of Eqs. (2.62) and (2.63a to c).

The momentum equations for an inviscid flow [such as Eqs. (2.63a to c)] are called the *Euler equations*. The momentum equations for a viscous flow [such as Eqs. (2.61a to c)] are called the *Navier-Stokes equations*. We will encounter this terminology in subsequent chapters.

2.6 AN APPLICATION OF THE MOMENTUM EQUATION: DRAG OF A TWO-DIMENSIONAL BODY

We briefly interrupt our orderly development of the fundamental equations of fluid dynamics in order to examine an important application of the integral form of the momentum equation. During the 1930s and 1940s, the National Advisory Committee for Aeronautics (NACA) measured the lift and drag characteristics of a series of systematically designed airfoil shapes (discussed in detail in Chap. 4). These measurements were carried out in a specially designed wind tunnel where the wing models spanned the entire test section; i.e., the wing tips were butted against both sidewalls of the wind tunnel. This was done in order to establish two-dimensional (rather than three-dimensional) flow over the wing, thus allowing the properties of an airfoil (rather than a finite wing) to be measured.

The distinction between the aerodynamics of airfoils and that of finite wings is made in Chaps. 4 and 5. The important point here is that because the wings were mounted against both sidewalls of the wind tunnel, the NACA did not use a conventional force balance to measure the lift and drag. Rather, the lift was obtained from the pressure distributions on the *ceiling and floor* of the tunnel (above and below the wing), and the drag was obtained from measurements of the flow velocity *downstream* of the wing. These measurements may appear to be a strange way to measure the aerodynamic force on a wing. Indeed, how are these measurements related to lift and drag? What is going on here? The answers to these questions are addressed in this section; they involve an application of the fundamental momentum equation in integral form, and they illustrate a basic technique that is frequently used in aerodynamics.

Consider a two-dimensional body in a flow, as sketched in Fig. 2.16*a*. A control volume is drawn around this body, as given by the dashed lines in Fig. 2.16*a*. The control volume is bounded by:

1. The upper and lower streamlines far above and below the body (*ab* and *hi*, respectively)
2. Lines perpendicular to the flow velocity far ahead of and behind the body (*ai* and *bh*, respectively)
3. A cut that surrounds and wraps the surface of the body (*cdefg*)

The entire control volume is *abcdefghia*. The width of the control volume in the *z* direction (perpendicular to the page) is unity. Stations 1 and 2 are inflow and outflow stations, respectively.

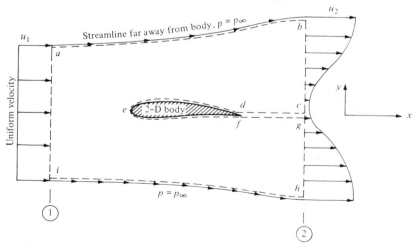

FIGURE 2.16*a*
Control volume for obtaining drag on a two-dimensional body.

Assume that the contour *abhi* is far enough from the body such that the pressure is everywhere the same on *abhi* and equal to the freestream pressure $p = p_\infty$. Also, assume that the inflow velocity u_1 is uniform across *ai* (as it would be in a freestream, or a test section of a wind tunnel). The outflow velocity u_2 is *not* uniform across *bh*, because the presence of the body has created a wake at the outflow station. However, assume that both u_1 and u_2 are in the x direction; hence, $u_1 = $ constant and $u_2 = f(y)$.

An actual photograph of the velocity profiles in a wake downstream of an airfoil is shown in Fig. 2.16*b*.

Consider the surface forces on the control volume. They stem from two contributions:

1. The pressure distribution over the surface *abhi*,

$$-\iint_{abhi} p \, \mathbf{dS}$$

2. The surface force on *def* created by the presence of the body

In the list above, the surface shear stress on *ab* and *hi* has been neglected. Also, note that in Fig. 2.16*a* the cuts *cd* and *fg* are taken adjacent to each other; hence,

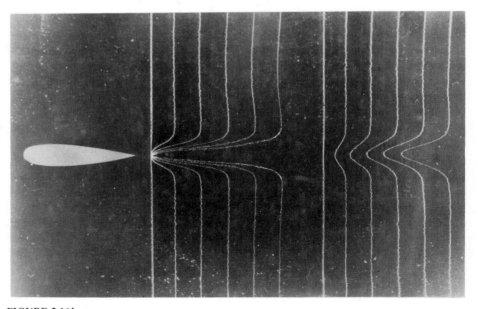

FIGURE 2.16*b*
Photograph of the velocity profiles downstream of an airfoil. The profiles are made visible in water flow by pulsing a voltage through a straight wire perpendicular to the flow, thus creating small bubbles of hydrogen that subsequently move downstream with the flow. (*Courtesy of Yasuki Nakayama, Tokai University, Japan.*)

any shear stress or pressure distribution on one is equal and opposite to that on the other; i.e., the surface forces on *cd* and *fg* cancel each other. Also, note that the surface force on *def* is the *equal and opposite reaction* to the shear stress and pressure distribution created by the flow over the surface of the body. To see this more clearly, examine Fig. 2.17. On the left is shown the flow over the body. As explained in Sec. 1.5, the moving fluid exerts pressure and shear stress distributions over the body surface which create a resultant aerodynamic force per unit span $\mathbf{R}'$ on the body. In turn, by Newton's third law, the body exerts equal and opposite pressure and shear stress distributions on the flow, i.e., on the part of the control surface bounded by *def*. Hence, the body exerts a force $-\mathbf{R}'$ on the control surface, as shown on the right of Fig. 2.17. With the above in mind, the total surface force on the entire control volume is

$$\text{Surface force} = -\iint_{abhi} p \, d\mathbf{S} - \mathbf{R}' \tag{2.64}$$

Moreover, this is the *total* force on the control volume shown in Fig. 2.16a because the volumetric body force is negligible.

Consider the integral form of the momentum equation as given by Eq. (2.55). The right-hand side of this equation is physically the force on the fluid moving through the control volume. For the control volume in Fig. 2.16a, this force is simply the expression given by Eq. (2.64). Hence, using Eq. (2.55), with the right-hand side given by Eq. (2.64), we have

$$\frac{\partial}{\partial t} \iiint_{\mathcal{V}} \rho \mathbf{V} \, d\mathcal{V} + \oiint_{S} (\rho \mathbf{V} \cdot d\mathbf{S})\mathbf{V} = -\iint_{abhi} p \, d\mathbf{S} - \mathbf{R}' \tag{2.65}$$

Assuming steady flow, Eq. (2.65) becomes

$$\mathbf{R}' = -\oiint_{S} (\rho \mathbf{V} \cdot d\mathbf{S})\mathbf{V} - \iint_{abhi} p \, d\mathbf{S} \tag{2.66}$$

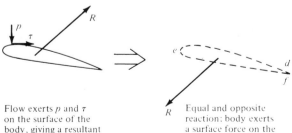

Flow exerts p and τ on the surface of the body, giving a resultant aerodynamic force $\mathbf{R}$

Equal and opposite reaction: body exerts a surface force on the section of the control volume *def* that equals $-\mathbf{R}$

FIGURE 2.17
Equal and opposite reactions on a body and adjacent section of control surface.

Equation (2.66) is a vector equation. Consider again the control volume in Fig. 2.16a. Take the x component of Eq. (2.66), noting that the inflow and outflow velocities u_1 and u_2 are in the x direction and the x component of $\mathbf{R}'$ is the aerodynamic drag per unit span D':

$$D' = -\oiint_S (\rho \mathbf{V} \cdot d\mathbf{S})u - \iint_{abhi} (p\, d\mathbf{S})_x \tag{2.67}$$

In Eq. (2.67), the last term is the component of the pressure force in the x direction. [The expression $(p\, d\mathbf{S})_x$ is the x component of the pressure force exerted on the elemental area $d\mathbf{S}$ of the control surface.] Recall that the boundaries of the control volume $abhi$ are chosen far enough from the body such that p is constant along these boundaries. For a constant pressure,

$$\iint_{abhi} (p\, d\mathbf{S})_x = 0 \tag{2.68}$$

because, looking along the x direction in Fig. 2.16a, the pressure force on $abhi$ pushing toward the right exactly balances the pressure force pushing toward the left. This is true no matter what the shape of $abhi$ is, as long as p is constant along the surface (for proof of this statement, see Prob. 2.3). Therefore, substituting Eq. (2.68) into (2.67), we obtain

$$D' = -\oiint_S (\rho \mathbf{V} \cdot d\mathbf{S})u \tag{2.69}$$

Evaluating the surface integral in Eq. (2.69), we note from Fig. 2.16a that:

1. The sections ab, hi, and def are streamlines of the flow. Since by definition $\mathbf{V}$ is parallel to the streamlines and $d\mathbf{S}$ is perpendicular to the control surface, along these sections $\mathbf{V}$ and $d\mathbf{S}$ are perpendicular vectors, and hence $\mathbf{V} \cdot d\mathbf{S} = 0$. As a result, the contributions of ab, hi, and def to the integral in Eq. (2.69) are zero.

2. The cuts cd and fg are adjacent to each other. The mass flux out of one is identically the mass flux into the other. Hence, the contributions of cd and fg to the integral in Eq. (2.69) cancel each other.

As a result, the only contributions to the integral in Eq. (2.69) come from sections ai and bh. These sections are oriented in the y direction. Also, the control volume has unit depth in the z direction (perpendicular to the page). Hence, for these sections, $d\mathbf{S} = dy(1)$. The integral in Eq. (2.69) becomes

$$\oiint_S (\rho \mathbf{V} \cdot d\mathbf{S})u = -\int_i^a \rho_1 u_1^2 \, dy + \int_h^b \rho_2 u_2^2 \, dy \tag{2.70}$$

Note that the minus sign in front of the first term on the right-hand side of Eq. (2.70) is due to $\mathbf{V}$ and $d\mathbf{S}$ being in opposite directions along ai (station 1 is an inflow boundary); in contrast, **V** and **dS** are in the same direction over hb (station 2 is an outflow boundary), and hence the second term has a positive sign.

Before going further with Eq. (2.70), consider the integral form of the continuity equation for steady flow, Eq. (2.44). Applied to the control volume in Fig. 2.16a, Eq. (2.44) becomes

$$-\int_i^a \rho_1 u_1 \, dy + \int_h^b \rho_2 u_2 \, dy = 0$$

or

$$\int_i^a \rho_1 u_1 \, dy = \int_h^b \rho_2 u_2 \, dy \tag{2.71}$$

Multiplying Eq. (2.71) by u_1, which is a constant, we obtain

$$\int_i^a \rho_1 u_1^2 \, dy = \int_h^b \rho_2 u_2 u_1 \, dy \tag{2.72}$$

Substituting Eq. (2.72) into Eq. (2.70), we have

$$\oiint_S (\rho \mathbf{V} \cdot \mathbf{dS}) u = -\int_h^b \rho_2 u_2 u_1 \, dy + \int_h^b \rho_2 u_2^2 \, dy$$

or

$$\oiint_S (\rho \mathbf{V} \cdot \mathbf{dS}) u = -\int_h^b \rho_2 u_2 (u_1 - u_2) \, dy \tag{2.73}$$

Substituting Eq. (2.73) into Eq. (2.69) yields

$$\boxed{D' = \int_h^b \rho_2 u_2 (u_1 - u_2) \, dy} \tag{2.74}$$

Equation (2.74) is the desired result of this section; it expresses the drag of a body in terms of the known freestream velocity u_1 and the flow-field properties ρ_2 and u_2, across a vertical station downstream of the body. These downstream properties can be measured in a wind tunnel, and the drag per unit span of the body, D', can be obtained by evaluating the integral in Eq. (2.74) numerically, using the measured data for ρ_2 and u_2 as a function of y.

Examine Eq. (2.74) more closely. The quantity $u_1 - u_2$ is the velocity decrement at a given y location. That is, because of the drag on the body, there is a wake that trails downstream of the body. In this wake, there is a loss in flow velocity $u_1 - u_2$. The quantity $\rho_2 u_2$ is simply the mass flux; when multiplied by $u_1 - u_2$, it gives the decrement in momentum. Therefore, the integral in Eq. (2.74) is physically the decrement in momentum flow that exists across the wake, and from Eq. (2.74), this wake momentum decrement is equal to the drag on the body.

For incompressible flow, ρ = constant and is known. For this case, Eq. (2.74) becomes

$$D' = \rho \int_h^b u_2 (u_1 - u_2) \, dy \tag{2.75}$$

Equation (2.75) is the answer to the questions posed at the beginning of this section. It shows how a measurement of the velocity distribution across the wake of a body can yield the drag. These velocity distributions are conventionally measured with a Pitot rake, such as shown in Fig. 2.18. This is nothing more than a series of Pitot tubes attached to a common stem, which allows the simultaneous measurement of velocity across the wake. (The principle of the Pitot tube as a velocity-measuring instrument is discussed in Chap. 3. See also pages 94–108 of Ref. 2 for an introductory discussion on Pitot tubes.)

The result embodied in Eq. (2.75) illustrates the power of the integral form of the momentum equation; it relates drag on a body located at some position in the flow to the flow-field variables at a completely different location.

At the beginning of this section, it was mentioned that lift on a two-dimensional body can be obtained by measuring the pressures on the ceiling and floor of a wind tunnel, above and below the body. This relation can be established from the integral form of the momentum equation in a manner analogous to that used to establish the drag relation; the derivation is left as a homework problem.

Example 2.1. Consider an incompressible flow, laminar boundary layer growing along the surface of a flat plate, with chord length c, as sketched in Fig. 2.19. The definition of a boundary layer was discussed in Sec. 1.10 and illustrated in Fig. 1.28. The significance of a laminar flow is discussed in Chap. 15; it is not relevant for this example. For an incompressible, laminar, flat plate boundary layer, the boundary-layer thickness, δ, at the trailing edge of the plate is

$$\frac{\delta}{c} = \frac{5}{\sqrt{\text{Re}_c}}$$

and the skin friction drag coefficient for the plate is

$$C_f \equiv \frac{D'}{q_\infty c(1)} = \frac{1.328}{\sqrt{\text{Re}_c}}$$

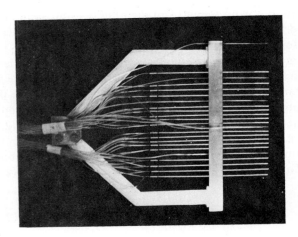

FIGURE 2.18
A Pitot rake for wake surveys.
(*Courtesy of the University of Maryland Aerodynamic Laboratory.*)

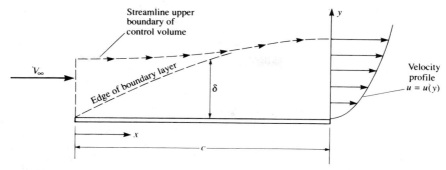

FIGURE 2.19
Sketch of a boundary layer and the velocity profile at $x = c$. The boundary-layer thickness, δ, is exaggerated here for clarity.

where the Reynolds number is based on chord length

$$\mathrm{Re}_c = \frac{\rho_\infty V_\infty c}{\mu_\infty}$$

[*Note*: Both δ/c and C_f are functions of the Reynolds number—just another demonstration of the power of the similarity parameters. Since we are dealing with a low-speed, incompressible flow, the Mach number is not a relevant parameter here.] Let us *assume* that the velocity profile through the boundary layer is given by a power-law variation

$$u = V_\infty \left(\frac{y}{\delta}\right)^n$$

Calculate the value of n, consistent with the information given above.

Solution. From Eq. (2.75),

$$C_f = \frac{D'}{q_\infty c} = \frac{\rho_\infty}{\frac{1}{2}\rho_\infty V_\infty^2 c} \int_0^\delta u_2(u_1 - u_2)\, dy$$

where the integral is evaluated at the trailing edge of the plate. Hence,

$$C_f = 2 \int_0^{\delta/c} \frac{u_2}{V_\infty}\left(\frac{u_1}{V_\infty} - \frac{u_2}{V_\infty}\right) d\left(\frac{y}{c}\right)$$

However, in Eq. (2.75), applied to the control volume in Fig. 1.28, $u_1 = V_\infty$. Thus,

$$C_f = 2 \int_0^{\delta/c} \frac{u_2}{V_\infty}\left(1 - \frac{u_2}{V_\infty}\right) d\left(\frac{y}{c}\right)$$

Inserting the laminar boundary-layer result for C_f as well as the assumed variation of velocity, both given above, we can write this integral as

$$\frac{1.328}{\sqrt{\mathrm{Re}_c}} = 2 \int_0^{\delta/c} \left[\left(\frac{y/c}{\delta/c}\right)^n - \left(\frac{y/c}{\delta/c}\right)^{2n}\right] d\left(\frac{y}{c}\right)$$

Carrying out the integration, we obtain

$$\frac{1.328}{\sqrt{\mathrm{Re}_c}} = \frac{2}{n+1}\left(\frac{\delta}{c}\right) - \frac{2}{2n+1}\left(\frac{\delta}{c}\right)$$

Since $\delta/c = 5/\sqrt{\mathrm{Re}_c}$, then

$$\frac{1.328}{\sqrt{\mathrm{Re}_c}} = \frac{10}{n+1}\left(\frac{1}{\sqrt{\mathrm{Re}_c}}\right) - \frac{10}{2n+1}\left(\frac{1}{\sqrt{\mathrm{Re}_c}}\right)$$

or

$$\frac{1}{n+1} - \frac{1}{2n+1} = \frac{1.328}{10}$$

or

$$0.2656n^2 - 0.6016n + 0.1328 = 0$$

Using the quadratic formula, we have

$$n = 2 \quad \text{or} \quad 0.25$$

By *assuming* a power-law velocity profile in the form of $u/V_\infty = (y/\delta)^n$, we have found two different velocity profiles that satisfy the momentum principle applied to a finite control volume. Both of these profiles are shown in Fig. 2.20 and are compared with an exact velocity profile obtained by means of a solution of the incompressible, laminar boundary-layer equations for a flat plate. (This boundary-layer solution is discussed in Chap. 17.) Note that the result $n = 2$ gives a concave velocity profile which is essentially nonphysical when compared to the convex profiles always observed in boundary layers. The result $n = 0.25$ gives a convex velocity profile which is qualitatively physically correct. However, this profile is quantitatively inaccurate, as can be seen in comparison to the exact profile. Hence,

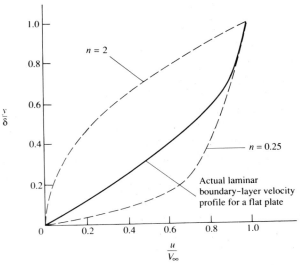

FIGURE 2.20
Comparison of the actual laminar boundary-layer profile with those calculated from Example 2.1.

our original *assumption* of a power-law velocity profile for the laminar boundary layer in the form of $u/V_\infty = (y/\delta)^n$ is not very good, in spite of the fact that when $n = 2$ or 0.25, this assumed velocity profile does satisfy the momentum principle, applied over a large, finite control volume.

2.6.1   Comment

In this section, we have applied the momentum principle (Newton's second law) to large, fixed control volumes in flows. On one hand, we demonstrated that, by knowing the detailed flow properties along the control surface, this application led to an accurate result for an overall quantity such as drag on a body, namely, Eq. (2.74) for a compressible flow and Eq. (2.75) for an incompressible flow. On the other hand, in Example 2.1, we have shown that, by knowing an overall quantity such as the net drag on a flat plate, the finite control volume concept by itself does not necessarily provide an accurate calculation of detailed flow-field properties along the control surface (in this case, the velocity profile), although the momentum principle is certainly satisfied in the aggregate. Example 2.1 is designed specifically to demonstrate this fact. The weakness here is the need to *assume* some form for the variation of flow properties over the control surface; in Example 2.1, the assumption of the particular power-law profile proved to be unsatisfactory.

2.7   ENERGY EQUATION

For an incompressible flow, where ρ is constant, the primary flow-field variables are p and $\mathbf{V}$. The continuity and momentum equations obtained earlier are two equations in terms of the two unknowns p and $\mathbf{V}$. Hence, for a study of incompressible flow, the continuity and momentum equations are sufficient tools to do the job.

However, for a compressible flow, ρ is an additional variable, and therefore we need an additional fundamental equation to complete the system. This fundamental relation is the energy equation, to be derived in this section. In the process, two additional flow-field variables arise, namely, the internal energy e and temperature T. Additional equations must also be introduced for these variables, as will be mentioned later in this section.

The material discussed in this section is germane to the study of compressible flow. For those readers interested only in the study of incompressible flow for the time being, you may bypass this section and return to it at a later stage.

Physical principle   Energy can be neither created nor destroyed; it can only change in form.

This physical principle is embodied in the first law of thermodynamics. A brief review of thermodynamics is given in Chap. 7. Thermodynamics is essential to the study of compressible flow; however, at this stage, we will only introduce the first law, and we defer any substantial discussion of thermodynamics until Chap. 7, where we begin to concentrate on compressible flow.

Consider a fixed amount of matter contained within a closed boundary. This matter defines the *system*. Because the molecules and atoms within the system are constantly in motion, the system contains a certain amount of energy. For simplicity, let the system contain a unit mass; in turn, denote the internal energy per unit mass by e.

The region outside the system defines the *surroundings*. Let an incremental amount of heat, δq, be added to the system from the surroundings. Also, let δw be the work done on the system by the surroundings. (The quantities δq and δw are discussed in more detail in Chap. 7.) Both heat and work are forms of energy, and when added to the system, they change the amount of internal energy in the system. Denote this change of internal energy by de. From our physical principle that energy is conserved, we have for the system

$$\delta q + \delta w = de \tag{2.76}$$

Equation (2.76) is a statement of the first law of thermodynamics.

Let us apply the first law to the fluid flowing through the fixed control volume shown in Fig. 2.15. Let

B_1 = rate of heat added to fluid inside control volume from surroundings

B_2 = rate of work done on fluid inside control volume

B_3 = rate of change of energy of fluid as it flows through control volume

From the first law,

$$B_1 + B_2 = B_3 \tag{2.77}$$

Note that each term in Eq. (2.77) involves the *time rate* of energy change; hence, Eq. (2.77) is, strictly speaking, a *power* equation. However, because it is a statement of the fundamental principle of conservation of energy, the equation is conventionally termed the "energy equation." We continue this convention here.

First, consider the rate of heat transferred to or from the fluid. This can be visualized as volumetric heating of the fluid inside the control volume due to absorption of radiation originating outside the system or the local emission of radiation by the fluid itself, if the temperature inside the control volume is high enough. In addition, there may be chemical combustion processes taking place inside the control volume, such as fuel-air combustion in a jet engine. Let this volumetric rate of heat addition per unit mass be denoted by $\dot{q}$. Typical units for $\dot{q}$ are J/s · kg or ft · lb/s · slug. Examining Fig. 2.15, the mass contained within an elemental volume is $\rho \, d\mathcal{V}$; hence, the rate of heat addition to this mass is $\dot{q}(\rho \, d\mathcal{V})$. Summing over the complete control volume, we obtain

$$\text{Rate of volumetric heating} = \iiint_{\mathcal{V}} \dot{q}\rho \, d\mathcal{V} \tag{2.78}$$

In addition, if the flow is viscous, heat can be transferred into the control volume by means of thermal conduction and mass diffusion across the control surface. At this stage, a detailed development of these viscous heat-addition terms is not warranted; they are considered in detail in Chap. 15. Rather, let us denote the rate of heat addition to the control volume due to viscous effects simply by $\dot{Q}_{viscous}$. Therefore, in Eq. (2.77), the total rate of heat addition is given by Eq. (2.78) plus $\dot{Q}_{viscous}$:

$$B_1 = \iiint\limits_{\mathcal{V}} \dot{q}\rho \, d\mathcal{V} + \dot{Q}_{viscous} \qquad (2.79)$$

Before considering the rate of work done on the fluid inside the control volume, consider a simpler case of a solid object in motion, with a force **F** being exerted on the object, as sketched in Fig. 2.21. The position of the object is measured from a fixed origin by the radius vector **r**. In moving from position $\mathbf{r}_1$ to $\mathbf{r}_2$ over an interval of time dt, the object is displaced through $d\mathbf{r}$. By definition, the work done on the object in time dt is $\mathbf{F} \cdot d\mathbf{r}$. Hence, the time rate of doing work is simply $\mathbf{F} \cdot d\mathbf{r}/dt$. However, $d\mathbf{r}/dt = \mathbf{V}$, the velocity of the moving object. Hence, we can state that

Rate of doing work on moving body $= \mathbf{F} \cdot \mathbf{V}$

In words, the rate of work done on a moving body is equal to the product of its velocity and the component of force in the direction of the velocity.

This result leads to an expression for B_2, as follows. Consider the elemental area dS of the control surface in Fig. 2.15. The pressure force on this elemental area is $-p \, d\mathbf{S}$. From the above result, the rate of work done on the fluid passing through dS with velocity **V** is $(-p \, d\mathbf{S}) \cdot \mathbf{V}$. Hence, summing over the complete control surface, we have

$$\begin{array}{c} \text{Rate of work done on fluid inside} \\ \mathcal{V} \text{ due to pressure force on } S \end{array} = -\oiint\limits_{S} (p \, d\mathbf{S}) \cdot \mathbf{V} \qquad (2.80)$$

In addition, consider an elemental volume, $d\mathcal{V}$, inside the control volume, as shown in Fig. 2.15. Recalling that **f** is the body force per unit mass, the rate of work done on the elemental volume due to the body force is $(\rho \mathbf{f} \, d\mathcal{V}) \cdot \mathbf{V}$. Summing

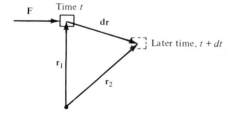

FIGURE 2.21
Schematic for the rate of doing work by a force **F** exerted on a moving body.

over the complete control volume, we obtain

$$\text{Rate of work done on fluid inside } \mathcal{V} \text{ due to body forces} = \oiint_{\mathcal{V}} (\rho \mathbf{f} \, d\mathcal{V}) \cdot \mathbf{V} \qquad (2.81)$$

If the flow is viscous, the shear stress on the control surface will also perform work on the fluid as it passes across the surface. Once again, a detailed development of this term is deferred until Chap. 15. Let us denote this contribution simply by $\dot{W}_{\text{viscous}}$. Then the total rate of work done on the fluid inside the control volume is the sum of Eqs. (2.80) and (2.81) and $\dot{W}_{\text{viscous}}$:

$$B_2 = -\oiint_S p\mathbf{V} \cdot \mathbf{dS} + \oiint_{\mathcal{V}} \rho(\mathbf{f} \cdot \mathbf{V}) \, d\mathcal{V} + \dot{W}_{\text{viscous}} \qquad (2.82)$$

To visualize the energy inside the control volume, recall that in the first law of thermodynamics as stated in Eq. (2.76), the internal energy e is due to the random motion of the atoms and molecules inside the system. Equation (2.76) is written for a stationary system. However, the fluid inside the control volume in Fig. 2.15 is not stationary; it is moving at the local velocity $\mathbf{V}$ with a consequent kinetic energy per unit mass of $V^2/2$. Hence, the energy per unit mass of the moving fluid is the sum of both internal and kinetic energies, $e + V^2/2$. This sum is called the *total energy* per unit mass.

We are now ready to obtain an expression for B_3, the rate of change of total energy of the fluid as it flows through the control volume. Keep in mind that mass flows into the control volume of Fig. 2.15 bringing with it a certain total energy; at the same time mass flows out of the control volume taking with it a generally different amount of total energy. The elemental mass flow across dS is $\rho\mathbf{V} \cdot \mathbf{dS}$, and therefore the elemental flow of total energy across dS is $(\rho\mathbf{V} \cdot \mathbf{dS})(e + V^2/2)$. Summing over the complete control surface, we obtain

$$\text{Net rate of flow of total energy across control surface} = \oiint_S (\rho\mathbf{V} \cdot \mathbf{dS})\left(e + \frac{V^2}{2}\right) \qquad (2.83)$$

In addition, if the flow is unsteady, there is a time rate of change of total energy inside the control volume due to the transient fluctuations of the flow-field variables. The total energy contained in the elemental volume $d\mathcal{V}$ is $\rho(e + V^2/2) \, d\mathcal{V}$, and hence the total energy inside the complete control volume at any instant in time is

$$\oiint_{\mathcal{V}} \rho\left(e + \frac{V^2}{2}\right) d\mathcal{V}$$

Therefore,

$$\text{Time rate of change of total energy inside } \mathcal{V} \text{ due to transient variations of flow-field variables} = \frac{\partial}{\partial t} \oiint_{\mathcal{V}} \rho\left(e + \frac{V^2}{2}\right) d\mathcal{V} \qquad (2.84)$$

In turn, B_3 is the sum of Eqs. (2.83) and (2.84):

$$B_3 = \frac{\partial}{\partial t} \iiint_{\mathcal{V}} \rho \left(e + \frac{V^2}{2} \right) d\mathcal{V} + \oiint_{S} (\rho \mathbf{V} \cdot d\mathbf{S}) \left(e + \frac{V^2}{2} \right) \tag{2.85}$$

Repeating the physical principle stated at the beginning of this section, the rate of heat added to the fluid plus the rate of work done on the fluid is equal to the rate of change of total energy of the fluid as it flows through the control volume; i.e., *energy is conserved.* In turn, these words can be directly translated into an equation by combining Eqs. (2.77), (2.79), (2.82), and (2.85):

$$\iiint_{\mathcal{V}} \dot{q}\rho \, d\mathcal{V} + \dot{Q}_{\text{viscous}} - \oiint_{S} p\mathbf{V} \cdot d\mathbf{S} + \iiint_{\mathcal{V}} \rho(\mathbf{f} \cdot \mathbf{V}) \, d\mathcal{V} + \dot{W}_{\text{viscous}}$$

$$= \frac{\partial}{\partial t} \iiint_{\mathcal{V}} \rho \left(e + \frac{V^2}{2} \right) d\mathcal{V} + \oiint_{S} \rho \left(e + \frac{V^2}{2} \right) \mathbf{V} \cdot d\mathbf{S}$$

$$\tag{2.86}$$

Equation (2.86) is the energy equation in integral form; it is essentially the first law of thermodynamics applied to a fluid flow.

For the sake of completeness, note that if a shaft penetrates the control surface in Fig. 2.15, driving some power machinery located inside the control volume (say, a compressor of a jet engine), then the rate of work delivered by the shaft, $\dot{W}_{\text{shaft}}$, must be added to the left side of Eq. (2.86). In addition, if the size of the control volume were so large that changes in height z were important, then the potential energy per unit mass, gz, must be added to the total energy; i.e., total energy would be $e + V^2/2 + gz$, rather than just $e + V^2/2$, as appears in Eq. (2.86). However, for the aerodynamic problems considered in this book, shaft work is not treated, and changes in potential energy are always negligible. Therefore, these effects have been intentionally not included in Eq. (2.86).

Following the approach established in Secs. 2.4 and 2.5, we can obtain a partial differential equation for total energy from the integral form given in Eq. (2.86). Applying the divergence theorem to the surface integrals in Eq. (2.86), collecting all terms inside the same volume integral, and setting the integrand equal to zero, we obtain

$$\frac{\partial}{\partial t} \left[\rho \left(e + \frac{V^2}{2} \right) \right] + \nabla \cdot \left[\rho \left(e + \frac{V^2}{2} \right) \mathbf{V} \right] = \rho \dot{q} - \nabla \cdot (p\mathbf{V}) + \rho(\mathbf{f} \cdot \mathbf{V})$$

$$+ \dot{Q}'_{\text{viscous}} + \dot{W}'_{\text{viscous}} \tag{2.87}$$

where $\dot{Q}'_{\text{viscous}}$ and $\dot{W}'_{\text{viscous}}$ represent the proper forms of the viscous terms, to be obtained in Chap. 15. Equation (2.87) is a partial differential equation which relates the flow-field variables at a given point in space.

If the flow is steady ($\partial/\partial t = 0$), inviscid ($\dot{Q}_{viscous} = 0$ and $\dot{W}_{viscous} = 0$), adiabatic (no heat addition, $\dot{q} = 0$), without body forces ($\mathbf{f} = 0$), then Eqs. (2.86) and (2.87) reduce to

$$\oiint_S \rho\left(e + \frac{V^2}{2}\right)\mathbf{V} \cdot \mathbf{dS} = -\oiint_S p\mathbf{V} \cdot \mathbf{dS} \qquad (2.88)$$

and

$$\nabla \cdot \left[\rho\left(e + \frac{V^2}{2}\right)\mathbf{V}\right] = -\nabla \cdot (p\mathbf{V}) \qquad (2.89)$$

Equations (2.88) and (2.89) are discussed and applied at length beginning with Chap. 7.

With the energy equation, we have introduced another unknown flow-field variable, e. We now have three equations, continuity, momentum, and energy, which involve four dependent variables, ρ, p, $\mathbf{V}$, and e. A fourth equation can be obtained from a thermodynamic state relation for e (see Chap. 7). If the gas is calorically perfect, then

$$e = c_v T \qquad (2.90)$$

where c_v is the specific heat at constant volume. Equation (2.90) introduces temperature as yet another dependent variable. However, the system can be completed by using the perfect gas equation of state

$$p = \rho R T \qquad (2.91)$$

where R is the specific gas constant. Therefore, the continuity, momentum, and energy equations, along with Eqs. (2.90) and (2.91) are five independent equations for the five unknowns, ρ, p, $\mathbf{V}$, e, and T. The matter of a perfect gas and related equations of state are reviewed in detail in Chap. 7; Eqs. (2.90) and (2.91) are presented here only to round out our development of the fundamental equations of fluid flow.

2.8 INTERIM SUMMARY

At this stage, let us pause and think about the various equations we have developed. Do not fall into the trap of seeing these equations as just a jumble of mathematical symbols that, by now, might look all the same to you. Quite the contrary, these equations speak words: e.g., Eqs. (2.39), (2.43), (2.44), and (2.45) all say that mass is conserved; Eqs. (2.55), (2.61a to c), (2.62), and (2.63a to c) are statements of Newton's second law applied to a fluid flow; Eqs. (2.86) to (2.89) say that energy is conserved. It is very important to be able to see the physical principles behind these equations. When you look at an equation, try to develop the ability to see past a collection of mathematical symbols and, instead, to read the physics that the equation represents.

The equations listed above are fundamental to all of aerodynamics. Take the time to go back over them. Become familiar with the way they are developed, and make yourself comfortable with their final forms. In this way, you will find our subsequent aerodynamic applications that much easier to understand.

Also, note our location on the road map shown in Fig. 2.1. We have finished the items on the left branch of the map—we have obtained the basic flow equations containing the fundamental physics of fluid flow. We now start with the branch on the right, which is a collection of useful concepts helpful in the application of the basic flow equations.

2.9   SUBSTANTIAL DERIVATIVE

Consider a small fluid element moving through a flow field, as shown in Fig. 2.22. The velocity field is given by $V = u\mathbf{i} + v\mathbf{j} + w\mathbf{k}$, where

$$u = u(x, y, z, t)$$

$$v = v(x, y, z, t)$$

$$w = w(x, y, z, t)$$

In addition, the density field is given by

$$\rho = \rho(x, y, z, t)$$

At time t_1, the fluid element is located at point 1 in the flow (see Fig. 2.22), and its density is

$$\rho_1 = \rho(x_1, y_1, z_1, t_1)$$

At a later time, t_2, the same fluid element has moved to a different location in the flow field, such as point 2 in Fig. 2.22. At this new time and location, the density of the fluid element is

$$\rho_2 = \rho(x_2, y_2, z_2, t_2)$$

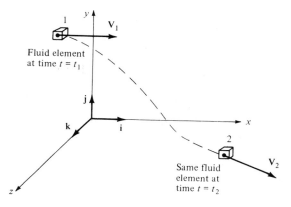

FIGURE 2.22
Fluid element moving in a flow field—illustration for the substantial derivative.

Since $\rho = \rho(x, y, z, t)$, we can expand this function in a Taylor series about point 1 as follows:

$$\rho_2 = \rho_1 + \left(\frac{\partial \rho}{\partial x}\right)_1 (x_2 - x_1) + \left(\frac{\partial \rho}{\partial y}\right)_1 (y_2 - y_1) + \left(\frac{\partial \rho}{\partial z}\right)_1 (z_2 - z_1)$$

$$+ \left(\frac{\partial \rho}{\partial t}\right)_1 (t_2 - t_1) + \text{higher-order terms}$$

Dividing by $t_2 - t_1$, and ignoring the higher-order terms, we have

$$\frac{\rho_2 - \rho_1}{t_2 - t_1} = \left(\frac{\partial \rho}{\partial x}\right)_1 \frac{x_2 - x_1}{t_2 - t_1} + \left(\frac{\partial \rho}{\partial y}\right)_1 \left(\frac{y_2 - y_1}{t_2 - t_1}\right) + \left(\frac{\partial \rho}{\partial z}\right)_1 \frac{z_2 - z_1}{t_2 - t_1} + \left(\frac{\partial \rho}{\partial t}\right)_1 \quad (2.92)$$

Consider the physical meaning of the left side of Eq. (2.92). The term $(\rho_2 - \rho_1)/(t_2 - t_1)$ is the *average* time rate of change in density of the fluid element as it moves from point 1 to point 2. In the limit, as t_2 approaches t_1, this term becomes

$$\lim_{t_2 \to t_1} \frac{\rho_2 - \rho_1}{t_2 - t_1} = \frac{D\rho}{Dt}$$

Here, $D\rho/Dt$ is a symbol for the *instantaneous* time rate of change of density of the fluid element as it moves through point 1. By definition, this symbol is called the *substantial derivative D/Dt*. Note that $D\rho/Dt$ is the time rate of change of density of a *given fluid element* as it moves through space. Here, our eyes are locked on the fluid element as it is moving, and we are watching the density of the element change as it moves through point 1. This is different from $(\partial \rho/\partial t)_1$, which is physically the time rate of change of density at the *fixed* point 1. For $(\partial \rho/\partial t)_1$, we fix our eyes on the stationary point 1, and watch the density change due to transient fluctuations in the flow field. Thus, $D\rho/Dt$ and $\partial \rho/\partial t$ are physically and numerically different quantities.

Returning to Eq. (2.92), note that

$$\lim_{t_2 \to t_1} \frac{x_2 - x_1}{t_2 - t_1} \equiv u$$

$$\lim_{t_2 \to t_1} \frac{y_2 - y_1}{t_2 - t_1} \equiv v$$

$$\lim_{t_2 \to t_1} \frac{z_2 - z_1}{t_2 - t_1} \equiv w$$

Thus, taking the limit of Eq. (2.92) as $t_2 \to t_1$, we obtain

$$\frac{D\rho}{Dt} = u \frac{\partial \rho}{\partial x} + v \frac{\partial \rho}{\partial y} + w \frac{\partial \rho}{\partial z} + \frac{\partial \rho}{\partial t} \quad (2.93)$$

Examine Eq. (2.93) closely. From it, we can obtain an expression for the substantial derivative in cartesian coordinates:

$$\frac{D}{Dt} \equiv \frac{\partial}{\partial t} + u \frac{\partial}{\partial x} + v \frac{\partial}{\partial y} + w \frac{\partial}{\partial z} \quad (2.94)$$

Furthermore, in cartesian coordinates, the vector operator ∇ is defined as

$$\nabla \equiv i\frac{\partial}{\partial x} + j\frac{\partial}{\partial y} + k\frac{\partial}{\partial z}$$

Hence, Eq. (2.94) can be written as

$$\frac{D}{Dt} \equiv \frac{\partial}{\partial t} + (\mathbf{V} \cdot \nabla) \tag{2.95}$$

Equation (2.95) represents a definition of the substantial derivative in vector notation; thus, it is valid for any coordinate system.

Focusing on Eq. (2.95), we once again emphasize that D/Dt is the substantial derivative, which is physically the time rate of change following a moving fluid element; $\partial/\partial t$ is called the *local derivative*, which is physically the time rate of change at a fixed point; $\mathbf{V} \cdot \nabla$ is called the *convective derivative*, which is physically the time rate of change due to the movement of the fluid element from one location to another in the flow field where the flow properties are spatially different. The substantial derivative applies to any flow-field variable, e.g., Dp/Dt, DT/Dt, Du/Dt. For example,

$$\frac{DT}{Dt} \equiv \underset{\substack{\text{local}\\\text{derivative}}}{\frac{\partial T}{\partial t}} + \underset{\substack{\text{convective}\\\text{derivative}}}{(\mathbf{V} \cdot \nabla) T} \equiv \frac{\partial T}{\partial t} + u\frac{\partial T}{\partial x} + v\frac{\partial T}{\partial y} + w\frac{\partial T}{\partial z} \tag{2.96}$$

Again, Eq. (2.96) states physically that the temperature of the fluid element is changing as the element sweeps past a point in the flow because at that point the flow-field temperature itself may be fluctuating with time (the local derivative) and because the fluid element is simply on its way to another point in the flow field where the temperature is different (the convective derivative).

Consider an example which will help to reinforce the physical meaning of the substantial derivative. Imagine that you are hiking in the mountains, and you are about to enter a cave. The temperature inside the cave is cooler than outside. Thus, as you walk through the mouth of the cave, you feel a temperature decrease—this is analogous to the convective derivative in Eq. (2.96). However, imagine that, at the same time, a friend throws a snowball at you such that the snowball hits you just at the same instant you pass through the mouth of the cave. You will feel an additional, but momentary, temperature drop when the snowball hits you—this is analogous to the local derivative in Eq. (2.96). The net temperature drop you feel as you walk through the mouth of the cave is therefore a combination of both the act of moving into the cave, where it is cooler, and being struck by the snowball at the same instant—this net temperature drop is analogous to the substantial derivative in Eq. (2.96).

2.10 FUNDAMENTAL EQUATIONS IN TERMS OF THE SUBSTANTIAL DERIVATIVE

In this section, we express the continuity, momentum, and energy equations in terms of the substantial derivative. In the process, we make use of the following vector identity:

$$\nabla \cdot (\rho \mathbf{V}) \equiv \rho \nabla \cdot \mathbf{V} + \mathbf{V} \cdot \nabla \rho \qquad (2.97)$$

In words, this identity states that the divergence of a scalar times a vector is equal to the scalar times the divergence of the vector plus the dot product of the vector and the gradient of the scalar.

First, consider the continuity equation given in the form of Eq. (2.43):

$$\frac{\partial \rho}{\partial t} + \nabla \cdot (\rho \mathbf{V}) = 0 \qquad (2.43)$$

Using the vector identity given by Eq. (2.97), Eq. (2.43) becomes

$$\frac{\partial \rho}{\partial t} + \mathbf{V} \cdot \nabla \rho + \rho \nabla \cdot \mathbf{V} = 0 \qquad (2.98)$$

However, the sum of the first two terms of Eq. (2.98) is the substantial derivative of ρ [see Eq. (2.95)]. Thus, from Eq. (2.98),

$$\boxed{\frac{D\rho}{Dt} + \rho \nabla \cdot \mathbf{V} = 0} \qquad (2.99)$$

Equation (2.99) is the continuity equation written in terms of the substantial derivative.

Next, consider the x component of the momentum equation given in the form of Eq. (2.61a):

$$\frac{\partial(\rho u)}{\partial t} + \nabla \cdot (\rho u \, \mathbf{V}) = -\frac{\partial p}{\partial x} + \rho f_x + (\mathscr{F}_x)_{\text{viscous}} \qquad (2.61a)$$

The first terms can be expanded as

$$\frac{\partial(\rho u)}{\partial t} = \rho \frac{\partial u}{\partial t} + u \frac{\partial \rho}{\partial t} \qquad (2.100)$$

In the second term of Eq. (2.61a), treat the scalar quantity as u and the vector quantity as $\rho \mathbf{V}$. Then the term can be expanded using the vector identity in Eq. (2.97):

$$\nabla \cdot (\rho u \, \mathbf{V}) \equiv \nabla \cdot [u(\rho \mathbf{V})] = u \nabla \cdot (\rho \mathbf{V}) + (\rho \mathbf{V}) \cdot \nabla u \qquad (2.101)$$

Substituting Eqs. (2.100) and (2.101) into (2.61a), we obtain

$$\rho\frac{\partial u}{\partial t}+u\frac{\partial \rho}{\partial t}+u\nabla\cdot(\rho\mathbf{V})+(\rho\mathbf{V})\cdot\nabla u=-\frac{\partial p}{\partial x}+\rho f_x+(\mathscr{F}_x)_{\text{viscous}}$$

or $\qquad \rho\dfrac{\partial u}{\partial t}+u\left[\dfrac{\partial \rho}{\partial t}+\nabla\cdot(\rho\mathbf{V})\right]+(\rho\mathbf{V})\cdot\nabla u=-\dfrac{\partial p}{\partial x}+\rho f_x+(\mathscr{F}_x)_{\text{viscous}}$ $\qquad$ (2.102)

Examine the two terms inside the square brackets; they are precisely the left side of the continuity equation, Eq. (2.43). Since the right side of Eq. (2.43) is zero, the sum inside the square brackets is zero. Hence, Eq. (2.102) becomes

$$\rho\frac{\partial u}{\partial t}+\rho\mathbf{V}\cdot\nabla u=-\frac{\partial p}{\partial x}+\rho f_x+(\mathscr{F}_x)_{\text{viscous}}$$

or $\qquad \rho\left(\dfrac{\partial u}{\partial t}+\mathbf{V}\cdot\nabla u\right)=-\dfrac{\partial p}{\partial x}+\rho f_x+(\mathscr{F}_x)_{\text{viscous}}$ $\qquad$ (2.103)

Examine the two terms inside the parentheses in Eq. (2.103); their sum is precisely the substantial derivative Du/Dt. Hence, Eq. (2.103) becomes

$$\rho\frac{Du}{Dt}=-\frac{\partial p}{\partial x}+\rho f_x+(\mathscr{F}_x)_{\text{viscous}} \qquad (2.104a)$$

In a similar manner, Eqs. (2.61b and c) yield

$$\rho\frac{Dv}{Dt}=-\frac{\partial p}{\partial y}+\rho f_y+(\mathscr{F}_y)_{\text{viscous}} \qquad (2.104b)$$

$$\rho\frac{Dw}{Dt}=-\frac{\partial p}{\partial z}+\rho f_z+(\mathscr{F}_z)_{\text{viscous}} \qquad (2.104c)$$

Equations (2.104a to c) are the x, y, and z components of the *momentum equation* written in terms of the substantial derivative. Compare these equations with Eqs. (2.61a to c). Note that the right sides of both sets of equations are unchanged; only the left sides are different.

In an analogous fashion, the energy equation given in the form of Eq. (2.87) can be expressed in terms of the substantial derivative. The derivation is left as a homework problem; the result is

$$\rho\frac{D(e+V^2/2)}{Dt}=\rho\dot{q}-\nabla\cdot(p\mathbf{V})+\rho(\mathbf{f}\cdot\mathbf{V})+\dot{Q}'_{\text{viscous}}+\dot{W}'_{\text{viscous}} \qquad (2.105)$$

Again, the right-hand sides of Eqs. (2.87) and (2.105) are the same; only the form of the left sides is different.

In modern aerodynamics, it is conventional to call the form of Eqs. (2.43), (2.61a to c), and (2.87) the *conservation* form of the fundamental equations (sometimes these equations are labeled as the *divergence* form because of the divergence terms on the left side). In contrast, the form of Eqs. (2.99), (2.104a to c), and (2.105), which deals with the substantial derivative on the left side, is called the *nonconservation* form. Both forms are equally valid statements of the fundamental principles, and in most cases, there is no particular reason to choose one form over the other. The nonconservation form is frequently found in textbooks and in aerodynamic theory. However, for the numerical solution of some aerodynamic problems, the conservation form sometimes leads to more accurate results. Hence, the distinction between the conservation form and the nonconservation form has become important in the modern discipline of computational fluid dynamics. (See Ref. 7 for more details.)

2.11 PATHLINES AND STREAMLINES OF A FLOW

In addition to knowing the density, pressure, temperature, and velocity fields, in aerodynamics we like to draw pictures of "where the flow is going." To accomplish this, we construct diagrams of pathlines and/or streamlines of the flow. The distinction between pathlines and streamlines is described in this section.

Consider an unsteady flow with a velocity field given by $\mathbf{V} = \mathbf{V}(x, y, z, t)$. Also, consider an infinitesimal fluid element moving through the flow field, say, element A as shown in Fig. 2.23a. Element A passes through point 1. Let us trace the path of element A as it moves downstream from point 1, as given by the dashed line in Fig. 2.23a. Such a path is defined as the *pathline* for element A. Now, trace the path of another fluid element, say, element B as shown in Fig. 2.23b. Assume that element B also passes through point 1, but at some different time from element A. The pathline of element B is given by the dashed line in Fig. 2.23b. Because the flow is unsteady, the velocity at point 1 (and at all other

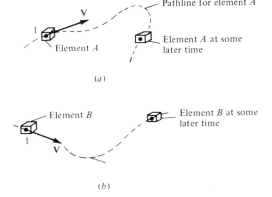

FIGURE 2.23
Pathlines for two different fluid elements passing through the same point in space: unsteady flow.

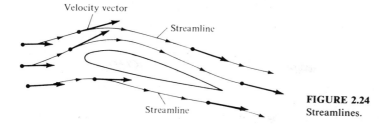

FIGURE 2.24
Streamlines.

points of the flow) changes with time. Hence, the pathlines of elements A and B are different curves in Fig. 2.23a and b. In general, for *unsteady* flow, the pathlines for different fluid elements passing through the same point are not the same.

In Sec. 1.4, the concept of a streamline was introduced in a somewhat heuristic manner. Let us be more precise here. By definition, a *streamline* is a curve whose tangent at any point is in the direction of the velocity vector at that point. Streamlines are illustrated in Fig. 2.24. The streamlines are drawn such that their tangents at every point along the streamline are in the same direction as the velocity vectors at those points. If the flow is unsteady, the streamline pattern is different at different times because the velocity vectors are fluctuating with time in both magnitude and direction.

In general, streamlines are different from pathlines. You can visualize a pathline as a time-exposure photograph of a given fluid element, whereas a streamline pattern is like a single frame of a motion picture of the flow. In an unsteady flow, the streamline pattern changes; hence, each "frame" of the motion picture is different.

However, for the case of *steady flow* (which applies to most of the applications in this book), the magnitude and direction of the velocity vectors at all points are fixed, invariant with time. Hence, the pathlines for different fluid elements going through the same point are the same. Moreover, the pathlines and streamlines are identical. Therefore, in steady flow, there is no distinction between pathlines and streamlines; they are the same curves in space. This fact is reinforced in Fig. 2.25, which illustrates the fixed, time-invariant streamline

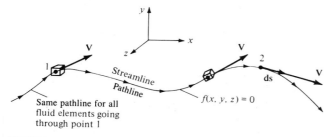

FIGURE 2.25
For steady flow, streamlines and pathlines are the same.

(pathline) through point 1. In Fig. 2.25, a given fluid element passing through point 1 traces a pathline downstream. All subsequent fluid elements passing through point 1 at later times trace the same pathline. Since the velocity vector is tangent to the pathline at all points on the pathline for all times, the pathline is also a streamline. For the remainder of this book, we deal mainly with the concept of streamlines rather than pathlines; however, always keep in mind the distinction described above.

Question: Given the velocity field of a flow, how can we obtain the mathematical equation for a streamline? Obviously, the streamline illustrated in Fig. 2.25 is a curve in space, and hence it can be described by the equation $f(x, y, z) = 0$. How can we obtain this equation? To answer this question, let **ds** be a directed element of the streamline, such as shown at point 2 in Fig. 2.25. The velocity at point 2 is **V**, and by definition of a streamline, **V** is parallel to **ds**. Hence, from the definition of the vector cross product [see Eq. (2.4)],

$$\mathbf{ds} \times \mathbf{V} = 0 \tag{2.106}$$

Equation (2.106) is a valid equation for a streamline. To put it in a more recognizable form, expand Eq. (2.106) in cartesian coordinates:

$$\mathbf{ds} = dx\,\mathbf{i} + dy\,\mathbf{j} + dz\,\mathbf{k}$$

$$\mathbf{V} = u\mathbf{i} + v\mathbf{j} + w\mathbf{k}$$

$$\mathbf{ds} \times \mathbf{V} = \begin{vmatrix} \mathbf{i} & \mathbf{j} & \mathbf{k} \\ dx & dy & dz \\ u & v & w \end{vmatrix}$$

$$= \mathbf{i}(w\,dy - v\,dz) + \mathbf{j}(u\,dz - w\,dx) + \mathbf{k}(v\,dx - u\,dy) = 0 \tag{2.107}$$

Since the vector given by Eq. (2.107) is zero, its components must each be zero:

$$w\,dy - v\,dz = 0 \tag{2.108a}$$

$$u\,dz - w\,dx = 0 \tag{2.108b}$$

$$v\,dx - u\,dy = 0 \tag{2.108c}$$

Equations (2.108a to c) are differential equations for the streamline. Knowing u, v, and w as functions of x, y, and z, Eqs. (2.108a to c) can be integrated to yield the equation for the streamline: $f(x, y, z) = 0$.

To reinforce the physical meaning of Eqs. (2.108a to c), consider a streamline in two dimensions, as sketched in Fig. 2.26a. The equation of this streamline is $y = f(x)$. Hence, at point 1 on the streamline, the slope is dy/dx. However, **V** with x and y components u and v, respectively, is tangent to the streamline at

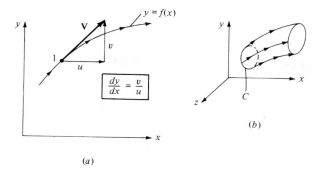

FIGURE 2.26
(*a*) Equation of a streamline in two-dimensional cartesian space.
(*b*) Sketch of a streamtube in three-dimensional space.

point 1. Thus, the slope of the streamline is also given by v/u, as shown in Fig. 2.24. Therefore,

$$\frac{dy}{dx} = \frac{v}{u} \qquad (2.109)$$

Equation (2.109) is a differential equation for a streamline in two dimensions. From Eq. (2.109),

$$v\,dx - u\,dy = 0$$

which is precisely Eq. (2.108c). Therefore, Eqs. (2.108a to c) and (2.109) simply state mathematically that the velocity vector is tangent to the streamline.

A concept related to streamlines is that of a streamtube. Consider an arbitrary closed curve C in three-dimensional space, as shown in Fig. 2.26b. Consider the streamlines which pass through all points on C. These streamlines form a tube in space as sketched in Fig. 2.26b; such a tube is called a *streamtube*. For example, the walls of an ordinary garden hose form a streamtube for the water flowing through the hose. For a steady flow, a direct application of the integral form of the continuity equation [Eq. (2.44)] proves that the mass flow across all cross sections of a streamtube is constant. (Prove this yourself.)

Example 2.2. Consider the velocity field given by $u = y/(x^2 + y^2)$ and $v = -x/(x^2 + y^2)$. Calculate the equation of the streamline passing through the point $(0, 5)$.

Solution. From Eq. (2.109), $dy/dx = v/u = -x/y$, and

$$y\,dy = -x\,dx$$

Integrating, we obtain

$$y^2 = -x^2 + c$$

where c is a constant of integration.

For the streamline through $(0, 5)$, we have

$$5^2 = 0 + c \qquad \text{or} \qquad c = 25$$

Thus, the equation of the streamline is

$$\boxed{x^2 + y^2 = 25}$$

Note that the streamline is a circle with its center at the origin and a radius of 5 units.

2.12 ANGULAR VELOCITY, VORTICITY, AND STRAIN

In several of our previous discussions, we made use of the concept of a fluid element moving through the flow field. In this section, we examine this motion more closely, paying particular attention to the orientation of the element and its change in shape as it moves along a streamline. In the process, we introduce the concept of vorticity, one of the most powerful quantities in theoretical aerodynamics.

Consider an infinitesimal fluid element moving in a flow field. As it translates along a streamline, it may also *rotate*, and in addition its shape may become *distorted*, as sketched in Fig. 2.27. The amount of rotation and distortion depends on the velocity field; the purpose of this section is to quantify this dependency.

Consider a two-dimensional flow in the xy plane. Also, consider an infinitesimal fluid element in this flow. Assume that at time t the shape of this fluid element is rectangular, as shown at the left of Fig. 2.28. Assume that the fluid element is moving upward and to the right; its position and shape at time $t + \Delta t$ are shown at the right in Fig. 2.28. Note that during the time increment Δt, the sides AB and AC have rotated through the angular displacements $-\Delta\theta_1$ and $\Delta\theta_2$, respectively. (Counterclockwise rotations by convention are considered positive; since line AB is shown with a clockwise rotation in Fig. 2.28, the angular displacement is negative, $-\Delta\theta_1$.) At present, consider just the line AC. It has rotated because during the time increment Δt, point C has moved differently from point A. Consider the velocity in the y direction. At point A at time t, this

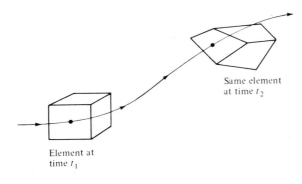

Same element
at time t_2

Element at
time t_1

FIGURE 2.27
The motion of a fluid element along a streamline is a combination of translation and rotation; in addition, the shape of the element can become distorted.

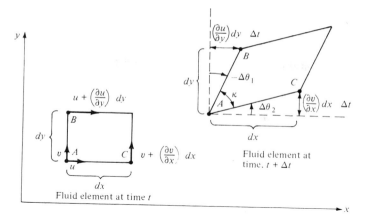

FIGURE 2.28
Rotation and distortion of a fluid element.

velocity is v, as shown in Fig. 2.28. Point C is a distance dx from point A; hence, at time t the vertical component of velocity of point C is given by $v + (\partial v/\partial x)\, dx$. Hence,

$$\text{Distance in } y \text{ direction that } A \text{ moves during time increment } \Delta t = v\Delta t$$

$$\text{Distance in } y \text{ direction that } C \text{ moves during time increment } \Delta t = \left(v + \frac{\partial v}{\partial x}\, dx\right)\Delta t$$

$$\text{Net displacement in } y \text{ direction of } C \text{ relative to } A = \left(v + \frac{\partial v}{\partial x}\, dx\right)\Delta t - v\Delta t$$

$$= \left(\frac{\partial v}{\partial x}\, dx\right)\Delta t$$

This net displacement is shown at the right of Fig. 2.28. From the geometry of Fig. 2.28,

$$\tan \Delta\theta_2 = \frac{[(\partial v/\partial x)\, dx]\, \Delta t}{dx} = \frac{\partial v}{\partial x}\Delta t \qquad (2.110)$$

Since $\Delta\theta_2$ is a small angle, $\tan \Delta\theta_2 \approx \Delta\theta_2$. Hence, Eq. (2.110) reduces to

$$\Delta\theta_2 = \frac{\partial v}{\partial x}\Delta t \qquad (2.111)$$

Now consider line AB. The x component of the velocity at point A at time t is u, as shown in Fig. 2.28. Because point B is a distance dy from point A, the horizontal component of velocity of point B at time t is $u + (\partial u/\partial y)\, dy$. By reasoning similar to that above, the net displacement in the x direction of B

relative to A over the time increment Δt is $[(\partial u/\partial y)\,dy]\,\Delta t$, as shown in Fig. 2.28. Hence,

$$\tan(-\Delta\theta_1) = \frac{[(\partial u/\partial y)\,dy]\,\Delta t}{dy} = \frac{\partial u}{\partial y}\Delta t \tag{2.112}$$

Since $-\Delta\theta_1$ is small, Eq. (2.112) reduces to

$$\Delta\theta_1 = -\frac{\partial u}{\partial y}\Delta t \tag{2.113}$$

Consider the angular velocities of lines AB and AC, defined as $d\theta_1/dt$ and $d\theta_2/dt$, respectively. From Eq. (2.113), we have

$$\frac{d\theta_1}{dt} \equiv \lim_{\Delta t \to 0}\frac{\Delta\theta_1}{\Delta t} = -\frac{\partial u}{\partial y} \tag{2.114}$$

From Eq. (2.111), we have

$$\frac{d\theta_2}{dt} = \lim_{\Delta t \to 0}\frac{\Delta\theta_2}{\Delta t} = \frac{\partial v}{\partial x} \tag{2.115}$$

By definition, the angular velocity of the fluid element as seen in the xy plane is the average of the angular velocities of lines AB and AC. Let ω_z denote this angular velocity. Therefore, by definition,

$$\omega_z = \frac{1}{2}\left(\frac{d\theta_1}{dt} + \frac{d\theta_2}{dt}\right) \tag{2.116}$$

Combining Eqs. (2.114) to (2.116) yields

$$\omega_z = \frac{1}{2}\left(\frac{\partial v}{\partial x} - \frac{\partial u}{\partial y}\right) \tag{2.117}$$

In the above discussion, we have considered motion in the xy plane only. However, the fluid element is generally moving in three-dimensional space, and its angular velocity is a vector $\boldsymbol{\omega}$ that is orientated in some general direction, as shown in Fig. 2.29. In Eq. (2.117), we have obtained only the component of $\boldsymbol{\omega}$ in the z direction; this explains the subscript z in Eqs. (2.116) and (2.117). The

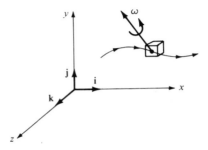

FIGURE 2.29
Angular velocity of a fluid element in three-dimensional space.

x and y components of $\boldsymbol{\omega}$ can be obtained in a similar fashion. The resulting angular velocity of the fluid element in three-dimensional space is

$$\boldsymbol{\omega} = \omega_x \mathbf{i} + \omega_y \mathbf{j} + \omega_z \mathbf{k}$$

$$\boxed{\boldsymbol{\omega} = \frac{1}{2}\left[\left(\frac{\partial w}{\partial y} - \frac{\partial v}{\partial z}\right)\mathbf{i} + \left(\frac{\partial u}{\partial z} - \frac{\partial w}{\partial x}\right)\mathbf{j} + \left(\frac{\partial v}{\partial x} - \frac{\partial u}{\partial y}\right)\mathbf{k}\right]} \qquad (2.118)$$

Equation (2.118) is the desired result; it expresses the angular velocity of the fluid element in terms of the velocity field, or more precisely, in terms of derivatives of the velocity field.

The angular velocity of a fluid element plays an important role in theoretical aerodynamics, as we shall soon see. However, the expression $2\boldsymbol{\omega}$ appears frequently, and therefore we define a new quantity, *vorticity*, which is simply twice the angular velocity. Denote vorticity by the vector $\boldsymbol{\xi}$:

$$\boldsymbol{\xi} \equiv 2\boldsymbol{\omega}$$

Hence, from Eq. (2.118),

$$\boxed{\boldsymbol{\xi} = \left(\frac{\partial w}{\partial y} - \frac{\partial v}{\partial z}\right)\mathbf{i} + \left(\frac{\partial u}{\partial z} - \frac{\partial w}{\partial x}\right)\mathbf{j} + \left(\frac{\partial v}{\partial x} - \frac{\partial u}{\partial y}\right)\mathbf{k}} \qquad (2.119)$$

Recall Eq. (2.22) for $\nabla \times \mathbf{V}$ in cartesian coordinates. Since u, v, and w denote the x, y, and z components of velocity, respectively, note that the right sides of Eqs. (2.22) and (2.119) are identical. Hence, we have the important result that

$$\boxed{\boldsymbol{\xi} = \nabla \times \mathbf{V}} \qquad (2.120)$$

In a velocity field, the curl of the velocity is equal to the vorticity.
The above leads to two important definitions:

1. If $\nabla \times \mathbf{V} \neq 0$ at every point in a flow, the flow is called *rotational*. This implies that the fluid elements have a finite angular velocity.
2. If $\nabla \times \mathbf{V} = 0$ at every point in a flow, the flow is called *irrotational*. This implies that the fluid elements have no angular velocity; rather, their motion through space is a pure translation.

The case of rotational flow is illustrated in Fig. 2.30. Here, fluid elements moving along two different streamlines are shown in various modes of rotation. In contrast, the case of irrotational flow is illustrated in Fig. 2.31. Here, the upper streamline shows a fluid element where the angular velocities of its sides are zero. The lower streamline shows a fluid element where the angular velocities of two intersecting sides are finite but equal and opposite to each other, and so their sum is identically

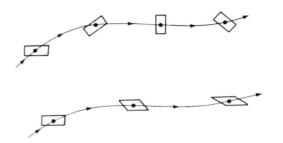

FIGURE 2.30
Fluid elements in a rotational flow.

zero. In both cases, the angular velocity of the fluid element is zero; i.e., the flow is irrotational.

If the flow is two-dimensional (say, in the xy plane), then from Eq. (2.119),

$$\boldsymbol{\xi} = \xi_z \mathbf{k} = \left(\frac{\partial v}{\partial x} - \frac{\partial u}{\partial y}\right)\mathbf{k} \tag{2.121}$$

Also, if the flow is irrotational, $\boldsymbol{\xi} = 0$. Hence, from Eq. (2.121),

$$\boxed{\frac{\partial v}{\partial x} - \frac{\partial u}{\partial y} = 0} \tag{2.122}$$

Equation (2.122) is the *condition of irrotationality for two-dimensional flow.* We will have frequent occasion to use Eq. (2.122).

Why is it so important to make a distinction between rotational and irrotational flows? The answer becomes blatantly obvious as we progress in our study of aerodynamics; we find that irrotational flows are much easier to analyze than rotational flows. However, irrotational flow may at first glance appear to be so special that its applications are limited. Amazingly enough, such is not the case. There are a large number of practical aerodynamic problems where the flow field is essentially irrotational, e.g., the subsonic flow over airfoils, the supersonic flow over slender bodies at small angle of attack, and the subsonic-supersonic flow through nozzles. For such cases, there is generally a thin boundary layer of viscous

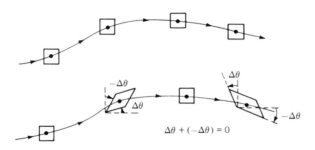

$\Delta\theta + (-\Delta\theta) = 0$

FIGURE 2.31
Fluid elements in an irrotational flow.

flow immediately adjacent to the surface; in this viscous region the flow is highly rotational. However, outside this boundary layer, the flow is frequently irrotational. As a result, the study of irrotational flow is an important aspect of aerodynamics.

Return to the fluid element shown in Fig. 2.28. Let the angle between sides AB and AC be denoted by κ. As the fluid element moves through the flow field, κ will change. In Fig. 2.28, at time t, κ is initially 90°. At time $t + \Delta t$, κ has changed by the amount $\Delta\kappa$, where

$$\Delta\kappa = -\Delta\theta_2 - (-\Delta\theta_1) \tag{2.123}$$

By definition, the *strain* of the fluid element as seen in the xy plane is the change in κ, where positive strain corresponds to a *decreasing* κ. Hence, from Eq. (2.123),

$$\text{Strain} = -\Delta\kappa = \Delta\theta_2 - \Delta\theta_1 \tag{2.124}$$

In viscous flows (to be discussed in Chaps. 15 to 17), the time rate of strain is an important quantity. Denote the time rate of strain by ε_{xy}, where in conjunction with Eq. (2.124)

$$\varepsilon_{xy} \equiv -\frac{d\kappa}{dt} = \frac{d\theta_2}{dt} - \frac{d\theta_1}{dt} \tag{2.125}$$

Substituting Eqs. (2.114) and (2.115) into (2.125), we have

$$\varepsilon_{xy} = \frac{\partial v}{\partial x} + \frac{\partial u}{\partial y} \tag{2.126a}$$

In the yz and zx planes, by a similar derivation the strain is, respectively,

$$\varepsilon_{yz} = \frac{\partial w}{\partial y} + \frac{\partial v}{\partial z} \tag{2.126b}$$

and

$$\varepsilon_{zx} = \frac{\partial u}{\partial z} + \frac{\partial w}{\partial x} \tag{2.126c}$$

Note that angular velocity (hence, vorticity) and time rate of strain depend solely on the velocity derivatives of the flow field. These derivatives can be displayed in a matrix as follows:

$$\begin{bmatrix} \dfrac{\partial u}{\partial x} & \dfrac{\partial u}{\partial y} & \dfrac{\partial u}{\partial z} \\[2mm] \dfrac{\partial v}{\partial x} & \dfrac{\partial v}{\partial y} & \dfrac{\partial v}{\partial z} \\[2mm] \dfrac{\partial w}{\partial x} & \dfrac{\partial w}{\partial y} & \dfrac{\partial w}{\partial z} \end{bmatrix}$$

The sum of the diagonal terms is simply equal to $\nabla \cdot \mathbf{V}$, which from Sec. 2.3 is equal to the time rate of change of volume of a fluid element; hence, the diagonal terms represent the *dilatation* of a fluid element. The off-diagonal terms are cross derivatives which appear in Eqs. (2.118), (2.119), and (2.126a to c). Hence, the off-diagonal terms are associated with rotation and strain of a fluid element.

In summary, in this section, we have examined the rotation and deformation of a fluid element moving in a flow field. The angular velocity of a fluid element and the corresponding vorticity at a point in the flow are concepts which are useful in the analysis of both inviscid and viscous flows; in particular, the absence of vorticity—irrotational flow—greatly simplifies the analysis of the flow, as we will see. We take advantage of this simplification in much of our treatment of inviscid flows in subsequent chapters. On the other hand, we do not make use of the time rate of strain until we discuss viscous flow, beginning with Chap. 15.

Example 2.3. For the velocity field given in Example 2.2, calculate the vorticity.

Solution

$$\boldsymbol{\xi} = \nabla \times \mathbf{V} = \begin{vmatrix} \mathbf{i} & \mathbf{j} & \mathbf{k} \\ \dfrac{\partial}{\partial x} & \dfrac{\partial}{\partial y} & \dfrac{\partial}{\partial z} \\ u & v & w \end{vmatrix} = \begin{vmatrix} \mathbf{i} & \mathbf{j} & \mathbf{k} \\ \dfrac{\partial}{\partial x} & \dfrac{\partial}{\partial y} & \dfrac{\partial}{\partial z} \\ \dfrac{y}{x^2+y^2} & \dfrac{-x}{x^2+y^2} & 0 \end{vmatrix}$$

$$= \mathbf{i}[0-0] - \mathbf{j}[0-0]$$

$$+ \mathbf{k}\left[\frac{(x^2+y^2)(-1)+x(2x)}{x^2+y^2} - \frac{(x^2+y^2)-y(2y)}{x^2+y^2} \right]$$

$$= 0\mathbf{i} + 0\mathbf{j} + 0\mathbf{k} = 0$$

The flow field is irrotational at every point except at the origin, where $x^2 + y^2 = 0$.

Example 2.4. Consider the boundary-layer velocity profile used in Example 2.1, namely, $u/V_\infty = (y/\delta)^{0.25}$. Is this flow rotational or irrotational?

Solution. For a two-dimensional flow, the irrotationality condition is given by Eq. (2.122), namely,

$$\frac{\partial v}{\partial x} - \frac{\partial u}{\partial y} = 0$$

Does this relation hold for the viscous boundary-layer flow in Example 2.1? Let us examine this question. From the boundary-layer velocity profile given by

$$\frac{u}{V_\infty} = \left(\frac{y}{\delta}\right)^{0.25}$$

we obtain

$$\frac{\partial u}{\partial y} = 0.25 \frac{V_\infty}{\delta} \left(\frac{y}{\delta}\right)^{-0.75} \tag{E.1}$$

What can we say about $\partial v/\partial x$? In Example 2.1, the flow was incompressible. From the continuity equation for a steady flow given by Eq. (2.45), repeated below,

$$\nabla \cdot (\rho \mathbf{V}) = \frac{\partial(\rho u)}{\partial x} + \frac{\partial(\rho v)}{\partial y} = 0$$

we have for an incompressible flow, where $\rho = $ constant,

$$\frac{\partial u}{\partial x} + \frac{\partial v}{\partial y} = 0 \qquad \text{(E.2)}$$

Equation (E.2) will provide an expression for v as follows:

$$\frac{\partial u}{\partial x} = \frac{\partial}{\partial x}\left[V_\infty \left(\frac{y}{\delta}\right)^{0.25} \right] \qquad \text{(E.3)}$$

However, from Example 2.1, we stated that

$$\frac{\delta}{c} = \frac{5}{\sqrt{\text{Re}_c}}$$

This equation holds at any x station along the plate, not just at $x = c$. Therefore, we can write

$$\frac{\delta}{x} = \frac{5}{\sqrt{\text{Re, } x}}$$

where

$$\text{Re, } x = \frac{\rho_\infty V_\infty x}{\mu_\infty}$$

Thus, δ is a function of x given by

$$\delta = 5 \sqrt{\frac{\mu_\infty x}{\rho_\infty V_\infty}}$$

and

$$\frac{d\delta}{dx} = \frac{5}{2} \sqrt{\frac{\mu_\infty}{\rho_\infty V_\infty}}\, x^{-1/2}$$

Substituting into Eq. (E.3), we have

$$\frac{\partial u}{\partial x} = \frac{\partial}{\partial x}\left[V_\infty \left(\frac{y}{\delta}\right)^{0.25} \right] = V_\infty y^{0.25}(-0.25)\delta^{-1.25}\frac{d\delta}{dx}$$

$$= -V_\infty y^{0.25}\delta^{-1.25}\left(\frac{5}{8}\right) \sqrt{\frac{\mu_\infty}{\rho_\infty V_\infty}}\, x^{-1/2}$$

$$= -\frac{5}{8} V_\infty y^{0.25}\left(\frac{1}{5}\right)^{1.25}\left(\frac{\mu_\infty}{\rho_\infty V_\infty}\right)^{-1/8} x^{-9/8}$$

Hence,

$$\frac{\partial u}{\partial x} = -C y^{1/4} x^{-9/8}$$

where C is a constant. Inserting this into Eq. (E.2), we have

$$\frac{\partial v}{\partial y} = C_y^{1/4} x^{-9/8}$$

Integrating with respect to y, we have

$$v = C_1 y^{5/4} x^{-9/8} + C_2 \tag{E.4}$$

where C_1 is a constant and C_2 can be a function of x. Evaluating Eq. (E.4) at the wall, where $v = 0$ and $y = 0$, we obtain $C_2 = 0$. Hence,

$$v = C_1 y^{5/4} x^{-9/8}$$

In turn, we obtain by differentiation

$$\frac{\partial v}{\partial x} = C_3 y^{5/4} x^{-17/8} \tag{E.5}$$

(*Note*: v is finite inside a boundary layer; the streamlines within a boundary are deflected upward. However, this "displacement" effect is usually small compared to the running length in the x direction, and v is of small magnitude in comparison to u. Both of these statements will be verified in Chaps. 16 and 17.) Recasting Eq. (E.1) in the same general form as Eq. (E.5), we have

$$\frac{\partial u}{\partial y} = 0.25 V_\infty y^{-0.75} \left(\frac{1}{\delta}\right)^{0.25}$$

$$= 0.25 V_\infty y^{-0.75} \left(\frac{1}{5\sqrt{\mu_\infty x / \rho_\infty V_\infty}}\right)^{0.25}$$

Hence,

$$\frac{\partial u}{\partial y} = C_4 y^{-3/4} x^{-1/8} \tag{E.6}$$

From Eqs. (E.5) and (E.6), we can write

$$\frac{\partial v}{\partial x} - \frac{\partial u}{\partial y} = C_3 y^{5/4} x^{-17/8} - C_4 y^{-3/4} x^{-1/9} \neq 0$$

Therefore, the irrotationality condition does *not* hold; the flow is *rotational.*

In Example 2.4, we demonstrated a basic result which holds in general for viscous flows, namely, viscous flows are *rotational.* This is almost intuitive. For example, consider an infinitesimally small fluid element moving along a streamline, as sketched in Fig. 2.32. If this is a viscous flow, and assuming that the velocity increases in the upward direction (i.e., the velocity is higher on the neighboring streamline above and lower on the neighboring streamline below), then the shear stresses on the upper and lower faces of the fluid element will be in the directions shown. Such shear stresses will be discussed at length in Chap. 15. Examining Fig. 2.32, we see clearly that the shear stresses exert a rotational moment about the center of the element, thus providing a mechanism for setting the fluid element into rotation. Although this picture is overly simplistic, it serves

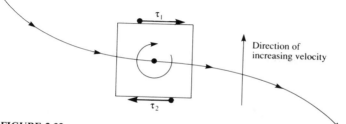

FIGURE 2.32
Shear stress and the consequent rotation of a fluid element.

to emphasize that viscous flows are rotational flows. On the other hand, as stated earlier in this section, there are numerous inviscid flow problems that are irrotational, with the attendant simplifications to be explained later. Some inviscid flows are rotational, but there exists such a large number of practical aerodynamic problems described by inviscid, irrotational flows that the distinction between rotational and irrotational flow is an important consideration.

2.13 CIRCULATION

You are reminded again that this is a tool-building chapter. Taken individually, each aerodynamic tool we have developed so far may not be particularly exciting. However, taken collectively, these tools allow us to obtain solutions for some very practical and exciting aerodynamic problems.

In this section, we introduce a tool which is fundamental to the calculation of aerodynamic lift, namely, *circulation*. This tool was used independently by Frederick Lanchester (1878–1946) in England, Wilhelm Kutta (1867–1944) in Germany, and Nikolai Joukowski (1847–1921) in Russia to create a breakthrough in the theory of aerodynamic lift at the turn of the twentieth century. The relationship between circulation and lift and the historical circumstances surrounding this breakthrough are discussed in Chaps. 3 and 4. The purpose of this section is only to define circulation and relate it to vorticity.

Consider a closed curve C in a flow field, as sketched in Fig. 2.33. Let $\mathbf{V}$ and $\mathbf{ds}$ be the velocity and directed line segment, respectively, at a point on C.

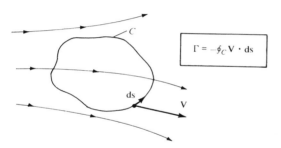

$$\Gamma = -\oint_C \mathbf{V} \cdot \mathbf{ds}$$

FIGURE 2.33
Definition of circulation.

The circulation, denoted by Γ, is defined as

$$\boxed{\Gamma \equiv -\oint_C \mathbf{V} \cdot \mathbf{ds}}$$

(2.127)

The circulation is simply the negative of the line integral of velocity around a closed curve in the flow; it is a kinematic property depending only on the velocity field and the choice of the curve C. As discussed in Sec. 2.2.8, Line Integrals, by mathematical convention the positive sense of the line integral is counterclockwise. However, in aerodynamics, it is convenient to consider a positive circulation as being clockwise. Hence, a minus sign appears in the definition given by Eq. (2.127) to account for the positive-counterclockwise sense of the integral and the positive-clockwise sense of circulation.†

The use of the word "circulation" to label the integral in Eq. (2.127) may be somewhat misleading because it leaves a general impression of something moving around in a loop. Indeed, according to the *American Heritage Dictionary of the English Language*, the first definition given to the word "circulation" is "movement in a circle or circuit." However, in aerodynamics, circulation has a very precise technical meaning, namely, Eq. (2.127). It does *not* necessarily mean that the fluid elements are moving around in circles within this flow field—a common early misconception of new students of aerodynamics. Rather, when circulation exists in a flow, it simply means that the line integral in Eq. (2.127) is finite. For example, if the airfoil in Fig. 2.24 is generating lift, the circulation taken around a closed curve enclosing the airfoil will be finite, although the fluid elements are by no means executing circles around the airfoil (as clearly seen from the streamlines sketched in Fig. 2.24).

Circulation is also related to vorticity as follows. Refer back to Fig. 2.9, which shows an open surface bounded by the closed curve C. Assume that the surface is in a flow field and the velocity at point P is $\mathbf{V}$, where P is any point on the surface (including any point on curve C). From Stokes' theorem [Eq. (2.25)],

$$\boxed{\Gamma \equiv -\oint_C \mathbf{V} \cdot \mathbf{ds} = -\iint_S (\nabla \times \mathbf{V}) \cdot \mathbf{dS}}$$

(2.128)

Hence, the circulation about a curve C is equal to the vorticity integrated over any open surface bounded by C. This leads to the immediate result that if the

---

† Some books do not use the minus sign in the definition of circulation. In such cases, the positive sense of both the line integral and Γ is in the same direction. This causes no problem as long as the reader is aware of the convention used in a particular book or paper.

flow is *irrotational everywhere within the contour of integration* (i.e., if $\nabla \times \mathbf{V} = 0$ over any surface bounded by C), *then* $\Gamma = 0$. A related result is obtained by letting the curve C shrink to an infinitesimal size, and denoting the circulation around this infinitesimally small curve by $d\Gamma$. Then, in the limit as C becomes infinitesimally small, Eq. (2.128) yields

$$d\Gamma = -(\nabla \times \mathbf{V}) \cdot \mathbf{dS} = -(\nabla \times \mathbf{V}) \cdot \mathbf{n} \, dS$$

or

$$\boxed{(\nabla \times \mathbf{V}) \cdot \mathbf{n} = -\frac{d\Gamma}{dS}}$$

(2.129)

where dS is the infinitesimal area enclosed by the infinitesimal curve C. Referring to Fig. 2.34, Eq. (2.129) states that at a point P in a flow, the component of vorticity normal to dS is equal to the negative of the "circulation per unit area," where the circulation is taken around the boundary of dS.

Example 2.5. For the velocity field given in Example 2.2, calculate the circulation around a circular path of radius 5 m. Assume that u and v given in Example 2.2 are in units of meters per second.

Solution. Since we are dealing with a circular path, it is easier to work this problem in polar coordinates, where $x = r \cos \theta$, $y = r \sin \theta$, $x^2 + y^2 = r^2$, $V_r = u \cos \theta + v \sin \theta$, and $V_\theta = -u \sin \theta + v \cos \theta$. Therefore,

$$u = \frac{y}{x^2 + y^2} = \frac{r \sin \theta}{r^2} = \frac{\sin \theta}{r}$$

$$v = -\frac{x}{x^2 + y^2} = -\frac{r \cos \theta}{r^2} = -\frac{\cos \theta}{r}$$

$$V_r = \frac{\sin \theta}{r} \cos \theta + \left(-\frac{\cos \theta}{r}\right) \sin \theta = 0$$

$$V_\theta = -\frac{\sin \theta}{r} \sin \theta + \left(-\frac{\cos \theta}{r}\right) \cos \theta = -\frac{1}{r}$$

$$\mathbf{V} \cdot \mathbf{dS} = (V_r \, \mathbf{e}_r + V_\theta \, \mathbf{e}_\theta) \cdot (dr \, \mathbf{e}_r + r \, d\theta \, \mathbf{e}_\theta)$$

$$= V_r \, dr + r V_\theta \, d\theta = 0 + r\left(-\frac{1}{r}\right) d\theta = -d\theta$$

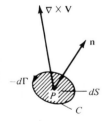

$$(\nabla \times \mathbf{V}) \cdot \mathbf{n} = -\frac{d\Gamma}{dS}$$

FIGURE 2.34
Relation between vorticity and circulation.

Hence,

$$\Gamma = -\oint_C \mathbf{V} \cdot \mathbf{ds} = -\int_0^{2\pi} -d\theta = \boxed{2\pi \quad \text{m}^2/\text{s}}$$

Note that we never used the 5-m diameter of the circular path; in this case, the value of Γ is independent of the diameter of the path.

2.14   STREAM FUNCTION

In this section, we consider two-dimensional steady flow. Recall from Sec. 2.11 that the differential equation for a streamline in such a flow is given by Eq. (2.109), repeated below

$$\frac{dy}{dx} = \frac{v}{u} \tag{2.109}$$

If u and v are known functions of x and y, then Eq. (2.109) can be integrated to yield the algebraic equation for a streamline:

$$f(x, y) = c \tag{2.130}$$

where c is an arbitrary constant of integration, with different values for different streamlines. In Eq. (2.130), denote the function of x and y by the symbol $\bar{\psi}$. Hence, Eq. (2.130) is written as

$$\bar{\psi}(x, y) = c \tag{2.131}$$

The function $\bar{\psi}(x, y)$ is called the *stream function*. From Eq. (2.131) we see that the equation for a streamline is given by *setting the stream function equal to a constant*, i.e., c_1, c_2, c_3, etc. Two different streamlines are illustrated in Fig. 2.35; streamlines ab and cd are given by $\bar{\psi} = c_1$ and $\bar{\psi} = c_2$, respectively.

There is a certain arbitrariness in Eqs. (2.130) and (2.131) via the arbitrary constant of integration c. Let us define the stream function more precisely in order to reduce this arbitrariness. Referring to Fig. 2.35, let us define the numerical value of $\bar{\psi}$ such that the *difference* $\Delta\bar{\psi}$ between $\bar{\psi} = c_2$ for streamline cd and $\bar{\psi} = c_1$ for streamline ab is equal to the *mass* flow between the two streamlines. Since Fig. 2.35 is a two-dimensional flow, the mass flow between two streamlines is defined *per unit depth perpendicular to the page*. That is, in Fig. 2.35 we are considering the mass flow inside a streamtube bounded by streamlines ab and cd, with a rectangular cross-sectional area equal to Δn times a unit depth perpendicular to the page. Here, Δn is the normal distance between ab and cd,

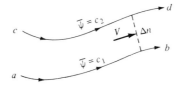

FIGURE 2.35
Different streamlines are given by different values of the stream function.

as shown in Fig. 2.35. Hence, mass flow between streamlines ab and cd per unit depth perpendicular to the page is

$$\Delta\bar{\psi} = c_2 - c_1 \tag{2.132}$$

The above definition does not completely remove the arbitrariness of the constant of integration in Eqs. (2.130) and (2.131), but it does make things a bit more precise. For example, consider a given two-dimensional flow field. Choose one streamline of the flow, and give it an arbitrary value of the stream function, say, $\bar{\psi} = c_1$. Then, the value of the stream function for any other streamline in the flow, say, $\bar{\psi} = c_2$, is fixed by the definition given in Eq. (2.132). Which streamline you choose to designate as $\bar{\psi} = c_1$ and what numerical value you give c_1 usually depend on the geometry of the given flow field, as we see in Chap. 3.

The equivalence between $\bar{\psi} = $ constant designating a streamline, and $\Delta\bar{\psi}$ equaling mass flow (per unit depth) between streamlines, is natural. For a steady flow, the mass flow inside a given streamtube is constant along the tube; the mass flow across any cross section of the tube is the same. Since by definition $\Delta\bar{\psi}$ is equal to this mass flow, then $\Delta\bar{\psi}$ itself is constant for a given streamtube. In Fig. 2.35, if $\bar{\psi}_1 = c_1$ designates the streamline on the bottom of the streamtube, then $\psi_2 = c_2 = c_1 + \Delta\bar{\psi}$ is also constant along the top of the streamtube. Since by definition of a streamtube (see Sec. 2.11) the upper boundary of the streamtube is a streamline itself, then $\psi_2 = c_2 = $ constant must designate this streamline.

We have yet to develop the most important property of the stream function, namely, derivatives of $\bar{\psi}$ yield the flow-field velocities. To obtain this relationship, consider again the streamlines ab and cd in Fig. 2.35. Assume that these streamlines are close together (i.e., assume n is small), such that the flow velocity V is a constant value across Δn. The mass flow through the streamtube per unit depth perpendicular to the page is

$$\Delta\bar{\psi} \equiv \rho V\,\Delta n (1)$$

or

$$\frac{\Delta\bar{\psi}}{\Delta n} = \rho V \tag{2.133}$$

Consider the limit of Eq. (2.133) as $\Delta n \to 0$:

$$\rho V = \lim_{\Delta n \to 0} \frac{\Delta\bar{\psi}}{\Delta n} \equiv \frac{\partial\bar{\psi}}{\partial n} \tag{2.134}$$

Equation (2.134) states that if we know $\bar{\psi}$, then we can obtain the product (ρV) by differentiating $\bar{\psi}$ in the direction *normal* to V. To obtain a practical form of Eq. (2.134) for cartesian coordinates, consider Fig. 2.36. Notice that the directed normal distance Δn is equivalent first to moving upward in the y direction by the amount Δy and then to the left in the negative x direction by the amount $-\Delta x$. Due to conservation of mass, the mass flow through Δn (per unit depth) is equal to the sum of the mass flows through Δy and $-\Delta x$ (per unit depth):

$$\text{Mass flow} = \Delta\bar{\psi} = \rho V\,\Delta n = \rho u\,\Delta y + \rho v(-\Delta x) \tag{2.135}$$

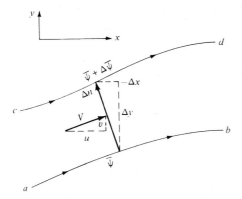

FIGURE 2.36
Mass flow through Δn is the sum of the mass flows through Δy and $-\Delta x$.

Letting cd approach ab, Eq. (2.135) becomes in the limit

$$d\bar{\psi} = \rho u\, dy - \rho v\, dx \qquad (2.136)$$

However, since $\bar{\psi} = \bar{\psi}(x, y)$, the chain rule of calculus states

$$d\bar{\psi} = \frac{\partial \bar{\psi}}{\partial x}\, dx + \frac{\partial \bar{\psi}}{\partial y}\, dy \qquad (2.137)$$

Comparing Eqs. (2.136) and (2.137), we have

$$\rho u = \frac{\partial \bar{\psi}}{\partial y} \qquad (2.138a)$$

$$\rho v = -\frac{\partial \bar{\psi}}{\partial x} \qquad (2.138b)$$

Equations (2.138a and b) are important. If $\bar{\psi}(x, y)$ is known for a given flow field, then at any point in the flow the products ρu and ρv can be obtained by differentiating $\bar{\psi}$ in the directions normal to u and v, respectively.

If Fig. 2.36 were to be redrawn in terms of polar coordinates, then a similar derivation yields

$$\rho V_r = \frac{1}{r}\frac{\partial \bar{\psi}}{\partial \theta} \qquad (2.139a)$$

$$\rho V_\theta = -\frac{\partial \bar{\psi}}{\partial r} \qquad (2.139b)$$

Such a derivation is left as a homework problem.

Note that the dimensions of $\bar{\psi}$ are equal to mass flow per unit depth perpendicular to the page. That is, in SI units, $\bar{\psi}$ is in terms of kilograms per second per meter perpendicular to the page, or simply kg/(s · m).

The stream function $\bar{\psi}$ defined above applies to both compressible and incompressible flow. Now consider the case of incompressible flow only, where $\rho = $ constant. Equation (2.134) can be written as

$$V = \frac{\partial(\bar{\psi}/\rho)}{\partial n} \qquad (2.140)$$

We define a new stream function, for incompressible flow only, as $\psi \equiv \bar{\psi}/\rho$. Then Eq. (2.140) becomes

$$V = \frac{\partial \psi}{\partial n}$$

and Eqs. (2.138) and (2.139) become

$$u = \frac{\partial \psi}{\partial y} \qquad (2.141a)$$

$$v = -\frac{\partial \psi}{\partial x} \qquad (2.141b)$$

and

$$V_r = \frac{1}{r}\frac{\partial \psi}{\partial \theta} \qquad (2.142a)$$

$$V_\theta = -\frac{\partial \psi}{\partial r} \qquad (2.142b)$$

The incompressible stream function ψ has characteristics analogous to its more general compressible counterpart $\bar{\psi}$. For example, since $\bar{\psi}(x, y) = c$ is the equation of a streamline, and since ρ is a constant for incompressible flow, then $\psi(x, y) \equiv \bar{\psi}/\rho = $ constant is also the equation for a streamline (for incompressible flow only). In addition, since $\Delta\bar{\psi}$ is mass flow between two streamlines (per unit depth perpendicular to the page), and since ρ is mass per unit volume, then physically $\Delta\psi = \Delta\bar{\psi}/\rho$ represents the *volume flow* (per unit depth) between two streamlines. In SI units, $\Delta\psi$ is expressed as cubic meters per second per meter perpendicular to the page, or simply m^2/s.

In summary, the concept of the stream function is a powerful tool in aerodynamics, for two primary reasons. Assuming that $\bar{\psi}(x, y)$ [or $\psi(x, y)$] is known through the two-dimensional flow field, then:

1. $\bar{\psi} = $ constant (or $\psi = $ constant) gives the equation of a streamline.
2. The flow velocity can be obtained by differentiating $\bar{\psi}$ (or ψ), as given by Eqs. (2.138) and (2.139) for compressible flow and Eqs. (2.141) and (2.142) for incompressible flow.

We have not yet discussed how $\bar{\psi}(x, y)$ [or $\psi(x, y)$] can be obtained in the first place; we are assuming that it is known. The actual determination of the stream function for various problems is discussed in Chap. 3.

2.15 VELOCITY POTENTIAL

Recall from Sec. 2.12 that an irrotational flow is defined as a flow where the vorticity is zero at every point. From Eq. (2.120), for an irrotational flow,

$$\boldsymbol{\xi} = \nabla \times \mathbf{V} = 0 \tag{2.143}$$

Consider the following vector identity: if ϕ is a scalar function, then

$$\nabla \times (\nabla \phi) = 0 \tag{2.144}$$

i.e., the curl of the gradient of a scalar function is identically zero. Comparing Eqs. (2.143) and (2.144), we see that

$$\boxed{\mathbf{V} = \nabla \phi} \tag{2.145}$$

Equation (2.145) states that for an *irrotational* flow, there exists a scalar function ϕ such that the velocity is given by the gradient of ϕ. We denote ϕ as the *velocity potential*. ϕ is a function of the spatial coordinates; i.e., $\phi = \phi(x, y, z)$, or $\phi = \phi(r, \theta, z)$, or $\phi = \phi(r, \theta, \Phi)$. From the definition of the gradient in cartesian coordinates given by Eq. (2.16), we have, from Eq. (2.145),

$$u\mathbf{i} + v\mathbf{j} + w\mathbf{k} = \frac{\partial \phi}{\partial x}\mathbf{i} + \frac{\partial \phi}{\partial y}\mathbf{j} + \frac{\partial \phi}{\partial z}\mathbf{k} \tag{2.146}$$

The coefficients of like unit vectors must be the same on both sides of Eq. (2.146). Thus, in cartesian coordinates,

$$\boxed{u = \frac{\partial \phi}{\partial x} \qquad v = \frac{\partial \phi}{\partial y} \qquad w = \frac{\partial \phi}{\partial z}} \tag{2.147}$$

In a similar fashion, from the definition of the gradient in cylindrical and spherical coordinates given by Eqs. (2.17) and (2.18), we have, in cylindrical coordinates,

$$V_r = \frac{\partial \phi}{\partial r} \qquad V_\theta = \frac{1}{r}\frac{\partial \phi}{\partial \theta} \qquad V_z = \frac{\partial \phi}{\partial z} \tag{2.148}$$

and in spherical coordinates,

$$V_r = \frac{\partial \phi}{\partial r} \qquad V_\theta = \frac{1}{r}\frac{\partial \phi}{\partial \theta} \qquad V_\Phi = \frac{1}{r \sin \theta}\frac{\partial \phi}{\partial \Phi} \tag{2.149}$$

The velocity potential is analogous to the stream function in the sense that derivatives of ϕ yield the flow-field velocities. However, there are distinct differences between ϕ and $\bar{\psi}$ (or ψ):

1. The flow-field velocities are obtained by differentiating ϕ in the same direction as the velocities [see Eqs. (2.147) to (2.149)], whereas $\bar{\psi}$ (or ψ) is differentiated normal to the velocity direction [see Eqs. (2.138) and (2.139), or Eqs. (2.141) and (2.142)].
2. The velocity potential is defined for irrotational flow only. In contrast, the stream function can be used in either rotational or irrotational flows.
3. The velocity potential applies to three-dimensional flows, whereas the stream function is defined for two-dimensional flows only.†

When a flow field is irrotational, hence allowing a velocity potential to be defined, there is a tremendous simplification. Instead of dealing with the velocity components (say, u, v, and w) as unknowns, hence requiring three equations for these three unknowns, we can deal with the velocity potential as one unknown, therefore requiring the solution of only one equation for the flow field. Once ϕ is known for a given problem, the velocities are obtained directly from Eqs. (2.147) to (2.149). This is why, in theoretical aerodynamics, we make a distinction between irrotational and rotational flows and why the analysis of irrotational flows is simpler than that of rotational flows.

Because irrotational flows can be described by the velocity potential ϕ, such flows are called *potential flows.*

In this section, we have not yet discussed how ϕ can be obtained in the first place; we are assuming that it is known. The actual determination of ϕ for various problems is discussed in Chaps. 3, 6, 11, and 12.

2.16 RELATIONSHIP BETWEEN THE STREAM FUNCTION AND VELOCITY POTENTIAL

In Sec. 2.15, we demonstrated that for an irrotational flow, $\mathbf{V} = \nabla\phi$. At this stage, take a moment and review some of the nomenclature introduced in Sec. 2.2.5 for the gradient of a scalar field. We see that a line of constant ϕ is an isoline of ϕ; since ϕ is the velocity potential, we give this isoline a specific name, *equipotential line.* In addition, a line drawn in space such that $\nabla\phi$ is tangent at every point is defined as a gradient line; however, since $\nabla\phi = \mathbf{V}$, this gradient line is a *streamline.* In turn, from Sec. 2.14, a streamline is a line of constant $\bar{\psi}$ (for a two-dimensional flow). Because gradient lines and isolines are perpendicular (see Sec. 2.2.5,

---

† $\bar{\psi}$ (or ψ) can be defined for axisymmetric flows, such as the flow over a cone at zero degrees angle of attack. However, for such flows, only two spatial coordinates are needed to describe the flow field (see Chap. 6).

Gradient of a Scalar Field), then equipotential lines (ϕ = constant) and stream-lines ($\bar{\psi}$ = constant) are mutually perpendicular.

To illustrate this result more clearly, consider a two-dimensional, irrotational, incompressible flow in cartesian coordinates. For a streamline, $\psi(x, y)$ = constant. Hence, the differential of ψ along the streamline is zero; i.e.,

$$d\psi = \frac{\partial \psi}{\partial x} dx + \frac{\partial \psi}{\partial y} dy = 0 \tag{2.150}$$

From Eqs. (2.141a and b), Eq. (2.150) can be written as

$$d\psi = -v \, dx + u \, dy = 0 \tag{2.151}$$

Solve Eq. (2.151) for dy/dx, which is the slope of the ψ = constant line, i.e., the slope of the streamline:

$$\left(\frac{dy}{dx}\right)_{\psi=\text{const}} = \frac{v}{u} \tag{2.152}$$

Similarly, for an equipotential line, $\phi(x, y)$ = constant. Along this line,

$$d\phi = \frac{\partial \phi}{\partial x} dx + \frac{\partial \phi}{\partial y} dy = 0 \tag{2.153}$$

From Eq. (2.147), Eq. (2.153) can be written as

$$d\phi = u \, dx + v \, dy = 0 \tag{2.154}$$

Solving Eq. (2.154) for dy/dx, which is the slope of the ϕ = constant line, i.e., the slope of the equipotential line, we obtain

$$\left(\frac{dy}{dx}\right)_{\phi=\text{const}} = -\frac{u}{v} \tag{2.155}$$

Combining Eqs. (2.152) and (2.155), we have

$$\left(\frac{dy}{dx}\right)_{\psi=\text{const}} = -\frac{1}{(dy/dx)_{\phi=\text{const}}} \tag{2.156}$$

Equation (2.156) shows that the slope of a ψ = constant line is the negative reciprocal of the slope of a ϕ = constant line, i.e., streamlines and equipotential lines are mutually perpendicular.

2.17 SUMMARY

Return to the road map for this chapter, as given in Fig. 2.1. We have now covered both the left and right branches of this map and are ready to launch into the solution of practical aerodynamic problems in subsequent chapters. Look at each block in Fig. 2.1; let your mind flash over the important equations and concepts represented by each block. If the flashes are dim, return to the appropriate sections of this chapter and review the material until you feel comfortable with these aerodynamic tools.

For your convenience, the most important results are summarized below:

Basic Flow Equations

Continuity equation

$$\frac{\partial}{\partial t} \iiint_{\mathcal{V}} \rho \, d\mathcal{V} + \oiint_{S} \rho \mathbf{V} \cdot \mathbf{dS} = 0 \qquad (2.39)$$

or

$$\frac{\partial \rho}{\partial t} + \nabla \cdot (\rho \mathbf{V}) = 0 \qquad (2.43)$$

or

$$\frac{D\rho}{Dt} + \rho \nabla \cdot \mathbf{V} = 0 \qquad (2.99)$$

Momentum equation

$$\frac{\partial}{\partial t} \iiint_{\mathcal{V}} \rho \mathbf{V} \, d\mathcal{V} + \oiint_{S} (\rho \mathbf{V} \cdot \mathbf{dS}) \mathbf{V} = - \oiint_{S} p \, \mathbf{dS} + \iiint_{\mathcal{V}} \rho \mathbf{f} \, d\mathcal{V} + F_{\text{viscous}} \qquad (2.55)$$

or

$$\frac{\partial(\rho u)}{\partial t} + \nabla \cdot (\rho u \mathbf{V}) = -\frac{\partial p}{\partial x} + \rho f_x + (\mathscr{F}_x)_{\text{viscous}} \qquad (2.61a)$$

$$\frac{\partial(\rho v)}{\partial t} + \nabla \cdot (\rho v \mathbf{V}) = -\frac{\partial p}{\partial y} + \rho f_y + (\mathscr{F}_y)_{\text{viscous}} \qquad (2.61b)$$

$$\frac{\partial(\rho w)}{\partial t} + \nabla \cdot (\rho w \mathbf{V}) = -\frac{\partial p}{\partial z} + \rho f_z + (\mathscr{F}_z)_{\text{viscous}} \qquad (2.61c)$$

or

$$\rho \frac{Du}{Dt} = -\frac{\partial p}{\partial x} + \rho f_x + (\mathscr{F}_x)_{\text{viscous}} \qquad (2.104a)$$

$$\rho \frac{Dv}{Dt} = -\frac{\partial p}{\partial y} + \rho f_y + (\mathscr{F}_y)_{\text{viscous}} \qquad (2.104b)$$

$$\rho \frac{Dw}{Dt} = -\frac{\partial p}{\partial z} + \rho f_z + (\mathscr{F}_z)_{\text{viscous}} \qquad (2.104c)$$

Energy equation

$$\frac{\partial}{\partial t} \iiint_{\mathcal{V}} \rho \left(e + \frac{V^2}{2} \right) d\mathcal{V} + \oiint_{S} \rho \left(e + \frac{V^2}{2} \right) \mathbf{V} \cdot \mathbf{dS}$$

$$= \iiint_{\mathcal{V}} \dot{q} \rho \, d\mathcal{V} + \dot{Q}_{\text{viscous}} - \oiint_{S} p \mathbf{V} \cdot \mathbf{dS}$$

$$+ \iiint_{\mathcal{V}} \rho (\mathbf{f} \cdot \mathbf{V}) \, d\mathcal{V} + \dot{W}_{\text{viscous}} \qquad (2.86)$$

or $\quad \dfrac{\partial}{\partial t}\left[\rho\left(e+\dfrac{V^2}{2}\right)\right]+\nabla\cdot\left[\rho\left(e+\dfrac{V^2}{2}\right)\mathbf{V}\right]=\rho\dot{q}-\nabla\cdot(p\mathbf{V})+\rho(\mathbf{f}\cdot\mathbf{V})$

$$+\,\dot{Q}'_{\text{viscous}}+\dot{W}'_{\text{viscous}} \qquad (2.87)$$

or $\quad \rho\dfrac{D(e+V^2/2)}{Dt}=\rho\dot{q}-\nabla\cdot(p\mathbf{V})+\rho(\mathbf{f}\cdot\mathbf{V})+\dot{Q}'_{\text{viscous}}+\dot{W}'_{\text{viscous}} \qquad (2.105)$

Substantial derivative

$$\dfrac{D}{Dt}\equiv\dfrac{\partial}{\partial t}+(\mathbf{V}\cdot\nabla) \qquad (2.95)$$

local   convective
derivative derivative

A streamline is a curve whose tangent at any point is in the direction of the velocity vector at that point. The equation of a streamline is given by

$$\mathbf{ds}\times\mathbf{V}=0 \qquad (2.106)$$

or, in cartesian coordinates,

$$w\,dy-v\,dz=0 \qquad (2.108a)$$

$$u\,dz-w\,dx=0 \qquad (2.108b)$$

$$v\,dx-u\,dy=0 \qquad (2.108c)$$

The vorticity $\boldsymbol{\xi}$ at any given point is equal to twice the angular velocity of a fluid element $\boldsymbol{\omega}$, and both are related to the velocity field by

$$\boldsymbol{\xi}=2\boldsymbol{\omega}=\nabla\times\mathbf{V} \qquad (2.120)$$

When $\nabla\times\mathbf{V}\neq 0$, the flow is rotational. When $\nabla\times\mathbf{V}=0$, the flow is irrotational.

Circulation Γ is related to lift and is defined as

$$\Gamma\equiv-\oint_C\mathbf{V}\cdot\mathbf{ds} \qquad (2.127)$$

Circulation is also related to vorticity via

$$\Gamma\equiv-\oint_C\mathbf{V}\cdot\mathbf{ds}=-\iint_S(\nabla\times\mathbf{V})\cdot\mathbf{dS} \qquad (2.128)$$

or
$$(\nabla \times \mathbf{V}) \cdot \mathbf{n} = -\frac{d\Gamma}{dS} \qquad (2.129)$$

The stream function $\bar{\psi}$ is defined such that $\bar{\psi}(x, y) = $ constant is the equation of a streamline, and the difference in the stream function between two streamlines, $\Delta\bar{\psi}$, is equal to the mass flow between the streamlines. As a consequence of this definition, in cartesian coordinates,

$$\rho u = \frac{\partial \bar{\psi}}{\partial y} \qquad (2.138a)$$

$$\rho v = -\frac{\partial \bar{\psi}}{\partial x} \qquad (2.138b)$$

and in cylindrical coordinates,

$$\rho V_r = \frac{1}{r}\frac{\partial \bar{\psi}}{\partial \theta} \qquad (2.139a)$$

$$\rho V_\theta = -\frac{\partial \bar{\psi}}{\partial r} \qquad (2.139b)$$

For incompressible flow, $\psi \equiv \bar{\psi}/\rho$ is defined such that $\psi(x, y) = $ constant denotes a streamline and $\Delta\psi$ between two streamlines is equal to the volume flow between these streamlines. As a consequence of this definition, in cartesian coordinates,

$$u = \frac{\partial \psi}{\partial y} \qquad (2.141a)$$

$$v = -\frac{\partial \psi}{\partial x} \qquad (2.141b)$$

and in cylindrical coordinates,

$$V_r = \frac{1}{r}\frac{\partial \psi}{\partial \theta} \qquad (2.142a)$$

$$V_\theta = -\frac{\partial \psi}{\partial r} \qquad (2.142b)$$

The stream function is valid for both rotational and irrotational flows, but it is restricted to two-dimensional flows only.

The velocity potential ϕ is defined for irrotational flows only, such that

$$\mathbf{V} = \nabla \phi \qquad (2.145)$$

In cartesian coordinates,

$$u = \frac{\partial \phi}{\partial x} \qquad v = \frac{\partial \phi}{\partial y} \qquad w = \frac{\partial \phi}{\partial z} \qquad (2.147)$$

In cylindrical coordinates,

$$V_r = \frac{\partial \phi}{\partial r} \qquad V_\theta = \frac{1}{r}\frac{\partial \phi}{\partial \theta} \qquad V_z = \frac{\partial \phi}{\partial z} \qquad (2.148)$$

In spherical coordinates,

$$V_r = \frac{\partial \phi}{\partial r} \qquad V_\theta = \frac{1}{r}\frac{\partial \phi}{\partial \theta} \qquad V_\Phi = \frac{1}{r \sin \theta}\frac{\partial \phi}{\partial \Phi} \qquad (2.149)$$

An irrotational flow is called a potential flow.

A line of constant ϕ is an equipotential line. Equipotential lines are perpendicular to streamlines (for two-dimensional irrotational flows).

PROBLEMS

2.1. Consider a body of arbitrary shape. If the pressure distribution over the surface of the body is constant, prove that the resultant pressure force on the body is zero. [Recall that this fact was used in Eq. (2.68).]

2.2. Consider an airfoil in a wind tunnel (i.e., a wing that spans the entire test section). Prove that the lift per unit span can be obtained from the pressure distributions on the top and bottom walls of the wind tunnel (i.e., from the pressure distributions on the walls above and below the airfoil).

2.3. Consider a velocity field where the x and y components of velocity are given by $u = cx/(x^2+y^2)$ and $v = cy/(x^2+y^2)$, where c is a constant. Obtain the equations of the streamlines.

2.4. Consider a velocity field where the x and y components of velocity are given by $u = cy/(x^2+y^2)$ and $v = -cx/(x^2+y^2)$, where c is a constant. Obtain the equations of the streamlines.

2.5. Consider a velocity field where the radial and tangential components of velocity are $V_r = 0$ and $V_\theta = cr$, respectively, where c is a constant. Obtain the equations of the streamlines.

2.6. Consider a velocity field where the x and y components of velocity are given by $u = cx$ and $v = -cy$, where c is a constant. Obtain the equations of the streamlines.

2.7. The velocity field given in Prob. 2.3 is called *source flow*. For source flow, calculate:
(a) The time rate of change of the volume of a fluid element per unit volume
(b) The vorticity
Hint: It is simpler to convert the velocity components to polar coordinates and deal with a polar coordinate system.

2.8. The velocity field given in Prob. 2.4 is called *vortex flow*. For vortex flow, calculate:
(*a*) The time rate of change of the volume of a fluid element per unit volume
(*b*) The vorticity
Hint: Again, for convenience use polar coordinates.

2.9. Is the flow field given in Prob. 2.5 irrotational? Prove your answer.

2.10. Consider a flow field in polar coordinates, where the stream function is given as $\psi = \psi(r, \theta)$. Starting with the concept of mass flow between two streamlines, derive Eqs. (2.139*a* and *b*).

2.11. Assuming the velocity field given in Prob. 2.6 pertains to an incompressible flow, calculate the stream function and velocity potential. Using your results, show that lines of constant ϕ are perpendicular to lines of constant ψ.

PART
II

INVISCID, INCOMPRESSIBLE FLOW

In Part II, we deal with the flow of a fluid which has constant density—incompressible flow. This applies to the flow of liquids, such as water flow, and to low-speed flow of gases. The material covered here is applicable to low-speed flight through the atmosphere—flight at a Mach number of about 0.3 or less.

CHAPTER
3

FUNDAMENTALS OF INVISCID, INCOMPRESSIBLE FLOW

Theoretical fluid dynamics, being a difficult subject, is for convenience, commonly divided into two branches, one treating of frictionless or perfect fluids, the other treating of viscous or imperfect fluids. The frictionless fluid has no existence in nature, but is hypothesized by mathematicians in order to facilitate the investigation of important laws and principles that may be approximately true of viscous or natural fluids.

Albert F. Zahm, 1912
(Professor of aeronautics, and developer of the first aeronautical laboratory in a U.S. university, The Catholic University of America)

3.1 INTRODUCTION AND ROAD MAP

The world of practical aviation was born on December 17, 1903, when, at 10:35 A.M., and in the face of cold, stiff, dangerous winds, Orville Wright piloted the Wright Flyer on its historic 12-s, 120-ft first flight. Figure 3.1 shows a photograph of the Wright Flyer at the instant of lift-off, with Wilbur Wright running along the right side of the machine, supporting the wing tip so that it will not drag the sand. This photograph is the most important picture in aviation history; the event it depicts launched the profession of aeronautical engineering into the mainstream of the twentieth century.†

The flight velocity of the Wright Flyer was about 30 mi/h. Over the ensuing decades, the flight velocities of airplanes steadily increased. By means of more powerful engines and attention to drag reduction, the flight velocities of airplanes

---

† See Ref. 2 for historical details leading to the first flight by the Wright brothers.

FIGURE 3.1
Historic photograph of the first successful heavier-than-air powered manned flight, achieved by the Wright brothers on December 17, 1903.

rose to approximately 300 mi/h just prior to World War II. Figure 3.2 shows a typical fighter airplane of the immediate pre-World War II era. From an aerodynamic point of view, at air velocities between 0 and 300 mi/h the air density remains essentially constant, varying by only a few percent. Hence, the aerodynamics of the family of airplanes spanning the period between the two photographs shown in Figs. 3.1 and 3.2 could be described by *incompressible flow*. As a result, a huge bulk of experimental and theoretical aerodynamic results was

FIGURE 3.2
The Seversky P-35. (*Courtesy of the U.S. Air Force.*)

acquired over the 40-year period beginning with the Wright Flyer—results which applied to incompressible flow. Today, we are still very interested in incompressible aerodynamics because most modern general aviation aircraft still fly at speeds below 300 mi/h; a typical light general aviation airplane is shown in Fig. 3.3. In addition to low-speed aeronautical applications, the principles of incompressible flow apply to the flow of fluids, e.g., water flow through pipes, the motion of submarines and ships through the ocean, the design of wind turbines (the modern term for windmills), and many other important applications.

For all the above reasons, the study of incompressible flow is as relevant today as it was at the time of the Wright brothers. Therefore, Chaps. 3 to 6 deal exclusively with incompressible flow. Moreover, for the most part, we ignore any effects of friction, thermal conduction, or diffusion; i.e., we deal with *inviscid* incompressible flow in these chapters.† Looking at our spectrum of aerodynamic flows as shown in Fig. 1.31, the material contained in Chaps. 3 to 6 falls within the combined blocks *D* and *E*.

The purpose of this chapter is to establish some fundamental relations applicable to inviscid, incompressible flows and to discuss some simple but important flow fields and applications. The material in this chapter is then used as a launching pad for the airfoil theory of Chap. 4 and the finite wing theory of Chap. 5.

A road map for this chapter is given in Fig. 3.4. There are three main avenues: (1) a development of Bernoulli's equation, with some straightforward

FIGURE 3.3
The Beechcraft Bonanza F33A. (*Courtesy of Beechcraft.*)

---

† An inviscid, incompressible fluid is sometimes called an *ideal fluid*, or *perfect fluid*. This terminology will not be used here because of the confusion it sometimes causes with "ideal gases" or "perfect gases" from thermodynamics. This author prefers to use the more precise descriptor "inviscid, incompressible flow," rather than ideal fluid or perfect fluid.

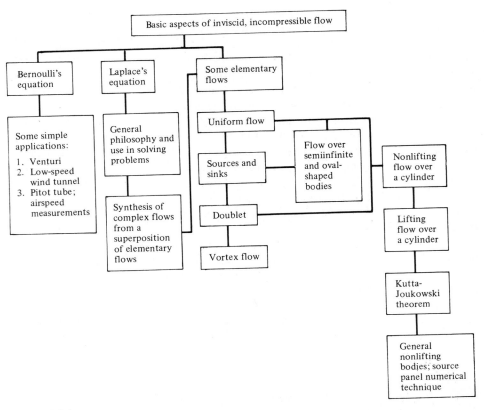

FIGURE 3.4
Road map for Chap. 3.

applications; (2) a discussion of Laplace's equation, which is the governing equation for inviscid, incompressible, irrotational flow; (3) the presentation of some elementary flow patterns, how they can be superimposed to synthesize both the nonlifting and lifting flow over a circular cylinder, and how they form the basis of a general numerical technique, called the *panel technique*, for the solution of flows over bodies of general shape. As you progress through this chapter, occasionally refer to this road map so that you can maintain your orientation and see how the various sections are related.

3.2 BERNOULLI'S EQUATION

As will be portrayed in Sec. 3.19, the early part of the eighteenth century saw the flowering of theoretical fluid dynamics, paced by the work of Johann and Daniel Bernoulli and, in particular, by Leonhard Euler. It was at this time that the relation between pressure and velocity in an inviscid, incompressible flow was first understood. The resulting equation is

$$p + \tfrac{1}{2}\rho V^2 = \text{const}$$

This equation is called *Bernoulli's equation*, although it was first presented in the above form by Euler (see Sec. 3.19). Bernoulli's equation is probably the most famous equation in fluid dynamics, and the purpose of this section is to derive it from the general equations discussed in Chap. 2.

Consider the x component of the momentum equation given by Eq. (2.104a). For an inviscid flow with no body forces, this equation becomes

$$\rho \frac{Du}{Dt} = -\frac{\partial p}{\partial x}$$

or

$$\rho \frac{\partial u}{\partial t} + \rho u \frac{\partial u}{\partial x} + \rho v \frac{\partial u}{\partial y} + \rho w \frac{\partial u}{\partial z} = -\frac{\partial p}{\partial x} \qquad (3.1)$$

For steady flow, $\partial u / \partial t = 0$. Equation (3.1) is then written as

$$u \frac{\partial u}{\partial x} + v \frac{\partial u}{\partial y} + w \frac{\partial u}{\partial z} = -\frac{1}{\rho} \frac{\partial p}{\partial x} \qquad (3.2)$$

Multiply Eq. (3.2) by dx:

$$u \frac{\partial u}{\partial x} dx + v \frac{\partial u}{\partial y} dx + w \frac{\partial u}{\partial z} dx = -\frac{1}{\rho} \frac{\partial p}{\partial x} dx \qquad (3.3)$$

Consider the flow along a streamline in three-dimensional space. The equation of a streamline is given by Eqs. (2.108a to c). In particular, substituting

$$u\, dz - w\, dx = 0 \qquad (2.108b)$$

and

$$v\, dx - u\, dy = 0 \qquad (2.108c)$$

into Eq. (3.3), we have

$$u \frac{\partial u}{\partial x} dx + u \frac{\partial u}{\partial y} dy + u \frac{\partial u}{\partial z} dz = -\frac{1}{\rho} \frac{\partial p}{\partial x} dx \qquad (3.4)$$

or

$$u \left(\frac{\partial u}{\partial x} dx + \frac{\partial u}{\partial y} dy + \frac{\partial u}{\partial z} dz \right) = -\frac{1}{\rho} \frac{\partial p}{\partial x} dx \qquad (3.5)$$

Recall from calculus that given a function $u = u(x, y, z)$, the differential of u is

$$du = \frac{\partial u}{\partial x} dx + \frac{\partial u}{\partial y} dy + \frac{\partial u}{\partial z} dz$$

This is exactly the term in parentheses in Eq. (3.5). Hence, Eq. (3.5) is written as

$$u\, du = -\frac{1}{\rho} \frac{\partial p}{\partial x} dx$$

or

$$\frac{1}{2} d(u^2) = -\frac{1}{\rho} \frac{\partial p}{\partial x} dx \qquad (3.6)$$

In a similar fashion, starting from the y component of the momentum equation given by Eq. (2.104b), specializing to an inviscid, steady flow, and applying the result to flow along a streamline, Eqs. (2.108a and c), we have

$$\frac{1}{2} d(v^2) = -\frac{1}{\rho} \frac{\partial p}{\partial y} \, dy \tag{3.7}$$

Similarly, from the z component of the momentum equation, Eq. (2.104c), we obtain

$$\frac{1}{2} d(w^2) = -\frac{1}{\rho} \frac{\partial p}{\partial z} \, dz \tag{3.8}$$

Adding Eqs. (3.6) through (3.8) yields

$$\frac{1}{2} d(u^2 + v^2 + w^2) = -\frac{1}{\rho} \left(\frac{\partial p}{\partial x} \, dx + \frac{\partial p}{\partial y} \, dy + \frac{\partial p}{\partial z} \, dz \right) \tag{3.9}$$

However,

$$u^2 + v^2 + w^2 = V^2 \tag{3.10}$$

and

$$\frac{\partial p}{\partial x} \, dx + \frac{\partial p}{\partial y} \, dy + \frac{\partial p}{\partial z} \, dz = dp \tag{3.11}$$

Substituting Eqs. (3.10) and (3.11) into (3.9), we have

$$\frac{1}{2} d(V^2) = -\frac{dp}{\rho}$$

or

$$\boxed{dp = -\rho V \, dV} \tag{3.12}$$

Equation (3.12) is called *Euler's equation.* It applies to an inviscid flow with no body forces, and it relates the change in velocity along a streamline dV to the change in pressure dp along the same streamline.

Equation (3.12) takes on a very special and important form for incompressible flow. In such a case, $\rho = \text{constant}$, and Eq. (3.12) can be easily integrated between any two points 1 and 2 along a streamline. From Eq. (3.12), with $\rho = \text{constant}$, we have

$$\int_{p_1}^{p_2} dp = -\rho \int_{V_1}^{V_2} V \, dV$$

or

$$p_2 - p_1 = -\rho \left(\frac{V_2^2}{2} - \frac{V_1^2}{2} \right)$$

or

$$\boxed{p_1 + \tfrac{1}{2}\rho V_1^2 = p_2 + \tfrac{1}{2}\rho V_2^2} \tag{3.13}$$

Equation (3.13) is *Bernoulli's equation*, which relates p_1 and V_1 at point 1 on a streamline to p_2 and V_2 at another point 2 on the same streamline. Equation (3.13) can also be written as

$$\boxed{p + \tfrac{1}{2}\rho V^2 = \text{const} \qquad \text{along a streamline}} \tag{3.14}$$

In the derivation of Eqs. (3.13) and (3.14), no stipulation has been made as to whether the flow is rotational or irrotational—these equations hold along a streamline in either case. For a general, rotational flow, the value of the constant in Eq. (3.14) will change from one streamline to the next. However, if the flow is irrotational, then Bernoulli's equation holds between *any* two points in the flow, not necessarily just on the same streamline. For an irrotational flow, the constant in Eq. (3.14) is the same for all streamlines, and

$$\boxed{p + \tfrac{1}{2}\rho V^2 = \text{const} \qquad \text{throughout the flow}} \tag{3.15}$$

The proof of this statement is given as Prob. 3.1.

The physical significance of Bernoulli's equation is obvious from Eqs. (3.13) to (3.15); namely, *when the velocity increases, the pressure decreases, and when the velocity decreases, the pressure increases.*

Note that Bernoulli's equation was derived from the momentum equation; hence, it is a statement of Newton's second law for an inviscid, incompressible flow with no body forces. However, note that the dimensions of Eqs. (3.13) to (3.15) are energy per unit volume ($\tfrac{1}{2}\rho V^2$ is the kinetic energy per unit volume). Hence, Bernoulli's equation is also a relation for mechanical energy in an incompressible flow; it states that the work done on a fluid by pressure forces is equal to the change in kinetic energy of the flow. Indeed, Bernoulli's equation can be derived from the general energy equation, such as Eq. (2.105). This derivation is left to the reader. The fact that Bernoulli's equation can be interpreted as either Newton's second law or an energy equation simply illustrates that the energy equation is redundant for the analysis of inviscid, incompressible flow. For such flows, the continuity and momentum equations suffice. (You may wish to review the opening comments of Sec. 2.7 on this same subject.)

The strategy for solving most problems in inviscid, incompressible flow is as follows:

1. Obtain the velocity field from the governing equations. These equations, appropriate for an inviscid, incompressible flow, are discussed in Secs. 3.6 and 3.7.
2. Once the velocity field is known, obtain the pressure field from Bernoulli's equation.

However, before treating the general approach to the solution of such flows (Sec. 3.7), several applications of the continuity equation and Bernoulli's equation are made to flows in ducts (Sec. 3.3) and to the measurement of airspeed using a Pitot tube (Sec. 3.4).

Example 3.1. Consider an airfoil in a flow at standard sea level conditions with a freestream velocity of 50 m/s. At a given point on the airfoil, the pressure is 0.9×10^5 N/m². Calculate the velocity at this point.

Solution. At standard sea level conditions, $\rho_\infty = 1.23$ kg/m³ and $p_\infty = 1.01 \times 10^5$ N/m². Hence,

$$p_\infty + \tfrac{1}{2}\rho V_\infty^2 = p + \tfrac{1}{2}\rho V^2$$

$$V = \sqrt{\frac{2(p_\infty - p)}{\rho} + V_\infty^2} = \sqrt{\frac{2(1.01 - 0.9) \times 10^5}{1.23} + (50)^2}$$

$$\boxed{V = 142.8 \text{ m/s}}$$

3.3 INCOMPRESSIBLE FLOW IN A DUCT: THE VENTURI AND LOW-SPEED WIND TUNNEL

Consider the flow through a duct, such as that sketched in Fig. 3.5. In general, the duct will be a three-dimensional shape, such as a tube with elliptical or rectangular cross sections which vary in area from one location to another. The flow through such a duct is three-dimensional and, strictly speaking, should be analyzed by means of the full three-dimensional conservation equations derived in Chap. 2. However, in many applications, the variation of area $A = A(x)$ is moderate, and for such cases it is reasonable to *assume* that the flow-field properties are uniform across any cross section, and hence vary only in the x direction. In Fig. 3.5, uniform flow is sketched at station 1, and another but different uniform flow is shown at station 2. Such flow, where the area changes as a function of x and all the flow-field variables are assumed to be functions of x only, i.e., $A = A(x)$, $V = V(x)$, $p = p(x)$, etc., is called *quasi-one-dimensional flow*. Although such flow is only an approximation of the truly three-dimensional flow in ducts, the results are sufficiently accurate for many aerodynamic

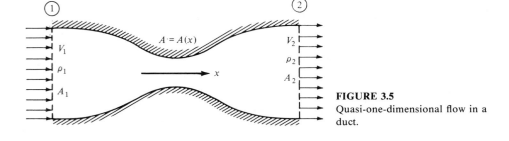

FIGURE 3.5
Quasi-one-dimensional flow in a duct.

applications. Such quasi-one-dimensional flow calculations are frequently used in engineering. They are the subject of this section.

Consider the integral form of the continuity equation written below:

$$\frac{\partial}{\partial t} \iiint_{\mathscr{V}} \rho \, d\mathscr{V} + \iint_{S} \rho \mathbf{V} \cdot \mathbf{dS} = 0 \qquad (2.39)$$

For steady flow, this becomes

$$\iint_{S} \rho \mathbf{V} \cdot \mathbf{dS} = 0 \qquad (3.16)$$

Apply Eq. (3.16) to the duct shown in Fig. 3.5, where the control volume is bounded by A_1 on the left, A_2 on the right, and the upper and lower walls of the duct. Hence, Eq. (3.16) is

$$\iint_{A_1} \rho \mathbf{V} \cdot \mathbf{dS} + \iint_{A_2} \rho \mathbf{V} \cdot \mathbf{dS} + \iint_{\text{wall}} \rho \mathbf{V} \cdot \mathbf{dS} = 0 \qquad (3.17)$$

Along the walls, the flow velocity is tangent to the wall. Since by definition $\mathbf{dS}$ is perpendicular to the wall, then along the wall, $\mathbf{V} \cdot \mathbf{dS} = 0$, and the integral over the wall surface is zero; i.e., in Eq. (3.17),

$$\iint_{\text{wall}} \rho \mathbf{V} \cdot \mathbf{dS} = 0 \qquad (3.18)$$

At station 1, the flow is uniform across A_1. Noting that $\mathbf{dS}$ and $\mathbf{V}$ are in opposite directions at station 1 ($\mathbf{dS}$ always points *out* of the control volume by definition), we have in Eq. (3.17)

$$\iint_{A_1} \rho \mathbf{V} \cdot \mathbf{dS} = -\rho_1 A_1 V_1 \qquad (3.19)$$

At station 2, the flow is uniform across A_2, and since $\mathbf{dS}$ and $\mathbf{V}$ are in the same direction, we have, in Eq. (3.17),

$$\iint_{A_2} \rho \mathbf{V} \cdot \mathbf{dS} = \rho_2 A_2 V_2 \qquad (3.20)$$

Substituting Eqs. (3.18) to (3.20) into (3.17), we obtain

$$-\rho_1 A_1 V_1 + \rho_2 A_2 V_2 + 0 = 0$$

or

$$\boxed{\rho_1 A_1 V_1 = \rho_2 A_2 V_2} \qquad (3.21)$$

Equation (3.21) is the quasi-one-dimensional continuity equation; it applies to both compressible and incompressible flow.† In physical terms, it states that the mass flow through the duct is constant (i.e., what goes in must come out). Compare Eq. (3.21) with Eq. (2.34) for mass flow.

Consider *incompressible* flow only, where ρ = constant. In Eq. (3.21), $\rho_1 = \rho_2$, and we have

$$A_1 V_1 = A_2 V_2 \tag{3.22}$$

Equation (3.22) is the quasi-one-dimensional continuity equation for incompressible flow. In physical terms, it states that the volume flow (cubic feet per second or cubic meters per second) through the duct is constant. From Eq. (3.22), we see that if the area decreases along the flow (convergent duct), the velocity increases; conversely, if the area increases (divergent duct), the velocity decreases. These variations are shown in Fig. 3.6; they are fundamental consequences of the incompressible continuity equation, and you should fully understand them. Moreover, from Bernoulli's equation, Eq. (3.15), we see that when the velocity increases in a convergent duct, the pressure decreases; conversely, when the velocity decreases in a divergent duct, the pressure increases. These pressure variations are also shown in Fig. 3.6.

Consider the incompressible flow through a convergent-divergent duct, shown in Fig. 3.7. The flow enters the duct with velocity V_1 and pressure p_1. The velocity increases in the convergent portion of the duct, reaching a maximum value V_2 at the minimum area of the duct. This minimum area is called the *throat*. Also, in the convergent section, the pressure decreases, as sketched in Fig. 3.7. At the throat, the pressure reaches a minimum value, p_2. In the divergent section downstream of the throat, the velocity decreases and the pressure increases. The duct shown in Fig. 3.7 is called a *venturi*; it is a device which finds many applications in engineering, and its use dates back more than a century. Its primary characteristic is that the pressure p_2 is lower at the throat than the ambient

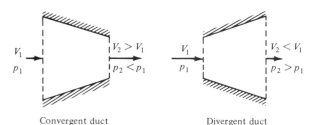

| Convergent duct | Divergent duct |

FIGURE 3.6
Incompressible flow in a duct.

---

† For a simpler, more rudimentary derivation of Eq. (3.21), see chap. 4 of Ref. 2. In the present discussion, we have established a more rigorous derivation of Eq. (3.21), consistent with the general integral form of the continuity equation.

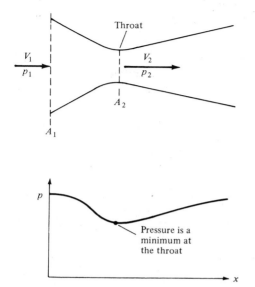

FIGURE 3.7
Flow through a venturi.

pressure p_1 outside the venturi. This pressure difference $p_1 - p_2$ is used to advantage in several applications. For example, in the carburetor of your automobile engine, there is a venturi through which the incoming air is mixed with fuel. The fuel line opens into the venturi at the throat. Because p_2 is less than the surrounding ambient pressure p_1, the pressure difference $p_1 - p_2$ helps to force the fuel into the airstream and mix it with the air downstream of the throat.

In an application closer to aerodynamics, a venturi can be used to measure airspeeds. Consider a venturi with a given inlet-to-throat area ratio A_1/A_2, as shown in Fig. 3.7. Assume that the venturi is inserted into an airstream that has an unknown velocity V_1. We wish to use the venturi to measure this velocity. With regard to the venturi itself, the most direct quantity that can be measured is the pressure difference $p_1 - p_2$. This can be accomplished by placing a small hole (a pressure tap) in the wall of the venturi at both the inlet and the throat and connecting the pressure leads (tubes) from these holes across a differential pressure gage, or to both sides of a U-tube manometer (see Sec. 1.9). In such a fashion, the pressure difference $p_1 - p_2$ can be obtained directly. This measured pressure difference can be related to the unknown velocity V_1 as follows. From Bernoulli's equation, Eq. (3.13), we have

$$V_1^2 = \frac{2}{\rho}(p_2 - p_1) + V_2^2 \tag{3.23}$$

From the continuity equation, Eq. (3.22), we have

$$V_2 = \frac{A_1}{A_2} V_1 \tag{3.24}$$

Substituting Eq. (3.24) into (3.23), we obtain

$$V_1^2 = \frac{2}{\rho}(p_2 - p_1) + \left(\frac{A_1}{A_2}\right)^2 V_1^2 \qquad (3.25)$$

Solving Eq. (3.25) for V_1, we obtain

$$V_1 = \sqrt{\frac{2(p_1 - p_2)}{\rho[(A_1/A_2)^2 - 1]}} \qquad (3.26)$$

Equation (3.26) is the desired result; it gives the inlet air velocity V_1 in terms of the measured pressure difference $p_1 - p_2$ and the known density ρ and area ratio A_1/A_2. In this fashion, a venturi can be used to measure airspeeds. Indeed, historically the first practical airspeed indicator on an airplane was a venturi used by the French Captain A. Eteve in January 1911, more than 7 years after the Wright brothers' first powered flight. Today, the most common airspeed-measuring instrument is the Pitot tube (to be discussed in Sec. 3.4); however, the venturi is still found on some general aviation airplanes, including home-built and simple experimental aircraft.

Another application of incompressible flow in a duct is the low-speed wind tunnel. The desire to build ground-based experimental facilities designed to produce flows of air in the laboratory which simulate actual flight in the atmosphere dates back to 1871, when Francis Wenham in England built and used the first wind tunnel in history.† From that date to the mid-1930s, almost all wind tunnels were designed to produce airflows with velocities from 0 to 250 mi/h. Such low-speed wind tunnels are still much in use today, along with a complement of transonic, supersonic, and hypersonic tunnels. The principles developed in this section allow us to examine the basic aspects of low-speed wind tunnels, as follows.

In essence, a low-speed wind tunnel is a large venturi where the airflow is driven by a fan connected to some type of motor drive. The wind-tunnel fan blades are similar to airplane propellers and are designed to draw the airflow through the tunnel circuit. The wind tunnel may be open circuit, where the air is drawn in the front directly from the atmosphere and exhausted out the back, again directly to the atmosphere, as shown in Fig. 3.8a; or the wind tunnel may be closed circuit, where the air from the exhaust is returned directly to the front of the tunnel via a closed duct forming a loop, as shown in Fig. 3.8b. In either case, the airflow with pressure p_1 enters the nozzle at a low velocity V_1, where the area is A_1. The nozzle converges to a smaller area A_2 at the test section, where the velocity has increased to V_2 and the pressure has decreased to p_2. After flowing over an aerodynamic model (which may be a model of a complete airplane or part of an airplane such as a wing, tail, engine, or nacelle), the air

---

† For a discussion on the history of wind tunnels, see chap. 4 of Ref. 2.

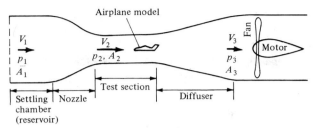

Airplane model

(a) Open-circuit tunnel

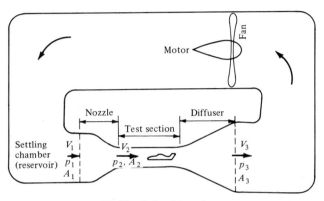

(b) Closed-circuit tunnel

FIGURE 3.8
(a) Open-circuit tunnel. (b) Closed-circuit tunnel.

passes into a diverging duct called a *diffuser*, where the area increases to A_3, the velocity decreases to V_3, and the pressure increases to p_3. From the continuity equation (3.22), the test-section air velocity is

$$V_2 = \frac{A_1}{A_2} V_1 \tag{3.27}$$

In turn, the velocity at the exit of the diffuser is

$$V_3 = \frac{A_2}{A_3} V_2 \tag{3.28}$$

The pressure at various locations in the wind tunnel is related to the velocity by Bernoulli's equation:

$$p_1 + \tfrac{1}{2}\rho V_1^2 = p_2 + \tfrac{1}{2}\rho V_2^2 = p_3 + \tfrac{1}{2}\rho V_3^2 \tag{3.29}$$

The basic factor that controls the air velocity in the test section of a given low-speed wind tunnel is the pressure difference $p_1 - p_2$. To see this more clearly, rewrite Eq. (3.29) as

$$V_2^2 = \frac{2}{\rho}(p_2 - p_1) + V_1^2 \tag{3.30}$$

From Eq. (3.27), $V_1 = (A_2/A_1)V_2$. Substituting into the right-hand side of Eq. (3.30), we have

$$V_2^2 = \frac{2}{\rho}(p_1 - p_2) + \left(\frac{A_2}{A_1}\right)^2 V_2^2 \qquad (3.31)$$

Solving Eq. (3.31) for V_2, we obtain

$$V_2 = \sqrt{\frac{2(p_1 - p_2)}{\rho[1 - (A_2/A_1)^2]}} \qquad (3.32)$$

The area ratio A_2/A_1 is a fixed quantity for a wind tunnel of given design. Moreover, the density is a known constant for incompressible flow. Therefore, Eq. (3.32) demonstrates conclusively that the test-section velocity V_2 is governed by the pressure difference $p_1 - p_2$. The fan driving the wind-tunnel flow creates this pressure difference by doing work on the air. When the wind-tunnel operator turns the "control knob" of the wind tunnel and adjusts the power to the fan, he or she is essentially adjusting the pressure difference $p_1 - p_2$ and, in turn, adjusting the velocity via Eq. (3.32).

In low-speed wind tunnels, the most convenient method of measuring the pressure difference $p_1 - p_2$, hence of measuring V_2 via Eq. (3.32), is by means of a manometer as discussed in Sec. 1.9. In Eq. (1.49), the density is the density of the liquid in the manometer (*not* the density of the air in the tunnel). The product of density and the acceleration of gravity g in Eq. (1.49) is the weight per unit volume of the manometer fluid. Denote this weight per unit volume by w. Referring to Eq. (1.49), if the side of the manometer associated with p_a is connected to a pressure tap in the settling chamber of the wind tunnel, where the pressure is p_1, and if the other side of the manometer (associated with p_b) is connected to a pressure tap in the test section, where the pressure is p_2, then, from Eq. (1.49),

$$p_1 - p_2 = w\,\Delta h$$

where Δh is the difference in heights of the liquid between the two sides of the manometer. In turn, Eq. (3.32) can be expressed as

$$V_2 = \sqrt{\frac{2w\,\Delta h}{\rho[1 - (A_2/A_1)^2]}}$$

In many low-speed wind tunnels, the test section is vented to the surrounding atmosphere by means of slots in the wall; in others, the test section is not a duct at all, but rather, an open area between the nozzle exit and the diffuser inlet. In both cases, the pressure in the surrounding atmosphere is impressed on the test-section flow; hence, $p_2 = 1$ atm. (In subsonic flow, a jet which is dumped freely into the surrounding air takes on the same pressure as the surroundings; in contrast, a supersonic free jet may have completely different pressures than the surrounding atmosphere, as we see in Chap. 10.)

Keep in mind that the basic equations used in this section have certain limitations—we are assuming a quasi-one-dimensional inviscid flow. Such equations can sometimes lead to misleading results when the neglected phenomena are in reality important. For example, if $A_3 = A_1$ (inlet area of the tunnel is equal to the exit area), then Eqs. (3.27) and (3.28) yield $V_3 = V_1$. In turn, from Eq. (3.29), $p_3 = p_1$; i.e., there is no pressure difference across the entire tunnel circuit. If this were true, the tunnel would run without the application of any power—we would have a perpetual motion machine. In reality, there are losses in the airflow due to friction at the tunnel walls and drag on the aerodynamic model in the test section. Bernoulli's equation, Eq. (3.29), does not take such losses into account. (Review the derivation of Bernoulli's equation in Sec. 3.2; note that viscous effects are neglected.) Thus, in an actual wind tunnel, there is a pressure loss due to viscous and drag effects, and $p_3 < p_1$. The function of the wind-tunnel motor and fan is to add power to the airflow in order to increase the pressure of the flow coming out of the diffuser so that it can be exhausted into the atmosphere (Fig. 3.8a) or returned to the inlet of the nozzle at the higher pressure p_1 (Fig. 3.8b). Photographs of a typical subsonic wind tunnel are shown in Fig. 3.9a and b.

(a)

FIGURE 3.9
(a) Test section of a large subsonic wind tunnel; the Glenn L. Martin Wind Tunnel at the University of Maryland. (b) The power fan drive section of the Glenn L. Martin Wind Tunnel. (*Courtesy of Dr. Jewel Barlow, University of Maryland.*)

(b)

FIGURE 3.9
(continued)

Example 3.2. Consider a venturi with a throat-to-inlet area ratio of 0.8 mounted in a flow at standard sea level conditions. If the pressure difference between the inlet and the throat is 7 lb/ft², calculate the velocity of the flow at the inlet.

Solution. At standard sea level conditions, $\rho = 0.002377$ slug/ft³. Hence,

$$V_1 = \sqrt{\frac{2(p_1 - p_2)}{\rho[(A_1/A_2)^2 - 1]}} = \sqrt{\frac{2(7)}{(0.002377)[(\frac{1}{0.8})^2 - 1]}} = \boxed{102.3 \text{ ft/s}}$$

Example 3.3. Consider a low-speed subsonic wind tunnel with a 12/1 contraction ratio for the nozzle. If the flow in the test section is at standard sea level conditions with a velocity of 50 m/s, calculate the height difference in a U-tube mercury manometer with one side connected to the nozzle inlet and the other to the test section.

Solution. At standard sea level, $\rho = 1.23$ kg/m³. From Eq. (3.32),

$$p_1 - p_2 = \frac{1}{2}\rho V_2^2\left[1 - \left(\frac{A_2}{A_1}\right)^2\right] = \frac{1}{2}(50)^2(1.23)\left[1 - \left(\frac{1}{12}\right)^2\right] = 1527 \text{ N/m}^2$$

However, $p_1 - p_2 = w\,\Delta h$. The density of liquid mercury is 1.36×10^4 kg/m³. Hence,

$$w = (1.36 \times 10^4 \text{ kg/m}^3)(9.8 \text{ m/s}^2) = 1.33 \times 10^5 \text{ N/m}^2$$

$$\Delta h = \frac{p_1 - p_2}{w} = \frac{1527}{1.33 \times 10^5} = \boxed{0.01148 \text{ m}}$$

Example 3.4. Consider a model of an airplane mounted in a subsonic wind tunnel, such as shown in Fig. 3.10. The wind-tunnel nozzle has a 12-to-1 contraction ratio. The maximum lift coefficient of the airplane model is 1.3. The wing planform area of the model is 6 ft². The lift is measured with a mechanical balance which is rated at a maximum force of 1000 lb; i.e., if the lift of the airplane model exceeds 1000 lb, the balance will be damaged. During a given test of this airplane model, the plan is to rotate the model through its whole range of angle of attack, including up to that for maximum C_L. Calculate the maximum pressure difference allowable between the wind-tunnel settling chamber and the test section, assuming standard sea level density in the test section, i.e., $\rho_\infty = 0.002377$ slug/ft³.

Solution. Maximum lift occurs when the model is at its maximum lift coefficient. Since the maximum allowable lift force is 1000 lb, the freestream velocity at which this occurs is obtained from

$$L_{\max} = \frac{1}{2}\rho_\infty V_\infty^2 S C_{L,\max}$$

or $$V_\infty = \sqrt{\frac{2L_{\max}}{\rho_\infty S C_{L,\max}}} = \sqrt{\frac{(2)(1000)}{(0.002377)(6)(1.3)}} = 328.4 \text{ ft/s}$$

From Eq. (3.32),

$$p_1 - p_2 = \frac{1}{2}\rho_\infty V_\infty^2\left[1 - \left(\frac{A_2}{A_1}\right)^2\right]$$

$$= \frac{1}{2}(0.002377)(328.4)^2\left[1 - \left(\frac{1}{12}\right)^2\right] = \boxed{117.5 \text{ lb/ft}^2}$$

FIGURE 3.10
Typical model installation in the test section of a large wind tunnel. The Glenn L. Martin Wind Tunnel at the University of Maryland.

3.4 PITOT TUBE: MEASUREMENT OF AIRSPEED

In 1732, the Frenchman Henri Pitot was busy trying to measure the flow velocity of the Seine River in Paris. One of the instruments he used was his own invention—a strange-looking tube bent into an L shape, as shown in Fig. 3.11. Pitot oriented one of the open ends of the tube so that it faced directly into the flow. In turn, he used the pressure inside this tube to measure the water flow velocity. This was the first time in history that a proper measurement of fluid velocity was made, and Pitot's invention has carried through to the present day as the *Pitot tube*—one of the most common and frequently used instruments in any modern aerodynamic laboratory. Moreover, a Pitot tube is the most common device for measuring flight velocities of airplanes. The purpose of this section is to describe the basic principle of the Pitot tube.†

---

† See chap. 4 of Ref. 2 for a detailed discussion of the history of the Pitot tube, how Pitot used it to overturn a basic theory in civil engineering, how it created some controversy in engineering, and how it finally found application in aeronautics.

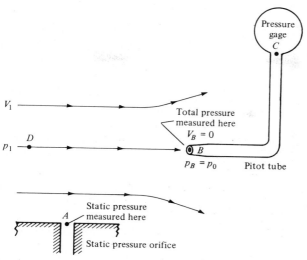

FIGURE 3.11
Pitot tube and a static pressure orifice.

Consider a flow with pressure p_1 moving with velocity V_1, as sketched at the left of Fig. 3.11. Let us consider the significance of the pressure p_1 more closely. In Sec. 1.4, the pressure is associated with the time rate of change of momentum of the gas molecules impacting on or crossing a surface; i.e., pressure is clearly related to the motion of the molecules. This motion is very random, with molecules moving in all directions with various velocities. Now imagine that you hop on a fluid element of the flow and ride with it at the velocity V_1. The gas molecules, because of their random motion, will still bump into you, and you will feel the pressure p_1 of the gas. We now give this pressure a specific name: the *static* pressure. Static pressure is a measure of the purely random motion of molecules in the gas; it is the pressure you feel when you ride along with the gas at the local flow velocity. All pressures used in this book so far have been static pressures; the pressure p appearing in all our previous equations has been the static pressure. In engineering, whenever a reference is made to "pressure" without further qualification, that pressure is always interpreted as the *static* pressure. Furthermore, consider a boundary of the flow, such as a wall, where a small hole is drilled perpendicular to the surface. The plane of the hole is parallel to the flow, as shown at point A in Fig. 3.11. Because the flow moves over the opening, the pressure felt at point A is due only to the random motion of the molecules; i.e., at point A, the static pressure is measured. Such a small hole in the surface is called a *static pressure orifice*, or a *static pressure tap*.

In contrast, consider that a Pitot tube is now inserted into the flow, with an open end facing directly into the flow. That is, the plane of the opening of the tube is perpendicular to the flow, as shown at point B in Fig. 3.11. The other end of the Pitot tube is connected to a pressure gage, such as point C in Fig.

3.11; i.e., the Pitot tube is closed at point C. For the first few milliseconds after the Pitot tube is inserted into the flow, the gas will rush into the open end and will fill the tube. However, the tube is closed at point C; there is no place for the gas to go, and hence after a brief period of adjustment, the gas inside the tube will stagnate; i.e., the gas velocity inside the tube will go to zero. Indeed, the gas will eventually pile up and stagnate *everywhere* inside the tube, including at the open mouth at point B. As a result, the streamline of the flow that impinges directly at the open face of the tube (streamline DB in Fig. 3.11) sees this face as an obstruction to the flow. The fluid elements along streamline DB slow down as they get closer to the Pitot tube and go to zero velocity right at point B. Any point in a flow where $V = 0$ is called a *stagnation point* of the flow; hence, point B at the open face of the Pitot tube is a stagnation point, where $V_B = 0$. In turn, from Bernoulli's equation we know the pressure increases as the velocity decreases. Hence, $p_B > p_1$. The pressure at a stagnation point is called the *stagnation* pressure, or *total* pressure, denoted by p_0. Hence, at point B, $p_B = p_0$.

From the above discussion, we see that two types of pressure can be defined for a given flow: static pressure, which is the pressure you feel by moving with the flow at its local velocity V_1, and total pressure, which is the pressure that the flow achieves when the velocity is reduced to zero. In aerodynamics, the distinction between total and static pressure is important; we have discussed this distinction at some length, and you should make yourself comfortable with the above paragraphs before proceeding further. (Further elaboration on the meaning and significance of total and static pressure will be made in Chap. 7.)

How is the Pitot tube used to measure flow velocity? To answer this question, first note that the total pressure p_0 exerted by the flow at the tube inlet (point B) is impressed throughout the tube (there is no flow inside the tube; hence, the pressure everywhere inside the tube is p_0). Therefore, the pressure gage at point C reads p_0. This measurement, in conjunction with a measurement of the static pressure p_1 at point A, yields the difference between total and static pressure, $p_0 - p_1$, and it is this pressure *difference* which allows the calculation of V_1 via Bernoulli's equation. In particular, apply Bernoulli's equation between point A, where the pressure and velocity are p_1 and V_1, respectively, and point B, where the pressure and velocity are p_0 and $V = 0$, respectively:

$$p_A + \tfrac{1}{2}\rho V_A^2 = p_B + \tfrac{1}{2}\rho V_B^2$$

or
$$p_1 + \tfrac{1}{2}\rho V_1^2 = p_0 + 0 \tag{3.33}$$

Solving Eq. (3.33) for V_1, we have

$$V_1 = \sqrt{\frac{2(p_0 - p_1)}{\rho}} \tag{3.34}$$

Equation (3.34) allows the calculation of velocity simply from the measured difference between total and static pressure. The total pressure p_0 is obtained from the Pitot tube, and the static pressure p_1 is obtained from a suitably placed static pressure tap.

It is possible to combine the measurement of both total and static pressure in one instrument, a *Pitot-static probe*, as sketched in Fig. 3.12. A Pitot-static probe measures p_0 at the nose of the probe and p_1 at a suitably placed static pressure tap on the probe surface downstream of the nose.

In Eq. (3.33), the term $\frac{1}{2}\rho V_1^2$ is called the *dynamic pressure* and is denoted by the symbol q_1. The grouping $\frac{1}{2}\rho V^2$ is called the dynamic pressure by definition and is used in all flows, incompressible to hypersonic:

$$q \equiv \tfrac{1}{2}\rho V^2$$

However, for incompressible flow, the dynamic pressure has special meaning; it is precisely the difference between total and static pressure. Repeating Eq. (3.33), we obtain

$$\underset{\substack{\text{static} \\ \text{pressure}}}{p_1} \quad + \quad \underset{\substack{\text{dynamic} \\ \text{pressure}}}{\tfrac{1}{2}\rho V_1^2} \quad = \quad \underset{\substack{\text{total} \\ \text{pressure}}}{p_0}$$

or

$$p_1 + q_1 = p_0$$

or

$$\boxed{q_1 = p_0 - p_1} \qquad (3.35)$$

It is important to keep in mind that Eq. (3.35) comes from Bernoulli's equation, and thus holds for *incompressible flow only*. For compressible flow, where Bernoulli's equation is not valid, the pressure difference $p_0 - p_1$ is *not* equal to q_1. Moreover, Eq. (3.34) is valid for incompressible flow only. The velocities of compressible flows, both subsonic and supersonic, can be measured by means of a Pitot tube, but the equations are different from Eq. (3.34). (Velocity measurements in subsonic and supersonic compressible flows are discussed in Chap. 8.)

At this stage, it is important to repeat that Bernoulli's equation holds for incompressible flow only, and therefore any result derived from Bernoulli's equation also holds for incompressible flow only, such as Eqs. (3.26), (3.32), (3.34). Experience has shown that some students when first introduced to aerodynamics seem to adopt Bernoulli's equation as the gospel and tend to use it for all applications, including many cases where it is not valid. Hopefully, the repetitive warnings given above will squelch such tendencies.

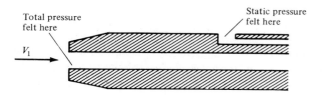

FIGURE 3.12
Pitot-static probe.

Example 3.5. An airplane is flying at standard sea level. The measurement obtained from a Pitot tube mounted on the wing tip reads 2190 lb/ft^2. What is the velocity of the airplane?

Solution. Standard sea level pressure is 2116 lb/ft^2. From Eq. (3.34), we have

$$V_1 = \sqrt{\frac{2(p_0 - p_1)}{\rho}} = \sqrt{\frac{2(2190 - 2116)}{0.002377}} = \boxed{250 \text{ ft/s}}$$

Example 3.6. In the wind-tunnel flow described in Example 3.4, a small Pitot tube is mounted in the flow just upstream of the model. Calculate the pressure measured by the Pitot tube for the same flow conditions as in Example 3.4.

Solution. From Eq. (3.35),

$$p_0 = p_\infty + q_\infty = p_\infty + \tfrac{1}{2}\rho_\infty V_\infty^2$$

$$= 2116 + \tfrac{1}{2}(0.002377)(328.4)^2$$

$$= 2116 + 128.2 = \boxed{2244 \text{ lb/ft}^2}$$

Note in this example that the dynamic pressure is $\frac{1}{2}\rho_\infty V_\infty^2 = 128.2$ lb/ft^2. This is only 8 percent larger than the pressure difference $(p_1 - p_2)$, calculated in Example 3.4, that is required to produce the test-section velocity in the wind tunnel. Why is $(p_1 - p_2)$ so close to the test-section dynamic pressure? *Answer:* Because the velocity in the settling chamber, V_1, is so small that p_1 is close to the total pressure of the flow. Indeed, from Eq. (3.22),

$$V_1 = \frac{A_2}{A_1} V_2 = \left(\frac{1}{12}\right)(328.4) = 27.3 \text{ ft/s}$$

Compared to the test-section velocity of 328.4 ft/s, V_1 is seen to be small. In regions of a flow where the velocity is finite but small, the local static pressure is close to the total pressure. (Indeed, in the limiting case of a fluid with zero velocity, the local static pressure is the same as the total pressure; here, the concepts of static pressure and total pressure are redundant. For example, consider the air in the room around you. Assuming the air is motionless, and assuming standard sea level conditions, the pressure is 2116 lb/ft^2, namely, 1 atm. Is this pressure a static pressure or a total pressure? *Answer:* It is *both.* By the definition of total pressure given in the present section, when the local flow velocity is itself zero, then the local static pressure and the local total pressure are exactly the same.)

3.5   PRESSURE COEFFICIENT

Pressure, by itself, is a dimensional quantity, e.g., pounds per square foot, newtons per square meter. However, in Secs. 1.7 and 1.8, we established the usefulness of certain dimensionless parameters such as M, Re, C_L. It makes sense, therefore, that a dimensionless pressure would also find use in aerodynamics. Such a quantity is the *pressure coefficient* C_p, first introduced in Sec. 1.5 and defined as

$$C_p \equiv \frac{p - p_\infty}{q_\infty}$$

(3.36)

where

$$q_\infty = \tfrac{1}{2}\rho_\infty V_\infty^2$$

The definition given in Eq. (3.36) is just that—a definition. It is used throughout aerodynamics, from incompressible to hypersonic flow. In the aerodynamic literature, it is very common to find pressures given in terms of C_p rather than the pressure itself. Indeed, the pressure coefficient is another similarity parameter that can be added to the list started in Secs. 1.7 and 1.8.

For *incompressible flow*, C_p can be expressed in terms of velocity only. Consider the flow over an aerodynamic body immersed in a freestream with pressure p_∞ and velocity V_∞. Pick an arbitrary point in the flow where the pressure and velocity are p and V, respectively. From Bernoulli's equation,

$$p_\infty + \tfrac{1}{2}\rho V_\infty^2 = p + \tfrac{1}{2}\rho V^2$$

or

$$p - p_\infty = \tfrac{1}{2}\rho(V_\infty^2 - V^2)$$

(3.37)

Substituting Eq. (3.37) into (3.36), we have

$$C_p = \frac{p - p_\infty}{q_\infty} = \frac{\tfrac{1}{2}\rho(V_\infty^2 - V^2)}{\tfrac{1}{2}\rho V_\infty^2}$$

or

$$C_p = 1 - \left(\frac{V}{V_\infty}\right)^2$$

(3.38)

Equation (3.38) is a useful expression for the pressure coefficient; however, note that the form of Eq. (3.38) holds for incompressible flow only.

Note from Eq. (3.38) that the pressure coefficient at a stagnation point (where $V = 0$) in an incompressible flow is always equal to 1.0. This is the highest allowable value of C_p anywhere in the flow field. (For compressible flows, C_p at a stagnation point is greater than 1.0, as shown in Chap. 14.) Also, keep in mind that in regions of the flow where $V > V_\infty$ or $p < p_\infty$, C_p will be a negative value.

Another interesting property of the pressure coefficient can be seen by rearranging the definition given by Eq. (3.36), as follows:

$$p = p_\infty + q_\infty C_p$$

Clearly, the value of C_p tells us how much p differs from p_∞ in multiples of the dynamic pressure. That is, if $C_p = 1$ (the value at a stagnation point in an incompressible flow), then $p = p_\infty + q_\infty$, or the local pressure is "one times" the dynamic pressure above freestream static pressure. If $C_p = -3$, then $p = p_\infty - 3q_\infty$, or the local pressure is three times the dynamic pressure below freestream static pressure.

Example 3.7. Consider an airfoil in a flow with a freestream velocity of 150 ft/s. The velocity at a given point on the airfoil is 225 ft/s. Calculate the pressure coefficient at this point.

Solution

$$C_p = 1 - \left(\frac{V}{V_\infty}\right)^2 = 1 - \left(\frac{225}{150}\right)^2 = \boxed{-1.25}$$

Example 3.8. Consider the airplane model in Example 3.4. When it is at a high angle of attack, slightly less than that when C_L becomes a maximum, the peak (negative) pressure coefficient which occurs at a certain point on the airfoil surface is -5.3. Assuming inviscid, incompressible flow, calculate the velocity at this point when (*a*) $V_\infty = 80$ ft/s and (*b*) $V_\infty = 300$ ft/s.

Solution. Using Eq. (3.38), we have

(*a*) $\quad V = \sqrt{V_\infty^2(1 - C_p)} = \sqrt{(80)^2[1 - (-5.3)]} = \boxed{200.8 \text{ ft/s}}$

(*b*) $\quad V = \sqrt{V_\infty^2(1 - C_p)} = \sqrt{(300)^2[1 - (-5.3)]} = \boxed{753 \text{ ft/s}}$

The above example illustrates two aspects of such a flow, as follows:

1. Consider a given point on the airfoil surface. The C_p is *given* at this point and, from the statement of the problem, C_p is obviously *unchanged* when the velocity is increased from 80 to 300 ft/s. Why? The answer underscores part of our discussion on dimensional analysis in Sec. 1.7, namely, C_p should depend only on the Mach number, Reynolds number, shape and orientation of the body, and location on the body. For the low-speed inviscid flow considered here, the Mach number and Reynolds number are not in the picture. For this type of flow, the variation of C_p is a function *only* of location on the surface of the body, and the body shape and orientation. Hence, C_p will *not* change with V_∞ or ρ_∞ as long as the flow can be considered inviscid and incompressible. For such a flow, once the C_p distribution over the body has been determined by some means, the *same* C_p distribution will exist for all freestream values of V_∞ and ρ_∞.

2. In part (*b*) of Example 3.8, the velocity at the point where C_p is a peak (negative) value is a large value, namely, 753 ft/s. Is Eq. (3.38) valid for this case? The answer is essentially *no*. Equation (3.38) assumes incompressible flow. The speed of sound at standard sea level is 1117 ft/s; hence, the freestream Mach number is $300/1117 = 0.269$. A flow where the local Mach number is less than 0.3 can be assumed to be essentially incompressible. Hence, the freestream Mach number satisfies this criterion. On the other hand, the flow rapidly expands over the top surface of the airfoil and accelerates to a velocity of 753 ft/s at the point of minimum pressure (the point of peak negative C_p). In the expansion, the speed of sound *decreases*. (We will find out why in

Part III.) Hence, at the point of minimum pressure, the local Mach number is *greater* than $\frac{753}{1117} = 0.674$. That is, the flow has expanded to such a high local Mach number that it is no longer incompressible. Therefore, the answer given in part (*b*) of Example 3.8 is not correct. (We will learn how to calculate the correct value in Part III.) There is an interesting point to be made here. Just because a model is being tested in a low-speed, subsonic wind tunnel, it does not mean that the assumption of incompressible flow will hold for all aspects of the flow field. As we see here, in some regions of the flow field around a body, the flow can achieve such high local Mach numbers that it must be considered as compressible.

3.6 CONDITION ON VELOCITY FOR INCOMPRESSIBLE FLOW

Consulting our chapter road map in Fig. 3.4, we have completed the left branch dealing with Bernoulli's equation. We now begin a more general consideration of incompressible flow, given by the center branch in Fig. 3.4. However, before introducing Laplace's equation, it is important to establish a basic condition on velocity in an incompressible flow, as follows.

First, consider the physical definition of incompressible flow, namely, $\rho =$ constant. Since ρ is the mass per unit volume and ρ is constant, then a fluid element of fixed mass moving through an incompressible flow field must also have a fixed, constant volume. Recall Eq. (2.32), which shows that $\nabla \cdot \mathbf{V}$ is physically the time rate of change of the volume of a moving fluid element per unit volume. However, for an incompressible flow, we have just stated that the volume of a fluid element is constant [e.g., in Eq. (2.32), $D(\delta \mathcal{V})/Dt \equiv 0$]. Therefore, for an incompressible flow,

$$\boxed{\nabla \cdot \mathbf{V} = 0} \tag{3.39}$$

The fact that the divergence of velocity is zero for an incompressible flow can also be shown directly from the continuity equation, Eq. (2.43):

$$\frac{\partial \rho}{\partial t} + \nabla \cdot \rho \mathbf{V} = 0 \tag{2.43}$$

For incompressible flow, $\rho =$ constant. Hence, $\partial \rho / \partial t = 0$ and $\nabla \cdot (\rho \mathbf{V}) = \rho \nabla \cdot \mathbf{V}$. Equation (2.43) then becomes

$$0 + \rho \nabla \cdot \mathbf{V} = 0$$

or $$\nabla \cdot \mathbf{V} = 0$$

which is precisely Eq. (3.39).

3.7 GOVERNING EQUATION FOR IRROTATIONAL, INCOMPRESSIBLE FLOW: LAPLACE'S EQUATION

We have seen in Sec. 3.6 that the principle of mass conservation for an incompressible flow can take the form of Eq. (3.39):

$$\nabla \cdot \mathbf{V} = 0 \tag{3.39}$$

In addition, for an irrotational flow we have seen in Sec. 2.15 that a velocity potential ϕ can be defined such that [from Eq. (2.145)]

$$\mathbf{V} = \nabla \phi \tag{2.145}$$

Therefore, for a flow that is both incompressible and irrotational, Eqs. (3.39) and (2.145) can be combined to yield

$$\nabla \cdot (\nabla \phi) = 0$$

or

$$\boxed{\nabla^2 \phi = 0} \tag{3.40}$$

Equation (3.40) is *Laplace's equation*—one of the most famous and extensively studied equations in mathematical physics. Solutions of Laplace's equation are called *harmonic functions*, for which there is a huge bulk of existing literature. Therefore, it is most fortuitous that incompressible, irrotational flow is described by Laplace's equation, for which numerous solutions exist and are well understood.

For convenience, Laplace's equation is written below in terms of the three common orthogonal coordinate systems employed in Sec. 2.2:

Cartesian coordinates: $\phi = \phi(x, y, z)$

$$\nabla^2 \phi = \frac{\partial^2 \phi}{\partial x^2} + \frac{\partial^2 \phi}{\partial y^2} + \frac{\partial^2 \phi}{\partial z^2} = 0 \tag{3.41}$$

Cylindrical coordinates: $\phi = \phi(r, \theta, z)$

$$\nabla^2 \phi = \frac{1}{r} \frac{\partial}{\partial r} \left(r \frac{\partial \phi}{\partial r} \right) + \frac{1}{r^2} \frac{\partial^2 \phi}{\partial \theta^2} + \frac{\partial^2 \phi}{\partial z^2} = 0 \tag{3.42}$$

Spherical coordinates: $\phi = \phi(r, \theta, \Phi)$

$$\nabla^2 \phi = \frac{1}{r^2 \sin \theta} \left[\frac{\partial}{\partial r} \left(r^2 \sin \theta \frac{\partial \phi}{\partial r} \right) + \frac{\partial}{\partial \theta} \left(\sin \theta \frac{\partial \phi}{\partial \theta} \right) + \frac{\partial}{\partial \Phi} \left(\frac{1}{\sin \theta} \frac{\partial \phi}{\partial \Phi} \right) \right] = 0 \tag{3.43}$$

Recall from Sec. 2.14 that, for a two-dimensional incompressible flow, a stream function ψ can be defined such that, from Eqs. (2.141a and b),

$$u = \frac{\partial \psi}{\partial y} \tag{2.141a}$$

$$v = -\frac{\partial \psi}{\partial x} \tag{2.141b}$$

The continuity equation, $\nabla \cdot \mathbf{V} = 0$, expressed in cartesian coordinates, is

$$\nabla \cdot \mathbf{V} = \frac{\partial u}{\partial x} + \frac{\partial v}{\partial y} = 0 \tag{3.44}$$

Substituting Eqs. (2.141a and b) into (3.44), we obtain

$$\frac{\partial}{\partial x}\left(\frac{\partial \psi}{\partial y}\right) + \frac{\partial}{\partial y}\left(-\frac{\partial \psi}{\partial x}\right) = \frac{\partial^2 \psi}{\partial x \, \partial y} - \frac{\partial^2 \psi}{\partial y \, \partial x} = 0 \tag{3.45}$$

Since mathematically $\partial^2 \psi / \partial x \, \partial y = \partial^2 \psi / \partial y \, \partial x$, we see from Eq. (3.45) that ψ automatically satisfies the continuity equation. *Indeed, the very definition and use of ψ is a statement of the conservation of mass, and therefore Eqs. (2.141a and b) can be used in place of the continuity equation itself.* If, in addition, the incompressible flow is irrotational, we have, from the irrotationality condition stated in Eq. (2.122),

$$\frac{\partial v}{\partial x} - \frac{\partial u}{\partial y} = 0 \tag{2.122}$$

Substituting Eqs. (2.141a and b) into (2.122), we have

$$\frac{\partial}{\partial x}\left(-\frac{\partial \psi}{\partial x}\right) - \frac{\partial}{\partial y}\left(\frac{\partial \psi}{\partial y}\right) = 0$$

or

$$\boxed{\frac{\partial^2 \psi}{\partial x^2} + \frac{\partial^2 \psi}{\partial y^2} = 0} \tag{3.46}$$

which is Laplace's equation. Therefore, the stream function also satisfies Laplace's equation, along with ϕ.

From Eqs. (3.40) and (3.46), we make the following obvious and important conclusions:

1. Any irrotational, incompressible flow has a velocity potential and stream function (for two-dimensional flow) that both satisfy Laplace's equation.
2. Conversely, any solution of Laplace's equation represents the velocity potential or stream function (two-dimensional) for an irrotational, incompressible flow.

Note that Laplace's equation is a second-order linear partial differential equation. The fact that it is *linear* is particularly important, because the sum of any particular solutions of a linear differential equation is also a solution of the equation. For example, if $\phi_1, \phi_2, \phi_3, \ldots, \phi_n$ represent n separate solutions of Eq. (3.40), then the sum

$$\phi = \phi_1 + \phi_2 + \cdots + \phi_n$$

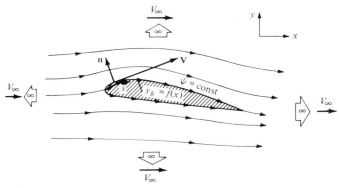

FIGURE 3.13
Boundary conditions at infinity and on a body; inviscid flow.

is also a solution of Eq. (3.40). Since irrotational, incompressible flow is governed by Laplace's equation and Laplace's equation is linear, we conclude that a *complicated flow pattern for an irrotational, incompressible flow can be synthesized by adding together a number of elementary flows which are also irrotational and incompressible.* Indeed, this establishes the grand strategy for the remainder of our discussions on inviscid, incompressible flow. We develop flow-field solutions for several different elementary flows, which by themselves may not seem to be practical flows in real life. However, we then proceed to add (i.e., superimpose) these elementary flows in different ways such that the resulting flow fields do pertain to practical problems.

Before proceeding further, consider the irrotational, incompressible flow fields over different aerodynamic shapes, such as a sphere, cone, or airplane wing. Clearly, each flow is going to be distinctly different; the streamlines and pressure distribution over a sphere are quite different from those over a cone. However, these different flows are all governed by the same equation, namely, $\nabla^2\phi = 0$. How, then, do we obtain different flows for the different bodies? The answer is found in the *boundary conditions.* Although the governing equation for the different flows is the same, the boundary conditions for the equation must conform to the different geometric shapes, and hence yield different flow-field solutions. Boundary conditions are therefore of vital concern in aerodynamic analysis. Let us examine the nature of boundary conditions further.

Consider the external aerodynamic flow over a stationary body, such as the airfoil sketched in Fig. 3.13. The flow is bounded by (1) the freestream flow which occurs (theoretically) an infinite distance away from the body and (2) the surface of the body itself. Therefore, two sets of boundary conditions apply, as follows.

3.7.1 Infinity Boundary Conditions

Far away from the body (toward infinity), in all directions, the flow approaches the uniform freestream conditions. Let V_∞ be aligned with the x direction as shown in Fig. 3.13. Hence, at infinity,

$$u = \frac{\partial \phi}{\partial x} = \frac{\partial \psi}{\partial y} = V_\infty \qquad (3.47a)$$

$$v = \frac{\partial \phi}{\partial y} = -\frac{\partial \psi}{\partial x} = 0 \qquad (3.47b)$$

Equations (3.47a and b) are the *boundary conditions on velocity at infinity*. They apply at an infinite distance from the body in all directions, above and below, and to the left and right of the body, as indicated in Fig. 3.13.

3.7.2 Wall Boundary Conditions

If the body in Fig. 3.13 has a solid surface, then it is impossible for the flow to penetrate the surface. Instead, if the flow is viscous, the influence of friction between the fluid and the solid surface creates a zero velocity at the surface. Such viscous flows are discussed in Chaps. 15 to 17. In contrast, for inviscid flows the velocity at the surface can be finite, but because the flow cannot penetrate the surface, the velocity vector must be *tangent* to the surface. This "wall tangency" condition is illustrated in Fig. 3.13, which shows **V** tangent to the body surface. If the flow is tangent to the surface, then the component of velocity *normal* to the surface must be zero. Let **n** be a unit vector normal to the surface as shown in Fig. 3.13. The wall boundary condition can be written as

$$\mathbf{V} \cdot \mathbf{n} = (\nabla \phi) \cdot \mathbf{n} = 0 \qquad (3.48a)$$

or

$$\frac{\partial \phi}{\partial n} = 0 \qquad (3.48b)$$

Equation (3.48a or b) gives the boundary condition for velocity at the wall; it is expressed in terms of ϕ. If we are dealing with ψ rather than ϕ, then the wall boundary condition is

$$\frac{\partial \psi}{\partial s} = 0 \qquad (3.48c)$$

where s is the distance measured along the body surface, as shown in Fig. 3.13. Note that the body contour is a streamline of the flow, as also shown in Fig. 3.13. Recall that $\psi = $ constant is the equation of a streamline. Thus, if the shape of the body in Fig. 3.13 is given by $y_b = f(x)$, then

$$\psi_{\text{surface}} = \psi_{y = y_b} = \text{const} \qquad (3.48d)$$

is an alternative expression for the boundary condition given in Eq. (3.48c).

If we are dealing with neither ϕ nor ψ, but rather with the velocity components u and v themselves, then the wall boundary condition is obtained from the equation of a streamline, Eq. (2.109), evaluated at the body surface; i.e.,

$$\frac{dy_b}{dx} = \left(\frac{v}{u}\right)_{surface} \tag{3.48e}$$

Equation (3.48e) states simply that the body surface is a streamline of the flow. The form given in Eq. (3.48e) for the flow tangency condition at the body surface is used for all inviscid flows, incompressible to hypersonic, and does not depend on the formulation of the problem in terms of ϕ or ψ (or $\bar{\psi}$).

3.8   INTERIM SUMMARY

Reflecting on our previous discussions, the general approach to the solution of irrotational, incompressible flows can be summarized as follows:

1. Solve Laplace's equation for ϕ [Eq. (3.40)] or ψ [Eq. (3.46)] along with the proper boundary conditions [such as Eqs. (3.47) and (3.48)]. These solutions are usually in the form of a sum of elementary solutions (to be discussed in the following sections).
2. Obtain the flow velocity from $\mathbf{V} = \nabla\phi$ or $u = \partial\psi/\partial y$ and $v = -\partial\psi/\partial x$.
3. Obtain the pressure from Bernoulli's equation, $p + \frac{1}{2}\rho V^2 = p_\infty + \frac{1}{2}\rho V_\infty^2$, where p_∞ and V_∞ are known freestream conditions.

Since $\mathbf{V}$ and p are the primary dependent variables for an incompressible flow, steps 1 to 3 are all that we need to solve a given problem as long as the flow is incompressible and irrotational.

3.9   UNIFORM FLOW: OUR FIRST ELEMENTARY FLOW

In this section, we present the first of a series of elementary incompressible flows which later will be superimposed to synthesize more complex incompressible flows. For the remainder of this chapter and in Chap. 4, we deal with two-dimensional steady flows; three-dimensional steady flows are treated in Chaps. 5 and 6.

Consider a uniform flow with velocity V_∞ oriented in the positive x direction, as sketched in Fig. 3.14. It is easily shown (see Prob. 3.8) that a uniform flow is a physically possible incompressible flow (i.e., it satisfies $\nabla \cdot \mathbf{V} = 0$) and that it is irrotational (i.e., it satisfies $\nabla \times \mathbf{V} = 0$). Hence, a velocity potential for uniform flow can be obtained such that $\nabla\phi = \mathbf{V}$. Examining Fig. 3.14, and recalling Eq. (2.147), we have

$$\frac{\partial\phi}{\partial x} = u = V_\infty \tag{3.49a}$$

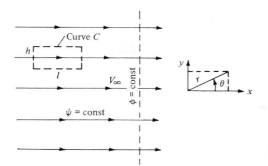

FIGURE 3.14
Uniform flow.

and
$$\frac{\partial \phi}{\partial y} = v = 0 \qquad (3.49b)$$

Integrating Eq. (3.49a) with respect to x, we have

$$\phi = V_\infty x + f(y) \qquad (3.50)$$

where $f(y)$ is a function of y only. Integrating Eq. (3.49b) with respect to y, we obtain

$$\phi = \text{const} + g(x) \qquad (3.51)$$

where $g(x)$ is a function of x only. In Eqs. (3.50) and (3.51), ϕ is the same function; hence, by comparing these equations, $g(x)$ must be $V_\infty x$, and $f(y)$ must be constant. Thus,

$$\phi = V_\infty x + \text{const} \qquad (3.52)$$

Note that in a practical aerodynamic problem, the actual value of ϕ is not significant; rather, ϕ is always used to obtain the velocity by differentiation; i.e., $\nabla \phi = \mathbf{V}$. Since the derivative of a constant is zero, we can drop the constant from Eq. (3.52) without any loss of rigor. Hence, Eq. (3.52) can be written as

$$\boxed{\phi = V_\infty x} \qquad (3.53)$$

Equation (3.53) is the velocity potential for a uniform flow with velocity V_∞ oriented in the positive x direction. Note that the derivation of Eq. (3.53) does not depend on the assumption of incompressibility; it applies to any uniform flow, compressible or incompressible.

Consider the incompressible stream function ψ. From Fig. 3.14 and Eqs. (2.141a and b), we have

$$\frac{\partial \psi}{\partial y} = u = V_\infty \qquad (3.54a)$$

and
$$\frac{\partial \psi}{\partial x} = -v = 0 \qquad (3.54b)$$

Integrating Eq. (3.54*a*) with respect to *y* and Eq. (3.54*b*) with respect to *x*, and comparing the results, we obtain

$$\psi = V_\infty y \qquad (3.55)$$

Equation (3.55) is the stream function for an incompressible uniform flow oriented in the positive *x* direction.

From Sec. 2.14, the equation of a streamline is given by $\psi = $ constant. Therefore, from Eq. (3.55), the streamlines for the uniform flow are given by $\psi = V_\infty y = $ constant. Because V_∞ is itself constant, the streamlines are thus given mathematically as $y = $ constant, i.e., as lines of constant *y*. This result is consistent with Fig. 3.14, which shows the streamlines as horizontal lines, i.e., as lines of constant *y*. Also, note from Eq. (3.53) that the equipotential lines are lines of constant *x*, as shown by the dashed line in Fig. 3.14. Consistent with our discussion in Sec. 2.16, note that the lines of $\psi = $ constant and $\phi = $ constant are mutually perpendicular.

Equations (3.53) and (3.55) can be expressed in terms of polar coordinates, where $x = r \cos \theta$ and $y = r \sin \theta$, as shown in Fig. 3.14. Hence,

$$\phi = V_\infty r \cos \theta \qquad (3.56)$$

and

$$\psi = V_\infty r \sin \theta \qquad (3.57)$$

Consider the circulation in a uniform flow. The definition of circulation is given by

$$\Gamma \equiv - \oint_C \mathbf{V} \cdot \mathbf{ds} \qquad (2.127)$$

Let the closed curve *C* in Eq. (2.127) be the rectangle shown at the left of Fig. 3.14; *h* and *l* are the lengths of the vertical and horizontal sides, respectively, of the rectangle. Then

$$\oint_C \mathbf{V} \cdot \mathbf{ds} = - V_\infty l - 0(h) + V_\infty l + 0(h) = 0$$

or

$$\Gamma = 0 \qquad (3.58)$$

Equation (3.58) is true for any arbitrary closed curve in the uniform flow. To show this, note that $\mathbf{V}_\infty$ is constant in both magnitude and direction, and hence

$$\Gamma = - \oint_C \mathbf{V} \cdot \mathbf{ds} = - \mathbf{V}_\infty \cdot \oint_C \mathbf{ds} = \mathbf{V}_\infty \cdot 0 = 0$$

because the line integral of **ds** around a closed curve is identically zero. Therefore, from Eq. (3.58), we state that *circulation around any closed curve in a uniform flow is zero.*

The above result is consistent with Eq. (2.128), which states that

$$\Gamma = -\iint\limits_{S} (\nabla \times \mathbf{V}) \cdot d\mathbf{S} \qquad (2.128)$$

We stated earlier that a uniform flow is irrotational; i.e., $\nabla \times \mathbf{V} = 0$ everywhere. Hence, Eq. (2.128) yields $\Gamma = 0$.

Note that Eqs. (3.53) and (3.55) satisfy Laplace's equation [see Eq. (3.41)], which can be easily proved by simple substitution. Therefore, uniform flow is a viable elementary flow for use in building more complex flows.

3.10 SOURCE FLOW: OUR SECOND ELEMENTARY FLOW

Consider a two-dimensional, incompressible flow where all the streamlines are straight lines emanating from a central point, *O*, as shown at the left of Fig. 3.15. Moreover, let the velocity along each of the streamlines vary inversely with distance from point *O*. Such a flow is called a *source flow.* Examining Fig. 3.15, we see that the velocity components in the radial and tangential directions are V_r and V_θ, respectively, where $V_\theta = 0$. The coordinate system in Fig. 3.15 is a cylindrical coordinate system, with the *z* axis perpendicular to the page. (Note that polar coordinates are simply the cylindrical coordinates *r* and θ confined to a single plane given by $z = \text{constant}$.) It is easily shown (see Prob. 3.9) that (1) source flow is a physically possible incompressible flow, i.e., $\nabla \cdot \mathbf{V} = 0$, at every point *except* the origin, where $\nabla \cdot \mathbf{V}$ becomes infinite, and (2) source flow is *irrotational* at every point.

In a source flow, the streamlines are directed *away from* the origin, as shown at the left of Fig. 3.15. The opposite case is that of a *sink flow,* where by definition

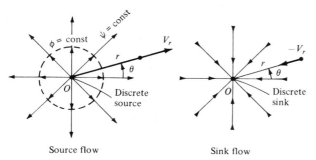

Source flow   Sink flow

FIGURE 3.15
Source and sink flows.

the streamlines are directed *toward* the origin, as shown at the right of Fig. 3.15. For sink flow, the streamlines are still radial lines from a common origin, along which the flow velocity varies inversely with distance from point O. Indeed, a sink flow is simply a negative source flow.

The flows in Fig. 3.15 have an alternate, somewhat philosophical interpretation. Consider the origin, point O, as a *discrete* source or sink. Moreover, interpret the radial flow surrounding the origin as simply being *induced* by the presence of the discrete source or sink at the origin (much like a magnetic field is induced in the space surrounding a current-carrying wire). Recall that, for a source flow, $\nabla \cdot \mathbf{V} = 0$ everywhere except at the origin, where it is infinite. Thus, the origin is a *singular point*, and we can interpret this singular point as a discrete source or sink of a given strength, with a corresponding induced flow field about the point. This interpretation is very convenient and is used frequently. Other types of singularities, such as doublets and vortices, are introduced in subsequent sections. Indeed, the irrotational, incompressible flow field about an arbitrary body can be visualized as a flow induced by a proper distribution of such singularities over the surface of the body. This concept is fundamental to many theoretical solutions of incompressible flow over airfoils and other aerodynamic shapes, and it is the very heart of modern numerical techniques for the solution of such flows. You will obtain a greater appreciation for the concept of distributed singularities for the solution of incompressible flow in Chaps. 4 through 6. At this stage, however, simply visualize a discrete source (or sink) as a singularity that induces the flows shown in Fig. 3.15.

Let us look more closely at the velocity field induced by a source or sink. By definition, the velocity is inversely proportional to the radial distance r. As stated earlier, this velocity variation is a physically possible flow, because it yields $\nabla \cdot \mathbf{V} = 0$. Moreover, it is the *only* such velocity variation for which the relation $\nabla \cdot \mathbf{V} = 0$ is satisfied for the radial flows shown in Fig. 3.15. Hence,

$$V_r = \frac{c}{r} \tag{3.59a}$$

and
$$V_\theta = 0 \tag{3.59b}$$

where c is a constant. The value of the constant is related to the volume flow from the source, as follows. In Fig. 3.15, consider a depth of length l perpendicular to the page, i.e., a length l along the z axis. This is sketched in three-dimensional perspective in Fig. 3.16. In Fig. 3.16, we can visualize an entire line of sources along the z axis, of which the source at O is just part. Therefore, in a two-dimensional flow, the discrete source, sketched in Fig. 3.15, is simply a single point on the line source shown in Fig. 3.16. The two-dimensional flow shown in Fig. 3.15 is the same in any plane perpendicular to the z axis, i.e., for any plane given by $z = $ constant. Consider the mass flow across the surface of the cylinder of radius r and height l as shown in Fig. 3.16. The elemental mass flow across the surface element $\mathbf{dS}$ shown in Fig. 3.16 is $\rho \mathbf{V} \cdot \mathbf{dS} = \rho V_r (r \, d\theta)(l)$. Hence, noting that V_r is the same value at any θ location for the fixed radius r, the total mass

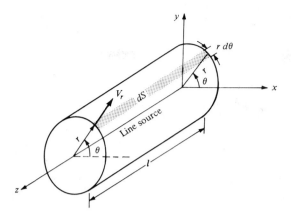

FIGURE 3.16
Volume flow rate from a line source.

flow across the surface of the cylinder is

$$\dot{m} = \int_0^{2\pi} \rho V_r (r\, d\theta) l = \rho r l V_r \int_0^{2\pi} d\theta = 2\pi r l \rho V_r \tag{3.60}$$

Since ρ is defined as the mass per unit volume and $\dot{m}$ is mass per second, then $\dot{m}/\rho$ is the volume flow per second. Denote this rate of volume flow by $\dot{v}$. Thus, from Eq. (3.60), we have

$$\dot{v} = \frac{\dot{m}}{\rho} = 2\pi r l V_r \tag{3.61}$$

Moreover, the rate of volume flow per unit length along the cylinder is $\dot{v}/l$. Denote this volume flow rate per unit length (which is the same as per unit depth perpendicular to the page in Fig. 3.15) as Λ. Hence, from Eq. (3.61), we obtain

$$\Lambda = \frac{\dot{v}}{l} = 2\pi r V_r$$

or

$$\boxed{V_r = \frac{\Lambda}{2\pi r}} \tag{3.62}$$

Hence, comparing Eqs. (3.59a) and (3.62), we see that the constant in Eq. (3.59a) is $c = \Lambda/2\pi$. In Eq. (3.62), Λ defines the *source strength*; it is physically the rate of volume flow from the source, per unit depth perpendicular to the page of Fig. 3.15. Typical units of Λ are square meters per second or square feet per second. In Eq. (3.62), a positive value of Λ represents a source, whereas a negative value represents a sink.

The velocity potential for a source can be obtained as follows. From Eqs. (2.148), (3.59b), and (3.62),

$$\frac{\partial \phi}{\partial r} = V_r = \frac{\Lambda}{2\pi r} \tag{3.63}$$

and
$$\frac{1}{r}\frac{\partial \phi}{\partial \theta} = V_\theta = 0 \tag{3.64}$$

Integrating Eq. (3.63) with respect to r, we have

$$\phi = \frac{\Lambda}{2\pi} \ln r + f(\theta) \tag{3.65}$$

Integrating Eq. (3.64) with respect to θ, we have

$$\phi = \text{const} + f(r) \tag{3.66}$$

Comparing Eqs. (3.65) and (3.66), we see that $f(r) = (\Lambda/2\pi) \ln r$ and $f(\theta) =$ constant. As explained in Sec. 3.9, the constant can be dropped without loss of rigor, and hence Eq. (3.65) yields

$$\boxed{\phi = \frac{\Lambda}{2\pi} \ln r} \tag{3.67}$$

Equation (3.67) is the velocity potential for a two-dimensional source flow.

The stream function can be obtained as follows. From Eqs. (2.142a and b), (3.59b), and (3.62),

$$\frac{1}{r}\frac{\partial \psi}{\partial \theta} = V_r = \frac{\Lambda}{2\pi r} \tag{3.68}$$

and
$$-\frac{\partial \psi}{\partial r} = V_\theta = 0 \tag{3.69}$$

Integrating Eq. (3.68) with respect to θ, we obtain

$$\psi = \frac{\Lambda}{2\pi} \theta + f(r) \tag{3.70}$$

Integrating Eq. (3.69) with respect to r, we have

$$\psi = \text{const} + f(\theta) \tag{3.71}$$

Comparing Eqs. (3.70) and (3.71) and dropping the constant, we obtain

$$\boxed{\psi = \frac{\Lambda}{2\pi} \theta} \tag{3.72}$$

Equation (3.72) is the stream function for a two-dimensional source flow.

The equation of the streamlines can be obtained by setting Eq. (3.72) equal to a constant:

$$\psi = \frac{\Lambda}{2\psi} \theta = \text{const} \tag{3.73}$$

From Eq. (3.73), we see that $\theta = $ constant, which, in polar coordinates, is the equation of a straight line from the origin. Hence, Eq. (3.73) is consistent with the picture of the source flow sketched in Fig. 3.15. Moreover, Eq. (3.67) gives an equipotential line as $r = $ constant, i.e., a circle with its center at the origin, as shown by the dashed line in Fig. 3.15. Once again, we see that streamlines and equipotential lines are mutually perpendicular.

To evaluate the circulation for source flow, recall that $\nabla \times \mathbf{V} = 0$ everywhere. In turn, from Eq. (2.128),

$$\Gamma = -\int\int_S (\nabla \times \mathbf{V}) \cdot d\mathbf{S} = 0$$

for any closed curve C chosen in the flow field. Hence, as in the case of uniform flow discussed in Sec. 3.9, there is no circulation associated with the source flow.

It is straightforward to show that Eqs. (3.67) and (3.72) satisfy Laplace's equation, simply by substitution into $\nabla^2 \phi = 0$ and $\nabla^2 \psi = 0$ written in terms of cylindrical coordinates [see Eq. (3.42)]. Therefore, source flow is a viable elementary flow for use in building more complex flows.

3.11 COMBINATION OF A UNIFORM FLOW WITH A SOURCE AND SINK

Consider a polar coordinate system with a source of strength Λ located at the origin. Superimpose on this flow a uniform stream with velocity V_∞ moving from left to right, as sketched in Fig. 3.17. The stream function for the resulting flow is the sum of Eqs. (3.57) and (3.72):

$$\psi = V_\infty r \sin \theta + \frac{\Lambda}{2\pi} \theta \tag{3.74}$$

Since both Eqs. (3.57) and (3.72) are solutions of Laplace's equation, we know that Eq. (3.74) also satisfies Laplace's equation; i.e., Eq. (3.74) describes a viable irrotational, incompressible flow. The streamlines of the combined flow are

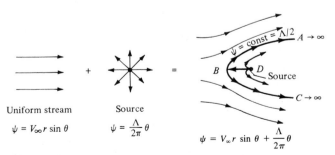

FIGURE 3.17
Superposition of a uniform flow and a source; flow over a semi-infinite body.

obtained from Eq. (3.74) as

$$\psi = V_\infty r \sin \theta + \frac{\Lambda}{2\pi} \theta = \text{const} \tag{3.75}$$

The resulting streamline shapes from Eq. (3.75) are sketched at the right of Fig. 3.17. The source is located at point D. The velocity field is obtained by differentiating Eq. (3.75):

$$V_r = \frac{1}{r} \frac{\partial \psi}{\partial \theta} = V_\infty \cos \theta + \frac{\Lambda}{2\pi r} \tag{3.76}$$

and

$$V_\theta = -\frac{\partial \psi}{\partial r} = -V_\infty \sin \theta \tag{3.77}$$

Note from Sec. 3.10 that the radial velocity from a source is $\Lambda/2\pi r$, and from Sec. 3.9 the component of the freestream velocity in the radial direction is $V_\infty \cos \theta$. Hence, Eq. (3.76) is simply the direct sum of the two velocity fields—a result which is consistent with the linear nature of Laplace's equation. Therefore, not only can we add the values of ϕ or ψ to obtain more complex solutions, we can add their derivatives, i.e., the velocities, as well.

The stagnation points in the flow can be obtained by setting Eqs. (3.76) and (3.77) equal to zero:

$$V_\infty \cos \theta + \frac{\Lambda}{2\pi r} = 0 \tag{3.78}$$

and

$$V_\infty \sin \theta = 0 \tag{3.79}$$

Solving for r and θ, we find that one stagnation point exists, located at $(r, \theta) = (\Lambda/2\pi V_\infty, \pi)$, which is labeled as point B in Fig. 3.17. That is, the stagnation point is a distance $(\Lambda/2\pi V_\infty)$ directly upstream of the source. From this result, the distance DB clearly grows smaller if V_∞ is increased and larger if Λ is increased—trends that also make sense based on intuition. For example, looking at Fig. 3.17, you would expect that as the source strength is increased, keeping V_∞ the same, the stagnation point B will be blown further upstream. Conversely, if V_∞ is increased, keeping the source strength the same, the stagnation point will be blown further downstream.

If the coordinates of the stagnation point at B are substituted into Eq. (3.75), we obtain

$$\psi = V_\infty \frac{\Lambda}{2\pi V_\infty} \sin \pi + \frac{\Lambda}{2\pi} \pi = \text{const}$$

$$\psi = \frac{\Lambda}{2} = \text{const}$$

Hence, the streamline that goes through the stagnation point is described by $\psi = \Lambda/2$. This streamline is shown as curve ABC in Fig. 3.17.

Examining Fig. 3.17, we now come to an important conclusion. Since we are dealing with inviscid flow, where the velocity at the surface of a solid body is tangent to the body, then *any* streamline of the combined flow at the right of Fig. 3.17 could be replaced by a solid surface of the same shape. In particular, consider the streamline *ABC*. Because it contains the stagnation point at *B*, the streamline *ABC* is a *dividing* streamline; i.e., it separates the fluid coming from the freestream and the fluid emanating from the source. All the fluid outside *ABC* is from the freestream, and all the fluid inside *ABC* is from the source. Therefore, as far as the freestream is concerned, the entire region inside *ABC* could be replaced with a solid body of the same shape, and the external flow, i.e., the flow from the freestream, would not feel the difference. The streamline $\psi = \Lambda/2$ extends downstream to infinity, forming a semi-infinite body. Therefore, we are led to the following important interpretation. If we want to construct the flow over a solid semi-infinite body described by the curve *ABC* as shown in Fig. 3.17, then all we need to do is take a uniform stream with velocity V_∞ and add to it a source of strength Λ at point *D*. The resulting superposition will then represent the flow over the prescribed solid semi-infinite body of shape *ABC*. *This illustrates the practicality of adding elementary flows to obtain a more complex flow over a body of interest.*

The superposition illustrated in Fig. 3.17 results in the flow over the semi-infinite body *ABC*. This is a half-body that stretches to infinity in the downstream direction; i.e., the body is not closed. However, if we take a sink of equal strength as the source and add it to the flow downstream of point *D*, then the resulting body shape will be closed. Let us examine this flow in more detail.

Consider a polar coordinate system with a source and sink placed a distance b to the left and right of the origin, respectively, as sketched in Fig. 3.18. The strengths of the source and sink are Λ and $-\Lambda$, respectively (equal and opposite). In addition, superimpose a uniform stream with velocity V_∞, as shown in Fig. 3.18. The stream function for the combined flow at any point *P* with coordinates (r, θ) is obtained from Eqs. (3.57) and (3.72):

$$\psi = V_\infty r \sin \theta + \frac{\Lambda}{2\pi} \theta_1 - \frac{\Lambda}{2\pi} \theta_2$$

or
$$\psi = V_\infty r \sin \theta + \frac{\Lambda}{2\pi} (\theta_1 - \theta_2) \tag{3.80}$$

The velocity field is obtained by differentiating Eq. (3.80) according to Eqs. (2.142a and b). Note from the geometry of Fig. 3.18 that θ_1 and θ_2 in Eq. (3.80) are functions of r, θ, and b. In turn, by setting $V = 0$, two stagnation points are found, namely, points $\bar{A}$ and *B* in Fig. 3.18. These stagnation points are located such that (see Prob. 3.13)

$$OA = OB = \sqrt{b^2 + \frac{\Lambda b}{\pi V_\infty}} \tag{3.81}$$

The equation of the streamlines is given by Eq. (3.80) as

$$\psi = V_\infty r \sin \theta + \frac{\Lambda}{2\pi} (\theta_1 - \theta_2) = \text{const} \tag{3.82}$$

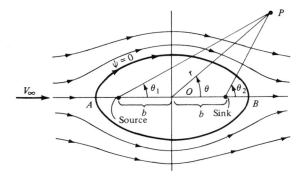

FIGURE 3.18
Superposition of a uniform flow and a source-sink pair; flow over a Rankine oval.

The equation of the specific streamline going through the stagnation points is obtained from Eq. (3.82) by noting that $\theta = \theta_1 = \theta_2 = \pi$ at point A and $\theta = \theta_1 = \theta_2 = 0$ at point B. Hence, for the stagnation streamline, Eq. (3.82) yields a value of zero for the constant. Thus, the stagnation streamline is given by $\psi = 0$, i.e.,

$$V_\infty r \sin\theta + \frac{\Lambda}{2\pi}(\theta_1 - \theta_2) = 0 \tag{3.83}$$

the equation of an oval, as sketched in Fig. 3.18. Equation (3.83) is also the dividing streamline; all the flow from the source is consumed by the sink and is contained entirely inside the oval, whereas the flow outside the oval has originated with the uniform stream only. Therefore, in Fig. 3.18, the region inside the oval can be replaced by a solid body with the shape given by $\psi = 0$, and the region outside the oval can be interpreted as the inviscid, potential (irrotational), incompressible flow over the solid body. This problem was first solved in the nineteenth century by the famous Scottish engineer W. J. M. Rankine; hence, the shape given by Eq. (3.83) and sketched in Fig. 3.18 is called a *Rankine oval*.

3.12 DOUBLET FLOW: OUR THIRD ELEMENTARY FLOW

There is a special, degenerate case of a source-sink pair that leads to a singularity called a *doublet*. The doublet is frequently used in the theory of incompressible flow; the purpose of this section is to describe its properties.

Consider a source of strength Λ and a sink of equal (but opposite) strength $-\Lambda$ separated by a distance l, as shown in Fig. 3.19a. At any point P in the flow, the stream function is

$$\psi = \frac{\Lambda}{2\pi}(\theta_1 - \theta_2) = -\frac{\Lambda}{2\pi}\Delta\theta \tag{3.84}$$

where $\Delta\theta = \theta_2 - \theta_1$ as seen from Fig. 3.19a. Equation (3.84) is the stream function for a source-sink pair separated by the distance l.

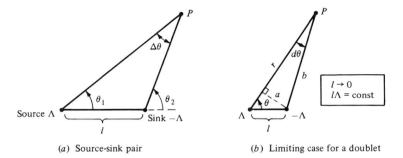

(a) Source-sink pair   (b) Limiting case for a doublet

FIGURE 3.19
How a source-sink pair approaches a doublet in the limiting case.

Now, in Fig. 3.19a, let the distance l approach zero while the absolute magnitudes of the strengths of the source and sink increase in such a fashion that the product $l\Lambda$ remains constant. This limiting process is shown in Fig. 3.19b. In the limit, as $l \to 0$ while $l\Lambda$ remains constant, we obtain a special flow pattern defined as a *doublet*. The *strength* of the doublet is denoted by κ and is defined as $\kappa \equiv l\Lambda$. The stream function for a doublet is obtained from Eq. (3.84) as follows:

$$\psi = \lim_{\substack{l \to 0 \\ \kappa = l\Lambda = \text{const}}} \left(-\frac{\Lambda}{2\pi} \, d\theta \right) \tag{3.85}$$

where in the limit $\Delta\theta \to d\theta \to 0$. (Note that the source strength Λ approaches an infinite value in the limit.) In Fig. 3.19b, let r and b denote the distances to point P from the source and sink, respectively. Draw a line from the sink perpendicular to r, and denote the length along this line by a. For an infinitesimal $d\theta$, the geometry of Fig. 3.19b yields

$$a = l \sin \theta$$

$$b = r - l \cos \theta$$

$$d\theta = \frac{a}{b}$$

Hence,

$$d\theta = \frac{a}{b} = \frac{l \sin \theta}{r - l \cos \theta} \tag{3.86}$$

Substituting Eq. (3.86) into (3.85), we have

$$\psi = \lim_{\substack{l \to 0 \\ \kappa = \text{const}}} \left(-\frac{\Lambda}{2\pi} \frac{l \sin \theta}{r - l \cos \theta} \right)$$

or

$$\psi = \lim_{\substack{l \to 0 \\ \kappa = \text{const}}} \left(-\frac{\kappa}{2\pi} \frac{\sin \theta}{r - l \cos \theta} \right)$$

or

$$\psi = -\frac{\kappa}{2\pi} \frac{\sin \theta}{r}$$ (3.87)

Equation (3.87) is the stream function for a doublet. In a similar fashion, the velocity potential for a doublet is given by (see Prob. 3.14)

$$\phi = \frac{\kappa}{2\pi} \frac{\cos \theta}{r}$$ (3.88)

The streamlines of a doublet flow are obtained from Eq. (3.87):

$$\psi = -\frac{\kappa}{2\pi} \frac{\sin \theta}{r} = \text{const} = c$$

or

$$r = -\frac{\kappa}{2\pi c} \sin \theta$$ (3.89)

Equation (3.89) gives the equation for the streamlines in doublet flow. Recall from analytic geometry that the following equation in polar coordinates

$$r = d \sin \theta$$ (3.90)

is a circle with a diameter d on the vertical axis and with the center located $d/2$ directly above the origin. Comparing Eqs. (3.89) and (3.90), we see that the streamlines for a doublet are a family of circles with diameter $\kappa/2\pi c$, as sketched in Fig. 3.20. The different circles correspond to different values of the parameter c. Note that in Fig. 3.19 we placed the source to the left of the sink; hence, in Fig. 3.20 the direction of flow is out of the origin to the left and back into the origin from the right. In Fig. 3.19, we could just as well have placed the sink to the left of the source. In such a case, the signs in Eqs. (3.87) and (3.88) would be reversed, and the flow in Fig. 3.20 would be in the opposite direction. Therefore, a doublet has associated with it a sense of direction—the direction with which

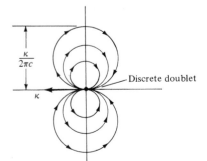

FIGURE 3.20
Doublet flow with strength κ.

the flow moves around the circular streamlines. By convention, we designate the direction of the doublet by an arrow drawn from the sink to the source, as shown in Fig. 3.20. In Fig. 3.20, the arrow points to the left, which is consistent with the form of Eqs. (3.87) and (3.88). If the arrow would point to the right, the sense of rotation would be reversed, Eq. (3.87) would have a positive sign, and Eq. (3.88) would have a negative sign.

Returning to Fig. 3.19, note that in the limit as $l \to 0$, the source and sink fall on top of each other. However, they do not extinguish each other, because the absolute magnitude of their strengths becomes infinitely large in the limit, and we have a singularity of strength $(\infty - \infty)$; this is an indeterminate form which can have a finite value.

As in the case of a source or sink, it is useful to interpret the doublet flow shown in Fig. 3.20 as being *induced* by a discrete doublet of strength κ placed at the origin. Therefore, a doublet is a singularity that induces about it the double-lobed circular flow pattern shown in Fig. 3.20.

3.13 NONLIFTING FLOW OVER A CIRCULAR CYLINDER

Consulting our road map given in Fig. 3.4, we see that we are well into the third column, having already discussed uniform flow, sources and sinks, and doublets. Along the way, we have seen how the flow over a semi-infinite body can be simulated by the combination of a uniform flow with a source, and the flow over an oval-shaped body can be constructed by superimposing a uniform flow and a source-sink pair. In this section, we demonstrate that the combination of a uniform flow and a doublet produces the flow over a circular cylinder. A circular cylinder is one of the most basic geometric shapes available, and the study of the flow around such a cylinder is a classic problem in aerodynamics.

Consider the addition of a uniform flow with velocity V_∞ and a doublet of strength κ, as shown in Fig. 3.21. The direction of the doublet is upstream, facing into the uniform flow. From Eqs. (3.57) and (3.87), the stream function for the combined flow is

$$\psi = V_\infty r \sin \theta - \frac{\kappa}{2\pi} \frac{\sin \theta}{r}$$

or
$$\psi = V_\infty r \sin \theta \left(1 - \frac{\kappa}{2\pi V_\infty r^2} \right) \qquad (3.91)$$

Let $R^2 \equiv \kappa/2\pi V_\infty$. Then Eq. (3.91) can be written as

$$\psi = (V_\infty r \sin \theta) \left(1 - \frac{R^2}{r^2} \right) \qquad (3.92)$$

Equation (3.92) is the stream function for a uniform flow–doublet combination;

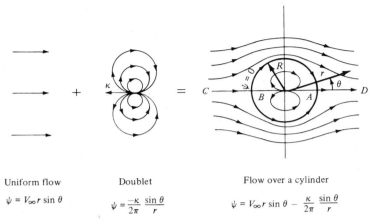

Uniform flow   Doublet   Flow over a cylinder

$\psi = V_\infty r \sin \theta$ $\psi = \dfrac{-\kappa}{2\pi} \dfrac{\sin \theta}{r}$ $\psi = V_\infty r \sin \theta - \dfrac{\kappa}{2\pi} \dfrac{\sin \theta}{r}$

FIGURE 3.21
Superposition of a uniform flow and a doublet; nonlifting flow over a circular cylinder.

it is also the stream function for the flow over a circular cylinder of radius R as shown in Fig. 3.21 and as demonstrated below.

The velocity field is obtained by differentiating Eq. (3.92), as follows:

$$V_r = \frac{1}{r}\frac{\partial \psi}{\partial \theta} = \frac{1}{r}(V_\infty r \cos \theta)\left(1 - \frac{R^2}{r^2}\right)$$

$$V_r = \left(1 - \frac{R^2}{r^2}\right) V_\infty \cos \theta \tag{3.93}$$

$$V_\theta = -\frac{\partial \psi}{\partial r} = -\left[(V_\infty r \sin \theta)\frac{2R^2}{r^3} + \left(1 - \frac{R^2}{r^2}\right)(V_\infty \sin \theta)\right]$$

$$V_\theta = -\left(1 + \frac{R^2}{r^2}\right) V_\infty \sin \theta \tag{3.94}$$

To locate the stagnation points, set Eqs. (3.93) and (3.94) equal to zero:

$$\left(1 - \frac{R^2}{r^2}\right) V_\infty \cos \theta = 0 \tag{3.95}$$

$$\left(1 + \frac{R^2}{r^2}\right) V_\infty \sin \theta = 0 \tag{3.96}$$

Simultaneously solving Eqs. (3.95) and (3.96) for r and θ, we find that there are two stagnation points, located at $(r, \theta) = (R, 0)$ and (R, π). These points are denoted as A and B, respectively, in Fig. 3.21.

The equation of the streamline that passes through the stagnation point B is obtained by inserting the coordinates of B into Eq. (3.92). For $r = R$ and $\theta = \pi$, Eq. (3.92) yields $\psi = 0$. Similarly, inserting the coordinates of point A into Eq.

(3.92), we also find that $\psi = 0$. Hence, the same streamline goes through both stagnation points. Moreover, the equation of this streamline, from Eq. (3.92), is

$$\psi = (V_\infty r \sin \theta)\left(1 - \frac{R^2}{r^2}\right) = 0 \qquad (3.97)$$

Note that Eq. (3.97) is satisfied by $r = R$ for all values of θ. However, recall that $R^2 \equiv \kappa/2\pi V_\infty$, which is a constant. Moreover, in polar coordinates, $r = $ constant $= R$ is the equation of a circle of radius R with its center at the origin. Therefore, Eq. (3.97) describes a circle with radius R, as shown in Fig. 3.21. Moreover, Eq. (3.97) is satisfied by $\theta = \pi$ and $\theta = 0$ for all values of r; hence, the entire horizontal axis through points A and B, extending infinitely far upstream and downstream, is part of the stagnation streamline.

Note that the $\psi = 0$ streamline, since it goes through the stagnation points, is the dividing streamline. That is, all the flow inside $\psi = 0$ (inside the circle) comes from the doublet, and all the flow outside $\psi = 0$ (outside the circle) comes from the uniform flow. Therefore, we can replace the flow inside the circle by a solid body, and the external flow will not know the difference. Consequently, *the inviscid irrotational, incompressible flow over a circular cylinder of radius R can be synthesized by adding a uniform flow with velocity V_∞ and a doublet of strength κ, where R is related to V_∞ and κ through*

$$R = \sqrt{\frac{\kappa}{2\pi V_\infty}} \qquad (3.98)$$

Note from Eqs. (3.92) to (3.94) that the entire flow field is symmetrical about both the horizontal and vertical axes through the center of the cylinder, as clearly seen by the streamline pattern sketched in Fig. 3.21. Hence, the pressure distribution is also symmetrical about both axes. As a result, the pressure distribution over the top of the cylinder is exactly balanced by the pressure distribution over the bottom of the cylinder; i.e., *there is no net lift*. Similarly, the pressure distribution over the front of the cylinder is exactly balanced by the pressure distribution over the back of the cylinder; i.e., *there is no net drag*. In real life, the result of zero lift is easy to accept, but the result of zero drag makes no sense. We know that any aerodynamic body immersed in a real flow will experience a drag. This paradox between the theoretical result of zero drag, and the knowledge that in real life the drag is finite, was encountered in the year 1744 by the Frenchman Jean Le Rond d'Alembert—and it has been known as *d'Alembert's paradox* ever since. For d'Alembert and other fluid dynamic researchers during the eighteenth and nineteenth centuries, this paradox was unexplained and perplexing. Of course, today we know that the drag is due to viscous effects which generate frictional shear stress at the body surface and which cause the flow to separate from the surface on the back of the body, thus creating a large wake downstream of the body and destroying the symmetry of the flow about the vertical axis through the cylinder. These viscous effects are discussed in detail in Chaps. 15 and 16. However, such viscous effects are not included in our present analysis of the inviscid flow over the cylinder. As a result, the inviscid theory

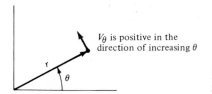

V_θ is positive in the
direction of increasing θ

FIGURE 3.22
Sign convention for V_θ in polar coordinates.

predicts that the flow closes smoothly and completely behind the body, as sketched in Fig. 3.21. It predicts no wake, and no asymmetries, resulting in the theoretical result of zero drag.

Let us quantify the above discussion. The velocity distribution on the surface of the cylinder is given by Eqs. (3.93) and (3.94) with $r = R$, resulting in

$$V_r = 0 \tag{3.99}$$

and
$$V_\theta = -2V_\infty \sin \theta \tag{3.100}$$

Note that at the surface of the cylinder, V_r is geometrically normal to the surface; hence, Eq. (3.99) is consistent with the physical boundary condition that the component of velocity normal to a stationary solid surface must be zero. Equation (3.100) gives the tangential velocity, which is the full magnitude of velocity on the surface of the cylinder, i.e., $V = V_\theta = -2V_\infty \sin \theta$ on the surface. The minus sign in Eq. (3.100) is consistent with the sign convention in polar coordinates that V_θ is positive in the direction of increasing θ, i.e., in the counterclockwise direction as shown in Fig. 3.22. However, in Fig. 3.21, the surface velocity for $0 \le \theta \le \pi$ is obviously in the opposite direction of increasing θ; hence, the minus sign in Eq. (3.100) is proper. For $\pi \le \theta \le 2\pi$, the surface flow is in the same direction as increasing θ, but $\sin \theta$ is itself negative; hence, once again the minus sign in Eq. (3.100) is proper. Note from Eq. (3.100) that the velocity at the surface reaches a maximum value of $2V_\infty$ at the top and the bottom of the cylinder (where $\theta = \pi/2$ and $3\pi/2$, respectively), as shown in Fig. 3.23. Indeed, these are the points of maximum velocity for the entire flow field around the cylinder, as can be seen from Eqs. (3.93) and (3.94).

The pressure coefficient is given by Eq. (3.38):

$$C_p = 1 - \left(\frac{V}{V_\infty}\right)^2 \tag{3.38}$$

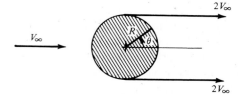

FIGURE 3.23
Maximum velocity in the flow over a circular cylinder.

Combining Eqs. (3.100) and (3.38), we find that the surface pressure coefficient over a circular cylinder is

$$C_p = 1 - 4\sin^2\theta$$

(3.101)

Note that C_p varies from 1.0 at the stagnation points to -3.0 at the points of maximum velocity. The pressure coefficient distribution over the surface is sketched in Fig. 3.24. The regions corresponding to the top and bottom halves of the cylinder are identified at the top of Fig. 3.24. Clearly, the pressure distribution over the top half of the cylinder is equal to the pressure distribution over the bottom half, and hence the lift must be zero, as discussed earlier. Moreover, the regions corresponding to the front and rear halves of the cylinder are identified at the bottom of Fig. 3.24. Clearly, the pressure distributions over the front and rear halves are the same, and hence the drag is theoretically zero, as also discussed previously. These results are confirmed by Eqs. (1.15) and (1.16). Since $c_f = 0$ (we are dealing with an inviscid flow), Eqs. (1.15) and (1.16) become, respectively,

$$c_n = \frac{1}{c}\int_0^c (C_{p,l} - C_{p,u})\, dx$$

(3.102)

$$c_a = \frac{1}{c}\int_{LE}^{TE} (C_{p,u} - C_{p,l})\, dy$$

(3.103)

For the circular cylinder, the chord c is the horizontal diameter. From Fig. 3.24, $C_{p,l} = C_{p,u}$ for corresponding stations measured along the chord, and hence the integrands in Eqs. (3.102) and (3.103) are identically zero, yielding $c_n = c_a = 0$. In turn, the lift and drag are zero, thus again confirming our previous conclusions.

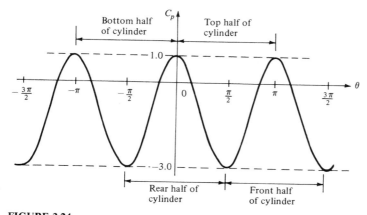

FIGURE 3.24
Pressure coefficient distribution over the surface of a circular cylinder; theoretical results for inviscid, incompressible flow.

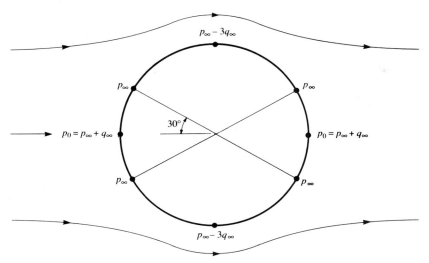

FIGURE 3.25
Values of pressure at various locations on the surface of a circular cylinder; nonlifting case.

Example 3.9. Consider the nonlifting flow over a circular cylinder. Calculate the locations on the surface of the cylinder where the surface pressure equals the freestream pressure.

Solution. When $p = p_\infty$, then $C_p = 0$. From Eq. (3.101),

$$C_p = 0 = 1 - 4 \sin^2 \theta$$

Hence,

$$\sin \theta = \pm \tfrac{1}{2}$$

$$\boxed{\theta = 30°, \ 150°, \ 210°, \ 330°}$$

These points, as well as the stagnation points and points of minimum pressure, are illustrated in Fig. 3.25. Note that at the stagnation point, where $C_p = 1$, the pressure is $p_\infty + q_\infty$; the pressure decreases to p_∞ in the first 30° of expansion around the body, and the minimum pressure at the top and bottom of the cylinder, consistent with $C_p = -3$, is $p_\infty - 3q_\infty$.

3.14 VORTEX FLOW: OUR FOURTH ELEMENTARY FLOW

Again, consulting our chapter road map in Fig. 3.4, we have discussed three elementary flows—uniform flow, source flow, and doublet flow—and have superimposed these elementary flows to obtain the nonlifting flow over several body shapes, such as ovals and circular cylinders. In this section, we introduce our fourth, and last, elementary flow: vortex flow. In turn, in Secs. 3.15 and 3.16, we see how the superposition of flows involving such vortices leads to cases with finite lift.

Consider a flow where all the streamlines are concentric circles about a given point, as sketched in Fig. 3.26. Moreover, let the velocity along any given circular streamline be constant, but let it vary from one streamline to another inversely with distance from the common center. Such a flow is called a *vortex flow*. Examine Fig. 3.26; the velocity components in the radial and tangential directions are V_r and V_θ, respectively, where $V_r = 0$ and $V_\theta = \text{constant}/r$. It is easily shown (try it yourself) that (1) vortex flow is a physically possible incompressible flow, i.e., $\nabla \cdot \mathbf{V} = 0$ at every point, and (2) vortex flow is irrotational, i.e., $\nabla \times \mathbf{V} = 0$, at every point except the origin.

From the definition of vortex flow, we have

$$V_\theta = \frac{\text{const}}{r} = \frac{C}{r} \tag{3.104}$$

To evaluate the constant C, take the circulation around a given circular streamline of radius r:

$$\Gamma = -\oint_C \mathbf{V} \cdot \mathbf{ds} = -V_\theta(2\pi r)$$

or

$$\boxed{V_\theta = -\frac{\Gamma}{2\pi r}} \tag{3.105}$$

Comparing Eqs. (3.104) and (3.105), we see that

$$C = -\frac{\Gamma}{2\pi} \tag{3.106}$$

Therefore, for vortex flow, Eq. (3.106) demonstrates that the circulation taken

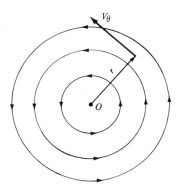

FIGURE 3.26
Vortex flow.

about all streamlines is the same value, namely, $\Gamma = -2\pi C$. By convention, Γ is called the *strength* of the vortex flow, and Eq. (3.105) gives the velocity field for a vortex flow of strength Γ. Note from Eq. (3.105) that V_θ is negative when Γ is positive; i.e., a vortex of positive strength rotates in the *clockwise* direction. (This is a consequence of our sign convention on circulation defined in Sec. 2.13, namely, positive circulation is clockwise.)

We stated earlier that vortex flow is irrotational except at the origin. What happens at $r = 0$? What is the value of $\nabla \times \mathbf{V}$ at $r = 0$? To answer these questions, recall Eq. (2.128) relating circulation to vorticity:

$$\Gamma = -\iint_S (\nabla \times \mathbf{V}) \cdot \mathbf{dS} \qquad (2.128)$$

Combining Eqs. (3.106) and (2.128), we obtain

$$2\pi C = \iint_S (\nabla \times \mathbf{V}) \cdot \mathbf{dS} \qquad (3.107)$$

Since we are dealing with two-dimensional flow, the flow sketched in Fig. 3.26 takes place in the plane of the paper. Hence, in Eq. (3.107), both $\nabla \times \mathbf{V}$ and $\mathbf{dS}$ are in the same direction, both perpendicular to the plane of the flow. Thus, Eq. (3.107) can be written as

$$2\pi C = \iint_S (\nabla \times \mathbf{V}) \cdot \mathbf{dS} = \iint_S |\nabla \times \mathbf{V}| \, dS \qquad (3.108)$$

In Eq. (3.108), the surface integral is taken over the circular area inside the streamline along which the circulation $\Gamma = -2\pi C$ is evaluated. However, Γ is the same for all the circular streamlines. In particular, choose a circle as close to the origin as we wish; i.e., let $r \to 0$. The circulation will still remain $\Gamma = -2\pi C$. However, the area inside this small circle around the origin will become infinitesimally small, and

$$\iint_S |\nabla \times \mathbf{V}| \, dS \to |\nabla \times \mathbf{V}| \, dS \qquad (3.109)$$

Combining Eqs. (3.108) and (3.109), in the limit as $r \to 0$, we have

$$2\pi C = |\nabla \times \mathbf{V}| \, dS$$

or

$$|\nabla \times \mathbf{V}| = \frac{2\pi C}{dS} \qquad (3.110)$$

However, as $r \to 0$, $dS \to 0$. Therefore, in the limit as $r \to 0$, from Eq. (3.110), we have

$$|\nabla \times \mathbf{V}| \to \infty$$

Conclusion: *Vortex flow is irrotational everywhere except at the point $r = 0$, where the vorticity is infinite.* Therefore, the origin, $r = 0$, is a singular point in the flow field. We see that, along with sources, sinks, and doublets, the vortex flow contains a singularity. Hence, we can interpret the singularity itself, i.e., point O in Fig. 3.26, to be a point vortex which induces about it the circular vortex flow shown in Fig. 3.26.

The velocity potential for vortex flow can be obtained as follows:

$$\frac{\partial \phi}{\partial r} = V_r = 0 \tag{3.111a}$$

$$\frac{1}{r} \frac{\partial \phi}{\partial \theta} = V_\theta = -\frac{\Gamma}{2\pi r} \tag{3.111b}$$

Integrating Eqs. (3.111a and b), we find

$$\boxed{\phi = -\frac{\Gamma}{2\pi} \theta} \tag{3.112}$$

Equation (3.112) is the velocity potential for vortex flow.

The stream function is determined in a similar manner:

$$\frac{1}{r} \frac{\partial \psi}{\partial \theta} = V_r = 0 \tag{3.113a}$$

$$-\frac{\partial \psi}{\partial r} = V_\theta = -\frac{\Gamma}{2\pi r} \tag{3.113b}$$

TABLE 3.1

Type of flow	Velocity	ϕ	ψ
Uniform flow in x direction	$u = V_\infty$	$V_\infty x$	$V_\infty y$
Source	$V_r = \dfrac{\Lambda}{2\pi r}$	$\dfrac{\Lambda}{2\pi} \ln r$	$\dfrac{\Lambda}{2\pi} \theta$
Vortex	$V_\theta = -\dfrac{\Gamma}{2\pi r}$	$-\dfrac{\Gamma}{2\pi} \theta$	$\dfrac{\Gamma}{2\pi} \ln r$
Doublet	$V_r = -\dfrac{\kappa}{2\pi} \dfrac{\cos \theta}{r^2}$	$\dfrac{\kappa}{2\pi} \dfrac{\cos \theta}{r}$	$-\dfrac{\kappa}{2\pi} \dfrac{\sin \theta}{r}$
	$V_\theta = -\dfrac{\kappa}{2\pi} \dfrac{\sin \theta}{r^2}$		

Integrating Eqs. (3.113a and b), we have

$$\psi = \frac{\Gamma}{2\pi} \ln r \qquad (3.114)$$

Equation (3.114) is the stream function for vortex flow. Note that since $\psi =$ constant is the equation of a streamline, Eq. (3.114) states that the streamlines of vortex flow are given by $r =$ constant; i.e., the streamlines are circles. Thus, Eq. (3.114) is consistent with our definition of vortex flow. Also, note from Eq. (3.112) that equipotential lines are given by $\theta =$ constant, i.e., straight radial lines from the origin. Once again, we see that equipotential lines and streamlines are mutually perpendicular.

At this stage, we summarize the pertinent results for our four elementary flows in Table 3.1.

3.15   LIFTING FLOW OVER A CYLINDER

In Sec. 3.13, we superimposed a uniform flow and a doublet to synthesize the flow over a circular cylinder, as shown in Fig. 3.21. In addition, we proved that both the lift and drag were zero for such a flow. However, the streamline pattern shown at the right of Fig. 3.21 is not the only flow that is theoretically possible around a circular cylinder. It *is* the only flow that is consistent with zero lift. However, there are other possible flow patterns around a circular cylinder— different flow patterns which result in a nonzero lift on the cylinder. Such lifting flows are discussed in this section.

Now you might be hesitant at this moment, perplexed by the question as to how a lift could possibly be exerted on a circular cylinder. Is not the body perfectly symmetric, and would not this geometry always result in a symmetric flow field with a consequent zero lift, as we have already discussed? You might be so perplexed that you run down to the laboratory, place a stationary cylinder in a low-speed tunnel, and measure the lift. To your satisfaction, you measure no lift, and you walk away muttering that the subject of this section is ridiculous— there is no lift on the cylinder. However, go back to the wind tunnel, and this time run a test with the cylinder *spinning* about its axis at relatively high revolutions per minute. This time you measure a *finite* lift. Also, by this time you might be thinking about other situations: spin on a baseball causes it to curve, and spin on a golf ball causes it to hook or slice. Clearly, in real life there are nonsymmetric aerodynamic forces acting on these symmetric, spinning bodies. So, maybe the subject matter of this section is not so ridiculous after all. Indeed, as you will soon appreciate, the concept of lifting flow over a cylinder will start us on a journey which leads directly to the theory of the lift generated by airfoils, as discussed in Chap. 4.

Consider the flow synthesized by the addition of the nonlifting flow over a cylinder and a vortex of strength Γ, as shown in Fig. 3.27. The stream function

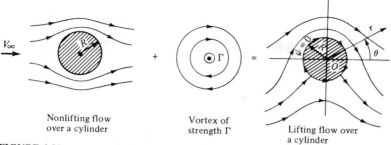

Nonlifting flow over a cylinder + Vortex of strength Γ = Lifting flow over a cylinder

FIGURE 3.27
The synthesis of lifting flow over a circular cylinder.

for nonlifting flow over a circular cylinder of radius R is given by Eq. (3.92):

$$\psi_1 = (V_\infty r \sin \theta)\left(1 - \frac{R^2}{r^2}\right) \tag{3.92}$$

The stream function for a vortex of strength Γ is given by Eq. (3.114). Recall that the stream function is determined within an arbitrary constant; hence, Eq. (3.114) can be written as

$$\psi_2 = \frac{\Gamma}{2\pi} \ln r + \text{const} \tag{3.115}$$

Since the value of the constant is arbitrary, let

$$\text{Const} = -\frac{\Gamma}{2\pi} \ln R \tag{3.116}$$

Combining Eqs. (3.115) and (3.116), we obtain

$$\psi_2 = \frac{\Gamma}{2\pi} \ln \frac{r}{R} \tag{3.117}$$

Equation (3.117) is the stream function for a vortex of strength Γ and is just as valid as Eq. (3.114) obtained earlier; the only difference between these two equations is a constant of the value given by Eq. (3.116).

The resulting stream function for the flow shown at the right of Fig. 3.27 is

$$\psi = \psi_1 + \psi_2$$

or

$$\psi = (V_\infty r \sin \theta)\left(1 - \frac{R^2}{r^2}\right) + \frac{\Gamma}{2\pi} \ln \frac{r}{R} \tag{3.118}$$

From Eq. (3.118), if $r = R$, then $\psi = 0$ for all values of θ. Since $\psi = \text{constant}$ is the equation of a streamline, $r = R$ is therefore a streamline of the flow, but

$r = R$ is the equation of a circle of radius R. Hence, Eq. (3.118) is a valid stream function for the inviscid, incompressible flow over a circular cylinder of radius R, as shown at the right of Fig. 3.27. Indeed, our previous result given by Eq. (3.92) is simply a special case of Eq. (3.118) with $\Gamma = 0$.

The resulting streamline pattern given by Eq. (3.118) is sketched at the right of Fig. 3.27. Note that the streamlines are no longer symmetrical about the horizontal axis through point O, and you might suspect (correctly) that the cylinder will experience a resulting finite normal force. However, the streamlines are symmetrical about the vertical axis through O, and as a result the drag will be zero, as we prove shortly. Note also that because a vortex of strength Γ has been added to the flow, the circulation about the cylinder is now finite and equal to Γ.

The velocity field can be obtained by differentiating Eq. (3.118). An equally direct method of obtaining the velocities is to add the velocity field of a vortex to the velocity field of the nonlifting cylinder. (Recall that because of the linearity of the flow, the velocity components of the superimposed elementary flows add directly.) Hence, from Eqs. (3.93) and (3.94) for nonlifting flow over a cylinder of radius R, and Eqs. (3.111a and b) for vortex flow, we have, for the lifting flow over a cylinder of radius R,

$$V_r = \left(1 - \frac{R^2}{r^2}\right) V_\infty \cos\theta \tag{3.119}$$

$$V_\theta = -\left(1 + \frac{R^2}{r^2}\right) V_\infty \sin\theta - \frac{\Gamma}{2\pi r} \tag{3.120}$$

To locate the stagnation points in the flow, set $V_r = V_\theta = 0$ in Eqs. (3.119) and (3.120) and solve for the resulting coordinates (r, θ):

$$V_r = \left(1 - \frac{R^2}{r^2}\right) V_\infty \cos\theta = 0 \tag{3.121}$$

$$V_\theta = -\left(1 + \frac{R^2}{r^2}\right) V_\infty \sin\theta - \frac{\Gamma}{2\pi r} = 0 \tag{3.122}$$

From Eq. (3.121), $r = R$. Substituting this result into Eq. (3.122) and solving for θ, we obtain

$$\theta = \arc\sin\left(-\frac{\Gamma}{4\pi V_\infty R}\right) \tag{3.123}$$

Since Γ is a positive number, from Eq. (3.123) θ must be in the third and fourth quadrants. That is, there can be two stagnation points on the bottom half of the circular cylinder, as shown by points 1 and 2 in Fig. 3.28a. These points are located at (R, θ), where θ is given by Eq. (3.123). However, this result is valid only when $\Gamma/4\pi V_\infty R < 1$. If $\Gamma/4\pi V_\infty R > 1$, then Eq. (3.123) has no meaning. If $\Gamma/4\pi V_\infty R = 1$, there is only one stagnation point on the surface of the cylinder, namely, point $(R, -\pi/2)$ labeled as point 3 in Fig. 3.28b. For the case of

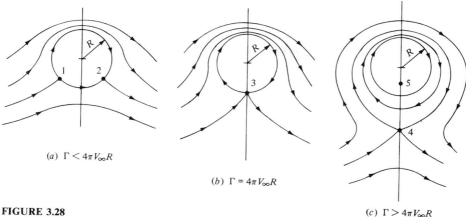

(a) $\Gamma < 4\pi V_\infty R$

(b) $\Gamma = 4\pi V_\infty R$

FIGURE 3.28

(c) $\Gamma > 4\pi V_\infty R$

Stagnation points for the lifting flow over a circular cylinder.

$\Gamma/4\pi V_\infty R > 1$, return to Eq. (3.121). We saw earlier that it is satisfied by $r = R$; however, it is also satisfied by $\theta = \pi/2$ or $-\pi/2$. Substituting $\theta = -\pi/2$ into Eq. (3.122), and solving for r, we have

$$r = \frac{\Gamma}{4\pi V_\infty} \pm \sqrt{\left(\frac{\Gamma}{4\pi V_\infty}\right)^2 - R^2} \qquad (3.124)$$

Hence, for $\Gamma/4\pi V_\infty R > 1$, there are two stagnation points, one inside and the other outside the cylinder, and both on the vertical axis, as shown by points 4 and 5 in Fig. 3.28c. [How does one stagnation point fall *inside* the cylinder? Recall that $r = R$, or $\psi = 0$, is just one of the allowed streamlines of the flow. There is a theoretical flow inside the cylinder—flow that is issuing from the doublet at the origin superimposed with the vortex flow for $r < R$. The circular streamline $r = R$ is the dividing streamline between this flow and the flow from the freestream. Therefore, as before, we can replace the dividing streamline by a solid body—our circular cylinder—and the *external* flow will not know the difference. Hence, although one stagnation point falls inside the body (point 5), we are not realistically concerned about it. Instead, from the point of view of flow over a solid cylinder of radius R, point 4 is the only meaningful stagnation point for the case $\Gamma/4\pi V_\infty R > 1$.]

The results shown in Fig. 3.28 can be visualized as follows. Consider the inviscid incompressible flow of given freestream velocity V_∞ over a cylinder of given radius R. If there is no circulation, i.e., if $\Gamma = 0$, the flow is given by the sketch at the right of Fig. 3.21, with horizontally opposed stagnation points A and B. Now assume that a circulation is imposed on the flow, such that $\Gamma = 4\pi V_\infty R$. The flow sketched in Fig. 3.28a will result; the two stagnation points will move to the lower surface of the cylinder as shown by points 1 and 2. Assume that Γ is further increased until $\Gamma = 4\pi V_\infty R$. The flow sketched in Fig. 3.28b will result, with only one stagnation point at the bottom of the cylinder, as shown by point 3. When Γ is increased still further such that $\Gamma > 4\pi V_\infty R$, the flow sketched in Fig. 3.28c will result. The stagnation point will lift from the cylinder's surface and will appear in the flow directly below the cylinder, as shown by point 4.

From the above discussion, Γ is clearly a parameter that can be chosen freely. There is no single value of Γ that "solves" the flow over a circular cylinder; rather, the circulation can be any value. Therefore, *for the incompressible flow over a circular cylinder, there are an infinite number of possible potential flow solutions, corresponding to the infinite choice for values of* Γ. This statement is not limited to flow over circular cylinders, but rather, it is a general statement that holds for the incompressible potential flow over all smooth two-dimensional bodies. We return to these ideas in subsequent sections.

From the symmetry, or lack of it, in the flows sketched in Figs. 3.27 and 3.28, we intuitively concluded earlier that a finite normal force (lift) must exist on the body but that the drag is zero; i.e., d'Alembert's paradox still prevails. Let us quantify these statements by calculating expressions for lift and drag, as follows.

The velocity on the surface of the cylinder is given by Eq. (3.120) with $r = R$.

$$V = V_\theta = -2V_\infty \sin\theta - \frac{\Gamma}{2\pi R} \tag{3.125}$$

In turn, the pressure coefficient is obtained by substituting Eq. (3.125) into Eq. (3.38):

$$C_p = 1 - \left(\frac{V}{V_\infty}\right)^2 = 1 - \left(-2\sin\theta - \frac{\Gamma}{2\pi R V_\infty}\right)^2$$

or

$$C_p = 1 - \left[4\sin^2\theta + \frac{2\Gamma\sin\theta}{\pi R V_\infty} + \left(\frac{\Gamma}{2\pi R V_\infty}\right)^2\right] \tag{3.126}$$

In Sec. 1.5, we discussed in detail how the aerodynamic force coefficients can be obtained by integrating the pressure coefficient and skin friction coefficient over the surface. For inviscid flow, $c_f = 0$. Hence, the drag coefficient c_d is given by Eq. (1.16) as

$$c_d = c_a = \frac{1}{c} \int_{LE}^{TE} (C_{p,u} - C_{p,l})\, dy$$

or

$$c_d = \frac{1}{c} \int_{LE}^{TE} C_{p,u}\, dy - \frac{1}{c} \int_{LE}^{TE} C_{p,l}\, dy \tag{3.127}$$

Converting Eq. (3.127) to polar coordinates, we note that

$$y = R\sin\theta \qquad dy = R\cos\theta\, d\theta \tag{3.128}$$

Substituting Eq. (3.128) into (3.127), and noting that $c = 2R$, we have

$$c_d = \frac{1}{2} \int_\pi^0 C_{p,u} \cos\theta\, d\theta - \frac{1}{2} \int_\pi^{2\pi} C_{p,l} \cos\theta\, d\theta \tag{3.129}$$

The limits of integration in Eq. (3.129) are explained as follows. In the first integral, we are integrating from the leading edge (the front point of the cylinder), moving over the *top* surface of the cylinder. Hence, θ is equal to π at the leading edge and, moving over the top surface, *decreases* to 0 at the trailing edge. In the second integral, we are integrating from the leading edge to the trailing edge while moving over the *bottom* surface of the cylinder. Hence, θ is equal to π at the leading edge and, moving over the bottom surface, *increases* to 2π at the trailing edge. In Eq. (3.129), both $C_{p,u}$ and $C_{p,l}$ are given by the same analytic expression for C_p, namely, Eq. (3.126). Hence, Eq. (3.129) can be written as

$$c_d = -\frac{1}{2}\int_0^\pi C_p \cos\theta \, d\theta - \frac{1}{2}\int_\pi^{2\pi} C_p \cos\theta \, d\theta$$

or

$$c_d = -\frac{1}{2}\int_0^{2\pi} C_p \cos\theta \, d\theta \qquad (3.130)$$

Substituting Eq. (3.126) into (3.130), and noting that

$$\int_0^{2\pi} \cos\theta \, d\theta = 0 \qquad (3.131a)$$

$$\int_0^{2\pi} \sin^2\theta \cos\theta \, d\theta = 0 \qquad (3.131b)$$

$$\int_0^{2\pi} \sin\theta \cos\theta \, d\theta = 0 \qquad (3.131c)$$

we immediately obtain

$$\boxed{c_d = 0} \qquad (3.132)$$

Equation (3.132) confirms our intuitive statements made earlier. The drag on a cylinder in an inviscid, incompressible flow is zero, regardless of whether or not the flow has circulation about the cylinder.

The lift on the cylinder can be evaluated in a similar manner as follows. From Eq. (1.15) with $c_f = 0$,

$$c_l = c_n = \frac{1}{c}\int_0^c C_{p,l} \, dx - \frac{1}{c}\int_0^c C_{p,u} \, dx \qquad (3.133)$$

Converting to polar coordinates, we obtain

$$x = R\cos\theta \qquad dx = -R\sin\theta \, d\theta \qquad (3.134)$$

Substituting Eq. (3.134) into (3.133), we have

$$c_l = -\frac{1}{2}\int_\pi^{2\pi} C_{p,l} \sin\theta \, d\theta + \frac{1}{2}\int_\pi^0 C_{p,u} \sin\theta \, d\theta \qquad (3.135)$$

Again, noting that $C_{p,l}$ and $C_{p,u}$ are both given by the same analytic expression, namely, Eq. (3.126), Eq. (3.135) becomes

$$c_l = -\frac{1}{2} \int_0^{2\pi} C_p \sin \theta \, d\theta \tag{3.136}$$

Substituting Eq. (3.126) into (3.136), and noting that

$$\int_0^{2\pi} \sin \theta \, d\theta = 0 \tag{3.137a}$$

$$\int_0^{2\pi} \sin^3 \theta \, d\theta = 0 \tag{3.137b}$$

$$\int_0^{2\pi} \sin^2 \theta \, d\theta = \pi \tag{3.137c}$$

we immediately obtain

$$c_l = \frac{\Gamma}{R V_\infty} \tag{3.138}$$

From the definition of c_l (see Sec. 1.5), the lift per unit span L' can be obtained from

$$L' = q_\infty S c_l = \tfrac{1}{2} \rho_\infty V_\infty^2 S c_l \tag{3.139}$$

Here, the planform area $S = 2R(1)$. Therefore, combining Eqs. (3.138) and (3.139), we have

$$L' = \frac{1}{2} \rho_\infty V_\infty^2 2R \frac{\Gamma}{R V_\infty}$$

or

$$\boxed{L' = \rho_\infty V_\infty \Gamma} \tag{3.140}$$

Equation (3.140) gives the lift per unit span for a circular cylinder with circulation Γ. It is a remarkably simple result, and it states that *the lift per unit span is directly proportional to circulation.* Equation (3.140) is a powerful relation in theoretical aerodynamics. It is called the *Kutta-Joukowski theorem,* named after the German mathematician M. Wilheim Kutta (1867-1944) and the Russian physicist Nikolai E. Joukowski (1847-1921), who independently obtained it during the first decade of this century. We will have more to say about the Kutta-Joukowski theorem in Sec. 3.16.

What are the connections between the above theoretical results and real life? As stated earlier, the prediction of zero drag is totally erroneous—viscous effects cause skin friction and flow separation which always produce a finite drag, as will be discussed in Chaps. 15 to 17. The inviscid flow treated in this chapter

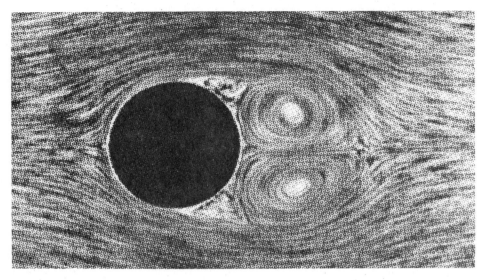

(a)

(b)

FIGURE 3.29
These flow-field pictures were obtained in water, where aluminum filings were scattered on the surface to show the direction of the streamlines. (*From Prandtl and Tietjens, Ref. 8.*) (*a*) Shown above is the case for the nonspinning cylinder. (*b*) Spinning cylinder: peripheral velocity of the surface = $3V_\infty$.

[*Continues*]

simply does not model the proper physics for drag calculations. On the other hand, the prediction of lift via Eq. (3.140) is quite realistic. Let us return to the wind-tunnel experiments mentioned at the beginning of this chapter. If a stationary, nonspinning cylinder is placed in a low-speed wind tunnel, the flow field will appear as shown in Fig. 3.29a. The streamlines over the front of the cylinder are similar to theoretical predictions, as sketched at the right of Fig. 3.21. However, because of viscous effects, the flow separates over the rear of the cylinder, creating a recirculating flow in the wake downstream of the body. This separated flow greatly contributes to the finite drag measured for the cylinder. On the other hand, Fig. 3.29a shows a reasonably symmetric flow about the horizontal axis, and the measurement of lift is essentially zero. Now let us *spin* the cylinder in a clockwise direction about its axis. The resulting flow fields are shown in Fig. 3.29b and c. For a moderate amount of spin (Fig. 3.29b), the stagnation points move to the lower part of the cylinder, similar to the theoretical flow sketched in Fig. 3.28a. If the spin is sufficiently increased (Fig. 3.29c), the stagnation point lifts off the surface, similar to the theoretical flow sketched in Fig. 3.28c. And what is most important, a *finite lift* is measured for the spinning cylinder in the wind tunnel. What is happening here? Why does spinning the cylinder produce lift? In actuality, the friction between the fluid and the surface of the cylinder tends to drag the fluid near the surface in the same direction as the rotational motion. Superimposed on top of the usual nonspinning flow, this "extra" velocity contribution creates a higher-than-usual velocity at the top of the cylinder and a lower-than-usual velocity at the bottom, as sketched in Fig. 3.30. These velocities are assumed to be just outside the viscous boundary layer on the surface. Recall

(c)

FIGURE 3.29
(c) Spinning cylinder: peripheral velocity of the surface $= 6V_\infty$.

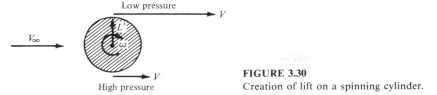

FIGURE 3.30
Creation of lift on a spinning cylinder.

from Bernoulli's equation that as the velocity increases, the pressure decreases. Hence, from Fig. 3.30, the pressure on the top of the cylinder is lower than on the bottom. This pressure imbalance creates a net upward force, i.e., a finite lift. Therefore, the theoretical prediction embodied in Eq. (3.140) that the flow over a circular cylinder can produce a finite lift is verified by experimental observation.

The general ideas discussed above concerning the generation of lift on a spinning circular cylinder in a wind tunnel also apply to a spinning sphere. This explains why a baseball pitcher can throw a curve and how a golfer can hit a hook or slice—all of which are due to nonsymmetric flows about the spinning bodies, and hence the generation of an aerodynamic force perpendicular to the body's angular velocity vector. This phenomenon is called the *Magnus effect*, named after the German engineer who first observed and explained it in Berlin in 1852.

It is interesting to note that a rapidly spinning cylinder can produce a much higher lift than an airplane wing of the same planform area; however, the drag on the cylinder is also much higher than a well-designed wing. As a result, the Magnus effect is not employed for powered flight. On the other hand, in the 1920s, the German engineer Anton Flettner replaced the sail on a boat with a rotating circular cylinder with its axis vertical to the deck. In combination with the wind, this spinning cylinder provided propulsion for the boat. Moreover, by the action of two cylinders in tandem and rotating in opposite directions, Flettner was able to turn the boat around. Flettner's device was a technical success, but an economic failure because the maintenance on the machinery to spin the cylinders at the necessary high rotational speeds was too costly. Today, the Magnus effect has an important influence on the performance of spinning missiles; indeed, a certain amount of modern high-speed aerodynamic research has focused on the Magnus forces on spinning bodies for missile applications.

Example 3.10. Consider the lifting flow over a circular cylinder. The lift coefficient is 5. Calculate the peak (negative) pressure coefficient.

Solution. Examining Fig. 3.27, note that the maximum velocity for the *nonlifting* flow over a cylinder is $2V_\infty$ and that it occurs at the top and bottom points on the cylinder. When the vortex in Fig. 3.27 is added to the flow field, the direction of the vortex velocity is in the *same* direction as the flow on the top of the cylinder, but opposes the flow on the bottom of the cylinder. Hence, the maximum velocity for the lifting case occurs at the *top* of the cylinder and is equal to the sum of the nonlifting value, $-2V_\infty$, and the vortex, $-\Gamma/2\pi R$. (*Note:* We are still following the usual sign convention; since the velocity on the top of the cylinder is in the opposite direction of increasing θ for the polar coordinate system, the velocity magnitudes here are negative.) Hence,

$$V = -2V_\infty - \frac{\Gamma}{2\pi R} \tag{E.1}$$

The lift coefficient and Γ are related through Eq. (3.138):

$$c_l = \frac{\Gamma}{RV_\infty} = 5$$

Hence,

$$\frac{\Gamma}{R} = 5V_\infty \tag{E.2}$$

Substituting Eq. (E.2) into (E.1), we have

$$V = -2V_\infty - \frac{5}{2\pi}V_\infty = -2.796\,V_\infty \tag{E.3}$$

Substituting Eq. (E.3) into Eq. (3.38), we obtain

$$C_p = 1 - \left(\frac{V}{V_\infty}\right) = 1 - (2.796)^2 = \boxed{-6.82}$$

This example is designed in part to make the following point. Recall that, for an inviscid, incompressible flow, the distribution of C_p over the surface of a body depends only on the shape and orientation of the body—the flow properties such as velocity, density, etc., are irrelevant here. Recall Eq. (3.101), which gives C_p as a function of θ only, namely, $C_p = 1 - 4\sin^2\theta$. However, for the case of lifting flow, the distribution of C_p over the surface is a function of one additional parameter—namely, the lift coefficient. Clearly, in this example, only the value of c_l is given. But this is powerful enough to define the flow uniquely; the value of C_p at any point on the surface follows directly from the value of lift coefficient, as demonstrated in the above problem.

Example 3.11. For the flow field in Example 3.10, calculate the location of the stagnation points and the points on the cylinder where the pressure equals freestream static pressure.

Solution. From Eq. (3.123), the stagnation points are given by

$$\theta = \arcsin\left(-\frac{\Gamma}{4\pi V_\infty R}\right)$$

From Example 3.10,

$$\frac{\Gamma}{RV_\infty} = 5$$

Thus,

$$\theta = \arcsin\left(-\frac{5}{4\pi}\right) = \boxed{203.4° \quad \text{and} \quad 336.6°}$$

To find the locations where $p = p_\infty$, first construct a formula for the pressure coefficient on the cylinder surface:

$$C_p = 1 - \left(\frac{V}{V_\infty}\right)^2$$

where

$$V = -2V_\infty \sin\theta - \frac{\Gamma}{2\pi R}$$

Thus,

$$C_p = 1 - \left(-2\sin\theta - \frac{\Gamma}{2\pi R}\right)^2$$

$$= 1 - 4\sin^2\theta - \frac{2\Gamma\sin\theta}{\pi R V_\infty} - \left(\frac{\Gamma}{2\pi R V_\infty}\right)^2$$

From Example 3.10, $\Gamma/R V_\infty = 5$. Thus,

$$C_p = 1 - 4\sin^2\theta - \frac{10}{\pi}\sin\theta - \left(\frac{5}{2\pi}\right)^2$$

$$= 0.367 - 3.183\sin\theta - 4\sin^2\theta$$

A check on this equation can be obtained by calculating C_p at $\theta = 90°$ and seeing if it agrees with the result obtained in Example 3.10. For $\theta = 90°$, we have

$$C_p = 0.367 - 3.183 - 4 = \boxed{-6.82}$$

This is the same result as in Example 3.10; the equation checks.

To find the values of θ where $p = p_\infty$, set $C_p = 0$:

$$0 = 0.367 - 3.183\sin\theta - 4\sin^2\theta$$

From the quadratic formula,

$$\sin\theta = \frac{3.183 \pm \sqrt{(3.183)^2 + 5.872}}{-8} = \boxed{-0.897 \quad \text{or} \quad 0.102}$$

Hence,

$$\theta = 243.8° \quad \text{and} \quad 296.23°$$

Also,

$$\theta = 5.85° \quad \text{and} \quad 174.1°$$

There are four points on the circular cylinder where $p = p_\infty$. These are sketched in Fig. 3.31, along with the stagnation point locations. As shown in Example 3.10, the minimum pressure occurs at the top of the cylinder and is equal to $p_\infty - 6.82 q_\infty$. A *local* minimum pressure occurs at the bottom of the cylinder, where $\theta = 3\pi/2$. This local minimum is given by

$$C_p = 0.367 - 3.183\sin\frac{3\pi}{2} - 4\sin^2\frac{3\pi}{2}$$

$$= 0.367 + 3.183 - 4 = \boxed{-0.45}$$

Hence, at the bottom of the cylinder, $p = p_\infty - 0.45 q_\infty$.

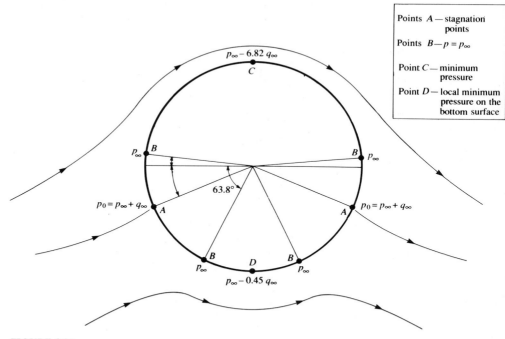

Points A — stagnation points

Points B — $p = p_\infty$

Point C — minimum pressure

Point D — local minimum pressure on the bottom surface

FIGURE 3.31
Values of pressure at various locations on the surface of a circular cylinder; lifting case with finite circulation. The values of pressure correspond to the case discussed in Example 3.10.

3.16 THE KUTTA-JOUKOWSKI THEOREM AND THE GENERATION OF LIFT

Although the result given by Eq. (3.140) was derived for a circular cylinder, it applies in general to cylindrical bodies of arbitrary cross section. For example, consider the incompressible flow over an airfoil section, as sketched in Fig. 3.32. Let curve A be any curve in the flow *enclosing* the airfoil. If the airfoil is producing lift, the velocity field around the airfoil will be such that the line integral of velocity around A will be finite, i.e., the circulation

$$\Gamma \equiv \oint_A \mathbf{V} \cdot \mathbf{ds}$$

is finite. In turn, the lift per unit span L' on the airfoil will be given by the *Kutta-Joukowski theorem*, as embodied in Eq. (3.140):

$$L' = \rho_\infty V_\infty \Gamma \qquad (3.140)$$

This result underscores the importance of the concept of circulation, defined in Sec. 2.13. The Kutta-Joukowski theorem states that lift per unit span on a

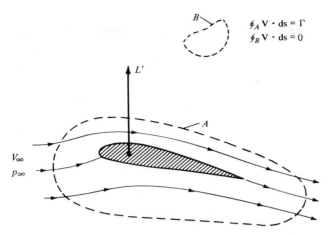

$$\oint_A \mathbf{V} \cdot d\mathbf{s} = \Gamma$$
$$\oint_B \mathbf{V} \cdot d\mathbf{s} = 0$$

FIGURE 3.32
Circulation around a lifting airfoil.

two-dimensional body is directly proportional to the circulation around the body. Indeed, the concept of circulation is so important at this stage of our discussion that you should reread Sec. 2.13 before proceeding further.

The general derivation of Eq. (3.140) for bodies of arbitrary cross section can be carried out using the method of complex variables. Such mathematics is beyond the scope of this book. (It can be shown that arbitrary functions of complex variables are general solutions of Laplace's equation, which in turn governs incompressible potential flow. Hence, more advanced treatments of such flows utilize the mathematics of complex variables as an important tool. See Ref. 9 for a particularly lucid treatment of inviscid, incompressible flow at a more advanced level.)

In Sec. 3.15, the lifting flow over a circular cylinder was synthesized by superimposing a uniform flow, a doublet, and a vortex. Recall that all three elementary flows are irrotational at all points, except for the vortex, which has infinite vorticity at the origin. Therefore, the lifting flow over a cylinder as shown in Fig. 3.28 is irrotational at every point except at the origin. If we take the circulation around any curve *not* enclosing the origin, we obtain from Eq. (2.128) the result that $\Gamma = 0$. It is only when we choose a curve that encloses the origin, where $\nabla \times \mathbf{V}$ is infinite, that Eq. (2.128) yields a finite Γ, equal to the strength of the vortex. The same can be said about the flow over the airfoil in Fig. 3.32. As we show in Chap. 4, the flow *outside* the airfoil is irrotational, and the circulation around any closed curve *not* enclosing the airfoil (such as curve *B* in Fig. 3.32) is consequently zero. On the other hand, we also show in Chap. 4 that the flow over an airfoil is synthesized by distributing vortices either on the surface or inside the airfoil. These vortices have the usual singularities in $\nabla \times \mathbf{V}$, and therefore, if we choose a curve that encloses the airfoil (such as curve *A* in Fig. 3.32),

Eq. (2.128) yields a finite value of Γ, equal to the *sum* of the vortex strengths distributed on or inside the airfoil. The important point here is that, in the Kutta-Joukowski theorem, the value of Γ used in Eq. (3.140) must be evaluated around a closed curve that *encloses the body*; the curve can be otherwise arbitrary, but it must have the body inside it.

At this stage, let us pause and assess our thoughts. The approach we have discussed above—the definition of circulation and the use of Eq. (3.140) to obtain the lift—is the essence of the *circulation theory of lift* in aerodynamics. Its development at the turn of the twentieth century created a breakthrough in aerodynamics. However, let us keep things in perspective. The circulation theory of lift is an *alternative* way of thinking about the generation of lift on an aerodynamic body. Keep in mind that the true physical sources of the aerodynamic force on a body are the pressure and shear stress distributions exerted on the surface of the body, as explained in Sec. 1.5. The Kutta-Joukowski theorem is simply an alternate way of expressing the *consequences* of the surface pressure distribution; it is a mathematical expression that is consistent with the special tools we have developed for the analysis of inviscid, incompressible flow. Indeed, recall that Eq. (3.140) was derived in Sec. 3.15 by integrating the pressure distribution over the surface. Therefore, it is not quite proper to say that circulation "causes" lift. Rather, lift is "caused" by the net imbalance of the surface pressure distribution, and circulation is simply a defined quantity determined from the same pressures. The relation between the surface pressure distribution (which produces lift L') and circulation is given by Eq. (3.140). However, in the theory of incompressible, potential flow, it is generally much easier to determine the circulation around the body rather than calculate the detailed surface pressure distribution. Therein lies the power of the circulation theory of lift.

Consequently, the theoretical analysis of lift on two-dimensional bodies in incompressible, inviscid flow focuses on the calculation of the circulation about the body. Once Γ is obtained, then the lift per unit span follows directly from the Kutta-Joukowski theorem. As a result, in subsequent sections we constantly address the question: How can we calculate the circulation for a given body in a given incompressible, inviscid flow?

3.17 NONLIFTING FLOWS OVER ARBITRARY BODIES: THE NUMERICAL SOURCE PANEL METHOD

In this section, we return to the consideration of nonlifting flows. Recall that we have already dealt with the nonlifting flows over a semi-infinite body and a Rankine oval and both the nonlifting and the lifting flows over a circular cylinder. For those cases, we added our elementary flows in certain ways and discovered that the dividing streamlines turned out to fit the shapes of such special bodies. However, this indirect method of starting with a given combination of elementary flows and seeing what body shape comes out of it can hardly be used in a practical sense for bodies of arbitrary shape. For example, consider the airfoil in Fig. 3.32.

Do we know in advance the correct combination of elementary flows to synthesize the flow over this specific body? The answer is no. Rather, what we want is a direct method; i.e., let us *specify* the shape of an arbitrary body and *solve* for the distribution of singularities which, in combination with a uniform stream, produce the flow over the given body. The purpose of this section is to present such a direct method, limited for the present to nonlifting flows. We consider a numerical method appropriate for solution on a high-speed digital computer. The technique is called the *source panel method*, which, since the late 1960s, has become a standard aerodynamic tool in industry and most research laboratories. In fact, the numerical solution of potential flows by both source and vortex panel techniques has revolutionized the analysis of low-speed flows. We return to various numerical panel techniques in Chaps. 4 through 6. As a modern student of aerodynamics, it is necessary for you to become familiar with the fundamentals of such panel methods. The purpose of the present section is to introduce the basic ideas of the source panel method, which is a technique for the numerical solution of nonlifting flows over arbitrary bodies.

First, let us extend the concept of a source or sink introduced in Sec. 3.10. In that section, we dealt with a single line source, as sketched in Fig. 3.16. Now imagine that we have an infinite number of such line sources side by side, where the strength of each line source is infinitesimally small. These side-by-side line sources form a *source sheet*, as shown in perspective in the upper left of Fig. 3.33. If we look along the series of line sources (looking along the z axis in Fig. 3.33), the source sheet will appear as sketched at the lower right of Fig. 3.33. Here, we are looking at an edge view of the sheet; the line sources are all perpendicular to the page. Let s be the distance measured along the source sheet in the edge view. Define $\lambda = \lambda(s)$ to be the *source strength per unit length along s*. [To keep things in perspective, recall from Sec. 3.10 that the strength of a single line source

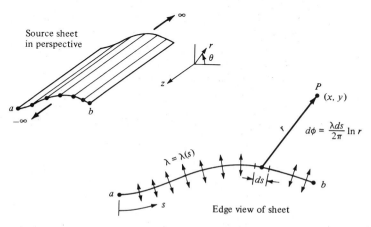

FIGURE 3.33
Source sheet.

Λ was defined as the volume flow rate per unit depth, i.e., per unit length in the z direction. Typical units for Λ are square meters per second or square feet per second. In turn, the strength of a source sheet $\lambda(s)$ is the volume flow rate per unit depth (in the z direction) and per unit length (in the s direction). Typical units for λ are meters per second or feet per second.] Therefore, the strength of an infinitesimal portion ds of the sheet, as shown in Fig. 3.33, is $\lambda\,ds$. This small section of the source sheet can be treated as a distinct source of strength $\lambda\,ds$. Now consider point P in the flow, located a distance r from ds; the cartesian coordinates of P are (x, y). The small section of the source sheet of strength $\lambda\,ds$ induces an infinitesimally small potential, $d\phi$, at point P. From Eq. (3.67), $d\phi$ is given by

$$d\phi = \frac{\lambda\,ds}{2\pi}\ln r \tag{3.141}$$

The complete velocity potential at point P, induced by the entire source sheet from a to b, is obtained by integrating Eq. (3.141):

$$\phi(x, y) = \int_a^b \frac{\lambda\,ds}{2\pi}\ln r \tag{3.142}$$

Note that, in general, $\lambda(s)$ can change from positive to negative along the sheet; i.e., the "source" sheet is really a combination of line sources and line sinks.

Next, consider a given body of arbitrary shape in a flow with freestream velocity V_∞, as shown in Fig. 3.34. Let us cover the surface of the prescribed body with a source sheet, where the strength $\lambda(s)$ varies in such a fashion that the combined action of the uniform flow and the source sheet makes the airfoil surface a streamline of the flow. Our problem now becomes one of finding the appropriate $\lambda(s)$. The solution of this problem is carried out numerically, as follows.

Let us approximate the source sheet by a series of straight panels, as shown in Fig. 3.35. Moreover, let the source strength λ per unit length be constant over a given panel, but allow it to vary from one panel to the next. That is, if there

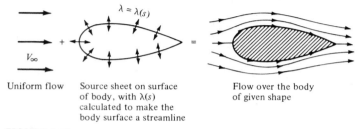

Uniform flow Source sheet on surface Flow over the body
of body, with $\lambda(s)$ of given shape
calculated to make the
body surface a streamline

FIGURE 3.34
Superposition of a uniform flow and a source sheet on a body of given shape, to produce the flow over the body.

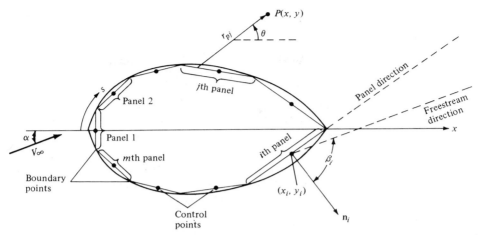

FIGURE 3.35
Source panel distribution over the surface of a body of arbitrary shape.

are a total of n panels, the source panel strengths per unit length are λ_1, $\lambda_2, \ldots, \lambda_j, \ldots, \lambda_n$. These panel strengths are unknown; the main thrust of the panel technique is to solve for λ_j, $j = 1$ to n, such that the body surface becomes a streamline of the flow. This boundary condition is imposed numerically by defining the midpoint of each panel to be a *control point* and by determining the λ_j's such that the normal component of the flow velocity is zero at each control point. Let us now quantify this strategy.

Let P be a point located at (x, y) in the flow, and let r_{pj} be the distance from any point on the jth panel to P, as shown in Fig. 3.35. The velocity potential induced at P due to the jth panel $\Delta \phi_j$ is, from Eq. (3.142),

$$\Delta \phi_j = \frac{\lambda_j}{2\pi} \int_j \ln r_{pj} \, ds_j \tag{3.143}$$

In Eq. (3.143), λ_j is constant over the jth panel, and the integral is taken over the jth panel only. In turn, the *potential at P due to all the panels* is Eq. (3.143) summed over all the panels:

$$\phi(P) = \sum_{j=1}^{n} \Delta \phi_j = \sum_{j=1}^{n} \frac{\lambda_j}{2\pi} \int_j \ln r_{pj} \, ds_j \tag{3.144}$$

In Eq. (3.144), the distance r_{pj} is given by

$$r_{pj} = \sqrt{(x - x_j)^2 + (y - y_j)^2} \tag{3.145}$$

where (x_j, y_j) are coordinates along the surface of the jth panel. Since point P is just an arbitrary point in the flow, let us put P at the control point of the ith panel. Let the coordinates of this control point be given by (x_i, y_i), as shown in

Fig. 3.35. Then Eqs. (3.144) and (3.145) become

$$\phi(x_i, y_i) = \sum_{j=1}^{n} \frac{\lambda_j}{2\pi} \int_j \ln r_{ij} \, ds_j \tag{3.146}$$

and

$$r_{ij} = \sqrt{(x_i - x_j)^2 + (y_i - y_j)^2} \tag{3.147}$$

Equation (3.146) is physically the contribution of *all* the panels to the potential at the control point of the *i*th panel.

Recall that the boundary condition is applied at the control points; i.e., the normal component of the flow velocity is zero at the control points. To evaluate this component, first consider the component of freestream velocity perpendicular to the panel. Let $\mathbf{n}_i$ be the unit vector normal to the *i*th panel, directed out of the body, as shown in Fig. 3.35. Also, note that the slope of the *i*th panel is $(dy/dx)_i$. In general, the freestream velocity will be at some incidence angle α to the x axis, as shown in Fig. 3.35. Therefore, inspection of the geometry of Fig. 3.35 reveals that the component of V_∞ normal to the *i*th panel is

$$V_{\infty,n} = \mathbf{V}_\infty \cdot \mathbf{n}_i = V_\infty \cos \beta_i \tag{3.148}$$

where β_i is the angle between $\mathbf{V}_\infty$ and $\mathbf{n}_i$. Note that $V_{\infty,n}$ is positive when directed away from the body, and negative when directed toward the body.

The normal component of velocity induced at (x_i, y_i) by the source panels is, from Eq. (3.146),

$$V_n = \frac{\partial}{\partial n_i} [\phi(x_i, y_i)] \tag{3.149}$$

where the derivative is taken in the direction of the outward unit normal vector, and hence, again, V_n is positive when directed away from the body. When the derivative in Eq. (3.149) is carried out, r_{ij} appears in the denominator. Consequently, a singular point arises on the *i*th panel because when $j = i$, at the control point itself $r_{ij} = 0$. It can be shown that when $j = i$, the contribution to the derivative is simply $\lambda_i/2$. Hence, Eq. (3.149) combined with Eq. (3.146) becomes

$$V_n = \frac{\lambda_i}{2} + \sum_{\substack{j=1 \\ (j \neq i)}}^{n} \frac{\lambda_j}{2\pi} \int_j \frac{\partial}{\partial n_i} (\ln r_{ij}) \, ds_j \tag{3.150}$$

In Eq. (3.150), the first term $\lambda_i/2$ is the normal velocity induced at the *i*th control point by the *i*th panel itself, and the summation is the normal velocity induced at the *i*th control point by all the other panels.

The normal component of the flow velocity at the *i*th control point is the sum of that due to the freestream [Eq. (3.148)] and that due to the source panels [Eq. (3.150)]. The boundary condition states that this sum must be zero:

$$V_{\infty,n} + V_n = 0 \tag{3.151}$$

Substituting Eqs. (3.148) and (3.150) into (3.151), we obtain

$$\frac{\lambda_i}{2} + \sum_{\substack{j=1 \\ (j \neq i)}}^{n} \frac{\lambda_j}{2\pi} \int_j \frac{\partial}{\partial n_i} (\ln r_{ij}) \, ds_j + V_\infty \cos \beta_i = 0 \tag{3.152}$$

Equation (3.152) is the crux of the source panel method. The values of the integrals in Eq. (3.152) depend simply on the panel geometry; they are not properties of the flow. Let $I_{i,j}$ be the value of this integral when the control point is on the ith panel and the integral is over the jth panel. Then Eq. (3.152) can be written as

$$\frac{\lambda_i}{2} + \sum_{\substack{j=1 \\ (j \neq i)}}^{n} \frac{\lambda_j}{2\pi} I_{i,j} + V_\infty \cos \beta_i = 0 \tag{3.153}$$

Equation (3.153) is a linear *algebraic* equation with n unknowns $\lambda_1, \lambda_2, \ldots, \lambda_n$. It represents the flow boundary condition evaluated at the control point of the ith panel. Now apply the boundary condition to the control points of *all* the panels; i.e., in Eq. (3.153), let $i = 1, 2, \ldots, n$. The results will be a system of n linear algebraic equations with n unknowns $(\lambda_1, \lambda_2, \ldots, \lambda_n)$, which can be solved simultaneously by conventional numerical methods.

Look what has happened! After solving the system of equations represented by Eq. (3.153) with $i = 1, 2, \ldots n$, we now have the distribution of source panel strengths which, in an appropriate fashion, cause the body surface in Fig. 3.35 to be a streamline of the flow. This approximation can be made more accurate by increasing the number of panels, hence more closely representing the source sheet of continuously varying strength $\lambda(s)$ shown in Fig. 3.34. Indeed, the accuracy of the source panel method is amazingly good; a circular cylinder can be accurately represented by as few as 8 panels, and most airfoil shapes, by 50 to 100 panels. (For an airfoil, it is desirable to cover the leading-edge region with a number of small panels to represent accurately the rapid surface curvature and to use larger panels over the relatively flat portions of the body. Note that, in general, all the panels in Fig. 3.35 can be different lengths.)

Once the λ_i's $(i = 1, 2, \ldots, n)$ are obtained, the velocity *tangent* to the surface at each control point can be calculated as follows. Let s be the distance along the body surface, measured positive from front to rear, as shown in Fig. 3.35. The component of freestream velocity tangent to the surface is

$$V_{\infty,s} = V_\infty \sin \beta_i \tag{3.154}$$

The tangential velocity V_s at the control point of the ith panel induced by all the panels is obtained by differentiating Eq. (3.146) with respect to s:

$$V_s = \frac{\partial \phi}{\partial s} = \sum_{j=1}^{n} \frac{\lambda_j}{2\pi} \int_j \frac{\partial}{\partial s} (\ln r_{ij}) \, ds_j \tag{3.155}$$

[The tangential velocity on a flat source panel induced by the panel itself is zero; hence, in Eq. (3.155), the term corresponding to $j = i$ is zero. This is easily seen by intuition, because the panel can only emit volume flow from its surface in a direction perpendicular to the panel itself.] The total surface velocity at the ith control point V_i is the sum of the contribution from the freestream [Eq. (3.154)] and from the source panels [Eq. (3.155)]:

$$V_i = V_{\infty,s} + V_s = V_\infty \sin \beta_i + \sum_{j=1}^{n} \frac{\lambda_j}{2\pi} \int_j \frac{\partial}{\partial s} (\ln r_{ij}) \, ds_j \qquad (3.156)$$

In turn, the pressure coefficient at the ith control point is obtained from Eq. (3.38):

$$C_{p,i} = 1 - \left(\frac{V_i}{V_\infty} \right)^2$$

In this fashion, the source panel method gives the pressure distribution over the surface of a nonlifting body of arbitrary shape.

When you carry out a source panel solution as described above, the accuracy of your results can be tested as follows. Let S_j be the length of the jth panel. Recall that λ_j is the strength of the jth panel *per unit length*. Hence, the strength of the jth panel itself is $\lambda_j S_j$. For a closed body, such as in Fig. 3.35, the *sum* of all the source and sink strengths must be zero, or else the body itself would be adding or absorbing mass from the flow—an impossible situation for the case we are considering here. Hence, the values of the λ_j's obtained above should obey the relation

$$\sum_{j=1}^{n} \lambda_j S_j = 0 \qquad (3.157)$$

Equation (3.157) provides an independent check on the accuracy of the numerical results.

Example 3.12. Calculate the pressure coefficient distribution around a circular cylinder using the source panel technique.

Solution. We choose to cover the body with eight panels of equal length, as shown in Fig. 3.36. This choice is arbitrary; however, experience has shown that, for the

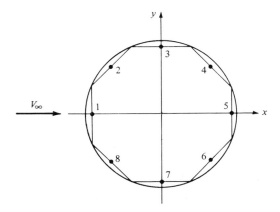

FIGURE 3.36
Source panel distribution around a circular cylinder.

case of a circular cylinder, the arrangement shown in Fig. 3.36 provides sufficient accuracy. The panels are numbered from 1 to 8, and the control points are shown by the dots in the center of each panel.

Let us evaluate the integrals $I_{i,j}$ which appear in Eq. (3.153). Consider Fig. 3.37, which illustrates two arbitrarily chosen panels. In Fig. 3.37, (x_i, y_i) are the coordinates of the control point of the ith panel and (x_j, y_j) are the running coordinates over the entire jth panel. The coordinates of the boundary points for the ith panel are (X_i, Y_i) and (X_{i+1}, Y_{i+1}); similarly, the coordinates of the boundary points for the jth panel are (X_j, Y_j) and (X_{j+1}, Y_{j+1}). In this problem, $\mathbf{V}_\infty$ is in the x direction; hence, the angles between the x axis and the unit vectors $\mathbf{n}_i$ and $\mathbf{n}_j$ are β_i and β_j, respectively. Note that, in general, both β_i and β_j vary from 0 to 2π. Recall that the integral $I_{i,j}$ is evaluated at the ith control point and the integral is taken over the complete jth panel:

$$I_{i,j} = \int_j \frac{\partial}{\partial n_i} (\ln r_{ij})\, ds_j \tag{3.158}$$

Since

$$r_{ij} = \sqrt{(x_i - x_j)^2 + (y_i - y_j)^2}$$

then

$$\frac{\partial}{\partial n_i} (\ln r_{ij}) = \frac{1}{r_{ij}} \frac{\partial r_{ij}}{\partial n_i}$$

$$= \frac{1}{r_{ij}} \frac{1}{2} [(x_i - x_j)^2 + (y_i - y_j)^2]^{-1/2}$$

$$\cdot \left[2(x_i - x_j)\frac{dx_i}{dn_i} + 2(y_i - y_j)\frac{dy_i}{dn_i} \right]$$

or

$$\frac{\partial}{\partial n_i} (\ln r_{ij}) = \frac{(x_i - x_j)\cos \beta_i + (y_i - y_j)\sin \beta_i}{(x_i - x_j)^2 + (y_i - y_j)^2} \tag{3.159}$$

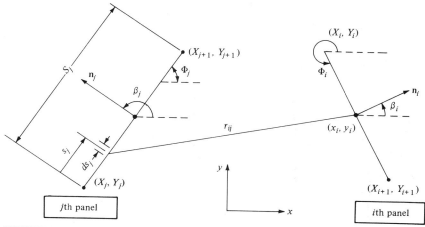

FIGURE 3.37
Geometry required for the evaluation of I_{ij}.

Note in Fig. 3.37 that Φ_i and Φ_j are angles measured in the counterclockwise direction from the x axis to the bottom of each panel. From this geometry,

$$\beta_i = \Phi_i + \frac{\pi}{2}$$

hence,

$$\sin \beta_i = \cos \Phi_i \qquad (3.160a)$$

$$\cos \beta_i = -\sin \Phi_i \qquad (3.160b)$$

Also, from the geometry of Fig. 3.33, we have

$$x_j = X_j + s_j \cos \Phi_j \qquad (3.161a)$$

and

$$y_j = Y_j + s_j \sin \Phi_j \qquad (3.161b)$$

Substituting Eqs. (3.159) to (3.161) into (3.158), we obtain

$$I_{i,j} = \int_0^{S_j} \frac{Cs_j + D}{s_j^2 + 2As_j + B} \, ds_j \qquad (3.162)$$

where

$$A = -(x_i - X_j) \cos \Phi_j - (y_i - Y_j) \sin \Phi_j$$

$$B = (x_i - X_j)^2 + (y_i - Y_j)^2$$

$$C = \sin(\Phi_i - \Phi_j)$$

$$D = (y_i - Y_j) \cos \Phi_i - (x_i - X_j) \sin \Phi_i$$

$$S_j = \sqrt{(X_{j+1} - X_j)^2 + (Y_{j+1} - Y_j)^2}$$

Letting

$$E = \sqrt{B - A^2} = (x_i - X_j) \sin \Phi_j - (y_i - Y_j) \cos \Phi_j$$

we obtain an expression for Eq. (3.162) from any standard table of integrals:

$$I_{i,j} = \frac{C}{2} \ln \left(\frac{S_j^2 + 2AS_j + B}{B} \right)$$

$$+ \frac{D - AC}{E} \left(\tan^{-1} \frac{S_j + A}{E} - \tan^{-1} \frac{A}{E} \right) \qquad (3.163)$$

Equation (3.163) is a general expression for two arbitrarily oriented panels; it is not restricted to the case of a circular cylinder.

We now apply Eq. (3.163) to the circular cylinder shown in Fig. 3.36. For purposes of illustration, let us choose panel 4 as the ith panel and panel 2 as the jth panel; i.e., let us calculate $I_{4,2}$. From the geometry of Fig. 3.36, assuming a unit radius for the cylinder, we see that

$$X_j = -0.9239 \qquad X_{j+1} = -0.3827 \qquad Y_j = 0.3827$$

$$Y_{j+1} = 0.9239 \qquad \Phi_i = 315° \qquad \Phi_j = 45°$$

$$x_i = 0.6533 \qquad y_i = 0.6533$$

Hence, substituting these numbers into the above formulas, we obtain

$$A = -1.3065 \qquad B = 2.5607 \qquad C = -1 \qquad D = 1.3065$$

$$S_j = 0.7654 \qquad E = 0.9239$$

Inserting the above values into Eq. (3.163), we obtain

$$I_{4,2} = 0.4018$$

Return to Figs. 3.36 and 3.37. If we now choose panel 1 as the jth panel, keeping panel 4 as the ith panel, we obtain, by means of a similar calculation, $I_{4,1} = 0.4074$. Similarly, $I_{4,3} = 0.3528$, $I_{4,5} = 0.3528$, $I_{4,6} = 0.4018$, $I_{4,7} = 0.4074$, and $I_{4,8} = 0.4084$.

Return to Eq. (3.153), which is evaluated for the ith panel. Written for panel 4, Eq. (3.153) becomes (after multiplying each term by 2 and noting that $\beta_i = 45°$ for panel 4)

$$0.4074\lambda_1 + 0.4018\lambda_2 + 0.3528\lambda_3 + \pi\lambda_4 + 0.3528\lambda_5$$

$$+ 0.4018\lambda_6 + 0.4074\lambda_7 + 0.4084\lambda_8 = -0.7071\, 2\pi V_\infty \qquad (3.164)$$

Equation (3.164) is a linear algebraic equation in terms of the eight unknowns, $\lambda_1, \lambda_2, \ldots, \lambda_8$. If we now evaluate Eq. (3.153) for each of the seven other panels, we obtain a total of eight equations, including Eq. (3.164), which can be solved simultaneously for the eight unknown λ's. The results are

$$\lambda_1/2\pi V_\infty = 0.3765 \qquad \lambda_2/2\pi V_\infty = 0.2662 \qquad \lambda_3/2\pi V_\infty = 0$$

$$\lambda_4/2\pi V_\infty = -0.2662 \qquad \lambda_5/2\pi V_\infty = -0.3765 \qquad \lambda_6/2\pi V_\infty = -0.2662$$

$$\lambda_7/2\pi V_\infty = 0 \qquad \lambda_8/2\pi V_\infty = 0.2662$$

Note the symmetrical distribution of the λ's, which is to be expected for the nonlifting circular cylinder. Also, as a check on the above solution, return to Eq. (3.157). Since each panel in Fig. 3.36 has the same length, Eq. (3.157) can be written simply as

$$\sum_{j=1}^{n} \lambda_j = 0$$

Substituting the values for the λ's obtained above into Eq. (3.157), we see that the equation is identically satisfied.

The velocity at the control point of the ith panel can be obtained from Eq. (3.156). In that equation, the integral over the jth panel is a geometric quantity which is evaluated in a similar manner as before. The result is

$$\int_j \frac{\partial}{\partial s} (\ln r_{ij})\, ds_j = \frac{D - AC}{2E} \ln \frac{S_j^2 + 2AS_j + B}{B}$$

$$- C\left(\tan^{-1}\frac{S_j + A}{E} - \tan^{-1}\frac{A}{E}\right) \qquad (3.165)$$

With the integrals in Eq. (3.156) evaluated by Eq. (3.165), and with the values for $\lambda_1, \lambda_2, \ldots, \lambda_8$ obtained above inserted into Eq. (3.156), we obtain the velocities $V_1, V_2, \ldots, V_8$. In turn, the pressure coefficients $C_{p,1}, C_{p,2}, \ldots, C_{p,8}$ are obtained directly from

$$C_{p,i} = 1 - \left(\frac{V_i}{V_\infty}\right)^2$$

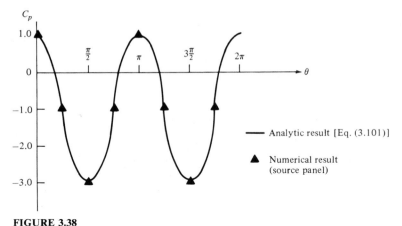

FIGURE 3.38
Pressure distribution over a circular cylinder; comparison of the source panel results and theory.

Results for the pressure coefficients obtained from this calculation are compared with the exact analytical result, Eq. (3.101) in Fig. 3.38. Amazingly enough, in spite of the relatively crude paneling shown in Fig. 3.36, the numerical pressure coefficient results are excellent.

3.18 APPLIED AERODYNAMICS: THE FLOW OVER A CIRCULAR CYLINDER—THE REAL CASE

The inviscid, incompressible flow over a circular cylinder was treated in Sec. 3.13. The resulting theoretical streamlines are sketched in Fig. 3.21, characterized by a symmetrical pattern where the flow "closes in" behind the cylinder. As a result, the pressure distribution over the front of the cylinder is the same as that over the rear (see Fig. 3.24). This leads to the theoretical result that the pressure drag is zero—d'Alembert's paradox.

The real flow over a circular cylinder is quite different from that studied in Sec. 3.13, the difference due to the influence of friction. Moreover, the drag coefficient for the real flow over a cylinder is certainly not zero. For a *viscous* incompressible flow, the results of dimensional analysis (Sec. 1.7) clearly demonstrate that the drag coefficient is a function of the Reynolds number. The variation of $C_D = f(\text{Re})$ for a circular cylinder is shown in Fig. 3.39, which is based on a wealth of experimental data. Here, $\text{Re} = (\rho_\infty V_\infty d)/\mu_\infty$, where d is the diameter of the cylinder. Note that C_D is very large for the extremely small values of $\text{Re} < 1$, but decreases monotonically until $\text{Re} \approx 300,000$. At this Reynolds number, there is a precipitous drop of C_D from a value near 1 to about 0.3, then a slight recovery to about 0.6 for $\text{Re} = 10^7$. (*Note:* These results are consistent with the comparison shown in Fig. 1.32d and e, contrasting C_D for a circular cylinder at low and high Re.) What causes this precipitous drop in C_D when the Reynolds number reaches about 300,000? A detailed answer must await our discussion of

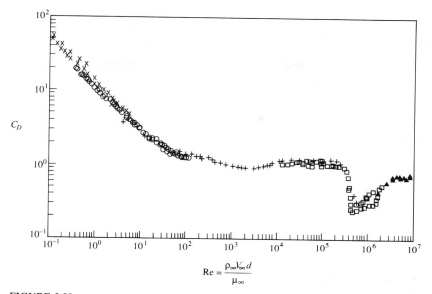

FIGURE 3.39

Variation of cylinder-drag coefficient with Reynolds number. (*Experimental data as compiled in Panton, Ronald*, Incompressible Flow, *Wiley-Interscience, New York, 1984.*)

viscous flow in Part IV. However, we state now that the phenomenon is caused by a sudden transition of laminar flow within the boundary layer at the lower values of Re to a turbulent boundary layer at the higher values of Re. Why does a turbulent boundary layer result in a smaller C_D for this case? Stay tuned; the answer is given in Part IV.

The variation of C_D shown in Fig. 3.39 across a range of Re from 10^{-1} to 10^7 is accompanied by tremendous variations in the qualitative aspects of the flow field, as itemized below, and as sketched in Fig. 3.40.

1. For very low values of Re, say, $0 < Re < 4$, the streamlines are almost (but not exactly) symmetrical, and the flow is attached, as sketched in Fig. 3.40*a*. This regime of viscous flow is called *Stokes flow*; it is characterized by a near balance of pressure forces with friction forces acting on any given fluid element; the flow velocity is so low that inertia effects are very small. A photograph of this type of flow is shown in Fig. 3.41, which shows the flow of water around a circular cylinder where Re = 1.54. The streamlines are made visible by aluminum powder on the surface, along with a time exposure of the film.

2. For $4 < Re < 40$, the flow becomes separated on the back of the cylinder, forming two distinct, stable vortices that remain in the position shown in Fig. 3.40*b*. A photograph of this type of flow is given in Fig. 3.42, where Re = 26.

3. As Re is increased above 40, the flow behind the cylinder becomes unstable; the vortices which were in a fixed position in Fig. 3.40*b* now are alternately

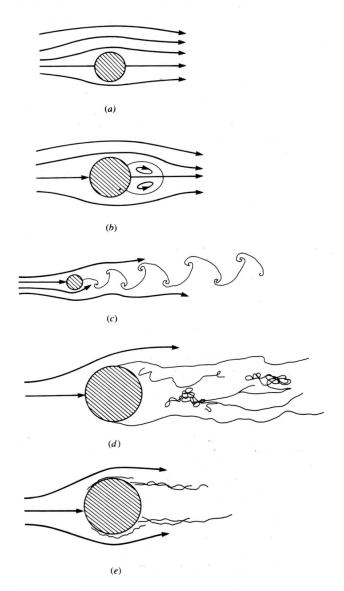

FIGURE 3.40
Various types of flow over a circular cylinder. (*From Panton, Ronald,* Incompressible Flow, *Wiley-Interscience, New York, 1984.*)

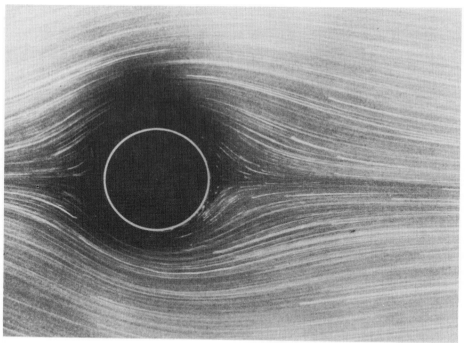

FIGURE 3.41
Flow over a circular cylinder. Re = 1.54. (*Photograph by Sadatoshi Taneda, from Van Dyke, Milton,* An Album of Fluid Motion, *The Parabolic Press, Stanford, Calif., 1982.*)

shed from the body in a regular fashion and flow downstream. This flow is sketched in Fig. 3.40c. A photograph of this type of flow is shown in Fig. 3.43, where Re = 140. This is a water flow where the streaklines are made visible by the electrolytic precipitation method. (In this method, metal plating on the cylinder surface acts as an anode, white particles are precipitated by electrolysis near the anode, and these particles subsequently flow downstream, forming a *streakline*. The definition of a streakline is given in Sec. 2.11.) The alternately shed vortex pattern shown in Figs. 3.40c and 3.43 is called a *Karman vortex street*, named after Theodore von Karman, who began to study and analyze this pattern in 1911 while at Göttingen University in Germany. (von Karman subsequently had a long and very distinguished career in aerodynamics, moving to the California Institute of Technology in 1933, and becoming America's best-known aerodynamicist in the mid-twentieth century. An autobiography of von Karman was published in 1967; see Ref. 49. This reference is "must" reading for anyone interested in a riveting perspective on the history of aerodynamics in the twentieth century.)

4. As the Reynolds number is increased to large numbers, the Karman vortex street becomes turbulent and begins to metamorphose into a distinct wake. The laminar boundary layer on the cylinder separates from the surface on the

FIGURE 3.42
Flow over a circular cylinder. Re = 26. (*Photograph by Sadatoshi Taneda, from Van Dyke, Milton,* An Album of Fluid Motion, *The Parabolic Press, Stanford, Calif., 1982.*)

FIGURE 3.43
Flow over a circular cylinder. Re = 140. A Karman vortex street exists behind the cylinder at this Reynolds number. (*Photograph by Sadatoshi Taneda, from Van Dyke, Milton,* An Album of Fluid Motion, *The Parabolic Press, Stanford, Calif., 1982.*)

forward face, at a point about 80° from the stagnation point. This is sketched in Fig. 3.40d. The value of the Reynolds number for this flow is on the order of 10^5. Note, from Fig. 3.39, that C_D is a relatively constant value near unity for $10^3 < \text{Re} < 3 \times 10^5$.

5. For $3 \times 10^5 < \text{Re} < 3 \times 10^6$, the separation of the laminar boundary layer still takes place on the forward face of the cylinder. However, in the free shear layer over the top of the separated region, transition to turbulent flow takes place. The flow then reattaches on the back face of the cylinder, but separates again at about 120° around the body measured from the stagnation point. This flow is sketched in Fig. 3.40e. This transition to turbulent flow, and the correspondingly thinner wake (comparing Fig. 3.40e with Fig. 3.40d), reduces the pressure drag on the cylinder and is responsible for the precipitous drop in C_D at $\text{Re} = 3 \times 10^5$ shown in Fig. 3.39. (More details on this phenomenon are covered in Part IV.)

6. For $\text{Re} < 3 < 10^6$, the boundary layer transits directly to turbulent flow at some point on the forward face, and the boundary layer remains totally attached over the surface until it separates at an angular location slightly less than 120° on the back surface. For this regime of flow, C_D actually increases slightly with increasing Re because the separation points on the back surface begin to move closer to the top and bottom of the cylinder, producing a fatter wake, and hence larger pressure drag.

In summary, from the photographs and sketches in this section, we see that the real flow over a circular cylinder is dominated by friction effects, namely, the separation of the flow over the rearward face of the cylinder. In turn, a finite pressure drag is created on the cylinder, and d'Alembert's paradox is resolved.

Let us examine the production of drag more closely. The theoretical pressure distribution over the surface of a cylinder in an inviscid, incompressible flow was given in Fig. 3.24. In contrast, several real pressure distributions based on experimental measurements for different Reynolds numbers are shown in Fig. 3.44, and are compared with the theoretical inviscid flow results obtained in Sec. 3.13. Note that theory and experiment agree well on the forward face of the cylinder, but that dramatic differences occur over the rearward face. The theoretical results show the pressure decreasing around the forward face from the initial total pressure at the stagnation point, reaching a minimum pressure at the top and bottom of the cylinder ($\theta = 90°$ and 270°), and then increasing again over the rearward face, recovering to the total pressure at the rear stagnation point. In contrast, in the real case where flow separation occurs, the pressures are relatively constant in the separated region over the rearward face and have values slightly less than freestream pressure. (In regions of separated flow, the pressure frequently exhibits a nearly constant value.) In the separated region over the rearward face, the pressure clearly does not recover to the higher values that exist on the front face. There is a net imbalance of the pressure distribution between the front and back faces, with the pressures on the front being higher than on the back, and this imbalance produces the drag on the cylinder.

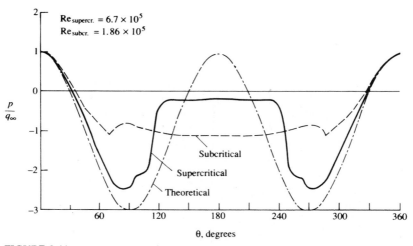

FIGURE 3.44
Pressure distribution over a circular cylinder in low-speed flow. Comparison of the theoretical pressure distribution with two experimental pressure distributions—one for a subcritical Re and the other for a supercritical Re.

Return to Fig. 3.39, and examine again the variation of C_D as a function of Re. The regimes associated with the very low Reynolds numbers, such as Stokes flow for Re ≈ 1, are usually of no interest to aeronautical applications. For example, consider a circular cylinder in an airflow of 30 m/s (about 100 ft/s, or 68 mi/h) at standard sea level conditions, where $\rho_\infty = 1.23$ kg/m^3 and $\mu_\infty = 1.79 \times 10^{-5}$ kg/(m · s). The smaller the diameter of the cylinder, the smaller will be the Reynolds number. *Question*: What is the required cylinder diameter in order to have Re $= 1$? The answer is obtained from

$$\text{Re} = \frac{\rho_\infty V_\infty d}{\mu_\infty} = 1$$

Hence,

$$d = \frac{\mu_\infty}{\rho_\infty V_\infty} = \frac{1.79 \times 10^{-5}}{(1.23)(30)} = 4 \times 10^{-7} \text{ m}$$

To have Re $= 1$ for the above conditions, the diameter of the cylinder would have to be extremely small; note that the value of $d = 4 \times 10^{-7}$ m is only slightly larger than the mean free path at standard sea level, which is 6.6×10^{-8} m. (See Sec. 1.10 for the definition of the mean free path.) Clearly, Reynolds numbers on the order of unity are of little practical aerodynamic importance.

If this is so, then what values of Re for the flow over cylinders are of practical importance? For one such example, consider the wing wires between the upper and lower wings on a World War I biplane, such as the SPAD XIII shown in Fig. 3.45. The diameter of these wires is about $\frac{3}{32}$ in, or 0.0024 m. The

FIGURE 3.45
The French SPAD XIII, an example of a strut-and-wire biplane from World War I. Captain Eddie Rickenbacker is shown at the front of the airplane. (*Courtesy of U.S. Air Force.*)

top speed of the SPAD was 130 mi/h, or 57.8 m/s. For this velocity at standard sea level, we have

$$\text{Re} = \frac{\rho_\infty V_\infty d}{\mu_\infty} = \frac{(1.23)(57.8)(0.0024)}{1.79 \times 10^{-5}} = 9532$$

With this value of Re, we are beginning to enter the world of practical aerodynamics for the flow over cylinders. It is interesting to note that, from Fig. 3.39, $C_D = 1$ for the wires on the SPAD. In terms of airplane aerodynamics, this is a high drag coefficient for any component of an aircraft. Indeed, the bracing wires used on biplanes of the World War I era were a source of high drag for the aircraft, so much so that early in the war, bracing wire with a symmetric airfoil-like cross section was utilized to help reduce this drag. Such wire was developed by the British at the Royal Aircraft Factory at Farnborough, and was first tested experimentally as early as 1914 on an SE-4 biplane. Interestingly enough, the SPAD used ordinary round wire, and in spite of this was the fastest of all World War I aircraft.

This author was struck by another example of the effect of cylinder drag while traveling in Charleston, South Carolina, shortly after hurricane Hugo devastated the area on September 28, 1989. Traveling north out of Charleston on U.S. Route 17, near the small fishing town of McClellanville, one passes through the Francis Marion National Forest. This forest was virtually destroyed by the hurricane; 60-ft pine trees were snapped off near their base, and approximately 8 out of every 10 trees were down. The sight bore an eerie resemblance to scenes from the battlefields in France during World War I. What type of force can destroy an entire forest in this fashion? To answer this question, we note that the Weather Bureau measured wind gusts as high as 175 mi/h during the hurricane. Let us approximate the wind force on a typical 60-ft pine tree by the aerodynamic drag on a cylinder of a length of 60 ft and a diameter of 5 ft. Since $V = 175$ mi/h $= 256.7$ ft/s, $\rho_\infty = 0.002377$ slug/ft^3, and $\mu_\infty = 3.7373 \times 10^{-7}$ slug/(ft · s), then the Reynolds number is

$$\text{Re} = \frac{\rho_\infty V_\infty d}{\mu_\infty} = \frac{(0.002377)(256.7)(5)}{3.7373 \times 10^{-7}} = 8.16 \times 10^6$$

Examining Fig. 3.39, we see that $C_D = 0.7$. Since C_D is based on the drag per unit length of the cylinder as well as the projected frontal area, we have for the total drag exerted on an entire tree that is 60 ft tall

$$D = q_\infty S C_D = \tfrac{1}{2}\rho_\infty V_\infty^2 (d)(60) C_D$$

$$= \tfrac{1}{2}(0.002377)(256.7)^2(5)(60)(0.7) = 16{,}446 \text{ lb}$$

a 16,000-lb force on the tree—it is no wonder a whole forest was destroyed. (In the above analysis, we neglected the end effects of the flow over the end of the vertical cylinder. Moreover, we did not correct the standard sea level density for the local reduction in barometric pressure experienced inside a hurricane. However, these are relatively small effects in comparison to the overall force on the cylinder.) The aerodynamics of a tree, and especially that of a forest, is more sophisticated than discussed here. Indeed, the aerodynamics of trees have been studied experimentally with trees actually mounted in a wind tunnel.[†]

3.19 HISTORICAL NOTE: BERNOULLI AND EULER—THE ORIGINS OF THEORETICAL FLUID DYNAMICS

Bernoulli's equation, expressed by Eqs. (3.14) and (3.15), is historically the most famous equation in fluid dynamics. Moreover, we derived Bernoulli's equation from the general momentum equation in partial differential equation form. The momentum equation is just one of the three fundamental equations of fluid dynamics—the others being continuity and energy. These equations are derived and discussed in Chap. 2 and applied to an incompressible flow in Chap. 3. Where did these equations first originate? How old are they, and who is responsible for them? Considering the fact that all of fluid dynamics in general, and

---

[†] For more details, see the interesting discussion on forest aerodynamics in the book by John E. Allen entitled *Aerodynamics, The Science of Air in Motion*, McGraw-Hill, New York, 1982.

aerodynamics in particular, is built on these fundamental equations, it is important to pause for a moment and examine their historical roots.

As discussed in Sec. 1.1, Isaac Newton, in his *Principia* of 1687, was the first to establish on a rational basis the relationships between force, momentum, and acceleration. Although he tried, he was unable to apply these concepts properly to a moving fluid. The real foundations of theoretical fluid dynamics were not laid until the next century—developed by a triumvirate consisting of David Bernoulli, Leonhard Euler, and Jean Le Rond d'Alembert.

First, consider Bernoulli. Actually, we must consider the whole family of Bernoulli's because Daniel Bernoulli was a member of a prestigious family that dominated European mathematics and physics during the early part of the eighteenth century. Figure 3.46 is a portion of the Bernoulli family tree. It starts with Nikolaus Bernoulli, who was a successful merchant and druggist in Basel, Switzerland, during the seventeenth century. With one eye on this family tree, let us simply list some of the subsequent members of this highly accomplished family:

1. Jakob—Daniel's uncle. Mathematician and physicist, he was professor of mathematics at the University of Basel. He made major contributions to the development of calculus and coined the term "integral."
2. Johann—Daniel's father. He was a professor of mathematics at Groningen, Netherlands, and later at the University of Basel. He taught the famous French mathematician L'Hospital the elements of calculus, and after the death of Newton in 1727 he was considered Europe's leading mathematician at that time.
3. Nikolaus—Daniel's cousin. He studied mathematics under his uncles and held a master's degree in mathematics and a doctor of jurisprudence.

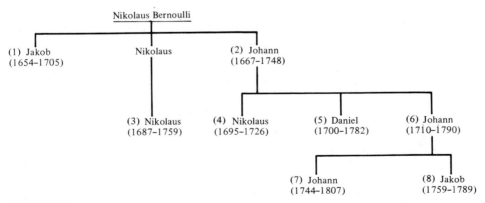

FIGURE 3.46
Bernoulli's family tree.

4. Nikolaus—Daniel's brother. He was Johann's favorite son. He held a master of arts degree, and assisted with much of Johann's correspondence to Newton and Liebniz concerning the development of calculus.

5. Daniel himself—to be discussed below.

6. Johann—Daniel's other brother. He succeeded his father in the Chair of Mathematics at Basel and won the prize of the Paris Academy four times for his work.

7. Johann—Daniel's nephew. A gifted child, he earned the master of jurisprudence at the age of 14. When he was 20, he was invited by Frederick II to reorganize the astronomical observatory at the Berlin Academy.

8. Jakob—Daniel's other nephew. He graduated in jurisprudence but worked in mathematics and physics. He was appointed to the Academy in St. Petersburg, Russia, but he had a promising career prematurely ended when he drowned in the river Neva at the age of 30.

With such a family pedigree, Daniel Bernoulli was destined for success.

Daniel Bernoulli was born in Groningen, Netherlands, on February 8, 1700. His father, Johann, was a professor at Groningen but returned to Basel, Switzerland, in 1705 to occupy the Chair of Mathematics which had been vacated by the death of Jacob Bernoulli. At the University of Basel, Daniel obtained a master's degree in 1716 in philosophy and logic. He went on to study medicine in Basel, Heidelburg, and Strasbourg, obtaining his Ph.D. in anatomy and botany in 1721. During these studies, he maintained an active interest in mathematics. He followed this interest by moving briefly to Venice, where he published an important work entitled *Exercitationes Mathematicae* in 1724. This earned him much attention and resulted in his winning the prize awarded by the Paris Academy—the first of 10 he was eventually to receive. In 1725, Daniel moved to St. Petersburg, Russia, to join the academy. The St. Petersburg Academy had gained a substantial reputation for scholarship and intellectual accomplishment at that time. During the next 8 years, Bernoulli experienced his most creative period. While at St. Petersburg, he wrote his famous book *Hydrodynamica*, completed in 1734, but not published until 1738. In 1733, Daniel returned to Basel to occupy the Chair of Anatomy and Botany, and in 1750 moved to the Chair of Physics created exclusively for him. He continued to write, give very popular and well-attended lectures in physics, and make contributions to mathematics and physics until his death in Basel on March 17, 1782.

Daniel Bernoulli was famous in his own time. He was a member of virtually all the existing learned societies and academies, such as Bologna, St. Petersburg, Berlin, Paris, London, Bern, Turin, Zurich, and Mannheim. His importance to fluid dynamics is centered on his book *Hydrodynamica* (1738). (With this book, Daniel introduced the term "hydrodynamics" to the literature.) In this book, he ranged over such topics as jet propulsion, manometers, and flow in pipes. Of most importance, he attempted to obtain a relationship between pressure and velocity. Unfortunately, his derivation was somewhat obscure, and Bernoulli's equation, ascribed by history to Daniel via his *Hydrodynamica*, is not to be found in this book, at least not in the form we see it today [such as Eqs. (3.14) and (3.15)]. The propriety of Eqs. (3.14) and (3.15) is further complicated by his father, Johann, who also published a book in 1743 entitled *Hydraulica*. It is clear from this latter book that the father understood Bernoulli's theorem better than his son; Daniel thought of pressure strictly in terms of the height of a manometer column, whereas Johann had the more fundamental understanding that pressure was a force acting on the fluid. (It is interesting to note that Johann Bernoulli was a person of some sensitivity and irritability, with an overpowering drive for recognition. He tried to undercut the impact of Daniel's *Hydrodynamica* by predating the publication date of *Hydraulica* to 1728, to make it appear to have been the first of the two. There was little love lost between son and father.)

During Daniel Bernoulli's most productive years, partial differential equations had not yet been introduced into mathematics and physics; hence, he could not approach the derivation of Bernoulli's equation in the same fashion as we have in Sec. 3.2. The introduction of partial differential equations to mathematical physics was due to d'Alembert in 1747. d'Alembert's role in fluid mechanics is detailed in Sec. 3.20. Suffice it to say here that his contributions were equally if not more important than Bernoulli's, and d'Alembert represents the second member of the triumvirate which molded the foundations of theoretical fluid dynamics in the eighteenth century.

The third and probably pivotal member of this triumvirate was Leonhard Euler. He was a giant among eighteenth-century mathematicians and scientists. As a result of his contributions, his name is associated with numerous equations and techniques, e.g., the Euler numerical solution of ordinary differential equations, eulerian angles in geometry, and the momentum equations for inviscid fluid flow [see Eq. (3.12)].

Leonhard Euler was born on April 15, 1707, in Basel, Switzerland. His father was a Protestant minister who enjoyed mathematics as a pastime. Therefore, Euler grew up in a family atmosphere that encouraged intellectual activity. At the age of 13, Euler entered the University of Basel which at that time had about 100 students and 19 professors. One of those professors was Johann Bernoulli, who tutored Euler in mathematics. Three years later, Euler received his master's degree in philosophy.

It is interesting that three of the people most responsible for the early development of theoretical fluid dynamics—Johann and Daniel Bernoulli and Euler—lived in the same town of Basel, were associated with the same university, and were contemporaries. Indeed, Euler and the Bernoulli's were close and respected friends—so much that, when Daniel Bernoulli moved to teach and study at the St. Petersburg Academy in 1725, he was able to convince the academy to hire Euler as well. At this invitation, Euler left Basel for Russia; he never returned to Switzerland, although he remained a Swiss citizen throughout his life.

Euler's interaction with Daniel Bernoulli in the development of fluid mechanics grew strong during these years at St. Petersburg. It was here that Euler conceived of pressure as a point property that can vary from point to point throughout a fluid and obtained a differential equation relating pressure and velocity, i.e., *Euler's equation* given by Eq. (3.12). In turn, Euler integrated the differential equation to obtain, for the first time in history, Bernoulli's equation in the form of Eqs. (3.14) and (3.15). Hence, we see that Bernoulli's equation is really a misnomer; credit for it is legitimately shared by Euler.

When Daniel Bernoulli returned to Basel in 1733, Euler succeeded him at St. Petersburg as a professor of physics. Euler was a dynamic and prolific man; by 1741 he had prepared 90 papers for publication and written the two-volume book *Mechanica.* The atmosphere surrounding St. Petersburg was conducive to such achievement. Euler wrote in 1749: "I and all others who had the good fortune to be for some time with the Russian Imperial Academy cannot but acknowledge that we owe everything which we are and possess to the favorable conditions which we had there."

However, in 1740, political unrest in St. Petersburg caused Euler to leave for the Berlin Society of Sciences, at that time just formed by Frederick the Great. Euler lived in Berlin for the next 25 years, where he transformed the society into a major academy. In Berlin, Euler continued his dynamic mode of working, preparing at least 380 papers for publication. Here, as a competitor with d'Alembert (see Sec. 3.20), Euler formulated the basis for mathematical physics.

In 1766, after a major disagreement with Frederick the Great over some financial aspects of the academy, Euler moved back to St. Petersburg. This second period of his life in Russia became one of physical suffering. In that same year, he became blind in one eye after a short illness. An operation in 1771 resulted in restoration of his sight, but only for a few days. He did not take proper precautions after the operation, and within a few days, he was completely blind. However, with the help of others, he continued his work. His mind was as sharp as ever, and his spirit did not diminish. His literary output even increased—about half of his total papers were written after 1765!

On September 18, 1783, Euler conducted business as usual—giving a mathematics lesson, making calculations of the motion of balloons, and discussing with friends the planet of Uranus, which had recently been discovered. At about 5 P.M., he suffered a brain hemorrhage. His only words before losing consciousness were "I am dying." By 11 P.M., one of the greatest minds in history had ceased to exist.

With the lives of Bernoulli, Euler, and d'Alembert (see Sec. 3.20) as background, let us now trace the geneology of the basic equations of fluid dynamics. For example, consider the continuity equation in the form of Eq. (2.43). Although Newton had postulated the obvious fact that the mass of a specified object was constant, this principle was not appropriately applied to fluid mechanics until 1749. In this year, d'Alembert gave a paper in Paris, entitled "Essai d'une nouvelle theorie de la resistance des fluides," in which he formulated differential equations for the conservation of mass in special applications to plane and axisymmetric flows. Euler took d'Alembert's results and, 8 years later, generalized them in a series of three basic papers on fluid mechanics. In these papers, Euler published, for the first time in history, the continuity equation in the form of Eq. (2.43) and the momentum equations in the form of Eqs. (2.104a to c), without the viscous terms. Hence, two of the three basic conservation equations used today in modern fluid dynamics were well established long before the American Revolutionary War—such equations were contemporary with the time of George Washington and Thomas Jefferson!

The origin of the energy equation in the form of Eq. (2.87) without viscous terms has its roots in the development of thermodynamics in the nineteenth century. Its precise first use is obscure and is buried somewhere in the rapid development of physical science in the nineteenth century.

The purpose of this section has been to give you some feeling for the historical development of the fundamental equations of fluid dynamics. Maybe we can appreciate these equations more when we recognize that they have been with us for quite some time and that they are the product of much thought from some of the greatest minds of the eighteenth century.

3.20 HISTORICAL NOTE: d'ALEMBERT AND HIS PARADOX

You can well imagine the frustration that Jean le Rond d'Alembert felt in 1744 when, in a paper entitled "Traite de l'equilibre et des mouvements de fluids pour servir de siute au traite de dynamique," he obtained the result of zero drag for the inviscid, incompressible flow over a closed two-dimensional body. Using different approaches, d'Alembert encountered this result again in 1752 in his paper entitled "Essai sur la resistance" and again in 1768 in his "Opuscules mathematiques." In this last paper can be found the quote given at the beginning of Chap. 15; in essence, he had given up trying to explain the cause of this paradox. Even though the prediction of fluid-dynamic drag was a very important problem in d'Alembert's time, and in spite of the number of great minds that addressed it, the fact that viscosity is responsible for drag was not appreciated. Instead, d'Alembert's analyses used momentum principles in a frictionless flow, and quite naturally he found that the flow field closed smoothly around the downstream portion of the bodies, resulting in zero drag. Who was this man, d'Alembert? Considering the role his paradox played in the development of fluid dynamics, it is worth our time to take a closer look at the man himself.

d'Alembert was born illegitimately in Paris on November 17, 1717. His mother was Madame De Tenun, a famous salon hostess of that time, and his father was Chevalier Destouches-Canon, a cavalry officer. d'Alembert was immediately abandoned by his mother (she was an ex-nun who was afraid of being forcibly returned to the convent). However, his father quickly arranged for a home for d'Alembert—with a family of modest means named Rousseau. d'Alembert lived with this family for the next 47 years. Under the support of his father, d'Alembert was educated at the College de Quatre-Nations, where he studied law and medicine, and later turned to mathematics. For the remainder of his life, d'Alembert would consider himself a mathematician. By a program of self-study, d'Alembert learned the works of Newton and the Bernoulli's. His early mathematics caught the attention of the Paris Academy of Sciences, of which he became a member in 1741. d'Alembert published frequently and sometimes rather hastily, in order to be in print before his competition. However, he made substantial contributions to the science of his time. For example, he was (1) the first to formulate the wave equation of classical physics, (2) the first to express the concept of a partial differential equation, (3) the first to solve a partial differential equation—he used separation of variables—and (4) the first to express the differential equations of fluid dynamics in terms of a field. His contemporary, Leonhard Euler (see Secs. 1.1 and 3.18) later expanded greatly on these equations and was responsible for developing them into a truly rational approach for fluid-dynamic analysis.

During the course of his life, d'Alembert became interested in many scientific and mathematical subjects, including vibrations, wave motion, and celestial mechanics. In the 1750s, he had the honored position of science editor for the *Encyclopedia*—a major French intellectual endeavor of the eighteenth century which attempted to compile all existing knowledge into a large series of books. As he grew older, he also wrote papers on nonscientific subjects, mainly musical structure, law, and religion.

In 1765, d'Alembert became very ill. He was helped to recover by the nursing of Mlle. Julie de Lespinasse, the woman who was d'Alembert's only love throughout his life. Although he never married, d'Alembert lived with Julie de Lespinasse until she died in 1776. d'Alembert had always been a charming gentleman, renowned for his intelligence, gaiety, and considerable conversational ability. However, after Mlle. de Lespinasse's death, he became frustrated and morose—living a life of despair. He died in this condition on October 29, 1783, in Paris.

d'Alembert was one of the great mathematicians and physicists of the eighteenth century. He maintained active communications and dialogue with both Bernoulli and Euler and ranks with them as one of the founders of modern fluid dynamics. This, then, is the man behind the paradox, which has existed as an integral part of fluid dynamics for the past two centuries.

3.21 SUMMARY

Return to the road map given in Fig. 3.4. Examine each block of the road map to remind yourself of the route we have taken in this discussion of the fundamentals of inviscid, incompressible flow. Before proceeding further, make certain that you feel comfortable with the detailed material represented by each block, and how each block is related to the overall flow of ideas and concepts.

For your convenience, some of the highlights of this chapter are summarized below:

Bernoulli's equation

$$p + \tfrac{1}{2}\rho V^2 = \text{const}$$

(a) Applies to inviscid, incompressible flows only.
(b) Holds along a streamline for a rotational flow.

(c) Holds at every point throughout an irrotational flow.
(d) In the form given above, body forces (such as gravity) are neglected, and steady flow is assumed.

Quasi-one-dimensional continuity equation

$$\rho A V = \text{const} \qquad \text{(for compressible flow)}$$

$$A V = \text{const} \qquad \text{(for incompressible flow)}$$

From a measurement of the Pitot pressure p_0 and static pressure p_1, the velocity of an incompressible flow is given by

$$V_1 = \sqrt{\frac{2(p_0 - p_1)}{\rho}} \tag{3.34}$$

Pressure coefficient

Definition:
$$C_p = \frac{p - p_\infty}{q_\infty} \tag{3.36}$$

where dynamic pressure is $q_\infty \equiv \frac{1}{2}\rho_\infty V_\infty^2$.

For incompressible steady flow with no friction:

$$C_p = 1 - \left(\frac{V}{V_\infty}\right)^2 \tag{3.38}$$

Governing equations

$$\nabla \cdot \mathbf{V} = 0 \qquad \text{(condition of incompressibility)} \tag{3.39}$$

$$\nabla^2 \phi = 0 \qquad \text{(Laplace's equation; holds for irrotational, incompressible flow)} \tag{3.40}$$

or
$$\nabla^2 \psi = 0 \tag{3.46}$$

Boundary conditions

$$u = \frac{\partial \phi}{\partial x} = \frac{\partial \psi}{\partial y} = V_\infty$$

at infinity

$$v = \frac{\partial \phi}{\partial y} = -\frac{\partial \psi}{\partial x} = 0$$

$$\mathbf{V} \cdot \mathbf{n} = 0 \qquad \text{at body (flow tangency condition)}$$

Elementary flows

(a) Uniform flow:

$$\phi = V_\infty x = V_\infty r \cos \theta \qquad (3.53)$$

$$\psi = V_\infty y = V_\infty r \sin \theta \qquad (3.55)$$

(b) Source flow:

$$\phi = \frac{\Lambda}{2\pi} \ln r \qquad (3.67)$$

$$\psi = \frac{\Lambda}{2\pi} \theta \qquad (3.72)$$

$$V_r = \frac{\Lambda}{2\pi r} \qquad V_\theta = 0 \qquad (3.62)$$

(c) Doublet flow:

$$\phi = \frac{\kappa}{2\pi} \frac{\cos \theta}{r} \qquad (3.88)$$

$$\psi = -\frac{\kappa}{2\pi} \frac{\sin \theta}{r} \qquad (3.87)$$

(d) Vortex flow:

$$\phi = -\frac{\Gamma}{2\pi} \theta \qquad (3.112)$$

$$\psi = \frac{\Gamma}{2\pi} \ln r \qquad (3.114)$$

$$V_\theta = -\frac{\Gamma}{2\pi r} \qquad V_r = 0 \qquad (3.105)$$

Inviscid flow over a cylinder

(a) Nonlifting (uniform flow and doublet)

$$\psi = (V_\infty r \sin \theta)\left(1 - \frac{R^2}{r^2}\right) \qquad (3.92)$$

where R = radius of cylinder = $\kappa / 2\pi V_\infty$.

Surface velocity: $\qquad V_\theta = -2 V_\infty \sin \theta \qquad (3.100)$

Surface pressure coefficient: $\qquad C_p = 1 - 4 \sin^2 \theta \qquad (3.101)$

$$L = D = 0$$

(b) Lifting (uniform flow + doublet + vortex)

$$\psi = (V_\infty r \sin \theta)\left(1 - \frac{R^2}{r^2}\right) + \frac{\Gamma}{2\pi} \ln \frac{r}{R} \qquad (3.118)$$

Surface velocity: $\qquad V_\theta = -2V_\infty \sin\theta - \dfrac{\Gamma}{2\pi R}$ (3.125)

$L' = \rho_\infty V_\infty \Gamma \qquad$ (lift per unit span) (3.140)

$D = 0$

Kutta-Joukowski theorem

For a closed two-dimensional body of arbitrary shape, the lift per unit span is $L' = \rho_\infty V_\infty \Gamma$.

Source panel method

This is a numerical method for calculating the nonlifting flow over bodies of arbitrary shape. Governing equations:

$$\frac{\lambda_i}{2} + \sum_{\substack{j=1 \\ (j \neq i)}}^{n} \frac{\lambda_j}{2\pi} \int_j \frac{\partial}{\partial n_i} (\ln r_{ij}) \, ds_j + V_\infty \cos\beta_i = 0 \qquad (i = 1, 2, \ldots, n) \quad (3.152)$$

PROBLEMS

Note: All the following problems assume an inviscid, incompressible flow. Also, standard sea level density and pressure are 1.23 kg/m^3 $(0.002377 \text{ slug/ft}^3)$ and $1.01 \times 10^5 \text{ N/m}^2$ (2116 lb/ft^2), respectively.

3.1. For an irrotational flow, show that Bernoulli's equation holds between *any* points in the flow, not just along a streamline.

3.2. Consider a venturi with a throat-to-inlet area ratio of 0.8, mounted on the side of an airplane fuselage. The airplane is in flight at standard sea level. If the static pressure at the throat is 2100 lb/ft^2, calculate the velocity of the airplane.

3.3. Consider a venturi with a small hole drilled in the side of the throat. This hole is connected via a tube to a closed reservoir. The purpose of the venturi is to create a vacuum in the reservoir when the venturi is placed in an airstream. (The *vacuum* is defined as the pressure difference *below* the outside ambient pressure.) The venturi has a throat-to-inlet area ratio of 0.85. Calculate the maximum vacuum obtainable in the reservoir when the venturi is placed in an airstream of 90 m/s at standard sea level conditions.

3.4. Consider a low-speed open-circuit subsonic wind tunnel with an inlet-to-throat area ratio of 12. The tunnel is turned on, and the pressure difference between the inlet (the settling chamber) and the test section is read as a height difference of 10 cm on a U-tube mercury manometer. (The density of liquid mercury is $1.36 \times 10^4 \text{ kg/m}^3$.) Calculate the velocity of the air in the test section.

3.5. Assume that a Pitot tube is inserted into the test-section flow of the wind tunnel in Prob. 3.4. The tunnel test section is completely sealed from the outside ambient pressure. Calculate the pressure measured by the Pitot tube, assuming the static pressure at the tunnel inlet is atmospheric.

3.6. A Pitot tube on an airplane flying at standard sea level reads 1.07×10^5 N/m^2. What is the velocity of the airplane?

3.7. At a given point on the surface of the wing of the airplane in Prob. 3.6, the flow velocity is 130 m/s. Calculate the pressure coefficient at this point.

3.8. Consider a uniform flow with velocity V_∞. Show that this flow is a physically possible incompressible flow and that it is irrotational.

3.9. Show that a source flow is a physically possible incompressible flow everywhere except at the origin. Also show that it is irrotational everywhere.

3.10. Prove that the velocity potential and the stream function for a uniform flow, Eqs. (3.53) and (3.55), respectively, satisfy Laplace's equation.

3.11. Prove that the velocity potential and the stream function for a source flow, Eqs. (3.67) and (3.72), respectively, satisfy Laplace's equation.

3.12. Consider the flow over a semi-infinite body as discussed in Sec. 3.11. If V_∞ is the velocity of the uniform stream, and the stagnation point is 1 ft upstream of the source:
(*a*) Draw the resulting semi-infinite body to scale on graph paper.
(*b*) Plot the pressure coefficient distribution over the body; i.e., plot C_p versus distance along the centerline of the body.

3.13. Derive Eq. (3.81). *Hint*: Make use of the symmetry of the flow field shown in Fig. 3.18; i.e., start with the knowledge that the stagnation points must lie on the axis aligned with the direction of V_∞.

3.14. Derive the velocity potential for a doublet; i.e., derive Eq. (3.88). *Hint*: The easiest method is to start with Eq. (3.87) for the stream function and extract the velocity potential.

3.15. Consider the nonlifting flow over a circular cylinder. Derive an expression for the pressure coefficient at an arbitrary point (r, θ) in this flow, and show that it reduces to Eq. (3.101) on the surface of the cylinder.

3.16. Consider the nonlifting flow over a circular cylinder of a given radius, where $V_\infty = 20$ ft/s. If V_∞ is doubled, i.e., $V_\infty = 40$ ft/s, does the shape of the streamlines change? Explain.

3.17. Consider the lifting flow over a circular cylinder of a given radius and with a given circulation. If V_∞ is doubled, keeping the circulation the same, does the shape of the streamlines change? Explain.

3.18. The lift on a spinning circular cylinder in a freestream with a velocity of 30 m/s and at standard sea level conditions is 6 N/m of span. Calculate the circulation around the cylinder.

3.19. A typical World War I biplane fighter (such as the French SPAD shown in Fig. 3.45) has a number of vertical interwing struts and diagonal bracing wires. Assume for a given airplane that the total length for the vertical struts (summed together) is 25 ft, and that the struts are cylindrical with a diameter of 2 in. Assume also that the total length of the bracing wires is 80 ft, with a cylindrical diameter of $\frac{3}{32}$ in. Calculate the drag (in pounds) contributed by these struts and bracing wires when the airplane is flying at 120 mi/h at standard sea level. Compare this component of drag with the total zero-lift drag for the airplane, for which the total wing area is 230 ft^2 and the zero-lift drag coefficient is 0.036.

CHAPTER
4

INCOMPRESSIBLE FLOW OVER AIRFOILS

Of the many problems now engaging attention, the following are considered of immediate importance and will be considered by the committee as rapidly as funds can be secured for the purpose.... The evolution of more efficient wing sections of practical form, embodying suitable dimensions for an economical structure, with moderate travel of the center-of-pressure and still affording a large range of angle-of-attack combined with efficient action.

*From the First Annual Report of the
NACA, 1915*

4.1 INTRODUCTION

With the advent of successful powered flight at the turn of the twentieth century, the importance of aerodynamics ballooned almost overnight. In turn, interest grew in the understanding of the aerodynamic action of such lifting surfaces as fixed wings on airplanes and, later, rotors on helicopters. In the period 1912–1918, the analysis of airplane wings took a giant step forward when Ludwig Prandtl and his colleagues at Göttingen, Germany, showed that the aerodynamic consideration of wings could be split into two parts: (1) the study of the *section* of a wing—an airfoil—and (2) the modification of such airfoil properties to account for the complete, finite wing. This approach is still used today; indeed, the theoretical calculation and experimental measurement of modern airfoil properties have been a major part of the aeronautics research carried out by the National Aeronautics and Space Administration (NASA) in the 1970s and 1980s. (See chap. 5 of Ref. 2 for a historical sketch on airfoil development and Ref. 10 for a description of modern airfoil research.) Following Prandtl's philosophy, the present chapter deals exclusively with airfoils, whereas Chap. 5 treats the case of a complete, finite wing. Therefore, in this chapter and Chap. 5, we make a major excursion into aerodynamics as applied to airplanes.

What is an airfoil? Consider a wing as drawn in perspective in Fig. 4.1. The wing extends in the y direction (the span direction). The freestream velocity V_∞

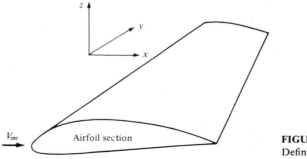

FIGURE 4.1
Definition of an airfoil.

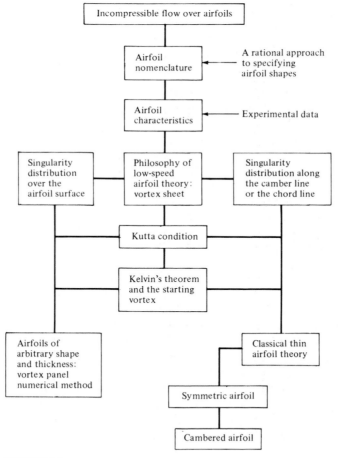

FIGURE 4.2
Road map for Chap. 4.

is parallel to the xz plane. Any section of the wing cut by a plane parallel to the xz plane is called an *airfoil*. The purpose of this chapter is to present theoretical methods for the calculation of airfoil properties. Since we are dealing with inviscid flow, we are not able to predict airfoil drag; indeed, d'Alembert's paradox says that the drag on an airfoil is zero—clearly not a realistic answer. We will have to wait until Chap. 15 and a discussion of viscous flow before predictions of drag can be made. However, the lift and moments on the airfoil are due mainly to the pressure distribution, which (below the stall) is dictated by inviscid flow. Therefore, this chapter concentrates on the theoretical prediction of airfoil lift and moments.

The road map for this chapter is given in Fig. 4.2. After some initial discussion on airfoil nomenclature and characteristics, we present two approaches to low-speed airfoil theory. One is the classical thin airfoil theory developed during the period 1910–1920 (the right-hand branch of Fig. 4.2). The other is the modern numerical approach for arbitrary airfoils using vortex panels (the left-hand branch of Fig. 4.2). Please refer to this road map as you work your way through this chapter.

4.2 AIRFOIL NOMENCLATURE

The first patented airfoil shapes were developed by Horatio F. Phillips in 1884. Phillips was an Englishman who carried out the first serious wind-tunnel experiments on airfoils. In 1902, the Wright brothers conducted their own airfoil tests in a wind tunnel, developing relatively efficient shapes which contributed to their successful first flight on December 17, 1903 (see Sec. 1.1). Clearly, in the early days of powered flight, airfoil design was basically customized and personalized. However, in the early 1930s, the National Advisory Committee for Aeronautics (NACA)—the forerunner of NASA—embarked on a series of definitive airfoil experiments using airfoil shapes that were constructed rationally and systematically. Many of these NACA airfoils are in common use today. Therefore, in this chapter we follow the nomenclature established by the NACA; such nomenclature is now a well-known standard.

Consider the airfoil sketched in Fig. 4.3. The *mean camber line* is the locus of points halfway between the upper and lower surfaces as measured perpendicular to the mean camber line itself. The most forward and rearward points of

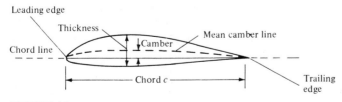

FIGURE 4.3
Airfoil nomenclature.

the mean camber line are the *leading* and *trailing edges*, respectively. The straight line connecting the leading and trailing edges is the *chord line* of the airfoil, and the precise distance from the leading to the trailing edge measured along the chord line is simply designated the *chord c* of the airfoil. The *camber* is the maximum distance between the mean camber line and the chord line, measured perpendicular to the chord line. The *thickness* is the distance between the upper and lower surfaces, also measured perpendicular to the chord line. The shape of the airfoil at the leading edge is usually circular, with a leading-edge radius of approximately $0.02c$. The shapes of all standard NACA airfoils are generated by specifying the shape of the mean camber line and then wrapping a specified symmetrical thickness distribution around the mean camber line.

The force-and-moment system on an airfoil was discussed in Sec. 1.5, and the relative wind, angle of attack, lift, and drag were defined in Fig. 1.10. You should review these considerations before proceeding further.

The NACA identified different airfoil shapes with a logical numbering system. For example, the first family of NACA airfoils, developed in the 1930s, was the "four-digit" series, such as the NACA 2412 airfoil. Here, the first digit is the maximum camber in hundredths of chord, the second digit is the location of maximum camber along the chord from the leading edge in tenths of chord, and the last two digits give the maximum thickness in hundredths of chord. For the NACA 2412 airfoil, the maximum camber is $0.02c$ located at $0.4c$ from the leading edge, and the maximum thickness is $0.12c$. It is common practice to state these numbers in percent of chord, i.e., 2 percent camber at 40 percent chord, with 12 percent thickness. An airfoil with no camber, i.e., with the camber line and chord line coincident, is called a *symmetric airfoil.* Clearly, the shape of a symmetric airfoil is the same above and below the chord line. For example, the NACA 0012 airfoil is a symmetric airfoil with a maximum thickness of 12 percent.

The second family of NACA airfoils was the "five-digit" series, such as the NACA 23012 airfoil. Here, the first digit when multiplied by $\frac{3}{2}$ gives the design lift coefficient† in tenths, the next two digits when divided by 2 give the location of maximum camber along the chord from the leading edge in hundredths of chord, and the final two digits give the maximum thickness in hundredths of chord. For the NACA 23012 airfoil, the design lift coefficient is 0.3, the location of maximum camber is at $0.15c$, and the airfoil has 12 percent maximum thickness.

One of the most widely used family of NACA airfoils is the "6-series" laminar flow airfoils, developed during World War II. An example is the NACA 65-218. Here, the first digit simply identifies the series, the second gives the location of minimum pressure in tenths of chord from the leading edge (for the basic symmetric thickness distribution at zero lift), the third digit is the design lift coefficient in tenths, and the last two digits give the maximum thickness in hundredths of chord. For the NACA 65-218 airfoil, the 6 is the series designation,

---

† The design lift coefficient is the theoretical lift coefficient for the airfoil when the angle of attack is such that the slope of the mean camber line at the leading edge is parallel to the freestream velocity.

the minimum pressure occurs at $0.5c$ for the basic symmetric thickness distribution at zero lift, the design lift coefficient is 0.2, and the airfoil is 18 percent thick.

The complete NACA airfoil numbering system is given in Ref. 11. Indeed, Ref. 11 is a definitive presentation of the classic NACA airfoil work up to 1949. It contains a discussion of airfoil theory, its application, coordinates for the shape of NACA airfoils, and a huge bulk of experimental data for these airfoils. This author strongly encourages you to read Ref. 11 for a thorough presentation of airfoil characteristics.

As a matter of interest, the following is a short partial listing of airplanes in service in 1982 which use standard NACA airfoils.

Airplane	Airfoil
Beechcraft Sundowner	NACA 63A415
Beechcraft Bonanza	NACA 23016.5 (at root)
	NACA 23012 (at tip)
Cessna 150	NACA 2412
Fairchild A-10	NACA 6716 (at root)
	NACA 6713 (at tip)
Gates Learjet 24D	NACA 64A109
General Dynamics F-16	NACA 64A204
Lockheed C-5 Galaxy	NACA 0012 (modified)

In addition, many of the large aircraft companies today design their own special-purpose airfoils; e.g., the Boeing 727, 737, 747, 757, and 767 all have specially designed Boeing airfoils. Such capability is made possible by modern airfoil design computer programs utilizing either panel techniques or direct numerical finite-difference solutions of the governing partial differential equations for the flow field. (Such equations are developed in Chap. 2.)

4.3 AIRFOIL CHARACTERISTICS

Before discussing the theoretical calculation of airfoil properties, let us examine some typical results. During the 1930s and 1940s, the NACA carried out numerous measurements of the lift, drag, and moment coefficients on the standard NACA airfoils. These experiments were performed at low speeds in a wind tunnel where the constant-chord wing spanned the entire test section from one sidewall to the other. In this fashion, the flow "sees" a wing without wing tips—a so-called infinite wing, which theoretically stretches to infinity along the span (in the y direction in Fig. 4.1). Because the airfoil section is the same at any spanwise location along the infinite wing, the properties of the airfoil and the infinite wing are identical. Hence, airfoil data are frequently called infinite wing data. (In contrast, we see in Chap. 5 that the properties of a finite wing are somewhat different from its airfoil properties.)

The typical variation of lift coefficient with angle of attack for an airfoil is sketched in Fig. 4.4. At low-to-moderate angles of attack, c_l varies *linearly* with α; the slope of this straight line is denoted by a_0 and is called the *lift slope*. In this region, the flow moves smoothly over the airfoil and is attached over most of the surface, as shown in the streamline picture at the left of Fig. 4.4. However, as α becomes large, the flow tends to separate from the top surface of the airfoil, creating a large wake of relatively "dead air" behind the airfoil as shown at the right of Fig. 4.4. Inside this separated region, the flow is recirculating, and part of the flow is actually moving in a direction opposite to the freestream—so-called reversed flow. (Refer also to Fig. 1.29.) This separated flow is due to viscous effects and is discussed in Chap. 15. The consequence of this separated flow at high α is a precipitous decrease in lift and a large increase in drag; under such conditions the airfoil is said to be *stalled*. The maximum value of c_l, which occurs just prior to the stall, is denoted by $c_{l,max}$; it is one of the most important aspects of airfoil performance, because it determines the stalling speed of an airplane. The higher is $c_{l,max}$, the lower is the stalling speed. A great deal of modern airfoil research has been directed toward increasing $c_{l,max}$. Again examining Fig. 4.4, we see that c_l increases linearly with α until flow separation begins to have an effect. Then the curve becomes nonlinear, c_l reaches a maximum value, and finally the airfoil stalls. At the other extreme of the curve, noting Fig. 4.4, the lift at $\alpha = 0$ is finite; indeed, the lift goes to zero only when the airfoil is pitched to some negative angle of attack. The value of α when lift equals zero is called the *zero-lift angle of attack* and is denoted by $\alpha_{L=0}$. For a symmetric airfoil, $\alpha_{L=0} = 0$, whereas for all airfoils with positive camber (camber *above* the chord line), $\alpha_{L=0}$ is a negative value, usually on the order of -2 or $-3°$.

The inviscid flow airfoil theory discussed in this chapter allows us to predict the lift slope a_0 and $\alpha_{l=0}$ for a given airfoil. It does not allow us to calculate $c_{l,max}$, which is a difficult viscous flow problem, to be discussed in Chaps. 15 to 17.

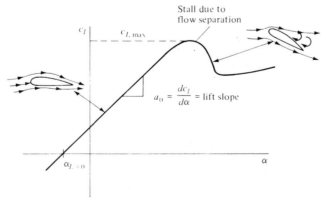

FIGURE 4.4
Schematic of lift-coefficient variation with angle of attack for an airfoil.

Experimental results for lift and moment coefficients for the NACA 2412 airfoil are given in Fig. 4.5. Here, the moment coefficient is taken about the quarter-chord point. Recall from Sec. 1.6 that the force-and-moment system on an airfoil can be transferred to any convenient point; however, the quarter-chord point is commonly used. (Refresh your mind on this concept by reviewing Sec. 1.6, especially Fig. 1.19.) Also shown in Fig. 4.5 are theoretical results to be discussed later. Note that the experimental data are given for two different Reynolds numbers. The lift slope a_0 is not influenced by Re; however, $c_{l,max}$ is dependent upon Re. This makes sense, because $c_{l,max}$ is governed by viscous effects, and Re is a similarity parameter that governs the strength of inertia forces relative to viscous forces in the flow. [See Sec. 1.7 and Eq. (1.35).] The moment coefficient is also insensitive to Re except at large α. The NACA 2412 airfoil is a commonly used airfoil, and the results given in Fig. 4.5 are quite typical of airfoil characteristics. For example, note from Fig. 4.5 that $\alpha_{L=0} = -2.1°$, $c_{l,max} \approx$ 1.6, and the stall occurs at $\alpha \approx 16°$.

This chapter deals with airfoil theory for an inviscid, incompressible flow; such theory is incapable of predicting airfoil drag, as noted earlier. However, for the sake of completeness, experimental data for the drag coefficient c_d for the

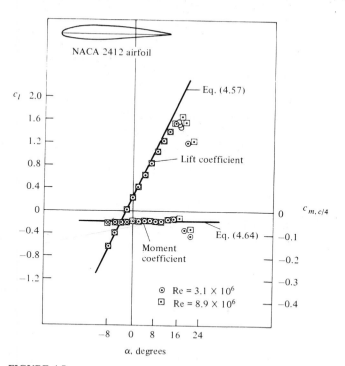

FIGURE 4.5

Experimental data for lift coefficient and moment coefficient about the quarter-chord point for an NACA 2412 airfoil. (*Data obtained from Abbott and von Doenhoff, Ref. 11.*) Also shown is a comparison with theory described in Sec. 4.8.

NACA 2412 airfoil are given in Fig. 4.6 as a function of the angle of attack.[†] The physical source of this drag coefficient is both skin friction drag and pressure drag due to flow separation (so-called form drag). The sum of these two effects yields the *profile* drag coefficient c_d for the airfoil, which is plotted in Fig. 4.6. Note that c_d is sensitive to Re, which is to be expected since both skin friction and flow separation are viscous effects. Again, we must wait until Chaps. 15 to 18 to obtain some tools for theoretically predicting c_d.

Also plotted in Fig. 4.6 is the moment coefficient about the aerodynamic center $c_{m,ac}$. In general, moments on an airfoil are a function of α. However, there is one point on the airfoil about which the moment is independent of angle of attack; such a point is defined as the *aerodynamic center*. Clearly, the data in Fig. 4.6 illustrate a constant value for $c_{m,ac}$ over a wide range of α.

For an elementary but extensive discussion of airfoil and wing properties, see chap. 5 of Ref. 2.

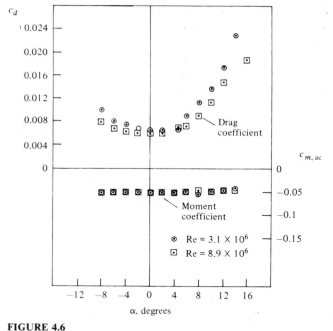

FIGURE 4.6
Experimental data for profile drag coefficient and moment coefficient about the aerodynamic center for the NACA 2412 airfoil. (*Data from Abbott and von Doenhoff, Ref. 11.*)

---

† In many references, such as Ref. 11, it is common to plot c_d versus c_l, rather than versus α. A plot of c_d versus c_l is called a *drag polar*. For the sake of consistency with Fig. 4.5, we choose to plot c_d versus α here.

Example 4.1. Consider an NACA 2412 airfoil with a chord of 0.64 m in an airstream at standard sea level conditions. The freestream velocity is 70 m/s. The lift per unit span is 1254 N/m. Calculate the angle of attack and the drag per unit span.

Solution. At standard sea level, $\rho = 1.23$ kg/m^3:

$$q_\infty = \tfrac{1}{2}\rho_\infty V_\infty^2 = \tfrac{1}{2}(1.23)(70)^2 = 3013.5 \text{ N/m}^2$$

$$c_l = \frac{L'}{q_\infty S} = \frac{L'}{q_\infty c(1)} = \frac{1254}{3013.5(0.64)} = 0.65$$

From Fig. 4.5, for $c_l = 0.65$, we obtain $\boxed{\alpha = 4°}$.

To obtain the drag per unit span, we must use the data in Fig. 4.6. However, since $c_d = f(\text{Re})$, let us calculate Re. At standard sea level, $\mu = 1.789 \times 10^{-5}$ kg/(m · s). Hence,

$$\text{Re} = \frac{\rho_\infty V_\infty c}{\mu_\infty} = \frac{1.23(70)(0.64)}{1.789 \times 10^{-5}} = 3.08 \times 10^6$$

Therefore, using the data for Re $= 3.1 \times 10^6$ in Fig. 4.6, we find $c_d = 0.0068$. Thus,

$$D' = q_\infty S c_d = q_\infty c(1) c_d = 3013.5(0.64)(0.0068) = \boxed{13.1 \text{ N/m}}$$

4.4 PHILOSOPHY OF THEORETICAL SOLUTIONS FOR LOW-SPEED FLOW OVER AIRFOILS: THE VORTEX SHEET

In Sec. 3.14, the concept of vortex flow was introduced; refer to Fig. 3.26 for a schematic of the flow induced by a point vortex of strength Γ located at a given point O. (Recall that Fig. 3.26, with its counterclockwise flow, corresponds to a negative value of Γ. By convention, a positive Γ induces a clockwise flow.) Let us now expand our concept of a point vortex. Referring to Fig. 3.26, imagine a straight line perpendicular to the page, going through point O, and extending to infinity both out of and into the page. This line is a straight *vortex filament* of strength Γ. A straight vortex filament is drawn in perspective in Fig. 4.7. (Here, we show a clockwise flow, which corresponds to a positive value of Γ.) The flow induced in any plane perpendicular to the straight vortex filament by the filament itself is identical to that induced by a point vortex of strength Γ; i.e., in Fig. 4.7, the flows in the planes perpendicular to the vortex filament at O and O' are identical to each other and are identical to the flow induced by a point vortex of strength Γ. Indeed, the point vortex described in Sec. 3.14 is simply a section of a straight vortex filament.

In Sec. 3.17, we introduced the concept of a source sheet, which is an infinite number of line sources side by side, with the strength of each line source being infinitesimally small. For vortex flow, consider an analogous situation. Imagine an infinite number of straight vortex filaments side by side, where the strength of each filament is infinitesimally small. These side-by-side vortex filaments form a *vortex sheet*, as shown in perspective in the upper left of Fig. 4.8. If we look along the series of vortex filaments (looking along the y axis in Fig. 4.8), the vortex sheet will appear as sketched at the lower right of Fig. 4.8. Here, we are

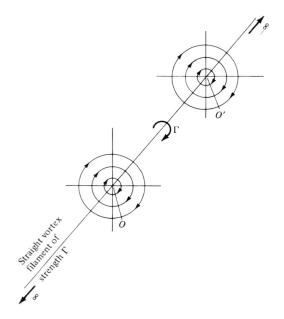

FIGURE 4.7
Vortex filament.

looking at an edge view of the sheet; the vortex filaments are all perpendicular to the page. Let s be the distance measured along the vortex sheet in the edge view. Define $\gamma = \gamma(s)$ as the strength of the vortex sheet, per unit length along s. Thus, the strength of an infinitesimal portion ds of the sheet is $\gamma\,ds$. This small section of the vortex sheet can be treated as a distinct vortex of strength $\gamma\,ds$.

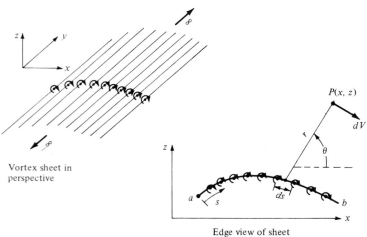

Vortex sheet in perspective

Edge view of sheet

FIGURE 4.8
Vortex sheet.

Now consider point P in the flow, located a distance r from ds; the cartesian coordinates of P are (x, z). The small section of the vortex sheet of strength $\gamma \, ds$ induces an infinitesimally small velocity, dV, at point P. From Eq. (3.105), dV is given by

$$dV = -\frac{\gamma \, ds}{2\pi r} \tag{4.1}$$

and is in a direction perpendicular to r, as shown in Fig. 4.8. The velocity at P induced by the entire vortex sheet is the summation of Eq. (4.1) from point a to point b. Note that dV, which is perpendicular to r, changes direction at point P as we sum from a to b; hence, the incremental velocities induced at P by different sections of the vortex sheet must be added vectorally. Because of this, it is sometimes more convenient to deal with the velocity potential. Again referring to Fig. 4.8, the increment in velocity potential, $d\phi$, induced at point P by the elemental vortex $\gamma \, ds$ is, from Eq. (3.112),

$$d\phi = -\frac{\gamma \, ds}{2\pi} \theta \tag{4.2}$$

In turn, the velocity potential at P due to the entire vortex sheet from a to b is

$$\boxed{\phi(x, z) = -\frac{1}{2\pi} \int_a^b \theta \gamma \, ds} \tag{4.3}$$

Equation (4.1) is particularly useful for our discussion of classical thin airfoil theory, whereas Eq. (4.3) is important for the numerical vortex panel method.

Recall from Sec. 3.14 that the circulation Γ around a point vortex is equal to the strength of the vortex. Similarly, the circulation around the vortex sheet in Fig. 4.8 is the sum of the strengths of the elemental vortices; i.e.,

$$\boxed{\Gamma = \int_a^b \gamma \, ds} \tag{4.4}$$

Recall that the source sheet introduced in Sec. 3.17 has a discontinuous change in the direction of the *normal* component of velocity across the sheet (from Fig. 3.33, note that the normal component of velocity changes direction by 180° in crossing the sheet), whereas the tangential component of velocity is the same immediately above and below the source sheet. In contrast, for a vortex sheet, there is a discontinuous change in the tangential component of velocity across the sheet, whereas the normal component of velocity is preserved across the sheet. This change in tangential velocity across the vortex sheet is related to the strength of the sheet as follows. Consider a vortex sheet as sketched in Fig. 4.9. Consider the rectangular dashed path enclosing a section of the sheet of

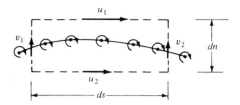

FIGURE 4.9
Tangential velocity jump across a vortex sheet.

length ds. The velocity components tangential to the top and bottom of this rectangular path are u_1 and u_2, respectively, and the velocity components tangential to the left and right sides are v_1 and v_2, respectively. The top and bottom of the path are separated by the distance dn. From the definition of circulation given by Eq. (2.127), the circulation around the dashed path is

$$\Gamma = -(v_2\, dn - u_1\, ds - v_1\, dn + u_2\, ds)$$

or

$$\Gamma = (u_1 - u_2)\, ds + (v_1 - v_2)\, dn \tag{4.5}$$

However, since the strength of the vortex sheet contained inside the dashed path is $\gamma\, ds$, we also have

$$\Gamma = \gamma\, ds \tag{4.6}$$

Therefore, from Eqs. (4.5) and (4.6),

$$\gamma\, ds = (u_1 - u_2)\, ds + (v_1 - v_2)\, dn \tag{4.7}$$

Let the top and bottom of the dashed line approach the vortex sheet; i.e., let $dn \to 0$. In the limit, u_1 and u_2 become the velocity components tangential to the vortex sheet immediately above and below the sheet, respectively, and Eq. (4.7) becomes

$$\gamma\, ds = (u_1 - u_2)\, ds$$

or

$$\boxed{\gamma = u_1 - u_2} \tag{4.8}$$

Equation (4.8) is important; it states that *the local jump in tangential velocity across the vortex sheet is equal to the local sheet strength.*

We have now defined and discussed the properties of a vortex sheet. The concept of a vortex sheet is instrumental in the analysis of the low-speed characteristics of an airfoil. A philosophy of airfoil theory of inviscid, incompressible flow is as follows. Consider an airfoil of arbitrary shape and thickness in a freestream with velocity V_∞, as sketched in Fig. 4.10. Replace the airfoil surface

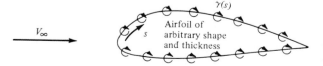

FIGURE 4.10
Simulation of an arbitrary airfoil by distributing a vortex sheet over the airfoil surface.

with a vortex sheet of variable strength $\gamma(s)$, as also shown in Fig. 4.10. Calculate the variation of γ as a function of s such that the induced velocity field from the vortex sheet when added to the uniform velocity of magnitude V_∞ will make the vortex sheet (hence the airfoil surface) a *streamline of the flow*. In turn, the circulation around the airfoil will be given by

$$\Gamma = \int \gamma \, ds$$

where the integral is taken around the complete surface of the airfoil. Finally, the resulting lift is given by the Kutta-Joukowski theorem:

$$L' = \rho_\infty V_\infty \Gamma$$

This philosophy is not new. It was first espoused by Ludwig Prandtl and his colleagues at Göttingen, Germany, during the period 1912-1922. However, no general analytical solution for $\gamma = \gamma(s)$ exists for an airfoil of arbitrary shape and thickness. Rather, the strength of the vortex sheet must be found numerically, and the practical implementation of the above philosophy had to wait until the 1960s with the advent of large digital computers. Today, the above philosophy is the foundation of the modern vortex panel method, to be discussed in Sec. 4.9.

The concept of replacing the airfoil surface in Fig. 4.10 with a vortex sheet is more than just a mathematical device; it also has physical significance. In real life, there is a thin boundary layer on the surface, due to the action of friction between the surface and the airflow (see Fig. 1.28). This boundary layer is a highly viscous region in which the large velocity gradients produce substantial vorticity; i.e., $\nabla \times \mathbf{V}$ is finite within the boundary layer. (Review Sec. 2.12 for a discussion of vorticity.) Hence, in real life, there is a distribution of vorticity along the airfoil surface due to viscous effects, and our philosophy of replacing the airfoil surface with a vortex sheet (such as in Fig. 4.10) can be construed as a way of modeling this effect in an inviscid flow.†

Imagine that the airfoil in Fig. 4.10 is made very thin. If you were to stand back and look at such a thin airfoil from a distance, the portions of the vortex sheet on the top and bottom surface of the airfoil would almost coincide. This gives rise to a method of approximating a thin airfoil by replacing it with a single vortex sheet distributed over the camber line of the airfoil, as sketched in Fig. 4.11. The strength of this vortex sheet, $\gamma(s)$, is calculated such that, in combination

---

† It is interesting to note that some recent research by NASA in 1982 is hinting that even as complex a problem as flow separation, heretofore thought to be a completely viscous-dominated phenomenon, may in reality be an inviscid-dominated flow which requires only a rotational flow. For example, some inviscid flow-field numerical solutions for flow over a circular cylinder, when vorticity is introduced either by means of a nonuniform freestream or a curved shock wave, are accurately predicting the separated flow on the rearward side of the cylinder. However, as exciting as these results may be, they are too preliminary to be emphasized in this book. We continue to talk about flow separation in Chaps. 15 to 18 as being a viscous-dominated effect, until definitely proved otherwise. This recent research is mentioned here only as another example of the physical connection between vorticity, vortex sheets, viscosity, and real life.

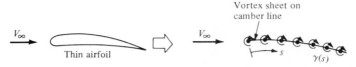

FIGURE 4.11
Thin airfoil approximation.

with the freestream, the camber line becomes a streamline of the flow. Although the approach shown in Fig. 4.11 is approximate in comparison with the case shown in Fig. 4.10, it has the advantage of yielding a closed-form analytical solution. This philosophy of thin airfoil theory was first developed by Max Munk, a colleague of Prandtl, in 1922 (see Ref. 12). It is discussed in Secs. 4.7 and 4.8.

4.5 THE KUTTA CONDITION

The lifting flow over a circular cylinder was discussed in Sec. 3.15, where we observed that an infinite number of potential flow solutions were possible, corresponding to the infinite choice of Γ. For example, Fig. 3.28 illustrates three different flows over the cylinder, corresponding to three different values of Γ. The same situation applies to the potential flow over an airfoil; for a given airfoil at a given angle of attack, there are an infinite number of valid theoretical solutions, corresponding to an infinite choice of Γ. For example, Fig. 4.12 illustrates two different flows over the same airfoil at the same angle of attack but with different values of Γ. At first, this may seem to pose a dilemma. We know from experience that a given airfoil at a given angle of attack produces a single value of lift (e.g., see Fig. 4.5). So, although there is an infinite number of possible potential flow solutions, nature knows how to pick a particular solution. Clearly, the philosophy discussed in the previous section is not complete—we need an additional condition that *fixes* Γ for a given airfoil at a given α.

To attempt to find this condition, let us examine some experimental results for the development of the flow field around an airfoil which is set into motion from an initial state of rest. Figure 4.13 shows a series of classic photographs of

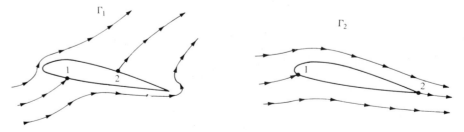

FIGURE 4.12
Effect of different values of circulation on the potential flow over a given airfoil at a given angle of attack. Points 1 and 2 are stagnation points.

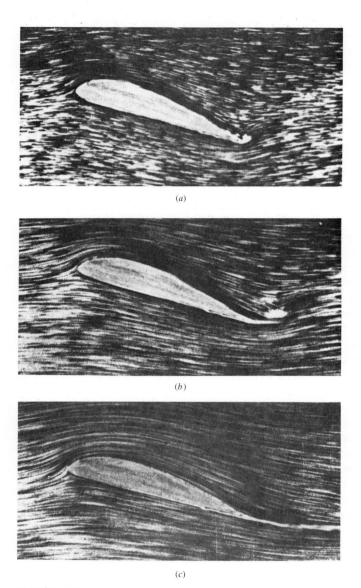

(a)

(b)

(c)

FIGURE 4.13
The development of steady flow over an airfoil; the airfoil is impulsively started from rest and attains a steady velocity through the fluid. (a) A moment just after starting. (b) An intermediate time. (c) The final steady flow. (*From Prandtl and Tietjens, Ref. 8.*)

the flow over an airfoil, taken from Prandtl and Tietjens (Ref. 8). In Fig. 4.13a, the flow has just started, and the flow pattern is just beginning to develop around the airfoil. In these early moments of development, the flow tries to curl around the sharp trailing edge from the bottom surface to the top surface, similar to the sketch shown at the left of Fig. 4.12. However, more advanced considerations of inviscid, incompressible flow (see, e.g., Ref. 9) show the theoretical result that the velocity becomes infinitely large at a sharp corner. Hence, the type of flow sketched at the left of Fig. 4.12, and shown in Fig. 4.13a, is not tolerated very long by nature. Rather, as the real flow develops over the airfoil, the stagnation point on the upper surface (point 2 in Fig. 4.12) moves toward the trailing edge. Figure 4.13b shows this intermediate stage. Finally, after the initial transient process dies out, the steady flow shown in Fig. 4.13c is reached. This photograph demonstrates that the flow is *smoothly leaving the top and the bottom surfaces of the airfoil at the trailing edge.* This flow pattern is sketched at the right of Fig. 4.12 and represents the type of pattern to be expected for the steady flow over an airfoil.

Reflecting on Figs. 4.12 and 4.13, we emphasize again that in establishing the steady flow over a given airfoil at a given angle of attack, nature adopts that particular value of circulation (Γ_2 in Fig. 4.12) which results in the flow leaving smoothly at the trailing edge. This observation was first made and used in a theoretical analysis by the German mathematician M. Wilhelm Kutta in 1902. Therefore, it has become known as the *Kutta condition.*

In order to apply the Kutta condition in a theoretical analysis, we need to be more precise about the nature of the flow at the trailing edge. The trailing edge can have a finite angle, as shown in Figs. 4.12 and 4.13 and as sketched at the left of Fig. 4.14, or it can be cusped, as shown at the right of Fig. 4.14. First, consider the trailing edge with a finite angle, as shown at the left of Fig. 4.14. Denote the velocities along the top surface and the bottom surface as V_1 and V_2, respectively. V_1 is parallel to the top surface at point a, and V_2 is parallel to the bottom surface at point a. For the finite-angle trailing edge, if these velocities were finite at point a, then we would have two velocities in two different directions at the same point, as shown at the left of Fig. 4.14. However, this is not physically possible, and the only recourse is for both V_1 and V_2 to be zero at point a. That is, for the finite trailing edge, point a is a stagnation point, where $V_1 = V_2 = 0$. In contrast, for the cusped trailing edge shown at the right of Fig. 4.14, V_1 and

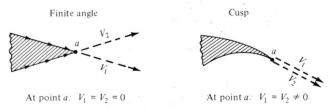

Finite angle   Cusp

At point a: $V_1 = V_2 = 0$ At point a: $V_1 = V_2 \neq 0$

FIGURE 4.14
Different possible shapes of the trailing edge and their relation to the Kutta condition.

V_2 are in the same direction at point a, and hence both V_1 and V_2 can be finite. However, the pressure at point a, p_2, is a single, unique value, and Bernoulli's equation applied at both the top and bottom surfaces immediately adjacent to point a yields

$$p_a + \tfrac{1}{2}\rho V_1^2 = p_a + \tfrac{1}{2}\rho V_2^2$$

or

$$V_1 = V_2$$

Hence, for the cusped trailing edge, we see that the velocities leaving the top and bottom surfaces of the airfoil at the trailing edge are finite and equal in magnitude and direction.

We can summarize the statement of the Kutta condition as follows:

1. For a given airfoil at a given angle of attack, the value of Γ around the airfoil is such that the flow leaves the trailing edge smoothly.
2. If the trailing-edge angle is finite, then the trailing edge is a stagnation point.
3. If the trailing edge is cusped, then the velocities leaving the top and bottom surfaces at the trailing edge are finite and equal in magnitude and direction.

Consider again the philosophy of simulating the airfoil with vortex sheets placed either on the surface or on the camber line, as discussed in Sec. 4.4. The strength of such a vortex sheet is variable along the sheet and is denoted by $\gamma(s)$. The statement of the Kutta condition in terms of the vortex sheet is as follows. At the trailing edge (TE), from Eq. (4.8), we have

$$\gamma(\text{TE}) = \gamma(a) = V_1 - V_2 \tag{4.9}$$

However, for the finite-angle trailing edge, $V_1 = V_2 = 0$; hence, from Eq. (4.9), $\gamma(\text{TE}) = 0$. For the cusped trailing edge, $V_1 = V_2 \neq 0$; hence, from Eq. (4.9), we again obtain the result that $\gamma(\text{TE}) = 0$. Therefore, the Kutta condition expressed in terms of the strength of the vortex sheet is

$$\boxed{\gamma(\text{TE}) = 0} \tag{4.10}$$

4.6 KELVIN'S CIRCULATION THEOREM AND THE STARTING VORTEX

In this section, we put the finishing touch to the overall philosophy of airfoil theory before developing the quantitative aspects of the theory itself in subsequent sections. This section also ties up a loose end introduced by the Kutta condition described in the previous section. Specifically, the Kutta condition states that the circulation around an airfoil is just the right value to ensure that the flow smoothly leaves the trailing edge. *Question*: How does nature generate this circulation?

Does it come from nowhere, or is circulation somehow conserved over the whole flow field? Let us examine these matters more closely.

Consider an arbitrary inviscid, incompressible flow as sketched in Fig. 4.15. Assume that all body forces **f** are zero. Choose an arbitrary curve, C_1, and identify the fluid elements that are on this curve at a given instant in time t_1. Also, by definition the circulation around curve C_1 is $\Gamma_1 = -\int_{C_1} \mathbf{V} \cdot \mathbf{ds}$. Now let these specific fluid elements move downstream. At some later time, t_2, these *same* fluid elements will form another curve C_2, around which the circulation is $\Gamma_2 = -\int_{C_2} \mathbf{V} \cdot \mathbf{ds}$. For the conditions stated above, we can readily show that $\Gamma_1 = \Gamma_2$. In fact, since we are following a set of specific fluid elements, we can state that circulation around a closed curve formed by a set of contiguous fluid elements remains constant as the fluid elements move throughout the flow. Recall from Sec. 2.9 that the substantial derivative gives the time rate of change following a given fluid element. Hence, a mathematical statement of the above discussion is simply

$$\frac{D\Gamma}{Dt} = 0 \tag{4.11}$$

which says that the time rate of change of circulation around a closed curve consisting of the same fluid elements is zero. Equation (4.11) along with its supporting discussion is called *Kelvin's circulation theorem*.† Its derivation from

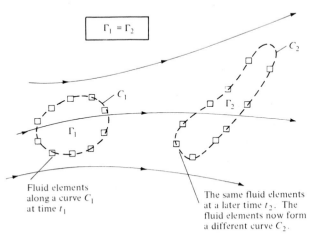

$$\Gamma_1 = \Gamma_2$$

Fluid elements along a curve C_1 at time t_1

The same fluid elements at a later time t_2. The fluid elements now form a different curve C_2.

FIGURE 4.15
Kelvin's theorem.

---

† Kelvin's theorem also holds for an inviscid compressible flow in the special case where $\rho = \rho(p)$; i.e., the density is some single-valued function of pressure. Such is the case for isentropic flow, to be treated in later chapters.

first principles is left as Prob. 4.3. Also, recall our definition and discussion of a vortex sheet in Sec. 4.4. An interesting consequence of Kelvin's circulation theorem is proof that a stream surface which is a vortex sheet at some instant in time remains a vortex sheet for all times.

Kelvin's theorem helps to explain the generation of circulation around an airfoil, as follows. Consider an airfoil in a fluid at rest, as shown in Fig. 4.16a. Because $V = 0$ everywhere, the circulation around curve C_1 is zero. Now start the flow in motion over the airfoil. Initially, the flow will tend to curl around the trailing edge, as explained in Sec. 4.5 and illustrated at the left of Fig. 4.12. In so doing, the velocity at the trailing edge theoretically becomes infinite. In real life, the velocity tends toward a very large finite number. Consequently, during the very first moments after the flow is started, a thin region of very large velocity gradients (and therefore high vorticity) is formed at the trailing edge. This high-vorticity region is fixed to the same fluid elements, and consequently it is flushed downstream as the fluid elements begin to move downstream from the trailing edge. As it moves downstream, this thin sheet of intense vorticity is unstable, and it tends to roll up and form a picture similar to a point vortex. This vortex is called the *starting vortex* and is sketched in Fig. 4.16b. After the flow around the airfoil has come to a steady state where the flow leaves the trailing edge smoothly (the Kutta condition), the high velocity gradients at the trailing edge disappear and vorticity is no longer produced at that point. However, the starting vortex has already been formed during the starting process, and it

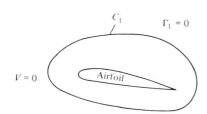

(a) Fluid at rest relative to the airfoil

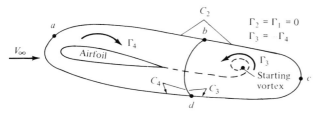

(b) Picture some moments after the start of the flow

FIGURE 4.16
The creation of the starting vortex and the resulting generation of circulation around the airfoil.

moves steadily downstream with the flow forever after. Figure 4.16b shows the flow field sometime after steady flow has been achieved over the airfoil, with the starting vortex somewhere downstream. The fluid elements that initially made up curve C_1 in Fig. 4.16a have moved downstream and now make up curve C_2, which is the complete circuit $abcda$ shown in Fig. 4.16b. Thus, from Kelvin's theorem, the circulation Γ_2 around curve C_2 (which encloses both the airfoil and the starting vortex) is the same as that around curve C_1, namely, zero. $\Gamma_2 = \Gamma_1 = 0$. Now let us subdivide C_2 into two loops by making the cut bd, thus forming curves C_3 (circuit $bcdb$) and C_4 (circuit $abda$). Curve C_3 encloses the starting vortex, and curve C_4 encloses the airfoil. The circulation Γ_3 around curve C_3 is due to the starting vortex; by inspecting Fig. 4.16b, we see that Γ_3 is in the counterclockwise direction, i.e., a negative value. The circulation around curve C_4 enclosing the airfoil is Γ_4. Since the cut bd is common to both C_3 and C_4, the sum of the circulations around C_3 and C_4 is simply equal to the circulation around C_2:

$$\Gamma_3 + \Gamma_4 = \Gamma_2$$

However, we have already established that $\Gamma_2 = 0$. Hence,

$$\Gamma_4 = -\Gamma_3$$

i.e., *the circulation around the airfoil is equal and opposite to the circulation around the starting vortex.*

This brings us to the summary as well as the crux of this section. As the flow over an airfoil is started, the large velocity gradients at the sharp trailing edge result in the formation of a region of intense vorticity which rolls up downstream of the trailing edge, forming the starting vortex. This starting vortex has associated with it a counterclockwise circulation. Therefore, as an equal-and-opposite reaction, a clockwise circulation around the airfoil is generated. As the starting process continues, vorticity from the trailing edge is constantly fed into the starting vortex, making it stronger with a consequent larger counterclockwise circulation. In turn, the clockwise circulation around the airfoil becomes stronger, making the flow at the trailing edge more closely approach the Kutta condition, thus weakening the vorticity shed from the trailing edge. Finally, the starting vortex builds up to just the right strength such that the equal-and-opposite clockwise circulation around the airfoil leads to smooth flow from the trailing edge (the Kutta condition is exactly satisfied). When this happens, the vorticity shed from the leading edge becomes zero, the starting vortex no longer grows in strength, and a steady circulation exists around the airfoil.

4.7 CLASSICAL THIN AIRFOIL THEORY: THE SYMMETRIC AIRFOIL

Some experimentally observed characteristics of airfoils and a philosophy for the theoretical prediction of these characteristics have been discussed in the preceding sections. Referring to our chapter road map in Fig. 4.2, we have now completed the central branch. In this section, we move to the right-hand branch of Fig. 4.2, namely, a quantitative development of thin airfoil theory. The basic equations necessary for the calculation of airfoil lift and moments are established in this section, with an application to symmetric airfoils. The case of cambered airfoils will be treated in Sec. 4.8.

For the time being, we deal with *thin* airfoils; for such a case, the airfoil can be simulated by a vortex sheet placed along the camber line, as discussed in Sec. 4.4. Our purpose is to calculate the variation of $\gamma(s)$ such that the camber line becomes a streamline of the flow and such that the Kutta condition is satisfied at the trailing edge; i.e., $\gamma(\text{TE}) = 0$ [see Eq. (4.10)]. Once we have found the particular $\gamma(s)$ which satisfies these conditions, then the total circulation Γ around the airfoil is found by integrating $\gamma(s)$ from the leading edge to the trailing edge. In turn, the lift is calculated from Γ via the Kutta-Joukowski theorem.

Consider a vortex sheet placed on the camber line of an airfoil, as sketched in Fig. 4.17a. The freestream velocity is V_∞, and the airfoil is at the angle of attack α. The x axis is oriented along the chord line, and the z axis is perpendicular to the chord. The distance measured along the camber line is denoted by s. The shape of the camber line is given by $z = z(x)$. The chord length is c. In Fig. 4.17a, w' is the component of velocity normal to the camber line induced by the vortex sheet; $w' = w'(s)$. For a thin airfoil, we rationalized in Sec. 4.4 that the distribution of a vortex sheet over the surface of the airfoil, when viewed from a distance, looks almost the same as a vortex sheet placed on the camber line. Let us stand back once again and view Fig. 4.17a from a distance. If the airfoil is thin, the

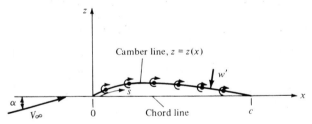

(a) Vortex sheet on the camber line

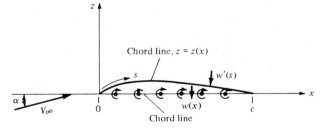

(b) Vortex sheet on the chord line

FIGURE 4.17
Placement of the vortex sheet for thin airfoil analysis.

camber line is close to the chord line, and viewed from a distance, the vortex sheet appears to fall approximately on the chord line. Therefore, once again, let us reorient our thinking and place the vortex sheet on the chord line, as sketched in Fig. 4.17b. Here, $\gamma = \gamma(x)$. We still wish the camber line to be a streamline of the flow, and $\gamma = \gamma(x)$ is calculated to satisfy this condition as well as the Kutta condition $\gamma(c) = 0$. That is, the strength of the vortex sheet on the chord line is determined such that the camber line (not the chord line) is a streamline.

For the camber line to be a streamline, the component of velocity normal to the camber line must be zero at all points along the camber line. The velocity at any point in the flow is the sum of the uniform freestream velocity and the velocity induced by the vortex sheet. Let $V_{\infty,n}$ be the component of the freestream velocity normal to the camber line. Thus, for the camber line to be a streamline,

$$V_{\infty,n} + w'(s) = 0 \tag{4.12}$$

at every point along the camber line.

An expression for $V_{\infty,n}$ in Eq. (4.12) is obtained by the inspection of Fig. 4.18. At any point P on. the camber line, where the slope of the camber line is dz/dx, the geometry of Fig. 4.18 yields

$$V_{\infty,n} = V_\infty \sin\left[\alpha + \tan^{-1}\left(-\frac{dz}{dx}\right)\right] \tag{4.13}$$

For a thin airfoil at small angle of attack, both α and $\tan^{-1}(-dz/dx)$ are small values. Using the approximation that $\sin\theta \approx \tan\theta \approx \theta$ for small θ, where θ is in radians, Eq. (4.13) reduces to

$$V_{\infty,n} = V_\infty\left(\alpha - \frac{dz}{dx}\right) \tag{4.14}$$

Equation (4.14) gives the expression for $V_{\infty,n}$ to be used in Eq. (4.12). Keep in mind that, in Eq. (4.14), α is in radians.

Returning to Eq. (4.12), let us develop an expression for $w'(s)$ in terms of the strength of the vortex sheet. Refer again to Fig. 4.17b. Here, the vortex sheet

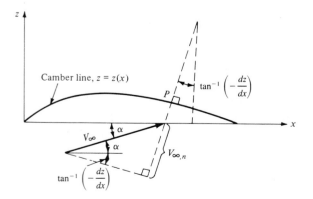

FIGURE 4.18
Determination of the component of freestream velocity normal to the camber line.

is along the chord line, and $w'(s)$ is the component of velocity normal to the camber line induced by the vortex sheet. Let $w(x)$ denote the component of velocity normal to the *chord line* induced by the vortex sheet, as also shown in Fig. 4.17b. If the airfoil is thin, the camber line is close to the chord line, and it is consistent with thin airfoil theory to make the approximation that

$$w'(s) \approx w(x) \qquad (4.15)$$

An expression for $w(x)$ in terms of the strength of the vortex sheet is easily obtained from Eq. (4.1), as follows. Consider Fig. 4.19, which shows the vortex sheet along the chord line. We wish to calculate the value of $w(x)$ at the location x. Consider an elemental vortex of strength $\gamma\,d\xi$ located at a distance ξ from the origin along the chord line, as shown in Fig. 4.19. The strength of the vortex sheet, γ, varies with the distance along the chord; i.e., $\gamma = \gamma(\xi)$. The velocity dw at point x induced by the elemental vortex at point ξ is given by Eq. (4.1) as

$$dw = -\frac{\gamma(\xi)\,d\xi}{2\pi(x-\xi)} \qquad (4.16)$$

In turn, the velocity $w(x)$ induced at point x by *all* the elemental vortices along the chord line is obtained by integrating Eq. (4.16) from the leading edge ($\xi = 0$) to the trailing edge ($\xi = c$):

$$w(x) = -\int_0^c \frac{\gamma(\xi)\,d\xi}{2\pi(x-\xi)} \qquad (4.17)$$

Combined with the approximation stated by Eq. (4.15), Eq. (4.17) gives the expression for $w'(s)$ to be used in Eq. (4.12).

Recall that Eq. (4.12) is the boundary condition necessary for the camber line to be a streamline. Substituting Eqs. (4.14), (4.15), and (4.17) into (4.12), we obtain

$$V_\infty\left(\alpha - \frac{dz}{dx}\right) - \int_0^c \frac{\gamma(\xi)\,d\xi}{2\pi(x-\xi)} = 0$$

or

$$\boxed{\frac{1}{2\pi}\int_0^c \frac{\gamma(\xi)\,d\xi}{x-\xi} = V_\infty\left(\alpha - \frac{dz}{dx}\right)} \qquad (4.18)$$

the *fundamental equation of thin airfoil theory*; it is simply a statement that the camber line is a streamline of the flow.

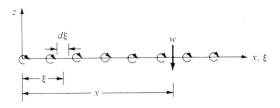

FIGURE 4.19
Calculation of the induced velocity at the chord line.

Note that Eq. (4.18) is written at a given point x on the chord line, and that dz/dx is evaluated at that point x. The variable ξ is simply a dummy variable of integration which varies from 0 to c along the chord line, as shown in Fig. 4.19. The vortex strength $\gamma = \gamma(\xi)$ is a variable along the chord line. For a given airfoil at a given angle of attack, both α and dz/dx are known values in Eq. (4.18). Indeed, the only unknown in Eq. (4.18) is the vortex strength $\gamma(\xi)$. Hence, Eq. (4.18) is an integral equation, the solution of which yields the variation of $\gamma(\xi)$ such that the camber line is a streamline of the flow. The central problem of thin airfoil theory is to solve Eq. (4.18) for $\gamma(\xi)$, subject to the Kutta condition, namely, $\gamma(c) = 0$.

In this section, we treat the case of a symmetric airfoil. As stated in Sec. 4.2, a symmetric airfoil has no camber; the camber line is coincident with the chord line. Hence, for this case, $dz/dx = 0$, and Eq. (4.18) becomes

$$\frac{1}{2\pi} \int_0^c \frac{\gamma(\xi)\, d\xi}{x - \xi} = V_\infty \alpha \tag{4.19}$$

In essence, within the framework of thin airfoil theory, a symmetric airfoil is treated the same as a flat plate; note that our theoretical development does not account for the airfoil thickness distribution. Equation (4.19) is an *exact* expression for the inviscid, incompressible flow over a flat plate at angle of attack.

To help deal with the integral in Eqs. (4.18) and (4.19), let us transform ξ into θ via the following transformation:

$$\xi = \frac{c}{2}(1 - \cos \theta) \tag{4.20}$$

Since x is a fixed point in Eqs. (4.18) and (4.19), it corresponds to a particular value of θ, namely, θ_0, such that

$$x = \frac{c}{2}(1 - \cos \theta_0) \tag{4.21}$$

Also, from Eq. (4.20),

$$d\xi = \frac{c}{2} \sin \theta\, d\theta \tag{4.22}$$

Substituting Eqs. (4.20) to (4.22) into (4.19), and noting that the limits of integration become $\theta = 0$ at the leading edge (where $\xi = 0$) and $\theta = \pi$ at the trailing edge (where $\xi = c$), we obtain

$$\frac{1}{2\pi} \int_0^\pi \frac{\gamma(\theta) \sin \theta\, d\theta}{\cos \theta - \cos \theta_0} = V_\infty \alpha \tag{4.23}$$

A rigorous solution of Eq. (4.23) for $\gamma(\theta)$ can be obtained from the mathematical theory of integral equations, which is beyond the scope of this book. Instead, we simply state that the solution is

$$\gamma(\theta) = 2\alpha V_\infty \frac{(1+\cos\theta)}{\sin\theta} \qquad (4.24)$$

We can verify this solution by substituting Eq. (4.24) into (4.23), yielding

$$\frac{1}{2\pi} \int_0^\pi \frac{\gamma(\theta)\sin\theta\,d\theta}{\cos\theta - \cos\theta_0} = \frac{V_\infty\alpha}{\pi} \int_0^\pi \frac{(1+\cos\theta)\,d\theta}{\cos\theta - \cos\theta_0} \qquad (4.25)$$

The following standard integral appears frequently in airfoil theory and is derived in appendix E of Ref. 9:

$$\int_0^\pi \frac{\cos n\theta\,d\theta}{\cos\theta - \cos\theta_0} \equiv \frac{\pi \sin n\theta_0}{\sin\theta_0} \qquad (4.26)$$

Using Eq. (4.26) in the right-hand side of Eq. (4.25), we find that

$$\frac{V_\infty\alpha}{\pi} \int_0^\pi \frac{(1+\cos\theta)\,d\theta}{\cos\theta - \cos\theta_0} = \frac{V_\infty\alpha}{\pi} \left(\int_0^\pi \frac{d\theta}{\cos\theta - \cos\theta_0} + \int_0^\pi \frac{\cos\theta\,d\theta}{\cos\theta - \cos\theta_0} \right)$$

$$= \frac{V_\infty\alpha}{\pi}(0+\pi) = V_\infty\alpha \qquad (4.27)$$

Substituting Eq. (4.27) into (4.25), we have

$$\frac{1}{2\pi} \int_0^\pi \frac{\gamma(\theta)\sin\theta\,d\theta}{\cos\theta - \cos\theta_0} = V_\infty\alpha$$

which is identical to Eq. (4.23). Hence, we have shown that Eq. (4.24) is indeed the solution to Eq. (4.23). Also, note that at the trailing edge, where $\theta = \pi$, Eq. (4.24) yields

$$\gamma(\pi) = 2\alpha V_\infty \frac{0}{0}$$

which is an indeterminant form. However, using L'Hospital's rule on Eq. (4.24),

$$\gamma(\pi) = 2\alpha V_\infty \frac{-\sin\pi}{\cos\pi} = 0$$

Thus, Eq. (4.24) also satisfies the Kutta condition.

We are now in a position to calculate the lift coefficient for a thin, symmetric airfoil. The total circulation around the airfoil is

$$\Gamma = \int_0^c \gamma(\xi)\,d\xi \qquad (4.28)$$

Using Eqs. (4.20) and (4.22), Eq. (4.28) transforms to

$$\Gamma = \frac{c}{2} \int_0^\pi \gamma(\theta)\sin\theta\,d\theta \qquad (4.29)$$

Substituting Eq. (4.24) into (4.29), we obtain

$$\Gamma = \alpha c V_\infty \int_0^\pi (1 + \cos\theta)\, d\theta = \pi \alpha c V_\infty \qquad (4.30)$$

Substituting Eq. (4.30) into the Kutta-Joukowski theorem, we find that the lift per unit span is

$$L' = \rho_\infty V_\infty \Gamma = \pi \alpha c \rho_\infty V_\infty^2 \qquad (4.31)$$

The lift coefficient is

$$c_l = \frac{L'}{q_\infty S} \qquad (4.32)$$

where

$$S = c(1)$$

Substituting Eq. (4.31) into (4.32), we have

$$c_l = \frac{\pi \alpha c \rho_\infty V_\infty^2}{\frac{1}{2}\rho_\infty V_\infty^2 c(1)}$$

or

$$\boxed{c_l = 2\pi\alpha} \qquad (4.33)$$

and

$$\boxed{\text{Lift slope} = \frac{dc_l}{d\alpha} = 2\pi} \qquad (4.34)$$

Equations (4.33) and (4.34) are important results; they state the theoretical result that the lift coefficient is *linearly proportional to angle of attack*, which is supported by the experimental results discussed in Sec. 4.3. They also state that the theoretical lift slope is equal to 2π rad^{-1}, which is 0.11 degree^{-1}. The experimental lift coefficient data for an NACA 0012 symmetric airfoil are given in Fig. 4.20; note that Eq. (4.33) accurately predicts c_l over a large range of angle of attack. (The NACA 0012 airfoil section is commonly used on airplane tails and helicopter blades.)

The moment about the leading edge can be calculated as follows. Consider the elemental vortex of strength $\gamma(\xi)\, d\xi$ located a distance ξ from the leading edge, as sketched in Fig. 4.21. The circulation associated with this elemental vortex is $d\Gamma = \gamma(\xi)\, d\xi$. In turn, the increment of lift, dL, contributed by the elemental vortex is $dL = \rho_\infty V_\infty\, d\Gamma$. This increment of lift creates a moment about the leading edge $dM = -\xi(dL)$. The total moment about the leading edge (LE) (per unit span) due to the entire vortex sheet is therefore

$$M'_{\text{LE}} = -\int_0^c \xi(dL) = -\rho_\infty V_\infty \int_0^c \xi \gamma(\xi)\, d\xi \qquad (4.35)$$

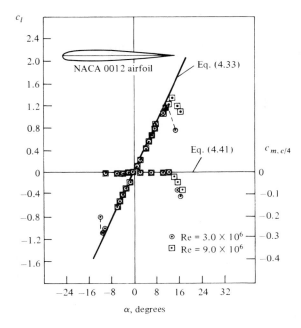

FIGURE 4.20
Comparison between theory and experiment for the lift and moment coefficients for an NACA 0012 airfoil. (*Experimental data are from Abbott and von Doenhoff, Ref. 11.*)

Transforming Eq. (4.35) via Eqs. (4.20) and (4.22), and performing the integration, we obtain (the details are left for Prob. 4.4):

$$M'_{LE} = -q_\infty c^2 \frac{\pi \alpha}{2} \tag{4.36}$$

The moment coefficient is

$$c_{m,le} = \frac{M'_{LE}}{q_\infty S c}$$

where $S = c(1)$. Hence,

$$c_{m,le} = \frac{M'_{LE}}{q_\infty c^2} = -\frac{\pi \alpha}{2} \tag{4.37}$$

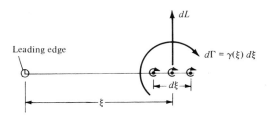

FIGURE 4.21
Calculation of moments about the leading edge.

However, from Eq. (4.33),

$$\pi\alpha = \frac{c_l}{2} \tag{4.38}$$

Combining Eqs. (4.37) and (4.38), we obtain

$$\boxed{c_{m,\text{le}} = -\frac{c_l}{4}} \tag{4.39}$$

From Eq. (1.22), the moment coefficient about the quarter-chord point is

$$c_{m,c/4} = c_{m,\text{le}} + \frac{c_l}{4} \tag{4.40}$$

Combining Eqs. (4.39) and (4.40), we have

$$\boxed{c_{m,c/4} = 0} \tag{4.41}$$

In Sec. 1.6, a definition is given for the center of pressure as that point about which the moments are zero. Clearly, Eq. (4.41) demonstrates the theoretical result that the *center of pressure is at the quarter-chord point for a symmetric airfoil.*

By the definition given in Sec. 4.3, that point on an airfoil where moments are independent of angle of attack is called the aerodynamic center. From Eq. (4.41), the moment about the quarter chord is zero for all values of α. Hence, for a symmetric airfoil, we have the theoretical result that the *quarter-chord point is both the center of pressure and the aerodynamic center.*

The above theoretical result for $c_{m,c/4} = 0$ is supported by the experimental data given in Fig. 4.20. Also, note that the experimental value of $c_{m,c/4}$ is constant over a wide range of α, thus demonstrating that the real aerodynamic center is essentially at the quarter chord.

Let us summarize the above results. The essence of thin airfoil theory is to find a distribution of vortex sheet strength along the chord line that will make the camber line a streamline of the flow while satisfying the Kutta condition $\gamma(\text{TE}) = 0$. Such a vortex distribution is obtained by solving Eq. (4.18) for $\gamma(\xi)$, or in terms of the transformed independent variable θ, solving Eq. (4.23) for $\gamma(\theta)$ [recall that Eq. (4.23) is written for a symmetric airfoil]. The resulting vortex distribution $\gamma(\theta)$ for a symmetric airfoil is given by Eq. (4.24). In turn, this vortex distribution, when inserted into the Kutta-Joukowski theorem, gives the following important theoretical results for a symmetric airfoil:

1. $c_l = 2\pi\alpha$.
2. Lift slope $= 2\pi$.
3. The center of pressure and the aerodynamic center are both located at the quarter-chord point.

4.8 THE CAMBERED AIRFOIL

Thin airfoil theory for a cambered airfoil is a generalization of the method for a symmetric airfoil discussed in Sec. 4.7. To treat the cambered airfoil, return to Eq. (4.18):

$$\frac{1}{2\pi} \int_0^c \frac{\gamma(\xi)\,d\xi}{x-\xi} = V_\infty \left(\alpha - \frac{dz}{dx} \right) \tag{4.18}$$

For a cambered airfoil, dz/dx is finite, and this makes the analysis more elaborate than in the case of a symmetric airfoil, where $dz/dx = 0$. Once again, let us transform Eq. (4.18) via Eqs. (4.20) to (4.22), obtaining

$$\frac{1}{2\pi} \int_0^\pi \frac{\gamma(\theta)\sin\theta\,d\theta}{\cos\theta - \cos\theta_0} = V_\infty \left(\alpha - \frac{dz}{dx} \right) \tag{4.42}$$

We wish to obtain a solution for $\gamma(\theta)$ from Eq. (4.42), subject to the Kutta condition $\gamma(\pi) = 0$. Such a solution for $\gamma(\theta)$ will make the camber line a streamline of the flow. However, as before, a rigorous solution of Eq. (4.42) for $\gamma(\theta)$ is beyond the scope of this book. Rather, the result is stated below:

$$\gamma(\theta) = 2V_\infty \left(A_0 \frac{1+\cos\theta}{\sin\theta} + \sum_{n=1}^\infty A_n \sin n\theta \right) \tag{4.43}$$

Note that the above expression for $\gamma(\theta)$ consists of a leading term very similar to Eq. (4.24) for a symmetric airfoil, plus a Fourier sine series with coefficients A_n. The values of A_n depend on the shape of the camber line, dz/dx, and A_0 depends on both dz/dx and α, as shown below.

The coefficients A_0 and A_n $(n = 1, 2, 3, \ldots)$ in Eq. (4.43) must be specific values in order that the camber line be a streamline of the flow. To find these specific values, substitute Eq. (4.43) into Eq. (4.42):

$$\frac{1}{\pi} \int_0^\pi \frac{A_0(1+\cos\theta)\,d\theta}{\cos\theta - \cos\theta_0} + \frac{1}{\pi} \sum_{n=1}^\infty \int_0^\pi \frac{A_n \sin n\theta \sin\theta\,d\theta}{\cos\theta - \cos\theta_0} = \alpha - \frac{dz}{dx} \tag{4.44}$$

The first integral can be evaluated from the standard form given in Eq. (4.26). The remaining integrals can be obtained from another standard form, which is derived in appendix E of Ref. 9, and which is given below:

$$\int_0^\pi \frac{\sin n\theta \sin\theta\,d\theta}{\cos\theta - \cos\theta_0} = -\pi \cos n\theta_0 \tag{4.45}$$

Hence, using Eqs. (4.26) and (4.45), we can reduce Eq. (4.44) to

$$A_0 - \sum_{n=1}^\infty A_n \cos n\theta_0 = \alpha - \frac{dz}{dx}$$

or

$$\frac{dz}{dx} = (\alpha - A_0) + \sum_{n=1}^\infty A_n \cos n\theta_0 \tag{4.46}$$

Recall that Eq. (4.46) was obtained directly from Eq. (4.42), which is the transformed version of the fundamental equation of thin airfoil theory, Eq. (4.18). Furthermore, recall that Eq. (4.18) is evaluated at a given point x along the chord line, as sketched in Fig. 4.19. Hence, Eq. (4.46) is also evaluated at the given point x; here, dz/dx and θ_0 correspond to the same point x on the chord line. Also, recall that dz/dx is a function of θ_0, where $x = (c/2)(1 - \cos \theta_0)$ from Eq. (4.21).

Examine Eq. (4.46) closely. It is in the form of a Fourier cosine series expansion for the function of dz/dx. In general, the Fourier cosine series representation of a function $f(\theta)$ over an interval $0 \le \theta \le \pi$ is given by

$$f(\theta) = B_0 + \sum_{n=1}^{\infty} B_n \cos n\theta \qquad (4.47)$$

where, from Fourier analysis, the coefficients B_0 and B_n are given by

$$B_0 = \frac{1}{\pi} \int_0^{\pi} f(\theta) \, d\theta \qquad (4.48)$$

and

$$B_n = \frac{2}{\pi} \int_0^{\pi} f(\theta) \cos n\theta \, d\theta \qquad (4.49)$$

(See, e.g., page 217 of Ref. 6.) In Eq. (4.46), the function dz/dx is analogous to $f(\theta)$ in the general form given in Eq. (4.47). Thus, from Eqs. (4.48) and (4.49), the coefficients in Eq. (4.46) are given by

$$\alpha - A_0 = \frac{1}{\pi} \int_0^{\pi} \frac{dz}{dx} \, d\theta_0$$

or

$$A_0 = \alpha - \frac{1}{\pi} \int_0^{\pi} \frac{dz}{dx} \, d\theta_0 \qquad (4.50)$$

and

$$A_n = \frac{2}{\pi} \int_0^{\pi} \frac{dz}{dx} \cos n\theta_0 \, d\theta_0 \qquad (4.51)$$

Keep in mind that in the above, dz/dx is a function of θ_0. Note from Eq. (4.50) that A_0 depends on both α and the shape of the camber line (through dz/dx), whereas from Eq. (4.51) the values of A_n depend only on the shape of the camber line.

Pause for a moment and think about what we have done. We are considering the flow over a cambered airfoil of given shape, dz/dx, at a given angle of attack α. In order to make the camber line a streamline of the flow, the strength of the vortex sheet along the chord line must have the distribution $\gamma(\theta)$ given by Eq. (4.43), where the coefficients A_0 and A_n are given by Eqs. (4.50) and (4.51), respectively. Also, note that Eq. (4.43) satisfies the Kutta condition $\gamma(\pi) = 0$. Actual numbers for A_0 and A_n can be obtained for a given shape airfoil at a given angle of attack simply by carrying out the integrations indicated in Eqs. (4.50) and (4.51). For an example of such calculations applied to an NACA 2412 airfoil, see pages 120–125 of Ref. 13. Also, note that when $dz/dx = 0$, Eq. (4.43) reduces to Eq. (4.24) for a symmetric airfoil. Hence, the symmetric airfoil is a special case of Eq. (4.43).

Let us now obtain expressions for the aerodynamic coefficients for a cambered airfoil. The total circulation due to the entire vortex sheet from the leading edge to the trailing edge is

$$\Gamma = \int_0^c \gamma(\xi) \, d\xi = \frac{c}{2} \int_0^\pi \gamma(\theta) \sin \theta \, d\theta \tag{4.52}$$

Substituting Eq. (4.43) for $\gamma(\theta)$ into Eq. (4.52), we obtain

$$\Gamma = cV_\infty \left[A_0 \int_0^\pi (1 + \cos \theta) \, d\theta + \sum_{n=1}^\infty A_n \int_0^\pi \sin n\theta \sin \theta \, d\theta \right] \tag{4.53}$$

From any standard table of integrals,

$$\int_0^\pi (1 + \cos \theta) \, d\theta = \pi$$

and

$$\int_0^\pi \sin n\theta \sin \theta \, d\theta = \begin{cases} \pi/2 & \text{for } n = 1 \\ 0 & \text{for } n \neq 1 \end{cases}$$

Hence, Eq. (4.53) becomes

$$\Gamma = cV_\infty \left(\pi A_0 + \frac{\pi}{2} A_1 \right) \tag{4.54}$$

From Eq. (4.54), the lift per unit span is

$$L' = \rho_\infty V_\infty \Gamma = \rho_\infty V_\infty^2 c \left(\pi A_0 + \frac{\pi}{2} A_1 \right) \tag{4.55}$$

In turn, Eq. (4.55) leads to the lift coefficient in the form

$$c_l = \frac{L'}{\frac{1}{2}\rho_\infty V_\infty^2 c(1)} = \pi(2A_0 + A_1) \tag{4.56}$$

Recall that the coefficients A_0 and A_1 in Eq. (4.56) are given by Eqs. (4.50) and (4.51), respectively. Hence, Eq. (4.56) becomes

$$\boxed{c_l = 2\pi \left[\alpha + \frac{1}{\pi} \int_0^\pi \frac{dz}{dx} (\cos \theta_0 - 1) \, d\theta_0 \right.} \tag{4.57}$$

and

$$\boxed{\text{Lift slope} \equiv \frac{dc_l}{d\alpha} = 2\pi} \tag{4.58}$$

Equations (4.57) and (4.58) are important results. Note that, as in the case of the symmetric airfoil, the theoretical lift slope for a cambered airfoil is 2π. It is a general result from thin airfoil theory that $dc_l/d\alpha = 2\pi$ *for any shape airfoil.* However, the expression for c_l itself differs between a symmetric and a cambered airfoil, the difference being the integral term in Eq. (4.57). This integral term has physical significance, as follows. Return to Fig. 4.4, which illustrates the lift curve for an airfoil. The angle of zero lift is denoted by $\alpha_{L=0}$ and is a negative value. From the geometry shown in Fig. 4.4, clearly

$$c_l = \frac{dc_l}{d\alpha}(\alpha - \alpha_{L=0}) \tag{4.59}$$

Substituting Eq. (4.58) into (4.59), we have

$$c_l = 2\pi(\alpha - \alpha_{L=0}) \tag{4.60}$$

Comparing Eqs. (4.60) and (4.57), we see that the integral term in Eq. (4.57) is simply the negative of the zero-lift angle; i.e.,

$$\alpha_{L=0} = -\frac{1}{\pi}\int_0^\pi \frac{dz}{dx}(\cos\theta_0 - 1)\,d\theta_0 \tag{4.61}$$

Hence, from Eq. (4.61), thin airfoil theory provides a means to predict the angle of zero lift. Note that Eq. (4.61) yields $\alpha_{L=0} = 0$ for a symmetric airfoil, which is consistent with the results shown in Fig. 4.20. Also, note that the more highly cambered the airfoil, the larger will be the absolute magnitude of $\alpha_{L=0}$.

Returning to Fig. 4.21, the moment about the leading edge can be obtained by substituting $\gamma(\theta)$ from Eq. (4.43) into the transformed version of Eq. (4.35). The details are left for Prob. 4.9. The result for the moment coefficient is

$$c_{m,\text{le}} = -\frac{\pi}{2}\left(A_0 + A_1 - \frac{A_2}{2}\right) \tag{4.62}$$

Substituting Eq. (4.56) into (4.62), we have

$$c_{m,\text{le}} = -\left[\frac{c_l}{4} + \frac{\pi}{4}(A_1 - A_2)\right] \tag{4.63}$$

Note that, for $dz/dx = 0$, $A_1 = A_2 = 0$ and Eq. (4.63) reduces to Eq. (4.39) for a symmetric airfoil.

The moment coefficient about the quarter chord can be obtained by substituting Eq. (4.63) into (4.40), yielding

$$c_{m,c/4} = \frac{\pi}{4}(A_2 - A_1) \tag{4.64}$$

Unlike the symmetric airfoil, where $c_{m,c/4} = 0$, Eq. (4.64) demonstrates that $c_{m,c/4}$ is finite for a cambered airfoil. Therefore, the quarter chord is *not* the center of pressure for a cambered airfoil. However, note that A_1 and A_2 depend only on the shape of the camber line and do not involve the angle of attack. Hence, from Eq. (4.64), $c_{m,c/4}$ is *independent* of α. Thus, the quarter-chord point is the *theoretical location* of the aerodynamic center for a cambered airfoil.

The location of the center of pressure can be obtained from Eq. (1.21):

$$x_{cp} = -\frac{M'_{LE}}{L'} = -\frac{c_{m,le} c}{c_l} \qquad (4.65)$$

Substituting Eq. (4.63) into (4.65), we obtain

$$\boxed{x_{cp} = \frac{c}{4}\left[1 + \frac{\pi}{c_l}(A_1 - A_2)\right]} \qquad (4.66)$$

Equation (4.66) demonstrates that the center of pressure for a cambered airfoil varies with the lift coefficient. Hence, as the angle of attack changes, the center of pressure also changes. Indeed, as the lift approaches zero, x_{cp} moves toward infinity; i.e., it leaves the airfoil. For this reason, the center of pressure is not always a convenient point at which to draw the force system on an airfoil. Rather, the force-and-moment system on an airfoil is more conveniently considered at the aerodynamic center. (Return to Fig. 1.19 and the discussion at the end of Sec. 1.6 for the referencing of the force-and-moment system on an airfoil.)

Example 4.2. Consider an NACA 23012 airfoil. The mean camber line for this airfoil is given by

$$\frac{z}{c} = 2.6595\left[\left(\frac{x}{c}\right)^3 - 0.6075\left(\frac{x}{c}\right)^2 + 0.1147\left(\frac{x}{c}\right)\right] \qquad \text{for } 0 \le \frac{x}{c} \le 0.2025$$

and

$$\frac{z}{c} = 0.02208\left(1 - \frac{x}{c}\right) \qquad \text{for } 0.2025 \le \frac{x}{c} \le 1.0$$

Calculate (*a*) the angle of attack at zero lift, (*b*) the lift coefficient when $\alpha = 4°$, (*c*) the moment coefficient about the quarter chord, and (*d*) the location of the center of pressure in terms of x_{cp}/c, when $\alpha = 4°$. Compare the results with experimental data.

Solution. We will need dz/dx. From the given shape of the mean camber line, this is

$$\frac{dz}{dx} = 2.6595\left[3\left(\frac{x}{c}\right)^2 - 1.215\left(\frac{x}{c}\right) + 0.1147\right] \qquad \text{for } 0 \le \frac{x}{c} \le 0.2025$$

and

$$\frac{dz}{dx} = -0.02208 \qquad \text{for } 0.2025 \le \frac{x}{c} \le 1.0$$

Transforming from x to θ, where $x = (c/2)(1 - \cos\theta)$, we have

$$\frac{dz}{dx} = 2.6595\left[\frac{3}{4}(1 - 2\cos\theta + \cos^2\theta) - 0.6075(1 - \cos\theta) + 0.1147\right]$$

or

$$= 0.6840 - 2.3736\cos\theta + 1.995\cos^2\theta \qquad \text{for } 0 \le \theta \le 0.9335 \text{ rad}$$

and

$$= -0.02208 \qquad \text{for } 0.9335 \le \theta \le \pi$$

(a) From Eq. (4.61),

$$\alpha_{L=0} = -\frac{1}{\pi} \int_0^\pi \frac{dz}{dx} (\cos\theta - 1)\, d\theta$$

(*Note*: For simplicity, we have dropped the subscript zero from θ; in Eq. (4.61), θ_0 is the variable of integration—it can just as well be symbolized as θ for the variable of integration.) Substituting the equation for dz/dx into Eq. (4.61), we have

$$\alpha_{L=0} = -\frac{1}{\pi} \int_0^{0.9335} (-0.6840 + 3.0576 \cos\theta - 4.3686 \cos^2\theta + 1.995 \cos^3\theta)\, d\theta$$

$$-\frac{1}{\pi} \int_{0.9335}^\pi (0.02208 - 0.02208 \cos\theta)\, d\theta \qquad\qquad (E.1)$$

From a table of integrals, we see that

$$\int \cos\theta\, d\theta = \sin\theta$$

$$\int \cos^2\theta\, d\theta = \tfrac{1}{2}\sin\theta\cos\theta + \tfrac{1}{2}\theta$$

$$\int \cos^3\theta\, d\theta = \tfrac{1}{3}\sin\theta(\cos^2\theta + 2)$$

Hence, Eq. (E.1) becomes

$$\alpha_{L=0} = -\frac{1}{\pi}[-2.8683\theta + 3.0576\sin\theta - 2.1843\sin\theta\cos\theta$$

$$+ 0.665\sin\theta(\cos^2\theta + 2)]_0^{0.9335}$$

$$-\frac{1}{\pi}[0.02208\theta - 0.02208\sin\theta]_{0.9335}^\pi$$

Hence,

$$\alpha_{L=0} = -\frac{1}{\pi}(-0.0065 + 0.0665) = -0.0191 \text{ rad}$$

or

$$\boxed{\alpha_{L=0} = -1.09^\circ}$$

(b) $\alpha = 4^\circ = 0.0698$ rad
From Eq. (4.60),

$$c_l = 2\pi(\alpha - \alpha_{L=0}) = 2\pi(0.0698 + 0.0191) = \boxed{0.559}$$

(c) The value of $c_{m,c/4}$ is obtained from Eq. (4.64). For this, we need the two Fourier coefficients, A_1 and A_2. From Eq. (4.51),

$$A_1 = \frac{2}{\pi} \int_0^\pi \frac{dz}{dx} \cos\theta\, d\theta$$

$$A_1 = \frac{2}{\pi} \int_0^{0.9335} (0.6840 \cos \theta - 2.3736 \cos^2 \theta + 1.995 \cos^3 \theta) \, d\theta$$

$$+ \frac{2}{\pi} \int_{0.9335}^{\pi} (-0.02208 \cos \theta) \, d\theta$$

$$= \frac{2}{\pi} [0.6840 \sin \theta - 1.1868 \sin \theta \cos \theta - 1.1868\theta + 0.665 \sin \theta (\cos^2 \theta + 2)]_0^{0.9335}$$

$$+ \frac{2}{\pi} [-0.02208 \sin \theta]_{0.9335}^{\pi}$$

$$= \frac{2}{\pi} (0.1322 + 0.0177) = 0.0954$$

From Eq. (4.51),

$$A_2 = \frac{2}{\pi} \int_0^{\pi} \frac{dz}{dx} \cos 2\theta \, d\theta = \frac{2}{\pi} \int_0^{\pi} \frac{dz}{dx} (2 \cos^2 \theta - 1) \, d\theta$$

$$= \frac{2}{\pi} \int_0^{0.9335} (-0.6840 + 2.3736 \cos \theta - 0.627 \cos^2 \theta$$

$$- 4.747 \cos^3 \theta + 3.99 \cos^4 \theta) \, d\theta$$

$$+ \frac{2}{\pi} \int_{0.9335}^{\pi} (0.02208 - 0.0446 \cos^2 \theta) \, d\theta$$

Note:

$$\int \cos^4 \theta \, d\theta = \frac{1}{4} \cos^3 \theta \sin \theta + \frac{3}{8} (\sin \theta \cos \theta + \theta)$$

Thus,

$$A_2 = \frac{2}{\pi} \left\{ -0.6840\theta + 2.3736 \sin \theta - 0.628 \left(\frac{1}{2}\right)(\sin \theta \cos \theta + \theta) \right.$$

$$\left. - 4.747 \left(\frac{1}{3}\right) \sin \theta (\cos^2 \theta + 2) + 3.99 \left[\frac{1}{4} \cos^3 \sin \theta + \frac{3}{8}(\sin \theta \cos \theta + \theta)\right] \right\}_0^{0.9335}$$

$$+ \frac{2}{\pi} \left[0.02208\theta - 0.0446 \left(\frac{1}{2}\right)(\sin \theta \cos \theta + \theta) \right]_{0.9335}^{\pi}$$

$$= \frac{2}{\pi} (0.11384 + 0.01056) = 0.0792$$

From Eq. (4.64),

$$c_{m,c/4} = \frac{\pi}{4} (A_2 - A_1) = \frac{\pi}{4} (0.0792 - 0.0954)$$

$$\boxed{c_{m,c/4} = -0.0127}$$

(*d*) From Eq. (4.66),

$$x_{cp} = \frac{c}{4} \left[1 + \frac{\pi}{c_l} (A_1 + A_2) \right]$$

Hence,

$$\frac{x_{cp}}{c} = \frac{1}{4} \left[1 + \frac{\pi}{0.559} (0.0954 - 0.0792) \right] = 0.273$$

COMPARISON WITH EXPERIMENTAL DATA. The data for the NACA 23012 airfoil are shown in Fig. 4.22. From this, we make the following tabulation:

	Calculated	Experiment
$\alpha_{L=0}$	$-1.09°$	$-1.1°$
c_l (at $\alpha = 4°$)	0.559	0.55
$c_{m,c/4}$	-0.0127	-0.01

Note that the results from thin airfoil theory for a cambered airfoil agree very well with the experimental data. Recall that excellent agreement between thin airfoil theory for a symmetric airfoil and experimental data has already been shown in Fig. 4.20. Hence, all of the work we have done in this section to develop thin airfoil theory is certainly worth the effort. Moreover, this illustrates that the development of thin airfoil theory in the early 1900s was a crowning achievement in the theoretical aerodynamics and validates the mathematical approach of replacing the chord line of the airfoil with a vortex sheet, with the flow tangency condition evaluated along the mean camber line.

This brings to an end our introduction to classical thin airfoil theory. Returning to our road map in Fig. 4.2, we have now completed the right-hand branch.

4.9 LIFTING FLOWS OVER ARBITRARY BODIES: THE VORTEX PANEL NUMERICAL METHOD

The thin airfoil theory described in Secs. 4.7 and 4.8 is just what it says—it applies only to thin airfoils at small angles of attack. (Make certain that you understand exactly where in the development of thin airfoil theory these assumptions are made and the reasons for making them.) The advantage of thin airfoil theory is that closed-form expressions are obtained for the aerodynamic coefficients. Moreover, the results compare favorably with experimental data for airfoils of about 12 percent thickness or less. However, the airfoils on many low-speed airplanes are thicker than 12 percent. Moreover, we are frequently interested in high angles of attack, such as occur during takeoff and landing.

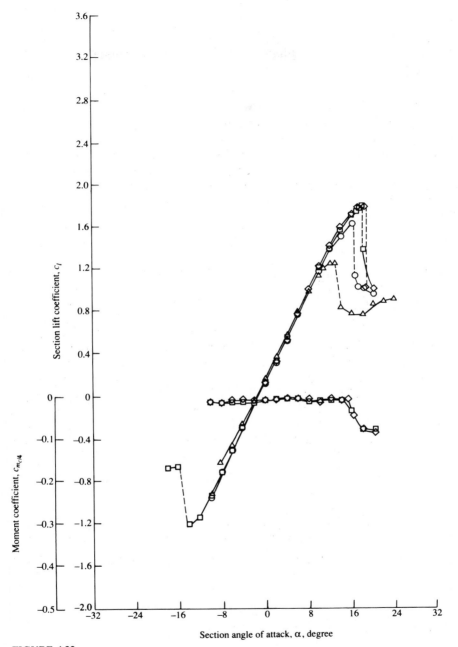

FIGURE 4.22
Lift- and moment-coefficient data for an NACA 23012 airfoil, for comparison with the theoretical results obtained in Example 4.2.

Finally, we are sometimes concerned with the generation of aerodynamic lift on other body shapes, such as automobiles or submarines. Hence, thin airfoil theory is quite restrictive when we consider the whole spectrum of aerodynamic applications. We need a method that allows us to calculate the aerodynamic characteristics of bodies of arbitrary shape, thickness, and orientation. Such a method is described in this section. Specifically, we treat the vortex panel method, which is a numerical technique that has come into widespread use since the early 1970s. In reference to our road map in Fig. 4.2, we now move to the left-hand branch. Also, since this chapter deals with airfoils, we limit our attention to two-dimensional bodies.

The vortex panel method is directly analogous to the source panel method described in Sec. 3.17. However, because a source has zero circulation, source panels are useful only for nonlifting cases. In contrast, vortices have circulation, and hence vortex panels can be used for lifting cases. (Because of the similarities between source and vortex panel methods, return to Sec. 3.17 and review the basic philosophy of the source panel method before proceeding further.)

The philosophy of covering a body *surface* with a vortex sheet of such a strength to make the surface a streamline of the flow was discussed in Sec. 4.4. We then went on to simplify this idea by placing the vortex sheet on the camber line of the airfoil as shown in Fig. 4.11, thus establishing the basis for thin airfoil theory. We now return to the original idea of wrapping the vortex sheet over the complete surface of the body, as shown in Fig. 4.10. We wish to find $\gamma(s)$ such that the body surface becomes a streamline of the flow. There exists no closed-form analytical solution for $\gamma(s)$; rather, the solution must be obtained numerically. This is the purpose of the vortex panel method.

Let us approximate the vortex sheet shown in Fig. 4.10 by a series of straight panels, as shown earlier in Fig. 3.35. (In Chap. 3, Fig. 3.35 was used to discuss source panels; here, we use the same sketch for discussion of vortex panels.) Let the vortex strength $\gamma(s)$ per unit length be constant over a given panel, but allow it to vary from one panel to the next. That is, for the n panels shown in Fig. 3.35, the vortex panel strengths per unit length are $\gamma_1, \gamma_2, \ldots, \gamma_j, \ldots, \gamma_n$. These panel strengths are unknowns; the main thrust of the panel technique is to solve for γ_j, $j = 1$ to n, such that the body surface becomes a streamline of the flow and such that the Kutta condition is satisfied. As explained in Sec. 3.17, the midpoint of each panel is a control point at which the boundary condition is applied; i.e., at each control point, the normal component of the flow velocity is zero.

Let P be a point located at (x, y) in the flow, and let r_{pj} be the distance from any point on the jth panel to P, as shown in Fig. 3.35. The radius r_{pj} makes the angle θ_{pj} with respect to the x axis. The velocity potential induced at P due to the jth panel, $\Delta\phi_j$, is, from Eq. (4.3),

$$\Delta\phi_j = -\frac{1}{2\pi}\int_j \theta_{pj}\gamma_j \, ds_j \qquad (4.67)$$

In Eq. (4.67), γ_j is constant over the jth panel, and the integral is taken over the *j*th panel only. The angle θ_{pj} is given by

$$\theta_{pj} = \tan^{-1} \frac{y - y_j}{x - x_j} \tag{4.68}$$

In turn, the potential at P due to *all* the panels is Eq. (4.67) summed over all the panels:

$$\phi(P) = \sum_{j=1}^{n} \phi_j = -\sum_{j=1}^{n} \frac{\gamma_j}{2\pi} \int_j \theta_{pj} \, ds_j \tag{4.69}$$

Since point P is just an arbitrary point in the flow, let us put P at the control point of the *i*th panel shown in Fig. 3.35. The coordinates of this control point are (x_i, y_i). Then Eqs. (4.68) and (4.69) become

$$\theta_{ij} = \tan^{-1} \frac{y_i - y_j}{x_i - x_j}$$

and

$$\phi(x_i, y_i) = -\sum_{j=1}^{n} \frac{\gamma_j}{2\pi} \int_j \theta_{ij} \, ds_j \tag{4.70}$$

Equation (4.70) is physically the contribution of *all* the panels to the potential at the control point of the *i*th panel.

At the control points, the normal component of the velocity is zero; this velocity is the superposition of the uniform flow velocity and the velocity induced by all the vortex panels. The component of V_∞ normal to the *i*th panel is given by Eq. (3.148):

$$V_{\infty,n} = V_\infty \cos \beta_i \tag{3.148}$$

The normal component of velocity induced at (x_i, y_i) by the vortex panels is

$$V_n = \frac{\partial}{\partial n_i} [\phi(x_i, y_i)] \tag{4.71}$$

Combining Eqs. (4.70) and (4.71), we have

$$V_n = -\sum_{j=1}^{n} \frac{\gamma_j}{2\pi} \int_j \frac{\partial \theta_{ij}}{\partial n_i} \, ds_j \tag{4.72}$$

where the summation is over all the panels. The normal component of the flow velocity at the *i*th control point is the sum of that due to the freestream [Eq. (3.148)] and that due to the vortex panels [Eq. (4.72)]. The boundary condition states that this sum must be zero:

$$V_{\infty,n} + V_n = 0 \tag{4.73}$$

Substituting Eqs. (3.148) and (4.72) into (4.73), we obtain

$$V_\infty \cos \beta_i - \sum_{j=1}^{n} \frac{\gamma_j}{2\pi} \int_j \frac{\partial \theta_{ij}}{\partial n_i} \, ds_j = 0 \tag{4.74}$$

Equation (4.74) is the crux of the vortex panel method. The values of the integrals in Eq. (4.74) depend simply on the panel geometry; they are not properties of the flow. Let $J_{i,j}$ be the value of this integral when the control point is on the ith panel. Then Eq. (4.74) can be written as

$$V_\infty \cos \beta_i - \sum_{j=1}^{n} \frac{\gamma_j}{2\pi} J_{i,j} = 0 \qquad (4.75)$$

Equation (4.75) is a linear algebraic equation with n unknowns, $\gamma_1, \gamma_2, \ldots, \gamma_n$. It represents the flow boundary condition evaluated at the control point of the ith panel. If Eq. (4.75) is applied to the control points of *all* the panels, we obtain a system of n linear equations with n unknowns.

To this point, we have been deliberately paralleling the discussion of the source panel method given in Sec. 3.17; however, the similarity stops here. For the source panel method, the n equations for the n unknown source strengths are routinely solved, giving the flow over a nonlifting body. In contrast, for the lifting case with vortex panels, in addition to the n equations given by Eq. (4.75) applied at all the panels, we must also satisfy the Kutta condition. This can be done in several ways. For example, consider Fig. 4.23, which illustrates a detail of the vortex panel distribution at the trailing edge. Note that the length of each panel can be different; their length and distribution over the body are up to your discretion. Let the two panels at the trailing edge (panels i and $i-1$ in Fig. 4.23) be very small. The Kutta condition is applied *precisely* at the trailing edge and is given by $\gamma(TE) = 0$. To approximate this numerically, if points i and $i-1$ are close enough to the trailing edge, we can write

$$\gamma_i = -\gamma_{i-1} \qquad (4.76)$$

such that the strengths of the two vortex panels i and $i-1$ exactly cancel at the point where they touch at the trailing edge. Thus, in order to impose the Kutta condition on the solution of the flow, Eq. (4.76) (or an equivalent expression) must be included. Note that Eq. (4.75) evaluated at all the panels and Eq. (4.76) constitute an *overdetermined* system of n unknowns with $n+1$ equations. Therefore, to obtain a determined system, Eq. (4.75) is not evaluated at one of the control points on the body. That is, we choose to ignore one of the control points, and we evaluate Eq. (4.75) at the other $n-1$ control points. This, in combination

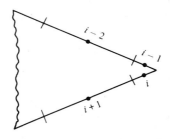

FIGURE 4.23
Vortex panels at the trailing edge.

with Eq. (4.76), now gives a system of n linear algebraic equations with n unknowns, which can be solved by standard techniques.

At this stage, we have conceptually obtained the values of $\gamma_1, \gamma_2, \ldots, \gamma_n$ which make the body surface a streamline of the flow and which also satisfy the Kutta condition. In turn, the flow velocity tangent to the surface can be obtained directly from γ. To see this more clearly, consider the airfoil shown in Fig. 4.24. We are concerned only with the flow outside the airfoil and on its surface. Therefore, let the velocity be zero at every point *inside* the body, as shown in Fig. 4.24. In particular, the velocity just inside the vortex sheet on the surface is zero. This corresponds to $u_2 = 0$ in Eq. (4.8). Hence, the velocity just outside the vortex sheet is, from Eq. (4.8),

$$\gamma = u_1 - u_2 = u_1 - 0 = u_1$$

In Eq. (4.8), u denotes the velocity tangential to the vortex sheet. In terms of the picture shown in Fig. 4.24, we obtain $V_a = \gamma_a$ at point a, $V_b = \gamma_b$ at point b, etc. Therefore, *the local velocities tangential to the airfoil surface are equal to the local values of γ*. In turn, the local pressure distribution can be obtained from Bernoulli's equation.

The total circulation and the resulting lift are obtained as follows. Let s_j be the length of the jth panel. Then the circulation due to the jth panel is $\gamma_j s_j$. In turn, the total circulation due to all the panels is

$$\Gamma = \sum_{j=1}^{n} \gamma_j s_j \tag{4.77}$$

Hence, the lift per unit span is obtained from

$$L' = \rho_\infty V_\infty \sum_{j=1}^{n} \gamma_j s_j \tag{4.78}$$

The presentation in this section is intended to give only the general flavor of the vortex panel method. There are many variations of the method in use today, and you are encouraged to read the modern literature, especially as it appears in the *AIAA Journal* and the *Journal of Aircraft* since 1970. The vortex panel method as described in this section is termed a "first-order" method because it assumes a constant value of γ over a given panel. Although the method may appear to be straightforward, its numerical implementation can sometimes be frustrating. For example, the results for a given body are sensitive to the number of panels used, their various sizes, and the way they are distributed over the body surface (i.e., it is usually advantageous to place a large number of small panels

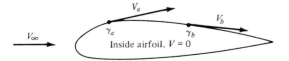

FIGURE 4.24
Airfoil as a solid body, with zero velocity inside the profile.

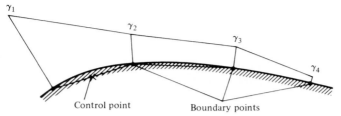

FIGURE 4.25
Linear distribution of γ over each panel—a second-order panel method.

near the leading and trailing edges of an airfoil and a smaller number of larger panels in the middle). The need to ignore one of the control points in order to have a determined system in n equations for n unknowns also introduces some arbitrariness in the numerical solution. Which control point do you ignore? Different choices sometimes yield different numerical answers for the distribution of γ over the surface. Moreover, the resulting numerical distributions for γ are not always smooth, but rather, they have oscillations from one panel to the next

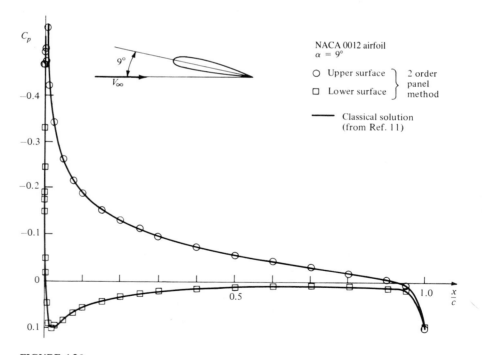

FIGURE 4.26
Pressure coefficient distribution over an NACA 0012 airfoil; comparison between second-order vortex panel method and NACA theoretical results from Ref. 11. The numerical panel results were obtained by one of the author's graduate students, Mr. Tae-Hwan Cho.

as a result of numerical inaccuracies. The problems mentioned above are usually overcome in different ways by different groups who have developed relatively sophisticated panel programs for practical use. Again, you are encouraged to consult the literature for more information.

Such accuracy problems have encouraged the development of higher-order panel techniques. For example, a "second-order" panel method assumes a *linear* variation of γ over a given panel, as sketched in Fig. 4.25. Here, the value of γ at the edges of each panel is matched to its neighbors, and the values γ_1, γ_2, γ_3, etc. at the *boundary points* become the unknowns to be solved. The flow-tangency boundary condition is still applied at the *control point* of each panel, as before. Some results using a second-order vortex panel technique are given in Fig. 4.26, which shows the distribution of pressure coefficients over the upper and lower surfaces of an NACA 0012 airfoil at a 9° angle of attack. The circles and squares are numerical results from a second-order vortex panel technique developed at the University of Maryland, and the solid lines are from NACA results given in Ref. 11. Excellent agreement is obtained.

Finally, many groups developing and using panel techniques use a combination of source panels and vortex panels for lifting bodies—source panels to accurately represent the thickness of the body and vortex panels to provide circulation. Again, you are encouraged to consult the literature. For example, Ref. 14 is a classic paper on panel methods, and Ref. 15 highlights many of the basic concepts of panel methods along with actual computer program statement listings for simple applications.

4.10 MODERN LOW-SPEED AIRFOILS

The nomenclature and aerodynamic characteristics of standard NACA airfoils are discussed in Secs. 4.2 and 4.3; before progressing further, you should review these sections in order to reinforce your knowledge of airfoil behavior, especially in light of our discussions on airfoil theory. Indeed, the purpose of this section is to provide a modern sequel to the airfoils discussed in Secs. 4.2 and 4.3.

During the 1970s, NASA designed a series of low-speed airfoils that have performance superior to the earlier NACA airfoils. The standard NACA airfoils were based almost exclusively on experimental data obtained during the 1930s and 1940s. In contrast, the new NASA airfoils were designed on a computer using a numerical technique similar to the source and vortex panel methods discussed earlier, along with numerical predictions of the viscous flow behavior (skin friction and flow separation). Wind-tunnel tests were then conducted to verify the computer-designed profiles and to obtain the definitive airfoil properties. Out of this work first came the general aviation—Whitcomb [GA(W) − 1] airfoil, which has since been redesignated the LS(1)-0417 airfoil. The shape of this airfoil is given in Fig. 4.27, obtained from Ref. 16. Note that it has a large leading-edge radius ($0.08c$ in comparison to the standard $0.02c$) in order to flatten the usual peak in pressure coefficient near the nose. Also, note that the bottom surface near the trailing edge is cusped in order to increase the camber and hence

FIGURE 4.27
Profile for the NASA LS(1)-0417 airfoil. When first introduced, this airfoil was labeled the GA (W)-1 airfoil, a nomenclature which has now been superseded. (*From Ref. 16.*)

the aerodynamic loading in that region. Both design features tend to discourage flow separation over the top surface at high angle of attack, hence yielding higher values of the maximum lift coefficient. The experimentally measured lift and moment properties (from Ref. 16) are given in Fig. 4.28, where they are compared with the properties for an NACA 2412 airfoil, obtained from Ref. 11. Note that $c_{l,max}$ for the NASA LS(1)-0417 is considerably higher than for the NACA 2412.

The NASA LS(1)-0417 airfoil has a maximum thickness of 17 percent and a design lift coefficient of 0.4. Using the same camber line, NASA has extended this airfoil into a family of low-speed airfoils of different thicknesses, e.g., the NASA LS(1)-0409 and the LS(1)-0413. (See Ref. 17 for more details.) In comparison with the standard NACA airfoils having the same thicknesses, these new LS(1)-04xx airfoils all have:

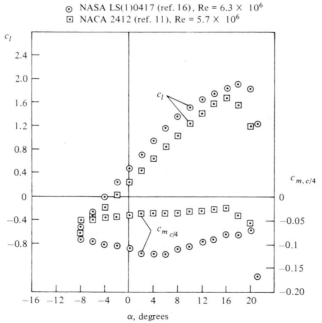

FIGURE 4.28
Comparison of the modern NASA LS(1)-0417 airfoil with the standard NACA 2412 airfoil.

1. Approximately 30 percent higher $c_{l,\max}$.
2. Approximately a 50 percent increase in the ratio of lift to drag (L/D) at a lift coefficient of 1.0. This value of $c_l = 1.0$ is typical of the climb lift coefficient for general aviation aircraft, and a high value of L/D greatly improves the climb performance. (See Ref. 2 for a general introduction to airplane performance and the importance of a high L/D ratio to airplane efficiency.)

It is interesting to note that the shape of the airfoil in Fig. 4.27 is very similar to the supercritical airfoils to be discussed in Chap. 11. The development of the supercritical airfoil by NASA aerodynamicist Richard Whitcomb in 1965 resulted in a major improvement in airfoil drag behavior at high subsonic speeds, near Mach 1. The supercritical airfoil was a major breakthrough in high-speed aerodynamics. The LS(1)-0417 low-speed airfoil shown in Fig. 4.27, first introduced as the GA(W)-1 airfoil, was a later spin-off from supercritical airfoil research. It is also interesting to note that the first production aircraft to use the NASA LS(1)-0417 airfoil was the Piper PA-38 Tomahawk, introduced in the late 1970s.

In summary, new airfoil development is alive and well in the aeronautics of the late twentieth century. Moreover, in contrast to the purely experimental development of the earlier airfoils, we now enjoy the benefit of powerful computer programs using panel methods and advanced viscous flow solutions for the design of new airfoils. Indeed, NASA has established an official Airfoil Design Center at The Ohio State University, which services the entire general aviation industry with over 30 different computer programs for airfoil design and analysis. For additional information on such new low-speed airfoil development, you are urged to read Ref. 16, which is the classic first publication dealing with these airfoils, as well as the concise review given in Ref. 17.

4.11 APPLIED AERODYNAMICS: THE FLOW OVER AN AIRFOIL— THE REAL CASE

In this chapter, we have studied the inviscid, incompressible flow over airfoils. When compared with actual experimental lift and moment data for airfoils in low-speed flows, we have seen that our theoretical results based on the assumption of inviscid flow are quite good—with one glaring exception. In the real case, flow separation occurs over the top surface of the airfoil when the angle of attack exceeds a certain value—the "stalling" angle of attack. As described in Sec. 4.3, this is a viscous effect. As shown in Fig. 4.4, the lift coefficient reaches a local maximum denoted by $c_{l,\max}$, and the angle of attack at which $c_{l,\max}$ is achieved is the stalling angle of attack. An increase in α beyond this value usually results in a (sometimes rather precipitous) drop in lift. At angles of attack well below the stalling angle, the experimental data clearly show a *linear* increase in c_l with increasing α—a result that is predicted by the theory presented in this chapter. Indeed, in this linear region, the inviscid flow theory is in excellent agreement with the experiment, as reflected in Fig. 4.5 and as demonstrated by Example 4.2. However, the inviscid theory does not predict flow separation, and consequently the prediction of $c_{l,\max}$ and the stalling angle of attack must be treated in some fashion by viscous flow theory. Such viscous flow analyses are the purview of Part IV. On the other hand, the purpose of this section is to examine the *physical* features of the real flow over an airfoil, and flow separation is an inherent part of this real flow. Therefore, let us take a more detailed look at how the flow field over an airfoil changes as the angle of attack is increased, and how the lift coefficient is affected by such changes.

The flow fields over an NACA 4412 airfoil at different angles of attack are shown in Fig. 4.29. Here, the streamlines are drawn to scale as obtained from the experimental results of Hikaru Ito given in Ref. 50. The experimental streamline patterns were made visible by a smoke wire technique, wherein metallic wires spread with oil over their surfaces were heated by an electric pulse and the resulting white smoke creates visible streaklines in the flow field. In Fig. 4.29, the angle of attack is progressively increased as we scan from Fig. 4.29a to e; to the right of each streamline picture is an arrow, the length of which is proportional to the value of the lift coefficient at the given angle of attack. The actual experimentally measured lift curve for the airfoil is given in Fig. 4.29f. Note that at low angle of attack, such as $\alpha = 2°$ in Fig. 4.29a, the streamlines are relatively undisturbed from their freestream shapes and c_l is small. As α is increased to 5°, as shown in Fig. 4.29b, and then to 10°, as shown in Fig. 4.29c, the streamlines exhibit a pronounced upward deflection in the region of the leading edge, and a subsequent downward deflection in the region of the trailing edge. Note that the stagnation point progressively moves downstream of the leading edge over the bottom surface of the airfoil as α is increased. Of course, c_l increases as α is increased, and, in this region, the increase is linear, as seen in Fig. 4.29f. When α is increased to slightly less than 15°, as shown in Fig. 4.29d, the curvature of the streamlines is particularly apparent. In Fig. 4.29d, the flow field is still attached over the top surface of the airfoil. However, as α is further increased slightly above 15°, massive flow-field separation occurs over the top surface, as shown in Fig. 4.29e. By slightly increasing α from that shown in Fig. 4.29d to that in Fig. 4.29e, the flow quite suddenly separates from the leading edge and the lift coefficient experiences a precipitous decrease, as seen in Fig. 4.29f.

The type of stalling phenomenon shown in Fig. 4.29 is called *leading-edge stall*; it is characteristic of relatively thin airfoils with thickness ratios between 10 and 16 percent of the chord length. As seen above, flow separation takes place rather suddenly and abruptly over the entire top surface of the airfoil, with the origin of this separation occurring at the leading edge. Note that the lift curve shown in Fig. 4.29f is rather sharp-peaked in the vicinity of $c_{l,\max}$ with a rapid decrease in c_l above the stall.

A second category of stall is the *trailing-edge stall*. This behavior is characteristic of thicker airfoils such as the NACA 4421 shown in Fig. 4.30. Here, we see a progressive and gradual movement of separation from the trailing edge toward the leading edge as α is increased. The lift curve for this case is shown

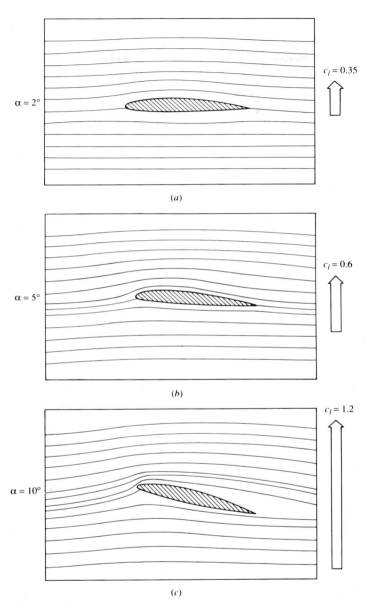

FIGURE 4.29

Example of leading-edge stall. Streamline patterns for an NACA 4412 airfoil at different angles of attack. (*The streamlines are drawn to scale from experimental data given by Hikaru Ito in Ref. 50.*) $Re = 2.1 \times 10^5$ and $V_\infty = 8$ m/s in air. The corresponding experimentally measured lift coefficients are indicated by arrows at the right of each streamline picture, where the length of each arrow indicates the relative magnitude of the lift. The lift coefficient is also shown in part (f).

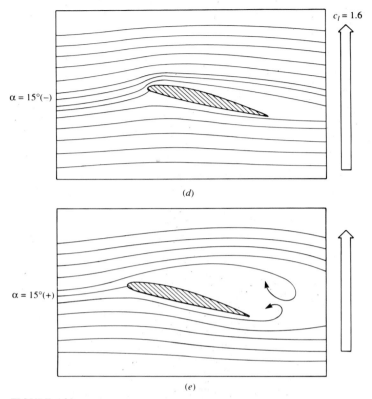

$\alpha = 15°(-)$

$c_l = 1.6$

(d)

$\alpha = 15°(+)$

(e)

FIGURE 4.29
(Continued.)

in Fig. 4.31. The solid curve in Fig. 4.31 is a repeat of the results for the NACA 4412 airfoil shown earlier in Fig. 4.29f—an airfoil with a leading-edge stall. The dot-dashed curve is the lift curve for the NACA 4421 airfoil—an airfoil with a trailing-edge stall. In comparing these two curves, note that:

1. The trailing-edge stall yields a gradual bending-over of the lift curve at maximum lift, in contrast to the sharp, precipitous drop in c_l for the leading-edge stall. The stall is "soft" for the trailing-edge stall.

2. The value of $c_{l,\max}$ is not so large for the trailing-edge stall.

3. For both the NACA 4412 and 4421 airfoils, the shape of the mean camber line is the same. From the thin airfoil theory discussed in this chapter, the linear lift slope and the zero-lift angle of attack should be the same for both airfoils; this is confirmed by the experimental data in Fig. 4.31. The only difference between the two airfoils is that one is thicker than the other. Hence, comparing results shown in Figs. 4.29 to 4.31, we conclude that the major effect of thickness of the airfoil is its effect on the value of $c_{l,\max}$, and this

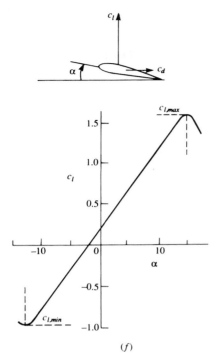

FIGURE 4.29

(f)　(Continued.)

effect is mirrored by the leading-edge stall behavior of the thinner airfoil versus the trailing-edge stall behavior of the thicker airfoil.

There is a third type of stall behavior, namely, behavior associated with the extreme thinness of an airfoil. This is sometimes labeled as "thin airfoil stall." An extreme example of a very thin airfoil is a flat plate; the lift curve for a flat plate is shown as the dashed curve in Fig. 4.31 labeled "thin airfoil stall." The streamline patterns for the flow over a flat plate at various angles of attack are given in Fig. 4.32. The thickness of the flat plate is 2 percent of the chord length. Inviscid, incompressible flow theory shows that the velocity becomes infinitely large at a sharp convex corner; the leading edge of a flat plate at an angle of attack is such a case. In the real flow over the plate as shown in Fig. 4.32, nature addresses this singular behavior by having the flow separate at the leading edge, even for very low values of α. Examining Fig. 4.32a, where $\alpha = 3°$, we observe a small region of separated flow at the leading edge. This separated flow reattaches to the surface further downstream, forming a *separation bubble* in the region near the leading edge. As α is increased, the reattachment point moves further downstream; i.e., the separation bubble becomes larger. This is illustrated in Fig. 4.32b where $\alpha = 7°$. At $\alpha = 9°$ (Fig. 4.32c), the separation bubble extends over almost the complete flat plate. Referring back to Fig. 4.31, we note that this angle of attack corresponds to $c_{l,\max}$ for the flat plate. When α is increased further, total

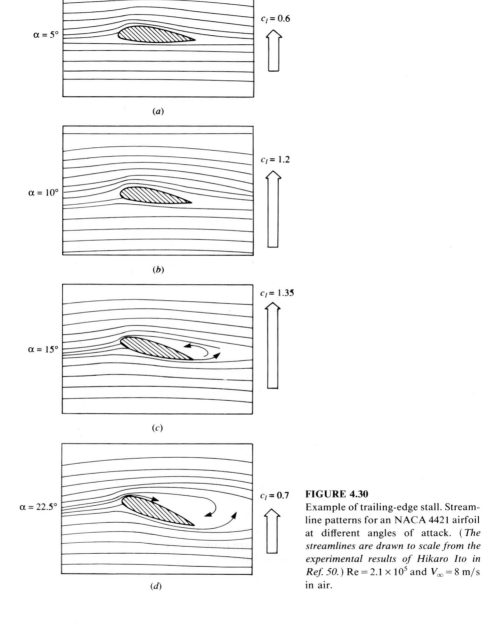

FIGURE 4.30

Example of trailing-edge stall. Streamline patterns for an NACA 4421 airfoil at different angles of attack. (*The streamlines are drawn to scale from the experimental results of Hikaro Ito in Ref. 50.*) $Re = 2.1 \times 10^5$ and $V_\infty = 8$ m/s in air.

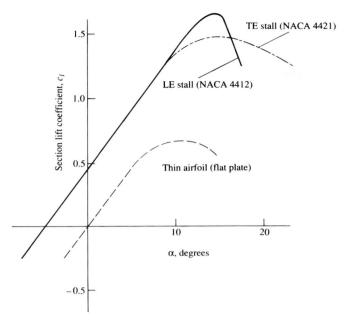

FIGURE 4.31
Lift-coefficient curves for three airfoils with different aerodynamic behavior: Trailing-edge stall (NACA 4421 airfoil), leading-edge stall (NACA 4412 airfoil), thin airfoil stall (flat plate).

flow separation is present, such as shown in Fig. 4.32d. The lift curve for the flat plate in Fig. 4.31 shows an early departure from its linear variation at about $\alpha = 3°$; this corresponds to the formation of the leading-edge separation bubble. The lift curve gradually bends over as α is increased further and exhibits a very gradual and "soft" stall. This is a trend similar to the case of the trailing-edge stall, although the physical aspects of the flow are quite different between the two cases. Of particular importance is the fact that $c_{l,\max}$ for the flat plate is considerably smaller than that for the two NACA airfoils compared in Fig. 4.31. Hence, we can conclude from Fig. 4.31 that the value of $c_{l,\max}$ is critically dependent on airfoil thickness. In particular, by comparing the flat plate with the two NACA airfoils, we see that *some* thickness is vital to obtaining a high value of $c_{l,\max}$. However, beyond that, the amount of thickness will influence the type of stall (leading-edge versus trailing-edge), and airfoils that are very thick tend to exhibit reduced values of $c_{l,\max}$ as the thickness increases. Hence, if we plot $c_{l,\max}$ versus thickness ratio, we expect to see a local minimum. Such is indeed the case, as shown in Fig. 4.33. Here, experimental data for $c_{l,\max}$ for the NACA 63-2XX series of airfoils is shown as a function of the thickness ratio. Note that as the thickness ratio increases from a small value, $c_{l,\max}$ first increases, reaches a maximum value at a thickness ratio of about 12 percent, and then decreases at larger thickness ratios. The experimental data in Fig. 4.33 is plotted with the Reynolds number as a parameter. Note that $c_{l,\max}$ for a given airfoil is clearly a

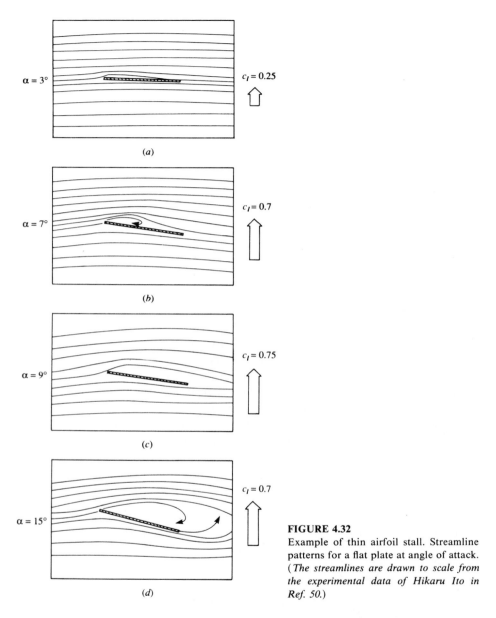

FIGURE 4.32
Example of thin airfoil stall. Streamline patterns for a flat plate at angle of attack. (*The streamlines are drawn to scale from the experimental data of Hikaru Ito in Ref. 50.*)

function of Re, with higher values of $c_{l,\mathrm{max}}$ corresponding to higher Reynolds numbers. Since flow separation is responsible for the lift coefficient exhibiting a local maximum, since flow separation is a viscous phenomenon, and since a viscous phenomenon is governed by a Reynolds number, it is no surprise that $c_{l,\mathrm{max}}$ exhibits some sensitivity to Re.

When was the significance of airfoil thickness first understood and appreciated? This question is addressed in the historical note in Sec. 4.12, where we will

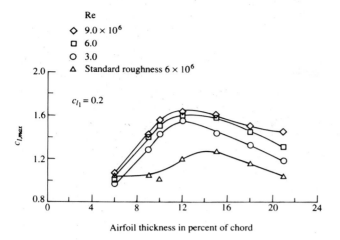

FIGURE 4.33
Effect of airfoil thickness on maximum lift coefficient for the NACA 63-2XX series of airfoils. (*From Abbott and von Doenhoff, Ref. 11.*)

see that the aerodynamic properties of thick airfoils even transcended technology during World War I and impacted the politics of the armistice.

Let us examine some other aspects of airfoil aerodynamics—aspects that are not always appreciated in a first study of the subject. The simple generation of lift by an airfoil is not the prime consideration in its design—even a barn door at an angle of attack produces lift. Rather, there are two figures of merit that are primarily used to judge the quality of a given airfoil:

1. *The lift-to-drag ratio, L/D.* An efficient airfoil produces lift with a minimum of drag; i.e., the ratio of lift-to-drag is a measure of the aerodynamic efficiency of an airfoil. The standard airfoils discussed in this chapter have high L/D ratios—much higher than that of a barn door. The L/D ratio for a complete flight vehicle has an important impact on its flight performance; e.g., the range of the vehicle is directly proportional to the L/D ratio. (See Ref. 2 for an extensive discussion of the role of L/D on flight performance of an airplane.)
2. *The maximum lift coefficient, $c_{l,max}$.* An effective airfoil produces a high value of $c_{l,max}$—much higher than that produced by a barn door.

The maximum lift coefficient is worth some additional discussion here. For a complete flight vehicle, the maximum lift coefficient, $c_{l,max}$, determines the stalling speed of the aircraft. This is easily seen by considering equilibrium, steady flight, where the lift L must equal the aircraft weight, W. Hence,

$$L = W = \tfrac{1}{2}\rho_\infty V_\infty^2 S C_L$$

or

$$V_\infty = \sqrt{\frac{2W}{\rho_\infty S C_L}}$$

In turn, the lowest possible speed of the aircraft—the stalling speed—occurs when the lift coefficient is a maximum value:

$$V_{\text{stall}} = \sqrt{\frac{2W}{\rho_\infty S C_{L,\max}}} \tag{4.79}$$

Therefore, a tremendous incentive exists to increase the maximum lift coefficient of an airfoil, in order to obtain either lower stalling speeds or higher payload weights at the same speed, as reflected in Eq. (4.79). Moreover, the maneuverability of an airplane (i.e., the smallest possible turn radius and the fastest possible turn rate) depends on a large value of $C_{L,\max}$ (see Sec. 6.17 of Ref. 2). On the other hand, for an airfoil at a given Reynolds number, the value of $c_{l,\max}$ is a function primarily of its shape. Once the shape is specified, the value of $c_{l,\max}$ is what nature dictates, as we have already seen. Therefore, to increase $c_{l,\max}$ beyond such a value, we must carry out some special measures. Such special measures include the use of flaps and/or leading-edge slats to increase $c_{l,\max}$ above that for the reference airfoil itself. These are called *high-lift devices,* and are discussed in more detail below.

A trailing-edge flap is simply a portion of the trailing-edge section of the airfoil that is hinged and which can be deflected upward or downward, as sketched in the insert in Fig. 4.34a. When the flap is deflected downward (a positive angle

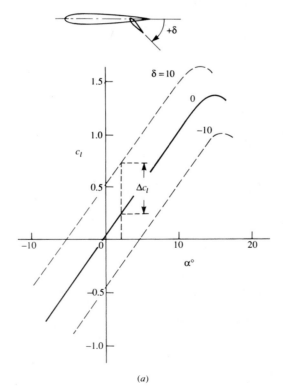

(a)

FIGURE 4.34
Effect of flap deflection on streamline shapes. (*The streamlines are drawn to scale from the experimental data of Hikaru Ito in Ref. 50.*) (a) Effect of flap deflection on lift coefficient. (b) Streamline pattern with no flap deflection. (c) Streamline pattern with a 15° flap deflection.

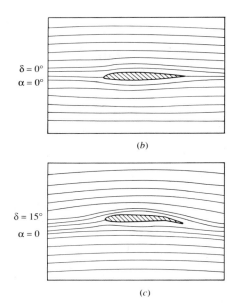

$\delta = 0°$
$\alpha = 0°$

(b)

$\delta = 15°$
$\alpha = 0$

(c)

FIGURE 4.34
(Continued.)

δ in Fig. 4.34a), the lift coefficient is increased, as shown in Fig. 4.34a. This increase is due to an effective increase in the camber of the airfoil as the flap is deflected downward. The thin airfoil theory presented in this chapter clearly shows that the zero-lift angle of attack is a function of the amount of camber [see Eq. (4.61)], with $\alpha_{L=0}$ becoming more negative as the camber is increased. In terms of the lift curve shown in Fig. 4.34a, the original curve for no flap deflection goes through the origin because the airfoil is symmetric; however, as the flap is deflected downward, this lift curve simply translates to the left because $\alpha_{L=0}$ is becoming more negative. In Fig. 4.34a, the results are given for flap deflections of $\pm 10°$. Comparing the case for $\delta = 10°$ with the no-deflection case, we see that, at a given angle of attack, the lift coefficient is increased by an amount Δc_l due to flap deflection. Moreover, the actual value of $c_{l,\max}$ is increased by flap deflection, although the angle of attack at which $c_{l,\max}$ occurs is slightly decreased. The change in the streamline pattern when the flap is deflected is shown in Fig. 4.34b and c. Figure 4.34b is the case for $\alpha = 0$ and $\delta = 0$, i.e., a symmetric flow. However, when α is held fixed at zero, but the flap is deflected by 15°, as shown in Fig. 4.34c, the flow field becomes unsymmetrical and resembles the lifting flows shown, e.g., in Fig. 4.29. That is, the streamlines in Fig. 4.34c are deflected upward in the vicinity of the leading edge and downward near the trailing edge, and the stagnation point moves to the lower surface of the airfoil— just by deflecting the flap downward.

High-lift devices can also be applied to the leading edge of the airfoil, as shown in the insert in Fig. 4.35. These can take the form of a leading-edge slat, leading-edge droop, or a leading-edge flap. Let us concentrate on the leading-edge

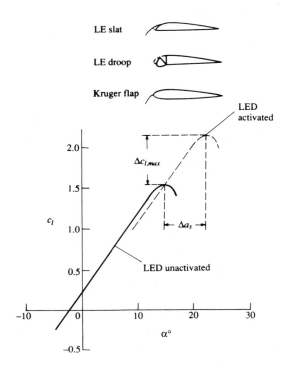

FIGURE 4.35
Effect of leading-edge flap on lift coefficient.

slat, which is simply a thin, curved surface that is deployed in front of the leading edge. In addition to the primary airflow over the airfoil, there is now a secondary flow that takes place through the gap between the slat and the airfoil leading edge. This secondary flow from the bottom to the top surface modifies the pressure distribution over the top surface; the adverse pressure gradient which would normally exist over much of the top surface is mitigated somewhat by this secondary flow, hence delaying flow separation over the top surface. Thus, a leading-edge slat increases the stalling angle of attack, and hence yields a higher $c_{l,max}$, as shown by the two lift curves in Fig. 4.35, one for the case without a leading-edge device and the other for the slat deployed. Note that the function of a leading-edge slat is inherently different from that of a trailing-edge flap. There is no change in $\alpha_{L=0}$; rather, the lift curve is simply extended to a higher stalling angle of attack, with the attendant increase in $c_{l,max}$. The streamlines of a flow field associated with an extended leading-edge slat are shown in Fig. 4.36. The airfoil is in an NACA 4412 section. (*Note*: The flows shown in Fig. 4.36 do not correspond exactly with the lift curves shown in Fig. 4.35, although the general behavior is the same.) The stalling angle of attack for the NACA 4412 airfoil without slat extension is about 15°, but increases to about 30° when the slat is extended. In Fig. 4.36*a*, the angle of attack is 10°. Note the flow through the gap between the slat and the leading edge. In Fig. 4.36*b*, the angle of attack

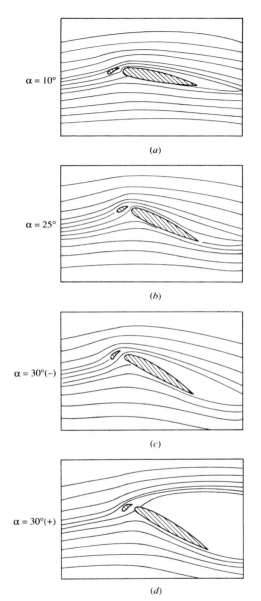

$\alpha = 10°$

(a)

$\alpha = 25°$

(b)

$\alpha = 30°(-)$

(c)

$\alpha = 30°(+)$

(d)

FIGURE 4.36
Effect of a leading-edge slat on the streamline pattern over an NACA 4412 airfoil. (*The streamlines are drawn to scale from the experimental data in Ref. 50.*)

is 25° and the flow is still attached. This prevails to an angle of attack slightly less than 30°, as shown in Fig. 4.36c. At slightly higher than 30°, flow separation suddenly occurs and the airfoil stalls.

The high-lift devices used on modern, high-performance aircraft are usually a combination of leading-edge slats (or flaps) and multi-element trailing-edge flaps. Typical airfoil configurations with these devices are sketched in Fig. 4.37. Three configurations including the high-lift devices are shown: *A*—the cruise configuration, with no deployment of the high-lift devices; *B*—a typical configuration at takeoff, with both the leading- and trailing-edge devices partially deployed; and *C*—a typical configuration at landing, with all devices fully extended. Note that for configuration *C*, there are a gap between the slat and the leading edge and several gaps between the different elements of the multi-element trailing-edge flap. The streamline pattern for the flow over such a configuration is shown in Fig. 4.38. Here, the leading-edge slat and the multi-element trailing-edge flap are fully extended. The angle of attack is 25°. Although the main flow over the top surface of the airfoil is essentially separated, the local flow through the gaps in the multi-element flap is locally attached to the top surface of the flap; because of this locally attached flow, the lift coefficient is still quite high, on the order of 4.5.

With this, we end our discussion of the real flow over airfoils. In retrospect, we can say that the real flow at high angles of attack is dominated by flow separation—a phenomenon that is not properly modeled by the inviscid theories presented in this chapter. On the other hand, at lower angles of attack, such as those associated with the cruise conditions of an airplane, the inviscid theories presented here do an excellent job of predicting both lift and moments on an airfoil. Moreover, in this section, we have clearly seen the importance of airfoil *thickness* in determining the angle of attack at which flow separation will occur, and hence greatly affecting the maximum lift coefficient.

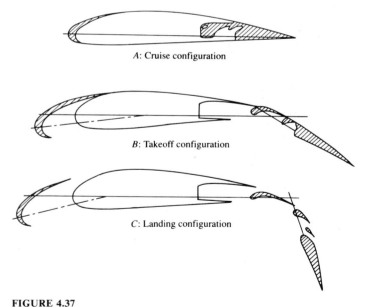

A: Cruise configuration

B: Takeoff configuration

C: Landing configuration

FIGURE 4.37
Airfoil with leading-edge and trailing-edge high-lift mechanisms. The trailing-edge device is a multi-element flap.

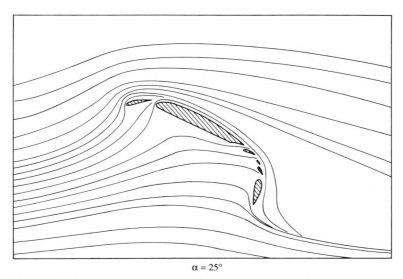

$$\alpha = 25°$$

FIGURE 4.38
Effect of leading-edge and multi-element trailing-edge flaps on the streamline pattern around an airfoil at angle of attack of 25°. (*The streamlines are drawn to scale from the experimental data of Ref. 50.*)

4.12 HISTORICAL NOTE: EARLY AIRPLANE DESIGN AND THE ROLE OF AIRFOIL THICKNESS

In 1804, the first modern configuration aircraft was conceived and built by Sir George Cayley in England—it was an elementary hand-launched glider, about a meter in length, and with a kitelike shape for a wing as shown in Fig. 4.39. (For the pivotal role played by George Cayley in the development of the airplane, see the extensive historical discussion in Chap. 1 of Ref. 2.) Note that right from the beginning of the modern configuration aircraft, the wing sections were very thin—whatever thickness was present, it was strictly for structural stiffness of the wing. Extremely thin airfoil sections were perpetuated by the work of Horatio Phillips in England. Phillips carried out the first serious wind-tunnel experiments in which the aerodynamic characteristics of a number of different airfoil shapes

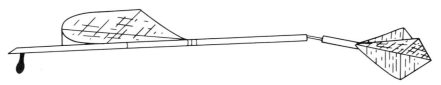

FIGURE 4.39
The first modern configuration airplane in history: George Cayley's model glider of 1804.

were measured. (See Sec. 5.20 of Ref. 2 for a presentation of the historical development of airfoils.) Some of Phillips airfoil sections are shown in Fig. 4.40—note that they are the epitome of exceptionally thin airfoils. The early pioneers of aviation such as Otto Lilienthal in Germany and Samuel Pierpont Langley in America (see Chap. 1 of Ref. 2) continued this thin airfoil tradition. This was especially true of the Wright brothers, who in the period of 1901–1902 tested hundreds of different wing sections and planform shapes in their wind tunnel in Dayton, Ohio (recall our discussion in Sec. 1.1 and the models shown in Fig. 1.2). A sketch of some of the Wrights' airfoil sections is given in Fig. 4.41—for the most part, very thin sections. Indeed, such a thin airfoil section was used on the 1903 Wright Flyer, as can be readily seen in the side view of the Flyer shown in Fig. 4.42. The important point here is that *all* of the early pioneering aircraft, and especially the Wright Flyer, incorporated very thin airfoil sections—airfoil sections that performed essentially like the flat plate results discussed in Sec. 4.11, and as shown in Fig. 4.31 (the dashed curve) and by the streamline pictures in Fig. 4.32. Conclusion: These early airfoil sections suffered flow-field separation at small angles of attack, and consequently had low values of $c_{l,\max}$. By the standards we apply today, these were simply very poor airfoil sections for the production of high lift.

This situation carried into the early part of World War I. In Fig. 4.43, we see four airfoil sections that were employed on World War I aircraft. The top three sections had thickness ratios of about 4 to 5 percent and are representative of the type of sections used on all aircraft until 1917. For example, the SPAD XIII (shown in Fig. 3.45), the fastest of all World War I fighters, had a thin airfoil section like the Eiffel section shown in Fig. 4.43. Why were such thin airfoil sections considered to be the best by most designers of World War I aircraft? The historical tradition described above might be part of the answer—a tradition that started with Cayley. Also, there was quite clearly a mistaken notion

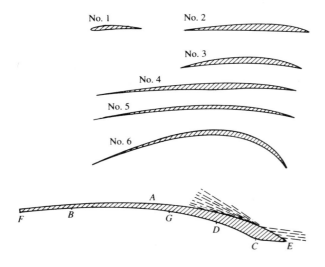

FIGURE 4.40
Double-surface airfoil sections by Horatio Phillips. The six upper shapes were patented by Phillips in 1884; the lower airfoil was patented in 1891. Note the thin profile shapes.

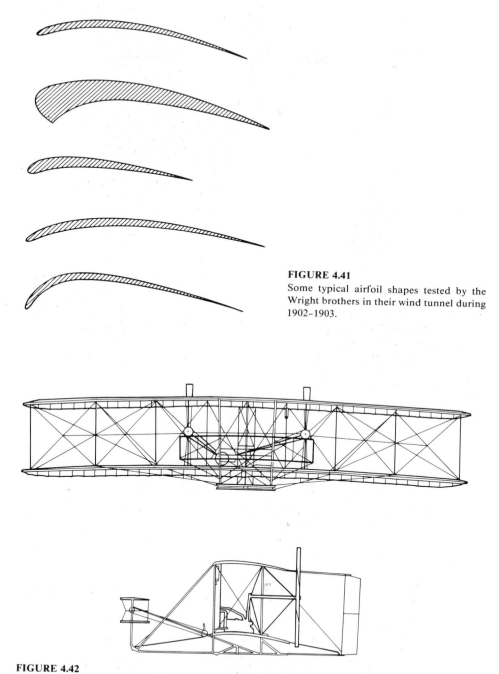

FIGURE 4.41
Some typical airfoil shapes tested by the Wright brothers in their wind tunnel during 1902–1903.

FIGURE 4.42
Front and side views of the 1903 Wright Flyer. Note the thin airfoil sections. (*Courtesy of the National Air and Space Museum.*)

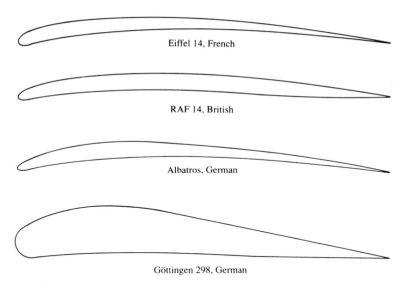

FIGURE 4.43
Some examples of different airfoil shapes used on World War I aircraft. (*From Loftin, Ref. 48.*)

at that time that *thick* airfoils would produce high drag. Of course, today we know the opposite to be true; review our discussion of streamlined shapes in Sec. 1.11 for this fact. Laurence Loftin in Ref. 48 surmises that the mistaken notion might have been fostered by early wind-tunnel tests. By the nature of the early wind tunnels in use—small sizes and very low speeds—the data were obtained at very low Reynolds numbers, less than 100,000 based on the airfoil-chord length. These Reynolds numbers are to be compared with typical values in the millions for actual airplane flight. Modern studies of low Reynolds number flows over conventional thick airfoils (e.g., see Ref. 51) clearly show high-drag coefficients, in contrast to the lower values that occur for the high Reynolds number associated with the flight of full-scale aircraft. Also, the reason for the World War I airplane designer's preference for thin airfoils might be as simple as the tendency to follow the example of the wings of birds, which are quite thin. In any event, the design of all English, French, and American World War I aircraft incorporated thin airfoils, and consequently suffered from poor high-lift performance. The fundamentals of airfoil aerodynamics as we know them today (and as being presented in this book) were simply not sufficiently understood by the designers during World War I. In turn, they never appreciated what they were losing.

This situation changed dramatically in 1917. Work carried out in Germany at the famous Göttingen aerodynamic laboratory of Ludwig Prandtl (see Sec. 5.7 for a biographical sketch of Prandtl) demonstrated the superiority of a thick airfoil section, such as the Göttingen 298 section shown at the bottom of Fig. 4.43. This revolutionary development was immediately picked up by the famous designer Anthony Fokker, who incorporated the 13-percent-thick Göttingen 298 profile in his new Fokker Dr-1—the famous triplane flown by the "Red Baron," Rittmeister Manfred Freiher von Richthofen. A photograph of the Fokker Dr-1 is shown in Fig. 4.44. The major benefits derived from Fokker's use of the thick airfoil were:

1. The wing structure could be completely internal; i.e., the wings of the Dr-1 were a cantilever design, which removed the need for the conventional wire bracing that was used in other aircraft. This, in turn, eliminated the high drag associated with these interwing wires, as discussed at the end of Sec. 1.11. For this reason, the Dr-1 had a zero-lift drag coefficient of 0.032, among the lowest of World War I airplanes. (By comparison the zero-lift drag coefficient of the French SPAD XIII was 0.037.)

2. The thick airfoil provided the Fokker Dr-1 with a high maximum lift coefficient. Its performance was analogous to the *upper* curves shown in Fig. 4.31. This, in turn, provided the Dr-1 with an exceptionally high rate-of-climb as well as enhanced maneuverability—characteristics that were dominant in dog-fighting combat.

Anthony Fokker continued the use of a thick airfoil in his design of the D-VII, as shown in Fig. 4.45. This gave the D-VII a much greater rate-of-climb than its two principal opponents at the end of the war—the English Sopwith Camel and the French SPAD XIII, both of which still used very thin airfoil sections. This rate-of-climb performance, as well as its excellent handling charac-

FIGURE 4.44
The World War I Fokker Dr-1 triplane, the first fighter aircraft to use a thick airfoil. (*From Loftin, Ref. 48.*)

FIGURE 4.45
The World War I Fokker D-VII, one of the most effective fighters of the war, due in part to its superior aerodynamic performance allowed by a thick airfoil section.

teristics, singled out the Fokker D-VII as the most effective of all German World War I fighters. The respect given by the Allies to this machine is no more clearly indicated than by a paragraph in article IV of the armistice agreement, which lists war material to be handed over to the Allies by Germany. In this article, the Fokker D-VII is *specifically* listed—the only airplane of any type to be explicitly mentioned in the armistice. To this author's knowledge, this is the one and only time where a breakthrough in airfoil technology is essentially reflected in any major political document, though somewhat implicitly.

4.13 HISTORICAL NOTE: KUTTA, JOUKOWSKI, AND THE CIRCULATION THEORY OF LIFT

Frederick W. Lanchester (1868–1946), an English engineer, automobile manufacturer, and self-styled aerodynamicist, was the first to connect the idea of circulation with lift. His thoughts were originally set forth in a presentation given before the Birmingham Natural History and Philosophical Society in 1894 and later contained in a paper submitted to the Physical Society, which turned it down. Finally, in 1907 and 1908, he published two books, entitled *Aerodynamics* and *Aero-*

donetics, where his thoughts on circulation and lift were described in detail. His books were later translated into German in 1909 and French in 1914. Unfortunately, Lanchester's style of writing was difficult to read and understand; this is partly responsible for the general lack of interest shown by British scientists in Lanchester's work. Consequently, little positive benefit was derived from Lanchester's writings. (See Sec. 5.6 for a more detailed portrait of Lanchester and his work.)

Quite independently, and with total lack of knowledge of Lanchester's thinking, M. Wilhelm Kutta (1867–1944) developed the idea that lift and circulation are related. Kutta was born in Pitschen, Germany, in 1867 and obtained a Ph.D. in mathematics from the University of Munich in 1902. After serving as professor of mathematics at several German technical schools and universities, he finally settled at the Technische Hochschule in Stuttgart in 1911 until his retirement in 1935. Kutta's interest in aerodynamics was initiated by the successful glider flights of Otto Lilienthal in Berlin during the period 1890–1896 (see chap. 1 of Ref. 2). Kutta attempted theoretically to calculate the lift on the curved wing surfaces used by Lilienthal. In the process, he surmised from experimental data that the flow left the trailing edge of a sharp-edged body smoothly and that this condition fixed the circulation around the body (the Kutta condition, described in Sec. 4.5). At the same time, he was convinced that circulation and lift were connected. Kutta was reluctant to publish these ideas, but after the strong insistence of his teacher, S. Finsterwalder, he wrote a paper entitled "Auftriebskrafte in Stromenden Flussigkecten" (Lift in Flowing Fluids). This was actually a short note abstracted from his longer graduation paper in 1902, but it represents the first time in history where the concepts of the Kutta condition as well as the connection of circulation with lift were officially published. Finsterwalder clearly repeated the ideas of his student in a lecture given on September 6, 1909, in which he stated:

> On the upper surface the circulatory motion increases the translatory one, therefore there is high velocity and consequently low pressure, while on the lower surface the two movements are opposite, therefore there is low velocity with high pressure, with the result of a thrust upward.

However, in his 1902 note, Kutta did not give the precise quantitative relation between circulation and lift. This was left to Nikolai Y. Joukowski (Zhukouski). Joukowski was born in Orekhovo in central Russia on January 5, 1847. The son of an engineer, he became an excellent student of mathematics and physics, graduating with a Ph.D. in applied mathematics from Moscow University in 1882. He subsequently held a joint appointment as a professor of mechanics at Moscow University and the Moscow Higher Technical School. It was at this latter institution that Joukowski built in 1902 the first wind tunnel in Russia. Joukowski was deeply interested in aeronautics, and he combined a rare gift for both experimental and theoretical work in the field. He expanded his wind tunnel into a major aerodynamics laboratory in Moscow. Indeed, during World War I, his laboratory was used as a school to train military pilots in the principles of aerodynamics and flight. When he died in 1921, Joukowski was by far the most noted aerodynamicist in Russia.

Much of Joukowski's fame was derived from a paper published in 1906, wherein he gives, for the first time in history, the relation $L' = \rho_\infty V_\infty \Gamma$—the Kutta-Joukowski theorem. In Joukowski's own words:

> If an irrotational two-dimensional fluid current, having at infinity the velocity V_∞, surrounds any closed contour on which the circulation of velocity is Γ, the force of the aerodynamic pressure acts on this contour in a direction perpendicular to the velocity and has the value
>
> $$L' = \rho_\infty V_\infty \Gamma$$
>
> The direction of this force is found by causing to rotate through a right angle the vector V_∞ around its origin in an inverse direction to that of the circulation.

Joukowski was unaware of Kutta's 1902 note, and developed his ideas on circulation and lift independently. However, in recognition of Kutta's contribution, the equation given above has propagated through the twentieth century as the "Kutta-Joukowski theorem."

Hence, by 1906—just 3 years after the first successful flight of the Wright brothers—the circulation theory of lift was in place, ready to aid aerodynamics in the design and understanding of lifting surfaces. In particular, this principle formed the cornerstone of the thin airfoil theory described in Secs. 4.7 and 4.8. Thin airfoil theory was developed by Max Munk, a colleague of Prandtl in Germany, during the first few years after World War I. However, the very existence of thin airfoil theory, as well as its amazingly good results, rests upon the foundation laid by Lanchester, Kutta, and Joukowski a decade earlier.

4.14 SUMMARY

Return to the road map given in Fig. 4.2. Make certain that you feel comfortable with the material represented by each box on the road map and that you understand the flow of ideas from one box to another. If you are uncertain about one or more aspects, review the pertinent sections before progressing further.

Some important results from this chapter are itemized below:

A vortex sheet can be used to synthesize the inviscid, incompressible flow over an airfoil. If the distance along the sheet is given by s and the strength of the sheet per unit length is $\gamma(s)$, then the velocity potential induced at point (x, y) by a vortex sheet that extends from point a to point b is

$$\phi(x, y) = -\frac{1}{2\pi} \int_a^b \theta \gamma(s) \, ds \tag{4.3}$$

The circulation associated with this vortex sheet is

$$\Gamma = \int_a^b \gamma(s) \, ds \tag{4.4}$$

Across the vortex sheet, there is a tangential velocity discontinuity, where

$$\gamma = u_1 - u_2 \tag{4.8}$$

The Kutta condition is an observation that for a lifting airfoil of given shape at a given angle of attack, nature adopts that particular value of circulation around the airfoil which results in the flow leaving smoothly at the trailing edge. If the trailing-edge angle is finite, then the trailing edge is a stagnation point. If the trailing edge is cusped, then the velocities leaving the top and bottom surfaces at the trailing edge are finite and equal in magnitude and direction. In either case,

$$\gamma(\text{TE}) = 0 \tag{4.10}$$

Thin airfoil theory is predicated on the replacement of the airfoil by the mean camber line. A vortex sheet is placed along the chord line, and its strength adjusted such that, in conjunction with the uniform freestream, the camber line becomes a streamline of the flow while at the same time satisfying the Kutta condition. The strength of such a vortex sheet is obtained from the fundamental equation of thin airfoil theory:

$$\frac{1}{2\pi} \int_0^c \frac{\gamma(\xi)\, d\xi}{x - \xi} = V_\infty \left(\alpha - \frac{dz}{dx} \right) \tag{4.18}$$

Results of thin airfoil theory:

Symmetric airfoil

1. $c_l = 2\pi\alpha$.
2. Lift slope = $dc_l/d\alpha = 2\pi$.
3. The center of pressure and the aerodynamic center are both at the quarter-chord point.
4. $c_{m,c/4} = c_{m,\text{ac}} = 0$.

Cambered airfoil

1. $c_l = 2\pi[\alpha + (1/\pi)\int_0^\pi (dz/dx)(\cos\theta_0 - 1)\, d\theta_0]$. $\tag{4.57}$
2. Lift slope = $dc_l/d\alpha = 2\pi$.
3. The aerodynamic center is at the quarter-chord point.
4. The center of pressure varies with the lift coefficient.

The vortex panel method is an important numerical technique for the solution of the inviscid, incompressible flow over bodies of arbitrary shape, thickness, and angle of attack. For panels of constant strength, the governing equations are

$$V_\infty \cos \beta_i - \sum_{j=1}^{n} \frac{\gamma_j}{2\pi} \int_j \frac{\partial \theta_{ij}}{\partial n_i} \, ds_j = 0 \qquad (i = 1, 2, \ldots, n)$$

and

$$\gamma_i = -\gamma_{i-1}$$

which is one way of expressing the Kutta condition for the panels immediately above and below the trailing edge.

PROBLEMS

4.1. Consider the data for the NACA 2412 airfoil given in Fig. 4.5. Calculate the lift and moment about the quarter chord (per unit span) for this airfoil when the angle of attack is 4° and the freestream is at standard sea level conditions with a velocity of 50 ft/s. The chord of the airfoil is 2 ft.

4.2. Consider an NACA 2412 airfoil with a 2-m chord in an airstream with a velocity of 50 m/s at standard sea level conditions. If the lift per unit span is 1353 N, what is the angle of attack?

4.3. Starting with the definition of circulation, derive Kelvin's circulation theorem, Eq. (4.11).

4.4. Starting with Eq. (4.35), derive Eq. (4.36).

4.5. Consider a thin, symmetric airfoil at 1.5° angle of attack. From the results of thin airfoil theory, calculate the lift coefficient and the moment coefficient about the leading edge.

4.6. The NACA 4412 airfoil has a mean camber line given by

$$\frac{z}{c} = \begin{cases} 0.25 \left[0.8 \dfrac{x}{c} - \left(\dfrac{x}{c} \right)^2 \right] & \text{for } 0 \le \dfrac{x}{c} \le 0.4 \\[2ex] 0.111 \left[0.2 + 0.8 \dfrac{x}{c} - \left(\dfrac{x}{c} \right)^2 \right] & \text{for } 0.4 \le \dfrac{x}{c} \le 1 \end{cases}$$

Using thin airfoil theory, calculate
(a) $\alpha_{L=0}$ (b) c_l when $\alpha = 3°$

4.7. For the airfoil given in Prob. 4.6, calculate $c_{m,c/4}$ and x_{cp}/c when $\alpha = 3°$.

4.8. Compare the results of Probs. 4.6 and 4.7 with experimental data for the NACA 4412 airfoil, and note the percentage difference between theory and experiment. (*Hint:* A good source of experimental airfoil data is Ref. 11.)

4.9. Starting with Eqs. (4.35) and (4.43), derive Eq. (4.62).

CHAPTER
5

INCOMPRESSIBLE FLOW OVER FINITE WINGS

The one who has most carefully watched the soaring birds of prey sees man with wings and the faculty of using them.

James Means, Editor of the Aeronautical Annual, 1895

5.1 INTRODUCTION: DOWNWASH AND INDUCED DRAG

In Chap. 4 we discussed the properties of airfoils, which are the same as the properties of a wing of infinite span; indeed, airfoil data are frequently denoted as "infinite wing" data. However, all real airplanes have wings of finite span, and the purpose of the present chapter is to apply our knowledge of airfoil properties to the analysis of such finite wings. This is the second step in Prandtl's philosophy of wing theory, as described in Sec. 4.1. You should review Sec. 4.1 before proceeding further.

Question: Why are the aerodynamic characteristics of a finite wing any different from the properties of its airfoil sections? Indeed, an airfoil is simply a section of a wing, and at first thought, you might expect the wing to behave exactly the same as the airfoil. However, as studied in Chap. 4, the flow over an airfoil is two-dimensional. In contrast, a finite wing is a three-dimensional body, and consequently the flow over the finite wing is three-dimensional; i.e., there is a component of flow in the spanwise direction. To see this more clearly, examine Fig. 5.1, which gives the top and front views of a finite wing. The physical mechanism for generating lift on the wing is the existence of a high pressure on the bottom surface and a low pressure on the top surface. The net imbalance of the pressure distribution creates the lift, as discussed in Sec. 1.5. However, as a

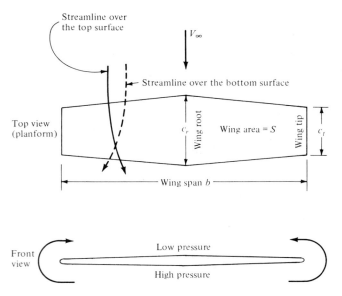

FIGURE 5.1
Finite wing. In this figure, the curvature of the streamlines over the top and bottom of the wing is exaggerated for clarity.

by-product of this pressure imbalance, the flow near the wing tips tends to curl around the tips, being forced from the high-pressure region just underneath the tips to the low-pressure region on top. This flow around the wing tips is shown in the front view of the wing in Fig. 5.1. As a result, on the top surface of the wing, there is generally a spanwise component of flow from the tip toward the wing root, causing the streamlines over the top surface to bend toward the root, as sketched on the top view shown in Fig. 5.1. Similarly, on the bottom surface of the wing, there is generally a spanwise component of flow from the root toward the tip, causing the streamlines over the bottom surface to bend toward the tip. Clearly, the flow over the finite wing is three-dimensional, and therefore you would expect the overall aerodynamic properties of such a wing to differ from those of its airfoil sections.

The tendency for the flow to "leak" around the wing tips has another important effect on the aerodynamics of the wing. This flow establishes a circulatory motion which trails downstream of the wing; i.e., a trailing *vortex* is created at each wing tip. These wing-tip vortices are sketched in Fig. 5.2 and are illustrated in Fig. 5.3. The tip vortices are essentially weak "tornadoes" that trail downstream of the finite wing. (For large airplanes such as a Boeing 747, these tip vortices can be powerful enough to cause light airplanes following too closely to go out of control. Such accidents have occurred, and this is one reason for large spacings between aircraft landing or taking off consecutively at airports.) These wing-tip vortices downstream of the wing induce a small downward component of air

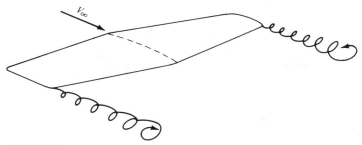

FIGURE 5.2
Schematic of wing-tip vortices.

velocity in the neighborhood of the wing itself. This can be seen by inspecting Fig. 5.3; the two vortices tend to drag the surrounding air around with them, and this secondary movement induces a small velocity component in the downward direction at the wing. This downward component is called *downwash*, denoted by the symbol w. In turn, the downwash combines with the freestream velocity

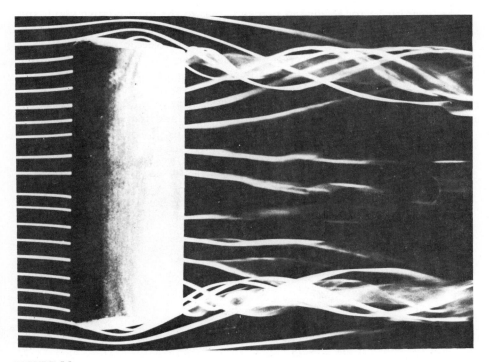

FIGURE 5.3
Wing-tip vortices from a rectangular wing. The wing is in a smoke tunnel, where individual streamtubes are made visible by means of smoke filaments. [*From Head, M. R., in* Flow Visualization II, *W. Merzkirch (ed.), Hemisphere Publishing Co., New York, 1982, pp. 399–403. Also available in Van Dyke, Milton,* An Album of Fluid Motion, *The Parabolic Press, Stanford, Calif., 1982.*]

V_∞ to produce a *local* relative wind which is canted downward in the vicinity of each airfoil section of the wing, as sketched in Fig. 5.4.

Examine Fig. 5.4 closely. The angle between the chord line and the direction of V_∞ is the angle of attack α, as defined in Sec. 1.5 and as used throughout our discussion of airfoil theory in Chap. 4. We now more precisely define α as the *geometric* angle of attack. In Fig. 5.4, the local relative wind is inclined below the direction of V_∞ by the angle α_i, called the *induced* angle of attack. The presence of downwash, and its effect on inclining the local relative wind in the downward direction, has two important effects on the local airfoil section, as follows:

1. The angle of attack actually seen by the local airfoil section is the angle between the chord line and the local relative wind. This angle is given by α_{eff} in Fig. 5.4 and is defined as the *effective* angle of attack. Hence, although the wing is at a geometric angle of attack α, the local airfoil section is seeing a smaller angle, namely, the effective angle of attack α_{eff}. From Fig. 5.4,

$$\alpha_{\text{eff}} = \alpha - \alpha_i \qquad (5.1)$$

2. The local lift vector is aligned perpendicular to the local relative wind, and hence is inclined behind the vertical by the angle α_i, as shown in Fig. 5.4. Consequently, there is a component of the local lift vector in the direction of V_∞; that is, there is a *drag* created by the presence of downwash. This drag is defined as *induced drag*, denoted by D_i in Fig. 5.4.

Hence, we see that the presence of downwash over a finite wing reduces the angle of attack that each section effectively sees, and moreover, it creates a

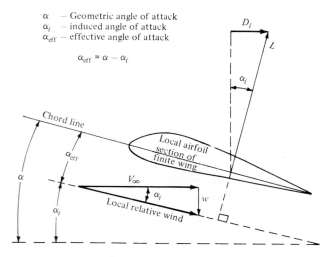

FIGURE 5.4
Effect of downwash on the local flow over a local airfoil section of a finite wing.

component of drag—the induced drag D_i. Keep in mind that we are still dealing with an inviscid, incompressible flow, where there is no skin friction or flow separation. For such a flow, there is a *finite* drag—the induced drag—on a finite wing. D'Alembert's paradox does *not* occur for a finite wing.

The tilting backward of the lift vector shown in Fig. 5.4 is one way of visualizing the physical generation of induced drag. Two alternate ways are as follows:

1. The three-dimensional flow induced by the wing-tip vortices shown in Figs. 5.2 and 5.3 simply alters the pressure distribution on the finite wing in such a fashion that a net pressure imbalance exists in the direction of V_∞; i.e., drag is created. In this sense, induced drag is a type of "pressure drag."

2. The wing-tip vortices contain a large amount of translational and rotational kinetic energy. This energy has to come from somewhere; indeed, it is ultimately provided by the aircraft engine, which is the only source of power associated with the airplane. Since the energy of the vortices serves no useful purpose, this power is essentially lost. In effect, the extra power provided by the engine that goes into the vortices is the extra power required from the engine to overcome the induced drag.

Clearly, from the discussion in this section, the characteristics of a finite wing are *not* identical to the characteristics of its airfoil sections. Therefore, let us proceed to develop a theory that will enable us to analyze the aerodynamic properties of finite wings. In the process, we follow the road map given in Fig. 5.5—keep in touch with this road map as we progress through the present chapter.

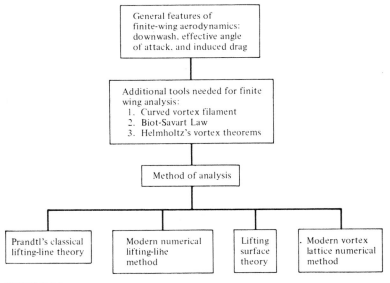

FIGURE 5.5
Road map for Chap. 5.

In this chapter, we note a difference in nomenclature. For the two-dimensional bodies considered in the previous chapters, the lift, drag, and moments per unit span have been denoted with primes, e.g., L', D', and M', and the corresponding lift, drag, and moment coefficients have been denoted by lowercase letters, e.g., c_l, c_d, and c_m. In contrast, the lift, drag, and moments on a complete three-dimensional body such as a finite wing are given without primes, e.g., L, D, and M, and the corresponding lift, drag, and moment coefficients are given by capital letters, e.g., C_L, C_D, and C_M. This distinction has already been mentioned in Sec. 1.5.

Finally, we note that the total drag on a subsonic finite wing in real life is the sum of the induced drag D_i, the skin friction drag D_f, and the pressure drag D_p due to flow separation. The latter two contributions are due to viscous effects, which are discussed in Chaps. 15 to 17. The sum of these two viscous-dominated drag contributions is called profile drag, as discussed in Sec. 4.3. The profile drag coefficient c_d for an NACA 2412 airfoil was given in Fig. 4.6. At moderate angle of attack, the profile drag coefficient for a finite wing is essentially the same as for its airfoil sections. Hence, defining the profile drag coefficient as

$$c_d = \frac{D_f + D_p}{q_\infty S} \tag{5.2}$$

and the induced drag coefficient as

$$C_{D,i} = \frac{D_i}{q_\infty S} \tag{5.3}$$

the total drag coefficient for the finite wing, C_D, is given by

$$C_D = c_d + C_{D,i} \tag{5.4}$$

In Eq. (5.4), the value of c_d is usually obtained from airfoil data, such as given in Fig. 4.6. The value of $C_{D,i}$ can be obtained from finite-wing theory as presented in this chapter. Indeed, one of the central objectives of the present chapter is to obtain an expression for induced drag and to study its variation with certain design characteristics of the finite wing. (See chap. 5 of Ref. 2 for an additional discussion of the characteristics of finite wings.)

5.2 THE VORTEX FILAMENT, THE BIOT-SAVART LAW, AND HELMHOLTZ'S THEOREMS

To establish a rational aerodynamic theory for a finite wing, we need to introduce a few additional aerodynamic tools. To begin with, we expand the concept of a vortex filament first introduced in Sec. 4.4. In Sec. 4.4, we discussed a *straight* vortex filament extending to $\pm\infty$. (Review the first paragraph of Sec. 4.4 before proceeding further.)

In general, a vortex filament can be *curved*, as shown in Fig. 5.6. Here, only a portion of the filament is illustrated. The filament induces a flow field in the

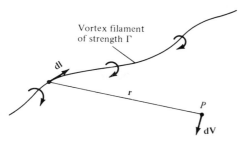

Vortex filament
of strength Γ

FIGURE 5.6
Vortex filament and illustration of the Biot-Savart law.

surrounding space. If the circulation is taken about any path enclosing the filament, a constant value, Γ, is obtained. Hence, the strength of the vortex filament is defined as Γ. Consider a directed segment of the filament **dl**, as shown in Fig. 5.6. The radius vector from **dl** to an arbitrary point P in space is **r**. The segment **dl** induces a velocity at P equal to

$$d\mathbf{V} = \frac{\Gamma}{4\pi} \frac{d\mathbf{l} \times \mathbf{r}}{|\mathbf{r}|^3} \tag{5.5}$$

Equation (5.5) is called the *Biot-Savart law* and is one of the most fundamental relations in the theory of inviscid, incompressible flow. Its derivation is given in more advanced books (see, e.g., Ref. 9). Here, we must accept it without proof. However, you might feel more comfortable if we draw an analogy with electromagnetic theory. If the vortex filament in Fig. 5.6 were instead visualized as a wire carrying an electrical current I, then the magnetic field strength **dB** induced at point P by a segment of the wire **dl** with the current moving in the direction of **dl** is

$$d\mathbf{B} = \frac{\mu I}{4\pi} \frac{d\mathbf{l} \times \mathbf{r}}{|\mathbf{r}|^3} \tag{5.6}$$

where μ is the permeability of the medium surrounding the wire. Equation (5.6) is identical in form to Eq. (5.5). Indeed, the Biot-Savart law is a general result of potential theory, and potential theory describes electromagnetic fields as well as inviscid, incompressible flows. In fact, our use of the word "induced" in describing velocities generated by the presence of vortices, sources, etc. is a carry-over from the study of electromagnetic fields induced by electrical currents. When developing their finite-wing theory during the period 1911–1918, Prandtl and his colleagues even carried the electrical terminology over to the generation of drag, hence the term "induced" drag.

Return again to our picture of the vortex filament in Fig. 5.6. Keep in mind that this single vortex filament and the associated Biot-Savart law [Eq. (5.5)] are simply conceptual aerodynamic tools to be used for synthesizing more complex flows of an inviscid, incompressible fluid. They are, for all practical purposes, a solution of the governing equation for inviscid, incompressible flow—Laplace's equation (see Sec. 3.7)—and, by themselves, are not of particular value. However, when a number of vortex filaments are used in conjunction with a uniform freestream, it is possible to synthesize a flow which has a practical application. The flow over a finite wing is one such example, as we will soon see.

Let us apply the Biot-Savart law to a straight vortex filament of infinite length, as sketched in Fig. 5.7. The strength of the filament is Γ. The velocity induced at point P by the directed segment of the vortex filament, dl, is given by Eq. (5.5). Hence, the velocity induced at P by the entire vortex filament is

$$V = \int_{-\infty}^{\infty} \frac{\Gamma}{4\pi} \frac{dl \times r}{|r|^3} \tag{5.7}$$

From the definition of the vector cross product (see Sec. 2.2), the direction of V is downward in Fig. 5.7. The magnitude of the velocity, $V = |V|$, is given by

$$V = \frac{\Gamma}{4\pi} \int_{-\infty}^{\infty} \frac{\sin\theta}{r^2} \, dl \tag{5.8}$$

In Fig. 5.7, let h be the perpendicular distance from point P to the vortex filament. Then, from the geometry shown in Fig. 5.7,

$$r = \frac{h}{\sin\theta} \tag{5.9a}$$

$$l = \frac{h}{\tan\theta} \tag{5.9b}$$

$$dl = -\frac{h}{\sin^2\theta} \, d\theta \tag{5.9c}$$

Substituting Eqs. (5.9a to c) in Eq. (5.8), we have

$$V = \frac{\Gamma}{4\pi} \int_{-\infty}^{\infty} \frac{\sin\theta}{r^2} \, dl = -\frac{\Gamma}{4\pi h} \int_{\pi}^{0} \sin\theta \, d\theta$$

or

$$V = \frac{\Gamma}{2\pi h} \tag{5.10}$$

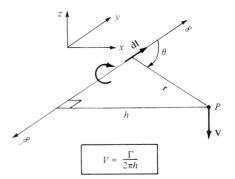

$$V = \frac{\Gamma}{2\pi h}$$

FIGURE 5.7
Velocity induced at point P by an infinite, straight vortex filament.

Thus, the velocity induced at a given point P by an infinite, straight vortex filament at a perpendicular distance h from P is simply $\Gamma/2\pi h$, which is precisely the result given by Eq. (3.105) for a point vortex in two-dimensional flow. [Note that the minus sign in Eq. (3.105) does not appear in Eq. (5.10); this is because V in Eq. (5.10) is simply the absolute magnitude of V, and hence it is positive by definition.]

Consider the *semi*-infinite vortex filament shown in Fig. 5.8. The filament extends from point A to ∞. Point A can be considered a boundary of the flow. Let P be a point in the plane through A perpendicular to the filament. Then, by an integration similar to that above (try it yourself), the velocity induced at P by the semi-infinite vortex filament is

$$V = \frac{\Gamma}{4\pi h} \tag{5.11}$$

We use Eq. (5.11) in the next section.

The great German mathematician, physicist, and physician Hermann von Helmholtz (1821–1894) was the first to make use of the vortex filament concept in the analysis of inviscid, incompressible flow. In the process, he established several basic principles of vortex behavior which have become known as Helmholtz's vortex theorems:

1. The strength of a vortex filament is constant along its length.
2. A vortex filament cannot end in a fluid; it must extend to the boundaries of the fluid (which can be $\pm\infty$) or form a closed path.

We make use of these theorems in the following sections.

Finally, let us introduce the concept of *lift distribution* along the span of a finite wing. Consider a given spanwise location y_1, where the local chord is c, the local geometric angle of attack is α, and the airfoil section is a given shape. The lift per unit span at this location is $L'(y_1)$. Now consider another location y_2 along the span, where c, α, and the airfoil shape may be different. (Most finite wings have a variable chord, with the exception of a simple rectangular wing. Also, many wings are geometrically twisted so that α is different at different spanwise locations—so-called geometric twist. If the tip is at a lower α than the

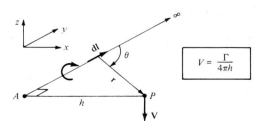

FIGURE 5.8
Velocity induced at point P by a semi-infinite straight vortex filament.

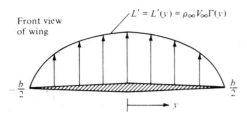

Front view
of wing

$L' = L'(y) = \rho_\infty V_\infty \Gamma(y)$

$-\dfrac{b}{2}$ $\dfrac{b}{2}$ y

FIGURE 5.9
Sketch of the lift distribution along the span of
a wing.

root, the wing is said to have *washout*; if the tip is at a higher α than the root, the wing has *washin*. In addition, the wings on a number of modern airplanes have different airfoil sections along the span, with different values of $\alpha_{L=0}$; this is called *aerodynamic twist*.) Consequently, the lift per unit span at this different location, $L'(y_2)$, will, in general, be different from $L'(y_1)$. Therefore, there is a distribution of lift per unit span along the wing, i.e., $L' = L'(y)$, as sketched in Fig. 5.9. In turn, the circulation is also a function of y, $\Gamma(y) = L'(y)/\rho_\infty V_\infty$. Note from Fig. 5.9 that the lift distribution goes to zero at the tips; this is because there is a pressure equalization from the bottom to the top of the wing precisely at $y = -b/2$ and $b/2$, and hence no lift is created at these points. The calculation of the lift distribution $L(y)$ [or the circulation distribution $\Gamma(y)$] is one of the central problems of finite-wing theory. It is addressed in the following sections.

In summary, we wish to calculate the induced drag, the total lift, and the lift distribution for a finite wing. This is the purpose of the remainder of this chapter.

5.3 PRANDTL'S CLASSICAL LIFTING-LINE THEORY

The first practical theory for predicting the aerodynamic properties of a finite wing was developed by Ludwig Prandtl and his colleagues at Göttingen, Germany, during the period 1911–1918, spanning World War I. The utility of Prandtl's theory is so great that it is still in use today for preliminary calculations of finite-wing characteristics. The purpose of this section is to describe Prandtl's theory and to lay the groundwork for the modern numerical methods described in subsequent sections.

Prandtl reasoned as follows. A vortex filament of strength Γ which is somehow bound to a fixed location in a flow—a so-called bound vortex—will experience a force $L = \rho_\infty V_\infty \Gamma$ from the Kutta-Joukowski theorem. This bound vortex is in contrast to a *free vortex*, which moves with the same fluid elements throughout a flow. Therefore, let us replace a finite wing of span b with a bound vortex, extending from $y = -b/2$ to $y = b/2$, as sketched in Fig. 5.10. However, due to Helmholtz's theorem, a vortex filament cannot end in the fluid. Therefore, assume the vortex filament continues as two free vortices trailing downstream

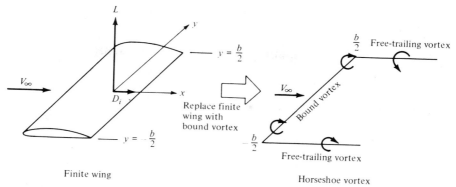

FIGURE 5.10
Replacement of the finite wing with a bound vortex.

from the wing tips to infinity, as also shown in Fig. 5.10. This vortex (the bound plus the two free) is in the shape of a horseshoe, and therefore is called a *horseshoe vortex*.

A single horseshoe vortex is shown in Fig. 5.11. Consider the downwash w induced along the bound vortex from $-b/2$ to $b/2$ by the horseshoe vortex. Examining Fig. 5.11, we see that the bound vortex induces no velocity along itself; however, the two trailing vortices both contribute to the induced velocity along the bound vortex, and both contributions are in the downward direction. Consistent with the *xyz* coordinate system in Fig. 5.11, such a downward velocity is negative; i.e., w (which is in the z direction) is a negative value when directed downward and a positive value when directed upward. If the origin is taken at the center of the bound vortex, then the velocity at any point y along the bound

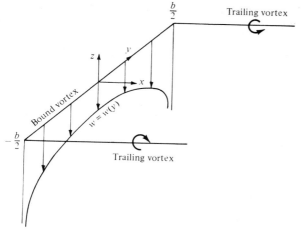

FIGURE 5.11
Downwash distribution along the y axis for a single horseshoe vortex.

vortex induced by the trailing semi-infinite vortices is, from Eq. (5.11),

$$w(y) = -\frac{\Gamma}{4\pi(b/2+y)} - \frac{\Gamma}{4\pi(b/2-y)} \tag{5.12}$$

In Eq. (5.12), the first term on the right-hand side is the contribution from the left trailing vortex (trailing from $-b/2$), and the second term is the contribution from the right trailing vortex (trailing from $b/2$). Equation (5.12) reduces to

$$w(y) = -\frac{\Gamma}{4\pi} \frac{b}{(b/2)^2 - y^2} \tag{5.13}$$

This variation of $w(y)$ is sketched in Fig. 5.11. Note that w approaches $-\infty$ as y approaches $-b/2$ or $b/2$.

The downwash distribution due to the single horseshoe vortex shown in Fig. 5.11 does not realistically simulate that of a finite wing; the downwash approaching an infinite value at the tips is especially disconcerting. During the early evolution of finite-wing theory, this problem perplexed Prandtl and his colleagues. After several years of effort, a resolution of this problem was obtained which, in hindsight, was simple and straightforward. Instead of representing the wing by a single horseshoe vortex, let us superimpose a large number of horseshoe vortices, each with a different length of the bound vortex, but with all the bound vortices coincident along a single line, called the *lifting line*. This concept is illustrated in Fig. 5.12, where only three horseshoe vortices are shown for the sake of clarity. In Fig. 5.12, a horseshoe vortex of strength $d\Gamma_1$ is shown, where the bound vortex spans the entire wing from $-b/2$ to $b/2$ (from point A to point F). Superimposed on this is a second horseshoe vortex of strength $d\Gamma_2$, where its bound vortex spans only part of the wing, from point B to point E. Finally, superimposed on this is a third horseshoe vortex of strength $d\Gamma_3$, where its bound vortex spans only the part of the wing from point C to point D. As a result, the circulation varies along the line of bound vortices—the lifting line defined above.

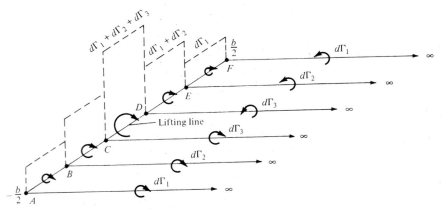

FIGURE 5.12
Superposition of a finite number of horseshoe vortices along the lifting line.

Along *AB* and *EF*, where only one vortex is present, the circulation is $d\Gamma_1$. However, along *BC* and *DE*, where two vortices are superimposed, the circulation is the sum of their strengths, $d\Gamma_1 + d\Gamma_2$. Along *CD*, three vortices are superimposed, and hence the circulation is $d\Gamma_1 + d\Gamma_2 + d\Gamma_3$. This variation of Γ along the lifting line is denoted by the vertical bars in Fig. 5.12. Also, note from Fig. 5.12 that we now have a series of trailing vortices distributed over the span, rather than just two vortices trailing downstream of the tips as shown in Fig. 5.11. The series of trailing vortices in Fig. 5.12 represents pairs of vortices, each pair associated with a given horseshoe vortex. Note that the strength of each trailing vortex is equal to the *change in circulation* along the lifting line.

Let us extrapolate Fig. 5.12 to the case where an *infinite number* of horseshoe vortices are superimposed along the lifting line, each with a vanishingly small strength, $d\Gamma$. This case is illustrated in Fig. 5.13. Note that the vertical bars in Fig. 5.12 have now become a continuous distribution of $\Gamma(y)$ along the lifting line in Fig. 5.13. The value of the circulation at the origin is Γ_0. Also, note that the finite number of trailing vortices in Fig. 5.12 have become a *continuous vortex sheet* trailing downstream of the lifting line in Fig. 5.13. This vortex sheet is parallel to the direction of V_∞. The total strength of the sheet integrated across the span of the wing is zero, because it consists of pairs of trailing vortices of equal strength but in opposite directions.

Let us single out an infinitesimally small segment of the lifting line, dy, located at the coordinate y as shown in Fig. 5.13. The circulation at y is $\Gamma(y)$, and the change in circulation over the segment dy is $d\Gamma = (d\Gamma/dy)\,dy$. In turn, the strength of the trailing vortex at y must equal the change in circulation $d\Gamma$ along the lifting line; this is simply an extrapolation of our result obtained for the strength of the finite trailing vortices in Fig. 5.12. Consider more closely the trailing vortex of strength $d\Gamma$ which intersects the lifting line at coordinate y, as shown in Fig. 5.13. Also consider the arbitrary location y_0 along the lifting line.

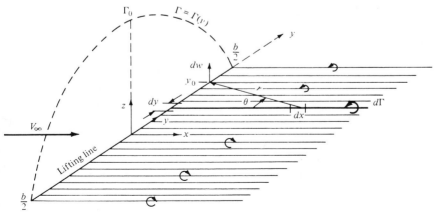

FIGURE 5.13
Superposition of an infinite number of horseshoe vortices along the lifting line.

Any segment of the trailing vortex, dx, will induce a velocity at y_0 with a magnitude and direction given by the Biot-Savart law, Eq. (5.5). In turn, the velocity dw at y_0 induced by the entire semi-infinite trailing vortex located at y is given by Eq. (5.11), which in terms of the picture given in Fig. 5.13 yields

$$dw = -\frac{(d\Gamma/dy)\, dy}{4\pi(y_0 - y)} \tag{5.14}$$

The minus sign in Eq. (5.14) is needed for consistency with the picture shown in Fig. 5.13; for the trailing vortex shown, the direction of dw at y_0 is upward and hence is a positive value, whereas Γ is decreasing in the y direction, making $d\Gamma/dy$ a negative quantity. The minus sign in Eq. (5.14) makes the positive dw consistent with the negative $d\Gamma/dy$.

The total velocity w induced at y_0 by the entire trailing vortex sheet is the summation of Eq. (5.14) over all the vortex filaments, i.e., the integral of Eq. (5.14) from $-b/2$ to $b/2$:

$$w(y_0) = -\frac{1}{4\pi} \int_{-b/2}^{b/2} \frac{(d\Gamma/dy)\, dy}{y_0 - y} \tag{5.15}$$

Equation (5.15) is important in that it gives the value of the downwash at y_0 due to all the trailing vortices. (Keep in mind that although we label w as downwash, w is treated as positive in the upward direction in order to be consistent with the normal convention in an xyz rectangular coordinate system.)

Pause for a moment and assess the status of our discussion so far. We have replaced the finite wing with the model of a lifting line along which the circulation $\Gamma(y)$ varies continuously, as shown in Fig. 5.13. In turn, we have obtained an expression for the downwash along the lifting line, given by Eq. (5.15). However, our central problem still remains to be solved; i.e., we want to *calculate* $\Gamma(y)$ for a given finite wing, along with its corresponding total lift and induced drag. Therefore, we must press on.

Return to Fig. 5.4, which shows the local airfoil section of a finite wing. Assume this section is located at the arbitrary spanwise station y_0. From Fig. 5.4, the induced angle of attack α_i is given by

$$\alpha_i(y_0) = \tan^{-1}\left(\frac{-w(y_0)}{V_\infty}\right) \tag{5.16}$$

[Note in Fig. 5.4 that w is downward, and hence is a negative quantity. Since α_i in Fig. 5.4 is positive, the negative sign in Eq. (5.16) is necessary for consistency.] Generally, w is much smaller than V_∞, and hence α_i is a small angle, on the order of a few degrees at most. For small angles, Eq. (5.16) yields

$$\alpha_i(y_0) = -\frac{w(y_0)}{V_\infty} \tag{5.17}$$

Substituting Eq. (5.15) into (5.17), we obtain

$$\alpha_i(y_0) = \frac{1}{4\pi V_\infty} \int_{-b/2}^{b/2} \frac{(d\Gamma/dy)\,dy}{y_0 - y} \tag{5.18}$$

i.e., an expression for the induced angle of attack in terms of the circulation distribution $\Gamma(y)$ along the wing.

Consider again the *effective* angle of attack α_{eff}, as shown in Fig. 5.4. As explained in Sec. 5.1, α_{eff} is the angle of attack actually seen by the local airfoil section. Since the downwash varies across the span, then α_{eff} is also variable; $\alpha_{\text{eff}} = \alpha_{\text{eff}}(y_0)$. The lift coefficient for the airfoil section located at $y = y_0$ is

$$c_l = a_0[\alpha_{\text{eff}}(y_0) - \alpha_{L=0}] = 2\pi[\alpha_{\text{eff}}(y_0) - \alpha_{L=0}] \tag{5.19}$$

In Eq. (5.19), the local section lift slope a_0 has been replaced by the thin airfoil theoretical value of 2π (rad^{-1}). Also, for a wing with aerodynamic twist, the angle of zero lift, $\alpha_{L=0}$, in Eq. (5.19) varies with y_0. If there is no aerodynamic twist, $\alpha_{L=0}$ is constant across the span. In any event, $\alpha_{L=0}$ is a known property of the local airfoil sections. From the definition of lift coefficient and from the Kutta-Joukowski theorem, we have, for the local airfoil section located at y_0,

$$L' = \tfrac{1}{2}\rho_\infty V_\infty^2 c(y_0) c_l = \rho_\infty V_\infty \Gamma(y_0) \tag{5.20}$$

From Eq. (5.20), we obtain

$$c_l = \frac{2\Gamma(y_0)}{V_\infty c(y_0)} \tag{5.21}$$

Substituting Eq. (5.21) into (5.19) and solving for α_{eff}, we have

$$\alpha_{\text{eff}} = \frac{\Gamma(y_0)}{\pi V_\infty c(y_0)} + \alpha_{L=0} \tag{5.22}$$

The above results come into focus if we refer to Eq. (5.1):

$$\alpha_{\text{eff}} = \alpha - \alpha_i \tag{5.1}$$

Substituting Eqs. (5.18) and (5.22) into (5.1), we obtain

$$\alpha(y_0) = \frac{\Gamma(y_0)}{\pi V_\infty c(y_0)} + \alpha_{L=0}(y_0) + \frac{1}{4\pi V_\infty} \int_{-b/2}^{b/2} \frac{(d\Gamma/dy)\,dy}{y_0 - y} \tag{5.23}$$

the *fundamental equation of Prandtl's lifting-line theory*; it simply states that the geometric angle of attack is equal to the sum of the effective angle plus the induced angle of attack. In Eq. (5.23), α_{eff} is expressed in terms of Γ, and α_i is expressed in terms of an integral containing $d\Gamma/dy$. Hence, Eq. (5.23) is an integro-differential equation, in which the only unknown is Γ; all the other quantities, α, c, V_∞, and $\alpha_{L=0}$, are known for a finite wing of given design at a given geometric angle of attack in a freestream with given velocity. Thus, a solution of Eq. (5.23) yields $\Gamma = \Gamma(y_0)$, where y_0 ranges along the span from $-b/2$ to $b/2$.

The solution $\Gamma = \Gamma(y_0)$ obtained from Eq. (5.23) gives us the three main aerodynamic characteristics of a finite wing, as follows:

1. The lift distribution is obtained from the Kutta-Joukowski theorem:

$$L'(y_0) = \rho_\infty V_\infty \Gamma(y_0) \tag{5.24}$$

2. The total lift is obtained by integrating Eq. (5.24) over the span:

$$L = \int_{-b/2}^{b/2} L'(y)\, dy$$

or
$$L = \rho_\infty V_\infty \int_{-b/2}^{b/2} \Gamma(y)\, dy \tag{5.25}$$

(Note that we have dropped the subscript on y, for simplicity.) The lift coefficient follows immediately from Eq. (5.25):

$$C_L = \frac{L}{q_\infty S} = \frac{2}{V_\infty S} \int_{-b/2}^{b/2} \Gamma(y)\, dy \tag{5.26}$$

3. The induced drag is obtained by inspection of Fig. 5.4. The induced drag per unit span is

$$D_i' = L_i' \sin \alpha_i$$

Since α_i is small, this relation becomes

$$D_i' = L_i' \alpha_i \tag{5.27}$$

The total induced drag is obtained by integrating Eq. (5.27) over the span:

$$D_i = \int_{-b/2}^{b/2} L'(y)\alpha_i(y)\, dy \tag{5.28}$$

or
$$D_i = \rho_\infty V_\infty \int_{-b/2}^{b/2} \Gamma(y)\alpha_i(y)\, dy \tag{5.29}$$

In turn, the induced drag coefficient is

$$C_{D,i} = \frac{D_i}{q_\infty S} = \frac{2}{V_\infty S} \int_{-b/2}^{b/2} \Gamma(y)\alpha_i(y)\, dy \tag{5.30}$$

In Eqs. (5.27) to (5.30), $\alpha_i(y)$ is obtained from Eq. (5.18).

Therefore, in Prandtl's lifting-line theory the solution of Eq. (5.23) for $\Gamma(y)$ is clearly the key to obtaining the aerodynamic characteristics of a finite wing. Before discussing the general solution of this equation, let us consider a special case, as outlined below.

5.3.1   Elliptical Lift Distribution

Consider a circulation distribution given by

$$\Gamma(y) = \Gamma_0 \sqrt{1 - \left(\frac{2y}{b}\right)^2} \tag{5.31}$$

In Eq. (5.31), note the following:

1. Γ_0 is the circulation at the origin, as shown in Fig. 5.13.
2. The circulation varies elliptically with distance y along the span; hence, it is designated as an *elliptical circulation distribution*. Since $L'(y) = \rho_\infty V_\infty \Gamma(y)$, we also have

$$L'(y) = \rho_\infty V_\infty \Gamma_0 \sqrt{1 - \left(\frac{2y}{b}\right)^2}$$

Hence, we are dealing with an *elliptical lift distribution*.
3. $\Gamma(b/2) = \Gamma(-b/2) = 0$. Thus, the circulation, hence lift, properly goes to zero at the wing tips, as shown in Fig. 5.13. We have not obtained Eq. (5.31) as a direct solution of Eq. (5.23); rather, we are simply stipulating a lift distribution that is elliptic. We now ask the question, What are the aerodynamic properties of a finite wing with such an elliptic lift distribution?

First, let us calculate the downwash. Differentiating Eq. (5.31), we obtain

$$\frac{d\Gamma}{dy} = -\frac{4\Gamma_0}{b^2} \frac{y}{(1 - 4y^2/b^2)^{1/2}} \tag{5.32}$$

Substituting Eq. (5.32) into (5.15), we have

$$w(y_0) = \frac{\Gamma_0}{\pi b^2} \int_{-b/2}^{b/2} \frac{y}{(1 - 4y^2/b^2)^{1/2}(y_0 - y)} \, dy \tag{5.33}$$

The integral can be evaluated easily by making the substitution

$$y = \frac{b}{2} \cos\theta \qquad dy = -\frac{b}{2} \sin\theta \, d\theta$$

Hence, Eq. (5.33) becomes

$$w(\theta_0) = -\frac{\Gamma_0}{2\pi b} \int_\pi^0 \frac{\cos\theta}{\cos\theta_0 - \cos\theta} \, d\theta$$

or

$$w(\theta_0) = -\frac{\Gamma_0}{2\pi b} \int_0^\pi \frac{\cos\theta}{\cos\theta - \cos\theta_0} \, d\theta \tag{5.34}$$

The integral in Eq. (5.34) is the standard form given by Eq. (4.26) for $n = 1$. Hence, Eq. (5.34) becomes

$$w(\theta_0) = -\frac{\Gamma_0}{2b} \tag{5.35}$$

which states the interesting and important result that the *downwash is constant over the span for an elliptical lift distribution*. In turn, from Eq. (5.17), we obtain, for the induced angle of attack,

$$\alpha_i = -\frac{w}{V_\infty} = \frac{\Gamma_0}{2bV_\infty} \tag{5.36}$$

For an elliptic lift distribution, the induced angle of attack is also constant along the span. Note from Eqs. (5.35) and (5.36) that both the downwash and induced angle of attack go to zero as the wing span becomes infinite—which is consistent with our previous discussions on airfoil theory.

A more useful expression for α_i can be obtained as follows. Substituting Eq. (5.31) into (5.25), we have

$$L = \rho_\infty V_\infty \Gamma_0 \int_{-b/2}^{b/2} \left(1 - \frac{4y^2}{b^2}\right)^{1/2} dy \tag{5.37}$$

Again, using the transformation $y = (b/2) \cos \theta$, Eq. (5.37) readily integrates to

$$L = \rho_\infty V_\infty \Gamma_0 \frac{b}{2} \int_0^\pi \sin^2 \theta \, d\theta = \rho_\infty V_\infty \Gamma_0 \frac{b}{4} \pi \tag{5.38}$$

Solving Eq. (5.38) for Γ_0, we have

$$\Gamma_0 = \frac{4L}{\rho_\infty V_\infty b \pi} \tag{5.39}$$

However, $L = \frac{1}{2}\rho_\infty V_\infty^2 S C_L$. Hence, Eq. (5.39) becomes

$$\Gamma_0 = \frac{2 V_\infty S C_L}{b\pi} \tag{5.40}$$

Substituting Eq. (5.40) into (5.36), we obtain

$$\alpha_i = \frac{2 V_\infty S C_L}{b\pi} \frac{1}{2bV_\infty}$$

or

$$\alpha_i = \frac{S C_L}{\pi b^2} \tag{5.41}$$

An important geometric property of a finite wing is the *aspect ratio*, denoted by AR and defined as

$$AR \equiv \frac{b^2}{S}$$

Hence, Eq. (5.41) becomes

$$\alpha_i = \frac{C_L}{\pi \text{AR}} \tag{5.42}$$

Equation (5.42) is a useful expression for the induced angle of attack, as shown below.

The induced drag coefficient is obtained from Eq. (5.30), noting that α_i is constant:

$$C_{D,i} = \frac{2\alpha_i}{V_\infty S} \int_{-b/2}^{b/2} \Gamma(y)\, dy = \frac{2\alpha_i \Gamma_0}{V_\infty S} \frac{b}{2} \int_0^\pi \sin^2 \theta\, d\theta = \frac{\pi \alpha_i \Gamma_0 b}{2 V_\infty S} \tag{5.42a}$$

Substituting Eqs. (5.40) and (5.42) into (5.42a), we obtain

$$C_{D,i} = \frac{\pi b}{2 V_\infty S} \left(\frac{C_L}{\pi \text{AR}} \right) \frac{2 V_\infty S C_L}{b\pi}$$

or

$$C_{D,i} = \frac{C_L^2}{\pi \text{AR}} \tag{5.43}$$

Equation (5.43) is an important result. It states that the induced drag coefficient is directly proportional to the square of the lift coefficient. The dependence of induced drag on the lift is not surprising, for the following reason. In Sec. 5.1, we saw that induced drag is a consequence of the presence of the wing-tip vortices, which in turn are produced by the difference in pressure between the lower and upper wing surfaces. The lift is produced by this same pressure difference. Hence, induced drag is intimately related to the production of lift on a finite wing; indeed, induced drag is frequently called the *drag due to lift*. Equation (5.43) dramatically illustrates this point. Clearly, an airplane cannot generate lift for free; the induced drag is the price for the generation of lift. The power required from an aircraft engine to overcome the induced drag is simply the power required to generate the lift of the aircraft. Also, note that because $C_{D,i} \propto C_L^2$, the induced drag coefficient increases rapidly as C_L increases and becomes a substantial part of the total drag coefficient when C_L is high, e.g., when the airplane is flying slowly such as on takeoff or landing. Even at relatively high cruising speeds, induced drag is typically 25 percent of the total drag.

Another important aspect of induced drag is evident in Eq. (5.43); i.e., $C_{D,i}$ is *inversely proportional to aspect ratio*. Hence, to reduce the induced drag, we want a finite wing with the highest possible aspect ratio. Wings with high and low aspect ratios are sketched in Fig. 5.14. Unfortunately, the design of very high aspect ratio wings with sufficient structural strength is difficult. Therefore, the aspect ratio of a conventional aircraft is a compromise between conflicting aerodynamic and structural requirements. It is interesting to note that the aspect

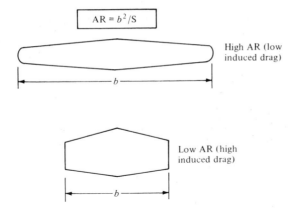

FIGURE 5.14
Schematic of high- and low-aspect-ratio wings.

ratio of the 1903 Wright Flyer was 6 and that today the aspect ratios of conventional subsonic aircraft range typically from 6 to 8. (Exceptions are the Lockheed U-2 high-altitude reconnaissance aircraft with $AR = 14.3$ and sailplanes with aspect ratios in the 10 to 22 range.)

Another property of the elliptical lift distribution is as follows. Consider a wing with no geometric twist (i.e., α is constant along the span) and no aerodynamic twist (i.e., $\alpha_{L=0}$ is constant along the span). From Eq. (5.42), we have seen that α_i is constant along the span. Hence, $\alpha_{\mathrm{eff}} = \alpha - \alpha_i$ is also constant along the span. Since the local section lift coefficient c_l is given by

$$c_l = a_0(\alpha_{\mathrm{eff}} - \alpha_{L=0})$$

then assuming that a_0 is the same for each section ($a_0 = 2\pi$ from thin airfoil theory), c_l must be constant along the span. The lift per unit span is given by

$$L'(y) = q_\infty c c_l \tag{5.44}$$

Solving Eq. (5.44) for the chord, we have

$$c(y) = \frac{L'(y)}{q_\infty c_l} \tag{5.45}$$

In Eq. (5.45), q_∞ and c_l are constant along the span. However, $L'(y)$ varies elliptically along the span. Thus, Eq. (5.45) dictates that for such an elliptic lift distribution, the *chord must vary elliptically along the span*; i.e., for the conditions given above, the *wing planform is elliptical*.

The related characteristics—the elliptic lift distribution, the elliptic planform, and the constant downwash—are sketched in Fig. 5.15. Although an elliptical lift distribution may appear to be a restricted, isolated case, in reality it gives a reasonable approximation for the induced drag coefficient for an arbitrary finite wing. The form of $C_{D,i}$ given by Eq. (5.43) is only slightly modified for the general case. Let us now consider the case of a finite wing with a general lift distribution.

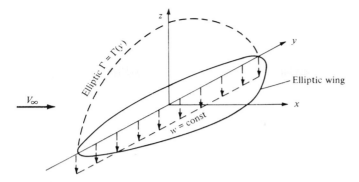

FIGURE 5.15
Illustration of the related quantities: an elliptic lift distribution, elliptic planform, and constant downwash.

5.3.2 General Lift Distribution

Consider the transformation

$$y = -\frac{b}{2}\cos\theta \qquad (5.46)$$

where the coordinate in the spanwise direction is now given by θ, with $0 \le \theta \le \pi$. In terms of θ, the elliptic lift distribution given by Eq. (5.31) is written as

$$\Gamma(\theta) = \Gamma_0 \sin\theta \qquad (5.47)$$

Equation (5.47) hints that a Fourier sine series would be an appropriate expression for the general circulation distribution along an arbitrary finite wing. Hence, assume for the general case that

$$\Gamma(\theta) = 2bV_\infty \sum_1^N A_n \sin n\theta \qquad (5.48)$$

where as many terms, N, in the series can be taken as we desire for accuracy. The coefficients A_n (where $n = 1, \ldots, N$) in Eq. (5.48) are unknowns; however, they must satisfy the fundamental equation of Prandtl's lifting-line theory; i.e., the A_n's must satisfy Eq. (5.23). Differentiating Eq. (5.48), we obtain

$$\frac{d\Gamma}{dy} = \frac{d\Gamma}{d\theta}\frac{d\theta}{dy} = 2bV_\infty \sum_1^N nA_n \cos n\theta \frac{d\theta}{dy} \qquad (5.49)$$

Substituting Eqs. (5.48) and (5.49) into (5.23), we obtain

$$\alpha(\theta_0) = \frac{2b}{\pi c(\theta_0)}\sum_1^N A_n \sin n\theta_0 + \alpha_{L=0}(\theta_0) + \frac{1}{\pi}\int_0^\pi \frac{\sum_1^N nA_n \cos n\theta}{\cos\theta - \cos\theta_0}\,d\theta \qquad (5.50)$$

The integral in Eq. (5.50) is the standard form given by Eq. (4.26). Hence, Eq.

(5.50) becomes

$$\alpha(\theta_0) = \frac{2b}{\pi c(\theta_0)} \sum_1^N A_n \sin n\theta_0 + \alpha_{L=0}(\theta_0) + \sum_1^N nA_n \frac{\sin n\theta_0}{\sin \theta_0} \qquad (5.51)$$

Examine Eq. (5.51) closely. It is evaluated at a given spanwise location; hence, θ_0 is specified. In turn, b, $c(\theta_0)$, and $\alpha_{L=0}(\theta_0)$ are known quantities from the geometry and airfoil section of the finite wing. The only unknowns in Eq. (5.51) are the A_n's. Hence, written at a given spanwise location (a specified θ_0), Eq. (5.51) is one algebraic equation with N unknowns, $A_1, A_2, \ldots, A_n$. However, let us choose N different spanwise stations, and let us evaluate Eq. (5.51) at each of these N stations. We then obtain a system of N independent algebraic equations with N unknowns, namely, $A_1, A_2, \ldots, A_N$. In this fashion, actual numerical values are obtained for the A_n's—numerical values which ensure that the general circulation distribution given by Eq. (5.48) satisfies the fundamental equation of finite-wing theory, Eq. (5.23).

Now that $\Gamma(\theta)$ is known via Eq. (5.48), the lift coefficient for the finite wing follows immediately from the substitution of Eq. (5.48) into (5.26):

$$C_L = \frac{2}{V_\infty S} \int_{-b/2}^{b/2} \Gamma(y) \, dy = \frac{2b^2}{S} \sum_1^N A_n \int_0^\pi \sin n\theta \sin \theta \, d\theta \qquad (5.52)$$

In Eq. (5.52), the integral is

$$\int_0^\pi \sin n\theta \sin \theta \, d\theta = \begin{cases} \pi/2 & \text{for } n = 1 \\ 0 & \text{for } n \neq 1 \end{cases}$$

Hence, Eq. (5.52) becomes

$$C_L = A_1 \pi \frac{b^2}{S} = A_1 \pi \text{AR} \qquad (5.53)$$

Note that C_L depends only on the leading coefficient of the Fourier series expansion. (However, although C_L depends on A_1 only, we must solve for all the A_n's simultaneously in order to obtain A_1.)

The induced drag coefficient is obtained from the substitution of Eq. (5.48) into Eq. (5.30) as follows:

$$C_{D,i} = \frac{2}{V_\infty S} \int_{-b/2}^{b/2} \Gamma(y) \alpha_i(y) \, dy$$

$$= \frac{2b^2}{S} \int_0^\pi \left(\sum_1^N A_n \sin n\theta \right) \alpha_i(\theta) \sin \theta \, d\theta \qquad (5.54)$$

The induced angle of attack $\alpha_i(\theta)$ in Eq. (5.54) is obtained from the substitution of Eqs. (5.46) and (5.49) into (5.18), which yields

$$\alpha_i(y_0) = \frac{1}{4\pi V_\infty} \int_{-b/2}^{b/2} \frac{(d\Gamma/dy)\, dy}{y_0 - y}$$

$$= \frac{1}{\pi} \sum_1^N nA_n \int_0^\pi \frac{\cos n\theta}{\cos \theta - \cos \theta_0}\, d\theta \qquad (5.55)$$

The integral in Eq. (5.55) is the standard form given by Eq. (4.26). Hence, Eq. (5.55) becomes

$$\alpha_i(\theta_0) = \sum_1^N nA_n \frac{\sin n\theta_0}{\sin \theta_0} \qquad (5.56)$$

In Eq. (5.56), θ_0 is simply a dummy variable which ranges from 0 to π across the span of the wing; it can therefore be replaced by θ, and Eq. (5.56) can be written as

$$\alpha_i(\theta) = \sum_1^N nA_n \frac{\sin n\theta}{\sin \theta} \qquad (5.57)$$

Substituting Eq. (5.57) into (5.54), we have

$$C_{D,i} = \frac{2b^2}{S} \int_0^\pi \left(\sum_1^N A_n \sin n\theta \right) \left(\sum_1^N nA_n \sin n\theta \right) d\theta \qquad (5.58)$$

Examine Eq. (5.58) closely; it involves the product of two summations. Also, note that, from the standard integral,

$$\int_0^\pi \sin m\theta \sin k\theta = \begin{cases} 0 & \text{for } m \neq k \\ \pi/2 & \text{for } m = k \end{cases} \qquad (5.59)$$

Hence, in Eq. (5.58), the mixed product terms involving unequal subscripts (such as A_1A_2, A_2A_4) are, from Eq. (5.59), equal to zero. Hence, Eq. (5.58) becomes

$$C_{D,i} = \frac{2b^2}{S} \left(\sum_1^N nA_n^2 \right) \frac{\pi}{2} = \pi \mathrm{AR} \sum_1^N nA_n^2$$

$$= \pi \mathrm{AR} \left(A_1^2 + \sum_2^N nA_n^2 \right)$$

$$= \pi \mathrm{AR}\, A_1^2 \left[1 + \sum_2^N n \left(\frac{A_n}{A_1} \right)^2 \right] \qquad (5.60)$$

Substituting Eq. (5.53) for C_L into Eq. (5.60), we obtain

$$\boxed{C_{D,i} = \frac{C_L^2}{\pi \mathrm{AR}} (1 + \delta)} \qquad (5.61)$$

where $\delta = \sum_2^N n(A_n/A_1)^2$. Note that $\delta \geq 0$; hence, the factor $1 + \delta$ in Eq. (5.61) is either greater than 1 or at least equal to 1. Let us define a span efficiency factor, e, as $e = (1 + \delta)^{-1}$. Then Eq. (5.61) can be written as

$$C_{D,i} = \frac{C_L^2}{\pi e \, \mathrm{AR}}$$ (5.62)

where $e \leq 1$. Comparing Eqs. (5.61) and (5.62) for the general lift distribution with Eq. (5.43) for the elliptical lift distribution, note that $\delta = 0$ and $e = 1$ for the elliptical lift distribution. Hence, the lift distribution which yields minimum induced drag is the *elliptical lift distribution.* This is why we have a practical interest in the elliptical lift distribution.

Recall that for a wing with no aerodynamic twist and no geometric twist, an elliptical lift distribution is generated by a wing with an elliptical planform, as sketched at the top of Fig. 5.16. Several aircraft have been designed in the past with elliptical wings; the most famous, perhaps, being the British Spitfire from World War II, shown in Fig. 5.17. However, elliptic planforms are more expensive to manufacture than, say, a simple rectangular wing as sketched in the middle of Fig. 5.16. On the other hand, a rectangular wing generates a lift distribution far from optimum. A compromise is the tapered wing shown at the bottom of Fig. 5.16. The tapered wing can be designed with a taper ratio, i.e.,

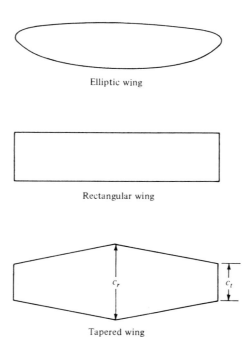

Elliptic wing

Rectangular wing

Tapered wing

FIGURE 5.16
Various planforms for straight wings.

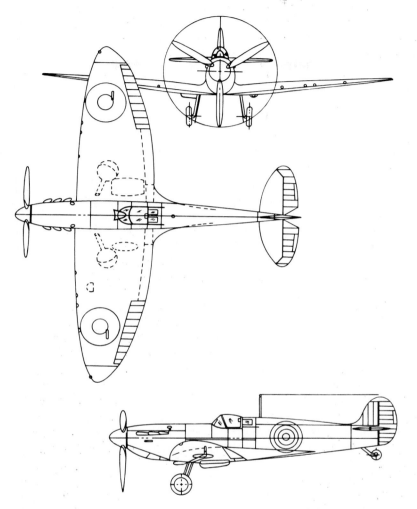

FIGURE 5.17
Three views of the Supermarine Spitfire, a famous British World War II fighter.

tip chord/root chord $\equiv c_t/c_r$, such that the lift distribution closely approximates the elliptic case. The variation of δ as a function of taper ratio for wings of different aspect ratio is illustrated in Fig. 5.18. Such calculations of δ were first performed by the famous English aerodynamicist, Hermann Glauert and published in Ref. 18 in the year 1926. Note from Fig. 5.18 that a tapered wing can be designed with an induced drag coefficient reasonably close to the minimum value. In addition, tapered wings with straight leading and trailing edges are considerably easier to manufacture than elliptic planforms. Therefore, most conventional aircraft employ tapered rather than elliptical wing planforms.

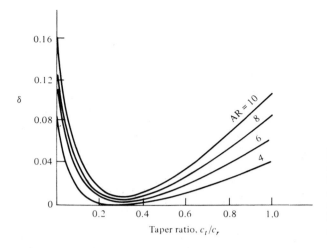

FIGURE 5.18
Induced drag factor δ as a function of taper ratio. (*From McCormick, B. W.,* Aerodynamics, Aeronautics, and Flight Mechanics, *John Wiley & Sons, New York, 1979.*)

5.3.3 Effect of Aspect Ratio

Returning to Eqs. (5.61) and (5.62), note that the induced drag coefficient for a finite wing with a general lift distribution is inversely proportional to the aspect ratio, as was discussed earlier in conjunction with the case of the elliptic lift distribution. Note that AR, which typically varies from 6 to 22 for standard subsonic airplanes and sailplanes, has a much stronger effect on $C_{D,i}$ than the value of δ, which from Fig. 5.18 varies only by about 10 percent over the practical range of taper ratio. Hence, the primary design factor for minimizing induced drag is not the closeness to an elliptical lift distribution, but rather, the ability to make the aspect ratio as large as possible. The determination that $C_{D,i}$ is inversely proportional to AR was one of the great victories of Prandtl's lifting-line theory. In 1915, Prandtl verified this result with a series of classic experiments wherein the lift and drag of seven rectangular wings with different aspect ratios were measured. The data are given in Fig. 5.19. Recall from Eq. (5.4) that the total drag of a finite wing is given by

$$C_D = c_d + \frac{C_L^2}{\pi e \, \mathrm{AR}} \tag{5.63}$$

The parabolic variation of C_D with C_L as expressed in Eq. (5.63) is reflected in the data of Fig. 5.19. If we consider two wings with different aspect ratios AR_1 and AR_2, Eq. (5.63) gives the drag coefficients $C_{D,1}$ and $C_{D,2}$ for the two wings as

$$C_{D,1} = c_d + \frac{C_L^2}{\pi e \, \mathrm{AR}_1} \tag{5.64a}$$

and

$$C_{D,2} = c_d + \frac{C_L^2}{\pi e \, \mathrm{AR}_2} \tag{5.64b}$$

Assume that the wings are at the same C_L. Also, since the airfoil section is the same for both wings, c_d is essentially the same. Moreover, the variation of e

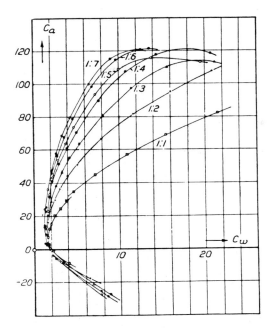

FIGURE 5.19
Prandtl's classic rectangular wing data for seven different aspect ratios from 1 to 7; variation of lift coefficient versus drag coefficient. For historical interest, we reproduce here Prandtl's actual graphs. Note that, in his nomenclature, C_a = lift coefficient and C_w = drag coefficient. Also, the numbers on both the ordinate and abscissa are 100 times the actual values of the coefficients. (*Taken from Prandtl, L., "Applications of Modern Hydrodynamics to Aeronautics," NACA Report No. 116, 1921.*)

between the wings is only a few percent and can be ignored. Hence, subtracting Eq. (5.64*b*) from (5.64*a*), we obtain

$$C_{D,1} = C_{D,2} + \frac{C_L^2}{\pi e}\left(\frac{1}{AR_1} - \frac{1}{AR_2}\right) \qquad (5.65)$$

Equation (5.65) can be used to scale the data of a wing with aspect ratio AR_2 to correspond to the case of another aspect ratio, AR_1. For example, Prandtl scaled the data of Fig. 5.19 to correspond to a wing with an aspect ratio of 5. For this case, Eq. (5.65) becomes

$$C_{D,1} = C_{D,2} + \frac{C_L^2}{\pi e}\left(\frac{1}{5} - \frac{1}{AR_2}\right) \qquad (5.66)$$

Inserting the respective values of $C_{D,2}$ and AR_2 from Fig. 5.19 into Eq. (5.66), Prandtl found that the resulting data for $C_{D,1}$ versus C_L collapsed to essentially the same curve, as shown in Fig. 5.20. Hence, the inverse dependence of $C_{D,i}$ on AR was substantially verified as early as 1915.

There are two primary differences between airfoil and finite-wing properties. We have discussed one difference, namely, a finite wing generates induced drag. However, a second major difference appears in the lift slope. In Fig. 4.4, the lift slope for an airfoil was defined as $a_0 \equiv dc_l / d\alpha$. Let us denote the lift slope for a finite wing as $a \equiv dC_L / d\alpha$. When the lift slope of a finite wing is compared with that of its airfoil section, we find that $a < a_0$. To see this more clearly, return to Fig. 5.4, which illustrates the influence of downwash on the flow over a local airfoil section of a finite wing. Note that although the geometric angle of attack

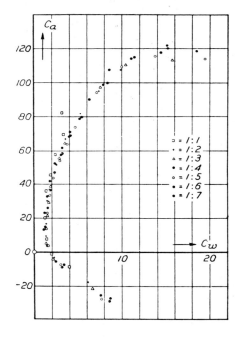

FIGURE 5.20
Data of Fig. 5.19 scaled by Prandtl to an aspect ratio of 5.

of the finite wing is α, the airfoil section effectively senses a smaller angle of attack, namely, α_{eff}, where $\alpha_{\text{eff}} = \alpha - \alpha_i$. For the time being, consider an elliptic wing with no twist; hence, α_i and α_{eff} are both constant along the span. Moreover, c_l is also constant along the span, and therefore $C_L = c_l$. Assume that we plot C_L for the finite wing versus α_{eff}, as shown at the top of Fig. 5.21. Because we are using α_{eff} the lift slope corresponds to that for an infinite wing, a_0. However, in real life, our naked eyes cannot see α_{eff}; instead, what we actually observe is a finite wing with a certain angle between the chord line and the relative wind; i.e., in practice, we always observe the geometric angle of attack α. Hence, C_L for a finite wing is generally given as a function of α, as sketched at the bottom of Fig. 5.21. Since $\alpha > \alpha_{\text{eff}}$, the bottom abscissa is stretched, and hence the bottom lift curve is less inclined; it has a slope equal to a, and Fig. 5.21 clearly shows that $a < a_0$. The effect of a finite wing is to *reduce* the lift slope. Also, recall that at zero lift, there are no induced effects; i.e., $\alpha_i = C_{D,i} = 0$. Thus, when $C_L = 0$, $\alpha = \alpha_{\text{eff}}$. As a result, $\alpha_{L=0}$ is the same for the finite and the infinite wings, as shown in Fig. 5.21.

The values of a_0 and a are related as follows. From the top of Fig. 5.21,

$$\frac{dC_L}{d(\alpha - \alpha_i)} = a_0$$

Integrating, we find

$$C_L = a_0(\alpha - \alpha_i) + \text{const} \tag{5.67}$$

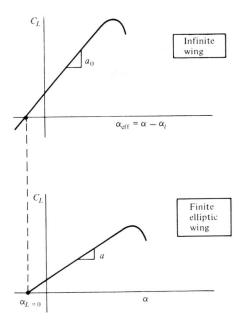

FIGURE 5.21
Lift curves for an infinite wing versus a finite elliptic wing.

Substituting Eq. (5.42) into (5.67), we obtain

$$C_L = a_0 \left(\alpha - \frac{C_L}{\pi \text{AR}} \right) + \text{const} \tag{5.68}$$

Differentiating Eq. (5.68) with respect to α, and solving for $dC_L/d\alpha$, we obtain

$$\frac{dC_L}{d\alpha} = a = \frac{a_0}{1 + a_0/\pi \text{AR}} \tag{5.69}$$

Equation (5.69) gives the desired relation between a_0 and a for an elliptic finite wing. For a finite wing of general planform, Eq. (5.69) is slightly modified, as given below:

$$a = \frac{a_0}{1 + (a_0/\pi \text{AR})(1 + \tau)} \tag{5.70}$$

In Eq. (5.70), τ is a function of the Fourier coefficients A_n. Values of τ were first calculated by Glauert in the early 1920s and were published in Ref. 18, which should be consulted for more details. Values of τ typically range between 0.05 and 0.25.

Of most importance in Eqs. (5.69) and (5.70) is the aspect-ratio variation. Note that for low-AR wings, a substantial difference can exist between a_0 and a. However, as $\text{AR} \to \infty$, $a \to a_0$. The effect of aspect ratio on the lift curve is dramatically shown in Fig. 5.22, which gives classic data obtained on rectangular

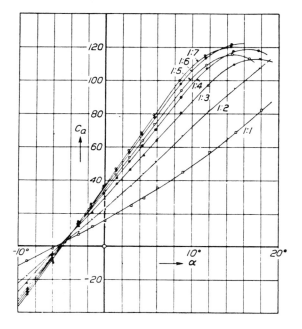

FIGURE 5.22
Prandtl's classic rectangular wing data. Variation of lift coefficient with angle of attack for seven different aspect ratios from 1 to 7. Nomenclature and scale are the same as given in Fig. 5.19.

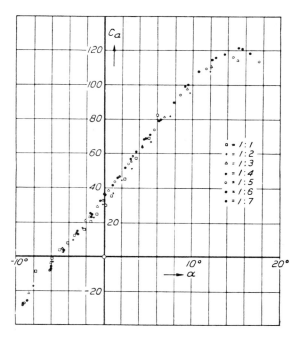

FIGURE 5.23
Data of Fig. 5.22 scaled by Prandtl to an aspect ratio of 5.

wings by Prandtl in 1915. Note the reduction in $dC_L/d\alpha$ as AR is reduced. Moreover, using the equations obtained above, Prandtl scaled the data in Fig. 5.22 to correspond to an aspect ratio of 5; his results collapsed to essentially the same curve, as shown in Fig. 5.23. In this manner, the aspect-ratio variation given in Eqs. (5.69) and (5.70) was confirmed as early as the year 1915.

5.3.4 Physical Significance

Consider again the basic model underlying Prandtl's lifting-line theory. Return to Fig. 5.13 and study it carefully. An infinite number of infinitesimally weak horseshoe vortices are superimposed in such a fashion as to generate a lifting line which spans the wing, along with a vortex sheet which trails downstream. This trailing-vortex sheet is the instrument which induces downwash at the lifting line. At first thought, you might consider this model to be somewhat abstract—a mathematical convenience which somehow produces surprisingly useful results. However, to the contrary, the model shown in Fig. 5.13 has real physical significance. To see this more clearly, return to Fig. 5.1. Note that in the three-dimensional flow over a finite wing, the streamlines leaving the trailing edge from the top and bottom surfaces are in different directions; i.e., there is a discontinuity in the tangential velocity at the trailing edge. We know from Chap. 4 that a discontinuous change in tangential velocity is theoretically allowed across a vortex sheet. In real life, such discontinuities do not exist; rather, the different velocities at the trailing edge generate a thin region of large velocity gradients—a thin region of shear flow with very large vorticity. Hence, a sheet of vorticity actually trails downstream from the trailing edge of a finite wing. This sheet tends to roll up at the edges and helps to form the wing-tip vortices sketched in Fig. 5.2. Thus, Prandtl's lifting-line model with its trailing-vortex sheet is physically consistent with the actual flow downstream of a finite wing.

Example 5.1. Consider a finite wing with an aspect ratio of 8 and a taper ratio of 0.8. The airfoil section is thin and symmetric. Calculate the lift and induced drag coefficients for the wing when it is at an angle of attack of 5°. Assume that $\delta = \tau$.

Solution. From Fig. 5.18, $\delta = 0.055$. Hence, from the stated assumption, τ also equals 0.055. From Eq. (5.70), assuming $a_0 = 2\pi$ from thin airfoil theory,

$$a = \frac{a_0}{1 + a_0/\pi AR(1+\tau)} = \frac{2\pi}{1 + 2\pi(1.055)/8\pi} = 4.97 \text{ rad}^{-1}$$

$$= 0.0867 \text{ degree}^{-1}$$

Since the airfoil is symmetric, $\alpha_{L=0} = 0°$. Thus,

$$C_L = a\alpha = (0.0867 \text{ degree}^{-1})(5°) = \boxed{0.4335}$$

From Eq. (5.61),

$$C_{D,i} = \frac{C_L^2}{\pi AR}(1+\delta) = \frac{(0.4335)^2(1+0.055)}{8\pi} = \boxed{0.00789}$$

Example 5.2. Consider a rectangular wing with an aspect ratio of 6, an induced drag factor $\delta = 0.055$, and a zero-lift angle of attack of $-2°$. At an angle of attack of 3.4°, the induced drag coefficient for this wing is 0.01. Calculate the induced drag coefficient for a similar wing (a rectangular wing with the same airfoil section) at the same angle of attack, but with an aspect ratio of 10. Assume that the induced factors for drag and the lift slope, δ and τ, respectively, are equal to each other; i.e., $\delta = \tau$. Also, for $AR = 10$, $\delta = 0.105$.

Solution. We must recall that although the angle of attack is the same for the two cases compared here ($AR = 6$ and 10), the value of C_L is different because of the aspect-ratio effect on the lift slope. First, let us calculate C_L for the wing with aspect ratio 6. From Eq. (5.61),

$$C_L^2 = \frac{\pi AR C_{D,i}}{1+\delta} = \frac{\pi(6)(0.01)}{1+0.055} = 0.1787$$

Hence,

$$C_L = 0.423$$

The lift slope of this wing is therefore

$$\frac{dC_L}{d\alpha} = \frac{0.423}{3.4° - (-2°)} = 0.078/\text{degree} = 4.485/\text{rad}$$

The lift slope for the airfoil (the infinite wing) can be obtained from Eq. (5.70):

$$\frac{dC_L}{d\alpha} = a = \frac{a_0}{1 + (a_0/\pi AR)(1+\tau)}$$

$$4.485 = \frac{a_0}{1 + [(1.055)a_0/\pi(6)]} = \frac{a_0}{1 + 0.056 a_0}$$

Solving for a_0, we find that this yields $a_0 = 5.989/\text{rad}$. Since the second wing (with $AR = 10$) has the same airfoil section, then a_0 is the same. The lift slope of the second wing is given by

$$a = \frac{a_0}{1 + (a_0/\pi AR)(1+\tau)} = \frac{5.989}{1 + [(5.989)(1.105)/\pi(10)]} = 4.95/\text{rad}$$

$$= 0.086/\text{degree}$$

The lift coefficient for the second wing is therefore

$$C_L = a(\alpha - \alpha_{L=0}) = 0.086[3.4° - (-2°)] = 0.464$$

In turn, the induced drag coefficient is

$$C_{D,i} = \frac{C_L^2}{\pi AR}(1+\delta) = \frac{(0.464)^2(1.105)}{\pi(10)} = \boxed{0.0076}$$

Note: This problem would have been more straightforward if the lift coefficients had been stipulated to be the same between the two wings rather than the angle of attack. Then Eq. (5.61) would have yielded the induced drag coefficient directly. A purpose of this example is to reinforce the rationale behind Eq. (5.65), which readily allows the scaling of drag coefficients from one aspect ratio to another, as long as the *lift coefficient is the same.* This allows the scaled drag-coefficient data to be plotted versus C_L (not the angle of attack) as in Fig. 5.20. However, in the present example where the angle of attack is the same between both cases, the effect of aspect ratio on the lift slope must be explicitly considered, as we have done above.

5.4 A NUMERICAL NONLINEAR LIFTING-LINE METHOD

The classical Prandtl lifting-line theory described in Sec. 5.3 assumes a linear variation of c_l versus α_{eff}. This is clearly seen in Eq. (5.19). However, as the angle of attack approaches and exceeds the stall angle, the lift curve becomes nonlinear, as shown in Fig. 4.4. This high-angle-of-attack regime is of interest to modern aerodynamicists. For example, when an airplane is in a spin, the angle of attack can range from 40 to 90°; an understanding of high-angle-of-attack aerodynamics is essential to the prevention of such spins. In addition, modern fighter airplanes achieve optimum maneuverability by pulling high angles of attack at subsonic speeds. Therefore, there are practical reasons for extending Prandtl's classical theory to account for a nonlinear lift curve. One simple extension is described in this section.

The classical theory developed in Sec. 5.4 is essentially closed form; i.e., the results are analytical equations as opposed to a purely numerical solution. Of course, in the end, the Fourier coefficients A_n for a given wing must come from a solution of a system of simultaneous linear algebraic equations. Until the advent of the modern digital computer, these coefficients were calculated by hand. Today, they are readily solved on a computer using standard matrix methods. However, the elements of the lifting-line theory lend themselves to a straightforward purely numerical solution which allows the treatment of nonlinear effects. Moreover, this numerical solution emphasizes the fundamental aspects of lifting-line theory. For these reasons, such a numerical solution is outlined in this section.

Consider the most general case of a finite wing of given planform and geometric twist, with different airfoil sections at different spanwise stations. Assume that we have experimental data for the lift curves of the airfoil sections, including the nonlinear regime (i.e., assume we have the conditions of Fig. 4.4 for all the given airfoil sections). A numerical iterative solution for the finite-wing properties can be obtained as follows:

1. Divide the wing into a number of spanwise stations, as shown in Fig. 5.24. Here $k + 1$ stations are shown, with n designating any specific station.
2. For the given wing at a given α, *assume* the lift distribution along the span; i.e., assume values for Γ at all the stations $\Gamma_1, \Gamma_2, \ldots, \Gamma_n, \ldots, \Gamma_{k+1}$. An elliptical lift distribution is satisfactory for such an assumed distribution.
3. With this assumed variation of Γ, calculate the induced angle of attack α_i

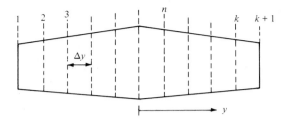

FIGURE 5.24
Stations along the span for a numerical solution.

from Eq. (5.18) at each of the stations:

$$\alpha_i(y_n) = \frac{1}{4\pi V_\infty} \int_{-b/2}^{b/2} \frac{(d\Gamma/dy)\,dy}{y_n - y} \tag{5.71}$$

The integral is evaluated numerically. If Simpson's rule is used, Eq. (5.71) becomes

$$\alpha_i(y_n) = \frac{1}{4\pi V_\infty} \frac{\Delta y}{3} \sum_{j=2,4,6}^{k} \frac{(d\Gamma/dy)_{j-1}}{(y_n - y_{j-1})} + 4\frac{(d\Gamma/dy)_j}{y_n - y_j} + \frac{(d\Gamma/dy)_{j+1}}{y_n - y_{j+1}} \tag{5.72}$$

where Δy is the distance between stations. In Eq. (5.72), when $y_n = y_{j-1}$, y_j, or y_{j+1}, a singularity occurs (a denominator goes to zero). When this singularity occurs, it can be avoided by replacing the given term by its average value based on the two adjacent sections.

4. Using α_i from step 3, obtain the effective angle of attack α_{eff} at each station from

$$\alpha_{\text{eff}}(y_n) = \alpha - \alpha_i(y_n)$$

5. With the distribution of α_{eff} calculated from step 4, obtain the section lift coefficient $(c_l)_n$ at each station. These values are read from the known lift curve for the airfoil.

6. From $(c_l)_n$ obtained in step 5, a *new* circulation distribution is calculated from the Kutta-Joukowski theorem and the definition of lift coefficient:

$$L'(y_n) = \rho_\infty V_\infty \Gamma(y_n) = \tfrac{1}{2}\rho_\infty V_\infty^2 c_n (c_l)_n$$

Hence,

$$\Gamma(y_n) = \tfrac{1}{2} V_\infty c_n (c_l)_n$$

where c_n is the local section chord. Keep in mind that in all the above steps, n ranges from 1 to $k+1$.

7. The new distribution of Γ obtained in step 6 is compared with the values that were initially fed into step 3. If the results from step 6 do not agree with the input to step 3, then a new input is generated. If the previous input to step 3 is designated as Γ_{old} and the result of step 6 is designated as Γ_{new}, then the

new input to step 3 is determined from

$$\Gamma_{input} = \Gamma_{old} + D(\Gamma_{new} - \Gamma_{old})$$

where D is a damping factor for the iterations. Experience has found that the iterative procedure requires heavy damping, with typical values of D on the order of 0.05.

8. Steps 3 to 7 are repeated a sufficient number of cycles until Γ_{new} and Γ_{old} agree at each spanwise station to within acceptable accuracy. If this accuracy is stipulated to be within 0.01 percent for a stretch of five previous iterations, then a minimum of 50 and sometimes as many as 150 iterations may be required for convergence.

9. From the converged $\Gamma(y)$, the lift and induced drag coefficients are obtained from Eqs. (5.26) and (5.30), respectively. The integrations in these equations can again be carried out by Simpson's rule.

The procedure outlined above generally works smoothly and quickly on a high-speed digital computer. Typical results are shown in Fig. 5.25, which shows the circulation distributions for rectangular wings with three different aspect ratios. The solid lines are from the classical calculations of Prandtl (Sec. 5.3), and the symbols are from the numerical method described above. Excellent agreement is obtained, thus verifying the integrity and accuracy of the numerical method. Also, Fig. 5.25 should be studied as an example of typical circulation distributions over general finite wings, with Γ reasonably high over the center section of the wing but rapidly dropping to zero at the tips.

An example of the use of the numerical method for the nonlinear regime is shown in Fig. 5.26. Here, C_L versus α is given for a rectangular wing up to an angle of attack of 50°—well beyond stall. The numerical results are compared with existing experimental data obtained at the University of Maryland (Ref.

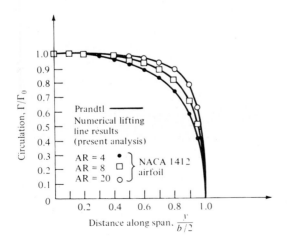

FIGURE 5.25
Lift distribution for a rectangular wing; comparison between Prandtl's classical theory and the numerical lifting-line method of Ref. 20.

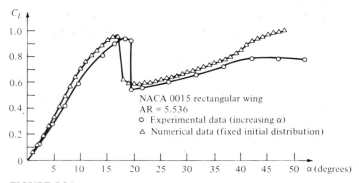

FIGURE 5.26
Lift coefficient versus angle of attack; comparison between experimental and numerical results.

19). The numerical lifting-line solution at high angle of attack agrees with experiment to within 20 percent, and much closer for many cases. Therefore, such solutions give reasonable preliminary engineering results for the high-angle-of-attack poststall region. However, it is wise not to stretch the applicability of lifting-line theory too far. At high angles of attack, the flow is highly three-dimensional. This is clearly seen in the surface oil pattern on a rectangular wing at high angle of attack shown in Fig. 5.27. At high α, there is a strong spanwise flow, in combination with mushroom-shaped flow separation regions. Clearly, the basic assumptions of lifting-line theory, classical or numerical, cannot properly account for such three-dimensional flows.

For more details and results on the numerical lifting-line method, please see Ref. 20.

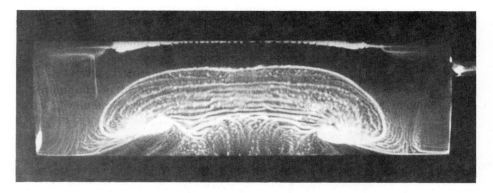

FIGURE 5.27
Surface oil flow pattern on a stalled, finite rectangular wing with a Clark Y-14 airfoil section. $AR = 3.5$, $\alpha = 22.8°$, $Re = 245,000$ (based on chord length). This pattern was established by coating the wing surface with pigmented mineral oil and inserting the model in a low-speed subsonic wind tunnel. In the photograph shown, flow is from top to bottom. Note the highly three-dimensional flow pattern. (*Courtesy of Allen E. Winkelmann, University of Maryland.*)

5.5 LIFTING-SURFACE THEORY; VORTEX LATTICE NUMERICAL METHOD

Prandtl's classical lifting-line theory (Sec. 5.3) gives reasonable results for straight wings at moderate to high aspect ratio. However, for low-aspect-ratio straight wings, swept wings, and delta wings, classical lifting-line theory is inappropriate. For such planforms, sketched in Fig. 5.28, a more sophisticated model must be used. The purpose of this section is to introduce such a model and to discuss its numerical implementation. However, it is beyond the scope of this book to elaborate on the details of such higher-order models; rather, only the flavor is given here. You are encouraged to pursue this subject by reading the literature and by taking more advanced studies in aerodynamics.

Return to Fig. 5.13. Here, a simple lifting line spans the wing, with its associated trailing vortices. The circulation Γ varies with y along the lifting line. Let us extend this model by placing a *series* of lifting lines on the plane of the wing, at different chordwise stations; i.e., consider a large number of lifting lines all parallel to the y axis, located at different values of x, as shown in Fig. 5.29. In the limit of an infinite number of lines of infinitesimal strength, we obtain a vortex sheet, where the vortex lines run parallel to the y axis. The strength of this sheet (per unit length in the x direction) is denoted by γ, where γ varies in the y direction, analogous to the variation of Γ for the single lifting line in Fig. 5.13. Moreover, each lifting line will have, in general, a different overall strength, so that γ varies with x also. Hence, $\gamma = \gamma(x, y)$ as shown in Fig. 5.29. In addition, recall that each lifting line has a system of trailing vortices; hence, the series of lifting lines is crossed by a series of superimposed trailing vortices parallel to the x axis. In the limit of an infinite number of infinitesimally weak vortices, these trailing vortices form another vortex sheet of strength δ (per unit length in the y direction). [Note that this δ is different from the δ used in Eq. (5.61); the use of the same symbol in both cases is standard, and there should be no confusion since the meanings and context are completely different.] To see this more clearly, consider a single line parallel to the x axis. As we move along this line from the leading edge to the trailing edge, we pick up an additional superimposed trailing vortex each time we cross a lifting line. Hence, δ must vary with x. Moreover, the trailing vortices are simply parts of the horseshoe vortex systems, the leading

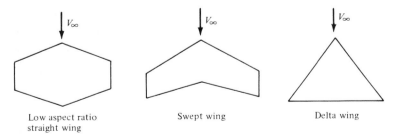

Low aspect ratio straight wing

Swept wing

Delta wing

FIGURE 5.28
Types of wing planforms for which classical lifting-line theory is not appropriate.

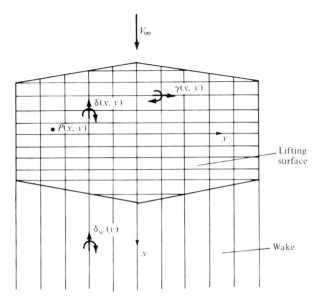

FIGURE 5.29
Schematic of a lifting surface.

edges of which make up the various lifting lines. Since the circulation about each lifting line varies in the y direction, the strengths of different trailing vortices will, in general, be different. Hence, δ also varies in the y direction, i.e., $\delta = \delta(x, y)$, as shown in Fig. 5.29. The two vortex sheets—the one with vortex lines running parallel to y with strength γ (per unit length in the x direction) and the other with vortex lines running parallel to x with strength δ (per unit length in the y direction)—result in a *lifting surface* distributed over the entire planform of the wing, as shown in Fig. 5.29. At any given point on the surface, the strength of the lifting surface is given by both γ and δ, which are functions of x and y. We denote $\gamma = \gamma(x, y)$ as the spanwise vortex strength distribution and $\delta = \delta(x, y)$ as the chordwise vortex strength distribution.

Note that downstream of the trailing edge we have no spanwise vortex lines, only trailing vortices. Hence, the wake consists of only chordwise vortices. The strength of this wake vortex sheet is given by δ_w (per unit length in the y direction). Since in the wake the trailing vortices do not cross any vortex lines, the strength of any given trailing vortex is constant with x. Hence, δ_w depends only on y and, throughout the wake, $\delta_w(y)$ is equal to its value at the trailing edge.

Now that we have defined the lifting surface, of what use is it? Consider point P located at (x, y) on the wing, as shown in Fig. 5.29. The lifting surface and the wake vortex sheet both induce a normal component of velocity at point P. Denote this normal velocity by $w(x, y)$. We want the wing planform to be a stream surface of the flow; i.e., we want the sum of the induced $w(x, y)$ and the normal component of the freestream velocity to be zero at point P and for all points on the wing—this is the flow-tangency condition on the wing surface. (Keep in mind that we are treating the wing as a flat surface in this discussion.)

The central theme of lifting-surface theory is to find $\gamma(x, y)$ and $\delta(x, y)$ such that the flow-tangency condition is satisfied at all points on the wing. [Recall that in the wake, $\delta_w(y)$ is fixed by the trailing-edge values of $\delta(x, y)$; hence, $\delta_w(y)$ is not, strictly speaking, one of the unknown dependent variables.]

Let us obtain an expression for the induced normal velocity $w(x, y)$ in terms of γ, δ, and δ_w. Consider the sketch given in Fig. 5.30, which shows a portion of the planview of a finite wing. Consider the point given by the coordinates (ξ, η). At this point, the spanwise vortex strength is $\gamma(\xi, \eta)$. Consider a thin ribbon, or filament, of the spanwise vortex sheet of incremental length $d\xi$ in the x direction. Hence, the strength of this filament is $\gamma\, d\xi$, and the filament stretches in the y (or η) direction. Also, consider point P located at (x, y) and removed a distance r from the point (ξ, η). From the Biot-Savart law, Eq. (5.5), the incremental velocity induced at P by a segment $d\eta$ of this vortex filament of strength $\gamma\, d\xi$ is

$$|\mathbf{dV}| = \left| \frac{\Gamma}{4\pi} \frac{\mathbf{dl} \times \mathbf{r}}{|\mathbf{r}|^3} \right| = \frac{\gamma\, d\xi}{4\pi} \frac{(d\eta)r \sin\theta}{r^3} \qquad (5.73)$$

Examining Fig. 5.30, and following the right-hand rule for the strength γ, note that $|\mathbf{dV}|$ is induced downward, into the plane of the wing, i.e., in the negative z direction. Following the usual sign convention that w is positive in the upward direction, i.e., in the positive z direction, we denote the contribution of Eq. (5.73)

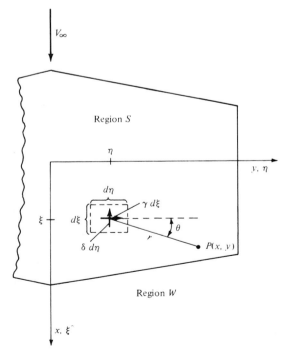

FIGURE 5.30
Velocity induced at point P by an infinitesimal segment of the lifting surface. The velocity is perpendicular to the plane of the paper.

to the induced velocity w as $(dw)_\gamma = -|d\mathbf{V}|$. Also, note that $\sin \theta = (x - \xi)/r$. Hence, Eq. (5.73) becomes

$$(dw)_\gamma = -\frac{\gamma}{4\pi} \frac{(x - \eta) \, d\xi \, d\eta}{r^3} \tag{5.74}$$

Considering the contribution of the elemental chordwise vortex of strength $\delta \, d\eta$ to the induced velocity at P, we find by an analogous argument that

$$(dw)_\delta = -\frac{\delta}{4\pi} \frac{(y - \eta) \, d\xi \, d\eta}{r^3} \tag{5.75}$$

To obtain the velocity induced at P by the entire lifting surface, Eqs. (5.74) and (5.75) must be integrated over the wing planform, designated as region S in Fig. 5.30. Moreover, the velocity induced at P by the complete wake is given by an equation analogous to Eq. (5.75), but with δ_w instead of δ, and integrated over the wake, designated as region W in Fig. 5.30. Noting that

$$r = \sqrt{(x - \xi)^2 + (y - \eta)^2}$$

the normal velocity induced at P by both the lifting surface and the wake is

$$
\begin{aligned}
w(x, y) = &-\frac{1}{4\pi} \iint\limits_{S} \frac{(x - \xi)\gamma(\xi, \eta) + (y - \eta)\delta(\xi, \eta)}{[(x - \xi)^2 + (y - \eta)^2]^{3/2}} \, d\xi \, d\eta \\
&-\frac{1}{4\pi} \iint\limits_{W} \frac{(y - \eta)\delta_w(\xi, \eta)}{[(x - \xi)^2 + (y - \eta)^2]^{3/2}} \, d\xi \, d\eta
\end{aligned}
\tag{5.76}
$$

The central problem of lifting-surface theory is to solve Eq. (5.76) for $\gamma(\xi, \eta)$ and $\delta(\xi, \eta)$ such that the sum of $w(x, y)$ and the normal component of the freestream is zero, i.e., such that the flow is tangent to the planform surface S. The details of various lifting-surface solutions are beyond the scope of this book; rather, our purpose here was simply to present the flavor of the basic model.

The advent of the high-speed digital computer has made possible the implementation of numerical solutions based on the lifting-surface concept. These solutions are similar to the panel solutions for two-dimensional flow discussed in Chaps. 3 and 4 in that the wing planform is divided into a number of panels, or elements. On each panel, either constant or prescribed variations of both γ and δ can be made. Control points on the panels can be chosen, where the net normal flow velocity is zero. The evaluation of equations like Eq. (5.76) at these control points results in a system of simultaneous algebraic equations which can be solved for the values of the γ's and δ's on all the panels.

A related but somewhat simpler approach is to superimpose a finite number of horseshoe vortices of different strengths Γ_n on the wing surface. For example, consider Fig. 5.31, which shows part of a finite wing. The dashed lines define a

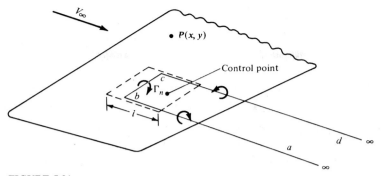

FIGURE 5.31
Schematic of a single horseshoe vortex, which is part of a vortex system on the wing.

panel on the wing planform, where l is the length of the panel in the flow direction. The panel is a trapezoid; it does not have to be a square, or even a rectangle. A horseshoe vortex, *abcd*, of strength Γ_n is placed on the panel such that the segment *bc* is a distance $l/4$ from the front of the panel. A control point is placed on the centerline of the panel at a distance $\frac{3}{4}l$ from the front. The velocity induced at an arbitrary point P only by the single horseshoe vortex can be calculated from the Biot-Savart law by treating each of the vortex filaments *ab*, *bc*, and *cd* separately. Now consider the entire wing covered by a finite number of panels, as sketched in Fig. 5.32. A series of horseshoe vortices is now superimposed. For example, on one panel at the leading edge, we have the horseshoe vortex *abcd*. On the panel behind it, we have the horseshoe vortex *aefd*. On the next panel, we have *aghd*, and on the next, *aijd*, etc. The entire wing is covered by this lattice of horseshoe vortices, each of different unknown strength Γ_n. At any control point P, the normal velocity induced by *all* the horseshoe vortices can be obtained from the Biot-Savart law. When the flow-tangency condition is applied at all the control points, a system of simultaneous algebraic equations results

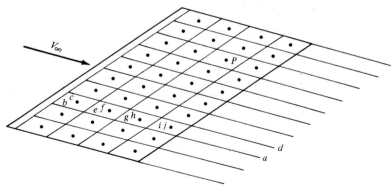

FIGURE 5.32
Vortex lattice system on a finite wing.

which can be solved for the unknown Γ_n's. This numerical approach is called the *vortex lattice method* and is in wide use today for the analysis of finite-wing properties. Once again, only the flavor of the method is given above; you are encouraged to read the volumes of literature that now exist on various versions of the vortex lattice method. In particular, Ref. 13 has an excellent introductory discussion on the vortex lattice method, including a worked example which clearly illustrates the salient points of the technique.

5.6 APPLIED AERODYNAMICS: THE DELTA WING

In Part III of this book, we will see that supersonic flow is dramatically different from subsonic flow in virtually all respects—the mathematics and physics of these two flow regimes are totally different. Such differences impact the design philosophy of aircraft for supersonic flight in comparison to aircraft for subsonic flight. In particular, supersonic airplanes usually have highly swept wings (the reasons for this are discussed in Part III). A special case of swept wings is those aircraft with a triangular planform—called delta wings. A comparison of the planform of a conventional swept wing and a delta wing was shown in Fig. 5.28. Two classic examples of aircraft with delta wings are the Convair F-102A, the first operational jet airplane in the United States to be designed with a delta wing, shown in Fig. 5.33a, and the space shuttle, basically a hypersonic airplane, shown in Fig. 5.33b. In reality, the planform of the space shuttle is more correctly denoted as a double-delta shape. Indeed, there are several variants of the basic delta wing used on modern aircraft; these are shown in Fig. 5.34. Delta wings are used on many different types of high-speed airplanes around the world; hence, the delta planform is an important aerodynamic configuration.

Question: Since delta-winged aircraft are high-speed vehicles, why are we discussing this topic in the present chapter, which deals with the low-speed, incompressible flow over finite wings? The obvious answer is that all high-speed aircraft fly at low speeds for takeoff and landing; moreover, in most cases, these aircraft spend the vast majority of their flight time at subsonic speeds, using their supersonic capability for short "supersonic dashes," depending on their mission. Several exceptions are, of course, the Concorde supersonic transport which cruises at supersonic speeds across oceans, and the space shuttle, which is hypersonic for most of its reentry into the earth's atmosphere. However, the vast majority of delta-winged aircraft spend a great deal of their flight time at subsonic speeds. For this reason, the low-speed aerodynamic characteristics of delta wings are of great importance; this is accentuated by the rather different and unique aerodynamic aspects associated with such delta wings. Therefore, the low-speed aerodynamics of delta wings has been a subject of much serious study over the past years, going back as far as the early work on delta wings by Alexander Lippisch in Germany during the 1930s. This is the answer to our question posed above—in the context of our discussion on finite wings, we must give the delta wing some special attention.

(a)

FIGURE 5.33
Some delta-winged vehicles. (a) The Convair F-102A. (*Courtesy of the U.S. Air Force.*) (b) The space shuttle. (*Courtesy of NASA.*)

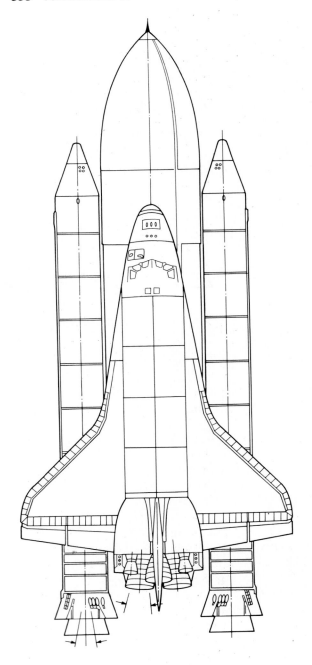

(b)

FIGURE 5.33
(Continued.)

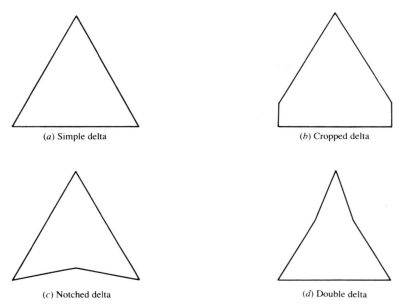

(a) Simple delta

(b) Cropped delta

(c) Notched delta

(d) Double delta

FIGURE 5.34
Four versions of a delta-wing planform. (*From Loftin, Ref. 48.*)

The subsonic flow pattern over the top of a delta wing at angle of attack is sketched in Fig. 5.35. The dominant aspect of this flow are the two vortex patterns that occur in the vicinity of the highly swept leading edges. These vortex patterns are created by the following mechanism. The pressure on the bottom surface of the wing at the angle of attack is higher than the pressure on the top surface. Thus, the flow on the bottom surface in the vicinity of the leading edge tries to curl around the leading edge from the bottom to the top. If the leading edge is sharp, the flow will separate along its entire length. (We have already mentioned several times that when low-speed, subsonic flow passes over a sharp convex corner, inviscid flow theory predicts an infinite velocity at the corner, and that nature copes with this situation by having the flow separate at the corner. The leading edge of a delta wing is such a case.) This separated flow curls into a primary vortex which exists above the wing just inboard of each leading edge, as sketched in Fig. 5.35. The stream surface which has separated at the leading edge (the primary separation line, S_1, in Fig. 5.35) loops above the wing and then reattaches along the primary attachment line (line A in Fig. 5.35). The primary vortex is contained within this loop. A secondary vortex is formed underneath the primary vortex, with its own separation line, denoted by S_2 in Fig. 5.35, and its own reattachment line, A_2. Notice that the surface streamlines flow *away* from the attachment lines A_1 and A_2 on both sides of these lines, whereas the surface streamlines tend to flow *toward* the separation lines S_1 and S_2 and then simply lift off the surface along these lines. Inboard of the leading-edge vortices, the surface streamlines are attached, and flow downstream virtually is

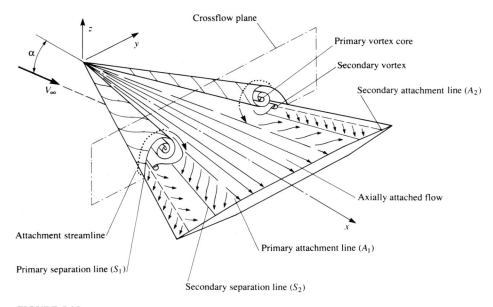

FIGURE 5.35
Schematic of the subsonic flow field over the top of a delta wing at angle of attack. (*Courtesy of John Stollery, Cranfield Institute of Technology, England.*)

undisturbed along a series of straight-line rays emanating from the vertex of the triangular shape. A graphic illustration of the leading-edge vortices is shown in both Figs. 5.36 and 5.37. In Fig. 5.36, we see a highly swept delta wing mounted in a water tunnel. Filaments of colored dye are introduced at two locations along each leading edge. This photograph, taken from an angle looking down on the top of the wing, clearly shows the entrainment of the colored dye in the vortices. Figure 5.37 is a photograph of the vortex pattern in the crossflow plane (the crossflow plane is shown in Fig. 5.35). From the photographs in Figs. 5.36 and 5.37, we clearly see that the leading-edge vortex is real and is positioned above and somewhat inboard of the leading edge itself.

The leading-edge vortices are strong and stable. Being a source of high energy, relatively high-vorticity flow, the local static pressure in the vicinity of the vortices is small. Hence, the surface pressure on the top surface of the delta wing is reduced near the leading edge and is higher and reasonably constant over the middle of the wing. The qualitative variation of the pressure coefficient in the spanwise direction (the y direction as shown in Fig. 5.35) is sketched in Fig. 5.38. The spanwise variation of pressure over the bottom surface is essentially constant and higher than the freestream pressure (a positive C_p). Over the top surface, the spanwise variation in the midsection of the wing is essentially constant and lower than the freestream pressure (a negative C_p). However, near the leading edges the static pressure drops considerably (the values of C_p become more negative). The leading-edge vortices are literally creating a strong "suction" on

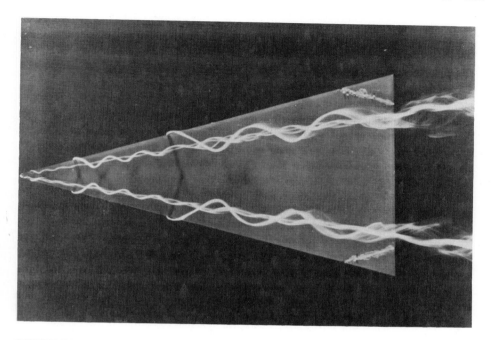

FIGURE 5.36
Leading-edge vortices over the top surface of a delta wing at angle of attack. The vortices are made
visible by dye streaks in water flow. (*Courtesy of H. Werle, ONERA, France. Also in Van Dyke,
Milton, An Album of Fluid Motion, The Parabolic Press, Stanford, Calif., 1982.*)

the top surface near the leading edges. In Fig. 5.38, vertical arrows are shown to
indicate further the effect on the spanwise lift distribution; the upward direction
of these arrows as well as their relative length show the local contribution of
each section of the wing to the normal force distribution. The suction effect of
the leading-edge vortices is clearly shown by these arrows.

The suction effect of the leading-edge vortices enhances the lift; for this
reason, the lift coefficient curve for a delta wing exhibits an increase in C_L for
values of α at which conventional wing planforms would be stalled. A typical
variation of C_L with α for a 60° delta wing is shown in Fig. 5.39. Note the
following characteristics:

1. The lift slope is small, on the order of 0.05/degree.
2. However, the lift continues to increase to large values of α; in Fig. 5.39, the
   stalling angle of attack is on the order of 35°. The net result is a reasonable
   value of $C_{L,\max}$, on the order of 1.3.

The next time you have an opportunity to watch a delta-winged airplane take
off or land, say, e.g., the televised landing of the space shuttle, note the large

FIGURE 5.37
The flow field in the crossflow plane above a delta wing at angle of attack, showing the two primary leading-edge vortices. The vortices are made visible by small air bubbles in water. (*Courtesy of H. Werle, ONERA, France. Also in Van Dyke, Milton, An Album of Fluid Motion, The Parabolic Press, Stanford, Calif., 1982.*)

angle of attack of the vehicle. Moreover, you will understand why the angle of attack is large—because the lift slope is small, and hence the angle of attack must be large enough to generate the high values of C_L required for low-speed flight.

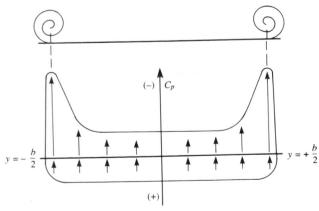

FIGURE 5.38
Schematic of the spanwise pressure coefficient distribution across a delta wing. (*Courtesy of John Stollery, Cranfield Institute of Technology, England.*)

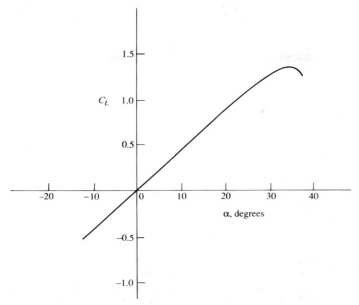

FIGURE 5.39
Variation of lift coefficient for a flat delta wing with angle of attack. (*Courtesy of John Stollery, Cranfield Institute of Technology, England.*)

The suction effect of the leading-edge vortices, in acting to increase the normal force, consequently, increases the drag at the same time it increases the lift. Hence, the aerodynamic effect of these vortices is not necessarily advantageous. In fact, the lift-to-drag ratio, L/D, for a delta planform is not so high as conventional wings. The typical variation of L/D with C_L for a delta wing is shown in Fig. 5.40; the results for the sharp leading edge, 60° delta wing are given by the lower curve. Note that the maximum value of L/D for this case is about 9.3—not a particularly exciting value for a low-speed aircraft.

There are two other phenomena that are reflected by the data in Fig. 5.40. The first is the effect of greatly rounding the leading edges of the delta wing. In our discussions above, we have treated the case of a sharp leading edge; such sharp edges cause the flow to separate at the leading edge, forming the leading-edge vortices. On the other hand, if the leading-edge radius is large, the flow separation will be minimized, or possibly will not occur. In turn, the drag penalty discussed above will not be present, and hence the L/D ratio will increase. The dashed curve in Fig. 5.40 is the case for a 60° delta wing with well-rounded leading edges. Note that $(L/D)_{max}$ for this case is about 16.5, almost a factor of 2 higher than the sharp leading-edge case. However, keep in mind that these are results for subsonic speeds. There is a major design compromise reflected in these results. At the beginning of this section, we mentioned that the delta-wing planform with sharp leading edges is advantageous for supersonic flight—its

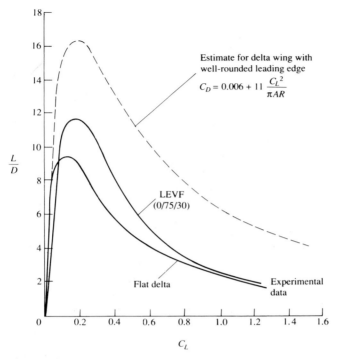

FIGURE 5.40
The effect of leading-edge shape on the lift-to-drag ratio for a delta wing of aspect ratio 2.31. The two solid curves apply to a sharp leading edge, and the dashed curve applies to a rounded leading edge. LEVF denotes a wing with a leading-edge vortex flap. (*Courtesy of John Stollery, Cranfield Institute of Technology, England.*)

highly swept shape in combination with sharp leading edges has a low supersonic drag. However, at supersonic speeds this advantage will be negated if the leading edges are rounded to any great extent. We will find in our study of supersonic flow in Part III that a blunt-nosed body creates very large values of wave drag. Therefore, leading edges with large radii are not appropriate for supersonic aircraft; indeed, it is desirable to have as sharp a leading edge as is practically possible for supersonic airplanes. A singular exception is the design of the space shuttle. The leading-edge radius of the space shuttle is large; this is due to three features that combine to make such blunt leading edges advantageous for the shuttle. First, the shuttle must slow down early during reentry into the earth's atmosphere to avoid massive aerodynamic heating (aspects of aerodynamic heating are discussed in Part IV). Therefore, in order to obtain this deceleration, a high drag is desirable for the space shuttle; indeed, the maximum L/D ratio of the space shuttle during reentry is about 2. A large leading-edge radius, with its attendant high drag, is therefore advantageous. Secondly, as we will see in Part IV, the rate of aerodynamic heating to the leading edge itself—a region of high heating—is inversely proportional to the square root of the leading-edge radius. Hence, the larger the radius, the smaller will be the heating rate to the leading edge. Thirdly, as already explained above, a highly rounded leading edge is certainly advantageous to the shuttle's subsonic aerodynamic characteristics. Hence, a well-rounded leading edge is an important design feature for the space shuttle on all accounts. However, we must be reminded that this is not the case for more conventional supersonic aircraft, which demand very sharp leading edges. For these aircraft, a delta wing with a sharp leading edge has relatively poor subsonic performance.

This leads to the second of the phenomena reflected in Fig. 5.40. The middle curve in Fig. 5.40 is labeled LEVF, which denotes the case for a leading-edge vortex flap. This pertains to a mechanical configuration where the leading edges can be deflected downward through a variable angle, analogous to the deflection of a conventional trailing-edge flap. The spanwise pressure-coefficient distribution for this case is sketched in Fig. 5.41; note that the direction of the suction due to the leading-edge vortice is now modified in comparison to the case with no leading-edge flap shown earlier in Fig. 5.38. Also, returning to Fig. 5.35, you can visualize what the wing geometry would look like with the leading edge drooped down; a front view of the downward deflected flap would actually show some projected frontal area. Since the pressure is low over this frontal area, the net drag can decrease. This phenomenon is illustrated by the middle curve in Fig. 5.40, which shows a generally higher L/D for the leading-edge vortex flap in comparison to the case with no flap (the flat delta wing).

In summary, the delta wing is a common planform for supersonic aircraft. In this section, we have examined the low-speed aerodynamic characteristics of such wings, and have found that these characteristics are in some ways quite different from a conventional planform.

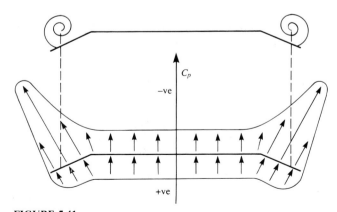

FIGURE 5.41
A schematic of the spanwise pressure coefficient distribution over the top of a delta wing as modified by leading-edge vortex flaps. (*Courtesy of John Stollery, Cranfield Institute of Technology, England.*)

5.7 HISTORICAL NOTE: LANCHESTER AND PRANDTL—THE EARLY DEVELOPMENT OF FINITE-WING THEORY

On June 27, 1866, in a paper entitled "Aerial Locomotion" given to the Aeronautical Society of Great Britain, the Englishman Francis Wenham expressed for the first time in history the effect of aspect ratio on finite-wing aerodynamics. He theorized (correctly) that most of the lift of a wing occurs from the portion near the leading edge, and hence a long, narrow wing would be most efficient. He suggested stacking a number of long thin wings above each other to generate the required lift, and he built two full-size gliders in 1858, both with five wings each, to demonstrate (successfully) his ideas. (Wenham is also known for designing and building the first wind tunnel in history, at Greenwich, England, in 1871.)

However, the true understanding of finite-wing aerodynamics, as well as ideas for the theoretical analysis of finite wings, did not come until 1907. In that year, Frederick W. Lanchester published his now famous book entitled *Aerodynamics*. We have met Lanchester before—in Sec. 4.11 concerning his role in the development of the circulation theory of lift. Here, we examine his contributions to finite-wing theory.

In Lanchester's *Aerodynamics*, we find the first mention of vortices that trail downstream of the wing tips. Figure 5.42 is one of Lanchester's own drawings from his 1907 book, showing the "vortex trunk" which forms at the wing tip. Moreover, he knew that a vortex filament could not end in space (see Sec. 5.2), and he theorized that the vortex filaments which constituted the two wing-tip vortices must cross the wing along its span—the first concept of bound vortices in the spanwise direction. Hence, the essence of the horseshoe vortex concept originated with Lanchester. In his own words:

> Thus the author regards the two trailed vortices as a definite proof of the existence of a cyclic component of equal strength in the motion surrounding the airfoil itself.

Considering the foresight and originality of Lanchester's thinking, let us pause for a moment and look at the man himself. Lanchester was born on October 23, 1868, in Lewisham, England. The son of an architect, Lanchester became

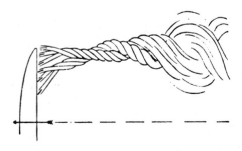

FIGURE 5.42
A figure from Lanchester's *Aerodynamics*, 1907; this is his own drawing of the wing-tip vortex on a finite wing.

interested in engineering at an early age. (He was told by his family that his mind was made up at the age of 4.) He studied engineering and mining during the years 1886–1889 at the Royal College of Science in South Kensington, London, but never officially graduated. He was a quick-minded and innovative thinker and became a designer at the Forward Gas Engine Company in 1889, specializing in internal combustion engines. He rose to the post of assistant works manager. In the early 1890s, Lanchester became very interested in aeronautics, and along with his development of high-speed engines, he also carried out numerous aerodynamics experiments. It was during this period that he formulated his ideas on both the circulation theory of lift and the finite-wing vortex concepts. A serious paper written by Lanchester first for the Royal Society, and then for the Physical Society, was turned down for publication—something Lanchester never forgot. Finally, his aeronautical concepts were published in his two books *Aerodynamics* and *Aerodonetics* in 1907 and 1908, respectively. To his detriment, Lanchester had a style of writing and a means of explanation which were not easy to follow, and his works were not immediately seized upon by other researchers. Lanchester's bitter feelings about the public's receipt of his papers and books are graphically seen in his letter to the Daniel Guggenheim Medal Fund decades later. In a letter dated June 6, 1931, Lanchester writes:

> So far as aeronautical science is concerned, I cannot say that I experienced anything but discouragement; in the early days my theoretical work (backed by a certain amount of experimental verification), mainly concerning the vortex theory of sustentation and the screw propeller, was refused by the two leading scientific societies in this country, and I was seriously warned that my profession as an engineer would suffer if I dabbled in a subject that was merely a dream of madmen! When I published my two volumes in 1907 and 1908 they were well received on the whole, but this was mainly due to the success of the brothers Wright, and the general interest aroused on the subject.

In 1899, he formed the Lanchester Motor Company, Limited, and sold automobiles of his own design. He married in 1919, but had no children. Lanchester maintained his interest in automobiles and related mechanical devices until his death on March 8, 1946, at the age of 77.

In 1908, Lanchester visited Göttingen, Germany, and fully discussed his wing theory with Ludwig Prandtl and his student Theodore von Karman. Prandtl spoke no English, Lanchester spoke no German, and in light of Lanchester's unclear way of explaining his ideas, there appeared to be little chance of understanding between the two parties. However, shortly after, Prandtl began to develop his own wing theory, using a bound vortex along the span and assuming that the vortex trails downstream from both wing tips. The first mention of Prandtl's work on finite-wing theory was made in a paper by O. Foppl in 1911, discussing some of Foppl's experimental work on finite wings. Commenting on his results, Foppl says:

> They agree very closely with the theoretical investigation by Professor Prandtl on the current around an airplane with a finite span wing. Already Lanchester in his

work, "Aerodynamics" (translated into German by C. and A. Runge), indicated that to the two extremities of an airplane wing are attached two vortex ropes (Wirbelzopfe) which make possible the transition from the flow around the airplane, which occurs nearly according to Kutta's theory, to the flow of the undisturbed fluid at both sides. These two vortex ropes continue the vortex which, according to Kutta's theory, takes place on the lamina.

We are led to admit this owing to the Helmholtz theorem that vortices cannot end in the fluid. At any rate these two vortex ropes have been made visible in the Göttingen Institute by emitting an ammonia cloud into the air. Prandtl's theory is constructed on the consideration of this current in reality existing.

In the same year, Prandtl expressed his own first published words on the subject. In a paper given at a meeting of the Representatives of Aeronautical Science in Göttingen in November 1911, entitled "Results and Purposes of the Model Experimental Institute of Göttingen," Prandtl states:

Another theoretical research relates to the conditions of the current which is formed by the air behind an airplane. The lift generated by the airplane is, on account of the principle of action and reaction, necessarily connected with a descending current behind the airplane. Now it seemed very useful to investigate this descending current in all its details. It appears that the descending current is formed by a pair of vortices, the vortex filaments of which start from the airplane wing tips. The distance of the two vortices is equal to the span of the airplane, their strength is equal to the circulation of the current around the airplane and the current in the vicinity of the airplane is fully given by the superposition of the uniform current with that of a vortex consisting of three rectilinear sections.

In discussing the results of his theory, Prandtl goes on to state in the same paper:

The same theory supplies, taking into account the variations of the current on the airplane which came from the lateral vortices, a relationship showing the dependence of the airplane lift on the aspect ratio; in particular it gives the possibility of extrapolating the results thus obtained experimentally to the airplane of infinite span wing. From the maximum aspect ratios measured by us (1:9 to that of $1:\infty$) the lifts increase further in marked degree—by some 30 or 40 percent. I would add here a remarkable result of this extrapolation, which is, that the results of Kutta's theory of the infinite wing, at least so far as we are dealing with small cambers and small angles of incidence, have been confirmed by these experimental results.

Starting from this line of thought we can attack the problem of calculating the surface of an airplane so that lift is distributed along its span in a determined manner, previously fixed. The experimental trial of these calculations has not yet been made, but it will be in the near future.

It is clear from the above comments that Prandtl was definitely following the model proposed earlier by Lanchester. Moreover, the major concern of the finite-wing theory was first in the calculation of lift—no mention is made of induced drag. It is interesting to note that Prandtl's theory first began with a single horseshoe vortex, such as sketched in Fig. 5.11. The results were not entirely satisfactory. During the period 1911–1918, Prandtl and his colleagues expanded and refined his finite-wing theory, which evolved to the concept of a lifting line consisting of an infinite number of horseshoe vortices, as sketched in Fig. 5.13. In 1918, the term "induced drag" was coined by Max Munk, a colleague of Prandtl at Göttingen. Much of Prandtl's development of finite-wing theory was classified secret by the German government during World War I. Finally, his lifting-line theory was released to the outside world, and his ideas were published in English in a special NACA report written by Prandtl and published in 1922, entitled "Applications of Modern Hydrodynamics to Aeronautics" (NACA TR 116). Hence, the theory we have outlined in Sec. 5.3 was well-established more than 60 years ago.

One of Prandtl's strengths was the ability to base his thinking on sound ideas, and to apply intuition that resulted in relatively straightforward theories that most engineers could understand and appreciate. This is in contrast to the difficult writings of Lanchester. As a result, the lifting theory for finite wings has come down through the years identified as *Prandtl's lifting-line theory*, although we have seen that Lanchester was the first to propose the basic model on which lifting-line theory is built.

In light of Lanchester's 1908 visit with Prandtl and Prandtl's subsequent development of the lifting-line theory, there has been some discussion over the years that Prandtl basically stole Lanchester's ideas. However, this is clearly not the case. We have seen in the above quotes that Prandtl's group at Göttingen was giving full credit to Lanchester as early as 1911. Moreover, Lanchester never gave the world a clear and practical theory with which results could be readily obtained—Prandtl did. Therefore, in this book we have continued the tradition of identifying the lifting-line theory with Prandtl's name. On the other hand, for very good reasons, in England and various places in western Europe, the theory is labeled the *Lanchester-Prandtl theory*.

To help put the propriety in perspective, Lanchester was awarded the Daniel Guggenheim Medal in 1936 (Prandtl had received this award some years earlier). In the medal citation, we find the following words:

Lanchester was the foremost person to propound the now famous theory of flight based on the Vortex theory, so brilliantly followed up by Prandtl and others. He first put forward his theory in a paper read before the Birmingham Natural History and Philosophical Society on 19th June, 1894. In a second paper in 1897, in his two books published in 1907 and 1908, and in his paper read before the Institution of Automobile Engineers in 1916, he further developed this doctrine.

Perhaps the best final words on Lanchester are contained in this excerpt from his obituary found in the British periodical *Flight* in March 1946:

And now Lanchester has passed from our ken but not from our thoughts. It is to be hoped that the nation which neglected him during much of his lifetime will at any rate perpetuate his work by a memorial worthy of the "Grand Old Man" of aerodynamics.

5.8 HISTORICAL NOTE: PRANDTL—THE MAN

The modern science of aerodynamics rests on a strong fundamental foundation, a large percentage of which was established in one place by one man—at the University of Göttingen by Ludwig Prandtl. Prandtl never received a Noble Prize, although his contributions to aerodynamics and fluid mechanics are felt by many to be of that caliber. Throughout this book, you will encounter his name in conjunction with major advances in aerodynamics: thin airfoil theory in Chap. 4, finite-wing theory in Chap. 5, supersonic shock- and expansion-wave theory in Chap. 9, compressibility corrections in Chap. 11, and what may be his most important contribution, namely, the boundary-layer concept in Chap. 16. Who was this man who has had such a major impact on fluid dynamics? Let us take a closer look.

Ludwig Prandtl was born on February 4, 1874, in Freising, Bavaria. His father was Alexander Prandtl, a professor of surveying and engineering at the agricultural college at Weihenstephan, near Freising. Although three children were born into the Prandtl family, two died at birth, and Ludwig grew up as an only child. His mother, the former Magdalene Ostermann, had a protracted illness, and partly as a result of this, Prandtl became very close to his father. At an early age, Prandtl became interested in his father's books on physics, machinery, and instruments. Much of Prandtl's remarkable ability to go intuitively to the heart of a physical problem can be traced to his environment at home as a child, where his father, a great lover of nature, induced Ludwig to observe natural phenomena and to reflect on them.

In 1894, Prandtl began his formal scientific studies at the Technische Hochschule in Munich, where his principal teacher was the well-known mechanics professor, August Foppl. Six years later, he graduated from the University of Munich with a Ph.D., with Foppl as his advisor. However, by this time Prandtl was alone, his father having died in 1896 and his mother in 1898.

By 1900, Prandtl had not done any work or shown any interest in fluid mechanics. Indeed, his Ph.D. thesis at Munich was in solid mechanics, dealing with unstable elastic equilibrium in which bending and distortion acted together. (It is not generally recognized by people in fluid dynamics that Prandtl continued his interest and research in solid mechanics through most of his life—this work is eclipsed, however, by his major contributions to the study of fluid flow.) However, soon after graduation from Munich, Prandtl had his first major encounter with fluid mechanics. Joining the Nuremburg works of the Maschinenfabrick Augsburg as an engineer, Prandtl worked in an office designing mechanical equipment for the new factory. He was made responsible for redesigning an apparatus for removing machine shavings by suction. Finding no reliable information in the scientific literature about the fluid mechanics of suction, Prandtl arranged his own experiments to answer a few fundamental questions about the flow. The result of this work was his new design for shavings' cleaners. The apparatus was modified with pipes of improved shape and size, and carried out satisfactory operation at one-third its original power consumption. Prandtl's contributions in fluid mechanics had begun.

One year later, in 1901, he became Professor of Mechanics in the Mathematical Engineering Department at the Technische Hochschule in Hanover. (Please note that in Germany a "technical high school" is equivalent to a technical university in the United States.) It was at Hanover that Prandtl enhanced and continued his new-found interest in fluid mechanics. It was here that Prandtl developed his boundary-layer theory and became interested in supersonic flow through nozzles. In 1904, Prandtl delivered his famous paper on the concept of the boundary layer to the Third Congress on Mathematicians at Heidelberg. Entitled "Über Flussigkeitsbewegung bei sehr kleiner Reibung," Prandtl's Heidelberg paper established the basis for most modern calculations of skin friction, heat transfer, and flow separation (see Chaps. 15 to 17). From that time on, the star of Prandtl was to rise meteorically. Later that year, he moved to the prestigious University of Göttingen to become Director of the Institute for Technical Physics, later to be renamed Applied Mechanics. Prandtl spent the remainder of his life at Göttingen, building his laboratory into the world's greatest aerodynamic research center of the 1904–1930 time period.

At Göttingen, during 1905–1908 Prandtl carried out numerous experiments on supersonic flow through nozzles and developed oblique shock- and expansion-wave theory (see Chap. 9). He took the first photographs of the supersonic flow through nozzles, using a special schlieren optical system (see chap. 4 of Ref. 21). From 1910 to 1920, he devoted most of his efforts to low-speed aerodynamics, principally airfoil and wing theory, developing the famous lifting-line theory for finite wings (see Sec. 5.3). Prandtl returned to high-speed flows in the 1920s, during which he contributed to the evolution of the famous Prandtl-Glauert compressibility correction (see Secs. 11.4 and 11.11).

By the 1930s, Prandtl was recognized worldwide as the "elder statesman" of fluid dynamics. Although he continued to do research in various areas, including structural mechanics and meteorology, his "Nobel Prize-level" contributions to fluid dynamics had all been made. Prandtl remained at Göttingen throughout the turmoil of World War II, engrossed in his work and seemingly insulated from the intense political and physical disruptions brought about by Nazi Germany. In fact, the German Air Ministry provided Prandtl's laboratory with new equipment and financial support. Prandtl's attitude at the end of the war is reflected in his comments to a U.S. Army interrogation team which swept through Göttingen in 1945; he complained about bomb damage to the roof of his house, and he asked how the Americans planned to support his current and future research. Prandtl was 70 at the time and was still going strong. However, the fate of Prandtl's laboratory at this time is summed up in the words of Irmgard Flugge-Lotz and Wilhelm Flugge, colleagues of Prandtl, who wrote 28 years later in the *Annual Review of Fluid Mechanics* (Vol. 5, 1973):

World War II swept over all of us. At its end some of the research equipment was dismantled, and most of the research staff was scattered with the winds. Many are now in this country (the United States) and in England, some have returned. The seeds sown by Prandtl have sprouted in many places, and there are now many "second growth" Göttingers who do not even know that they are.

What type of person was Prandtl? By all accounts he was a gracious man, studious, likable, friendly, and totally focused on those things that interested him. He enjoyed music and was an accomplished pianist. Figure 5.43 shows a rather introspective man busily at work. One of Prandtl's most famous students, Theodore von Karman, wrote in his autobiography *The Wind and Beyond* (Little, Brown and Company, Boston, 1967) that Prandtl bordered on being naive. A favorite story along these lines is that, in 1909, Prandtl decided that he should be married, but he did not know quite what to do. He finally wrote to Mrs. Foppl, the wife of his respected teacher, asking permission to marry one of her two

FIGURE 5.43
Ludwig Prandtl (1875–1953).

daughters. Prandtl and Foppl's daughters were acquainted, but nothing more than that. Moreover, Prandtl did not stipulate which daughter. The Foppl's made a family decision that Prandtl should marry the elder daughter, Gertrude. The marriage took place, leading to a happy relationship. The Prandtl's had two daughters, born in 1914 and 1917.

Prandtl was considered a tedious lecturer because he could hardly make a statement without qualifying it. However, he attracted excellent students who later went on to distinguish themselves in fluid mechanics—such as Jakob Ackeret in Zurich, Switzerland, Adolf Busemann in Germany, and Theodore von Karman at Aachen, Germany, and later at Cal Tech in the United States.

Prandtl died in 1953. He was clearly the father of modern aerodynamics—a monumental figure in fluid dynamics. His impact will be felt for centuries to come.

5.9   SUMMARY

Return to the chapter road map in Fig. 5.5, and review the straightforward path we have taken during the development of finite-wing theory. Make certain that you feel comfortable with the flow of ideas before proceeding further.

A brief summary of the important results of this chapter follows:

The wing-tip vortices from a finite wing induce a downwash which reduces the angle of attack effectively seen by a local airfoil section:

$$\alpha_{\text{eff}} = \alpha - \alpha_i \tag{5.1}$$

In turn, the presence of downwash results in a component of drag defined as induced drag D_i.

Vortex sheets and vortex filaments are useful in modeling the aerodynamics of finite wings. The velocity induced by a directed segment $\mathbf{dl}$ of a vortex filament is given by the Biot-Savart law:

$$\mathbf{dV} = \frac{\Gamma}{4\pi} \frac{\mathbf{dl} \times \mathbf{r}}{|\mathbf{r}|^3} \tag{5.2}$$

In Prandtl's classical lifting-line theory, the finite wing is replaced by a single spanwise lifting line along which the circulation $\Gamma(y)$ varies. A system of vortices trails downstream from the lifting line, which induces a downwash at the lifting line. The circulation distribution is determined from the fundamental equation

$$\alpha(y_0) = \frac{\Gamma(y_0)}{\pi V_\infty c(y_0)} + \alpha_{L=0}(y_0) + \frac{1}{4\pi V_\infty} \int_{-b/2}^{b/2} \frac{(d\Gamma/dy)\, dy}{y_0 - y} \tag{5.23}$$

Results from classical lifting-line theory:

Elliptic wing

Downwash is constant:

$$w = -\frac{\Gamma_0}{2b} \tag{5.35}$$

$$\alpha_i = \frac{C_L}{\pi AR} \tag{5.42}$$

$$C_{D,i} = \frac{C_L^2}{\pi AR} \tag{5.43}$$

$$a = \frac{a_0}{1 + a_0/\pi AR} \tag{5.69}$$

General wing:

$$C_{D,i} = \frac{C_L^2}{\pi AR}(1+\delta) = \frac{C_L^2}{\pi e AR} \qquad \text{(5.61) and (5.62)}$$

$$a = \frac{a_0}{1 + (a_0/\pi AR)(1+\tau)} \tag{5.70}$$

For low-aspect-ratio wings, swept wings, and delta wings, lifting-surface theory must be used. In modern aerodynamics, such lifting-surface theory is implemented by the vortex panel or the vortex lattice techniques.

PROBLEMS

5.1. Consider a vortex filament of strength Γ in the shape of a closed circular loop of radius R. Obtain an expression for the velocity induced at the center of the loop in terms of Γ and R.

5.2. Consider the same vortex filament as in Prob. 5.1. Consider also a straight line through the center of the loop, perpendicular to the plane of the loop. Let A be the distance along this line, measured from the plane of the loop. Obtain an expression for the velocity at distance A on the line, as induced by the vortex filament.

5.3. The measured lift slope for the NACA 23012 airfoil is 0.1080 degree^{-1}, and $\alpha_{L=0} = -1.3°$. Consider a finite wing using this airfoil, with $AR = 8$ and taper ratio $= 0.8$. Assume that $\delta = \tau$. Calculate the lift and induced drag coefficients for this wing at a geometric angle of attack $= 7°$.

5.4. The Piper Cherokee (a light, single-engine general aviation aircraft) has a wing area of 170 ft^2 and a wing span of 32 ft. Its maximum gross weight is 2450 lb. The wing uses an NACA 65-415 airfoil, which has a lift slope of 0.1033 degree^{-1} and $\alpha_{L=0} = -3°$. Assume $\tau = 0.12$. If the airplane is cruising at 120 mi/h at standard sea level at its maximum gross weight and is in straight-and-level flight, calculate the geometric angle of attack of the wing.

5.5. Consider the airplane and flight conditions given in Prob. 5.4. The span efficiency factor e for the complete airplane is generally much less than that for the finite wing alone. Assume $e = 0.64$. Calculate the induced drag for the airplane in Prob. 5.4.

CHAPTER
6

THREE-DIMENSIONAL
INCOMPRESSIBLE
FLOW

Treat nature in terms of the cylinder, the sphere, the cone, all in perspective.

Paul Cézanne, 1890

6.1 INTRODUCTION

To this point in our aerodynamic discussions, we have been working mainly in a two-dimensional world; the flows over the bodies treated in Chap. 3 and the airfoils in Chap. 4 involved only two dimensions in a single plane—so-called planar flows. In Chap. 5, the analyses of a finite wing were carried out in the plane of the wing, in spite of the fact that the detailed flow over a finite wing is truly three-dimensional. The relative simplicity of dealing with two dimensions, i.e., having only two independent variables, is self-evident and is the reason why a large bulk of aerodynamic theory deals with two-dimensional flows. Fortunately, the two-dimensional analyses go a long way toward understanding many practical flows, but they also have distinct limitations.

The real world of aerodynamic applications is three-dimensional. However, because of the addition of one more independent variable, the analyses generally become more complex. The accurate calculation of three-dimensional flow fields has been, and still is, one of the most active areas of aerodynamic research.

The purpose of this book is to present the fundamentals of aerodynamics. Therefore, it is important to recognize the predominance of three-dimensional flows, although it is beyond our scope to go into detail. Therefore, the purpose of this chapter is to introduce some very basic considerations of three-dimensional incompressible flow. This chapter is short; we do not even need a road map to guide us through it. Its function is simply to open the door to the analysis of three-dimensional flow.

The governing fluid flow equations have already been developed in three dimensions in Chaps. 2 and 3. In particular, if the flow is irrotational, Eq. (2.145) states that

$$\mathbf{V} = \nabla \phi \qquad (2.145)$$

where, if the flow is also incompressible, the velocity potential is given by Laplace's equation:

$$\nabla^2 \phi = 0 \qquad (3.40)$$

Solutions of Eq. (3.40) for flow over a body must satisfy the flow-tangency boundary condition on the body, i.e.,

$$\mathbf{V} \cdot \mathbf{n} = 0 \qquad (3.48a)$$

where **n** is a unit vector normal to the body surface. In all of the above equations, ϕ is, in general, a function of three-dimensional space; e.g., in spherical coordinates, $\phi = \phi(r, \theta, \Phi)$. Let us use these equations to treat some elementary three-dimensional incompressible flows.

6.2 THREE-DIMENSIONAL SOURCE

Return to Laplace's equation written in spherical coordinates, as given by Eq. (3.43). Consider the velocity potential given by

$$\phi = -\frac{C}{r} \qquad (6.1)$$

where C is a constant and r is the radial coordinate from the origin. Equation (6.1) satisfies Eq. (3.43), and hence it describes a physically possible incompressible, irrotational three-dimensional flow. Combining Eq. (6.1) with the definition of the gradient in spherical coordinates, Eq. (2.18), we obtain

$$\mathbf{V} = \nabla \phi = \frac{C}{r^2} \mathbf{e}_r \qquad (6.2)$$

In terms of the velocity components, we have

$$V_r = \frac{C}{r^2} \qquad (6.3a)$$

$$V_\theta = 0 \qquad (6.3b)$$

$$V_\Phi = 0 \qquad (6.3c)$$

Clearly, Eq. (6.2), or Eqs. (6.3a to c), describes a flow with straight streamlines emanating from the origin, as sketched in Fig. 6.1. Moreover, from Eq. (6.2) or (6.3a), the velocity varies inversely as the square of the distance from the origin. Such a flow is defined as a *three-dimensional source*. Sometimes it is called simply a *point source*, in contrast to the two-dimensional line source discussed in Sec. 3.10.

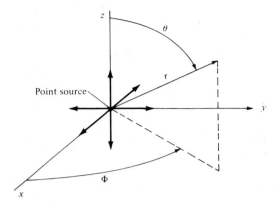

FIGURE 6.1
Three-dimensional (point) source.

To evaluate the constant C in Eq. (6.3a), consider a sphere of radius r and surface S centered at the origin. From Eq. (2.37), the mass flow across the surface of this sphere is

$$\text{Mass flow} = \oiint_S \rho \mathbf{V} \cdot \mathbf{dS}$$

Hence, the *volume* flow, denoted by λ, is

$$\lambda = \oiint_S \mathbf{V} \cdot \mathbf{dS} \tag{6.4}$$

On the surface of the sphere, the velocity is a constant value equal to $V_r = C/r^2$ and is normal to the surface. Hence, Eq. (6.4) becomes

$$\lambda = \frac{C}{r^2} 4\pi r^2 = 4\pi C$$

Hence,

$$C = \frac{\lambda}{4\pi} \tag{6.5}$$

Substituting Eq. (6.5) into (6.3a), we find

$$\boxed{V_r = \frac{\lambda}{4\pi r^2}} \tag{6.6}$$

Compare Eq. (6.6) with its counterpart for a two-dimensional source given by Eq. (3.62). Note that the three-dimensional effect is to cause an inverse *r-squared* variation and that the quantity 4π appears rather than 2π. Also, substituting Eq. (6.5) into (6.1), we obtain, for a point source,

$$\phi = -\frac{\lambda}{4\pi r} \qquad (6.7)$$

In the above equations, λ is defined as the *strength* of the source. When λ is a negative quantity, we have a point sink.

6.3   THREE-DIMENSIONAL DOUBLET

Consider a sink and source of equal but opposite strength located at points O and A, as sketched in Fig. 6.2. The distance between the source and sink is l. Consider an arbitrary point P located a distance r from the sink and a distance r_1 from the source. From Eq. (6.7), the velocity potential at P is

$$\phi = -\frac{\lambda}{4\pi} \left(\frac{1}{r_1} - \frac{1}{r} \right)$$

or
$$\phi = -\frac{\lambda}{4\pi} \frac{r - r_1}{rr_1} \qquad (6.8)$$

Let the source approach the sink as their strengths become infinite; i.e., let

$$l \to 0 \qquad \text{as} \qquad \lambda \to \infty$$

In the limit, as $l \to 0$, $r - r_1 \to OB = l \cos\theta$, and $rr_1 \to r^2$. Thus, in the limit, Eq. (6.8) becomes

$$\phi = -\lim_{\substack{l \to 0 \\ \lambda \to \infty}} \frac{\lambda}{4\pi} \frac{r - r_1}{rr_1} = -\frac{\lambda}{4\pi} \frac{l \cos\theta}{r^2}$$

or
$$\phi = -\frac{\mu}{4\pi} \frac{\cos\theta}{r^2} \qquad (6.9)$$

where $\mu = \lambda l$. The flow field produced by Eq. (6.9) is a *three-dimensional doublet*;

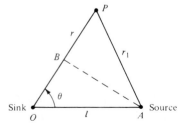

FIGURE 6.2
Source-sink pair. In the limit as $l \to 0$, a three-dimensional doublet is obtained.

μ is defined as the strength of the doublet. Compare Eq. (6.9) with its two-dimensional counterpart given in Eq. (3.88). Note that the three-dimensional effects lead to an inverse r-squared variation and introduce a factor 4π, versus 2π for the two-dimensional case.

From Eqs. (2.18) and (6.9), we find

$$\mathbf{V} = \nabla\phi = \frac{\mu}{2\pi}\frac{\cos\theta}{r^3}\mathbf{e}_r + \frac{\mu}{4\pi}\frac{\sin\theta}{r^3}\mathbf{e}_\theta + 0\mathbf{e}_\Phi \qquad (6.10)$$

The streamlines of this velocity field are sketched in Fig. 6.3. Shown are the streamlines in the zr plane; they are the same in all the zr planes, i.e., for all values of Φ. Hence, the flow induced by the three-dimensional doublet is a series of stream surfaces generated by revolving the streamlines in Fig. 6.3 about the z axis. Compare these streamlines with the two-dimensional case illustrated in Fig. 3.18; they are qualitatively similar but quantitatively different.

Note that the flow in Fig. 6.3 is independent of Φ; indeed, Eq. (6.10) clearly shows that the velocity field depends only on r and θ. Such a flow is defined as *axisymmetric flow*. Once again, we have a flow with two independent variables. For this reason, axisymmetric flow is sometimes labeled "two-dimensional" flow. However, it is quite different from the two-dimensional planar flows discussed earlier. In reality, axisymmetric flow is a degenerate three-dimensional flow, and it is somewhat misleading to refer to it as "two-dimensional." Mathematically, it has only two independent variables, but it exhibits some of the same physical characteristics as general three-dimensional flows, such as the three-dimensional relieving effect to be discussed later.

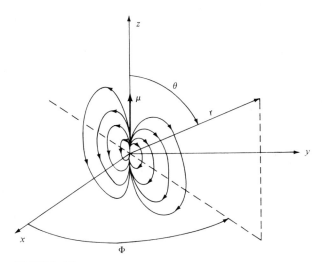

FIGURE 6.3
Sketch of the streamlines in the zr plane (Φ = constant plane) for a three-dimensional doublet.

6.4 FLOW OVER A SPHERE

Consider again the flow induced by the three-dimensional doublet illustrated in Fig. 6.3. Superimpose on this flow a uniform velocity field of magnitude V_∞ in the negative z direction. Since we are more comfortable visualizing a freestream which moves horizontally, say, from left to right, let us flip the coordinate system in Fig. 6.3 on its side. The picture shown in Fig. 6.4 results.

Examining Fig. 6.4, the spherical coordinates of the freestream are

$$V_r = -V_\infty \cos \theta \tag{6.11a}$$

$$V_\theta = V_\infty \sin \theta \tag{6.11b}$$

$$V_\Phi = 0 \tag{6.11c}$$

Adding V_r, V_θ, and V_Φ for the freestream, Eqs. (6.11a to c), to the representative components for the doublet given in Eq. (6.10), we obtain, for the combined flow,

$$V_r = -V_\infty \cos \theta + \frac{\mu}{2\pi} \frac{\cos \theta}{r^3} = -\left(V_\infty - \frac{\mu}{2\pi r^3}\right) \cos \theta \tag{6.12}$$

$$V_\theta = V_\infty \sin \theta + \frac{\mu}{4\pi} \frac{\sin \theta}{r^3} = \left(V_\infty + \frac{\mu}{4\pi r^3}\right) \sin \theta \tag{6.13}$$

$$V_\Phi = 0 \tag{6.14}$$

To find the stagnation points in the flow, set $V_r = V_\theta = 0$ in Eqs. (6.12) and (6.13). From Eq. (6.13), $V_\theta = 0$ gives $\sin \theta = 0$; hence, the stagnation points are located at $\theta = 0$ and π. From Eq. (6.12), with $V_r = 0$, we obtain

$$V_\infty - \frac{\mu}{2\pi R^3} = 0 \tag{6.15}$$

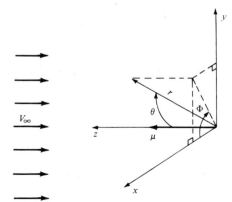

FIGURE 6.4
The superposition of a uniform flow and a three-dimensional doublet.

where $r = R$ is the radial coordinate of the stagnation points. Solving Eq. (6.15) for R, we obtain

$$R = \left(\frac{\mu}{2\pi V_\infty}\right)^{1/3}$$ (6.16)

Hence, there are two stagnation points, both on the z axis, with (r, θ) coordinates

$$\left[\left(\frac{\mu}{2\pi V_\infty}\right)^{1/3}, 0\right] \quad \text{and} \quad \left[\left(\frac{\mu}{2\pi V_\infty}\right)^{1/3}, \pi\right]$$

Insert the value of $r = R$ from Eq. (6.16) into the expression for V_r given by Eq. (6.12). We obtain

$$V_r = -\left(V_\infty - \frac{\mu}{2\pi R^3}\right)\cos\theta = -\left[V_\infty - \frac{\mu}{2\pi}\left(\frac{2\pi V_\infty}{\mu}\right)\right]\cos\theta$$

$$= -(V_\infty - V_\infty)\cos\theta = 0$$

Thus, $V_r = 0$ when $r = R$ for *all values* of θ and Φ. This is precisely the flow-tangency condition for flow over a sphere of radius R. Hence, the velocity field given by Eqs. (6.12) to (6.14) is the *incompressible flow over a sphere of radius R*. This flow is shown in Fig. 6.5; it is qualitatively similar to the flow over the cylinder shown in Fig. 3.19, but quantitatively the two flows are different.

On the surface of the sphere, where $r = R$, the tangential velocity is obtained from Eq. (6.13) as follows:

$$V_\theta = \left(V_\infty + \frac{\mu}{4\pi R^3}\right)\sin\theta$$ (6.17)

From Eq. (6.16),

$$\mu = 2\pi R^3 V_\infty$$ (6.18)

Substituting Eq. (6.18) into (6.17), we have

$$, = \left(V_\infty + \frac{1}{4\pi}\frac{2\pi R^3 V_\infty}{R^3}\right)\sin\theta$$

or

$$\boxed{V_\theta = \tfrac{3}{2}V_\infty \sin\theta}$$ (6.19)

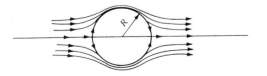

FIGURE 6.5
Schematic of the incompressible flow over a sphere.

The maximum velocity occurs at the top and bottom points of the sphere, and its magnitude is $\frac{3}{2}V_\infty$. Compare these results with the two-dimensional circular cylinder case given by Eq. (3.100). For the two-dimensional flow, the maximum velocity is $2V_\infty$. Hence, for the same V_∞, the maximum surface velocity on a sphere is *less* than that for a cylinder. The flow over a sphere is somewhat "relieved" in comparison with the flow over a cylinder. The flow over a sphere has an extra dimension in which to move out of the way of the solid body; the flow can move sideways as well as up and down. In contrast, the flow over a cylinder is more constrained; it can only move up and down. Hence, the maximum velocity on a sphere is less than that on a cylinder. This is an example of the *three-dimensional relieving effect*, which is a general phenomenon for all types of three-dimensional flows.

The pressure distribution on the surface of the sphere is given by Eqs. (3.38) and (6.19) as follows:

$$C_p = 1 - \left(\frac{V}{V_\infty}\right)^2 = 1 - \left(\frac{3}{2}\sin\theta\right)^2$$

or
$$\boxed{C_p = 1 - \tfrac{9}{4}\sin^2\theta}$$
(6.20)

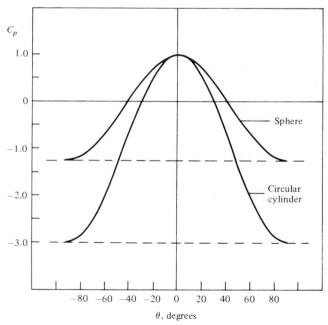

FIGURE 6.6
The pressure distribution over the surface of a sphere and a cylinder. Illustration of the three-dimensional relieving effect.

Compare Eq. (6.20) with the analogous result for a circular cylinder given by Eq. (3.101). Note that the absolute magnitude of the pressure coefficient on a sphere is less than that for a cylinder—again, an example of the three-dimensional relieving effect. The pressure distributions over a sphere and a cylinder are compared in Fig. 6.6, which dramatically illustrates the three-dimensional relieving effect.

6.5 GENERAL THREE-DIMENSIONAL FLOWS: PANEL TECHNIQUES

In modern aerodynamic applications, three-dimensional, inviscid, incompressible flows are almost always calculated by means of numerical panel techniques. The philosophy of the two-dimensional panel methods discussed in previous chapters is readily extended to three dimensions. The details are beyond the scope of this book—indeed, there are dozens of different variations, and the resulting computer programs are frequently long and sophisticated. However, the general idea behind all such panel programs is to cover the three-dimensional body with panels over which there is an unknown distribution of singularities (such as point sources, doublets, or vortices). Such paneling is illustrated in Fig. 6.7. These unknowns are solved through a system of simultaneous linear algebraic equations generated by calculating the induced velocity at control points on the panels and applying the flow-tangency condition. For a nonlifting body such as illustrated in Fig. 6.7, a distribution of source panels is sufficient. However, for a lifting body, both source and vortex panels (or their equivalent) are necessary. A striking example of the extent to which panel methods are now used for three-dimensional lifting

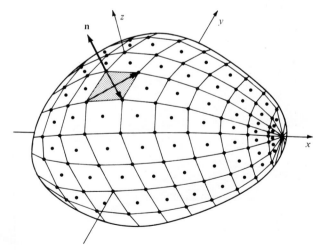

FIGURE 6.7
Distribution of three-dimensional source panels over a general nonlifting body (*Ref. 14*). (*Courtesy of the McDonnell-Douglas Corp.*)

bodies is shown in Fig. 6.8, which illustrates the paneling used for calculations made by the Boeing Company of the potential flow over a Boeing 747–space shuttle piggyback combination. Such applications are very impressive; moreover, they have become an industry standard and are today used routinely as part of the airplane design process by the major aircraft companies.

Examining Figs. 6.7 and 6.8, one aspect stands out, namely, the geometric complexity of distributing panels over the three-dimensional bodies. How do you get the computer to "see" the precise shape of the body? How do you distribute the panels over the body; i.e., do you put more at the wing leading edges and less on the fuselage, etc.? How many panels do you use? These are all nontrivial questions. It is not unusual for an aerodynamicist to spend weeks or even a few months determining the best geometric distribution of panels over a complex body.

We end this chapter on the following note. From the time they were introduced in the 1960s, panel techniques have revolutionized the calculation of three-dimensional potential flows. However, no matter how complex the application of these methods may be, the techniques are still based on the fundamentals we have discussed in this and all the preceding chapters. You are encouraged to pursue these matters further by reading the literature, particularly as it appears in such journals as the *Journal of Aircraft* and the *AIAA Journal.*

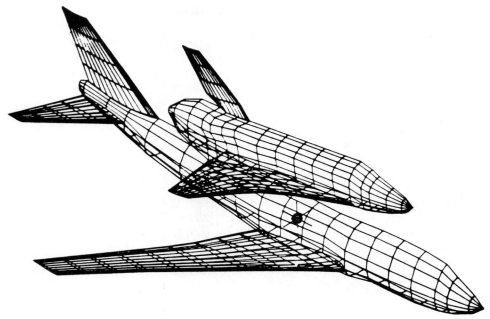

FIGURE 6.8
Panel distribution for the analysis of the Boeing 747 carrying the space shuttle orbiter. (*Courtesy of the Boeing Airplane Company.*)

6.6 APPLIED AERODYNAMICS: THE FLOW OVER A SPHERE— THE REAL CASE

The present section is a complement to Sec. 3.18, in which the real flow over a circular cylinder was discussed. Since the present chapter deals with three-dimensional flows, it is fitting at this stage to discuss the three-dimensional analog of the circular cylinder, namely, the sphere. The qualitative features of the real flow over a sphere are similar to those discussed for a cylinder in Sec. 3.18—the phenomenon of flow separation, the variation of drag coefficient with a Reynolds number, the precipitous drop in drag coefficient when the flow transits from laminar to turbulent ahead of the separation point at the critical Reynolds number, and the general structure of the wake. These items are similar for both cases. However, because of the three-dimensional relieving effect, the flow over a sphere is *quantitatively* different from that for a cylinder. These differences are the subject of the present section.

The laminar flow over a sphere is shown in Fig. 6.9. Here, the Reynolds number is 15,000, certainly low enough to maintain laminar flow over the spherical surface. However, in response to the adverse pressure gradient on the back surface of the sphere predicted by inviscid, incompressible flow theory (see Sec. 6.4 and Fig. 6.6), the laminar flow readily separates from the surface. Indeed, in Fig. 6.9,

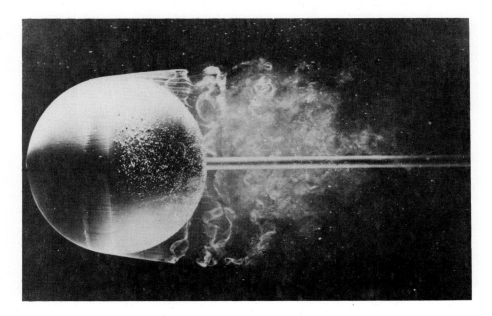

FIGURE 6.9
Laminar flow case: Instantaneous flow past a sphere in water. Re = 15,000. Flow is made visible by dye in the water. (*Courtesy of H. Werle, ONERA, France. Also in Van Dyke, Milton,* An Album of Fluid Motion, *The Parabolic Press, Stanford, Calif., 1982.*)

separation is clearly seen on the *forward* surface, slightly ahead of the vertical equator of the sphere. Thus, a large, fat wake trails downstream of the sphere, with a consequent large pressure drag on the body (analogous to that discussed in Sec. 3.18 for a cylinder.) In contrast, the turbulent flow case is shown in Fig. 6.10. Here, the Reynolds number is 30,000, still a low number normally conducive to laminar flow. However, in this case, turbulent flow is induced artificially by the presence of a wire loop in a vertical plane on the forward face. (Trip wires are frequently used in experimental aerodynamics to induce transition to turbulent flow; this is in order to study such turbulent flows under conditions where they would not naturally exist.) Because the flow is turbulent, separation takes place much farther over the back surface, resulting in a thinner wake, as can be seen by comparing Figs. 6.9 and 6.10. Consequently, the pressure drag is less for the turbulent case.

The variation of drag coefficient, C_D, with the Reynolds number for a sphere, is shown in Fig. 6.11. Compare this figure with Fig. 3.39 for a circular cylinder; the C_D variations are qualitatively similar, both with a precipitous decrease in C_D near a critical Reynolds number of 300,000, coinciding with

FIGURE 6.10
Turbulent flow case: Instantaneous flow past a sphere in water. Re = 30,000. The turbulent flow is forced by a trip wire hoop ahead of the equator, causing the laminar flow to become turbulent suddenly. The flow is made visible by air bubbles in water. (*Courtesy of H. Werle, ONERA, France. Also in Van Dyke, Milton, An Album of Fluid Motion, The Parabolic Press, Stanford, Calif., 1982.*)

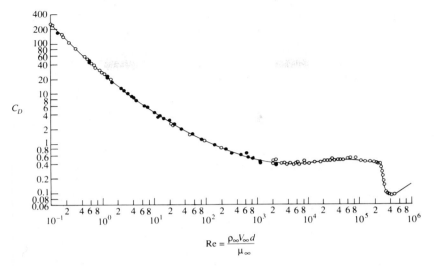

$$Re = \frac{\rho_\infty V_\infty d}{\mu_\infty}$$

FIGURE 6.11
Variation of drag coefficient with Reynolds number for a sphere. (*From Schlichting, Ref. 42.*)

natural transition from laminar to turbulent flow. However, *quantitatively* the two curves are quite different. In the Reynolds number range most appropriate to practical problems, i.e., for Re > 1000, the values of C_D for the sphere are considerably smaller than those for a cylinder—a classic example of the three-dimensional relieving effect. Reflecting on Fig. 3.39 for the cylinder, note that the value of C_D for Re slightly less than the critical value is about 1 and drops to 0.3 for Re slightly above the critical value. In contrast, for the sphere as shown in Fig. 6.11, C_D is about 0.4 in the Reynolds number range below the critical value and drops to about 0.1 for Reynolds numbers above the critical value. These variations in C_D for both the cylinder and sphere are classic results in aerodynamics; you should keep the actual C_D values in mind for future reference and comparisons.

As a final point in regard to both Figs. 3.39 and 6.11, the value of the critical Reynolds number at which transition to turbulent flow takes place upstream of the separation point is not a fixed, universal number. Quite the contrary, transition is influenced by many factors, as will be discussed in Part IV. Among these is the amount of turbulence in the freestream; the higher the freestream turbulence, the more readily transition takes place. In turn, the higher the freestream turbulence, the lower is the value of the critical Reynolds number. Because of this trend, calibrated spheres are used in wind-tunnel testing actually to assess the degree of freestream turbulence in the test section, simply by measuring the value of the critical Reynolds number on the sphere.

6.7 SUMMARY

For a three-dimensional (point) source,

$$V_r = \frac{\lambda}{4\pi r^2} \tag{6.6}$$

and

$$\phi = -\frac{\lambda}{4\pi r} \tag{6.7}$$

For a three-dimensional doublet,

$$\phi = -\frac{\mu}{4\pi} \frac{\cos\theta}{r^2} \tag{6.9}$$

and

$$V = \frac{\mu}{2\pi} \frac{\cos\theta}{r^3} e_r + \frac{\mu}{4\pi} \frac{\sin\theta}{r^3} e_\theta \tag{6.10}$$

The flow over a sphere is generated by superimposing a three-dimensional doublet and a uniform flow. The resulting surface velocity and pressure distributions are given by

$$V_\theta = \tfrac{3}{2} V_\infty \sin\theta \tag{6.19}$$

and

$$C_p = 1 - \tfrac{9}{4} \sin^2\theta \tag{6.20}$$

In comparison with flow over a cylinder, the surface velocity and magnitude of the pressure coefficient are smaller for the sphere—an example of the three-dimensional relieving effect.

In modern aerodynamic applications, inviscid, incompressible flows over complex three-dimensional bodies are usually computed via three-dimensional panel techniques.

PROBLEMS

6.1. Prove that three-dimensional source flow is irrotational.

6.2. Prove that three-dimensional source flow is a physically possible incompressible flow.

6.3. A sphere and a circular cylinder (with its axis perpendicular to the flow) are mounted in the same freestream. A pressure tap exists at the top of the sphere, and this is connected via a tube to one side of a manometer. The other side of the manometer is connected to a pressure tap on the surface of the cylinder. This tap is located on the cylindrical surface such that no deflection of the manometer fluid takes place. Calculate the location of this tap.

PART
III

INVISCID, COMPRESSIBLE FLOW

I n Part III, we deal with high-speed flows—subsonic, supersonic, and hypersonic. In such flows, the density is a variable—this is compressible flow.

CHAPTER
7

COMPRESSIBLE FLOW: SOME PRELIMINARY ASPECTS

With the realization of aeroplane and missile speeds equal to or even surpassing many times the speed of sound, thermodynamics has entered the scene and will never again leave our considerations.

Jakob Ackeret, 1962

7.1 INTRODUCTION

On September 30, 1935, the leading aerodynamicists from all corners of the world converged on Rome, Italy. Some of them arrived in airplanes which, in those days, lumbered along at speeds of 130 mi/h. Ironically, these people were gathering to discuss airplane aerodynamics not at 130 mi/h but rather at the unbelievable speeds of 500 mi/h and faster. By invitation only, such aerodynamic giants as Theodore von Karman and Eastman Jacobs from the United States, Ludwig Prandtl and Adolf Busemann from Germany, Jakob Ackeret from Switzerland, G. I. Taylor from England, Arturo Crocco and Enrico Pistolesi from Italy, and others assembled for the fifth Volta Conference, which had as its topic "High Velocities in Aviation." Although the jet engine had not yet been developed, these men were convinced that the future of aviation was "faster and higher." At that time, some aeronautical engineers felt that airplanes would never fly faster than the speed of sound—the myth of the "sound barrier" was propagating through the ranks of aviation. However, the people who attended the fifth Volta Conference knew better. For 6 days, inside an impressive Renaissance building that served as the city hall during the Holy Roman Empire, these individuals presented papers that discussed flight at high subsonic, supersonic, and even hypersonic speeds. Among these presentations was the first public revelation of the concept of a swept wing for high-speed flight; Adolf Busemann, who originated the concept, discussed the technical reasons why swept wings would have less drag at high speeds than conventional straight wings. (One year later, the swept-wing concept was classified by the German Luftwaffe as a military secret. The Germans went on to produce a large bulk of swept-wing research during World War II, resulting in the design of the first operational jet airplane—the Me 262—which had a moderate degree of sweep.) Many of the discussions at the Volta Conference centered on the effects of "compressibility" at high subsonic speeds, i.e., the effects of variable density, because this was clearly going to be the first problem to be encountered by future high-speed airplanes. For example, Eastman Jacobs presented wind-tunnel test results for compressibility effects on standard NACA four- and five-digit airfoils at high subsonic speeds and noted extraordinarily large increases in drag beyond certain freestream Mach numbers. In regard to supersonic flows, Ludwig Prandtl presented a series of photographs showing shock waves inside nozzles and on various bodies—with some of the photographs dating as far back as 1907, when Prandtl started serious work in supersonic aerodynamics. (Clearly, Ludwig Prandtl was busy with much more than just the development of his incompressible airfoil and finite-wing theory discussed in Chaps. 4 and 5.) Jakob Ackeret gave a paper on the design of supersonic wind tunnels, which, under his direction, were being established in Italy, Switzerland, and Germany. There were also presentations on propulsion techniques for high-speed flight, including rockets and ramjets. The atmosphere surrounding the participants in the Volta Conference was exciting and heady; the conference launched the world aerodynamic community into the area of high-speed subsonic and supersonic flight—an area which today is as common-place as the 130-mi/h flight speeds of 1935. Indeed, the purpose of the next eight chapters of this book is to present the fundamentals of such high-speed flight.

In contrast to the low-speed, incompressible flows discussed in Chaps. 3 to 6, the pivotal aspect of high-speed flow is that the density is a variable. Such flows are called *compressible flows* and are the subject of Chaps. 7 to 14. Return to Fig. 1.31, which gives a block diagram categorizing types of aerodynamic flows. In Chaps. 7 to 14, we discuss flows which fall into blocks D and F; i.e., we will deal with *inviscid compressible* flow. In the process, we touch all the flow regimes itemized in blocks G through J. These flow regimes are illustrated in Fig. 1.30; study Figs. 1.30 and 1.31 carefully, and review the surrounding discussion in Sec. 1.10 before proceeding further.

In addition to variable density, another pivotal aspect of high-speed compressible flow is *energy*. A high-speed flow is a high-energy flow. For example, consider the flow of air at standard sea level conditions moving at twice the speed of sound. The internal energy of 1 kg of this air is 2.07×10^5 J, whereas the kinetic energy is larger, namely 2.31×10^5 J. When the flow velocity is decreased, some of this kinetic energy is lost and reappears as an increase in internal energy, hence increasing the temperature of the gas. Therefore, in a high-speed flow, energy transformations and temperature changes are important considerations.

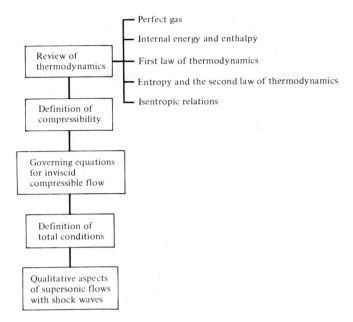

FIGURE 7.1
Road map for Chap. 7.

Such considerations come under the science of *thermodynamics*. For this reason, thermodynamics is a vital ingredient in the study of compressible flow. One purpose of the present chapter is to review briefly the particular aspects of thermodynamics which are essential to our subsequent discussions of compressible flow.

The road map for this chapter is given in Fig. 7.1. As our discussion proceeds, refer to this road map in order to provide an orientation for our ideas.

7.2 A BRIEF REVIEW OF THERMODYNAMICS

The importance of thermodynamics in the analysis and understanding of compressible flow was underscored in Sec. 7.1. Hence, the purpose of the present section is to review those aspects of thermodynamics which are important to compressible flows. This is in no way intended to be an exhaustive discussion of thermodynamics; rather, it is a review of only those fundamental ideas and equations which will be of direct use in subsequent chapters. If you have studied thermodynamics, this review should serve as a ready reminder of some important relations. If you are not familiar with thermodynamics, this section is somewhat self-contained so as to give you a feeling for the fundamental ideas and equations which we use frequently in subsequent chapters.

7.2.1 Perfect Gas

As described in Sec. 1.2, a gas is a collection of particles (molecules, atoms, ions, electrons, etc.) which are in more or less random motion. Due to the electronic structure of these particles, a force field pervades the space around them. The force field due to one particle reaches out and interacts with neighboring particles, and vice versa. Hence, these fields are called *intermolecular forces*. However, if the particles of the gas are far enough apart, the influence of the intermolecular forces is small and can be neglected. A gas in which the intermolecular forces are neglected is defined as a *perfect gas*. For a perfect gas, p, ρ, and T are related through the following *equation of state*:

$$p = \rho RT \tag{7.1}$$

where R is the specific gas constant, which is a different value for different gases. For air at standard conditions, $R = 287 \text{ J}/(\text{kg} \cdot \text{K}) = 1716 \text{ (ft} \cdot \text{lb)}/(\text{slug} \cdot \text{°R})$.

At the temperatures and pressures characteristic of many compressible flow applications, the gas particles are, on the average, more than 10 molecular diameters apart; this is far enough to justify the assumption of a perfect gas. Therefore, throughout the remainder of this book, we use the equation of state in the form of Eq. (7.1), or its counterpart,

$$pv = RT \tag{7.2}$$

where v is the specific volume, i.e., the volume per unit mass; $v = 1/\rho$. (*Please note*: Starting with this chapter, we use the symbol v to denote both specific volume and the y component of velocity. This usage is standard, and in all cases it should be obvious and cause no confusion.)

7.2.2 Internal Energy and Enthalpy

Consider an individual molecule of a gas, say, an O_2 molecule in air. This molecule is moving through space in a random fashion, occasionally colliding with a neighboring molecule. Because of its velocity through space, the molecule has translational kinetic energy. In addition, the molecule is made up of individual atoms which we can visualize as connected to each other along various axes; for example, we can visualize the O_2 molecule as a "dumbbell" shape, with an O atom at each end of a connecting axis. In addition to its translational motion, such a molecule can execute a rotational motion in space; the kinetic energy of this rotation contributes to the net energy of the molecule. Also, the atoms of a given molecule can vibrate back and forth along and across the molecular axis, thus contributing a potential and kinetic energy of vibration to the molecule. Finally, the motion of the electrons around each of the nuclei of the molecule contributes an "electronic" energy to the molecule. Hence, the energy of a given molecule is the sum of its translational, rotational, vibrational, and electronic energies.

Now consider a finite volume of gas consisting of a large number of molecules. The sum of the energies of all the molecules in this volume is defined as the *internal energy* of the gas. The internal energy per unit mass of gas is defined as the specific internal energy, denoted by e. A related quantity is the specific enthalpy, denoted by h and defined as

$$h = e + pv \tag{7.3}$$

For a perfect gas, both e and h are functions of temperature only:

$$e = e(T) \tag{7.4a}$$

$$h = h(T) \tag{7.4b}$$

Let de and dh represent differentials of e and h, respectively. Then, for a perfect gas,

$$de = c_v \, dT \tag{7.5a}$$

$$dh = c_p \, dT \tag{7.5b}$$

where c_v and c_p are the specific heats at constant volume and constant pressure, respectively. In Eqs. (7.5a and b), c_v and c_p can themselves be functions of T. However, for moderate temperatures (for air, for $T < 1000$ K), the specific heats are reasonably constant. A perfect gas where c_v and c_p are constants is defined as a *calorically perfect gas*, for which Eqs. (7.5a and b) become

$$\boxed{\begin{aligned} e &= c_v T \\ h &= c_p T \end{aligned}}$$

$$e = c_v T \tag{7.6a}$$

$$h = c_p T \tag{7.6b}$$

For a large number of practical compressible flow problems, the temperatures are moderate; for this reason, in this book we always treat the gas as calorically perfect; i.e., we assume that the specific heats are constant. For a discussion of compressible flow problems where the specific heats are not constant (such as the high-temperature chemically reacting flow over a high-speed atmospheric entry vehicle, i.e., the space shuttle), see Ref. 21.

Note that e and h in Eqs. (7.3) through (7.6) are thermodynamic state variables—they depend only on the state of the gas and are independent of any process. Although c_v and c_p appear in these equations, there is no restriction to just a constant volume or a constant pressure process. Rather, Eqs. (7.5a and b) and (7.6a and b) are relations for thermodynamic state variables, namely, e and h as functions of T, and have nothing to do with the process that may be taking place.

For a specific gas, c_p and c_v are related through the equation

$$c_p - c_v = R \tag{7.7}$$

Dividing Eq. (7.7) by c_p, we obtain

$$1 - \frac{c_v}{c_p} = \frac{R}{c_p} \tag{7.8}$$

Define $\gamma \equiv c_p/c_v$. For air at standard conditions, $\gamma = 1.4$. Then Eq. (7.8) becomes

$$1 - \frac{1}{\gamma} = \frac{R}{c_p}$$

or

$$\boxed{c_p = \frac{\gamma R}{\gamma - 1}} \tag{7.9}$$

Similarly, dividing Eq. (7.7) by c_v, we obtain

$$\boxed{c_v = \frac{R}{\gamma - 1}} \tag{7.10}$$

Equations (7.9) and (7.10) are particularly useful in our subsequent discussion of compressible flow.

7.2.3 First Law of Thermodynamics

Consider a fixed mass of gas, which we define as the *system*. (For simplicity, assume a unit mass, e.g., 1 kg or 1 slug.) The region outside the system is called the *surroundings*. The interface between the system and its surroundings is called the *boundary*, as shown in Fig. 7.2. Assume that the system is stationary. Let δq be an incremental amount of heat added to the system across the boundary, as sketched in Fig. 7.2. Examples of the source of δq are radiation from the surroundings which is absorbed by the mass in the system and thermal conduction due to temperature gradients across the boundary. Also, let δw denote the work done on the system by the surroundings (say, by a displacement of the boundary, squeezing the volume of the system to a smaller value). As discussed earlier, due

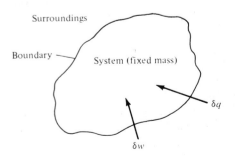

FIGURE 7.2
Thermodynamic system.

to the molecular motion of the gas, the system has an internal energy e. The heat added and work done on the system cause a change in energy, and since the system is stationary, this change in energy is simply de:

$$\boxed{\delta q + \delta w = de} \tag{7.11}$$

This is the *first law of thermodynamics*: It is an empirical result confirmed by experience. In Eq. (7.11), e is a state variable. Hence, de is an exact differential, and its value depends only on the initial and final states of the system. In contrast, δq and δw depend on the process in going from the initial to the final states.

For a given de, there are in general an infinite number of different ways (processes) by which heat can be added and work done on the system. We are primarily concerned with three types of processes:

1. *Adiabatic process.* One in which no heat is added to or taken away from the system
2. *Reversible process.* One in which no dissipative phenomena occur, i.e., where the effects of viscosity, thermal conductivity, and mass diffusion are absent
3. *Isentropic process.* One which is both adiabatic and reversible

For a reversible process, it can be easily shown that $\delta w = -p\,dv$, where dv is an incremental change in the volume due to a displacement of the boundary of the system. Thus, Eq. (7.11) becomes

$$\delta q - p\,dv = de \tag{7.12}$$

7.2.4 Entropy and the Second Law of Thermodynamics

Consider a block of ice in contact with a red-hot plate of steel. Experience tells us that the ice will warm up (and probably melt) and the steel plate will cool down. However, Eq. (7.11) does not necessarily say this will happen. Indeed, the first law allows that the ice may get cooler and the steel plate hotter—just as long as energy is conserved during the process. Obviously, in real life this does not happen; instead, nature imposes another condition on the process, a condition which tells us *which direction* a process will take. To ascertain the proper direction of a process, let us define a new state variable, the entropy, as follows:

$$ds = \frac{\delta q_{\text{rev}}}{T} \tag{7.13}$$

where s is the entropy of the system, δq_{rev} is an incremental amount of heat added reversibly to the system, and T is the system temperature. Do not be confused by the above definition. It defines a change in entropy in terms of a reversible addition of heat, δq_{rev}. However, entropy is a state variable, and it can be used in conjunction with any type of process, reversible or irreversible. The quantity δq_{rev} in Eq. (7.13) is just an artifice; an effective value of δq_{rev} can always be assigned to relate the initial and end points of an irreversible process, where the actual amount of heat added is δq. Indeed, an alternative and probably more lucid relation is

$$ds = \frac{\delta q}{T} + ds_{\text{irrev}} \tag{7.14}$$

In Eq. (7.14), δq is the actual amount of heat added to the system during an actual irreversible process, and ds_{irrev} is the generation of entropy due to the irreversible, dissipative phenomena of viscosity, thermal conductivity, and mass diffusion occurring *within* the system. These dissipative phenomena always increase the entropy:

$$ds_{\text{irrev}} \geq 0 \tag{7.15}$$

In Eq. (7.15), the equals sign denotes a reversible process, where by definition no dissipative phenomena occur within the system. Combining Eqs. (7.14) and (7.15), we have

$$ds \geq \frac{\delta q}{T} \tag{7.16}$$

Furthermore, if the process is adiabatic, $\delta q = 0$, and Eq. (7.16) becomes

$$ds \geq 0 \tag{7.17}$$

Equations (7.16) and (7.17) are forms of the *second law of thermodynamics*. The second law tells us in what direction a process will take place. A process will proceed in a direction such that the entropy of the system plus that of its surroundings always increases or, at best, stays the same. In our example of the ice in contact with hot steel, consider the system to be both the ice and steel plate combined. The simultaneous heating of the ice and cooling of the plate yield a net increase in entropy for the system. On the other hand, the impossible situation of the ice getting cooler and the plate hotter would yield a net decrease in entropy, a situation forbidden by the second law. In summary, the concept of entropy in combination with the second law allows us to predict the *direction* that nature takes.

The practical calculation of entropy is carried out as follows. In Eq. (7.12), assume that heat is added reversibly; then the definition of entropy, Eq. (7.13), substituted in Eq. (7.12) yields

$$T \, ds - p \, dv = de$$

or

$$T \, ds = de + p \, dv \qquad (7.18)$$

From the definition of enthalpy, Eq. (7.3), we have

$$dh = de + p \, dv + v \, dp \qquad (7.19)$$

Combining Eqs. (7.18) and (7.19), we obtain

$$T \, ds = dh - v \, dp \qquad (7.20)$$

Equations (7.18) and (7.20) are important; they are essentially alternate forms of the first law expressed in terms of entropy. For a perfect gas, recall Eqs. (7.5a and b), namely, $de = c_v \, dT$ and $dh = c_p \, dT$. Substituting these relations into Eqs. (7.18) and (7.20), we obtain

$$ds = c_v \frac{dT}{T} + \frac{p \, dv}{T} \qquad (7.21)$$

and

$$ds = c_p \frac{dT}{T} - \frac{v \, dp}{T} \qquad (7.22)$$

Working with Eq. (7.22), substitute the equation of state $pv = RT$, or $v/T = R/p$, into the last term:

$$ds = c_p \frac{dT}{T} - R \frac{dp}{p} \qquad (7.23)$$

Consider a thermodynamic process with initial and end states denoted by 1 and 2, respectively. Equation (7.23), integrated between states 1 and 2, becomes

$$s_2 - s_1 = \int_{T_1}^{T_2} c_p \frac{dT}{T} - \int_{p_1}^{p_2} R \frac{dp}{p} \qquad (7.24)$$

For a calorically perfect gas, both R and c_p are constants; hence, Eq. (7.24) becomes

$$s_2 - s_1 = c_p \ln \frac{T_2}{T_1} - R \ln \frac{p_2}{p_1} \qquad (7.25)$$

In a similar fashion, Eq. (7.21) leads to

$$s_2 - s_1 = c_v \ln \frac{T_2}{T_1} + R \ln \frac{v_2}{v_1} \qquad (7.26)$$

Equations (7.25) and (7.26) are practical expressions for the calculation of the entropy change of a calorically perfect gas between two states. Note from these equations that s is a function of two thermodynamic variables, e.g., $s = s(p, T)$, $s = s(v, T)$.

7.2.5   Isentropic Relations

We have defined an isentropic process as one which is both adiabatic and reversible. Consider Eq. (7.14). For an adiabatic process, $\delta q = 0$. Also, for a reversible process, $ds_{\text{irrev}} = 0$. Thus, for an adiabatic, reversible process, Eq. (7.14) yields $ds = 0$, or entropy is constant; hence, the word "isentropic." For such an isentropic process, Eq. (7.25) is written as

$$0 = c_p \ln \frac{T_2}{T_1} - R \ln \frac{p_2}{p_1}$$

$$\ln \frac{p_2}{p_1} = \frac{c_p}{R} \ln \frac{T_2}{T_1}$$

or

$$\frac{p_2}{p_1} = \left(\frac{T_2}{T_1} \right)^{c_p / R} \tag{7.27}$$

However, from Eq. (7.9),

$$\frac{c_p}{R} = \frac{\gamma}{\gamma - 1}$$

and hence Eq. (7.27) is written as

$$\frac{p_2}{p_1} = \left(\frac{T_2}{T_1} \right)^{\gamma/(\gamma - 1)} \tag{7.28}$$

In a similar fashion, Eq. (7.26) written for an isentropic process gives

$$0 = c_v \ln \frac{T_2}{T_1} + R \ln \frac{v_2}{v_1}$$

$$\ln \frac{v_2}{v_1} = -\frac{c_v}{R} \ln \frac{T_2}{T_1}$$

$$\frac{v_2}{v_1} = \left(\frac{T_2}{T_1} \right)^{-c_v / R} \tag{7.29}$$

From Eq. (7.10),

$$\frac{c_v}{R} = \frac{1}{\gamma - 1}$$

and hence Eq. (7.29) is written as

$$\frac{v_2}{v_1} = \left(\frac{T_2}{T_1} \right)^{-1/(\gamma - 1)} \tag{7.30}$$

Since $\rho_2/\rho_1 = v_1/v_2$, Eq. (7.30) becomes

$$\frac{\rho_2}{\rho_1} = \left(\frac{T_2}{T_1}\right)^{1/(\gamma-1)}$$

(7.31)

Combining Eqs. (7.28) and (7.31), we can summarize the isentropic relations as

$$\boxed{\frac{p_2}{p_1} = \left(\frac{\rho_2}{\rho_1}\right)^{\gamma} = \left(\frac{T_2}{T_1}\right)^{\gamma/(\gamma-1)}}$$

(7.32)

Equation (7.32) is very important; it relates pressure, density, and temperature for an isentropic process. We use this equation frequently, so make certain to brand it on your mind. Also, keep in mind the source of Eq. (7.32); it stems from the first law and the definition of entropy. Therefore, Eq. (7.32) is basically an energy relation for an isentropic process.

Why is Eq. (7.32) so important? Why is it frequently used? Why are we so interested in an isentropic process when it seems so restrictive—requiring both adiabatic and reversible conditions? The answers rest on the fact that a large number of practical compressible flow problems can be assumed to be isentropic—contrary to what you might initially think. For example, consider the flow over an airfoil or through a rocket engine. In the regions adjacent to the airfoil surface and the rocket nozzle walls, a boundary layer is formed wherein the dissipative mechanisms of viscosity, thermal conduction, and diffusion are strong. Hence, the entropy increases within these boundary layers. However, consider the fluid elements moving outside the boundary layer. Here, the dissipative effects of viscosity, etc., are very small and can be neglected. Moreover, no heat is being transferred to or from the fluid element (i.e., we are not heating the fluid element with a Bunsen burner or cooling it in a refrigerator); thus, the flow outside the boundary layer is adiabatic. Consequently, the fluid elements outside the boundary layer are experiencing an adiabatic reversible process—namely, isentropic flow. In the vast majority of practical applications, the viscous boundary layer adjacent to the surface is thin compared with the entire flow field, and hence large regions of the flow can be assumed isentropic. This is why a study of isentropic flow is directly applicable to many types of practical compressible flow problems. In turn, Eq. (7.32) is a powerful relation for such flows, valid for a calorically perfect gas.

This ends our brief review of thermodynamics. Its purpose has been to give a quick summary of ideas and equations which will be employed throughout our subsequent discussions of compressible flow. For a more thorough discussion of the power and beauty of thermodynamics, see any good thermodynamics text, such as Refs. 22 to 24.

Example 7.1. Consider a Boeing 747 flying at a standard altitude of 36,000 ft. The pressure at a point on the wing is $400 \, \text{lb/ft}^2$. Assuming isentropic flow over the wing, calculate the temperature at this point.

Solution. At a standard altitude of 36,000 ft, $p_\infty = 476$ lb/ft^2 and $T_\infty = 391$ °R. From Eq. (7.32),

$$\frac{p}{p_\infty} = \left(\frac{T}{T_\infty}\right)^{\gamma/(\gamma-1)}$$

or

$$T = T_\infty \left(\frac{p}{p_\infty}\right)^{(\gamma-1)/\gamma} = 391\left(\frac{400}{476}\right)^{0.4/1.4} = \boxed{372\ °R}$$

7.3 DEFINITION OF COMPRESSIBILITY

All real substances are compressible to some greater or lesser extent; i.e., when you squeeze or press on them, their density will change. This is particularly true of gases, much less so for liquids, and virtually unnoticeable for solids. The amount by which a substance can be compressed is given by a specific property of the substance called the *compressibility*, defined below.

Consider a small element of fluid of volume v, as sketched in Fig. 7.3. The pressure exerted on the sides of the element is p. Assume the pressure is now increased by an infinitesimal amount, dp. The volume of the element will change by a corresponding amount, dv; here, the volume will decrease; hence, dv shown in Fig. 7.3 is a negative quantity. By definition, the compressibility τ of the fluid is

$$\tau = -\frac{1}{v}\frac{dv}{dp} \tag{7.33}$$

Physically, the compressibility is the fractional change in volume of the fluid element per unit change in pressure. However, Eq. (7.33) is not precise enough. We know from experience that when a gas is compressed (say, in a bicycle pump), its temperature tends to increase, depending on the amount of heat transferred into or out of the gas through the boundaries of the system. If the temperature of the fluid element in Fig. 7.3 is held constant (due to some heat transfer mechanism), then τ is identified as the *isothermal compressibility* τ_T, defined from Eq. (7.33) as

$$\tau_T = -\frac{1}{v}\left(\frac{\partial v}{\partial p}\right)_T \tag{7.34}$$

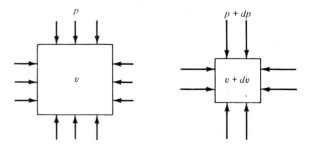

FIGURE 7.3
Definition of compressibility.

On the other hand, if no heat is added to or taken away from the fluid element, and if friction is ignored, the compression of the fluid element takes place isentropically, and τ is identified as the isentropic compressibility τ_s, defined from Eq. (7.33) as

$$\tau_s = -\frac{1}{v}\left(\frac{\partial v}{\partial p}\right)_s \tag{7.35}$$

where the subscript s denotes that the partial derivative is taken at constant entropy. Both τ_T and τ_s are precise thermodynamic properties of the fluid; their values for different gases and liquids can be obtained from various handbooks of physical properties. In general, the compressibility of gases is several orders of magnitude larger than that of liquids.

The role of the compressibility τ in determining the properties of a fluid in motion is seen as follows. Define v as the specific volume, i.e., the volume per unit mass. Hence, $v = 1/\rho$. Substituting this definition into Eq. (7.33), we obtain

$$\tau = \frac{1}{\rho}\frac{d\rho}{dp} \tag{7.36}$$

Thus, whenever the fluid experiences a change in pressure, dp, the corresponding change in density, $d\rho$, from Eq. (7.36) is

$$d\rho = \rho\tau\, dp \tag{7.37}$$

Consider a fluid flow, say, e.g., the flow over an airfoil. If the fluid is a *liquid*, where the compressibility τ is very small, then for a given pressure change dp from one point to another in the flow, Eq. (7.37) states that $d\rho$ will be negligibly small. In turn, we can reasonably assume that ρ is constant and that the flow of a liquid is incompressible. On the other hand, if the fluid is a *gas*, where the compressibility τ is large, then for a given pressure change dp from one point to another in the flow, Eq. (7.37) states that $d\rho$ can be large. Thus, ρ is *not* constant, and in general, the flow of a gas is a *compressible flow*. The exception to this is the *low-speed flow* of a gas; in such flows, the actual magnitude of the pressure changes throughout the flow field is small compared with the pressure itself. Thus, for a low-speed flow, dp in Eq. (7.37) is small, and even though τ is large, the value of $d\rho$ can be dominated by the small dp. In such cases, ρ can be assumed to be constant, hence allowing us to analyze low-speed gas flows as incompressible flows (such as discussed in Chaps. 3 to 6).

Later, we demonstrate that the most convenient index to gage whether a gas flow can be considered incompressible, or whether it must be treated as compressible, is the Mach number M, defined in Chap. 1 as the ratio of local flow velocity V to the local speed of sound, a:

$$M \equiv \frac{V}{a} \tag{7.38}$$

We show that, when $M > 0.3$, the flow should be considered compressible. Also, we show that the speed of sound in a gas is related to the isentropic compressibility τ_s, given by Eq. (7.35).

7.4 GOVERNING EQUATIONS FOR INVISCID, COMPRESSIBLE FLOW

In Chaps. 3 to 6, we studied inviscid, incompressible flow; recall that the primary dependent variables for such flows are p and $\mathbf{V}$, and hence we need only two basic equations, namely, the continuity and momentum equations, to solve for these two unknowns. Indeed, the basic equations are combined to obtain Laplace's equation and Bernoulli's equation, which are the primary tools used for the applications discussed in Chaps. 3 to 6. Note that both ρ and T are assumed to be constant throughout such inviscid, incompressible flows. As a result, no additional governing equations are required; in particular, there is no need for the energy equation or energy concepts in general. Basically, incompressible flow obeys purely mechanical laws and does not require thermodynamic considerations.

In contrast, for compressible flow, ρ is variable and becomes an unknown. Hence, we need an additional governing equation—the energy equation—which in turn introduces internal energy e as an unknown. Since e is related to temperature, then T also becomes an important variable. Therefore, the primary dependent variables for the study of compressible flow are p, $\mathbf{V}$, ρ, e, and T; to solve for these five variables, we need five governing equations. Let us examine this situation further.

To begin with, the flow of a compressible fluid is governed by the basic equations derived in Chap. 2. At this point in our discussion, it is most important for you to be familiar with these equations as well as their derivation. Therefore, before proceeding further, return to Chap. 2 and carefully review the basic ideas and relations contained therein. This is a serious study tip, and if you follow it, the material in our next seven chapters will flow much easier for you. In particular, review the integral and differential forms of the continuity equation (Sec. 2.4), the momentum equation (Sec. 2.5), and the energy equation (Sec. 2.7); indeed, pay particular attention to the energy equation because this is an important aspect which sets compressible flow apart from incompressible flow.

For convenience, some of the more important forms of the governing equations for an inviscid, compressible flow from Chap. 2 are repeated below:

Continuity: From Eq. (2.39),

$$\frac{\partial}{\partial t} \iiint_\mathscr{V} \rho \, d\mathscr{V} + \oiint_S \rho \mathbf{V} \cdot d\mathbf{S} = 0 \tag{7.39}$$

From Eq. (2.43),

$$\frac{\partial \rho}{\partial t} + \nabla \cdot \rho \mathbf{V} = 0 \tag{7.40}$$

Momentum: From Eq. (2.55),

$$\frac{\partial}{\partial t}\iiint_{\mathscr{V}}\rho\mathbf{V}\,d\mathscr{V}+\oiint_{S}(\rho\mathbf{V}\cdot\mathbf{dS})\mathbf{V}=-\oiint_{S}p\,\mathbf{dS}+\iiint_{\mathscr{V}}\rho\mathbf{f}\,d\mathscr{V} \qquad (7.41)$$

From Eqs. (2.104*a* to *c*),

$$\rho\frac{Du}{Dt}=-\frac{\partial p}{\partial x}+\rho f_x \qquad (7.42a)$$

$$\rho\frac{Dv}{Dt}=-\frac{\partial p}{\partial y}+\rho f_y \qquad (7.42b)$$

$$\rho\frac{Dw}{Dt}=-\frac{\partial p}{\partial z}+\rho f_z \qquad (7.42c)$$

Energy: From Eq. (2.86),

$$\frac{\partial}{\partial t}\iiint_{\mathscr{V}}\rho\left(e+\frac{V^2}{2}\right)d\mathscr{V}+\oiint_{S}\rho\left(e+\frac{V^2}{2}\right)\mathbf{V}\cdot\mathbf{dS}$$

$$=\iiint_{\mathscr{V}}\dot{q}\rho\,d\mathscr{V}-\oiint_{S}p\mathbf{V}\cdot\mathbf{dS}+\iiint_{\mathscr{V}}\rho(\mathbf{f}\cdot\mathbf{V})\,d\mathscr{V} \qquad (7.43)$$

From Eq. (2.105),

$$\rho\frac{D(e+V^2/2)}{Dt}=\rho\dot{q}-\nabla\cdot p\mathbf{V}+\rho(\mathbf{f}\cdot\mathbf{V}) \qquad (7.44)$$

The above continuity, momentum, and energy equations are three equations in terms of the five unknowns p, $\mathbf{V}$, ρ, T, and e. Assuming a calorically perfect gas, the additional two equations needed to complete the system are obtained from Sec. 7.2:

Equation of state: $p=\rho RT$ (7.1)

Internal energy: $e=c_v T$ (7.6a)

In regard to the basic equations for compressible flow, please note that Bernoulli's equation as derived in Sec. 3.2 and given by Eq. (3.13) does *not* hold for compressible flow; it clearly contains the assumption of constant density, and hence is invalid for compressible flow. This warning is necessary because experience shows that a certain number of students of aerodynamics, apparently attracted by the simplicity of Bernoulli's equation, attempt to use it for all situations, compressible as well as incompressible. Do not do it! *Always remember that Bernoulli's equation in the form of Eq. (3.13) holds for incompressible flow only, and we must dismiss it from our thinking when dealing with compressible flow.*

As a final note, we use both the integral and differential forms of the above equations in our subsequent discussions. Make certain that you feel comfortable with these equations before proceeding further.

7.5 DEFINITION OF TOTAL (STAGNATION) CONDITIONS

At the beginning of Sec. 3.4, the concept of static pressure p was discussed in some detail. Static pressure is a measure of the purely random motion of molecules in a gas; it is the pressure you feel when you ride along with the gas at the local flow velocity. In contrast, the total (or stagnation) pressure was defined in Sec. 3.4 as the pressure existing at a point (or points) in the flow where $V = 0$. Let us now define the concept of total conditions more precisely.

Consider a fluid element passing through a given point in a flow where the local pressure, temperature, density, Mach number, and velocity are p, T, ρ, M, and V, respectively. Here, p, T, and ρ are static quantities, i.e., static pressure, static temperature, and static density, respectively; they are the pressure, temperature, and density you feel when you ride along with the gas at the local flow velocity. Now imagine that you grab hold of the fluid element and *adiabatically* slow it down to zero velocity. Clearly, you would expect (correctly) that the values of p, T, and ρ would change as the fluid element is brought to rest. In particular, the value of the temperature of the fluid element after it has been brought to rest adiabatically is defined as the *total temperature*, denoted by T_0. The corresponding value of enthalpy is defined as the *total enthalpy* h_0, where $h_0 = c_p T_0$ for a calorically perfect gas. Keep in mind that we do not *actually* have to bring the flow to rest in real life in order to talk about the total temperature or total enthalpy; rather, they are *defined quantities* that would exist at a point in a flow *if* (in our imagination) the fluid element passing through that point were brought to rest adiabatically. Therefore, at a given point in a flow, where the static temperature and enthalpy are T and h, respectively, we can also assign a value of total temperature T_0 and a value of total enthalpy h_0 defined as above.

The energy equation, Eq. (7.44), provides some important information about total enthalpy and hence total temperature, as follows. Assume that the flow is adiabatic ($\dot{q} = 0$), and that body forces are negligible ($\mathbf{f} = 0$). For such a flow, Eq. (7.44) becomes

$$\rho \frac{D(e + V^2/2)}{Dt} = -\nabla \cdot p\mathbf{V} \tag{7.45}$$

Expand the right-hand side of Eq. (7.45) using the following vector identity:

$$\nabla \cdot p\mathbf{V} \equiv p\nabla \cdot \mathbf{V} + \mathbf{V} \cdot \nabla p \tag{7.46}$$

Also, note that the substantial derivative defined in Sec. 2.9 follows the normal laws of differentiation; e.g.,

$$\rho \frac{D(p/\rho)}{Dt} = \rho \frac{\rho Dp/Dt - pD\rho/Dt}{\rho^2} = \frac{Dp}{Dt} - \frac{p}{\rho}\frac{D\rho}{Dt} \tag{7.47}$$

Recall the form of the continuity equation given by Eq. (2.99):

$$\frac{D\rho}{Dt} + \rho\nabla \cdot \mathbf{V} = 0 \tag{2.99}$$

Substituting Eq. (2.99) into (7.47), we obtain

$$\rho \frac{D(p/\rho)}{Dt} = \frac{Dp}{Dt} + p\nabla \cdot \mathbf{V} = \frac{\partial p}{\partial t} + \mathbf{V} \cdot \nabla p + p\nabla \cdot \mathbf{V} \qquad (7.48)$$

Substituting Eq. (7.46) into (7.45), and adding Eq. (7.48) to the result, we obtain

$$\rho \frac{D}{Dt}\left(e + \frac{p}{\rho} + \frac{V^2}{2}\right) = -p\nabla \cdot \mathbf{V} - \mathbf{V} \cdot \nabla p + \frac{\partial p}{\partial t} + \mathbf{V} \cdot \nabla p + p\nabla \cdot \mathbf{V} \qquad (7.49)$$

Note that

$$e + \frac{p}{\rho} = e + pv \equiv h \qquad (7.50)$$

Substituting Eq. (7.50) into (7.49), and noting that some of the terms on the right-hand side of Eq. (7.49) cancel each other, we have

$$\rho \frac{D(h + V^2/2)}{Dt} = \frac{\partial p}{\partial t} \qquad (7.51)$$

If the flow is steady, $\partial p/\partial t = 0$, and Eq. (7.51) becomes

$$\rho \frac{D(h + V^2/2)}{Dt} = 0 \qquad (7.52)$$

From the definition of the substantial derivative given in Sec. 2.9, Eq. (7.52) states that the time rate of change of $h + V^2/2$ following a moving fluid element is zero; i.e.,

$$\boxed{h + \frac{V^2}{2} = \text{const}} \qquad (7.53)$$

along a streamline. Recall that the assumptions which led to Eq. (7.53) are that the flow is steady, adiabatic, and inviscid. In particular, since Eq. (7.53) holds for an adiabatic flow, it can be used to elaborate on our previous definition of total enthalpy. Since h_0 is defined as that enthalpy which would exist at a point if the fluid element were brought to rest adiabatically, we find from Eq. (7.53) with $V = 0$ and hence $h = h_0$ that the value of the constant in Eq. (7.53) is h_0. Hence, Eq. (7.53) can be written as

$$\boxed{h + \frac{V^2}{2} = h_0} \qquad (7.54)$$

Equation (7.54) is important; it states that at any point in a flow, the total enthalpy is given by the sum of the static enthalpy plus the kinetic energy, all per unit mass. Whenever we have the combination $h + V^2/2$ in any subsequent equations, it can be identically replaced by h_0. For example, Eq. (7.52), which was derived for a steady, adiabatic, inviscid flow, states that

$$\rho \frac{Dh_0}{Dt} = 0$$

i.e., the total enthalpy is constant along a streamline. Moreover, if all the streamlines of the flow originate from a common uniform freestream (as is usually the case), then h_0 is the same for each streamline. Consequently, we have for such a steady, adiabatic flow that

$$\boxed{h_0 = \text{const}} \qquad (7.55)$$

throughout the *entire* flow, and h_0 is equal to its freestream value. Equation (7.55), although simple in form, is a powerful tool. For steady, inviscid, adiabatic flow, Eq. (7.55) is a statement of the energy equation, and hence it can be used in *place of* the more complex partial differential equation given by Eq. (7.52). This is a great simplification, as we will see in subsequent discussions.

For a calorically perfect gas, $h_0 = c_p T_0$. Thus, the above results also state that the total temperature is constant throughout the steady, inviscid, adiabatic flow of a calorically perfect gas; i.e.,

$$\boxed{T_0 = \text{const}} \qquad (7.56)$$

For such a flow, Eq. (7.56) can be used as a form of the governing energy equation.

Keep in mind that the above discussion marbled two trains of thought: On the one hand, we dealt with the general concept of an adiabatic flow field [which led to Eqs. (7.51) to (7.53)], and on the other hand, we dealt with the definition of total enthalpy [which led to Eq. (7.54)]. These two trains of thought are really separate and should not be confused. Consider, for example, a general *non-adiabatic* flow, such as a viscous boundary layer with heat transfer. Clearly, Eqs. (7.51) to (7.53) do not hold for such a flow. However, Eq. (7.54) holds locally at each point in the flow, because the assumption of an adiabatic flow contained in Eq. (7.54) is made through the *definition* of h_0 and has nothing to do with the general overall flow field. For example, consider two different points, 1 and 2, in the general flow. At point 1, the local static enthalpy and velocity are h_1 and V_1, respectively. Hence, the local total enthalpy at point 1 is $h_{0,1} = h_1 + V_1^2/2$. At point 2, the local static enthalpy and velocity are h_2 and V_2, respectively. Hence, the local total enthalpy at point 2 is $h_{0,2} = h_2 + V_2^2/2$. If the flow between points 1 and 2 is nonadiabatic, then $h_{0,1} \neq h_{0,2}$. Only for the special case where the flow is adiabatic between the two points would $h_{0,1} = h_{0,2}$. Of course, this is the special case treated by Eqs. (7.55) and (7.56).

Return to the beginning of this section, where we considered a fluid element passing through a point in a flow where the local properties are p, T, ρ, M, and $\mathbf{V}$. Once again, imagine that you grab hold of the fluid element and slow it down to zero velocity, but this time, let us slow it down both adiabatically *and* reversibly. That is, let us slow the fluid element down to zero velocity *isentropically.* When the fluid element is brought to rest isentropically, the resulting pressure and density are defined as the *total pressure* p_0 and *total density* ρ_0. (Since an isentropic process is also adiabatic, the resulting temperature is the same total temperature T_0 as discussed earlier.) As before, keep in mind that we do not have to actually bring the flow to rest in real life in order to talk about total pressure and total density; rather, they are *defined* quantities that would exist at a point in a flow *if* (in our imagination) the fluid element passing through that point were brought to rest isentropically. Therefore, at a given point in a flow, where the static pressure and static density are p and ρ, respectively, we can also assign a value of total pressure p_0, and total density ρ_0 defined as above.

The definition of p_0 and ρ_0 deals with an isentropic compression to zero velocity. Keep in mind that the isentropic assumption is involved with the definition only. The concept of total pressure and density can be applied throughout any general *nonisentropic* flow. For example, consider two different points, 1 and 2, in a general flow field. At point 1, the local static pressure and static density are p_1 and ρ_1, respectively; also the local total pressure and total density are $p_{0,1}$ and $\rho_{0,1}$, respectively, defined as above. Similarly, at point 2, the local static pressure and static density are p_2 and ρ_2, respectively, and the local total pressure and total density are $p_{0,2}$ and $\rho_{0,2}$, respectively. If the flow is nonisentropic between points 1 and 2, then $p_{0,1} \neq p_{0,2}$ and $\rho_{0,1} \neq \rho_{0,2}$. On the other hand, if the flow is isentropic between points 1 and 2, then $p_{0,1} = p_{0,2}$ and $\rho_{0,1} = \rho_{0,2}$. Indeed, if the general flow field is isentropic throughout, then both p_0 and ρ_0 are constant values throughout the flow.

As a corollary to the above considerations, we need another defined temperature, denoted by T^*, and defined as follows. Consider a point in a subsonic flow where the local static temperature is T. At this point, imagine that the fluid element is speeded up to *sonic velocity, adiabatically.* The temperature it would have at such sonic conditions is denoted as T^*. Similarly, consider a point in a supersonic flow, where the local static temperature is T. At this point, imagine that the fluid element is slowed down to sonic velocity, adiabatically. Again, the temperature it would have at such sonic conditions is denoted as T^*. The quantity T^* is simply a defined quantity at a given point in a flow, in exactly the same vein as T_0, p_0, and ρ_0 are defined quantities. Also, $a^* = \sqrt{\gamma R T^*}$.

7.6 SOME ASPECTS OF SUPERSONIC FLOW: SHOCK WAVES

Return to the different regimes of flow sketched in Fig. 1.30. Note that subsonic compressible flow is qualitatively (but not quantitatively) the same as incompressible flow; Fig. 1.30*a* shows a subsonic flow with a smoothly varying streamline pattern, where the flow far ahead of the body is forewarned about the presence of the body and begins to adjust accordingly. In contrast, supersonic flow is quite different, as sketched in Fig. 1.30*d* and *e*. Here, the flow is dominated by shock waves, and the flow upstream of the body does not know about the presence of the body until it encounters the leading-edge shock wave. In fact, any flow with a supersonic region, such as those sketched in Fig. 1.30b to e, is subject to shock waves. Thus, an essential ingredient of a study of supersonic flow is the calculation of the shape and strength of shock waves. This is the main thrust of Chaps. 8 and 9.

A shock wave is an extremely thin region, typically on the order of 10^{-5} cm, across which the flow properties can change drastically. The shock wave is usually at an oblique angle to the flow, such as sketched in Fig. 7.4a; however, there are many cases where we are interested in a shock wave normal to the flow, as sketched in Fig. 7.4b. Normal shock waves are discussed at length in Chap. 8, whereas oblique shocks are considered in Chap. 9. In both cases, the shock wave is an almost explosive compression process, where the pressure increases almost discontinuously across the wave. Examine Fig. 7.4 closely. In region 1 ahead of the shock, the Mach number, flow velocity, pressure, density, temperature, entropy, total pressure, and total enthalpy are denoted by M_1, V_1, p_1, ρ_1, T_1, s_1, $p_{0,1}$, and $h_{0,1}$, respectively. The analogous quantities in region 2 behind the shock are M_2, V_2, p_2, ρ_2, T_2, s_2, $p_{0,2}$, and $h_{0,2}$, respectively. The qualitative changes across the wave are noted in Fig. 7.4. The pressure, density, temperature, and entropy increase across the shock, whereas the total pressure, Mach number, and velocity decrease. Physically, the flow across a shock wave is adiabatic (we are not heating the gas with a laser beam or cooling it in a refrigerator, for example). Therefore, recalling the discussion in Sec. 7.5, the total enthalpy is constant across the wave. In both the oblique shock and normal shock cases, the flow ahead of the shock wave must be supersonic; i.e., $M_1 > 1$. Behind the oblique

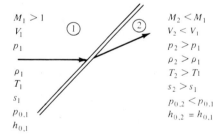

(a) Oblique shock wave

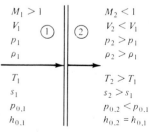

(b) Normal shock wave

FIGURE 7.4
Qualitative pictures of flow through oblique and normal shock waves.

shock, the flow usually remains supersonic; i.e., $M_2 > 1$, but at a reduced Mach number; i.e., $M_2 < M_1$. However, as discussed in Chap. 9, there are special cases where the oblique shock is strong enough to decelerate the downstream flow to a subsonic Mach number; hence, $M_2 < 1$ can occur behind an oblique shock. For the normal shock, as sketched in Fig. 7.4b, the downstream flow is always subsonic; i.e., $M_2 < 1$. Study the qualitative variations illustrated in Fig. 7.4 closely. They are important, and you should have them in mind for our subsequent discussions. One of the primary purposes of Chaps. 8 and 9 is to develop a shock-wave theory which allows the quantitative evaluation of these variations. We prove that pressure increases across the shock, that the upstream Mach number must be supersonic, etc. Moreover, we obtain equations that allow the direct calculation of changes across the shock.

Several photographs of shock waves are shown in Fig. 7.5. Since air is transparent, we cannot usually see shock waves with the naked eye. However, because the density changes across the shock wave, light rays propagating through the flow will be refracted across the shock. Special optical systems, such as shadowgraphs, schlieren, and interferometers, take advantage of this refraction and allow the visual imaging of shock waves on a screen or a photographic negative. For details of the design and characteristics of these optical systems, see Refs. 25 and 26. (Under certain conditions, you can see the refracted light from a shock wave with your naked eye. Recall from Fig. 1.30b that a shock wave can form in the locally supersonic region on the top surface of an airfoil

FIGURE 7.5
Schlieren photographs illustrating shock waves on various bodies. Apollo Command Module wind-tunnel model at Mach 8. (*Courtesy of the NASA Langley Research Center.*)

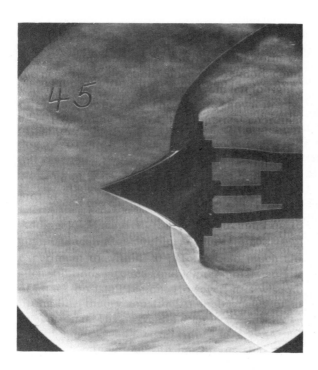

FIGURE 7.5
(*Continued*) "Tension shell" body once proposed for a Martian entry vehicle at Mach 8. (*Courtesy of the NASA Langley Research Center.*)

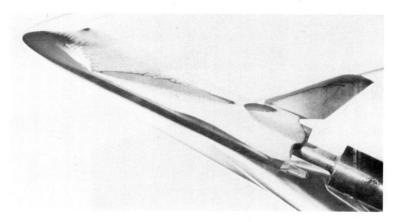

FIGURE 7.5
(*Continued*) Space Shuttle Orbiter model at Mach 6. This photo also shows regions of high aerodynamic heating on the model surface by means of the visible phase-change paint pattern. (*Courtesy of the NASA Langley Research Center.*)

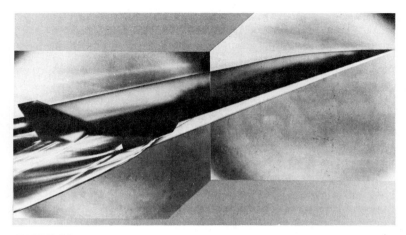

FIGURE 7.5
(*Continued*) A conceptual hypersonic aircraft at Mach 6. (*Courtesy of the NASA Langley Research Center.*)

if the freestream subsonic Mach number is high enough. The next time you are flying in a jet transport, and the sun is directly overhead, look out the window along the span of the wing. If you are lucky, you will see the shock wave dancing back and forth over the top of the wing.)

In summary, compressible flows introduce some very exciting physical phenomena into our aerodynamic studies. Moreover, as the flow changes from subsonic to supersonic, the complete nature of the flow changes, not the least of which is the occurrence of shock waves. The purpose of the next seven chapters is to describe and analyze these flows.

7.7 SUMMARY

As usual, examine the road map for this chapter (Fig. 7.1), and make certain that you feel comfortable with the material represented by this road map before continuing further.

Some of the highlights of this chapter are summarized below:

Thermodynamic relations:

Equation of state: $$p = \rho R T \tag{7.1}$$

For a calorically perfect gas,

$$e = c_v T \quad \text{and} \quad h = c_p T \tag{7.6a \text{ and } b}$$

$$c_p = \frac{\gamma R}{\gamma - 1} \tag{7.9}$$

$$c_v = \frac{R}{\gamma - 1} \tag{7.10}$$

Forms of the first law:

$$\delta q + \delta w = de \tag{7.11}$$

$$T\,ds = de + p\,dv \tag{7.18}$$

$$T\,ds = dh - v\,dp \tag{7.20}$$

Definition of entropy:

$$ds = \frac{\delta q_{\text{rev}}}{T} \tag{7.13}$$

Also,

$$ds = \frac{\delta q}{T} + ds_{\text{irrev}} \tag{7.14}$$

The second law:

$$ds \geq \frac{\delta q}{T} \tag{7.16}$$

or, for an adiabatic process,

$$ds \geq 0 \tag{7.17}$$

Entropy changes can be calculated from (for a calorically perfect gas)

$$s_2 - s_1 = c_p \ln \frac{T_2}{T_1} - R \ln \frac{p_2}{p_1} \tag{7.25}$$

and

$$s_2 - s_1 = c_v \ln \frac{T_2}{T_1} + R \ln \frac{v_2}{v_1} \tag{7.26}$$

For an isentropic flow,

$$\frac{p_2}{p_1} = \left(\frac{\rho_2}{\rho_1}\right)^{\gamma} = \left(\frac{T_2}{T_1}\right)^{\gamma/(\gamma-1)} \tag{7.32}$$

General definition of compressibility:

$$\tau = -\frac{1}{v}\frac{dv}{dp} \tag{7.33}$$

For an isothermal process,

$$\tau_T = -\frac{1}{v}\left(\frac{\partial v}{\partial p}\right)_T \tag{7.34}$$

For an isentropic process,

$$\tau_s = -\frac{1}{v}\left(\frac{\partial v}{\partial p}\right)_s \tag{7.35}$$

The governing equations for inviscid, compressible flow are

Continuity:

$$\frac{\partial}{\partial t} \iiint_{\mathcal{V}} \rho \, d\mathcal{V} + \iint_S \rho \mathbf{V} \cdot \mathbf{dS} = 0 \tag{7.39}$$

$$\frac{\partial \rho}{\partial t} + \nabla \cdot \rho \mathbf{V} = 0 \tag{7.40}$$

Momentum:

$$\frac{\partial}{\partial t} \iiint_{\mathcal{V}} \rho \mathbf{V} \, d\mathcal{V} + \iint_S (\rho \mathbf{V} \cdot \mathbf{dS}) \mathbf{V} = - \iint_S p \, \mathbf{dS} + \iiint_{\mathcal{V}} \rho \mathbf{f} \, d\mathcal{V} \tag{7.41}$$

$$\rho \frac{Du}{Dt} = -\frac{\partial p}{\partial x} + \rho f_x \tag{7.42a}$$

$$\rho \frac{Dv}{Dt} = -\frac{\partial p}{\partial y} + \rho f_y \tag{7.42b}$$

$$\rho \frac{Dw}{Dt} = -\frac{\partial p}{\partial z} + \rho f_z \tag{7.42c}$$

Energy:

$$\frac{\partial}{\partial t} \iiint_{\mathcal{V}} \rho \left(e + \frac{V^2}{2} \right) d\mathcal{V} + \iint_S \rho \left(e + \frac{V^2}{2} \right) \mathbf{V} \cdot \mathbf{dS}$$

$$= \iiint_{\mathcal{V}} \dot{q} \rho \, d\mathcal{V} - \iint_S p \mathbf{V} \cdot \mathbf{dS} + \iiint_{\mathcal{V}} \rho (\mathbf{f} \cdot \mathbf{V}) \, d\mathcal{V} \tag{7.43}$$

$$\rho \frac{D(e + V^2/2)}{Dt} = \rho \dot{q} - \nabla \cdot p \mathbf{V} + \rho (\mathbf{f} \cdot \mathbf{V}) \tag{7.44}$$

If the flow is steady and adiabatic, Eqs. (7.43) and (7.44) can be replaced by

$$h_0 = h + \frac{V^2}{2} = \text{const}$$

Equation of state (perfect gas):

$$p = \rho R T \tag{7.1}$$

Internal energy (calorically perfect gas):

$$e = c_v T \tag{7.6a}$$

Total temperature T_0 and total enthalpy h_0 are defined as the properties that would exist if (in our imagination) we slowed the fluid element at a point in the flow to zero velocity adiabatically. Similarly, total pressure p_0 and total density ρ_0 are defined as the properties that would exist if (in our imagination) we slowed the fluid element at a point in the flow to zero velocity isentropically. If a general flow field is adiabatic, h_0 is constant throughout the flow; in contrast, if the flow field is nonadiabatic, h_0 varies from one point to another. Similarly, if a general flow field is isentropic, p_0 and ρ_0 are constant throughout the flow; in contrast, if the flow field is nonisentropic, p_0 and ρ_0 vary from one point to another.

Shock waves are very thin regions in a supersonic flow across which the pressure, density, temperature, and entropy increase; the Mach number, flow velocity, and total pressure decrease; and the total enthalpy stays the same.

PROBLEMS

Note: In the following problems, you will deal with both the International System of Units (SI) (N, kg, m, s, K) and the English Engineering System (lb, slug, ft, s, °R). Which system to use will be self-evident in each problem. All problems deal with calorically perfect air as the gas, unless otherwise noted. Also, recall that $1\text{ atm} = 2116\text{ lb/ft}^2 = 1.01 \times 10^5\text{ N/m}^2$.

7.1. The temperature and pressure at the stagnation point of a high-speed missile are 934°R and 7.8 atm, respectively. Calculate the density at this point.

7.2. Calculate c_p, c_v, e, and h for
   (a) The stagnation point conditions given in Prob. 7.1
   (b) Air at standard sea level conditions
   (If you do not remember what standard sea level conditions are, find them in an appropriate reference, such as Ref. 2.)

7.3. Just upstream of a shock wave, the air temperature and pressure are 288 K and 1 atm, respectively; just downstream of the wave, the air temperature and pressure are 690 K and 8.656 atm, respectively. Calculate the changes in enthalpy, internal energy, and entropy across the wave.

7.4. Consider the isentropic flow over an airfoil. The freestream conditions are $T_\infty = 245$ K and $p_\infty = 4.35 \times 10^4$ N/m². At a point on the airfoil, the pressure is 3.6×10^4 N/m². Calculate the density at this point.

7.5. Consider the isentropic flow through a supersonic wind-tunnel nozzle. The reservoir properties are $T_0 = 500$ K and $p_0 = 10$ atm. If $p = 1$ atm at the nozzle exit, calculate the exit temperature and density.

7.6. Consider air at a pressure of 0.2 atm. Calculate the values of τ_T and τ_s. Express your answer in SI units.

7.7. Consider a point in a flow where the velocity and temperature are 1300 ft/s and 480°R, respectively. Calculate the total enthalpy at this point.

7.8. In the reservoir of a supersonic wind tunnel, the velocity is negligible, and the temperature is 1000 K. The temperature at the nozzle exit is 600 K. Assuming adiabatic flow through the nozzle, calculate the velocity at the exit.

7.9. An airfoil is in a freestream where $p_\infty = 0.61$ atm, $\rho_\infty = 0.819$ kg/m^3, and $V_\infty = 300$ m/s. At a point on the airfoil surface, the pressure is 0.5 atm. Assuming isentropic flow, calculate the velocity at that point.

7.10. Calculate the percentage error obtained if Prob. 7.9 is solved using (incorrectly) the incompressible Bernoulli equation.

7.11. Repeat Prob. 7.9, considering a point on the airfoil surface where the pressure is 0.3 atm.

7.12. Repeat Prob. 7.10, considering the flow of Prob. 7.11.

CHAPTER
8

NORMAL SHOCK
WAVES AND
RELATED
TOPICS

Shock wave: A large-amplitude compression wave, such as that produced by an explosion, caused by supersonic motion of a body in a medium.

From the American Heritage Dictionary of
the English Language, 1969

8.1 INTRODUCTION

The purpose of this chapter and Chap. 9 is to develop shock-wave theory, thus giving us the means to calculate the changes in the flow properties across a wave. These changes were discussed qualitatively in Sec. 7.6; make certain that you are familiar with these changes before continuing.

The focus of this chapter is on normal shock waves, as sketched in Fig. 7.4*b*. At first thought, a shock wave which is normal to the upstream flow may seem to be a very special case—and therefore a case of little practical interest—but nothing could be further from the truth. Normal shocks occur frequently in nature. Two such examples are sketched in Fig. 8.1; there are many more. The supersonic flow over a blunt body is shown at the left of Fig. 8.1. Here, a strong bow shock wave exists in front of the body. (We study such bow shocks in Chap. 9.) Although this wave is curved, the region of the shock closest to the nose is essentially normal to the flow. Moreover, the streamline that passes through this normal portion of the bow shock later impinges on the nose of the body and controls the values of stagnation (total) pressure and temperature at the nose. Since the nose region of high-speed blunt bodies is of practical interest in the calculation of drag and aerodynamic heating, the properties of the flow behind the normal portion of the shock wave take on some importance. In another example, shown at the right of Fig. 8.1, supersonic flow is established inside a nozzle (which can be a supersonic wind tunnel, a rocket engine, etc.) where the

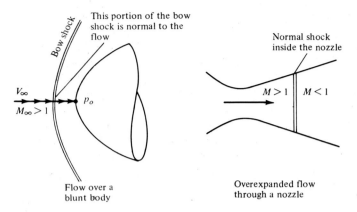

FIGURE 8.1
Two examples where normal shock waves are of interest.

back pressure is high enough to cause a normal shock wave to stand inside the nozzle. (We discuss such "overexpanded" nozzle flows in Chap. 10.) The conditions under which this shock wave will occur and the determination of flow properties at the nozzle exit downstream of the normal shock are both important questions to be answered. In summary, for these and many other applications, the study of normal shock waves is important.

Finally, we will find that many of the normal shock relations derived in this chapter carry over directly to the analysis of oblique shock waves, as discussed in Chap. 9. So once again, time spent on normal shock waves is time well spent.

The road map for this chapter is given in Fig. 8.2. As you can see, our objectives are fairly short and straightforward. We start with a derivation of the

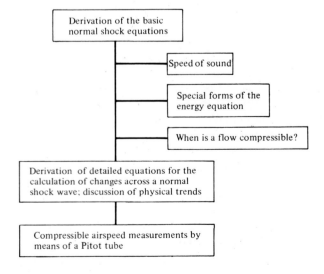

FIGURE 8.2
Road map for Chap. 8.

basic continuity, momentum, and energy equations for normal shock waves, and then we employ these basic relations to obtain detailed equations for the calculation of flow properties across the shock wave. In addition, we emphasize the physical trends indicated by the equations. On the way toward this objective, we take three side streets having to do with (1) the speed of sound, (2) special forms of the energy equation, and (3) a further discussion of the criteria used to judge when a flow must be treated as compressible. Finally, we apply the results of this chapter to the measurement of airspeed in a compressible flow using a Pitot tube. Keep the road map in Fig. 8.2 in mind as you progress through the chapter.

8.2   THE BASIC NORMAL SHOCK EQUATIONS

Consider the normal shock wave sketched in Fig. 8.3. Region 1 is a uniform flow upstream of the shock, and region 2 is a different uniform flow downstream of the shock. The pressure, density, temperature, Mach number, velocity, total pressure, total enthalpy, total temperature, and entropy in region 1 are p_1, ρ_1, T_1, M_1, u_1, $p_{0,1}$, $h_{0,1}$, $T_{0,1}$, and s_1, respectively. The corresponding variables in region 2 are denoted by p_2, ρ_2, T_2, M_2, u_2, $p_{0,2}$, $h_{0,2}$, $T_{0,2}$, and s_2. (Note that we are denoting the magnitude of the flow velocity by u rather than V; reasons for this will become obvious as we progress.) The problem of the normal shock wave is simply stated as follows: given the flow properties upstream of the wave (p_1, T_1, M_1, etc.), calculate the flow properties (p_2, T_2, M_2, etc.) downstream of the wave. Let us proceed.

Consider the rectangular control volume $abcd$ given by the dashed line in Fig. 8.3. The shock wave is inside the control volume, as shown. Side ab is the edge view of the left face of the control volume; this left face is perpendicular to the flow, and its area is A. Side cd is the edge view of the right face of the control volume; this right face is also perpendicular to the flow, and its area is

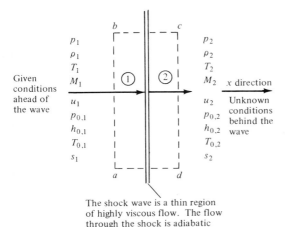

The shock wave is a thin region of highly viscous flow. The flow through the shock is adiabatic but nonisentropic

FIGURE 8.3

Sketch of a normal wave.

A. We apply the integral form of conservation equations to this control volume. In the process, we observe three important physical facts about the flow given in Fig. 8.3:

1. The flow is steady, i.e., $\partial/\partial t = 0$.
2. The flow is adiabatic, i.e., $\dot{q} = 0$. We are not adding or taking away heat from the control volume (we are not heating the shock wave with a Bunsen burner, for example). The temperature increases across the shock wave, not because heat is being added, but rather, because kinetic energy is converted to internal energy across the shock wave.
3. There are no viscous effects on the sides of the control volume. The shock wave itself is a thin region of extremely high velocity and temperature gradients; hence, friction and thermal conduction play an important role on the flow structure inside the wave. However, the wave itself is buried inside the control volume, and with the integral form of the conservation equations, we are not concerned about the details of what goes on inside the control volume.
4. There are no body forces; $\mathbf{f} = 0$.

Consider the continuity equation in the form of Eq. (7.39). For the conditions described above, Eq. (7.39) becomes

$$\oiint_S \rho \mathbf{V} \cdot \mathbf{dS} = 0 \tag{8.1}$$

To evaluate Eq. (8.1) over the face *ab*, note that $\mathbf{V}$ is pointing into the control volume whereas $\mathbf{dS}$ by definition is pointing out of the control volume, in the opposite direction of $\mathbf{V}$; hence, $\mathbf{V} \cdot \mathbf{dS}$ is negative. Moreover, ρ and $|\mathbf{V}|$ are uniform over the face *ab* and equal to ρ_1 and u_1, respectively. Hence, the contribution of face *ab* to the surface integral in Eq. (8.1) is simply $-\rho_1 u_1 A$. Over the right face, *cd*, both $\mathbf{V}$ and $\mathbf{dS}$ are in the same direction, and hence $\mathbf{V} \cdot \mathbf{dS}$ is positive. Moreover, ρ and $|\mathbf{V}|$ are uniform over the face *cd* and equal to ρ_2 and u_2, respectively. Thus, the contribution of face *cd* to the surface integral is $\rho_2 u_2 A$. On sides *bc* and *ad*, $\mathbf{V}$ and $\mathbf{dS}$ are always perpendicular; hence, $\mathbf{V} \cdot \mathbf{dS} = 0$, and these sides make no contribution to the surface integral. Hence, for the control volume shown in Fig. 8.3, Eq. (8.1) becomes

$$-\rho_1 u_1 A + \rho_2 u_2 A = 0$$

or

$$\boxed{\rho_1 u_1 = \rho_2 u_2} \tag{8.2}$$

Equation (8.2) is the continuity equation for normal shock waves.

Consider the momentum equation in the form of Eq. (7.41). For the flow we are treating here, Eq. (7.41) becomes

$$\oiint_S (\rho \mathbf{V} \cdot \mathbf{dS})\mathbf{V} = -\oiint_S p\,\mathbf{dS} \tag{8.3}$$

Equation (8.3) is a vector equation. Note that in Fig. 8.3, the flow is moving only in one direction, i.e., in the x direction. Hence, we need to consider only the scalar x component of Eq. (8.3), which is

$$\oiint_S (\rho \mathbf{V} \cdot \mathbf{dS}) u = -\oiint_S (p\, dS)_x \tag{8.4}$$

In Eq. (8.4), $(p\, dS)_x$ is the x component of the vector $(p\, \mathbf{dS})$. Note that over the face ab, $\mathbf{dS}$ points to the left, i.e., in the negative x direction. Hence, $(p\, dS)_x$ is negative over face ab. By similar reasoning, $(p\, dS)_x$ is positive over the face cd. Again noting that all the flow variables are uniform over the faces ab and cd, the surface integrals in Eq. (8.4) become

$$\rho_1(-u_1 A)u_1 + \rho_2(u_2 A)u_2 = -(-p_1 A + p_2 A) \tag{8.5}$$

or

$$\boxed{p_1 + \rho_1 u_1^2 = p_2 + \rho_2 u_2^2} \tag{8.6}$$

Equation (8.6) is the momentum equation for normal shock waves.

Consider the energy equation in the form of Eq. (7.43). For steady, adiabatic, inviscid flow with no body forces, this equation becomes

$$\oiint_S \rho \left(e + \frac{V^2}{2} \right) \mathbf{V} \cdot \mathbf{dS} = -\oiint_S p \mathbf{V} \cdot \mathbf{dS} \tag{8.7}$$

Evaluating Eq. (8.7) for the control surface shown in Fig. 8.3, we have

$$-\rho_1 \left(e_1 + \frac{u_1^2}{2} \right) u_1 A + \rho_2 \left(e_2 + \frac{u_2^2}{2} \right) u_2 A = -(-p_1 u_1 A + p_2 u_2 A)$$

Rearranging, we obtain

$$p_1 u_1 + \rho_1 \left(e_1 + \frac{u_1^2}{2} \right) u_1 = p_2 u_2 + \rho_2 \left(e_2 + \frac{u_2^2}{2} \right) u_2 \tag{8.8}$$

Dividing by Eq. (8.2), i.e., dividing the left-hand side of Eq. (8.8) by $\rho_1 u_1$ and the right-hand side by $\rho_2 u_2$, we have

$$\frac{p_1}{\rho_1} + e_1 + \frac{u_1^2}{2} = \frac{p_2}{\rho_2} + e_2 + \frac{u_2^2}{2} \tag{8.9}$$

From the definition of enthalpy, $h \equiv e + pv = e + p/\rho$. Hence, Eq. (8.9) becomes

$$\boxed{h_1 + \frac{u_1^2}{2} = h_2 + \frac{u_2^2}{2}} \tag{8.10}$$

Equation (8.10) is the energy equation for normal shock waves. Equation (8.10) should come as no surprise; the flow through a shock wave is adiabatic, and we derived in Sec. 7.5 the fact that for a steady, adiabatic flow, $h_0 = h + V^2/2 = \text{const.}$ Equation (8.10) simply states that h_0 (hence, for a calorically perfect gas, T_0) is constant across the shock wave. Therefore, Eq. (8.10) is consistent with the general results obtained in Sec. 7.5.

Repeating the above results for clarity, the basic normal shock equations are

Continuity:	$\rho_1 u_1 = \rho_2 u_2$	(8.2)
Momentum:	$p_1 + \rho_1 u_1^2 = p_2 + \rho_2 u_2^2$	(8.6)
Energy:	$h_1 + \dfrac{u_1^2}{2} = h_2 + \dfrac{u_2^2}{2}$	(8.10)

Examine these equations closely. Recall from Fig. 8.3 that all conditions upstream of the wave, ρ_1, u_1, p_1, etc., are known. Thus, the above equations are a system of three algebraic equations in four unknowns, ρ_2, u_2, p_2, and h_2. However, if we add the following thermodynamic relations

Enthalpy:	$h_2 = c_p T_2$
Equation of state:	$p_2 = \rho_2 R T_2$

we have five equations for five unknowns, namely, ρ_2, u_2, p_2, h_2, and T_2. In Sec. 8.6, we explicitly solve these equations for the unknown quantities behind the shock. However, rather than going directly to that solution, we first take three side trips as shown in the road map in Fig. 8.2. These side trips involve discussions of the speed of sound (Sec. 8.3), alternate forms of the energy equation (Sec. 8.4), and compressibility (Sec. 8.5)—all of which are necessary for a viable discussion of shock-wave properties in Sec. 8.6.

Finally, we note that Eqs. (8.2), (8.6), and (8.10) are not limited to normal shock waves; they describe the changes that take place in any steady, adiabatic, inviscid flow where only one direction is involved. That is, in Fig. 8.3, the flow is in the x direction only. This type of flow, where the flow-field variables are functions of x only [$p = p(x)$, $u = u(x)$, etc.], is defined as *one-dimensional flow*. Thus, Eqs. (8.2), (8.6), and (8.10) are governing equations for one-dimensional, steady, adiabatic, inviscid flow.

8.3 SPEED OF SOUND

Common experience tells us that sound travels through air at some finite velocity. For example, you see a flash of lightning in the distance, but you hear the corresponding thunder at some later moment. What is the physical mechanism of the propagation of sound waves? How can we calculate the speed of sound? What properties of the gas does it depend on? The speed of sound is an extremely important quantity which dominates the physical properties of compressible flow, and hence the answers to the above questions are vital to our subsequent discussions. The purpose of this section is to address these questions.

The physical mechanism of sound propagation in a gas is based on molecular motion. For example, imagine that you are sitting in a room, and suppose that a firecracker goes off in one corner. When the firecracker detonates, chemical energy (basically a form of heat release) is transferred to the air molecules adjacent to the firecracker. These energized molecules are moving about in a random fashion. They eventually collide with some of their neighboring molecules and transfer their high energy to these neighbors. In turn, these neighboring molecules eventually collide with their neighbors and transfer energy in the process. By means of this "domino" effect, the energy released by the firecracker is propagated through the air by molecular collisions. Moreover, because T, p, and ρ for a gas are macroscopic averages of the detailed microscopic molecular motion, the regions of energized molecules are also regions of slight variations in the local temperature, pressure, and density. Hence, as this energy wave from the firecracker passes over our eardrums, we "hear" the slight pressure changes in the wave. This is *sound*, and the propagation of the energy wave is simply the propagation of a *sound wave* through the gas.

Because a sound wave is propagated by molecular collisions, and because the molecules of a gas are moving with an average velocity of $\sqrt{8RT/\pi}$ given by kinetic theory, then we would expect the velocity of propagation of a sound wave to be approximately the average molecular velocity. Indeed, the speed of sound is about three-quarters of the average molecular velocity. In turn, because the kinetic theory expression given above for the average molecular velocity depends only on the *temperature* of the gas, we might expect the speed of sound to also depend on temperature only. Let us explore this matter further; indeed, let us now derive an equation for the speed of sound in a gas. Although the propagation of sound is due to molecular collisions, we do not use such a microscopic picture for our derivation. Rather, we take advantage of the fact that the macroscopic properties p, T, ρ, etc., change across the wave, and we use our macroscopic equations of continuity, momentum, and energy to analyze these changes.

Consider a sound wave propagating through a gas with velocity a, as sketched in Fig. 8.4a. Here, the sound wave is moving from right to left into a stagnant gas (region 1), where the local pressure, temperature, and density are p, T, and ρ, respectively. Behind the sound wave (region 2), the gas properties are slightly different and are given by $p + dp$, $T + dT$, and $\rho + d\rho$, respectively. Now imagine that you hop on the wave and ride with it. When you look upstream, into region 1, you see the gas moving toward you with a relative velocity a, as sketched in Fig. 8.4b. When you look downstream, into region 2, you see the gas receding away from you with a relative velocity $a + da$, as also shown in Fig. 8.4. (We have enough fluid-dynamic intuition by now to realize that because the pressure changes across the wave by the amount dp, then the relative flow velocity must also change across the wave by some amount da. Hence, the relative flow velocity behind the wave is $a + da$.) Consequently, in Fig. 8.4b, we have a picture of a stationary sound wave, with the flow ahead of it moving left to right with velocity a. The pictures in Fig. 8.4a and b are analogous; only the perspective is different. For purposes of analysis, we use Fig. 8.4b.

(*Note*: Figure 8.4b is similar to the picture of a normal shock wave shown in Fig. 8.3. In Fig. 8.3, the normal shock wave is stationary, and the upstream

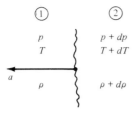

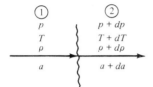

(a) A sound wave propagating with velocity a into a stagnant gas

(b) A stationary sound wave in a moving gas; the upstream velocity relative to the wave is a

FIGURE 8.4

Moving and stationary sound waves; two analogous pictures, only the perspective is different.

flow is moving left to right at a velocity u_1. If the upstream flow were to be suddenly shut off, then the normal shock wave in Fig. 8.3 would suddenly propagate to the left with a wave velocity of u_1, similar to the moving sound wave shown in Fig. 8.4a. The analysis of moving waves is slightly more subtle than the analysis of stationary waves; hence, it is simpler to begin a study of shock waves and sound waves with the pictures of stationary waves as shown in Figs. 8.3 and 8.4b. Also, please note that the sound wave in Fig. 8.4b is nothing more than an infinitely weak normal shock wave.)

Examine closely the flow through the sound wave sketched in Fig. 8.4b. The flow is one-dimensional. Moreover, it is adiabatic, because we have no source of heat transfer into or out of the wave (e.g., we are not "zapping" the wave with a laser beam or heating it with a torch). Finally, the gradients within the wave are very small—the changes dp, dT, $d\rho$, and da are infinitesimal. Therefore, the influence of dissipative phenomena (viscosity and thermal conduction) is negligible. As a result, the flow through the sound wave is both adiabatic and reversible—the flow is *isentropic*. Since we have now established that the flow is one-dimensional and isentropic, let us apply the appropriate governing equations to the picture shown in Fig. 8.4b.

Applying the continuity equation, Eq. (8.2), to Fig. 8.4b, we have

$$\rho a = (\rho + d\rho)(a + da)$$

or
$$\rho a = \rho a + a\, d\rho + \rho\, da + d\rho\, da \tag{8.11}$$

The product of two differentials, $dp\, da$, can be neglected in comparison with the other terms in Eq. (8.11). Hence, solving Eq. (8.11) for a, we obtain

$$a = -\rho \frac{da}{d\rho}$$
(8.12)

Now consider the one-dimensional momentum equation, Eq. (8.6), applied to Fig. 8.4b:

$$p + \rho a^2 = (p + dp) + (\rho + d\rho)(a + da)^2$$
(8.13)

Again ignoring products of differentials, Eq. (8.13) becomes

$$dp = -2a\rho\, da - a^2\, d\rho$$
(8.14)

Solving Eq. (8.14) for da, we have

$$da = \frac{dp + a^2\, d\rho}{-2a\rho}$$
(8.15)

Substituting Eq. (8.15) into (8.12), we obtain

$$a = -\rho \frac{dp/d\rho + a^2}{-2a\rho}$$
(8.16)

Solving Eq. (8.16) for a^2, we have

$$a^2 = \frac{dp}{d\rho}$$
(8.17)

As discussed above, the flow through a sound wave is isentropic; hence, in Eq. (8.17), the rate of change of pressure with respect to density, $dp/d\rho$, is an isentropic change. Hence, we can rewrite Eq. (8.17) as

$$\boxed{a = \sqrt{\left(\frac{\partial p}{\partial \rho}\right)_s}}$$
(8.18)

Equation (8.18) is a fundamental expression for the speed of sound in a gas.

Assume that the gas is calorically perfect. For such a case, the isentropic relation given by Eq. (7.32) holds, namely,

$$\frac{p_1}{p_2} = \left(\frac{\rho_1}{\rho_2}\right)^\gamma$$
(8.19)

From Eq. (8.19), we have

$$\frac{p}{\rho^\gamma} = \text{const} = c$$

or

$$p = c\rho^\gamma$$
(8.20)

Differentiating Eq. (8.20) with respect to ρ, we obtain

$$\left(\frac{\partial p}{\partial \rho}\right)_s = c\gamma\rho^{\gamma-1} \tag{8.21}$$

Substituting Eq. (8.20) for the constant c in Eq. (8.21), we have

$$\left(\frac{\partial p}{\partial \rho}\right)_s = \left(\frac{p}{\rho^\gamma}\right)\gamma\rho^{\gamma-1} = \frac{\gamma p}{\rho} \tag{8.22}$$

Substituting Eq. (8.22) into (8.18), we obtain

$$\boxed{a = \sqrt{\frac{\gamma p}{\rho}}} \tag{8.23}$$

Equation (8.23) is an expression for the speed of sound in a calorically perfect gas. At first glance, Eq. (8.23) seems to imply that the speed of sound would depend on both p and ρ. However, pressure and density are related through the perfect gas equation of state,

$$\frac{p}{\rho} = RT \tag{8.24}$$

Hence, substituting Eq. (8.24) into (8.23), we have

$$\boxed{a = \sqrt{\gamma RT}} \tag{8.25}$$

which is our final expression for the speed of sound; it clearly states that the *speed of sound in a calorically perfect gas is a function of temperature* only. This is consistent with our earlier discussion of the speed of sound being a molecular phenomenon, and therefore it is related to the average molecular velocity $\sqrt{8RT/\pi}$.

The speed of sound at standard sea level is a useful value to remember; it is

$$a_s = 340.9 \text{ m/s} = 1117 \text{ ft/s}$$

Recall the definition of compressibility given in Sec. 7.3. In particular, from Eq. (7.35) for the isentropic compressibility, repeated below,

$$\tau_s = -\frac{1}{v}\left(\frac{\partial v}{\partial p}\right)_s$$

and recalling that $v = 1/\rho$ (hence, $dv = -d\rho/\rho^2$), we have

$$\tau_s = -\rho\left[-\frac{1}{\rho^2}\left(\frac{\partial \rho}{\partial p}\right)_s\right] = \frac{1}{\rho(\partial p/\partial \rho)_s} \tag{8.26}$$

However, recall from Eq. (8.18) that $(\partial p/\partial \rho)_s = a^2$. Hence, Eq. (8.26) becomes

$$\tau_s = \frac{1}{\rho a^2}$$

or

$$a = \sqrt{\frac{1}{\rho \tau_s}} \tag{8.27}$$

Equation (8.27) relates the speed of sound to the compressibility of a gas. The lower the compressibility, the higher the speed of sound. Recall that for the limiting case of an incompressible fluid, $\tau_s = 0$. Hence, Eq. (8.27) states that the speed of sound in a theoretically incompressible fluid is infinite. In turn, for an incompressible flow with finite velocity V, the Mach number, $M = V/a$, is zero. Hence, the incompressible flows treated in Chaps. 3 to 6 are theoretically zero-Mach-number flows.

Finally, in regard to additional physical meaning of the Mach number, consider a fluid element moving along a streamline. The kinetic and internal energies per unit mass are $V^2/2$ and e, respectively. Their ratio is [recalling Eqs. (7.6a), (7.10), and (8.25)]

$$\frac{V^2/2}{e} = \frac{V^2/2}{c_v T} = \frac{V^2/2}{RT/(\gamma-1)} = \frac{(\gamma/2)V^2}{a^2/(\gamma-1)} = \frac{\gamma(\gamma-1)}{2}M^2$$

Hence, we see that the square of the Mach number is proportional to the ratio of kinetic energy to internal energy of a gas flow. In other words, the Mach number is a measure of the directed motion of the gas compared with the random thermal motion of the molecules.

8.4 SPECIAL FORMS OF THE ENERGY EQUATION

In this section, we elaborate upon the energy equation for adiabatic flow, as originally given by Eq. (7.44). In Sec. 7.5, we obtained for a steady, adiabatic, inviscid flow the result that

$$h_1 + \frac{V_1^2}{2} = h_2 + \frac{V_2^2}{2} \tag{8.28}$$

where V_1 and V_2 are velocities at any two points along a three-dimensional streamline. For the sake of consistency in our current discussion of one-dimensional flow, let us use u_1 and u_2 in Eq. (8.28):

$$h_1 + \frac{u_1^2}{2} = h_2 + \frac{u_2^2}{2} \tag{8.29}$$

However, keep in mind that all the subsequent results in this section hold in general along a streamline and are by no means limited to just one-dimensional flows.

Specializing Eq. (8.29) to a calorically perfect gas, where $h = c_p T$, we have

$$c_p T_1 + \frac{u_1^2}{2} = c_p T_2 + \frac{u_2^2}{2} \tag{8.30}$$

From Eq. (7.9), Eq. (8.30) becomes

$$\frac{\gamma R T_1}{\gamma - 1} + \frac{u_1^2}{2} = \frac{\gamma R T_2}{\gamma - 1} + \frac{u_2^2}{2} \tag{8.31}$$

Since $a = \sqrt{\gamma R T}$, Eq. (8.31) can be written as

$$\frac{a_1^2}{\gamma - 1} + \frac{u_1^2}{2} = \frac{a_2^2}{\gamma - 1} + \frac{u_2^2}{2} \tag{8.32}$$

If we consider point 2 in Eq. (8.32) to be a stagnation point, where the stagnation speed of sound is denoted by a_0, then, with $u_2 = 0$, Eq. (8.32) yields (dropping the subscript 1)

$$\frac{a^2}{\gamma - 1} + \frac{u^2}{2} = \frac{a_0^2}{\gamma - 1} \tag{8.33}$$

In Eq. (8.33), a and u are the speed of sound and flow velocity, respectively, at any given point in the flow, and a_0 is the stagnation (or total) speed of sound *associated* with that same point. Equivalently, if we have any two points along a streamline, Eq. (8.33) states that

$$\frac{a_1^2}{\gamma - 1} + \frac{u_1^2}{2} = \frac{a_2^2}{\gamma - 1} + \frac{u_2^2}{2} = \frac{a_0^2}{\gamma - 1} = \text{const} \tag{8.34}$$

Recalling the definition of a^* given at the end of Sec. 7.5, let point 2 in Eq. (8.32) represent sonic flow, where $u = a^*$. Then

$$\frac{a^2}{\gamma - 1} + \frac{u^2}{2} = \frac{a^{*2}}{\gamma - 1} + \frac{a^{*2}}{2}$$

or

$$\frac{a^2}{\gamma - 1} + \frac{u^2}{2} = \frac{\gamma + 1}{2(\gamma - 1)} a^{*2} \tag{8.35}$$

In Eq. (8.35), a and u are the speed of sound and flow velocity, respectively, at any given point in the flow, and a^* is a characteristic value *associated* with that same point. Equivalently, if we have any two points along a streamline, Eq. (8.35) states that

$$\frac{a_1^2}{\gamma-1}+\frac{u_1^2}{2}=\frac{a_2^2}{\gamma-1}+\frac{u_2^2}{2}=\frac{\gamma+1}{2(\gamma-1)}\,a^{*2}=\text{const} \tag{8.36}$$

Comparing the right-hand sides of Eqs. (8.34) and (8.36), the two properties a_0 and a^* associated with the flow are related by

$$\frac{\gamma+1}{2(\gamma-1)}\,a^{*2}=\frac{a_0^2}{\gamma-1}=\text{const} \tag{8.37}$$

Clearly, these defined quantities, a_0 and a^*, are both constants along a given streamline in a steady, adiabatic, inviscid flow. If all the streamlines emanate from the same uniform freestream conditions, then a_0 and a^* are constants throughout the entire flow field.

Recall the definition of total temperature T_0, as discussed in Sec. 7.5. In Eq. (8.30), let $u_2=0$; hence $T_2=T_0$. Dropping the subscript 1, we have

$$\boxed{c_p T+\frac{u^2}{2}=c_p T_0} \tag{8.38}$$

Equation (8.38) provides a formula from which the defined total temperature T_0 can be calculated from the given actual conditions of T and u at any given point in a general flow field. Equivalently, if we have any two points along a streamline in a steady, adiabatic, inviscid flow, Eq. (8.38) states that

$$c_p T_1+\frac{u_1^2}{2}=c_p T_2+\frac{u_2^2}{2}=c_p T_0=\text{const} \tag{8.39}$$

If all the streamlines emanate from the same uniform freestream, then Eq. (8.39) holds throughout the entire flow, not just along a streamline.

For a calorically perfect gas, the ratio of total temperature to static temperature, T_0/T, is a function of Mach number only, as follows. From Eqs. (8.38) and (7.9), we have

$$\frac{T_0}{T}=1+\frac{u^2}{2c_p T}=1+\frac{u^2}{2\gamma RT/(\gamma-1)}=1+\frac{u^2}{2a^2/(\gamma-1)}$$

$$=1+\frac{\gamma-1}{2}\left(\frac{u}{a}\right)^2$$

Hence,

$$\boxed{\frac{T_0}{T}=1+\frac{\gamma-1}{2}\,M^2} \tag{8.40}$$

Equation (8.40) is very important; it states that only M (and, of course, the value of γ) dictates the ratio of total temperature to static temperature.

Recall the definition of total pressure p_0 and total density ρ_0, as discussed in Sec. 7.5. These definitions involve an *isentropic* compression of the flow to zero velocity. From Eq. (7.32), we have

$$\frac{p_0}{p} = \left(\frac{\rho_0}{\rho}\right)^{\gamma} = \left(\frac{T_0}{T}\right)^{\gamma/(\gamma-1)} \tag{8.41}$$

Combining Eqs. (8.40) and (8.41), we obtain

$$\frac{p_0}{p} = \left(1 + \frac{\gamma-1}{2} M^2\right)^{\gamma/(\gamma-1)} \tag{8.42}$$

$$\frac{\rho_0}{\rho} = \left(1 + \frac{\gamma-1}{2} M^2\right)^{1/(\gamma-1)} \tag{8.43}$$

Similar to the case of T_0/T, we see from Eqs. (8.42) and (8.43) that the total-to-static ratios p_0/p and ρ_0/ρ are determined by M and γ only. Hence, for a given gas (i.e., given γ), the ratios T_0/T, p_0/p, and ρ_0/ρ depend only on Mach number.

Equations (8.40), (8.42), and (8.43) are very important; they should be branded on your mind. They provide formulas from which the defined quantities T_0, p_0, and ρ_0 can be calculated from the actual conditions of M, T, p, and ρ at a given point in a general flow field (assuming a calorically perfect gas). They are so important that values of T_0/T, p_0/p, and ρ_0/ρ obtained from Eqs. (8.40), (8.42), and (8.43), respectively, are tabulated as functions of M in App. A for $\gamma = 1.4$ (which corresponds to air at standard conditions).

Consider a point in a general flow where the velocity is exactly sonic, i.e., where $M = 1$. Denote the static temperature, pressure, and density at this sonic condition as T^*, p^*, and ρ^*, respectively. Inserting $M = 1$ into Eqs. (8.40), (8.42), and (8.43), we obtain

$$\frac{T^*}{T_0} = \frac{2}{\gamma+1} \tag{8.44}$$

$$\frac{p^*}{p_0} = \left(\frac{2}{\gamma+1}\right)^{\gamma/(\gamma-1)} \tag{8.45}$$

$$\frac{\rho^*}{\rho_0} = \left(\frac{2}{\gamma+1}\right)^{1/(\gamma-1)} \tag{8.46}$$

For $\gamma = 1.4$, these ratios are

$$\frac{T^*}{T_0} = 0.833 \qquad \frac{p^*}{p_0} = 0.528 \qquad \frac{\rho^*}{\rho_0} = 0.634$$

which are useful numbers to keep in mind for subsequent discussions.

We have one final item of business in this section. In Chap. 1, we defined the Mach number as $M = V/a$ (or, following the one-dimensional notation in this chapter, $M = u/a$). In turn, this allowed us to define several regimes of flow, among them being

$M < 1$ (subsonic flow)
$M = 1$ (sonic flow)
$M > 1$ (supersonic flow)

In the definition of M, a is the local speed of sound, $a = \sqrt{\gamma R T}$. In the theory of supersonic flow, it is sometimes convenient to introduce a "characteristic" Mach number, M^*, defined as

$$M^* \equiv \frac{u}{a^*}$$

where a^* is the value of the speed of sound at sonic conditions, *not* the actual local value. This is the same a^* introduced at the end of Sec. 7.5 and used in Eq. (8.35). The value of a^* is given by $a^* = \sqrt{\gamma R T^*}$. Let us now obtain a relation between the actual Mach number M and this defined characteristic Mach number M^*. Dividing Eq. (8.35) by u^2, we have

$$\frac{(a/u)^2}{\gamma - 1} + \frac{1}{2} = \frac{\gamma + 1}{2(\gamma - 1)}\left(\frac{a^*}{u}\right)^2$$

$$\frac{(1/M)^2}{\gamma - 1} = \frac{\gamma + 1}{2(\gamma - 1)}\left(\frac{1}{M^*}\right)^2 - \frac{1}{2}$$

$$M^2 = \frac{2}{(\gamma + 1)/M^{*2} - (\gamma - 1)} \tag{8.47}$$

Equation (8.47) gives M as a function of M^*. Solving Eq. (8.47) for M^{*2}, we have

$$M^{*2} = \frac{(\gamma + 1)M^2}{2 + (\gamma - 1)M^2} \tag{8.48}$$

which gives M^* as a function of M. As can be shown by inserting numbers into Eq. (8.48) (try some yourself),

$M^* = 1$ if $M = 1$
$M^* < 1$ if $M < 1$
$M^* > 1$ if $M > 1$

$$M^* \to \sqrt{\frac{\gamma + 1}{\gamma - 1}} \quad \text{if } M \to \infty$$

Therefore, M^* acts qualitatively in the same fashion as M except that M^* approaches a finite value when the actual Mach number approaches infinity.

In summary, a number of equations have been derived in this section, all of which stem in one fashion or another from the basic energy equation for steady, inviscid, adiabatic flow. Make certain that you understand these equations and become very familiar with them before progressing further. These equations are pivotal in the analysis of shock waves and in the study of compressible flow in general.

Example 8.1. Consider a point in an airflow where the local Mach number, static pressure, and static temperature are 3.5, 0.3 atm, and 180 K, respectively. Calculate the local values of p_0, T_0, T^*, a^*, and M^* at this point.

Solution. From App. A, for $M = 3.5$, $p_0/p = 76.27$ and $T_0/T = 3.45$. Hence,

$$p_0 = \left(\frac{p_0}{p}\right) p = 76.27(0.3 \text{ atm}) = \boxed{22.9 \text{ atm}}$$

$$T_0 = \frac{T_0}{T} T = 3.45(180) = \boxed{621 \text{ K}}$$

For $M = 1$, $T_0/T^* = 1.2$. Hence,

$$T^* = \frac{T_0}{1.2} = \frac{621}{1.2} = \boxed{517.5 \text{ K}}$$

$$a^* = \sqrt{\gamma R T^*} = \sqrt{1.4(287)(517.5)} = \boxed{456 \text{ m/s}}$$

$$a = \sqrt{\gamma R T} = \sqrt{1.4(287)(180)} = 268.9 \text{ m/s}$$

$$V = Ma = 3.5(268.9) = 941 \text{ m/s}$$

$$M^* = \frac{V}{a^*} = \frac{941}{456} = \boxed{2.06}$$

The above result for M^* can also be obtained directly from Eq. (8.48):

$$M^{*2} = \frac{(\gamma+1)M^2}{2+(\gamma-1)M^2} = \frac{2.4(3.5)^2}{2+0.4(3.5)^2} = 4.26$$

Hence, $M^* = \sqrt{4.26} = 2.06$, as obtained above.

8.5 WHEN IS A FLOW COMPRESSIBLE?

As a corollary to Sec. 8.4, we are now in a position to examine the question, When does a flow have to be considered compressible, i.e., when do we have to use analyses based on Chaps. 7 to 14 rather than the incompressible techniques discussed in Chaps. 3 to 6? There is no specific answer to this question; for subsonic flows, it is a matter of the degree of accuracy desired whether we treat ρ as a constant or as a variable, whereas for supersonic flow the qualitative aspects of the flow are so different that the density *must* be treated as variable. We have stated several times in the preceding chapters the rule of thumb that a flow can be reasonably assumed to be incompressible when $M < 0.3$, whereas it should be considered compressible when $M > 0.3$. There is nothing magic about the value 0.3, but it is a convenient dividing line. We are now in a position to add substance to this rule of thumb.

Consider a fluid element initially at rest, say, an element of the air around you. The density of this gas at rest is ρ_0. Let us now accelerate this fluid element isentropically to some velocity V and Mach number M, say, by expanding the air through a nozzle. As the velocity of the fluid element increases, the other flow properties will change according to the basic governing equations derived in Chap. 7 and in this chapter. In particular, the density ρ of the fluid element will change according to Eq. (8.43):

$$\frac{\rho_0}{\rho} = \left(1 + \frac{\gamma-1}{2} M^2\right)^{1/(\gamma-1)} \tag{8.43}$$

For $\gamma = 1.4$, this variation is illustrated in Fig. 8.5, where ρ/ρ_0 is plotted as a function of M from zero to sonic flow. Note that at low subsonic Mach numbers, the variation of ρ/ρ_0 is relatively flat. Indeed, for $M < 0.32$, the value of ρ deviates from ρ_0 by less than 5 percent, and for all practical purposes the flow can be treated as incompressible. However, for $M > 0.32$, the variation in ρ is larger than 5 percent, and its change becomes even more pronounced as M increases. As a result, many aerodynamicists have adopted the rule of thumb that the density variation should be accounted for at Mach numbers above 0.3; i.e., the flow should be treated as compressible. Of course, keep in mind that all flows, even at the lowest Mach numbers, are, strictly speaking, compressible. Incompressible flow is really a myth. However, as shown in Fig. 8.5, the *assumption* of incompressible flow is very reasonable at low Mach numbers. For this reason, the analyses in Chaps. 3 to 6 and the vast bulk of existing literature for incompressible flow are quite practical for many aerodynamic applications.

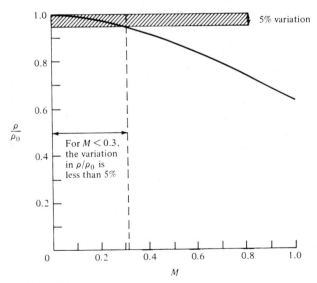

FIGURE 8.5
Isentropic variation of density with Mach number.

To obtain additional insight into the significance of Fig. 8.5, let us ask how the ratio ρ/ρ_0 affects the change in pressure associated with a given change in velocity. The differential relation between pressure and velocity for a compressible flow is given by Euler's equation, Eq. (3.12) repeated below:

$$dp = -\rho V \, dV \qquad\qquad (3.12)$$

This can be written as

$$\frac{dp}{p} = -\frac{\rho}{p} V^2 \frac{dV}{V}$$

This equation gives the fractional change in pressure for a given fractional change in velocity for a compressible flow with local density ρ. If we now *assume* that the density is constant, say, equal to ρ_0 as denoted in Fig. 8.5, then Eq. (3.12) yields

$$\left(\frac{dp}{p}\right)_0 = -\frac{\rho_0}{p} V^2 \frac{dV}{V}$$

where the subscript zero implies the assumption of constant density. Dividing the last two equations, and assuming the same dV/V and p, we have

$$\frac{dp/p}{(dp/p)_0} = \frac{\rho}{\rho_0}$$

Hence, the degree by which ρ/ρ_0 deviates from unity as shown in Fig. 8.5 is related to the same degree by which the fractional pressure change for a given dV/V is predicted. For example, if $\rho/\rho_0 = 0.95$, which occurs at about $M = 0.3$ in Fig. 8.5, then the fractional change in pressure for a compressible flow with local density ρ as compared to that for an incompressible flow with density ρ_0 will be about 5 percent different. Keep in mind that the above comparison is for the local fractional change in pressure; the actual integrated pressure change is less sensitive. For example, consider the flow of air through a nozzle starting in the reservoir at nearly zero velocity and standard sea level values of $p_0 = 2116$ lb/ft^2 and $T_0 = 510°$R, and expanding to a velocity of 350 ft/s at the nozzle exit. The pressure at the nozzle exit will be calculated assuming first incompressible flow and then compressible flow.

Incompressible flow: From Bernoulli's equation,

$$p = p_0 - \tfrac{1}{2}\rho V^2 = 2116 - \tfrac{1}{2}(0.002377)(350)^2 = \boxed{1970 \text{ lb/ft}^2}$$

Compressible flow: From the energy equation, Eq. (8.30), with $c_p = 6006$ [(ft)(lb)/slug°R] for air,

$$T = T_0 - \frac{V^2}{2c_p} = 519 - \frac{(350)^2}{2(6006)} = 508.8°\text{R}$$

From Eq. (7.32),

$$\frac{p}{p_0} = \left(\frac{T}{T_0}\right)^{\gamma/(\gamma-1)} = \left(\frac{508.8}{519}\right)^{3.5} = 0.9329$$

$$p = 0.9329 p_0 = 0.93299(2116) = \boxed{1974 \text{ lb/ft}^2}$$

Note that the two results are almost the same, with the compressible value of pressure only 0.2 percent higher than the incompressible value. Clearly, the assumption of incompressible flow (hence, the use of Bernoulli's equation) is certainly justified in this case. Also, note that the Mach number at the exit is 0.317 (work this out for yourself). Hence, we have shown that for a flow wherein the Mach number ranges from zero to about 0.3, Bernoulli's equation yields a reasonably accurate value for the pressure—another justification for the statement that flows wherein $M < 0.3$ are essentially incompressible flows. On the other hand, if this flow were to continue to expand to a velocity of 900 ft/s, a repeat of the above calculation yields the following results for the static pressure at the end of the expansion:

Incompressible (Bernoulli's equation): $p = 1153 \text{ lb/ft}^2$

Compressible: $p = 1300 \text{ lb/ft}^2$

Here, the difference between the two sets of results is considerable—a 13 percent difference. In this case, the Mach number at the end of the expansion is 0.86. Clearly, for such values of Mach number, the flow must be treated as compressible.

In summary, although it may be somewhat conservative, this author suggests on the strength of all the above information, including Fig. 8.5, that flows wherein the local Mach number exceeds 0.3 should be treated as compressible. Moreover, when $M < 0.3$, the assumption of incompressible flow is quite justified.

8.6 CALCULATION OF NORMAL SHOCK-WAVE PROPERTIES

Consider again the road map given in Fig. 8.2. We have finished our three side trips (Secs. 8.3 to 8.5) and are now ready to get back on the main road toward the calculation of changes of flow properties across a normal shock wave. Return again to Sec. 8.2, and recall the basic normal shock equations given by Eqs. (8.2), (8.6), and (8.10):

Continuity:
$$\rho_1 u_1 = \rho_2 u_2 \tag{8.2}$$

Momentum:
$$p_1 + \rho_1 u_1^2 = p_2 + \rho_2 u_2^2 \tag{8.6}$$

Energy:
$$h_1 + \frac{u_1^2}{2} = h_2 + \frac{u_2^2}{2} \tag{8.10}$$

In addition, for a calorically perfect gas, we have

$$h_2 = c_p T_2 \tag{8.49}$$

$$p_2 = \rho_2 R T_2 \tag{8.50}$$

Return again to Fig. 8.3, and recall the basic normal shock-wave problem: given the conditions in region 1 ahead of the shock, calculate the conditions in region 2 behind the shock. Examining the five equations given above, we see that they involve five unknowns, namely, ρ_2, u_2, p_2, h_2, and T_2. Hence, Eqs. (8.2), (8.6), (8.10), (8.49), and (8.50) are sufficient for determining the properties behind a normal shock wave in a calorically perfect gas. Let us proceed.

First, dividing Eq. (8.6) by (8.2), we obtain

$$\frac{p_1}{\rho_1 u_1} + u_1 = \frac{p_2}{\rho_2 u_2} + u_2$$

$$\frac{p_1}{\rho_1 u_1} - \frac{p_2}{\rho_2 u_2} = u_2 - u_1 \tag{8.51}$$

Recalling from Eq. (8.23) that $a = \sqrt{\gamma p / \rho}$, Eq. (8.51) becomes

$$\frac{a_1^2}{\gamma u_1} - \frac{a_2^2}{\gamma u_2} = u_2 - u_1 \tag{8.52}$$

Equation (8.52) is a combination of the continuity and momentum equations. The energy equation, Eq. (8.10), can be used in one of its alternate forms, namely, Eq. (8.35), rearranged below, and applied first in region 1 and then in region 2:

$$a_1^2 = \frac{\gamma+1}{2} a^{*2} - \frac{\gamma-1}{2} u_1^2 \tag{8.53}$$

and

$$a_2^2 = \frac{\gamma+1}{2} a^{*2} - \frac{\gamma-1}{2} u_2^2 \tag{8.54}$$

In Eqs. (8.53) and (8.54), a^* is the same constant value because the flow across the shock wave is adiabatic (see Secs. 7.5 and 8.4). Substituting Eqs. (8.53) and (8.54) into Eq. (8.52), we have

$$\frac{\gamma+1}{2} \frac{a^{*2}}{\gamma u_1} - \frac{\gamma-1}{2\gamma} u_1 - \frac{\gamma+1}{2} \frac{a^{*2}}{\gamma u_2} + \frac{\gamma-1}{2\gamma} u_2 = u_2 - u_1$$

or

$$\frac{\gamma+1}{2\gamma u_1 u_2} (u_2 - u_1) a^{*2} + \frac{\gamma-1}{2\gamma} (u_2 - u_1) = u_2 - u_1$$

Dividing by $u_2 - u_1$, we obtain

$$\frac{\gamma+1}{2\gamma u_1 u_2} a^{*2} + \frac{\gamma-1}{2\gamma} = 1$$

Solving for a^*, we obtain

$$\boxed{a^{*2} = u_1 u_2} \tag{8.55}$$

Equation (8.55) is called the *Prandtl relation* and is a useful intermediate relation for normal shock waves. For example, from Eq. (8.55),

$$1 = \frac{u_1}{a^*} \frac{u_2}{a^*} \tag{8.56}$$

Recall the definition of characteristic Mach number, $M^* = u/a^*$, given in Sec. 8.4. Hence, Eq. (8.56) becomes

$$1 = M_1^* M_2^*$$

or

$$M_2^* = \frac{1}{M_1^*} \tag{8.57}$$

Substituting Eq. (8.48) into (8.57), we have

$$\frac{(\gamma+1)M_2^2}{2+(\gamma-1)M_2^2} = \left[\frac{(\gamma+1)M_1^2}{2+(\gamma-1)M_1^2}\right]^{-1} \tag{8.58}$$

Solving Eq. (8.58) for M_2^2, we obtain

$$M_2^2 = \frac{1+[(\gamma-1)/2]M_1^2}{\gamma M_1^2 - (\gamma-1)/2} \tag{8.59}$$

Equation (8.59) is our first major result for a normal shock wave. Examine Eq. (8.59) closely; it states that *the Mach number behind the wave, M_2, is a function only of the Mach number ahead of the wave, M_1*. Moreover, if $M_1 = 1$, then $M_2 = 1$. This is the case of an infinitely weak normal shock wave, defined as a *Mach wave*. Furthermore, if $M_1 > 1$, then $M_2 < 1$; i.e., the Mach number behind the normal shock wave is *subsonic*. As M_1 increases above 1, the normal shock wave becomes stronger, and M_2 becomes progressively less than 1. However, in the limit as $M_1 \to \infty$, M_2 approaches a finite minimum value, $M_2 \to \sqrt{(\gamma-1)/2\gamma}$, which for air is 0.378.

Let us now obtain the ratios of the thermodynamic properties ρ_2/ρ_1, p_2/p_1, and T_2/T_1 across a normal shock wave. Rearranging Eq. (8.2) and using Eq. (8.55), we have

$$\frac{\rho_2}{\rho_1} = \frac{u_1}{u_2} = \frac{u_1^2}{u_2 u_1} = \frac{u_1^2}{a^{*2}} = M_1^{*2} \tag{8.60}$$

Substituting Eq. (8.48) into (8.60), we obtain

$$\frac{\rho_2}{\rho_1} = \frac{u_1}{u_2} = \frac{(\gamma+1)M_1^2}{2+(\gamma-1)M_1^2} \tag{8.61}$$

To obtain the pressure ratio, return to the momentum equation, Eq. (8.6), combined with the continuity equation, Eq. (8.2):

$$p_2 - p_1 = \rho_1 u_1^2 - \rho_2 u_2^2 = \rho_1 u_1(u_1 - u_2) = \rho_1 u_1^2 \left(1 - \frac{u_2}{u_1}\right) \tag{8.62}$$

Dividing Eq. (8.62) by p_1, and recalling that $a_1^2 = \gamma p_1 / \rho_1$, we obtain

$$\frac{p_2 - p_1}{p_1} = \frac{\gamma \rho_1 u_1^2}{\gamma p_1} \left(1 - \frac{u_2}{u_1}\right) = \frac{\gamma u_1^2}{a_1^2} \left(1 - \frac{u_2}{u_1}\right) = \gamma M_1^2 \left(1 - \frac{u_2}{u_1}\right) \tag{8.63}$$

For u_2 / u_1 in Eq. (8.63), substitute Eq. (8.61):

$$\frac{p_2 - p_1}{p_1} = \gamma M_1^2 \left[1 - \frac{2 + (\gamma - 1) M_1^2}{(\gamma + 1) M_1^2}\right] \tag{8.64}$$

Equation (8.64) simplifies to

$$\boxed{\frac{p_2}{p_1} = 1 + \frac{2\gamma}{\gamma + 1} (M_1^2 - 1)} \tag{8.65}$$

To obtain the temperature ratio, recall the equation of state $p = \rho R T$. Hence,

$$\frac{T_2}{T_1} = \left(\frac{p_2}{p_1}\right)\left(\frac{\rho_1}{\rho_2}\right) \tag{8.66}$$

Substituting Eqs. (8.61) and (8.65) into (8.66), and recalling that $h = c_p T$, we obtain

$$\boxed{\frac{T_2}{T_1} = \frac{h_2}{h_1} = \left[1 + \frac{2\gamma}{\gamma + 1} (M_1^2 - 1)\right] \frac{2 + (\gamma - 1) M_1^2}{(\gamma + 1) M_1^2}} \tag{8.67}$$

Equations (8.61), (8.65), and (8.67) are important. Examine them closely. Note that ρ_2/ρ_1, p_2/p_1, and T_2/T_1 are *functions of the upstream Mach number M_1 only*. Therefore, in conjunction with Eq. (8.59) for M_2, we see that the upstream Mach number M_1 is *the* determining parameter for changes across a normal shock wave in a calorically perfect gas. This is a dramatic example of the power of the Mach number as a governing parameter in compressible flows. In the above equations, if $M_1 = 1$, then $p_2/p_1 = \rho_2/\rho_1 = T_2/T_1 = 1$; i.e., we have the case of a normal shock wave of vanishing strength—a Mach wave. As M_1 increases above 1, p_2/p_1, ρ_2/ρ_1, and T_2/T_1 progressively increase above 1. In the limiting case of $M_1 \to \infty$ in Eqs. (8.59), (8.61), (8.65), and (8.67), we find, for $\gamma = 1.4$,

$$\lim_{M_1 \to \infty} M_2 = \sqrt{\frac{\gamma - 1}{2\gamma}} = 0.378 \qquad \text{(as discussed previously)}$$

$$\lim_{M_1 \to \infty} \frac{p_2}{p_1} = \frac{\gamma + 1}{\gamma - 1} = 6$$

$$\lim_{M_1 \to \infty} \frac{p_2}{p_1} = \infty \qquad \lim_{M_1 \to \infty} \frac{T_2}{T_1} = \infty$$

Note that, as the upstream Mach number increases toward infinity, the pressure and temperature increase without bound, whereas the density approaches a rather moderate finite limit.

We have stated earlier that shock waves occur in supersonic flows; a stationary normal shock such as shown in Fig. 8.3 does not occur in subsonic flow. That is, in Eqs. (8.59), (8.61), (8.65), and (8.67), the upstream Mach number is supersonic, $M_1 \geq 1$. However, on a *mathematical basis*, these equations also allow solutions for $M_1 \leq 1$. These equations embody the continuity, momentum, and energy equations, which in principle do not care whether the value of M_1 is subsonic or supersonic. Here is an ambiguity which can only be resolved by appealing to the second law of thermodynamics (see Sec. 7.2). Recall that the second law determines the *direction* which a given process can take. Let us apply the second law to the flow across a normal shock wave, and examine what it tells us about allowable values of M_1.

First, consider the entropy change across the normal shock wave. From Eq. (7.25),

$$s_2 - s_1 = c_p \ln \frac{T_2}{T_1} - R \ln \frac{p_2}{p_1} \tag{7.25}$$

with Eqs. (8.65) and (8.67), we have

$$s_2 - s_1 = c_p \ln \left\{ \left[1 + \frac{2\gamma}{\gamma+1}(M_1^2 - 1) \right] \frac{2 + (\gamma-1)M_1^2}{(\gamma+1)M_1^2} \right\}$$

$$- R \ln \left[1 + \frac{2\gamma}{\gamma+1}(M_1^2 - 1) \right] \tag{8.68}$$

From Eq. (8.68), we see that the entropy change $s_2 - s_1$ across the shock is a function of M_1 only. The second law dictates that

$$s_2 - s_1 \geq 0$$

In Eq. (8.68), if $M_1 = 1$, $s_2 = s_1$, and if $M_1 > 1$, then $s_2 - s_1 > 0$, both of which obey the second law. However, if $M_1 < 1$, then Eq. (8.68) gives $s_2 - s_1 < 0$, which is *not* allowed by the second law. Consequently, in nature, only cases involving $M_1 \geq 1$ are valid; i.e., normal shock waves can occur only in supersonic flow.

Why does the entropy increase across the shock wave? The second law tells us that it must, but what mechanism does nature use to accomplish this increase? To answer these questions, recall that a shock wave is a very thin region (on the order of 10^{-5} cm) across which some large changes occur almost discontinuously. Therefore, within the shock wave itself, large gradients in velocity and temperature occur; i.e., the mechanisms of friction and thermal conduction are strong. These are dissipative, irreversible mechanisms that always increase the entropy. Therefore, the precise entropy increase predicted by Eq. (8.68) for a given supersonic M_1 is appropriately provided by nature in the form of friction and thermal conduction within the interior of the shock wave itself.

In Sec. 7.5, we defined the total temperature T_0 and total pressure p_0. What happens to these total conditions across a shock wave? To help answer this question, consider Fig. 8.6, which illustrates the definition of total conditions ahead of and behind the shock. In region 1 ahead of the shock, a fluid element has the actual conditions of M_1, p_1, T_1, and s_1. Now imagine that we bring this fluid element to rest isentropically, creating the "imaginary" state 1a ahead of the shock. In state 1a, the fluid element at rest would have a pressure and temperature $p_{0,1}$ and $T_{0,1}$, respectively, i.e., the total pressure and total temperature, respectively, in region 1. The entropy in state 1a would still be s_1 because the fluid element is brought to rest isentropically; $s_{1a} = s_1$. Now consider region 2 behind the shock. Again consider a fluid element with the actual conditions of M_2, p_2, T_2, and s_2, as sketched in Fig. 8.6. And again let us imagine that we bring this fluid element to rest isentropically, creating the "imaginary" state 2a behind the shock. In state 2a, the fluid element at rest would have pressure and temperature $p_{0,2}$ and $T_{0,2}$, respectively, i.e., the total pressure and total temperature, respectively, in region 2. The entropy in state 2a would still be s_2 because the fluid element is brought to rest isentropically; $s_{2a} = s_2$. The questions are now asked: How does $T_{0,2}$ compare with $T_{0,1}$, and how does $p_{0,2}$ compare with $p_{0,1}$?

To answer the first of these questions, consider Eq. (8.30):

$$c_p T_1 + \frac{u_1^2}{2} = c_p T_2 + \frac{u_2^2}{2} \qquad (8.30)$$

From Eq. (8.38), the total temperature is given by

$$c_p T_0 = c_p T + \frac{u^2}{2} \qquad (8.38)$$

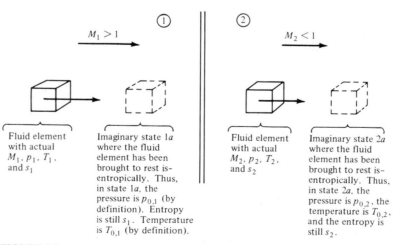

Fluid element with actual M_1, p_1, T_1, and s_1

Imaginary state 1a where the fluid element has been brought to rest isentropically. Thus, in state 1a, the pressure is $p_{0,1}$ (by definition). Entropy is still s_1. Temperature is $T_{0,1}$ (by definition).

Fluid element with actual M_2, p_2, T_2, and s_2

Imaginary state 2a where the fluid element has been brought to rest isentropically. Thus, in state 2a, the pressure is $p_{0,2}$, the temperature is $T_{0,2}$, and the entropy is still s_2.

FIGURE 8.6
Total conditions ahead of and behind a normal shock wave.

Combining Eqs. (8.30) and (8.38), we have

$$c_p T_{0,1} = c_p T_{0,2}$$

or

$$\boxed{T_{0,1} = T_{0,2}} \tag{8.69}$$

Equation (8.69) states that *total temperature is constant across a stationary normal shock wave.* This should come as no surprise; the flow across a shock wave is adiabatic, and in Sec. 7.5 we demonstrated that in a steady, adiabatic, inviscid flow of a calorically perfect gas, the total temperature is constant.

To examine the variation of total pressure across a normal shock wave, write Eq. (7.25) between the imaginary states 1a and 2a:

$$s_{2a} - s_{1a} = c_p \ln \frac{T_{2a}}{T_{1a}} - R \ln \frac{p_{2a}}{p_{1a}} \tag{8.70}$$

However, from the above discussion, as well as the sketch in Fig. 8.6, we have $s_{2a} = s_2$, $s_{1a} = s_1$, $T_{2a} = T_{0,2}$, $T_{1a} = T_{0,1}$, $p_{2a} = p_{0,2}$, and $p_{1a} = p_{0,1}$. Thus, Eq. (8.70) becomes

$$s_2 - s_1 = c_p \ln \frac{T_{0,2}}{T_{0,1}} - R \ln \frac{p_{0,2}}{p_{0,1}} \tag{8.71}$$

We have already shown that $T_{0,2} = T_{0,1}$; hence, Eq. (8.71) yields

$$\boxed{s_2 - s_1 = -R \ln \frac{p_{0,2}}{p_{0,1}}} \tag{8.72}$$

or

$$\boxed{\frac{p_{0,2}}{p_{0,1}} = e^{-(s_2 - s_1)/R}} \tag{8.73}$$

From Eq. (8.68), we know that $s_2 - s_1 > 0$ for a normal shock wave. Hence, Eq. (8.73) states that $p_{0,2} < p_{0,1}$. *The total pressure decreases across a shock wave.* Moreover, since $s_2 - s_1$ is a function of M_1 only [from Eq. (8.68)], then Eq. (8.73) clearly states that the total pressure ratio $p_{0,2}/p_{0,1}$ across a normal shock wave is a function of M_1 only.

In summary, we have now verified the qualitative changes across a normal shock wave as sketched in Fig. 7.4b and as originally discussed in Sec. 7.6. Moreover, we have obtained closed-form analytic expressions for these changes in the case of a calorically perfect gas. We have seen that p_2/p_1, ρ_2/ρ_1, T_2/T_1, M_2, and $p_{0,2}/p_{0,1}$ are functions of the upstream Mach number M_1 only. To help you obtain a stronger physical feeling of normal shock-wave properties, these variables are plotted in Fig. 8.7 as a function of M_1. Note that (as stated earlier) these curves show how, as M_1 becomes very large, T_2/T_1 and p_2/p_1 also become very large, whereas ρ_2/ρ_1 and M_2 approach finite limits. Examine Fig. 8.7 carefully, and become comfortable with the trends shown.

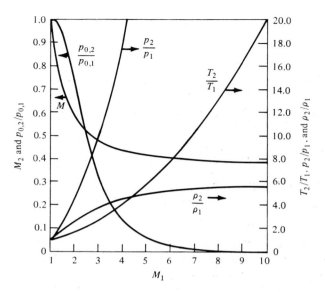

FIGURE 8.7
The variation of properties across a normal shock wave as a function of upstream Mach number: $\gamma = 1.4$.

The results given by Eqs. (8.59), (8.61), (8.65), (8.67), and (8.73) are so important that they are tabulated as a function of M_1 in App. B for $\gamma = 1.4$.

Example 8.2. Consider a normal shock wave in air where the upstream flow properties are $u_1 = 680$ m/s, $T_1 = 288$ K, and $p_1 = 1$ atm. Calculate the velocity, temperature, and pressure downstream of the shock.

Solution

$$a_1 = \sqrt{\gamma R T_1} = \sqrt{1.4(287)(288)} = 340 \text{ m/s}$$

$$M_1 = \frac{u_1}{a_1} = \frac{680}{340} = 2$$

From App. B, since $p_2/p_1 = 4.5$, $T_2/T_1 = 1.687$, $M_2 = 0.5774$, then

$$p_2 = \frac{p_2}{p_1} p_1 = 4.5(1 \text{ atm}) = \boxed{4.5 \text{ atm}}$$

$$T_2 = \frac{T_2}{T_1} T_1 = 1.687(288) = \boxed{486 \text{ K}}$$

$$a_2 = \sqrt{\gamma R T_2} = \sqrt{1.4(287)(486)} = 442 \text{ m/s}$$

$$u_2 = M_2 a_2 = 0.5774(486) = \boxed{255 \text{ m/s}}$$

8.7 MEASUREMENT OF VELOCITY IN A COMPRESSIBLE FLOW

The use of a Pitot tube for measuring the velocity of a low-speed, incompressible flow was discussed in Sec. 3.4. Before progressing further, return to Sec. 3.4, and review the principal aspects of a Pitot tube, as well as the formulas used to obtain the flow velocity from the Pitot pressure, assuming incompressible flow.

For low-speed, incompressible flow, we saw in Sec. 3.4 that the velocity can be obtained from a knowledge of both the total pressure and the static pressure at a point. The total pressure is measured by a Pitot tube, and the static pressure is obtained from a static pressure orifice or by some independent means. The important aspect of Sec. 3.4 is that the pressure sensed by a Pitot tube, along with the static pressure, is all that is necessary to extract the flow velocity for an incompressible flow. In the present section, we see that the same is true for a compressible flow, both subsonic and supersonic, if we consider the Mach number rather than the velocity. In both subsonic and supersonic compressible flows, a knowledge of the Pitot pressure and the static pressure is sufficient to calculate Mach number, although the formulas are different for each Mach-number regime. Let us examine this matter further.

8.7.1 Subsonic Compressible Flow

Consider a Pitot tube in a subsonic, compressible flow, as sketched in Fig. 8.8a. As usual, the mouth of the Pitot tube (point b) is a stagnation region. Hence, a fluid element moving along streamline ab is brought to rest isentropically at point b. In turn, the pressure sensed at point b is the total pressure of the freestream, $p_{0,1}$. This is the Pitot pressure read at the end of the tube. If, in addition, we know the freestream static pressure p_1, then the Mach number in region 1 can be obtained from Eq. (8.42),

$$\frac{p_{0,1}}{p_1} = \left(1 + \frac{\gamma-1}{2} M_1^2\right)^{\gamma/(\gamma-1)} \tag{8.42}$$

or solving for M_1^2,

$$M_1^2 = \frac{2}{\gamma-1}\left[\left(\frac{p_{0,1}}{p_1}\right)^{(\gamma-1)/\gamma} - 1\right] \tag{8.74}$$

Clearly, from Eq. (8.74), the Pitot pressure $p_{0,1}$ and the static pressure p_1 allow the direct calculation of Mach number.

The flow velocity can be obtained from Eq. (8.74) by recalling that $M_1 = u_1/a_1$. Hence,

$$u_1^2 = \frac{2a_1^2}{\gamma-1}\left[\left(\frac{p_{0,1}}{p_1}\right)^{(\gamma-1)/\gamma} - 1\right] \tag{8.75}$$

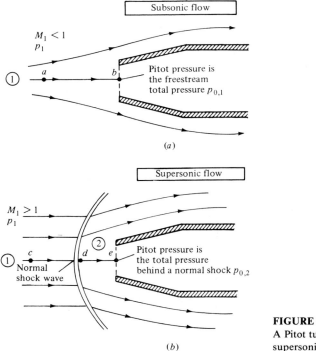

FIGURE 8.8
A Pitot tube in (*a*) subsonic flow and (*b*) supersonic flow.

From Eq. (8.75), we see that, unlike incompressible flow, a knowledge of $p_{0,1}$ and p_1 is not sufficient to obtain u_1; we also need the freestream speed of sound, a_1.

8.7.2 Supersonic Flow

Consider a Pitot tube in a supersonic freestream, as sketched in Fig. 8.8*b*. As usual, the mouth of the Pitot tube (point *e*) is a stagnation region. Hence, a fluid element moving along streamline *cde* is brought to rest at point *e*. However, because the freestream is supersonic and the Pitot tube presents an obstruction to the flow, there is a strong bow shock wave in front of the tube, much like the picture shown at the left of Fig. 8.1 for supersonic flow over a blunt body. Hence, streamline *cde* crosses the normal portion of the bow shock. A fluid element moving along streamline *cde* will first be decelerated *nonisentropically* to a subsonic velocity at point *d* just behind the shock. Then it is isentropically compressed to zero velocity at point *e*. As a result, the pressure at point *e* is *not* the total pressure of the freestream but rather the total pressure *behind a normal shock wave*, $p_{0,2}$. This is the Pitot pressure read at the end of the tube. Keep in mind that because of the entropy increase across the shock, there is a loss in total pressure across the shock, $p_{0,2} < p_{0,1}$. However, knowing $p_{0,2}$ and the free-stream static pressure p_1 is still sufficient to calculate the freestream Mach number M_1, as follows:

$$\frac{p_{0,2}}{p_1} = \frac{p_{0,2}}{p_2} \frac{p_2}{p_1} \tag{8.76}$$

Here, $p_{0,2}/p_2$ is the ratio of total pressure to static pressure in region 2 immediately behind the normal shock, and p_2/p_1 is the static pressure ratio across the shock. From Eq. (8.42),

$$\frac{p_{0,2}}{p_2} = \left(1 + \frac{\gamma-1}{2} M_2^2\right)^{\gamma/(\gamma-1)} \tag{8.77}$$

where, from Eq. (8.59),

$$M_2^2 = \frac{1 + [(\gamma-1)/2]M_1^2}{\gamma M_1^2 - (\gamma-1)/2} \tag{8.78}$$

Also, from Eq. (8.65),

$$\frac{p_2}{p_1} = 1 + \frac{2\gamma}{\gamma+1}(M_1^2 - 1) \tag{8.79}$$

Substituting Eq. (8.78) into (8.77), and substituting the result as well as Eq. (8.79) into Eq. (8.76), we obtain, after some algebraic simplification (see Prob. 8.14),

$$\boxed{\frac{p_{0,2}}{p_1} = \left(\frac{(\gamma+1)^2 M_1^2}{4\gamma M_1^2 - 2(\gamma-1)}\right)^{\gamma/(\gamma-1)} \frac{1 - \gamma + 2\gamma M_1^2}{\gamma+1}} \tag{8.80}$$

Equation (8.80) is called the *Rayleigh Pitot tube formula*. It relates the Pitot pressure $p_{0,2}$ and the freestream static pressure p_1 to the freestream Mach number M_1. Equation (8.80) gives M_1 as an implicit function of $p_{0,2}/p_1$ and allows the calculation of M_1 from a known $p_{0,2}/p_1$. For convenience in making calculations, the ratio $p_{0,2}/p_1$ is tabulated versus M_1 in App. B.

Example 8.3. A Pitot tube is inserted into an airflow where the static pressure is 1 atm. Calculate the flow Mach number when the Pitot tube measures (a) 1.276 atm, (b) 2.714 atm, (c) 12.06 atm.

Solution. First, we must assess whether the flow is subsonic or supersonic. At Mach 1, the Pitot tube would measure $p_0 = p/0.528 = 1.893p$. Hence, when $p_0 < 1.893$ atm, the flow is subsonic, and when $p_0 > 1.893$ atm, the flow is supersonic.

(a) Pitot tube measurement = 1.276 atm. The flow is subsonic. Hence, the Pitot tube is directly sensing the total pressure of the flow. From App. A, for $p_0/p = 1.276$, $\boxed{M = 0.6}$.

(b) Pitot tube measurement = 2.714 atm. The flow is supersonic. Hence, the Pitot tube is sensing the total pressure behind a normal shock wave. From App. B, for $p_{0,2}/p_1 = 2.714$, $\boxed{M_1 = 1.3}$.

(c) Pitot tube measurement = 12.06 atm. The flow is supersonic. From App. B, for $p_{0,2}/p_1 = 12.06$, $\boxed{M_1 = 3.0}$.

8.8 SUMMARY

Return to the road map given in Fig. 8.2, and make certain that you are comfortable with the areas we have covered in this chapter. A brief summary of the more important relations is given below.

The speed of sound in a gas is given by

$$a = \sqrt{\left(\frac{\partial p}{\partial \rho}\right)_s} \tag{8.18}$$

For a calorically perfect gas,

$$a = \sqrt{\frac{\gamma p}{\rho}} \tag{8.23}$$

or

$$a = \sqrt{\gamma R T} \tag{8.25}$$

The speed of sound depends only on the gas temperature.

For a steady, adiabatic, inviscid flow, the energy equation can be expressed as

$$h_1 + \frac{u_1^2}{2} = h_2 + \frac{u_2^2}{2} \tag{8.29}$$

$$c_p T_1 + \frac{u_1^2}{2} = c_p T_2 + \frac{u_2^2}{2} \tag{8.30}$$

$$\frac{a_1^2}{\gamma - 1} + \frac{u_1^2}{2} = \frac{a_2^2}{\gamma - 1} + \frac{u_2^2}{2} \tag{8.32}$$

$$\frac{a^2}{\gamma - 1} + \frac{u^2}{2} = \frac{a_0^2}{\gamma - 1} \tag{8.33}$$

$$\frac{a^2}{\gamma - 1} + \frac{u^2}{2} = \frac{\gamma + 1}{2(\gamma - 1)} a^{*2} \tag{8.35}$$

Total conditions in a flow are related to static conditions via

$$c_p T + \frac{u^2}{2} = c_p T_0 \tag{8.38}$$

$$\frac{T_0}{T} = 1 + \frac{\gamma - 1}{2} M^2 \tag{8.40}$$

$$\frac{p_0}{p} = \left(1 + \frac{\gamma - 1}{2} M^2\right)^{\gamma/(\gamma - 1)} \tag{8.42}$$

$$\frac{\rho_0}{\rho} = \left(1 + \frac{\gamma - 1}{2} M^2\right)^{1/(\gamma - 1)} \tag{8.43}$$

Note that the ratios of total to static properties are a function of local Mach number only. These functions are tabulated in App. A.

The basic normal shock equations are

Continuity: $\qquad\qquad\qquad \rho_1 u_1 = \rho_2 u_2 \tag{8.2}$

Momentum: $\qquad\qquad p_1 + \rho_1 u_1^2 = p_2 + \rho_2 u_2^2 \tag{8.6}$

Energy: $\qquad\qquad h_1 + \frac{u_1^2}{2} = h_2 + \frac{u_2^2}{2} \tag{8.10}$

These equations lead to relations for changes across a normal shock as a function of upstream Mach number M_1 only:

$$M_2^2 = \frac{1 + [(\gamma - 1)/2]M_1^2}{\gamma M_1^2 - (\gamma - 1)/2} \tag{8.59}$$

$$\frac{\rho_2}{\rho_1} = \frac{u_1}{u_2} = \frac{(\gamma + 1)M_1^2}{2 + (\gamma - 1)M_1^2} \tag{8.61}$$

$$\frac{p_2}{p_1} = 1 + \frac{2\gamma}{\gamma + 1}(M_1^2 - 1) \tag{8.65}$$

$$\frac{T_2}{T_1} = \frac{h_2}{h_1} = \left[1 + \frac{2\gamma}{\gamma + 1}(M_1^2 - 1)\right]\frac{2 + (\gamma - 1)M_1^2}{(\gamma + 1)M_1^2} \tag{8.67}$$

$$s_2 - s_1 = c_p \ln\left\{\left[1 + \frac{2\gamma}{\gamma + 1}(M_1^2 - 1)\right]\frac{2 + (\gamma - 1)M_1^2}{(\gamma + 1)M_1^2}\right\}$$

$$- R \ln\left[1 + \frac{2\gamma}{\gamma + 1}(M_1^2 - 1)\right] \tag{8.68}$$

$$\frac{p_{0,2}}{p_{0,1}} = e^{-(s_2 - s_1)/R} \tag{8.73}$$

The normal shock properties are tabulated versus M_1 in App. B.

For a calorically perfect gas, the total temperature is constant across a normal shock wave:

$$T_{0,2} = T_{0,1}$$

However, there is a loss in total pressure across the wave.

$$p_{0,2} < p_{0,1}$$

For subsonic and supersonic compressible flow, the freestream Mach number is determined by the ratio of Pitot pressure to freestream static pressure. However, the equations are different:

Subsonic flow:
$$M_1^2 = \frac{2}{\gamma - 1} \left[\left(\frac{p_{0,1}}{p_1} \right)^{(\gamma-1)/\gamma} - 1 \right] \tag{8.74}$$

Supersonic flow:
$$\frac{p_{0,2}}{p_1} = \left[\frac{(\gamma+1)^2 M_1^2}{4\gamma M_1^2 - 2(\gamma - 1)} \right]^{\gamma/(\gamma-1)} \frac{1 - \gamma + 2\gamma M_1^2}{\gamma + 1} \tag{8.80}$$

PROBLEMS

8.1. Consider air at a temperature of 230 K. Calculate the speed of sound.

8.2. The temperature in the reservoir of a supersonic wind tunnel is 519°R. In the test section, the flow velocity is 1385 ft/s. Calculate the test-section Mach number. Assume the tunnel flow is adiabatic.

8.3. At a given point in a flow, $T = 300$ K, $p = 1.2$ atm, and $V = 250$ m/s. At this point, calculate the corresponding values of p_0, T_0, p^*, T^*, and M^*.

8.4. At a given point in a flow, $T = 700°R$, $p = 1.6$ atm, and $V = 2983$ ft/s. At this point, calculate the corresponding values of p_0, T_0, p^*, T^*, and M^*.

8.5. Consider the isentropic flow through a supersonic nozzle. If the test-section conditions are given by $p = 1$ atm, $T = 230$ K, and $M = 2$, calculate the reservoir pressure and temperature.

8.6. Consider the isentropic flow over an airfoil. The freestream conditions correspond to a standard altitude of 10,000 ft and $M_\infty = 0.82$. At a given point on the airfoil, $M = 1.0$. Calculate p and T at this point. (*Note:* You will have to consult a standard atmosphere table for this problem, such as given in Ref. 2. If you do not have one, you can find such tables in any good technical library.)

8.7. The flow just upstream of a normal shock wave is given by $p_1 = 1$ atm, $T_1 = 288$ K, and $M_1 = 2.6$. Calculate the following properties just downstream of the shock: p_2, T_2, ρ_2, M_2, $p_{0,2}$, $T_{0,2}$, and the change in entropy across the shock.

8.8. The pressure upstream of a normal shock wave is 1 atm. The pressure and temperature downstream of the wave are 10.33 atm and 1390°R, respectively. Calculate the Mach number and temperature upstream of the wave and the total temperature and total pressure downstream of the wave.

8.9. The entropy increase across a normal shock wave is 199.5 J/(kg · K). What is the upstream Mach number?

8.10. The flow just upstream of a normal shock wave is given by $p_1 = 1800$ lb/ft^2, $T_1 = 480°$R, and $M_1 = 3.1$. Calculate the velocity and M^* behind the shock.

8.11. Consider a flow with a pressure and temperature of 1 atm and 288 K. A Pitot tube is inserted into this flow and measures a pressure of 1.555 atm. What is the velocity of the flow?

8.12. Consider a flow with a pressure and temperature of 2116 lb/ft^2 and 519°R, respectively. A Pitot tube is inserted into this flow and measures a pressure of 7712.8 lb/ft^2. What is the velocity of this flow?

8.13. Repeat Probs. 8.11 and 8.12 using (incorrectly) Bernoulli's equation for incompressible flow. Calculate the percent error induced by using Bernoulli's equation.

8.14. Derive the Rayleigh Pitot tube formula, Eq. (8.80).

8.15. On March 16, 1990, an Air Force SR-71 set a new continental speed record, averaging a velocity of 2112 mi/h at an altitude of 80,000 ft. Calculate the temperature (in degrees Fahrenheit) at a stagnation point on the vehicle. (Consult a standard altitude table for any additional information you need.)

8.16. In the test section of a supersonic wind tunnel, a Pitot tube in the flow reads a pressure of 1.13 atm. A static pressure measurement (from a pressure tap on the sidewall of the test section) yields 0.1 atm. Calculate the Mach number of the flow in the test section.

CHAPTER
9

OBLIQUE SHOCK AND EXPANSION WAVES

In the case of air (and the same is true for all gases) the shock wave is extremely thin so that calculations based on one-dimensional flow are still applicable for determining the changes in velocity and density on passing through it, even when the rest of the flow system is not limited to one dimension, provided that only the velocity component normal to the wave is considered.

G. I. Taylor and J. W. Maccoll, 1934

9.1 INTRODUCTION

In Chap. 8, we discussed normal shock waves, i.e., shock waves that make an angle of 90° with the upstream flow. The behavior of normal shock waves is important; moreover, the study of normal shock waves provides a relatively straightforward introduction to shock-wave phenomena. However, examining Fig. 7.4a and the photographs shown in Fig. 7.5, we see that, in general, a shock wave will make an oblique angle with respect to the upstream flow. These are called *oblique shock waves* and are the subject of part of this chapter. A normal shock wave is simply a special case of the general family of oblique shocks, namely, the case where the wave angle is 90°.

In addition to oblique shock waves, where the pressure increases discontinuously across the wave, supersonic flows are also characterized by oblique expansion waves, where the pressure *decreases continuously* across the wave. Let us examine these two types of waves further. Consider a supersonic flow over a wall with a corner at point *A*, as sketched in Fig. 9.1. In Fig. 9.1a, the wall is turned upward at the corner through the deflection angle θ; i.e., the corner is concave. The flow at the wall must be tangent to the wall; hence, the streamline at the wall is also deflected upward through the angle θ. The bulk of the gas is above the wall, and in Fig. 9.1a, the streamlines are turned upward, *into* the main bulk of the flow. Whenever a supersonic flow is "turned into itself" as shown in

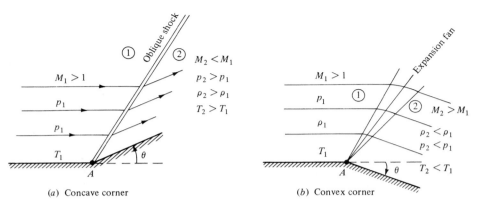

(a) Concave corner   (b) Convex corner

FIGURE 9.1
Supersonic flow over a corner.

Fig. 9.1*a*, an oblique shock wave will occur. The originally horizontal streamlines ahead of the wave are uniformly deflected in crossing the wave, such that the streamlines behind the wave are parallel to each other and inclined upward at the deflection angle θ. Across the wave, the Mach number discontinuously decreases, and the pressure, density, and temperature discontinuously increase. In contrast, Fig. 9.1*b* shows the case where the wall is turned downward at the corner through the deflection angle θ; i.e., the corner is convex. Again, the flow at the wall must be tangent to the wall; hence, the streamline at the wall is deflected downward through the angle θ. The bulk of the gas is above the wall, and in Fig. 9.1*b*, the streamlines are turned downward, *away from* the main bulk of the flow. Whenever a supersonic flow is "turned away from itself" as shown in Fig. 9.1*b*, an expansion wave will occur. This expansion wave is in the shape of a fan centered at the corner. The fan continuously opens in the direction away from the corner, as shown in Fig. 9.1*b*. The originally horizontal streamlines ahead of the expansion wave are deflected smoothly and continuously through the expansion fan such that the streamlines behind the wave are parallel to each other and inclined downward at the deflection angle θ. Across the expansion wave, the Mach number increases, and the pressure, temperature, and density decrease. Hence, an expansion wave is the direct antithesis of a shock wave.

Oblique shock and expansion waves are prevalent in two- and three-dimensional supersonic flows. These waves are inherently two-dimensional in nature, in contrast to the one-dimensional normal shock waves discussed in Chap. 8. That is, in Fig. 9.1*a* and *b*, the flow-field properties are a function of *x* and *y*. The purpose of the present chapter is to determine and study the properties of these oblique waves.

What is the physical mechanism that creates waves in a supersonic flow? To address this question, recall our picture of the propagation of a sound wave via molecular collisions, as portrayed in Sec. 8.3. If a slight disturbance takes place at some point in a gas, information is transmitted to other points in the gas by sound waves which propagate in all directions away from the source of the disturbance. Now consider a body in a flow, as sketched in Fig. 9.2. The gas molecules which impact the body surface experience a change in momentum. In turn, this change is transmitted to neighboring molecules by random molecular collisions. In this fashion, information about the presence of the body attempts to be transmitted to the surrounding flow via molecular collisions; i.e., the information is propagated upstream at approximately the local speed of sound. If the upstream flow is subsonic, as shown in Fig. 9.2a, the disturbances have no problem working their way far upstream, thus giving the incoming flow plenty of time to move out of the way of the body. On the other hand, if the upstream flow is supersonic, as shown in Fig. 9.2b, the disturbances cannot work their way upstream; rather, at some finite distance from the body, the disturbance waves

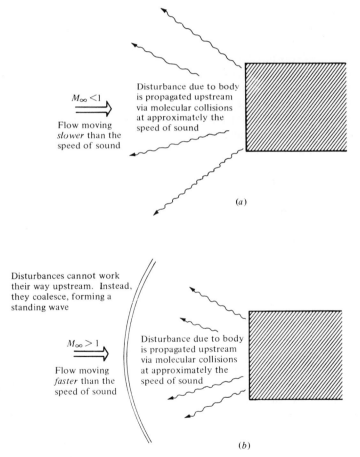

FIGURE 9.2
Propagation of disturbances. (a) Subsonic flow. (b) Supersonic flow.

pile up and coalesce, forming a standing wave in front of the body. Hence, the physical generation of waves in a supersonic flow—both shock and expansion waves—is due to the propagation of information via molecular collisions and due to the fact that such propagation cannot work its way into certain regions of the supersonic flow.

Why are most waves oblique rather than normal to the upstream flow? To answer this question, consider a small source of disturbance moving through a stagnant gas. For lack of anything better, let us call this disturbance source a "beeper," which periodically emits sound. First, consider the beeper moving at *subsonic* speed through the gas, as shown in Fig. 9.3a. The speed of the beeper is V, where $V < a$. At time $t = 0$, the beeper is located at point A; at this point, it emits a sound wave which propagates in all directions at the speed of sound, a. At a later time, t, this sound wave has propagated a distance at from point A and is represented by the circle of radius at shown in Fig. 9.3a. During the same time, the beeper has moved a distance Vt and is now at point B in Fig. 9.3a. Moreover, during its transit from A to B, the beeper has emitted several other sound waves, which at time t are represented by the smaller circles in Fig. 9.3a. Note that the beeper always stays *inside* the family of circular sound waves and that the waves continuously move ahead of the beeper. This is because the beeper is traveling at a subsonic speed, $V < a$. In contrast, consider the beeper moving at a *supersonic* speed, $V > a$, through the gas, as shown in Fig. 9.3b. At time $t = 0$, the beeper is located at point A, where it emits a sound wave. At a later time, t, this sound wave has propagated a distance at from point A and is represented by the circle of radius at shown in Fig. 9.3b. During the same time, the beeper has moved a distance Vt to point B. Moreover, during its transit from A to B, the beeper has emitted several other sound waves, which at time t are represented by the smaller circles in Fig. 9.3b. However, in contrast to the subsonic case, the beeper is now constantly *outside* the family of circular sound waves; i.e., it is moving ahead of the wave fronts because $V > a$. Moreover, something new is

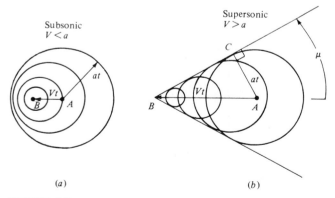

FIGURE 9.3
Another way of visualizing the propagation of disturbances in (a) subsonic and (b) supersonic flow.

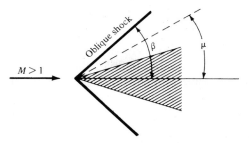

FIGURE 9.4
Relation between the oblique shock-wave angle and the Mach angle.

happening; these wave fronts form a disturbance envelope given by the straight line *BC*, which is tangent to the family of circles. This line of disturbances is defined as a *Mach wave*. In addition, the angle *ABC* which the Mach wave makes with respect to the direction of motion of the beeper is defined as the *Mach angle* μ. From the geometry of Fig. 9.3*b*, we readily find that

$$\sin \mu = \frac{at}{Vt} = \frac{a}{V} = \frac{1}{M}$$

Thus, the Mach angle is simply determined by the local Mach number as

$$\mu = \sin^{-1} \frac{1}{M} \tag{9.1}$$

Examining Fig. 9.3*b*, the Mach wave, i.e., the envelope of disturbances in the

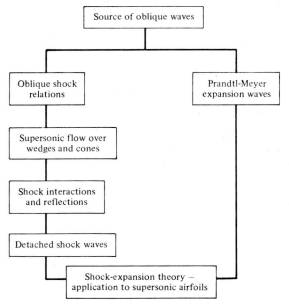

FIGURE 9.5
Road map for Chap. 9.

supersonic flow, is clearly *oblique* to the direction of motion. If the disturbances are stronger than a simple sound wave, then the wave front becomes stronger than a Mach wave, creating an oblique shock wave at an angle β to the freestream, where $\beta > \mu$. This comparison is shown in Fig. 9.4. However, the physical mechanism creating the oblique shock is essentially the same as that described above for the Mach wave. Indeed, a Mach wave is a limiting case for oblique shock; i.e., it is an infinitely weak oblique shock.

This finishes our discussion of the physical source of oblique waves in a supersonic flow. Let us now proceed to develop the equations which allow us to calculate the change in properties across these oblique waves, first for oblique shock waves, and then for expansion waves. In the process, we follow the road map given in Fig. 9.5.

9.2   OBLIQUE SHOCK RELATIONS

Consider the oblique shock wave sketched in Fig. 9.6. The angle between the shock wave and the upstream flow direction is defined as the *wave angle*, denoted by β. The upstream flow (region 1) is horizontal, with a velocity V_1 and Mach number M_1. The downstream flow (region 2) is inclined upward through the deflection angle θ and has velocity V_2 and Mach number M_2. The upstream velocity V_1 is split into components tangential and normal to the shock wave, w_1 and u_1, respectively, with the associated tangential and normal Mach numbers $M_{t,1}$ and $M_{n,1}$, respectively. Similarly, the downstream velocity is split into tangential and normal components w_2 and u_2, respectively, with the associated Mach numbers $M_{t,2}$ and $M_{n,2}$.

Consider the control volume shown by the dashed lines in the upper part of Fig. 9.6. Sides a and d are parallel to the shock wave. Segments b and c follow

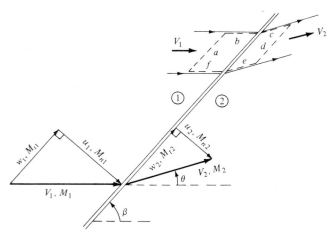

FIGURE 9.6
Oblique shock geometry.

the upper streamline, and segments e and f follow the lower streamline. Let us apply the integral form of the conservation equations to this control volume, keeping in mind that we are dealing with a steady, inviscid, adiabatic flow with no body forces. For these assumptions, the continuity equation, Eq. (2.39), becomes

$$\oiint_S \rho \mathbf{V} \cdot \mathbf{dS} = 0$$

This surface integral evaluated over faces a and d yields $-\rho_1 u_1 A_1 + \rho_2 u_2 A_2$, where $A_1 = A_2 =$ area of faces a and d. The faces b, c, e, and f are parallel to the velocity, and hence contribute nothing to the surface integral (i.e., $\mathbf{V} \cdot \mathbf{dS} = 0$ for these faces). Thus, the continuity equation for an oblique shock wave is

$$-\rho_1 u_1 A_1 + \rho_2 u_2 A_2 = 0$$

or

$$\boxed{\rho_1 u_1 = \rho_2 u_2} \qquad (9.2)$$

Keep in mind that u_1 and u_2 in Eq. (9.2) are *normal* to the shock wave.

The integral form of the momentum equation, Eq. (2.55), is a vector equation. Hence, it can be resolved into two components, tangential and normal to the shock wave. First, consider the tangential component, keeping in mind the type of flow we are considering:

$$\oiint_S (\rho \mathbf{V} \cdot \mathbf{dS}) w = -\oiint_S (p \, dS)_{\text{tangential}} \qquad (9.3)$$

In Eq. (9.3), w is the component of velocity tangential to the wave. Since $\mathbf{dS}$ is perpendicular to the control surface, then $(p \, dS)_{\text{tangential}}$ over faces a and d is zero. Also, since the vectors $p \, \mathbf{dS}$ on faces b and f are equal and opposite, the pressure integral in Eq. (9.3) involves two tangential forces that cancel each other over faces b and f. The same is true for faces c and e. Hence, Eq. (9.3) becomes

$$-(\rho_1 u_1 A_1) w_1 + (\rho_2 u_2 A_2) w_2 = 0 \qquad (9.4)$$

Dividing Eq. (9.4) by Eq. (9.2), we have

$$\boxed{w_1 = w_2} \qquad (9.5)$$

Equation (9.5) is an important result; it states that the *tangential component of the flow velocity is constant across an oblique shock.*

The normal component of the integral momentum equation is, from Eq. (2.55),

$$\oiint_S (\rho \mathbf{V} \cdot \mathbf{dS})u = -\oiint_S (p \, dS)_{\text{normal}} \tag{9.6}$$

Here, the pressure integral evaluated over faces a and d yields the net sum $-p_1 A_1 + p_2 A_2$. Once again, the equal and opposite pressure forces on b and f cancel, as do those on c and e. Hence, Eq. (9.6) becomes, for the control volume shown in Fig. 9.6,

$$-(\rho_1 u_1 A_1)u_1 + (\rho_2 u_2 A_2)u_2 = -(-p_1 A_1 + p_2 A_2)$$

Since $A_1 = A_2$, this becomes

$$\boxed{p_1 + \rho_1 u_1^2 = p_2 + \rho_2 u_2^2} \tag{9.7}$$

Again, note that the only velocities appearing in Eq. (9.7) are the components *normal* to the shock.

Finally, consider the integral form of the energy equation, Eq. (2.86). For our present case, this can be written as

$$\oiint_S \rho\left(e + \frac{V^2}{2}\right)\mathbf{V} \cdot \mathbf{dS} = -\oiint_S p\mathbf{V} \cdot \mathbf{dS} \tag{9.8}$$

Again noting that the flow is tangent to faces b, c, f, and e, and hence $\mathbf{V} \cdot \mathbf{dS} = 0$ on these faces, Eq. (9.8) becomes, for the control volume in Fig. 9.6,

$$-\rho_1\left(e_1 + \frac{V_1^2}{2}\right)u_1 A_1 + \rho_2\left(e_2 + \frac{V_2^2}{2}\right)u_2 A_2 = -(-p_1 u_1 A_1 + p_2 u_2 A_2) \tag{9.9}$$

Collecting terms in Eq. (9.9), we have

$$-\rho_1 u_1\left(e_1 + \frac{p_1}{\rho_1} + \frac{V_1^2}{2}\right) + \rho_2 u_2\left(e_2 + \frac{p_2}{\rho_2} + \frac{V_2^2}{2}\right) = 0$$

or

$$\rho_1 u_1\left(h_1 + \frac{V_1^2}{2}\right) = \rho_2 u_2\left(h_2 + \frac{V_2^2}{2}\right) \tag{9.10}$$

Dividing Eq. (9.10) by (9.2), we have

$$h_1 + \frac{V_1^2}{2} = h_2 + \frac{V_2^2}{2} \tag{9.11}$$

Since $h + V^2/2 = h_0$, we have again the familiar result that the *total enthalpy is constant across the shock wave.* Moreover, for a calorically perfect gas, $h_0 = c_p T_0$; hence, the *total temperature is constant across the shock wave.* Carrying Eq. (9.11) a bit further, note from Fig. 9.6 that $V^2 = u^2 + w^2$. Also, from Eq. (9.5), we know that $w_1 = w_2$. Hence,

$$V_1^2 - V_2^2 = (u_1^2 + w_1^2) - (u_2^2 + w_2^2) = u_1^2 - u_2^2$$

Thus, Eq. (9.11) becomes

$$h_1 + \frac{u_1^2}{2} = h_2 + \frac{u_2^2}{2}$$

(9.12)

Let us now gather our results. Look carefully at Eqs. (9.2), (9.7), and (9.12). They are the continuity, normal momentum, and energy equations, respectively, for an oblique shock wave. Note that they involve the *normal components only* of velocity, u_1 and u_2; the tangential component w does not appear in these equations. Hence, we deduce that *changes across an oblique shock wave are governed only by the component of velocity normal to the wave.*

Again, look hard at Eqs. (9.2), (9.7), and (9.12). *They are precisely the governing equations for a normal shock wave*, as given by Eqs. (8.2), (8.6), and (8.10). Hence, precisely the same algebra as applied to the normal shock equations in Sec. 8.6, when applied to Eqs. (9.2), (9.7), and (9.12), will lead to identical expressions for changes across an oblique shock in terms of the normal component of the upstream Mach number $M_{n,1}$. Note that

$$M_{n,1} = M_1 \sin \beta$$

(9.13)

Hence, for an oblique shock wave, with $M_{n,1}$ given by Eq. (9.13), we have, from Eqs. (8.59), (8.61), and (8.65),

$$M_{n,2}^2 = \frac{1 + [(\gamma - 1)/2] M_{n,1}^2}{\gamma M_{n,1}^2 - (\gamma - 1)/2}$$

(9.14)

$$\frac{\rho_2}{\rho_1} = \frac{(\gamma + 1) M_{n,1}^2}{2 + (\gamma - 1) M_{n,1}^2}$$

(9.15)

$$\frac{p_2}{p_1} = 1 + \frac{2\gamma}{\gamma + 1} (M_{n,1}^2 - 1)$$

(9.16)

The temperature ratio T_2/T_1 follows from the equation of state:

$$\frac{T_2}{T_1} = \frac{p_2}{p_1} \frac{\rho_1}{\rho_2}$$

(9.17)

Note that $M_{n,2}$ is the *normal* Mach number behind the shock wave. The downstream Mach number itself, M_2, can be found from $M_{n,2}$ and the geometry of Fig. 9.6 as

$$M_2 = \frac{M_{n,2}}{\sin (\beta - \theta)}$$

(9.18)

Examine Eqs. (9.14) to (9.17). They state that oblique shock-wave properties in a calorically perfect gas depend only on the normal component of the upstream Mach number $M_{n,1}$. However, note from Eq. (9.13) that $M_{n,1}$ depends on both M_1 and β. Recall from Sec. 8.6 that changes across a normal shock wave depend on one parameter only—the upstream Mach number M_1. In contrast, we now see that changes across an oblique shock wave depend on two parameters—say, M_1 and β. However, this distinction is slightly moot because in reality a normal shock wave is a special case of oblique shocks where $\beta = \pi/2$.

Equation (9.18) introduces the deflection angle θ into our oblique shock analysis; we need θ to be able to calculate M_2. However, θ is not an independent, third parameter; rather, θ is a function of M_1 and β, as derived below. From the geometry of Fig. 9.6,

$$\tan \beta = \frac{u_1}{w_1} \tag{9.19}$$

and

$$\tan (\beta - \theta) = \frac{u_2}{w_2} \tag{9.20}$$

Dividing Eq. (9.20) by (9.19), recalling that $w_1 = w_2$, and invoking the continuity equation, Eq. (9.2), we obtain

$$\frac{\tan (\beta - \theta)}{\tan \beta} = \frac{u_2}{u_1} = \frac{\rho_1}{\rho_2} \tag{9.21}$$

Combining Eq. (9.21) with Eqs. (9.13) and (9.15), we obtain

$$\frac{\tan (\beta - \theta)}{\tan \beta} = \frac{2 + (\gamma - 1) M_1^2 \sin^2 \beta}{(\gamma + 1) M_1^2 \sin^2 \beta} \tag{9.22}$$

which gives θ as an implicit function of M_1 and β. After some trigonometric substitutions and rearrangement, Eq. (9.22) can be cast explicitly for θ as

$$\boxed{\tan \theta = 2 \cot \beta \frac{M_1^2 \sin^2 \beta - 1}{M_1^2(\gamma + \cos 2\beta) + 2}} \tag{9.23}$$

Equation (9.23) is an important equation. It is called the θ-β-M relation, and it specifies θ as a unique function of M_1 and β. This relation is vital to the analysis of oblique shock waves, and results from it are plotted in Fig. 9.7 for $\gamma = 1.4$. Examine this figure closely. It is a plot of wave angle versus deflection angle, with the Mach number as a parameter. The results given in Fig. 9.7 are plotted in some detail—this is a chart which you will need to use for solving oblique shock problems.

Figure 9.7 illustrates a wealth of physical phenomena associated with oblique shock waves. For example:

1. For any given upstream Mach number M_1, there is a maximum deflection angle, θ_{max}. If the physical geometry is such that $\theta > \theta_{max}$, then no solution exists for a *straight* oblique shock wave. Instead, nature establishes a curved shock wave, detached from the corner or the nose of a body. This is illustrated in Fig. 9.8. Here, the left side of the figure illustrates flow over a wedge and

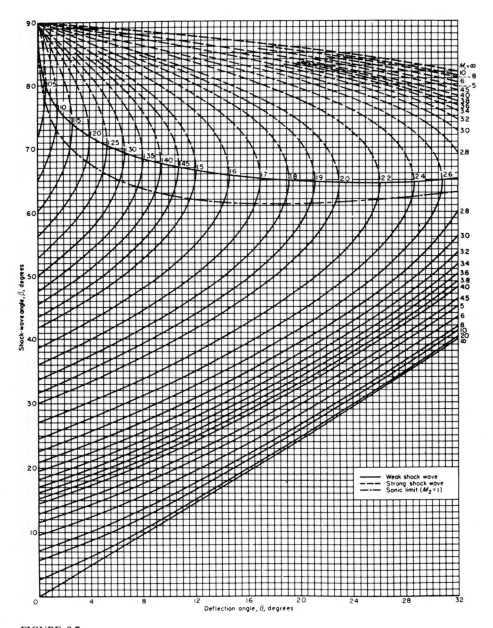

FIGURE 9.7

Oblique shock properties: $\gamma = 1.4$. The θ-β-M diagram. (*From NACA Report 1135, Ames Research Staff, "Equations, Tables and Charts for Compressible Flow," 1953.*)

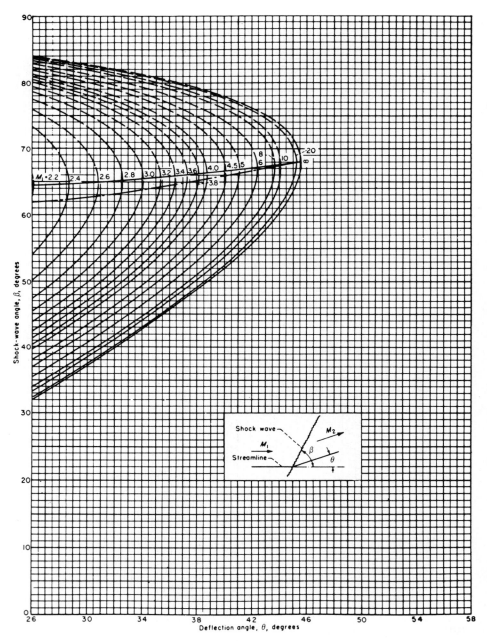

FIGURE 9.7
(Continued.)

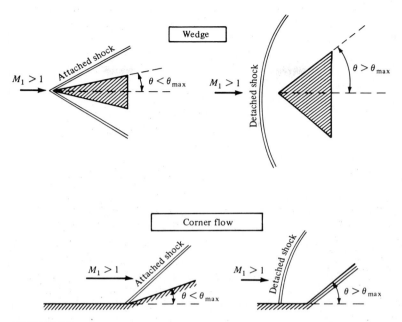

FIGURE 9.8
Attached and detached shocks.

a concave corner where the deflection angle is less than θ_{max} for the given upstream Mach number. Therefore, we see a straight oblique shock wave attached to the nose of the wedge and to the corner. The right side of Fig. 9.8 gives the case where the deflection angle is greater than θ_{max}; hence, there is no allowable straight oblique shock solution from the theory developed earlier in this section. Instead, we have a curved shock wave detached from the nose of the wedge or from the corner. Return to Fig. 9.7, and note that the value of θ_{max} increases with increasing M_1. Hence, at higher Mach numbers, the straight oblique shock solution can exist at higher deflection angles. However, there is a limit; as M_1 approaches infinity, θ_{max} approaches 45.5° (for $\gamma = 1.4$).

2. For any given θ *less than* θ_{max}, there are *two* straight oblique shock solutions for a given upstream Mach number. For example, if $M_1 = 2.0$ and $\theta = 15°$, then from Fig. 9.7, β can equal either 45.3 or 79.8°. The smaller value of β is called the *weak* shock solution, and the larger value of β is the *strong* shock solution. These two cases are illustrated in Fig. 9.9. The classifications "weak" and "strong" derive from the fact that for a given M_1, the larger the wave angle, the larger the normal component of upstream Mach number, $M_{n,1}$, and from Eq. (9.16) the larger the pressure ratio p_2/p_1. Thus, in Fig. 9.9, the higher-angle shock wave will compress the gas more than the lower-angle shock wave, hence the terms "strong" and "weak" solutions. In nature, the weak shock solution usually prevails. Whenever you see straight, attached oblique shock waves, such as sketched at the left of Fig. 9.8, they are almost

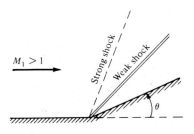

FIGURE 9.9
The weak and strong shock cases.

always the weak shock solution. It is safe to make this assumption, unless you have specific information to the contrary. Note in Fig. 9.7 that the locus of points connecting all the values of θ_{max} (the curve that sweeps approximately horizontally across the middle of Fig. 9.7) divides the weak and strong shock solutions. Above this curve, the strong shock solution prevails (as further indicated by the θ-β-M curves being dashed); below this curve, the weak shock solution prevails (where the θ-β-M curves are shown as solid lines). Note that slightly below this curve is another curve which also sweeps approximately horizontally across Fig. 9.7. This curve is the dividing line above which $M_2 < 1$ and below which $M_2 > 1$. For the strong shock solution, the downstream Mach number is always subsonic, $M_2 < 1$. For the weak shock solution very near θ_{max}, the downstream Mach number is also subsonic, but barely so. For the vast majority of cases involving the weak shock solution, the downstream Mach number is supersonic, $M_2 > 1$. Since the weak shock solution is almost always the case encountered in nature, we can readily state that the Mach number downstream of a straight, attached oblique shock is almost always supersonic.

3. If $\theta = 0$, then β equals either 90° or μ. The case of $\beta = 90°$ corresponds to a normal shock wave (i.e., the normal shocks discussed in Chap. 8 belong to the family of strong shock solutions). The case of $\beta = \mu$ corresponds to the Mach wave illustrated in Fig. 9.3b. In both cases, the flow streamlines experience no deflection across the wave.

4. (In all of the following discussions, we consider the weak shock solution exclusively, unless otherwise noted.) Consider an experiment where we have supersonic flow over a wedge of given semiangle θ, as sketched in Fig. 9.10. Now assume that we increase the freestream Mach number M_1. As M_1 increases, we observe that β decreases. For example, consider $\theta = 20°$ and $M_1 = 2.0$, as shown on the left of Fig. 9.10. From Fig. 9.7, we find that $\beta = 53.3°$. Now assume M_1 is increased to 5, keeping θ constant at 20°, as sketched on the right of Fig. 9.10. Here, we find that $\beta = 29.9°$. Interestingly enough, although this shock is at a lower wave angle, it is a stronger shock than the one on the left. This is because $M_{n,1}$ is larger for the case on the right. Although β is smaller, which decreases $M_{n,1}$, the upstream Mach number M_1 is larger, which increases $M_{n,1}$ by an amount which more than compensates for the decreased β. For example, note the values of $M_{n,1}$ and p_2/p_1 given in Fig.

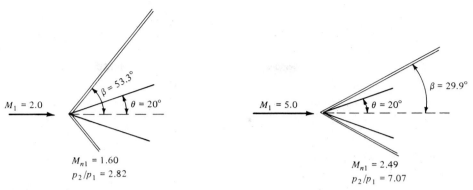

FIGURE 9.10
Effects of increasing the upstream Mach number.

9.10. Clearly, the Mach 5 case on the right yields the stronger shock wave. Hence, in general for attached shocks with a fixed deflection angle, as the upstream Mach number M_1 increases, the wave angle β decreases, and the shock wave becomes stronger. Going in the other direction, as M_1 decreases, the wave angle increases, and the shock wave becomes weaker. Finally, if M_1 is decreased enough, the shock wave will become detached. For the case of $\theta = 20°$ shown in Fig. 9.10, the shock will be detached for $M_1 < 1.84$.

5. Consider another experiment. Here, let us keep M_1 fixed and increase the deflection angle. For example, consider the supersonic flow over a wedge shown in Fig. 9.11. Assume that we have $M_1 = 2.0$ and $\theta = 10°$, as sketched at the left of Fig. 9.11. The wave angle will be 39.2° (from Fig. 9.7). Now assume that the wedge is hinged so that we can increase its deflection angle, keeping M_1 constant. In such a case, the wave angle will increase, as shown on the right of Fig. 9.11. Also, $M_{n,1}$ will increase, and hence the shock will become stronger. Therefore, in general for attached shocks with a fixed upstream Mach number, as the deflection angle increases, the wave angle β increases, and the

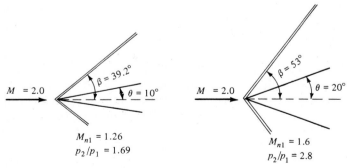

FIGURE 9.11
Effect of increasing the deflection angle.

shock becomes stronger. However, once θ exceeds θ_{max}, the shock wave will become detached. For the case of $M_1 = 2.0$ in Fig. 9.11, this will occur when $\theta > 23°$.

The physical properties of oblique shocks discussed above are very important. Before proceeding further, make certain to go over this discussion several times until you feel perfectly comfortable with these physical variations.

Example 9.1. Consider a supersonic flow with $M = 2$, $p = 1$ atm, and $T = 288$ K. This flow is deflected at a compression corner through 20°. Calculate M, p, T, p_0, and T_0 behind the resulting oblique shock wave.

Solution. From Fig. 9.7, for $M_1 = 2$ and $\theta = 20°$, $\beta = 53.4°$. Hence, $M_{n,1} = M_1 \sin \beta = 2 \sin 53.4° = 1.606$. From App. B, for $M_{n,1} = 1.60$ (rounded to the nearest table entry),

$$M_{n,2} = 0.6684 \qquad \frac{p_2}{p_1} = 2.82 \qquad \frac{T_2}{T_1} = 1.388 \qquad \frac{p_{0,2}}{p_{0,1}} = 0.8952$$

Hence,

$$M_2 = \frac{M_{n,2}}{\sin(\beta - \theta)} = \frac{0.6684}{\sin(53.4 - 20)} = \boxed{1.21}$$

$$p_2 = \frac{p_2}{p_1} p_1 = 2.82(1 \text{ atm}) = \boxed{2.82 \text{ atm}}$$

$$T_2 = \frac{T_2}{T_1} T_1 = 1.388(288) = \boxed{399.7 \text{ K}}$$

For $M_1 = 2$, from App. A, $p_{0,1}/p_1 = 7.824$ and $T_{0,1}/T_1 = 1.8$; thus,

$$p_{0,2} = \frac{p_{0,2}}{p_{0,1}} \frac{p_{0,1}}{p_1} p_1 = 0.8952(7.824)(1 \text{ atm}) = \boxed{7.00 \text{ atm}}$$

The total temperature is constant across the shock. Hence,

$$T_{0,2} = T_{0,1} = \frac{T_{0,1}}{T_1} T_1 = 1.8(288) = \boxed{518.4 \text{ K}}$$

Note: For oblique shocks, the entry for $p_{0,2}/p_1$ in App. B *cannot* be used to obtain $p_{0,2}$; this entry in App. B is for normal shocks only and is obtained directly from Eq. (8.80). In turn, Eq. (8.80) is derived using (8.77), where M_2 is the *actual* flow Mach number, not the normal component. Only in the case of a normal shock is this also the Mach number normal to the wave. Hence, Eq. (8.80) holds only for normal shocks; it *cannot* be used for oblique shocks with M_1 replaced by $M_{n,1}$. For example, an *incorrect* calculation would be to use $p_{0,2}/p_1 = 3.805$ for $M_{n,1} = 1.60$. This gives $p_{0,2} = 3.805$ atm, a totally incorrect result compared with the correct value of 7.00 atm obtained above.

Example 9.2. Consider an oblique shock wave with a wave angle of 30°. The upstream flow Mach number is 2.4. Calculate the deflection angle of the flow, the pressure and temperature ratios across the shock wave, and the Mach number behind the wave.

Solution. From Fig. 9.7, for $M_1 = 2.4$ and $\beta = 30°$, we have $\boxed{\theta = 6.5°}$. Also,

$$M_{n,1} = M_1 \sin \beta = 2.4 \sin 30° = 1.2$$

From App. B,

$$\frac{p_2}{p_1} = \boxed{1.513}$$

$$\frac{T_2}{T_1} = \boxed{1.128}$$

$$M_{n,2} = 0.8422$$

Thus,

$$M_2 = \frac{M_{n,2}}{\sin (\beta - \theta)} = \frac{0.8422}{\sin (30 - 6.5)} = \boxed{2.11}$$

Note: Two aspects are illustrated by this example:

1. This is a fairly weak shock wave—only a 51 percent increase in pressure across the wave. Indeed, examining Fig. 9.7, we find that this case is close to that of a Mach wave, where $\mu = \sin^{-1} (1/M) = \sin^{-1} (\frac{1}{2.4}) = 24.6°$. The shock-wave angle of 30° is not much larger than μ; the deflection angle of 6.5° is also small—consistent with the relative weakness of the shock wave.

2. Only two properties need to be specified in order to define uniquely a given oblique shock wave. In this example, M_1 and β were those two properties. In Example 9.1, the specified M_1 and θ were the two properties. Once any two properties about the oblique shock are specified, the shock is uniquely defined. This is analogous to the case of a normal shock wave studied in Chap. 8. There, we proved that all the changes across a normal shock wave were uniquely defined by specifying only *one* property, such as M_1. However, implicit in all of Chap. 8 was an additional property, namely, the wave angle of a normal shock wave is 90°. Of course, a normal shock is simply one example of the whole spectrum of oblique shocks, namely, a shock with $\beta = 90°$. An examination of Fig. 9.7 shows that the normal shock belongs to the family of strong shock solutions, as discussed earlier.

Example 9.3. Consider an oblique shock wave with $\beta = 35°$ and a pressure ratio $p_2/p_1 = 3$. Calculate the upstream Mach number.

Solution. From App. B, for $p_2/p_1 = 3$, $M_{n,1} = 1.64$ (nearest entry). Since

$$M_{n,1} = M_1 \sin \beta$$

then

$$M_1 = \frac{M_{n,1}}{\sin \beta} = \frac{1.66}{\sin 35°} = \boxed{2.86}$$

Note: Once again, the oblique shock is uniquely defined by two properties, in this case β and p_2/p_1.

Example 9.4. Consider a Mach 3 flow. It is desired to slow this flow to a subsonic speed. Consider two separate ways of achieving this: (1) the Mach 3 flow is slowed by passing directly through a normal shock wave; (2) the Mach 3 flow first passes through an oblique shock with a 40° wave angle, and then subsequently through a normal shock. These two cases are sketched in Fig. 9.12. Calculate the ratio of the final total pressure values for the two cases, i.e., the total pressure behind the normal shock for case 2 divided by the total pressure behind the normal shock for case 1. Comment on the significance of the result.

Solution. For case 1, at $M = 3$, we have, from App. B,

$$\left(\frac{p_{0_2}}{p_{0_1}}\right)_{\text{case 1}} = 0.3283$$

For case 2, we have $M_{n,1} = M_1 \sin \beta = 3 \sin 40° = 1.93$. From App. B,

$$\frac{p_{0_2}}{p_{0_1}} = 0.7535 \qquad \text{and} \qquad M_{n,2} = 0.588$$

From Fig. 9.7, for $M_1 = 3$ and $\beta = 40°$, we have the deflection angle $\theta = 22°$. Hence,

$$M_2 = \frac{M_{n,2}}{\sin(\beta - \theta)} = \frac{0.588}{\sin(40 - 22)} = 1.90$$

From App. B, for a normal shock with an upstream Mach number of 1.9, we have $p_{0_3}/p_{0_2} = 0.7674$. Thus, for case 2,

$$\left(\frac{p_{0_3}}{p_{0_1}}\right)_{\text{case 2}} = \left(\frac{p_{0_2}}{p_{0_1}}\right)\left(\frac{p_{0_3}}{p_{0_2}}\right) = (0.7535)(0.7674) = 0.578$$

Hence,

$$\left(\frac{p_{0_3}}{p_{0_1}}\right)_{\text{case 2}} \bigg/ \left(\frac{p_{0_2}}{p_{0_1}}\right)_{\text{case 1}} = \frac{0.578}{0.3283} = \boxed{1.76}$$

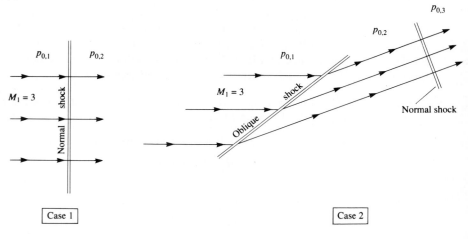

FIGURE 9.12
Illustration for Example 9.4.

The result of Example 9.4 shows that the final total pressure is 76 percent higher for the case of the multiple shock system (case 2) in comparison to the single normal shock (case 1). In principle, the total pressure is an indicator of how much useful work can be done by the gas; this is described later in Sec. 10.4. Everything else being equal, the higher the total pressure, the more useful is the flow. Indeed, losses of total pressure are an index of the *efficiency* of a fluid flow—the lower the total pressure loss, the more efficient is the flow process. In this example, case 2 is more efficient in slowing the flow to subsonic speeds than case 1 because the loss in total pressure across the multiple shock system of case 2 is actually less than that for case 1 with a single, strong, normal shock wave. The physical reason for this is straightforward. The loss in total pressure across a normal shock wave becomes particularly severe as the upstream Mach number increases; a glance at the $p_{0,2}/p_{0,1}$ column in App. B attests to this. If the Mach number of a flow can be reduced *before* passing through a normal shock, the loss in total pressure is much less because the normal shock is weaker. This is the function of the oblique shock in case 2, namely, to reduce the Mach number of the flow before passing through the normal shock. Although there is a total pressure loss across the oblique shock also, it is much less than across a

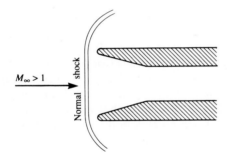

(a) Normal shock inlet

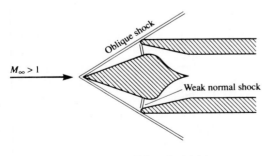

(b) Oblique shock inlet

FIGURE 9.13
Illustration of (a) normal shock inlet and (b) oblique shock inlet.

normal shock at the same upstream Mach number. The net effect of the oblique shock reducing the flow Mach number before passing through the normal shock more than makes up for the total pressure loss across the oblique shock, with the beneficial result that the multiple shock system in case 2 produces a *smaller* loss in total pressure than a single normal shock at the same freestream Mach number.

A practical application of these results is in the design of supersonic inlets for jet engines. A normal shock inlet is sketched in Fig. 9.13*a*. Here, a normal shock forms ahead of the inlet, with an attendant large loss in total pressure. In contrast, an oblique shock inlet is sketched in Fig. 9.13*b*. Here, a central cone creates an oblique shock wave, and the flow subsequently passes through a relatively weak normal shock at the lip of the inlet. For the same flight conditions (Mach number and altitude), the total pressure loss for the oblique shock inlet is less than for a normal shock inlet. Hence, everything else being equal, the resulting engine thrust will be higher for the oblique shock inlet. This, of course, is why most modern supersonic aircraft have oblique shock inlets.

9.3 SUPERSONIC FLOW OVER WEDGES AND CONES

For the supersonic flow over wedges, as shown in Figs. 9.10 and 9.11, the oblique shock theory developed in Sec. 9.2 is an *exact* solution of the flow field; no simplifying assumptions have been made. Supersonic flow over a wedge is characterized by an attached, straight oblique shock wave from the nose, a uniform flow downstream of the shock with streamlines parallel to the wedge surface, and a surface pressure equal to the static pressure behind the oblique shock, p_2. These properties are summarized in Fig. 9.14*a*. Note that the wedge is a two-dimensional profile; in Fig. 9.14*a*, it is a section of a body that stretches to plus

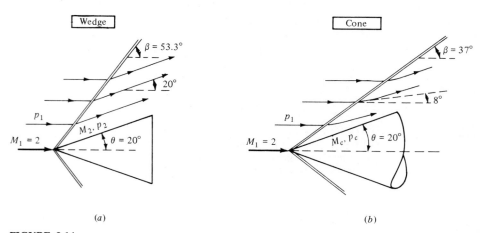

FIGURE 9.14
Relation between wedge and cone flow; illustration of the three-dimensional relieving effect.

or minus infinity in the direction perpendicular to the page. Hence, wedge flow is, by definition, two-dimensional flow, and our two-dimensional oblique shock theory fits this case nicely.

In contrast, consider the supersonic flow over a cone, as sketched in Fig. 9.14b. There is a straight oblique shock which emanates from the tip, just as in the case of a wedge, but the similarity stops there. Recall from Chap. 6 that flow over a three-dimensional body experiences a "three-dimensional relieving effect." That is, in comparing the wedge and cone in Fig. 9.14, both with the same 20° angle, the flow over the cone has an extra dimension in which to move, and hence it more easily adjusts to the presence of the conical body in comparison to the two-dimensional wedge. One consequence of this three-dimensional reliev-ing effect is that the shock wave on the cone is weaker than on the wedge; i.e., it has a smaller wave angle, as compared in Fig. 9.14. Specifically, the wave angles for the wedge and cone are 53.3 and 37°, respectively, for the same body angle of 20° and the same upstream Mach number of 2.0. In the case of the wedge (Fig. 9.14a), the streamlines are deflected by exactly 20° through the shock wave, and hence downstream of the shock the flow is exactly parallel to the wedge surface. In contrast, because of the weaker shock on the cone, the stream-lines are deflected by only 8° through the shock, as shown in Fig. 9.14b. Therefore, between the shock wave and the cone surface, the streamlines must gradually curve upward in order to accommodate the 20° cone. Also, as a consequence of the three-dimensional relieving effect, the pressure on the surface of the cone, p_c, is less than the wedge surface pressure p_2, and the cone surface Mach number M_c is greater than that on the wedge surface, M_2. In short, the main differences between the supersonic flow over a cone and wedge, both with the same body angle, are that (1) the shock wave on the cone is weaker, (2) the cone surface pressure is less, and (3) the streamlines above the cone surface are curved rather than straight.

The analysis of the supersonic flow over a cone is more sophisticated than the oblique shock theory given in this chapter and is beyond the scope of this book. For details concerning supersonic conical flow analysis, see chap. 10 of Ref. 21. However, it is important for you to recognize that conical flows are inherently different from wedge flows and to recognize in what manner they differ. This has been the purpose of the present section.

Example 9.5. Consider a wedge with a 15° half angle in a Mach 5 flow, as sketched in Fig. 9.15. Calculate the drag coefficient for this wedge. (Assume that the pressure over the base is equal to freestream static pressure, as shown in Fig. 9.15.)

Solution. Consider the drag on a unit span of the wedge, D'. Hence,

$$c_d = \frac{D'}{q_1 S} = \frac{D'}{q_1 c(1)} = \frac{D'}{q_1 c}$$

From Fig. 9.15,

$$D' = 2p_2 l \sin \theta - 2p_1 l \sin \theta = (2l \sin \theta)(p_2 - p_1)$$

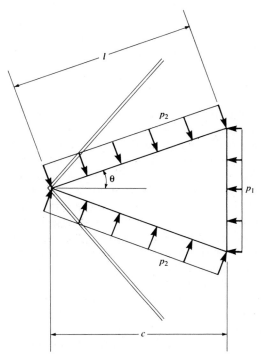

FIGURE 9.15
Illustration for Example 9.5.

However,

$$l = \frac{c}{\cos \theta}$$

Thus,

$$D' = (2c \tan \theta)(p_2 - p_1)$$

and

$$c_d = (2 \tan \theta)\left(\frac{p_2 - p_1}{q_1}\right)$$

Note that

$$q_1 \equiv \frac{1}{2}\rho_1 V_1^2 = \frac{1}{2}\rho_1 \frac{\gamma p_1}{\gamma p_1} V_1^2 = \frac{\gamma p_1}{2a_1^2} V_1^2 = \frac{\gamma}{2}p_1 M_1^2$$

Thus,

$$c_d = (2 \tan \theta)\left(\frac{p_2 - p_1}{(\gamma/2)p_1 M_1^2}\right) = \frac{4 \tan \theta}{\gamma M_1^2}\left(\frac{p_2}{p_1} - 1\right)$$

From Fig. 9.7, for $M_1 = 5$ and $\theta = 15°$, $\beta = 24.2°$. Hence,

$$M_{n,1} = M_1 \sin \beta = 5 \sin (24.2°) = 2.05$$

From App. B, for $M_{n,1} = 2.05$, we have

$$\frac{p_2}{p_1} = 4.736$$

Hence,

$$c_d = \frac{4 \tan \theta}{\gamma M_1^2} \left(\frac{p_2}{p_1} - 1 \right) = \frac{4 \tan 15°}{(1.4)(5)^2} (4.736 - 1) = \boxed{0.114}$$

(*Note*: The drag is *finite* for this case. In a supersonic or hypersonic inviscid flow over a two-dimensional body, the drag is always finite. D'Alembert's paradox does *not* hold for freestream Mach numbers such that shock waves appear in the flow. The fundamental reason for the generation of drag here is the presence of shock waves. Shocks are always a dissipative, drag-producing mechanism. For this reason, the drag in this case is called *wave drag*, and c_d is the wave-drag coefficient, more properly denoted as $c_{d,w}$.)

9.4 SHOCK INTERACTIONS AND REFLECTIONS

Return to the oblique shock wave illustrated in Fig. 9.1a. In this picture, we can imagine the shock wave extending unchanged above the corner to infinity. However, in real life this does not happen. In reality, the oblique shock in Fig. 9.1a will impinge somewhere on another solid surface and/or will intersect other waves, either shock or expansion waves. Such wave intersections and interactions are important in the practical design and analysis of supersonic airplanes, missiles, wind tunnels, rocket engines, etc. A perfect historical example of this, as well as the consequences that can be caused by not paying suitable attention to wave interactions, is a ramjet flight-test program conducted in the early 1960s. During this period, a ramjet engine was mounted underneath the X-15 hypersonic airplane for a series of flight tests at high Mach numbers, in the range from 4 to 7. (The X-15, shown in Fig. 9.16, was an experimental, rocket-powered airplane designed to probe the lower end of hypersonic manned flight.) During the first high-speed tests, the shock wave from the engine cowling impinged on the bottom surface of the X-15, and because of locally high aerodynamic heating in the impingement region, a hole was burned in the X-15 fuselage. Although this problem was later fixed, it is a graphic example of what shock-wave interactions can do to a practical configuration.

The purpose of this section is to present a mainly qualitative discussion of shock-wave interactions. For more details, see chap. 4 of Ref. 21.

First, consider an oblique shock wave generated by a concave corner, as shown in Fig. 9.17. The deflection angle at the corner is θ, thus generating an oblique shock at point A with a wave angle, β_1. Assume that a straight, horizontal wall is present above the corner, as also shown in Fig. 9.17. The shock wave generated at point A, called the *incident shock wave*, impinges on the upper wall at point B. *Question*: Does the shock wave simply disappear at point B? If not,

FIGURE 9.16
The X-15 hypersonic research vehicle. Designed and built during the late 1950s, it served as a test vehicle for the U.S. Air Force and NASA. (*Courtesy of Rockwell Intl., North America.*)

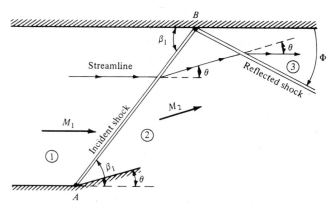

FIGURE 9.17
Regular reflection of a shock wave from a solid boundary.

what happens to it? To answer this question, we appeal to our knowledge of shock-wave properties. Examining Fig. 9.17, we see that the flow in region 2 behind the incident shock is inclined upward at the deflection angle θ. However, the flow must be tangent everywhere along the upper wall; if the flow in region 2 were to continue unchanged, it would run into the wall and have no place to go. Hence, the flow in region 2 must eventually be bent downward through the angle θ in order to maintain a flow tangent to the upper wall. Nature accomplishes this downward deflection via a second shock wave originating at the impingement point B in Fig. 9.17. This second shock is called the *reflected shock wave*. The purpose of the reflected shock is to deflect the flow in region 2 so that it is parallel to the upper wall in region 3, thus preserving the wall boundary condition.

The strength of the reflected shock wave is weaker than the incident shock. This is because $M_2 < M_1$, and M_2 represents the upstream Mach number for the reflected shock wave. Since the deflection angles are the same, whereas the reflected shock sees a lower upstream Mach number, we know from Sec. 9.2 that the reflected wave must be weaker. For this reason, the angle the reflected shock makes with the upper wall, Φ, is not equal to β_1; i.e., the wave reflection is not specular. The properties of the reflected shock are uniquely defined by M_2 and θ; since M_2 is in turn uniquely defined by M_1 and θ, then the properties in region 3 behind the reflected shock as well as the angle Φ are easily determined from the given conditions of M_1 and θ by using the results of Sec. 9.2 as follows:

1. Calculate the properties in region 2 from the given M_1 and θ. In particular, this gives us M_2.
2. Calculate the properties in region 3 from the value of M_2 calculated above and the known deflection angle θ.

An interesting situation can arise as follows. Assume that M_1 is only slightly above the minimum Mach number necessary for a straight, attached shock wave at the given deflection angle θ. For this case, the oblique shock theory from Sec. 9.2 allows a solution for a straight, attached incident shock. However, we know that the Mach number decreases across a shock, i.e., $M_2 < M_1$. This decrease may be enough such that M_2 is *not* above the minimum Mach number for the required deflection θ through the reflected shock. In such a case, our oblique shock theory does not allow a solution for a straight reflected shock wave. The regular reflection as shown in Fig. 9.17 is not possible. Nature handles this situation by creating the wave pattern shown in Fig. 9.18. Here, the originally straight incident shock becomes curved as it nears the upper wall and becomes a normal shock wave at the upper wall. This allows the streamline at the wall to continue parallel to the wall behind the shock intersection. In addition, a curved reflected shock branches from the normal shock and propagates downstream. This wave pattern, shown in Fig. 9.18, is called a *Mach reflection.* The calculation of the wave pattern and general properties for a Mach reflection requires numerical techniques such as those to be discussed in Chap. 13.

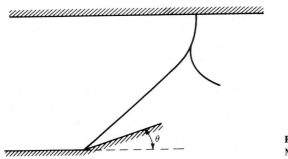

FIGURE 9.18
Mach reflection.

Another type of shock interaction is shown in Fig. 9.19. Here, a shock wave is generated by the concave corner at point G and propagates upward. Denote this wave as shock A. Shock A is a *left-running wave*, so-called because if you stand on top of the wave and look downstream, you see the shock wave running in front of you toward the left. Another shock wave is generated by the concave corner at point H, and propagates downward. Denote this wave as shock B. Shock B is a *right-running wave*, so-called because if you stand on top of the wave and look downstream, you see the shock running in front of you toward the right. The picture shown in Fig. 9.19 is the intersection of right- and left-running shock waves. The intersection occurs at point E. At the intersection, wave A is refracted and continues as wave D. Similarly, wave B is refracted and continues as wave C. The flow behind the refracted shock D is denoted by region 4; the flow behind the refracted shock C is denoted by region 4'. These two regions are divided by a slip line, EF. Across the slip line, the pressures are constant, i.e., $p_4 = p_{4'}$, and the direction (but not necessarily the magnitude) of velocity is the same, namely, parallel to the slip line. All other properties in regions 4 and 4' are different, most notably the entropy ($s_4 \neq s_{4'}$). The conditions

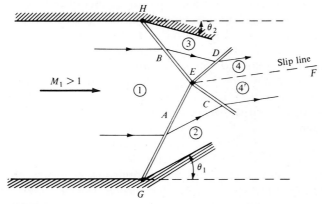

FIGURE 9.19
Intersection of right- and left-running shock waves.

which must hold across the slip line, along with the known M_1, θ_1, and θ_2, uniquely determine the shock-wave interaction shown in Fig. 9.19. (See chap. 4 of Ref. 21 for details concerning the calculation of this interaction.)

Figure 9.20 illustrates the intersection of two left-running shocks generated at corners A and B. The intersection occurs at point C, at which the two shocks merge and propagate as the stronger shock CD, usually along with a weak reflected wave CE. This reflected wave is necessary to adjust the flow so that the velocities in regions 4 and 5 are in the same direction. Again, a slip line CF trails downstream of the intersection point.

The above cases are by no means all the possible wave interactions in a supersonic flow. However, they represent some of the more common situations encountered frequently in practice.

Example 9.6. Consider an oblique shock wave generated by a compression corner with a 10° deflection angle. The Mach number of the flow ahead of the corner is 3.6; the flow pressure and temperature are standard sea level conditions. The oblique shock wave subsequently impinges on a straight wall opposite the compression corner. The geometry for this flow is given in Fig. 9.17. Calculate the angle of the reflected shock wave, Φ, relative to the straight wall. Also, obtain the pressure, temperature, and Mach number behind the reflected wave.

Solution. From the θ-β-M diagram, Fig. 9.7, for $M_1 = 3.6$ and $\theta = 10°$, $\beta_1 = 24°$. Hence,

$$M_{n,1} = M_1 \sin \beta_1 = 3.6 \sin 24° = 1.464$$

From App. B,

$$M_{n,2} = 0.7157, \qquad \frac{p_2}{p_1} = 2.32, \qquad \text{and} \qquad \frac{T_2}{T_1} = 1.294$$

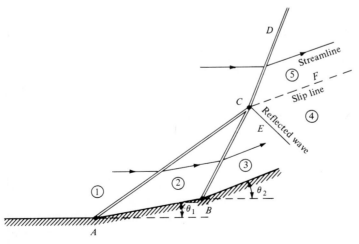

FIGURE 9.20
Intersection of two left-running shock waves.

Also,

$$M_2 = \frac{M_{n,2}}{\sin(\beta - \theta)} = \frac{0.7157}{\sin(24 - 10)} = 2.96$$

These are the conditions behind the incident shock wave. They constitute the upstream flow properties for the reflected shock wave. We know that the flow must be deflected again by $\theta = 10°$ in passing through the reflected shock. Thus, from the θ-β-M diagram, for $M_2 = 2.96$ and $\theta = 10°$, we have the wave angle for the reflected shock, $\beta_2 = 27.3°$. Note that β_2 is *not* the angle the reflected shock makes with respect to the upper wall; rather, by definition of the wave angle, β_2 is the angle between the reflected shock and the direction of the flow in region 2. The shock angle relative to the wall is, from the geometry shown in Fig. 9.17,

$$\Phi = \beta_2 - \theta = 27.3 - 10 = \boxed{17.3°}$$

Also, the normal component of the upstream Mach number relative to the reflected shock is $M_2 \sin \beta_2 = (2.96) \sin 27.3° = 1.358$. From App. B,

$$\frac{p_3}{p_2} = 1.991, \qquad \frac{T_3}{T_2} = 1.229, \qquad M_{n,3} = 0.7572$$

Hence,

$$M_3 = \frac{M_{n,3}}{\sin(\beta_2 - \theta)} = \frac{0.7572}{\sin(27.3 - 10)} = \boxed{2.55}$$

For standard sea level conditions, $p_1 = 2116 \text{ lb/ft}^3$ and $T_1 = 519°R$. Thus,

$$p_3 = \frac{p_3}{p_2}\frac{p_2}{p_1} p_1 = (1.991)(2.32)(2116) = \boxed{9774 \text{ lb/ft}^3}$$

$$T_3 = \frac{T_3}{T_2}\frac{T_2}{T_1} T_1 = (1.229)(1.294)(519) = \boxed{825°R}$$

Note that the reflected shock is weaker than the incident shock, as indicated by the smaller pressure ratio for the reflected shock, $p_3/p_2 = 1.991$ as compared to $p_2/p_1 = 2.32$ for the incident shock.

9.5 DETACHED SHOCK WAVE IN FRONT OF A BLUNT BODY

The curved bow shock which stands in front of a blunt body in a supersonic flow is sketched in Fig. 8.1. We are now in a position to better understand the properties of this bow shock, as follows.

The flow in Fig. 8.1 is sketched in more detail in Fig. 9.21. Here, the shock wave stands a distance δ in front of the nose of the blunt body; δ is defined as the *shock detachment distance*. At point a, the shock wave is normal to the upstream flow; hence, point a corresponds to a normal shock wave. Away from point a, the shock wave gradually becomes curved and weaker, eventually evolving into a Mach wave at large distances from the body (illustrated by point e in Fig. 9.21).

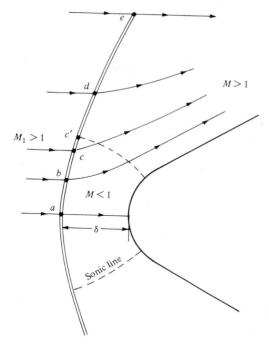

FIGURE 9.21
Flow over a supersonic blunt body.

A curved bow shock wave is one of the instances in nature when you can observe *all* possible oblique shock solutions at once for a given freestream Mach number, M_1. This takes place between points a and e. To see this more clearly, consider the θ-β-M diagram sketched in Fig. 9.22 in conjunction with Fig. 9.21. In Fig. 9.22, point a corresponds to the normal shock, and point e corresponds

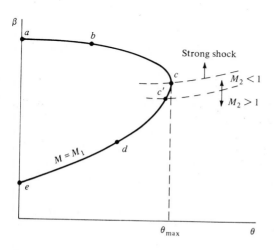

FIGURE 9.22
θ-β-M diagram for the sketch shown in Fig. 9.21.

to the Mach wave. Slightly above the centerline, at point b in Fig. 9.21, the shock is oblique but pertains to the strong shock-wave solution in Fig. 9.22. The flow is deflected slightly upward behind the shock at point b. As we move further along the shock, the wave angle becomes more oblique, and the flow deflection increases until we encounter point c. Point c on the bow shock corresponds to the maximum deflection angle shown in Fig. 9.22. Above point c, from c to e, all points on the shock correspond to the weak shock solution. Slightly above point c, at point c', the flow behind the shock becomes sonic. From a to c', the flow is subsonic behind the bow shock; from c' to e, it is supersonic. Hence, the flow field between the curved bow shock and the blunt body is a mixed region of both subsonic and supersonic flow. The dividing line between the subsonic and supersonic regions is called the *sonic line*, shown as the dashed line in Fig. 9.21.

The shape of the detached shock wave, its detachment distance δ, and the complete flow field between the shock and the body depend on M_1 and the size and shape of the body. The solution of this flow field is not trivial. Indeed, the supersonic blunt-body problem was a major focus for supersonic aerodynamicists during the 1950s and 1960s, spurred by the need to understand the high-speed flow over blunt-nosed missiles and reentry bodies. Indeed, it was not until the late 1960s that truly sufficient numerical techniques became available for satisfactory engineering solutions of supersonic blunt-body flows. These modern techniques are discussed in Chap. 13.

9.6 PRANDTL-MEYER EXPANSION WAVES

Oblique shock waves, as discussed in Secs. 9.2 to 9.5, occur when a supersonic flow is turned into itself (see again Fig. 9.1a). In contrast, when a supersonic flow is turned away from itself, an expansion wave is formed, as sketched in Fig. 9.1b. Examine this figure carefully, and review the surrounding discussion in Sec. 9.1 before progressing further. The purpose of the present section is to develop a theory which allows us to calculate the changes in flow properties across such expansion waves. To this stage in our discussion of oblique waves, we have completed the left-hand branch of the road map in Fig. 9.5. In this section, we cover the right-hand branch.

The expansion fan in Fig. 9.1b is a *continuous* expansion region which can be visualized as an infinite number of Mach waves, each making the Mach angle μ [see Eq. (9.1)] with the local flow direction. As sketched in Fig. 9.23, the expansion fan is bounded upstream by a Mach wave which makes the angle μ_1 with respect to the upstream flow, where $\mu_1 = \arcsin(1/M_1)$. The expansion fan is bounded downstream by another Mach wave which makes the angle μ_2 with respect to the downstream flow, where $\mu_2 = \arcsin(1/M_2)$. Since the expansion through the wave takes place across a continuous succession of Mach waves, and since $ds = 0$ for each Mach wave, the expansion is *isentropic*. This is in direct contrast to flow across an oblique shock, which always experiences an entropy increase. The fact that the flow through an expansion wave is isentropic is a greatly simplifying aspect, as we will soon appreciate.

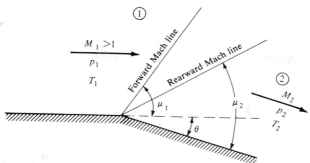

FIGURE 9.23
Prandtl-Meyer expansion.

An expansion wave emanating from a sharp convex corner as sketched in Figs. 9.1*b* and 9.23 is called a *centered* expansion wave. Ludwig Prandtl and his student Theodor Meyer first worked out a theory for centered expansion waves in 1907–1908, and hence such waves are commonly denoted as *Prandtl-Meyer expansion waves.*

The problem of an expansion wave is as follows: Referring to Fig. 9.23, given the upstream flow (region 1) and the deflection angle θ, calculate the downstream flow (region 2). Let us proceed.

Consider a very weak wave produced by an infinitesimally small flow deflection, $d\theta$, as sketched in Fig. 9.24. We consider the limit of this picture as $d\theta \to 0$; hence, the wave is essentially a Mach wave at the angle μ to the upstream flow. The velocity ahead of the wave is V. As the flow is deflected downward through the angle $d\theta$, the velocity is increased by the infinitesimal amount dV, and hence the flow velocity behind the wave is $V + dV$ inclined at the angle $d\theta$. Recall from the treatment of the momentum equation in Sec. 9.2 that any change

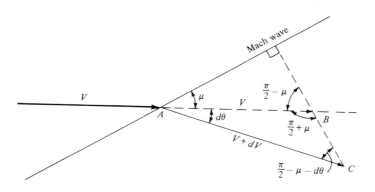

FIGURE 9.24
Geometrical construction for the infinitesimal changes across an infinitesimally weak wave (in the limit, a Mach wave).

in velocity across a wave takes place *normal* to the wave; the tangential component is unchanged across the wave. In Fig. 9.24, the horizontal line segment AB with length V is drawn behind the wave. Also, the line segment AC is drawn to represent the new velocity $V + dV$ behind the wave. Then line BC is normal to the wave because it represents the line along which the change in velocity occurs. Examining the geometry in Fig. 9.24, from the law of sines applied to triangle ABC, we see that

$$\frac{V + dV}{V} = \frac{\sin(\pi/2 + \mu)}{\sin(\pi/2 - \mu - d\theta)} \tag{9.24}$$

However, from trigonometric identities,

$$\sin\left(\frac{\pi}{2} + \mu\right) = \sin\left(\frac{\pi}{2} - \mu\right) = \cos\mu \tag{9.25}$$

$$\sin\left(\frac{\pi}{2} - \mu - d\theta\right) = \cos(\mu + d\theta) = \cos\mu\,\cos d\theta - \sin\mu\,\sin d\theta \tag{9.26}$$

Substituting Eqs. (9.25) and (9.26) into (9.24), we have

$$1 + \frac{dV}{V} = \frac{\cos\mu}{\cos\mu\,\cos d\theta - \sin\mu\,\sin d\theta} \tag{9.27}$$

For small $d\theta$, we can make the small-angle assumptions $\sin d\theta \approx d\theta$ and $\cos d\theta \approx 1$. Then, Eq. (9.27) becomes

$$1 + \frac{dV}{V} = \frac{\cos\mu}{\cos\mu - d\theta\,\sin\mu} = \frac{1}{1 - d\theta\,\tan\mu} \tag{9.28}$$

Note that the function $1/(1 - x)$ can be expanded in a power series (for $x < 1$) as

$$\frac{1}{1 - x} = 1 + x + x^2 + x^3 + \cdots$$

Hence, Eq. (9.28) can be expanded as (ignoring terms of second order and higher)

$$1 + \frac{dV}{V} = 1 + d\theta\,\tan\mu + \cdots \tag{9.29}$$

Thus, from Eq. (9.29),

$$d\theta = \frac{dV/V}{\tan\mu} \tag{9.30}$$

From Eq. (9.1), we know that $\mu = \arcsin(1/M)$. Hence, the right triangle in Fig. 9.25 demonstrates that

$$\tan\mu = \frac{1}{\sqrt{M^2 - 1}} \tag{9.31}$$

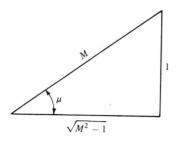

FIGURE 9.25
Right triangle associated with the Mach angle.

Substituting Eq. (9.31) into (9.30), we obtain

$$d\theta = \sqrt{M^2 - 1}\,\frac{dV}{V}$$

(9.32)

Equation (9.32) relates the infinitesimal change in velocity, dV, to the infinitesimal deflection $d\theta$ across a wave of vanishing strength. In the precise limit of a Mach wave, of course dV and hence $d\theta$ are zero. In this sense, Eq. (9.32) is an approximate equation for a finite $d\theta$, but it becomes a true equality as $d\theta \to 0$. Since the expansion fan illustrated in Figs. 9.1b and 9.23 is a region of an infinite number of Mach waves, Eq. (9.32) is a differential equation which precisely describes the flow inside the expansion wave.

Return to Fig. 9.23. Let us integrate Eq. (9.32) from region 1, where the deflection angle is zero and the Mach number is M_1, to region 2, where the deflection angle is θ and the Mach number is M_2:

$$\int_0^\theta d\theta = \theta = \int_{M_1}^{M_2} \sqrt{M^2 - 1}\,\frac{dV}{V}$$

(9.33)

To carry out the integral on the right-hand side of Eq. (9.33), dV/V must be obtained in terms of M, as follows. From the definition of Mach number, $M = V/a$, we have $V = Ma$, or

$$\ln V = \ln M + \ln a$$

(9.34)

Differentiating Eq. (9.34), we obtain

$$\frac{dV}{V} = \frac{dM}{M} + \frac{da}{a}$$

(9.35)

From Eqs. (8.25) and (8.40), we have

$$\left(\frac{a_0}{a}\right)^2 = \frac{T_0}{T} = 1 + \frac{\gamma - 1}{2}M^2$$

(9.36)

Solving Eq. (9.36) for a, we obtain

$$a = a_0\left(1 + \frac{\gamma - 1}{2}M^2\right)^{-\frac{1}{2}}$$

(9.37)

Differentiating Eq. (9.37), we obtain

$$\frac{da}{a} = -\left(\frac{\gamma-1}{2}\right) M \left(1+\frac{\gamma-1}{2} M^2\right)^{-1} dM \tag{9.38}$$

substituting Eq. (9.38) into (9.35), we have

$$\frac{dV}{V} = \frac{1}{1+[(\gamma-1)/2]M^2} \frac{dM}{M} \tag{9.39}$$

Equation (9.39) is a relation for dV/V strictly in terms of M—this is precisely what is desired for the integral in Eq. (9.33). Hence, substituting Eq. (9.39) into (9.33), we have

$$\theta = \int_{M_1}^{M_2} \frac{\sqrt{M^2-1}}{1+[(\gamma-1)/2]M^2} \frac{dM}{M} \tag{9.40}$$

In Eq. (9.40), the integral

$$\nu(M) \equiv \int \frac{\sqrt{M^2-1}}{1+[(\gamma-1)/2]M^2} \frac{dM}{M} \tag{9.41}$$

is called the *Prandtl-Meyer function*, denoted by ν. Carrying out the integration, Eq. (9.41) becomes

$$\boxed{\nu(M) = \sqrt{\frac{\gamma+1}{\gamma-1}} \tan^{-1} \sqrt{\frac{\gamma-1}{\gamma+1}(M^2-1)} - \tan^{-1}\sqrt{M^2-1}} \tag{9.42}$$

The constant of integration that would ordinarily appear in Eq. (9.42) is not important, because it drops out when Eq. (9.42) is used for the definite integral in Eq. (9.40). For convenience, it is chosen as zero, such that $\nu(M)=0$ when $M=1$. Finally, we can now write Eq. (9.40), combined with (9.41), as

$$\boxed{\theta = \nu(M_2) - \nu(M_1)} \tag{9.43}$$

where $\nu(M)$ is given by Eq. (9.42) for a calorically perfect gas. The Prandtl-Meyer function ν is very important; it is the key to the calculation of changes across an expansion wave. Because of its importance, ν is tabulated as a function of M in App. C. For convenience, values of μ are also tabulated in App. C.

How do the above results solve the problem stated in Fig. 9.23; i.e., how can we obtain the properties in region 2 from the known properties in region 1 and the known deflection angle θ? The answer is straightforward:

1. For the given M_1, obtain $\nu(M_1)$ from App. C.
2. Calculate $\nu(M_2)$ from Eq. (9.43), using the known θ and the value of $\nu(M_1)$ obtained in step 1.

3. Obtain M_2 from App. C corresponding to the value of $\nu(M_2)$ from step 2.
4. The expansion wave is isentropic; hence, p_0 and T_0 are constant through the wave. That is, $T_{0,2} = T_{0,1}$ and $p_{0,2} = p_{0,1}$. From Eq. (8.40), we have

$$\frac{T_2}{T_1} = \frac{T_2/T_{0,2}}{T_1/T_{0,1}} = \frac{1+[(\gamma-1)/2]M_1^2}{1+[(\gamma-1)/2]M_2^2} \tag{9.44}$$

From Eq. (8.42), we have

$$\frac{p_2}{p_1} = \frac{p_2/p_0}{p_1/p_0} = \left(\frac{1+[(\gamma-1)/2]M_1^2}{1+[(\gamma-1)/2]M_2^2}\right)^{\gamma/(\gamma-1)} \tag{9.45}$$

Since we know both M_1 and M_2, as well as T_1 and p_1, Eqs. (9.44) and (9.45) allow the calculation of T_2 and p_2 downstream of the expansion wave.

Example 9.7. A supersonic flow with $M_1 = 1.5$, $p_1 = 1$ atm, and $T_1 = 288$ K is expanded around a sharp corner (see Fig. 9.23) through a deflection angle of $15°$. Calculate $M_2, p_2, T_2, p_{0,2}, T_{0,2}$, and the angles that the forward and rearward Mach lines make with respect to the upstream flow direction.

Solution. From App. C, for $M_1 = 1.5$, $\nu_1 = 11.91°$. From Eq. (9.43), $\nu_2 = \nu_1 + \theta = 11.91 + 15 = 26.91°$. Thus, $\boxed{M_2 = 2.0}$ (rounding to the nearest entry in the table).
From App. A, for $M_1 = 1.5$, $p_{0,1}/p_1 = 3.671$ and $T_{0,1}/T_1 = 1.45$, and for $M_2 = 2.0$, $p_{0,2}/p_2 = 7.824$ and $T_{0,2}/T_2 = 1.8$.
Since the flow is isentropic, $T_{0,2} = T_{0,1}$ and $p_{0,2} = p_{0,1}$. Thus,

$$p_2 = \frac{p_2}{p_{0,2}}\frac{p_{0,2}}{p_{0,1}}\frac{p_{0,1}}{p_1} p_1 = \frac{1}{7.824}(1)(3.671)(1 \text{ atm}) = \boxed{0.469 \text{ atm}}$$

$$T_2 = \frac{T_2}{T_{0,2}}\frac{T_{0,2}}{T_{0,1}}\frac{T_{0,1}}{T_1} T_1 = \frac{1}{1.8}(1)(1.45)(288) = \boxed{232 \text{ K}}$$

$$p_{0,2} = p_{0,1} = \frac{p_{0,1}}{p_1} p_1 = 3.671(1 \text{ atm}) = \boxed{3.671 \text{ atm}}$$

$$T_{0,2} = T_{0,1} = \frac{T_{0,1}}{T_1} T_1 = 1.45(288) = \boxed{417.6 \text{ K}}$$

Returning to Fig. 9.23, we have

$$\text{Angle of forward Mach line} = \mu_1 = \boxed{41.81°}$$

$$\text{Angle of rearward Mach line} = \mu_2 - \theta = 30 - 15 = \boxed{15°}$$

9.7 SHOCK-EXPANSION THEORY: APPLICATIONS TO SUPERSONIC AIRFOILS

Consider a flat plate of length c at an angle of attack α in a supersonic flow, as sketched in Fig. 9.26. On the top surface, the flow is turned away from itself; hence, an expansion wave occurs at the leading edge, and the pressure on the

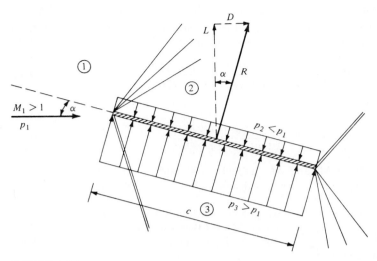

FIGURE 9.26
Flat plate at an angle of attack in a supersonic flow.

top surface, p_2, is less than the freestream pressure, $p_2 < p_1$. At the trailing edge, the flow must return to approximately (but not precisely) the freestream direction. Here, the flow is turned back into itself, and consequently a shock wave occurs at the trailing edge. On the bottom surface, the flow is turned into itself; an oblique shock wave occurs at the leading edge, and the pressure on the bottom surface, p_3, is greater than the freestream pressure $p_3 > p_1$. At the trailing edge, the flow is turned into approximately (but not precisely) the freestream direction by means of an expansion wave. Examining Fig. 9.26, note that the top and bottom surfaces of the flat plate experience uniform pressure distribution of p_2 and p_3, respectively, and that $p_3 > p_2$. This creates a net pressure imbalance which generates the resultant aerodynamic force R, shown in Fig. 9.26. Indeed, for a unit span, the resultant force and its components, lift and drag, per unit span are

$$R' = (p_3 - p_2)c \qquad (9.46)$$

$$L' = (p_3 - p_2)c \cos \alpha \qquad (9.47)$$

$$D' = (p_3 - p_2)c \sin \alpha \qquad (9.48)$$

In Eqs. (9.47) and (9.48), p_3 is calculated from oblique shock properties (Sec. 9.2), and p_2 is calculated from expansion-wave properties (Sec. 9.6). Moreover, these are *exact* calculations; no approximations have been made. The inviscid, supersonic flow over a flat plate at angle of attack is exactly given by the combination of shock and expansion waves sketched in Fig. 9.26.

The flat-plate case given above is the simplest example of a general technique called *shock-expansion theory*. Whenever we have a body made up of straight-line segments and the deflection angles are small enough so that no detached shock waves occur, the flow over the body goes through a series of distinct oblique shock and expansion waves, and the pressure distribution on the surface (hence the lift and drag) can be obtained *exactly* from both the shock- and expansion-wave theories discussed in this chapter.

As another example of the application of shock-expansion theory, consider the diamond-shape airfoil in Fig. 9.27. Assume the airfoil is at $0°$ angle of attack. The supersonic flow over the airfoil is first compressed and deflected through the angle ε by the oblique shock wave at the leading edge. At midchord, the flow is expanded through an angle 2ε, creating an expansion wave. At the trailing edge, the flow is turned back to the freestream direction through another oblique shock. The pressure distributions on the front and back faces of the airfoil are sketched in Fig. 9.27; note that the pressures on faces a and c are uniform and equal to p_2 and that the pressures on faces b and d are also uniform but equal to p_3, where $p_3 < p_2$. In the lift direction, perpendicular to the freestream, the pressure distributions on the top and bottom faces exactly cancel; i.e., $L' = 0$. In contrast, in the drag direction, parallel to the freestream, the pressure on the front faces a and c is larger than on the back faces b and d, and this results in a finite drag. To calculate this drag (per unit span), consider the geometry of the diamond airfoil in Fig. 9.27, where l is the length of each face and t is the airfoil thickness. Then,

$$D' = 2(p_2 l \sin \varepsilon - p_3 l \sin \varepsilon) = 2(p_2 - p_3)\frac{t}{2}$$

Hence,

$$D' = (p_2 - p_3)t \tag{9.49}$$

In Eq. (9.49), p_2 is calculated from oblique shock theory, and p_3 is obtained from

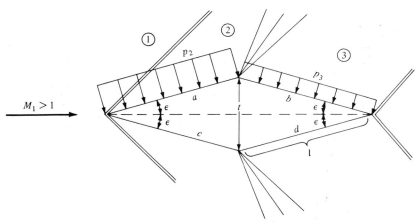

FIGURE 9.27
Diamond-wedge airfoil at zero angle of attack in a supersonic flow.

expansion-wave theory. Moreover, these pressures are the *exact* values for supersonic, inviscid flow over the diamond airfoil.

At this stage, it is worthwhile to recall our discussion in Sec. 1.5 concerning the source of aerodynamic force on a body. In particular, examine Eqs. (1.1), (1.2), (1.7), and (1.8). These equations give the means to calculate L' and D' from the pressure and shear stress distributions over the surface of a body of general shape. The results of the present section, namely, Eqs. (9.47) and (9.48) for a flat plate and Eq. (9.49) for the diamond airfoil, are simply specialized results from the more general formulas given in Sec. 1.5. However, rather than formally going through the integration indicated in Eqs. (1.7) and (1.8), we obtained our results for the simple bodies in Figs. 9.26 and 9.27 in a more direct fashion.

The results of this section illustrate a very important aspect of inviscid, supersonic flow. Note that Eq. (9.48) for the flat plate and Eq. (9.49) for the diamond airfoil predict a *finite drag* for these two-dimensional profiles. This is in direct contrast to our results for two-dimensional bodies in a low-speed, incompressible flow, as discussed in Chaps. 3 and 4, where the drag was theoretically zero. That is, in supersonic flow, d'Alembert's paradox does not occur. In a supersonic, inviscid flow, the drag per unit span on a two-dimensional body is finite. This new source of drag is called *wave drag*, and it represents a serious consideration in the design of all supersonic airfoils. The existence of wave drag is inherently related to the increase in entropy and consequently to the loss of total pressure across the oblique shock waves created by the airfoil.

Finally, the results of this section represent a merger of both the left- and right-hand branches of our road map shown in Fig. 9.5. As such, it brings us to a logical conclusion of our discussion of oblique waves in supersonic flows.

Example 9.8. Calculate the lift and drag coefficients for a flat plate at a 5° angle of attack in a Mach 3 flow.

Solution. Refer to Fig. 9.26. First, calculate p_2/p_1 on the upper surface. From Eq. (9.43),

$$\nu_2 = \nu_1 + \theta$$

where $\theta = \alpha$. From App. C, for $M_1 = 3$, $\nu_1 = 49.76°$. Thus,

$$\nu_2 = 49.76° + 5° = 54.76°$$

From App. C,

$$M_2 = 3.27$$

From App. A, for $M_1 = 3$, $p_{0_1}/p_1 = 36.73$; for $M_2 = 3.27$, $p_{0_2}/p_2 = 55$. Since $p_{0_1} = p_{0_2}$,

$$\frac{p_2}{p_1} = \frac{p_{0_1}}{p_1} \bigg/ \frac{p_{0_2}}{p_2} = \frac{36.73}{55} = 0.668$$

Next, calculate p_3/p_1 on the bottom surface. From the θ-β-M diagram (Fig. 9.7), for $M_1 = 3$ and $\theta = 5°$, $\beta = 23.1°$. Hence,

$$M_{n,1} = M_1 \sin \beta = 3 \sin 23.1° = 1.177$$

From App. B, for $M_{n,1} = 1.177$, $p_3/p_1 = 1.458$ (nearest entry).
Returning to Eq. (9.47), we have

$$L' = (p_3 - p_2)c \cos \alpha$$

The lift coefficient is obtained from

$$c_l = \frac{L'}{q_1 S} = \frac{L'}{(\gamma/2)p_1 M_1^2 c} = \frac{2}{\gamma M_1^2}\left(\frac{p_3}{p_1} - \frac{p_2}{p_1}\right) \cos \alpha$$

$$= \frac{2}{(1.4)(3)^2}(1.458 - 0.668) \cos 5° = \boxed{0.125}$$

From Eq. (9.48),

$$D' = (p_3 - p_2)c \sin \alpha$$

Hence,

$$c_d = \frac{D'}{q_1 S} = \frac{2}{\gamma M_1^2}\left(\frac{p_3}{p_1} - \frac{p_2}{p_1}\right) \sin \alpha$$

$$= \frac{2}{(1.4)(3^2)}(1.458 - 0.668) \sin 5° = \boxed{0.011}$$

A slightly simpler calculation for c_d is to recognize from Eqs. (9.47) and (9.48), or from the geometry of Fig. 9.26, that

$$\frac{c_d}{c_l} = \tan \alpha$$

Hence,

$$c_d = c_l \tan \alpha = 0.125 \tan 5° = 0.011$$

9.8 HISTORICAL NOTE: ERNST MACH— A BIOGRAPHICAL SKETCH

The Mach number is named in honor of Ernst Mach, an Austrian physicist and philosopher who was an illustrious and controversial figure in late nineteenth-century physics. Mach conducted the first meaningful experiments in supersonic flight, and his results triggered a similar interest in Ludwig Prandtl 20 years later. Who was Mach? What did he actually accomplish in supersonic aerodynamics? Let us look at this man further.

Mach was born at Turas, Moravia, in Austria, on February 18, 1838. His father, Johann, was a student of classical literature who settled with his family on a farm in 1840. An extreme individualist, Johann raised his family in an atmosphere of seclusion, working on various improved methods of farming, including silkworm cultivation. Ernst's mother, on the other hand, came from a family of lawyers and doctors and brought with her a love of poetry and music. Ernst seemed to thrive in this family atmosphere. Until the age of 14, his education came exclusively from instruction by his father, who read extensively in the Greek and Latin classics. In 1853, Mach entered public school, where he became interested in the world of science. He went on to obtain a Ph.D. in physics in 1860 at the University of Vienna, writing his dissertation on electrical discharge and induction. In 1864, he became a full professor of mathematics at the University of Graz and was given the title of Professor of Physics in 1866. Mach's work during this period centered on optics—a subject which was to interest him for the rest of his life. The year 1867 was important for Mach—during that year he married, and he also became a professor of experimental physics at the University of Prague, a position he held for the next 28 years. While at Prague, Mach published over 100 technical papers—work which was to constitute the bulk of his technical contributions.

Mach's contribution to supersonic aerodynamics involves a series of experiments covering the period from 1873 to 1893. In collaboration with his son, Ludwig, Mach studied the flow over supersonic projectiles, as well as the propagation of sound waves and shock waves. His work included the flow fields associated with meteorites, explosions, and gas jets. The main experimental data were photographic results. Mach combined his interest in optics and supersonic motion by designing several photographic techniques for making shock waves in air visible. He was the first to use the schlieren system in aerodynamics; this system senses density gradients and allows shock waves to appear on screens or photographic negatives. He also devised an interferometric technique which senses directly the change in density in a flow. A pattern of alternate dark and light bands are set up on a screen by the superposition of light rays passing through regions of different density. Shock waves are visible as a shift in this pattern along the shock. Mach's optical device still perpetuates today in the form of the Mach-Zehnder interferometer, an instrument present in many aerodynamic laboratories. Mach's major contributions in supersonic aerodynamics are contained in a paper given to the Academy of Sciences in Vienna in 1887. Here, for the first time in history, Mach shows a photograph of a weak wave on a slender cone moving at supersonic speed, and he demonstrates that the angle μ between this wave and the direction of flight is given by $\sin \mu = a/V$. This angle was later denoted as the Mach angle by Prandtl and his colleagues after their work on shock and expansion waves in 1907 and 1908. Also, Mach was the first person to point out the discontinuous and marked changes in a flow field as the ratio V/a changes from below 1 to above 1.

It is interesting to note that the ratio V/a was not denoted as Mach number by Mach himself. Rather, the term "Mach number" was coined by the Swiss engineer Jacob Ackeret in his inaugural lecture in 1929 as Privatdozent at the Eidgenossiche Technische Hochschule in Zurich. Hence, the term "Mach number" is of fairly recent usage, not being introduced into the English literature until the early 1930s.

In 1895, the University of Vienna established the Ernst Mach chair in the philosophy of inductive sciences. Mach moved to Vienna to occupy this chair. In 1897 he suffered a stroke which paralyzed the right side of his body. Although he eventually partially recovered, he officially retired in 1901. From that time until his death on February 19, 1916 near Munich, Mach continued to be an active thinker, lecturer, and writer.

In our time, Mach is most remembered for his early experiments on supersonic flow and, of course, through the Mach number itself. However, Mach's contemporaries, as well as Mach himself, viewed him more as a philosopher and historian of science. Coming at the end of the nineteenth century, when most physicists felt comfortable with newtonian mechanics, and many believed that virtually all was known about physics, Mach's outlook on science is summarized by the following passage from his book *Die Mechanik*:

> The most important result of our reflections is that precisely the apparently simplest mechanical theorems are of a very complicated nature; that they are founded on incomplete experiences, even on experiences that never can be fully completed; that in view of the tolerable stability of our environment they are, in fact, practically safeguarded to serve as the foundation of mathematical deduction; but that they by no means themselves can be regarded as mathematically established truths, but only as theorems that not only admit of constant control by experience but actually require it.

In other words, Mach was a staunch experimentalist who believed that the established laws of nature were simply theories and that only observations that are apparent to the senses are the fundamental truth. In particular, Mach could not accept the elementary ideas of atomic theory or the basis of relativity, both of which were beginning to surface during Mach's later years and, of course, were to form the basis of twentieth-century modern physics. As a result, Mach's philosophy did not earn him favor with most of the important physicists of his day. Indeed, at the time of his death, Mach was planning to write a book pointing out the flaws of Einstein's theory of relativity.

Although Mach's philosophy was controversial, he was respected for being a thinker. In fact, in spite of Mach's critical outlook on the theory of relativity, Albert Einstein had the following to say in the year of Mach's death: "I even believe that those who consider themselves to be adversaries of Mach scarcely know how much of Mach's outlook they have, so to speak, adsorbed with their mother's milk."

Hopefully, this section has given you a new dimension to think about whenever you encounter the term "Mach number." Maybe you will pause now and then to reflect on the man himself and to appreciate that the term "Mach number" is in honor of a man who devoted his life to experimental physics, but who at the same time was bold enough to view the physical world through the eyes of a self-styled philosopher.

9.9   SUMMARY

The road map given in Fig. 9.5 illustrates the flow of our discussion on oblique waves in supersonic flow. Review this road map, and make certain that you are familiar with all the ideas and results that are represented in Fig. 9.5.

Some of the more important results are summarized below:

An infinitesimal disturbance in a multidimensional supersonic flow creates a Mach wave which makes an angle μ with respect to the upstream velocity. This angle is defined as the Mach angle and is given by

$$\mu = \sin^{-1} \frac{1}{M} \qquad (9.1)$$

Changes across an oblique shock wave are determined by the normal component of velocity ahead of the wave. For a calorically perfect gas, the normal component of the upstream Mach number is the determining factor. Changes across an oblique shock can be determined from the normal shock relations derived in Chap. 8 by using $M_{n,1}$ in these relations, where

$$M_{n,1} = M_1 \sin \beta \qquad (9.13)$$

Changes across an oblique shock depend on two parameters, e.g., M_1 and β, or M_1 and θ. The relationship between M_1, β, and θ is given in Fig. 9.7, which should be studied closely.

Oblique shock waves incident on a solid surface reflect from that surface in such a fashion to maintain flow tangency on the surface. Oblique shocks also intersect each other, with the results of the intersection depending on the arrangement of the shocks.

The governing factor in the analysis of a centered expansion wave is the Prandtl-Meyer function $\nu(M)$. The key equation which relates the downstream Mach number M_2, the upstream Mach number M_1, and the deflection angle θ is

$$\theta = \nu(M_2) - \nu(M_1) \qquad (9.43)$$

The pressure distribution over a supersonic airfoil made up of straight-line segments can usually be calculated exactly from a combination of oblique and expansion waves—i.e., from exact shock-expansion theory.

PROBLEMS

9.1. A slender missile is flying at Mach 1.5 at low altitude. Assume the wave generated by the nose of the missile is a Mach wave. This wave intersects the ground 559 ft behind the nose. At what altitude is the missile flying?

9.2. Consider an oblique shock wave with a wave angle of 30° in a Mach 4 flow. The upstream pressure and temperature are 2.65×10^4 N/m^2 and 223.3 K, respectively (corresponding to a standard altitude of 10,000 m). Calculate the pressure, temperature, Mach number, total pressure, and total temperature behind the wave and the entropy increase across the wave.

9.3. Equation (8.80) does *not* hold for an oblique shock wave, and hence the column in App. B labeled $p_{0,2}/p_1$ *cannot* be used, in conjunction with the normal component of the upstream Mach number, to obtain the total pressure behind an oblique shock wave. On the other hand, the column labeled $p_{0,2}/p_{0,1}$ can be used for an oblique shock wave, using $M_{n,1}$. Explain why all this is so.

9.4. Consider an oblique shock wave with a wave angle of 36.87°. The upstream flow is given by $M_1 = 3$ and $p_1 = 1$ atm. Calculate the total pressure behind the shock using
(*a*) $p_{0,2}/p_{0,1}$ from App. B (the correct way)
(*b*) $p_{0,2}/p_1$ from App. B (the incorrect way)
Compare the results.

9.5. Consider the flow over a 22.2° half-angle wedge. If $M_1 = 2.5$, $p_1 = 1$ atm, and $T_1 = 300$ K, calculate the wave angle and p_2, T_2, and M_2.

9.6. Consider a flat plate at an angle of attack α to a Mach 2.4 airflow at 1 atm pressure. What is the maximum pressure that can occur on the plate surface and still have an attached shock wave at the leading edge? At what value of α does this occur?

9.7. A 30.2° half-angle wedge is inserted into a freestream with $M_\infty = 3.5$ and $p_\infty = 0.5$ atm. A Pitot tube is located above the wedge surface and behind the shock wave. Calculate the magnitude of the pressure sensed by the Pitot tube.

9.8. Consider a Mach 4 airflow at a pressure of 1 atm. We wish to slow this flow to subsonic speed through a system of shock waves with as small a loss in total pressure as possible. Compare the loss in total pressure for the following three shock systems:
(*a*) A single normal shock wave
(*b*) An oblique shock with a deflection angle of 25.3°, followed by a normal shock
(*c*) An oblique shock with a deflection angle of 25.3°, followed by a second oblique shock of deflection angle of 20°, followed by a normal shock
From the results of (*a*), (*b*), and (*c*), what can you induce about the efficiency of the various shock systems?

9.9. Consider an oblique shock generated at a compression corner with a deflection angle $\theta = 18.2°$. A straight horizontal wall is present above the corner, as shown in Fig. 9.14. If the upstream flow has the properties $M_1 = 3.2$, $p_1 = 1$ atm and $T_1 = 520°$R, calculate M_3, p_3, and T_3 behind the reflected shock from the upper wall. Also, obtain the angle Φ which the reflected shock makes with the upper wall.

9.10. Consider the supersonic flow over an expansion corner, such as given in Fig. 9.20. The deflection angle $\theta = 23.38°$. If the flow upstream of the corner is given by $M_1 = 2$, $p_1 = 0.7$ atm, $T_1 = 630°$R, calculate M_2, p_2, T_2, ρ_2, $p_{0,2}$, and $T_{0,2}$ downstream of the corner. Also, obtain the angles the forward and rearward Mach lines make with respect to the upstream direction.

9.11. A supersonic flow at $M_1 = 1.58$ and $p_1 = 1$ atm expands around a sharp corner. If the pressure downstream of the corner is 0.1306 atm, calculate the deflection angle of the corner.

9.12. A supersonic flow at $M_1 = 3$, $T_1 = 285$ K, and $p_1 = 1$ atm is deflected upward through a compression corner with $\theta = 30.6°$ and then is subsequently expanded around a corner of the same angle such that the flow direction is the same as its original direction. Calculate M_3, p_3, and T_3 downstream of the expansion corner. Since the resulting flow is in the same direction as the original flow, would you expect $M_3 = M_1$, $p_3 = p_1$, and $T_3 = T_1$? Explain.

9.13. Consider an infinitely thin flat plate at an angle of attack α in a Mach 2.6 flow. Calculate the lift and wave-drag coefficients for
(*a*) $\alpha = 5°$ (*b*) $\alpha = 15°$ (*c*) $\alpha = 30°$
(*Note*: Save the results of this problem for use in Chap. 12.)

9.14. Consider a diamond-wedge airfoil such as shown in Fig. 9.27, with a half-angle $\varepsilon = 10°$. The airfoil is at an angle of attack $\alpha = 15°$ to a Mach 3 freestream. Calculate the lift and wave-drag coefficients for the airfoil.

9.15. Consider sonic flow. Calculate the maximum deflection angle through which this flow can be expanded via a centered expansion wave.

9.16. Consider a circular cylinder (oriented with its axis perpendicular to the flow) and a symmetric diamond-wedge airfoil with a half-angle of 5° at zero angle of attack; both bodies are in the same Mach 5 freestream. The thickness of the airfoil and the diameter of the cylinder are the same. The drag coefficient (based on projected frontal area) of the cylinder is 4/3. Calculate the *ratio* of the cylinder drag to the diamond airfoil drag. What does this say about the aerodynamic performance of a blunt body compared to a sharp-nosed slender body in supersonic flow?

CHAPTER
10

COMPRESSIBLE FLOW THROUGH NOZZLES, DIFFUSERS, AND WIND TUNNELS

Having wondered from what source there is so much difficulty in successfully applying the principles of dynamics to fluids than to solids, finally, turning the matter over more carefully in my mind, I found the true origin of the difficulty; I discovered it to consist of the fact that a certain part of the pressing forces important in forming the <u>throat</u> (so called by me, not considered by others) was neglected, and moreover regarded as if of no importance, for no other reason than the throat is composed of a very small, or even an infinitely small, quantity of fluid, such as occurs whenever fluid passes from a wider place to a narrower, or vice versa, from a narrower to a wider.

Johann Bernoulli; from his Hydraulics, 1743

10.1 INTRODUCTION

Chapters 8 and 9 treated normal and oblique waves in supersonic flow. These waves are present on any aerodynamic vehicle in supersonic flight. Aeronautical engineers are concerned with observing the characteristics of such vehicles, especially the generation of lift and drag at supersonic speeds, as well as details of the flow field, including the shock- and expansion-wave patterns. To make such observations, we usually have two standard choices: (1) conduct flight tests using the actual vehicle, and (2) run wind-tunnel tests on a small-scale model of the vehicle. Flight tests, although providing the final answers in the full-scale environment, are costly and, not to say the least, dangerous if the vehicle is

499

unproven. Hence, the vast bulk of supersonic aerodynamic data have been obtained in wind tunnels on the ground. What do such supersonic wind tunnels look like? How do we produce a uniform flow of supersonic gas in a laboratory environment? What are the characteristics of supersonic wind tunnels? The answers to these and other questions are addressed in this chapter.

The first practical supersonic wind tunnel was built and operated by Adolf Busemann in Germany in the mid-1930s, although Prandtl had a small supersonic facility operating as early as 1905 for the study of shock waves. A photograph of Busemann's tunnel is shown in Fig. 10.1. Such facilities proliferated quickly during and after World War II. Today, all modern aerodynamic laboratories have one or more supersonic wind tunnels, and many are equipped with hypersonic tunnels as well. Such machines come in all sizes; an example of a moderately large hypersonic tunnel is shown in Fig. 10.2.

In this chapter, we discuss the aerodynamic fundamentals of compressible flow through ducts. Such fundamentals are vital to the proper design of high-speed wind tunnels, rocket engines, high-energy gas-dynamic and chemical lasers, and jet engines, to list just a few. Indeed, the material developed in this chapter is used almost daily by practicing aerodynamicists and is indispensable toward a full understanding of compressible flow.

The road map for this chapter is given in Fig. 10.3. After deriving the governing equations, we treat the cases of a nozzle and diffuser separately. Then we merge this information to examine the case of supersonic wind tunnels.

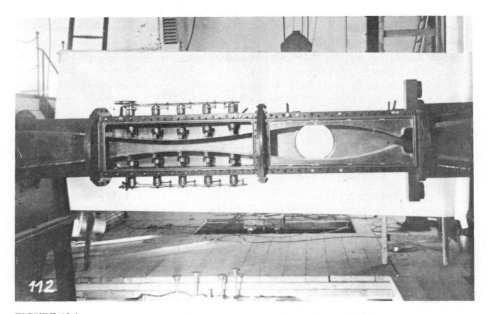

FIGURE 10.1
The first practical supersonic wind tunnel, built by A. Busemann in Germany in the mid-1930s. (*Courtesy of A. Busemann.*)

FIGURE 10.2
A large hypersonic wind tunnel at the U.S. Air Force Wright Aeronautical Laboratory, Dayton, Ohio.
(*Courtesy of the U.S. Air Force.*)

10.2 GOVERNING EQUATIONS FOR QUASI-ONE-DIMENSIONAL FLOW

Recall the one-dimensional flow treated in Chap. 8. There, we considered the flow-field variables to be a function of x only, i.e., $p = p(x)$, $u = u(x)$, etc. Strictly speaking, a streamtube for such a flow must be of constant area; i.e., the

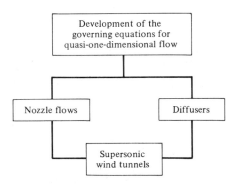

FIGURE 10.3
Road map for Chap. 10.

one-dimensional flow discussed in Chap. 8 is constant-area flow, as sketched in Fig. 10.4a.

In contrast, assume that the area of the streamtube changes as a function of x, i.e., $A = A(x)$, as sketched in Fig. 10.4b. Strictly speaking, this flow is three-dimensional; the flow-field variables are functions of x, y, and z, as can be seen simply by examining Fig. 10.4b. In particular, the velocity at the boundary of the streamtube must be tangent to the boundary, and hence it has components in the y and z directions as well as the axial x direction. However, if the area variation is moderate, the components in the y and z directions are small in comparison with the component in the x direction. In such a case, the flow-field variables can be *assumed* to vary with x only; i.e., the flow can be assumed to be uniform across any cross section at a given x station. Such a flow, where $A = A(x)$, but $p = p(x)$, $\rho = \rho(x)$, $u = u(x)$, etc., is defined as *quasi-one-dimensional flow*, as sketched in Fig. 10.4b. Such flow is the subject of this chapter. We have encountered quasi-one-dimensional flow earlier, in our discussion of incompressible flow through a duct in Sec. 3.3. Return to Sec. 3.3, and review the concepts presented there before progressing further.

Although the assumption of quasi-one-dimensional flow is an approximation to the actual flow in a variable-area duct, the integral forms of the conservation equations, namely, continuity [Eq. (2.39)], momentum [Eq. (2.55)], and energy [Eq. (2.86)], can be used to obtain governing equations for quasi-one-dimensional flow which are physically consistent, as follows. Consider the control volume given in Fig. 10.5. At station 1, the flow across area A_1 is assumed to be uniform with properties p_1, ρ_1, u_1, etc. Similarly, at station 2, the flow across area A_2 is assumed to be uniform with properties p_2, ρ_2, u_2, etc. The application of the integral form of the continuity equation was made to such a variable-area control volume in Sec. 3.3. The resulting continuity equation for steady, quasi-one-dimensional flow was obtained as Eq. (3.21), which in terms of the nomenclature in Fig. 10.5 yields

$$\rho_1 u_1 A_1 = \rho_2 u_2 A_2 \qquad (10.1)$$

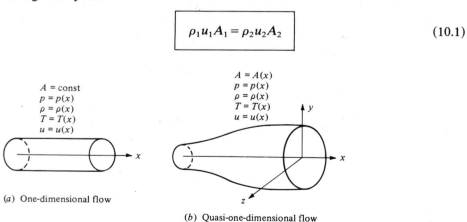

(a) One-dimensional flow

(b) Quasi-one-dimensional flow

FIGURE 10.4
One-dimensional and quasi-one-dimensional flows.

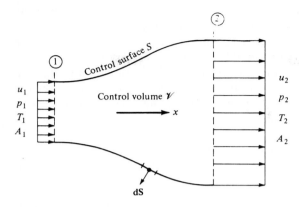

FIGURE 10.5
Finite control volume for quasi-one-dimensional flow.

Consider the integral form of the momentum equation, Eq. (2.55). For a steady, inviscid flow with no body forces, this equation becomes

$$\oiint_S (\rho \mathbf{V} \cdot d\mathbf{S})\mathbf{V} = -\oiint_S p \, d\mathbf{S} \tag{10.2}$$

Since Eq. (10.2) is a vector equation, let us examine its x component, given below:

$$\oiint_S (\rho \mathbf{V} \cdot d\mathbf{S})u = -\oiint_S (p \, d\mathbf{S})_x \tag{10.3}$$

where $(p \, d\mathbf{S})_x$ denotes the x component of the pressure force. Since Eq. (10.3) is a scalar equation, we must be careful about the sign of the x components when evaluating the surface integrals. All components pointing to the right in Fig. 10.5 are positive, and those pointing to the left are negative. The upper and lower surfaces of the control volume in Fig. 10.5 are streamlines; hence, $\mathbf{V} \cdot d\mathbf{S} = 0$ along these surfaces. Also, recall that across A_1, $\mathbf{V}$ and $d\mathbf{S}$ are in opposite directions; hence, $\mathbf{V} \cdot d\mathbf{S}$ is negative. Therefore, the integral on the left of Eq. (10.3) becomes $-\rho_1 u_1^2 A_1 + \rho_2 u_2^2 A_2$. The pressure integral on the right of Eq. (10.2), evaluated over the faces A_1 and A_2 of the control volume, becomes $-(-p_1 A_1 + p_2 A_2)$. (The negative sign in front of $p_1 A_1$ is because $d\mathbf{S}$ over A_1 points to the left, which is the negative direction for the x components.) Evaluated over the upper and lower surface of the control volume, the pressure integral can be expressed as

$$-\int_{A_1}^{A_2} -p \, dA = \int_{A_1}^{A_2} p \, dA \tag{10.4}$$

where dA is simply the x component of the vector $d\mathbf{S}$, i.e., the area dS projected on a plane perpendicular to the x axis. The negative sign *inside* the integral on the left of Eq. (10.4) is due to the direction of $d\mathbf{S}$ along the upper and lower surfaces; note that $d\mathbf{S}$ points in the backward direction along these surfaces, as shown in Fig. 10.5. Hence, the x component of $p \, d\mathbf{S}$ is to the left, and therefore appears in our equations as a negative component. [Recall from Sec. 2.5 that the negative sign *outside* the pressure integral, i.e., outside the integral on the left of Eq. (10.4), is always present to account for the physical fact that the pressure force $p\,\mathbf{dS}$ exerted on a control surface always acts in the opposite direction of $\mathbf{dS}$. If you are unsure about this, review the derivation of the momentum equation in Sec. 2.5. Also, do not let the signs in the above results confuse you; they are all quite logical if you keep track of the direction of the x components.] With the above results, Eq. (10.3) becomes

$$-\rho_1 u_1^2 A_1 + \rho_2 u_2^2 A_2 = -(-p_1 A_1 + p_2 A_2) + \int_{A_1}^{A_2} p\,dA$$

$$\boxed{p_1 A_1 + \rho_1 u_1^2 A_1 + \int_{A_1}^{A_2} p\,dA = p_2 A_2 + \rho_2 u_2^2 A_2} \tag{10.5}$$

Equation (10.5) is the momentum equation for steady, quasi-one-dimensional flow.

Consider the energy equation given by Eq. (2.86). For inviscid, adiabatic, steady flow with no body forces, this equation becomes

$$\oiint_S \rho \left(e + \frac{V^2}{2} \right) \mathbf{V} \cdot \mathbf{dS} = -\oiint_S p\mathbf{V} \cdot \mathbf{dS} \tag{10.6}$$

Applied to the control volume in Fig. 10.5, Eq. (10.6) yields

$$\rho_1 \left(e_1 + \frac{u_1^2}{2} \right)(-u_1 A_1) + \rho_2 \left(e_2 + \frac{u_2^2}{2} \right)(u_2 A_2) = -(-p_1 u_1 A_1 + p_2 u_2 A_2)$$

or

$$p_1 u_1 A_1 + \rho_1 u_1 A_1 \left(e_1 + \frac{u_1^2}{2} \right) = p_2 u_2 A_2 + \rho_2 u_2 A_2 \left(e_2 + \frac{u_2^2}{2} \right) \tag{10.7}$$

Dividing Eq. (10.7) by Eq. (10.1), we have

$$\frac{p_1}{\rho_1} + e_1 + \frac{u_1^2}{2} = \frac{p_2}{\rho_2} + e_2 + \frac{u_2^2}{2} \tag{10.8}$$

Recall that $h = e + pv = e + p/\rho$. Hence, Eq. (10.8) becomes

$$\boxed{h_1 + \frac{u_1^2}{2} = h_2 + \frac{u_2^2}{2}} \tag{10.9}$$

which is the energy equation for steady, adiabatic, inviscid quasi-one-dimensional flow. Examine Eq. (10.9) closely; it is a statement that the total enthalpy, $h_0 = h + u^2/2$, is a constant throughout the flow. Once again, this should come as no surprise; Eq. (10.9) is simply another example of the general result for steady, inviscid, adiabatic flow discussed in Sec. 7.5. Hence, we can replace Eq. (10.9) by

$$h_0 = \text{const} \tag{10.10}$$

Pause for a moment and examine our results given above. We have applied the integral forms of the conservation equations to the control volume in Fig. 10.5. We have obtained, as a result, Eqs. (10.1), (10.5), and (10.9) or (10.10) as the governing continuity, momentum, and energy equations, respectively, for quasi-one-dimensional flow. Examine these equations—they are *algebraic* equations (with the exception of the single integral term in the momentum equation). In Fig. 10.5, assume that the inflow conditions ρ_1, u_1, p_1, T_1, and h_1 are given and that the area distribution $A = A(x)$ is presented. Also, assume a calorically perfect gas, where

$$p_2 = \rho_2 R T_2 \tag{10.11}$$

and

$$h_2 = c_p T_2 \tag{10.12}$$

Equations (10.1), (10.5), (10.9) or (10.10), (10.11), and (10.12) constitute five equations for the five unknowns ρ_2, u_2, p_2, T_2, and h_2. We could, in principle, solve these equations directly for the unknown flow quantities at station 2 in Fig. 10.5. However, such a direct solution would involve substantial algebraic manipulations. Instead, we take a simpler tack, as described in Sec. 10.3.

Before moving on to a solution of the governing equations, let us examine some physical characteristics of a quasi-one-dimensional flow. To help this examination, we first obtain some *differential* expressions for the governing equations, in contrast to the algebraic equations obtained above. For example, consider Eq. (10.1), which states that

$$\rho u A = \text{const} \tag{10.13}$$

through a variable-area duct. Differentiating Eq. (10.13), we have

$$d(\rho u A) = 0 \tag{10.14}$$

which is the differential form of the continuity equation for quasi-one-dimensional flow.

To obtain a differential form of the momentum equation, apply Eq. (10.5) to the infinitesimal control volume sketched in Fig. 10.6. The flow going into the volume at station 1, where the area is A, has properties p, u, and ρ. In traversing the length dx, where the area changes by dA, the flow properties change by the corresponding amounts dp, $d\rho$, and du. Hence, the flow leaving at station 2 has the properties $p + dp$, $u + du$, and $\rho + d\rho$, as shown in Fig. 10.6. For this case, Eq. (10.5) becomes [recognizing that the integral in Eq. (10.5) can be replaced by its integrand for the differential volume in Fig. 10.6]

$$pA + \rho u^2 A + p\, dA = (p + dp)(A + dA) + (\rho + d\rho)(u + du)^2(A + dA) \tag{10.15}$$

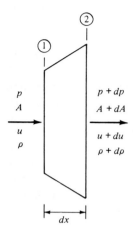

FIGURE 10.6
Incremental control volume.

In Eq. (10.15), all products of differentials, such as $dp\, dA$, $dp(du)^2$, are very small and can be ignored. Hence, Eq. (10.15) becomes

$$A\, dp + Au^2\, d\rho + \rho u^2\, dA + 2\rho uA\, du = 0 \tag{10.16}$$

Expanding the continuity equation, Eq. (10.14), and multiplying by u, we have

$$\rho u^2\, dA + \rho uA\, du + Au^2\, d\rho = 0 \tag{10.17}$$

Subtracting Eq. (10.17) from (10.16), we obtain

$$\boxed{dp = -\rho u\, du} \tag{10.18}$$

which is the differential form of the momentum equation for steady, inviscid, quasi-one-dimensional flow. Equation (10.18) is called *Euler's equation.* We have seen it before—as Eq. (3.12). In Sec. 3.2, it was derived from the differential form of the general momentum equation in three dimensions. (Make certain to review that derivation before progressing further.) In Sec. 3.2, we demonstrated that Eq. (3.12) holds along a streamline in a general three-dimensional flow. Now we see Euler's equation again, in Eq. (10.18), which was derived from the governing equations for quasi-one-dimensional flow.

A differential form of the energy equation follows directly from Eq. (10.9), which states that

$$h + \frac{u^2}{2} = \text{const}$$

Differentiating this equation, we have

$$\boxed{dh + u\, du = 0} \tag{10.19}$$

In summary, Eqs. (10.14), (10.18), and (10.19) are differential forms of the continuity, momentum, and energy equations, respectively, for a steady, inviscid, adiabatic, quasi-one-dimensional flow. We have obtained them from the algebraic forms of the equations derived earlier, applied essentially to the picture shown in Fig. 10.6. Now you might ask the question, Since we spent some effort obtaining partial differential equations for continuity, momentum, and energy in Chap. 2, applicable to a general three-dimensional flow, why would we not simply set $\partial/\partial y = 0$ and $\partial/\partial z = 0$ in those equations and obtain differential equations applicable to the one-dimensional flow treated in the present chapter? The answer is that we certainly could perform such a reduction, and we would obtain Eqs. (10.18) and (10.19) directly. [Return to the differential equations, Eqs. (2.104a) and (2.105), and prove this to yourself.] However, if we take the general continuity equation, Eq. (2.43), and reduce it to one-dimensional flow, we obtain $d(\rho u) = 0$. Comparing this result with Eq. (10.14) for quasi-one-dimensional flow, we see an inconsistency. This is another example of the physical inconsistency between the *assumption* of quasi-one-dimensional flow in a variable-area duct and the three-dimensional flow which actually occurs in such a duct. The result obtained from Eq. (2.43), namely, $d(\rho u) = 0$, is a *truly* one-dimensional result, which applies to *constant-area* flows such as considered in Chap. 8. [Recall in Chap. 8 that the continuity equation was used in the form $\rho u =$ constant, which is compatible with Eq. (2.43).] However, once we make the quasi-one-dimensional assumption, i.e., that uniform properties hold across a given cross section in a variable-area duct, then Eq. (10.14) is the only differential form of the continuity equation which insures mass conservation for such an assumed flow.

Let us now use the differential forms of the governing equations, obtained above, to study some physical characteristics of quasi-one-dimensional flow. Such physical information can be obtained from a particular combination of these equations, as follows. From Eq. (10.14),

$$\frac{d\rho}{\rho} + \frac{du}{u} + \frac{dA}{A} = 0 \tag{10.20}$$

We wish to obtain an equation which relates the change in velocity, du, to the change in area, dA. Hence, to eliminate $d\rho/\rho$ in Eq. (10.20), consider Eq. (10.18) written as

$$\frac{dp}{\rho} = \frac{dp}{d\rho}\frac{d\rho}{\rho} = -u\,du \tag{10.21}$$

Keep in mind that we are dealing with inviscid, adiabatic flow. Moreover, for the time being, we are assuming no shock waves in the flow. Hence, the flow is *isentropic*. In particular, any change in density, $d\rho$, with respect to a change in pressure, dp, takes place isentropically; i.e.,

$$\frac{dp}{d\rho} \equiv \left(\frac{\partial p}{\partial \rho}\right)_s \tag{10.22}$$

From Eq. (8.18) for the speed of sound, Eq. (10.22) becomes

$$\frac{dp}{d\rho} = a^2 \tag{10.23}$$

Substituting Eq. (10.23) into (10.21), we have

$$a^2 \frac{d\rho}{\rho} = -u\,du$$

or

$$\frac{d\rho}{\rho} = -\frac{u\,du}{a^2} = -\frac{u^2}{a^2}\frac{du}{u} = -M^2\frac{du}{u} \tag{10.24}$$

Substituting Eq. (10.24) into (10.20), we have

$$-M^2\frac{du}{u} + \frac{du}{u} + \frac{dA}{A} = 0$$

or

$$\boxed{\frac{dA}{A} = (M^2 - 1)\frac{du}{u}} \tag{10.25}$$

Equation (10.25) is the desired equation which relates dA to du; it is called the *area-velocity relation.*

Equation (10.25) is very important; study it closely. In the process, recall the standard convention for differentials; e.g., a positive value of du connotes an *increase* in velocity, a negative value of du connotes a *decrease* in velocity, etc. With this in mind, Eq. (10.25) tells us the following information:

1. For $0 \le M < 1$ (subsonic flow), the quantity in parentheses in Eq. (10.25) is negative. Hence, an *increase* in velocity (positive du) is associated with a *decrease* in area (negative dA). Likewise, a decrease in velocity (negative du) is associated with an increase in area (positive dA). Clearly, for a subsonic compressible flow, to increase the velocity, we must have a convergent duct, and to decrease the velocity, we must have a divergent duct. These results are illustrated at the top of Fig. 10.7. Also, these results are similar to the familiar trends for incompressible flow studied in Sec. 3.3. Once again we see that subsonic compressible flow is qualitatively (but not quantitatively) similar to incompressible flow.

2. For $M > 1$ (supersonic flow), the quantity in parentheses in Eq. (10.25) is positive. Hence, an *increase* in velocity (positive du) is associated with an *increase* in area (positive dA). Likewise, a decrease in velocity (negative du) is associated with a decrease in area (negative dA). For a supersonic flow, to increase the velocity, we must have a divergent duct, and to decrease the velocity, we must have a convergent duct. These results are illustrated at the bottom of Fig. 10.7; they are the *direct opposite* of the trends for subsonic flow.

3. For $M = 1$ (sonic flow), Eq. (10.25) shows that $dA = 0$ even though a finite du exists. Mathematically, this corresponds to a local maximum or minimum in

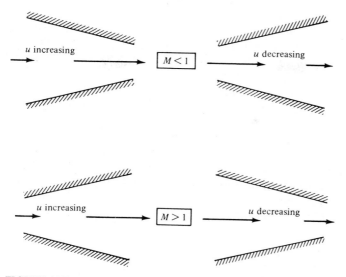

FIGURE 10.7
Compressible flow in converging and diverging ducts.

the area distribution. Physically, it corresponds to a minimum area, as discussed below.

Imagine that we want to take a gas at rest and isentropically expand it to supersonic speeds. The above results show that we must first accelerate the gas subsonically in a convergent duct. However, as soon as sonic conditions are achieved, we must further expand the gas to supersonic speeds by diverging the

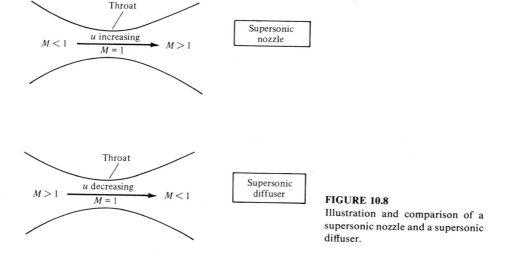

FIGURE 10.8
Illustration and comparison of a supersonic nozzle and a supersonic diffuser.

duct. Hence, a nozzle designed to achieve supersonic flow at its exit is a *convergent-divergent* duct, as sketched at the top of Fig. 10.8. The minimum area of the duct is called the *throat*. Whenever an isentropic flow expands from subsonic to supersonic speeds, the flow must pass through a throat; moreover, in such a case, $M = 1$ at the throat. The converse is also true; i.e., if we wish to take a supersonic flow and slow it down isentropically to subsonic speeds, we must first decelerate the gas in a convergent duct, and then as soon as sonic flow is obtained, we must further decelerate it to subsonic speeds in a divergent duct. Here, the convergent-divergent duct at the bottom of Fig. 10.8 is operating as a diffuser. Note that whenever an isentropic flow is slowed from supersonic to subsonic speeds, the flow must pass through a throat; moreover, in such a case, $M = 1$ at the throat.

As a final note on Eq. (10.25), consider the case when $M = 0$. Then we have $dA/A = -du/u$, which integrates to $Au = $ constant. This is the familiar continuity equation for incompressible flow in ducts as derived in Sec. 3.3 and as given by Eq. (3.22).

10.3 NOZZLE FLOWS

In this section, we move to the left-hand branch of the road map given in Fig. 10.3; i.e., we study in detail the compressible flow through nozzles. To expedite this study, we first derive an important equation which relates Mach number to the ratio of duct area to sonic throat area.

Consider the duct shown in Fig. 10.9. Assume that sonic flow exists at the throat, where the area is A^*. The Mach number and the velocity at the throat are denoted by M^* and u^*, respectively. Since the flow is sonic at the throat, $M^* = 1$ and $u^* = a^*$. (Note that the use of an asterisk to denote sonic conditions was introduced in Sec. 7.5; we continue this convention in our present discussion.) At any other section of this duct, the area, the Mach number, and the velocity are denoted by A, M, and u, respectively, as shown in Fig. 10.9. Writing Eq. (10.1) between A and A^*, we have

$$\rho^* u^* A^* = \rho u A \tag{10.26}$$

Since $u^* = a^*$, Eq. (10.26) becomes

$$\frac{A}{A^*} = \frac{\rho^*}{\rho} \frac{a^*}{u} = \frac{\rho^*}{\rho_0} \frac{\rho_0}{\rho} \frac{a^*}{u} \tag{10.27}$$

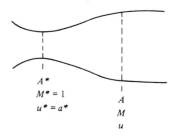

A^*
$M^* = 1$
$u^* = a^*$

A
M
u

FIGURE 10.9
Geometry for the derivation of the area–Mach number relation.

where ρ_0 is the stagnation density defined in Sec. 7.5 and is constant throughout an isentropic flow. From Eq. (8.46), we have

$$\frac{\rho^*}{\rho_0} = \left(\frac{2}{\gamma+1}\right)^{1/(\gamma-1)} \tag{10.28}$$

Also, from Eq. (8.43), we have

$$\frac{\rho_0}{\rho} = \left(1 + \frac{\gamma-1}{2} M^2\right)^{1/(\gamma-1)} \tag{10.29}$$

Also, recalling the definition of M^* in Sec. 8.4, as well as Eq. (8.48), we have

$$\left(\frac{u}{a^*}\right)^2 = M^{*2} = \frac{[(\gamma+1)/2]M^2}{1+[(\gamma-1)/2]M^2} \tag{10.30}$$

Squaring Eq. (10.27) and substituting Eqs. (10.28) to (10.30), we obtain

$$\left(\frac{A}{A^*}\right)^2 = \left(\frac{\rho^*}{\rho_0}\right)^2 \left(\frac{\rho_0}{\rho}\right)^2 \left(\frac{a^*}{u}\right)^2$$

or $\quad\left(\frac{A}{A^*}\right) = \left(\frac{2}{\gamma+1}\right)^{2/(\gamma-1)} \left(1 + \frac{\gamma-1}{2} M^2\right)^{2/(\gamma-1)} \frac{1+[(\gamma-1)/2]M^2}{[(\gamma+1)/2]M^2} \tag{10.31}$

Algebraically simplifying Eq. (10.31), we have

$$\boxed{\left(\frac{A}{A^*}\right)^2 = \frac{1}{M^2}\left[\frac{2}{\gamma+1}\left(1+\frac{\gamma-1}{2} M^2\right)\right]^{(\gamma+1)/(\gamma-1)}} \tag{10.32}$$

Equation (10.32) is very important; it is called the *area–Mach number relation*, and it contains a striking result. "Turned inside out," Eq. (10.32) tells us that $M = f(A/A^*)$; i.e., *the Mach number at any location in the duct is a function of the ratio of the local duct area to the sonic throat area.* Recall from our discussion of Eq. (10.25) that A must be greater than or at least equal to A^*; the case where $A < A^*$ is physically not possible in an isentropic flow. Thus, in Eq. (10.32), $A/A^* \geq 1$. Also, Eq. (10.32) yields *two* solutions for M at a given A/A^*—a subsonic value and a supersonic value. Which value of M that actually holds in a given case depends on the pressures at the inlet and exit of the duct, as explained later. The results for A/A^* as a function of M, obtained from Eq. (10.32), are tabulated in App. A. Examining App. A, we note that for subsonic values of M, as M increases, A/A^* decreases; i.e., the duct converges. At $M = 1$, $A/A^* = 1$ in App. A. Finally, for supersonic values of M, as M increases, A/A^* increases; i.e., the duct diverges. These trends in App. A are consistent with our physical discussion of convergent-divergent ducts at the end of Sec. 10.2. Moreover, App. A shows the double-valued nature of M as a function of A/A^*. For example, for $A/A^* = 2$, we have either $M = 0.31$ or $M = 2.2$.

Consider a given convergent-divergent nozzle, as sketched in Fig. 10.10a. Assume that the area ratio at the inlet, A_i/A^*, is very large and that the flow at the inlet is fed from a large gas reservoir where the gas is essentially stationary. The reservoir pressure and temperature are p_0 and T_0, respectively. Since A_i/A^* is very large, the subsonic Mach number at the inlet is very small, $M \approx 0$. Thus, the pressure and temperature at the inlet are essentially p_0 and T_0, respectively. The area distribution of the nozzle, $A = A(x)$, is specified, so that A/A^* is known at every station along the nozzle. The area of the throat is denoted by A_t, and the exit area is denoted by A_e. The Mach number and static pressure at the exit are denoted by M_e and p_e, respectively. Assume that we have an isentropic expansion of the gas through this nozzle to a supersonic Mach number $M_e = M_{e,6}$

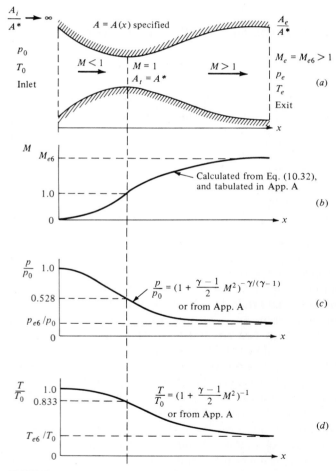

FIGURE 10.10
Isentropic supersonic nozzle flow.

at the exit (the reason for the subscript 6 will be apparent later). The corresponding exit pressure is $p_{e,6}$. For this expansion, the flow is sonic at the throat; hence, $M = 1$ and $A_t = A^*$ at the throat. The flow properties through the nozzle are a function of the local area ratio A/A^* and are obtained as follows:

1. The local Mach number as a function of x is obtained from Eq. (10.32), or more directly from the tabulated values in App. A. For the specified $A = A(x)$, we know the corresponding $A/A^* = f(x)$. Then read the related subsonic Mach numbers in the convergent portion of the nozzle from the first part of App. A (for $M < 1$) and the related supersonic Mach numbers in the divergent portion of the nozzle from the second part of App. A (for $M > 1$). The Mach number distribution through the complete nozzle is thus obtained and is sketched in Fig. 10.10b.

2. Once the Mach number distribution is known, then the corresponding variation of temperature, pressure, and density can be found from Eqs. (8.40), (8.42), and (8.43), respectively, or more directly from App. A. The distributions of p/p_0 and T/T_0 are sketched in Fig. 10.10c and d, respectively.

Examine the variations shown in Fig. 10.10. For the isentropic expansion of a gas through a convergent-divergent nozzle, the Mach number monotonically increases from near 0 at the inlet to $M = 1$ at the throat, and to the supersonic value $M_{e,6}$ at the exit. The pressure monotonically decreases from p_0 at the inlet to $0.528p_0$ at the throat and to the lower value $p_{e,6}$ at the exit. Similarly, the temperature monotonically decreases from T_0 at the inlet to $0.833T_0$ at the throat and to the lower value $T_{e,6}$ at the exit. Again, for the isentropic flow shown in Fig. 10.10, we emphasize that the distribution of M, and hence the resulting distributions of p and T, through the nozzle depends only on the local area ratio A/A^*. This is the key to the analysis of isentropic, supersonic, quasi-one-dimensional nozzle flows.

Imagine that you take a convergent-divergent nozzle, and simply place it on a table in front of you. What is going to happen? Is the air going to suddenly start flowing through the nozzle of its own accord? The answer is, of course not! Rather, by this stage in your study of aerodynamics, your intuition should tell you that we have to impose a force on the gas in order to produce any acceleration. Indeed, this is the essence of the momentum equation derived in Sec. 2.5. For the inviscid flows considered here, the only mechanism to produce an accelerating force on a gas is a pressure gradient. Thus, returning to the nozzle on the table, a pressure difference must be created between the inlet and exit; only then will the gas start to flow through the nozzle. The exit pressure must be less than the inlet pressure; i.e., $p_e < p_0$. Moreover, if we wish to produce the isentropic supersonic flow sketched in Fig. 10.10, the pressure p_e/p_0 must be *precisely* the value stipulated by App. A for the known exit Mach number $M_{e,6}$; i.e., $p_e/p_0 = p_{e,6}/p_0$. If the pressure ratio is different from the above isentropic value, the flow either inside or outside the nozzle will be different from that shown in Fig. 10.10.

Let us examine the type of nozzle flows that occur when p_e/p_0 is not equal to the precise isentropic value for $M_{e,6}$, i.e., when $p_e/p_0 \neq p_{e,6}/p_0$. To begin with, consider the convergent-divergent nozzle sketched in Fig. 10.11a. If $p_e = p_0$, no pressure difference exists, and no flow occurs inside the nozzle. Now assume that p_e is minutely reduced below p_0, say, $p_e = 0.999p_0$. This small pressure difference will produce a very low-speed subsonic flow inside the nozzle—essentially a gentle wind. The local Mach number will increase slightly through the convergent portion, reaching a maximum value at the throat, as shown by curve 1 in Fig. 10.11b. This Mach number at the throat will *not* be sonic; rather, it will be some small subsonic value. Downstream of the throat, the local Mach number will decrease in the divergent section, reaching a very small but finite value $M_{e,1}$ at the exit. Correspondingly, the pressure in the convergent section will gradually decrease from p_0 at the inlet to a minimum value at the throat, and then will gradually increase to the value $p_{e,1}$ at the exit. This variation is shown as curve 1 in Fig. 10.11c. Please note that because the flow is not sonic at the throat in this case, A_t is not equal to A^*. Recall that A^*, which appears in Eq. (10.32), is

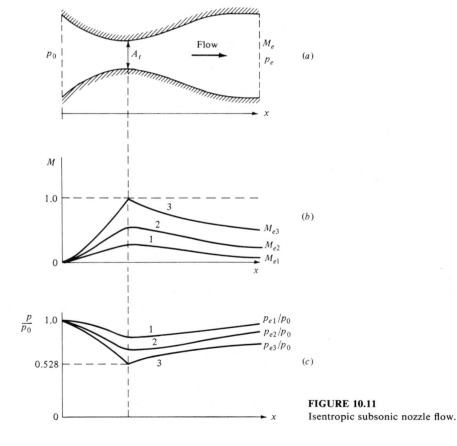

FIGURE 10.11
Isentropic subsonic nozzle flow.

the *sonic* throat area. In the case of purely subsonic flow through a convergent-divergent nozzle, A^* takes on the character of a reference area; it is not the same as the actual geometric area of the nozzle throat, A_t. Rather, A^* is the area the flow in Fig. 10.11 would have *if* it were somehow accelerated to sonic velocity. If this did happen, the flow area would have to be decreased further than shown in Fig. 10.11a. Hence, for a purely subsonic flow, $A_t > A^*$.

Assume that we further decrease the exit pressure in Fig. 10.11, say, to the value $p_e = p_{e,2}$. The flow is now illustrated by the curves labeled 2 in Fig. 10.11. The flow moves faster through the nozzle, and the maximum Mach number at the throat increases but remains less than 1. Now, let us reduce p_e to the value $p_e = p_{e,3}$, such that the flow just reaches sonic conditions at the throat. This is shown by curve 3 in Fig. 10.11. The throat Mach number is 1, and the throat pressure is $0.528p_0$. The flow downstream of the throat is subsonic.

Upon comparing Figs. 10.10 and 10.11, we are struck by an important physical difference. For a given nozzle shape, there is only *one* allowable isentropic flow solution for the supersonic case shown in Fig. 10.10. In contrast, there are an *infinite* number of possible isentropic subsonic solutions, each one corresponding to some value of p_e, where $p_0 \geq p_e \geq p_{e,3}$. Only three solutions of this infinite set of solutions are sketched in Fig. 10.11. Hence, the key factors for the analysis of purely subsonic flow in a convergent-divergent nozzle are both A/A^* and p_e/p_0.

Consider the mass flow through the convergent-divergent nozzle in Fig. 10.11. As the exit pressure is decreased, the flow velocity in the throat increases; hence, the mass flow increases. The mass flow can be calculated by evaluating Eq. (10.1) at the throat; i.e., $\dot{m} = \rho_t u_t A_t$. As p_e decreases, u_t increases and ρ_t decreases. However, the percentage increase in u_t is much greater than the decrease in ρ_t. As a result, $\dot{m}$ increases, as sketched in Fig. 10.12. When $p_e = p_{e,3}$, sonic flow is achieved at the throat, and $\dot{m} = \rho^* u^* A^* = \rho^* u^* A_t$. Now, if p_e is further reduced below $p_{e,3}$, the conditions at the throat take on a new behavior; they remain *unchanged*. From our discussion in Sec. 10.2, the Mach number at the throat cannot exceed 1; hence, as p_e is further reduced, M will remain equal to 1 at the throat. Consequently, the mass flow will remain constant as p_e is reduced below $p_{e,3}$, as shown in Fig. 10.12. In a sense, the flow at the throat, as well as upstream of the throat, becomes "frozen." Once the flow becomes sonic

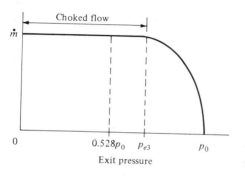

FIGURE 10.12
Variation of mass flow with exit pressure; illustration of choked flow.

at the throat, disturbances cannot work their way upstream of the throat. Hence, the flow in the convergent section of the nozzle no longer communicates with the exit pressure and has no way of knowing that the exit pressure is continuing to decrease. This situation—when the flow goes sonic at the throat, and the mass flow remains constant no matter how low p_e is reduced—is called *choked flow*. It is a vital aspect of the compressible flow through ducts, and we consider it further in our subsequent discussions.

Return to the subsonic nozzle flows sketched in Fig. 10.11. *Question*: What happens in the duct when p_e is reduced below $p_{e,3}$? In the convergent portion, as described above, nothing happens. The flow properties remain fixed at the conditions shown by curve 3 in the convergent section of the duct (the left side of Fig. 10.11b and c). However, a lot happens in the divergent section of the duct. As the exit pressure is reduced below $p_{e,3}$, a region of supersonic flow appears downstream of the throat. However, the exit pressure is too high to allow an isentropic supersonic flow throughout the entire divergent section. Instead, for p_e less than $p_{e,3}$ but substantially higher than the fully isentropic value $p_{e,6}$ (see Fig. 10.10c), a normal shock wave is formed downstream of the throat. This situation is sketched in Fig. 10.13.

In Fig. 10.13, the exit pressure has been reduced to $p_{e,4}$, where $p_{e,4} < p_{e,3}$, but where $p_{e,4}$ is also substantially higher than $p_{e,6}$. Here we observe a normal shock wave standing inside the nozzle at a distance d downstream of the throat.

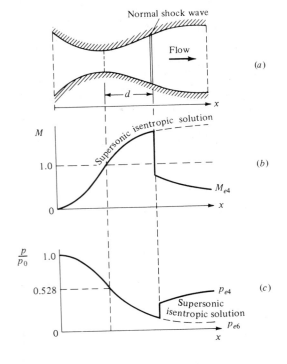

FIGURE 10.13
Supersonic nozzle flow with a normal shock inside the nozzle.

Between the throat and the normal shock wave, the flow is given by the supersonic isentropic solution, as shown in Fig. 10.13b and c. Behind the shock wave, the flow is subsonic. This subsonic flow sees the divergent duct and isentropically slows down further as it moves to the exit. Correspondingly, the pressure experiences a discontinuous increase across the shock wave and then is further increased as the flow slows down toward the exit. The flow on both the left and right sides of the shock wave is isentropic; however, the entropy increases across the shock wave. Hence, the flow on the left side of the shock wave is isentropic with one value of entropy, s_1, and the flow on the right side of the shock wave is isentropic with another value of entropy, s_2, where $s_2 > s_1$. The location of the shock wave inside the nozzle, given by d in Fig. 10.13a, is determined by the requirement that the increase in static pressure across the wave plus that in the divergent portion of the subsonic flow behind the shock be just right to achieve $p_{e,4}$ at the exit. As p_e is further reduced, the normal shock wave moves downstream, closer to the nozzle exit. At a certain value of exit pressure, $p_e = p_{e,5}$, the normal shock stands precisely at the exit. This is sketched in Fig. 10.14a to c. At this

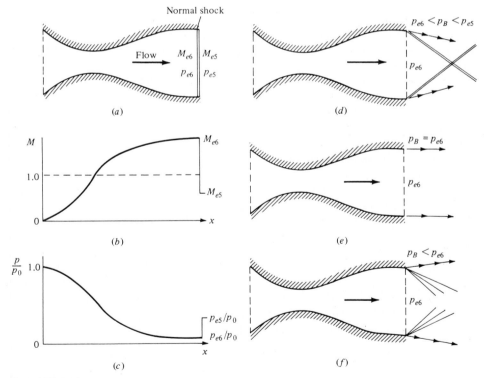

FIGURE 10.14
Supersonic nozzle flows with waves at the nozzle exit: (a), (b), and (c) pertain to a normal shock at the exit, (d) overexpanded nozzle, (e) isentropic expansion to the back pressure equal to the exit pressure, (f) underexpanded nozzle.

stage, when $p_e = p_{e,5}$, the flow through the entire nozzle, except precisely at the exit, is isentropic.

To this stage in our discussion, we have dealt with p_e, which is the pressure right at the nozzle exit. In Figs. 10.10, 10.11, 10.13, and 10.14a to c, we have not been concerned with the flow downstream of the nozzle exit. Now imagine that the nozzle in Fig. 10.14a exhausts directly into a region of surrounding gas downstream of the exit. These surroundings could be, e.g., the atmosphere. In any case, the pressure of the surroundings downstream of the exit is defined as the *back pressure*, denoted by p_B. When the flow at the nozzle exit is subsonic, the exit pressure must equal the back pressure, $p_e = p_B$, because a pressure discontinuity cannot be maintained in a steady subsonic flow. That is, when the exit flow is subsonic, the surrounding back pressure is impressed on the exit flow. Hence, in Fig. 10.11, $p_B = p_{e,1}$ for curve 1, $p_B = p_{e,2}$ for curve 2, and $p_B = p_{e,3}$ for curve 3. For the same reason, $p_B = p_{e,4}$ in Fig. 10.13, and $p_B = p_{e,5}$ in Fig. 10.14. Hence, in discussing these figures, instead of stating that we reduced the exit pressure p_e and observed the consequences, we could just as well have stated that we reduced the back pressure p_B. It would have amounted to the same thing.

For the remainder of our discussion in this section, let us now imagine that we have control over p_B and that we are going to continue to decrease p_B. Consider the case when the back pressure is reduced below $p_{e,5}$. When $p_{e,6} < p_B < p_{e,5}$, the back pressure is still above the isentropic pressure at the nozzle exit. Hence, in flowing out to the surroundings, the jet of gas from the nozzle must somehow be compressed such that its pressure is compatible with p_B. This compression takes place across oblique shock waves attached to the exit, as shown in Fig. 10.14d. When p_B is reduced to the value such that $p_B = p_{e,6}$, there is no mismatch of the exit pressure and the back pressure; the nozzle jet exhausts smoothly into the surroundings without passing through any waves. This is shown in Fig. 10.14e. Finally, as p_B is reduced below $p_{e,6}$ the jet of gas from the nozzle must expand further in order to match the lower back pressure. This expansion takes place across centered expansion waves attached to the exit, as shown in Fig. 10.14f.

When the situation in Fig. 10.14d exists, the nozzle is said to be *overexpanded*, because the pressure at the exit has expanded below the back pressure, $p_{e,6} < p_B$. That is, the nozzle expansion has gone too far, and the jet must pass through oblique shocks in order to come back up to the higher back pressure. Conversely, when the situation in Fig. 10.14f exists, the nozzle is said to be *underexpanded*, because the exit pressure is higher than the back pressure, $p_{e,6} > p_B$, and hence the flow is capable of additional expansion after leaving the nozzle.

Surveying Figs. 10.10 through 10.14, note that the purely isentropic super-sonic flow originally illustrated in Fig. 10.10 exists throughout the nozzle for all cases when $p_B \leq p_{e,5}$. For example, in Fig. 10.14a, the isentropic supersonic flow solution holds throughout the nozzle except right at the exit, where a normal shock exists. In Fig. 10.14d to f, the flow through the entire nozzle, including at the exit plane, is given by the isentropic supersonic flow solution.

Keep in mind that our entire discussion of nozzle flows in this section is predicated on having a duct of *given shape*. We assume that $A = A(x)$ is prescribed.

When this is the case, the quasi-one-dimensional theory of this chapter gives a reasonable prediction of the flow inside the duct, where the results are interpreted as mean properties averaged over each cross section. This theory does *not* tell us how to design the *contour* of the nozzle. In reality, if the walls of the nozzle are not curved just right, then oblique shocks occur inside the nozzle. To obtain the proper contour for a supersonic nozzle so that it produces isentropic shock-free flow inside the nozzle, we must account for the three-dimensionality of the actual flow. This is one purpose of the method of characteristics, a technique for analyzing two- and three-dimensional supersonic flow. A brief introduction to the method of characteristics is given in Chap. 13.

Example 10.1. Consider the isentropic supersonic flow through a convergent-divergent nozzle with an exit-to-throat area ratio of 10.25. The reservoir pressure and temperature are 5 atm and 600°R, respectively. Calculate M, p, and T at the nozzle exit.

Solution. From the supersonic portion of App. A, for $A_e/A^* = 10.25$,

$$M_e = \boxed{3.95}$$

Also,

$$\frac{p_e}{p_0} = \frac{1}{142} \quad \text{and} \quad \frac{T_e}{T_0} = \frac{1}{4.12}$$

Thus,

$$p_e = 0.007 p_0 = 0.007(5) = \boxed{0.035 \text{ atm}}$$

$$T_e = 0.2427 T_0 = 0.2427(600) = \boxed{145.6°R}$$

Example 10.2. Consider the isentropic flow through a convergent-divergent nozzle with an exit-to-throat area ratio of 2. The reservoir pressure and temperature are 1 atm and 288 K, respectively. Calculate the Mach number, pressure, and temperature at both the throat and the exit for the cases where (*a*) the flow is supersonic at the exit and (*b*) the flow is subsonic throughout the entire nozzle except at the throat, where $M = 1$.

Solution. (*a*) At the throat, the flow is sonic. Hence,

$$M_t = \boxed{1.0}$$

$$p_t = p^* = \frac{p^*}{p_0} p_0 = 0.528(1 \text{ atm}) = \boxed{0.528 \text{ atm}}$$

$$T_t = T^* = \frac{T^*}{T_0} T_0 = 0.833(288) = \boxed{240 \text{ K}}$$

At the exit, the flow is supersonic. Hence, from the supersonic portion of App. A, for $A_e/A^* = 2$,

$$M_e = \boxed{2.2}$$

$$p_e = \frac{p_e}{p_0} p_0 = \frac{1}{10.69} (1 \text{ atm}) = \boxed{0.0935 \text{ atm}}$$

$$T_e = \frac{T_e}{T_0} T_0 = \frac{1}{1.968} (288) = \boxed{146 \text{ K}}$$

(b) At the throat, the flow is still sonic. Hence, from above, $M_t = 1.0$, $p_t = 0.528$ atm, and $T_t = 240$ K. However, at all other locations in the nozzle, the flow is subsonic. At the exit, where $A_e/A^* = 2$, from the subsonic portion of App. A,

$$M_e = \boxed{0.3} \qquad \text{(rounded to the nearest entry in App. A)}$$

$$p_e = \frac{p_e}{p_0} p_0 = \frac{1}{1.064} (1 \text{ atm}) = \boxed{0.94 \text{ atm}}$$

$$T_e = \frac{T_e}{T_0} T_0 = \frac{1}{1.018} (288) = \boxed{282.9 \text{ K}}$$

Example 10.3. For the nozzle in Example 10.2, assume the exit pressure is 0.973 atm. Calculate the Mach numbers at the throat and the exit.

Solution. In Example 10.2, we saw that if $p_e = 0.94$ atm, the flow is sonic at the throat, but subsonic elsewhere. Hence, $p_e = 0.94$ atm corresponds to $p_{e,3}$ in Fig. 10.11. In the present problem, $p_e = 0.973$ atm, which is higher than $p_{e,3}$. Hence, in this case, the flow is subsonic throughout the nozzle, including at the throat. For this case, A^* takes on a reference value, and the actual geometric throat area is denoted by A_t. At the exit,

$$\frac{p_0}{p_e} = \frac{1}{0.973} = 1.028$$

From the subsonic portion of App. A, for $p_0/p_e = 1.028$, we have

$$M_e = \boxed{0.2} \qquad \text{and} \qquad \frac{A_e}{A^*} = 2.964$$

$$\frac{A_t}{A^*} = \frac{A_t}{A_e} \frac{A_e}{A^*} = 0.5(2.964) = 1.482$$

From the subsonic portion of App. A, for $A_t/A^* = 1.482$, we have

$$M_t = \boxed{0.44} \qquad \text{(nearest entry)}$$

10.4 DIFFUSERS

The role of a diffuser was first introduced in Sec. 3.3 in the context of a low-speed subsonic wind tunnel. There, a diffuser was a divergent duct downstream of the test section whose role was to slow the higher-velocity air from the test section down to a very low velocity at the diffuser exit (see Fig. 3.8). Indeed, in general, we can define a diffuser as any duct designed to slow an incoming gas flow to lower velocity at the exit of the diffuser. The incoming flow can be subsonic, as discussed in Fig. 3.8, or it can be supersonic, as discussed in the present section. However, the shape of the diffuser is drastically different, depending on whether the incoming flow is subsonic or supersonic.

Before pursuing this matter further, let us elaborate on the concept of total pressure, p_0, as discussed in Sec. 7.5. In a semiqualitative sense, the total pressure of a flowing gas is a measure of the capacity of the flow to perform useful work. Let us consider two examples:

1. A pressure vessel containing stagnant air at 10 atm
2. A supersonic flow at $M = 2.16$ and $p = 1$ atm

In case 1, the air velocity is zero; hence, $p_0 = p = 10$ atm. Now, imagine that we want to use air to drive a piston in a piston-cylinder arrangement, where useful work is performed by the piston being displaced through a distance. The air is ducted into the cylinder from a large manifold, in the same vein as the reciprocating internal combustion engine in our automobile. In case 1, the pressure vessel can act as the manifold; hence, the pressure on the piston is 10 atm, and a certain amount of useful work is performed, say, W_1. However, in case 2, the supersonic flow must be slowed to a low velocity before we can readily feed it into the manifold. If this slowing process can be achieved without loss of total pressure, then the pressure in the manifold is this case is also 10 atm (assuming $V \approx 0$), and the same amount of useful work, W_1, is performed. On the other hand, assume that in slowing down the supersonic stream, a loss of 3 atm takes place in the total pressure. Then the pressure in the manifold is only 7 atm, with the consequent generation of useful work, W_2, which is less than in the first case; i.e., $W_2 < W_1$. The purpose of this simple example is to indicate that the total pressure of a flowing gas is indeed a measure of its capability to perform useful work. On this basis, a loss of total pressure is always an inefficiency—a loss of the capability to do a certain amount of useful work.

In light of the above, let us expand our definition of a diffuser. A diffuser is a duct designed to slow an incoming gas flow to lower velocity at the exit of the diffuser *with as small a loss in total pressure as possible*. Consequently, an ideal diffuser would be characterized by an *isentropic* compression to lower velocities; this is sketched in Fig. 10.15a, where a supersonic flow enters the diffuser at M_1, is isentropically compressed in a convergent duct to Mach 1 at the throat, where the area is A^*, and then is further isentropically compressed in a divergent duct to a low subsonic Mach number at the exit. Because the flow is isentropic, $s_2 = s_1$, and from Eq. (8.73), $p_{0,2} = p_{0,1}$. Indeed, p_0 is constant throughout the entire diffuser—a characteristic of isentropic flow. However, common sense should tell you that the ideal diffuser in Fig. 10.15a can never be achieved. It is extremely difficult to slow a supersonic flow without generating shock waves in the process. For example, examine the convergent portion of the diffuser in Fig. 10.15a. Note that the supersonic flow is turned into itself; hence, the converging flow will inherently generate oblique shock waves, which will destroy the isentropic nature of the flow. Moreover, in real life, the flow is viscous; there will be an entropy increase within the boundary layers on the walls of the diffuser. For these reasons, an ideal isentropic diffuser can never be constructed;

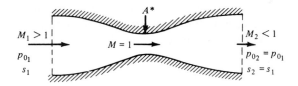

(a) Ideal (isentropic) supersonic diffuser

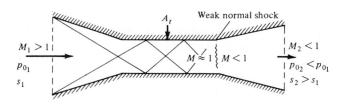

(b) Actual supersonic diffuser

FIGURE 10.15
The ideal (isentropic) diffuser compared with the actual situation.

an ideal diffuser is of the nature of a "perpetual motion machine"—only a utopian wish in the minds of engineers.

An actual supersonic diffuser is sketched in Fig. 10.15b. Here, the incoming flow is slowed by a series of reflected oblique shocks, first in a convergent section usually consisting of straight walls, and then in a constant-area throat. Due to the interaction of the shock waves with the viscous flow near the wall, the reflected shock pattern eventually weakens and becomes quite diffuse, sometimes ending in a weak normal shock wave at the end of the constant-area throat. Finally, the subsonic flow downstream of the constant-area throat is further slowed by moving through a divergent section. At the exit, clearly $s_2 > s_1$; hence $p_{0,2} < p_{0,1}$. The art of diffuser design is to obtain as small a total pressure loss as possible, i.e., to design the convergent, divergent, and constant-area throat sections so that $p_{0,2}/p_{0,1}$ is as close to unity as possible. Unfortunately, in most cases, we fall far short of that goal. For more details on supersonic diffusers, see chap. 5 of Ref. 21 and chap. 12 of Ref. 1.

Please note that due to the entropy increase across the shock waves and in the boundary layers, the real diffuser throat area A_t is larger than A^*, i.e., in Fig. 10.15, $A_t > A^*$.

10.5 SUPERSONIC WIND TUNNELS

Return to the road map given in Fig. 10.3. The material for the left and right branches is covered in Secs. 10.3 and 10.4, respectively. In turn, a mating of these two branches gives birth to the fundamental aspects of supersonic wind tunnels, to be discussed in this section.

Imagine that you want to create a Mach 2.5 uniform flow in a laboratory for the purpose of testing a model of a supersonic vehicle, say, a cone. How do you do it? Clearly, we need a convergent-divergent nozzle with an area ratio $A_e/A^* = 2.637$ (see App. A). Moreover, we need to establish a pressure ratio, $p_0/p_e = 17.09$, across the nozzle in order to obtain a shock-free expansion to $M_e = 2.5$ at the exit. Your first thought might be to exhaust the nozzle directly into the laboratory, as sketched in Fig. 10.16. Here, the Mach 2.5 flow passes into the surroundings as a "free jet." The test model is placed in the flow downstream of the nozzle exit. In order to make certain that the free jet does not have shock or expansion waves, the nozzle exit pressure p_e must equal the back pressure p_B, as originally sketched in Fig. 10.14e. Since the back pressure is simply that of the atmosphere surrounding the free jet, $p_B = p_e = 1$ atm. Consequently, to establish the proper isentropic expansion through the nozzle, you need a high-pressure reservoir with $p_0 = 17.09$ atm at the inlet to the nozzle. In this manner, you would be able to accomplish your objective, namely, to produce a uniform stream of air at Mach 2.5 in order to test a supersonic model, as sketched in Fig. 10.16.

In the above example, you may have a problem obtaining the high-pressure air supply at 17.09 atm. You need an air compressor or a bank of high-pressure air bottles—both of which can be expensive. It requires work, hence money, to create reservoirs of high-pressure air—the higher the pressure, the more the cost. So, can you accomplish your objective in a more efficient way, at less cost? The answer is yes, as follows. Instead of the free jet as sketched in Fig. 10.16, imagine that you have a long constant-area section downstream of the nozzle exit, with a normal shock wave standing at the end of the constant-area section; this is shown in Fig. 10.17. The pressure downstream of the normal shock wave is $p_2 = p_B = 1$ atm. At $M = 2.5$, the static pressure ratio across the normal shock is $p_2/p_e = 7.125$. Hence, the pressure upstream of the normal shock is 0.14 atm. Since the flow is uniform in the constant-area section, this pressure is also equal to the nozzle exit pressure; i.e., $p_e = 0.14$ atm. Thus, in order to obtain the proper isentropic flow through the nozzle, which requires a pressure ratio of $p_0/p_e = 17.09$, we need a reservoir with a pressure of only 2.4 atm. This is considerably more

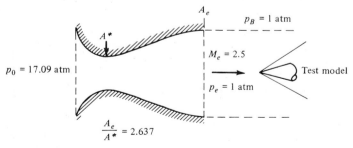

FIGURE 10.16
Nozzle exhausting directly to the atmosphere.

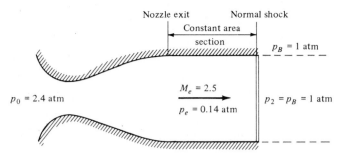

FIGURE 10.17
Nozzle exhausting into a constant-area duct, where a normal shock stands at the exit of the duct.

efficient than the 17.09 atm required in Fig. 10.16. Hence, we have created a uniform Mach 2.5 flow (in the constant-area duct) at a considerable reduction in cost compared with the scheme in Fig. 10.16.

In Fig. 10.17, the normal shock wave is acting as a diffuser, slowing the air originally at Mach 2.5 to the subsonic value of Mach 0.513 immediately behind the shock. Hence, by the addition of this "diffuser," we can more efficiently produce our uniform Mach 2.5 flow. This illustrates one of the functions of a diffuser. However, the "normal shock diffuser" sketched in Fig. 10.17 has several problems:

1. A normal shock is the strongest possible shock, hence creating the largest total pressure loss. If we could replace the normal shock in Fig. 10.17 with a weaker shock, the total pressure loss would be less, and the required reservoir pressure p_0 would be less than 2.4 atm.

2. It is extremely difficult to hold a normal shock wave stationary at the duct exit; in real life, flow unsteadiness and instabilities would cause the shock to move somewhere else and to fluctuate constantly in position. Thus, we could never be certain about the quality of the flow in the constant-area duct.

3. As soon as a test model is introduced into the constant-area section, the oblique waves from the model would propagate downstream, causing the flow to become two- or three-dimensional. The normal shock sketched in Fig. 10.17 could not exist in such a flow.

Hence, let us replace the normal shock in Fig. 10.17 with the oblique shock diffuser shown in Fig. 10.15b. The resulting duct would appear as sketched in Fig. 10.18. Examine this figure closely. We have a convergent-divergent nozzle feeding a uniform supersonic flow into the constant-area duct, which is called the *test section*. This flow is subsequently slowed to a low subsonic speed by means of a diffuser. This arrangement—namely, a convergent-divergent nozzle, a test section, and a convergent-divergent diffuser—is a *supersonic wind tunnel*. A test model, the cone in Fig. 10.18, is placed in the test section, where aerodynamic measurements such as lift, drag, and pressure distribution are made. The wave system from the model propagates downstream and interacts with the

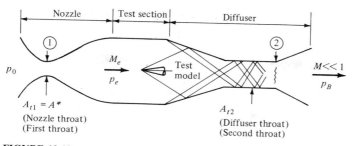

FIGURE 10.18
Sketch of a supersonic wind tunnel.

multireflected shocks in the diffuser. The pressure ratio required to run the supersonic tunnel is p_0/p_B. This can be obtained by making p_0 large via a high-pressure reservoir at the inlet to the nozzle or by making p_B small via a vacuum source at the exit of the diffuser, or a combination of both.

The main source of total pressure loss in a supersonic wind tunnel is the diffuser. How does the oblique shock diffuser in Fig. 10.18 compare with the hypothetical normal shock diffuser in Fig. 10.17? Is the total pressure loss across all the reflected oblique shocks in Fig. 10.18 greater or less than across the single normal shock wave in Fig. 10.17? This is an important question, since the smaller the total pressure loss in the diffuser, the smaller is the pressure ratio p_0/p_B required to run the supersonic tunnel. There is no pat answer to this question. However, it is usually true that progressively reducing the velocity of a supersonic flow through a series of oblique shocks to a low supersonic value, and then further reducing the flow to subsonic speeds across a weak normal shock, results in a *smaller* total pressure loss than simply reducing the flow to subsonic speeds across a single, strong normal shock wave at the initially high supersonic Mach number. This trend is illustrated by Example 9.4. Therefore, the oblique shock diffuser shown in Figs. 10.15*b* and 10.18 is usually more efficient than the simple normal shock diffuser shown in Fig. 10.17. This is not always true, however, because in an actual real-life oblique shock diffuser, the shock waves interact with the boundary layers on the walls, causing local thickening and even possible separation of the boundary layers. This creates an additional total pressure loss. Moreover, the simple aspect of skin friction exerted on the surface generates a total pressure loss. Hence, actual oblique shock diffusers may have efficiencies greater or less than a hypothetical normal shock diffuser. Nevertheless, virtually all supersonic wind tunnels use oblique shock diffusers qualitatively similar to that shown in Fig. 10.18.

Notice that the supersonic wind tunnel shown in Fig. 10.18 has two throats: the nozzle throat with area $A_{t,1}$ is called the *first throat*, and the diffuser throat with area $A_{t,2}$ is called the *second throat*. The mass flow through the nozzle can be expressed as $\dot{m} = \rho u A$ evaluated at the first throat. This station is denoted as station 1 in Fig. 10.18, and hence the mass flow through the nozzle is $\dot{m}_1 = \rho_1 u_1 A_{t,1} = \rho_1^* a_1^* A_{t,1}$. In turn, the mass flow through the diffuser can be expressed as $\dot{m} = \rho u A$ evaluated at station 2, namely, $\dot{m}_2 = \rho_2 u_2 A_{t,2}$. For steady flow through the wind tunnel, $\dot{m}_1 = \dot{m}_2$. Hence,

$$\rho_1^* a_1^* A_{t,1} = \rho_2 u_2 A_{t,2} \tag{10.33}$$

Since the thermodynamic state of the gas is irreversibly changed in going through the shock waves created by the test model and generated in the diffuser, clearly ρ_2 and possibly u_2 are different from ρ_1^* and a_1^*, respectively. Hence, from Eq. (10.33), *the second throat must have a different area from the first throat*; that is, $A_{t,2} \neq A_{t,1}$.

   Question: How does $A_{t,2}$ differ from $A_{t,1}$? Let us assume that sonic flow occurs at *both* stations 1 and 2 in Fig. 10.18. Thus, Eq. (10.33) can be written as

$$\frac{A_{t,2}}{A_{t,1}} = \frac{\rho_1^* a_1^*}{\rho_2^* a_2^*} \tag{10.34}$$

Recall from Sec. 8.4 that a^* is constant for an adiabatic flow. Also, recall that the flow across shock waves is adiabatic (but not isentropic). Hence, the flow throughout the wind tunnel sketched in Fig. 10.18 is adiabatic, and therefore $a_1^* = a_2^*$. In turn, Eq. (10.34) becomes

$$\frac{A_{t,2}}{A_{t,1}} = \frac{\rho_1^*}{\rho_2^*} \tag{10.35}$$

Recall from Sec. 8.4 that T^* is also constant throughout the adiabatic flow of a calorically perfect gas. Hence, from the equation of state,

$$\frac{\rho_1^*}{\rho_2^*} = \frac{p_1^*/RT_1^*}{p_2^*/RT_2^*} = \frac{p_1^*}{p_2^*} \tag{10.36}$$

Substituting Eq. (10.36) into (10.35), we have

$$\frac{A_{t,2}}{A_{t,1}} = \frac{p_1^*}{p_2^*} \tag{10.37}$$

From Eq. (8.45), we have

$$p_1^* = p_{0,1} \left(\frac{2}{\gamma + 1} \right)^{\gamma/(\gamma - 1)}$$

and

$$p_2^* = p_{0,2} \left(\frac{2}{\gamma + 1} \right)^{\gamma/(\gamma - 1)}$$

Substituting the above into Eq. (10.37), we obtain

$$\boxed{\frac{A_{t,2}}{A_{t,1}} = \frac{p_{0,1}}{p_{0,2}}} \tag{10.38}$$

Examining Fig. 10.18, the total pressure always decreases across shock waves;

therefore, $p_{0,2} < p_{0,1}$. In turn, from Eq. (10.38), $A_{t,2} > A_{t,1}$. Thus, the second throat must always be *larger* than the first throat. Only in the case of an ideal isentropic diffuser, where $p_0 = $ constant, would $A_{t,2} = A_{t,1}$, and we have already discussed the impossibility of such an ideal diffuser.

Equation (10.38) is a useful relation to size the second throat relative to the first throat *if* we know the total pressure ratio across the tunnel. In the absence of such information, for the preliminary design of supersonic wind tunnels, the total pressure ratio across a normal shock is assumed.

For a given wind tunnel, if $A_{t,2}$ is less than the value given by Eq. (10.38), the diffuser will "choke"; i.e., the diffuser cannot pass the mass flow coming from the isentropic, supersonic expansion through the nozzle. In this case, nature adjusts the flow through the wind tunnel by creating shock waves in the nozzle, which in turn reduce the Mach number in the test section, producing weaker shocks in the diffuser with an attendant overall reduction in the total pressure loss; i.e., nature adjusts the total pressure loss such that $p_{0,1}/p_{0,2} = p_{0,1}/p_B$ satisfies Eq. (10.38). Sometimes this adjustment is so severe that a normal shock stands inside the nozzle, and the flow through the test section and diffuser is totally subsonic. Obviously, this choked situation is not desirable because we no longer have uniform flow at the desired Mach number in the test section. In such a case, the supersonic wind tunnel is said to be *unstarted*. The only way to rectify this situation is to make $A_{t,2}/A_{t,1}$ large enough so that the diffuser can pass the mass flow from the isentropic expansion in the nozzle, i.e., so that Eq. (10.38) is satisfied along with a shock-free isentropic nozzle expansion.

As a general concluding comment, the basic concepts and relations discussed in this chapter are not limited to nozzles, diffusers, and supersonic wind tunnels. Rather, we have been discussing quasi-one-dimensional flow, which can be applied in many applications involving flow in a duct. For example, inlets on jet engines, which diffuse the flow to lower speeds before entering the engine compressor, obey the same principles. Also, a rocket engine is basically a supersonic nozzle designed to optimize the thrust from the expanded jet. The applications of the ideas presented in this chapter are numerous, and you should make certain that you understand these ideas before progressing further.

In Sec. 1.2, we subdivided aerodynamics into external and internal flows. You are reminded that the material in this chapter deals exclusively with internal flows.

Example 10.4. For the preliminary design of a Mach 2 supersonic wind tunnel, calculate the ratio of the diffuser throat area to the nozzle throat area.

Solution. Assuming a normal shock wave at the entrance of the diffuser (for starting), from App. B, $p_{0,2}/p_{0,1} = 0.7209$ for $M = 2.0$. Hence, from Eq. (10.38),

$$\frac{A_{t,2}}{A_{t,1}} = \frac{p_{0,1}}{p_{0,2}} = \frac{1}{0.7209} = \boxed{1.387}$$

10.6 SUMMARY

The results of this chapter are highlighted below:

Quasi-one-dimensional flow is an approximation to the actual three-dimensional flow in a variable-area duct; this approximation assumes that $p = p(x)$, $u = u(x)$, $T = T(x)$, etc., although the area varies as $A = A(x)$. Thus, we can visualize the quasi-one-dimensional results as giving the mean properties at a given station, averaged over the cross section. The quasi-one-dimensional flow assumption gives reasonable results for many internal flow problems; it is a "workhorse" in the everyday application of compressible flow. The governing equations for this are

Continuity:
$$\rho_1 u_1 A_1 = \rho_2 u_2 A_2 \tag{10.1}$$

Momentum:
$$p_1 A_1 + \rho_1 u_1^2 A_1 + \int_{A_1}^{A_2} p\, dA = p_2 A_2 + \rho_2 u_2^2 A_2 \tag{10.5}$$

Energy:
$$h_1 + \frac{u_1^2}{2} = h_2 + \frac{u_2^2}{2} \tag{10.9}$$

The area velocity relation
$$\frac{dA}{A} = (M^2 - 1)\frac{du}{u} \tag{10.25}$$

tells us that

1. To accelerate (decelerate) a subsonic flow, the area must decrease (increase).
2. To accelerate (decelerate) a supersonic flow, the area must increase (decrease).
3. Sonic flow can only occur at a throat or minimum area of the flow.

The isentropic flow of a calorically perfect gas through a nozzle is governed by the relation
$$\left(\frac{A}{A^*}\right)^2 = \frac{1}{M^2}\left[\frac{2}{\gamma+1}\left(1 + \frac{\gamma-1}{2}M^2\right)\right]^{(\gamma+1)/(\gamma-1)} \tag{10.32}$$

This tells us that the Mach number in a duct is governed by the ratio of local duct area to the sonic throat area; moreover, for a given area ratio, there are two values of Mach number that satisfy Eq. (10.32)—a subsonic value and a supersonic value.

For a given convergent-divergent duct, there is only one possible isentropic flow solution for supersonic flow; in contrast, there are an infinite number of subsonic isentropic solutions, each one associated with a different pressure ratio across the nozzle, $p_0/p_e = p_0/p_B$.

In a supersonic wind tunnel, the ratio of second throat area to first throat area should be approximately

$$\frac{A_{t,2}}{A_{t,1}} = \frac{p_{0,1}}{p_{0,2}} \qquad (10.38)$$

If $A_{t,2}$ is reduced much below this value, the diffuser will choke and the tunnel will unstart.

PROBLEMS

10.1. The reservoir pressure and temperature for a convergent-divergent nozzle are 5 atm and 520°R, respectively. The flow is expanded isentropically to supersonic speed at the nozzle exit. If the exit-to-throat area ratio is 2.193, calculate the following properties at the exit: M_e, p_e, T_e, ρ_e, u_e, $p_{0,e}$, $T_{0,e}$.

10.2. A flow is isentropically expanded to supersonic speeds in a convergent-divergent nozzle. The reservoir and exit pressures are 1 and 0.3143 atm, respectively. What is the value of A_e/A^*?

10.3. A Pitot tube inserted at the exit of a supersonic nozzle reads 8.92×10^4 N/m². If the reservoir pressure is 2.02×10^5 N/m², calculate the area ratio A_e/A^* of the nozzle.

10.4. For the nozzle flow given in Prob. 10.1, the throat area is 4 in². Calculate the mass flow through the nozzle.

10.5. A closed-form expression for the mass flow through a choked nozzle is

$$\dot{m} = \frac{p_0 A^*}{\sqrt{T_0}} \sqrt{\frac{\gamma}{R} \left(\frac{2}{\gamma+1} \right)^{(\gamma+1)/(\gamma-1)}}$$

Derive this expression.

10.6. Repeat Prob. 10.4, using the formula derived in Prob. 10.5, and check your answer from Prob. 10.4.

10.7. A convergent-divergent nozzle with an exit-to-throat area ratio of 1.616 has exit and reservoir pressures equal to 0.947 and 1.0 atm, respectively. Assuming isentropic flow through the nozzle, calculate the Mach number and pressure at the throat.

10.8. For the flow in Prob. 10.7, calculate the mass flow through the nozzle, assuming that the reservoir temperature is 288 K and the throat area is 0.3 m².

10.9. Consider a convergent-divergent nozzle with an exit-to-throat area ratio of 1.53. The reservoir pressure is 1 atm. Assuming isentropic flow, except for the possibility of a normal shock wave inside the nozzle, calculate the exit Mach number when the exit pressure p_e is
(*a*) 0.94 atm   (*b*) 0.886 atm   (*c*) 0.75 atm   (*d*) 0.154 atm

10.10. A 20° half-angle wedge is mounted at 0° angle of attack in the test section of a supersonic wind tunnel. When the tunnel is operating, the wave angle from the wedge leading edge is measured to be 41.8°. What is the exit-to-throat area ratio of the tunnel nozzle?

10.11. The nozzle of a supersonic wind tunnel has an exit-to-throat area ratio of 6.79. When the tunnel is running, a Pitot tube mounted in the test section measures 1.448 atm. What is the reservoir pressure for the tunnel?

10.12. We wish to design a supersonic wind tunnel which produces a Mach 2.8 flow at standard sea level conditions in the test section and has a mass flow of air equal to 1 slug/s. Calculate the necessary reservoir pressure and temperature, the nozzle throat and exit areas, and the diffuser throat area.

10.13. Consider a rocket engine burning hydrogen and oxygen. The total mass flow of the propellant plus oxidizer into the combustion chamber is 287.2 kg/s. The combustion chamber temperature is 3600 K. Assume that the combustion chamber is a low-velocity reservoir for the rocket engine. If the area of the rocket nozzle throat is 0.2 m^2, calculate the combustion chamber (reservoir) pressure. Assume that the gas that flows through the engine has a ratio of specific heats, $\gamma = 1.2$, and a molecular weight of 16.

10.14. For supersonic and hypersonic wind tunnels, a diffuser efficiency, η_D, can be defined as the ratio of the total pressures at the diffuser exit and nozzle reservoir, divided by the total pressure ratio across a normal shock at the test-section Mach number. This is a measure of the efficiency of the diffuser relative to normal shock pressure recovery. Consider a supersonic wind tunnel designed for a test-section Mach number of 3.0 which exhausts directly to the atmosphere. The diffuser efficiency is 1.2. Calculate the minimum reservoir pressure necessary for running the tunnel.

CHAPTER 11

SUBSONIC COMPRESSIBLE FLOW OVER AIRFOILS: LINEAR THEORY

During the war a British engineer named Frank Whittle invented the jet engine, and deHavilland built the first production-type model. He produced a jet plane named Vampire, the first to exceed 500 mph. Then he built the experimental DH 108, and released it to young Geoffrey for test. In the first cautious trials the new plane behaved beautifully; but as Geoffrey stepped up the speed he unsuspectingly drew closer to an invisible wall in the sky then unknown to anyone, later named the sound barrier, which can destroy a plane not designed to pierce it. One evening he hit the speed of sound, and the plane disintegrated. Young Geoffrey's body was not found for ten days.

From the Royal Air Force Flying
Review, as digested in Reader's
Digest, 1959

11.1 INTRODUCTION

The above quotation refers to an accident which took place on September 27, 1946, when Geoffrey deHavilland, son of the famed British airplane designer Sir Geoffrey deHavilland, took the D. H. 108 Swallow up for an attack on the world's speed record. At that time, no airplane had flown at or beyond the speed of sound. The Swallow was an experimental jet-propelled aircraft with swept wings and no tail. During its first high-speed, low-level run, the Swallow encountered major compressibility problems and broke up in the air. deHavilland was killed instantly. This accident strengthened the opinion of many that Mach 1 stood as a barrier to manned flight and that no airplane would ever fly faster than the speed of sound. This myth of the "sound barrier" originated in the early 1930s.

It was in full force by the time of the Volta Conference in 1935 (see Sec. 7.1). In light of the above quotation, the idea of a sound barrier was still being discussed in the popular literature as late as 1959, 12 years after the first successful supersonic flight by Captain Charles Yeager on October 14, 1947.

Of course, we know today that the sound barrier is indeed a myth; the supersonic transport Concorde flies at Mach 2, and some military aircraft are capable of Mach 3 and slightly beyond. The X-15 hypersonic research airplane has flown at Mach 7, and the Apollo lunar return capsule successfully reentered the earth's atmosphere at Mach 36. Supersonic flight is now an everyday occurrence. So, what caused the early concern about a sound barrier? In the present chapter, we develop a theory applicable to high-speed subsonic flight, and we see how the theory predicts a monotonically increasing drag going to infinity as $M_\infty \to 1$. It was this type of result which led some people in the early 1930s to believe that flight beyond the speed of sound was impossible. However, we also show in this chapter that the approximations made in the theory break down near Mach 1 and that in reality, although the drag coefficient at Mach 1 is large, it is still a manageable finite number.

Specifically, the purpose of this chapter is to examine the properties of two-dimensional airfoils at Mach numbers above 0.3, where we can no longer assume incompressible flow, but below Mach 1. That is, this chapter is an extension of the airfoil discussions in Chap. 4 (which applied to incompressible flow) to the high-speed subsonic regime.

In the process, we climb to a new tier in our study of compressible flow. If you survey our discussions so far of compressible flow, you will observe that they treat one-dimensional cases such as normal shock waves and flows in ducts. Even oblique shock waves, which are two- and three-dimensional in nature, depend only on the component of Mach number normal to the wave. Therefore, we have not been explicitly concerned with a multidimensional flow. As a consequence, note that the types of equations which allow an analysis of these flows are *algebraic equations*, and hence are relatively easy to solve in comparison with partial differential equations. In Chaps. 8 to 10, we have dealt primarily with such algebraic equations. These algebraic equations were obtained by applying the integral forms of the conservation equations [Eqs. (2.39), (2.55), and (2.86)] to appropriate control volumes where the flow properties were uniform over the inflow and outflow faces of the control volume. However, for general two- and three-dimensional flows, we are usually not afforded such a luxury. Instead, we must deal directly with the governing equations in their partial differential equation form (see Chap. 2). Such is the nature of the present chapter. Indeed, for the remainder of our aerodynamic discussions in this book, we appeal mainly to the differential forms of the continuity, momentum, and energy equations [such as Eqs. (2.43), (2.104a to c), and (2.105)].

The road map for this chapter is given in Fig. 11.1. We are going to return to the concept of a velocity potential, first introduced in Sec. 2.15. We are going to combine our governing equations so as to obtain a single equation simply in terms of the velocity potential; i.e., we are going to obtain for compressible flow

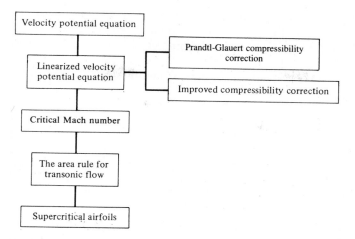

FIGURE 11.1
Road map for Chap. 11.

an equation analogous to Laplace's equation derived for incompressible flow in Sec. 3.7 [see Eq. (3.40)]. However, unlike Laplace's equation, which is linear, the exact velocity potential equation for compressible flow is nonlinear. By making suitable approximations, we are able to linearize this equation and apply it to thin airfoils at small angles of attack. The results enable us to correct incompressible airfoil data for the effects of compressibility—so-called *compressibility corrections*. Finally, we conclude this chapter by discussing several practical aspects of airfoil and general wing-body aerodynamics at speeds near Mach 1.

11.2 THE VELOCITY POTENTIAL EQUATION

The inviscid, compressible, subsonic flow over a body immersed in a uniform stream is *irrotational*; there is no mechanism in such a flow to start rotating the fluid elements (see Sec. 2.12). Thus, a velocity potential (see Sec. 2.15) can be defined. Since we are dealing with irrotational flow and the velocity potential, review Secs. 2.12 and 2.15 before progressing further.

Consider two-dimensional, steady, irrotational, isentropic flow. A velocity potential, $\phi = \phi(x, y)$, can be defined such that [from Eq. (2.145)]

$$\mathbf{V} = \nabla \phi \tag{11.1}$$

or in terms of the cartesian velocity components,

$$u = \frac{\partial \phi}{\partial x} \tag{11.2a}$$

$$v = \frac{\partial \phi}{\partial y} \tag{11.2b}$$

Let us proceed to obtain an equation for ϕ which represents a combination of the continuity, momentum, and energy equations. Such an equation would be very useful, because it would be simply one governing equation in terms of one unknown, namely ϕ.

The continuity equation for steady, two-dimensional flow is obtained from Eq. (2.43) as

$$\frac{\partial(\rho u)}{\partial x} + \frac{\partial(\rho v)}{\partial y} = 0 \tag{11.3}$$

or

$$\rho\frac{\partial u}{\partial x} + u\frac{\partial \rho}{\partial x} + v\frac{\partial \rho}{\partial y} + \rho\frac{\partial v}{\partial y} = 0 \tag{11.4}$$

Substituting Eqs. (11.2a and b) into (11.4), we have

$$\rho\frac{\partial^2 \phi}{\partial x^2} + \frac{\partial \phi}{\partial x}\frac{\partial \rho}{\partial x} + \frac{\partial \phi}{\partial y}\frac{\partial \rho}{\partial y} + \rho\frac{\partial^2 \phi}{\partial y^2} = 0$$

or

$$\rho\left(\frac{\partial^2 \phi}{\partial x^2} + \frac{\partial^2 \phi}{\partial y^2}\right) + \frac{\partial \phi}{\partial x}\frac{\partial \rho}{\partial x} + \frac{\partial \phi}{\partial y}\frac{\partial \rho}{\partial y} = 0 \tag{11.5}$$

We are attempting to obtain an equation completely in terms of ϕ; hence, we need to eliminate ρ from Eq. (11.5). To do this, consider the momentum equation in terms of Euler's equation:

$$dp = -\rho V\,dV \tag{3.12}$$

This equation holds for a steady, compressible, inviscid flow and relates p and V along a streamline. It can readily be shown that Eq. (3.12) holds in *any* direction throughout an irrotational flow, not just along a streamline (try it yourself). Therefore, from Eqs. (3.12) and (11.2a and b), we have

$$dp = -\rho V\,dV = -\frac{\rho}{2}\,d(V^2) = -\frac{\rho}{2}\,d(u^2 + v^2)$$

or

$$dp = -\frac{\rho}{2}\,d\left[\left(\frac{\partial \phi}{\partial x}\right)^2 + \left(\frac{\partial \phi}{\partial y}\right)^2\right] \tag{11.6}$$

Recall that we are also considering the flow to be isentropic. Hence, any change in pressure, dp, in the flow is automatically accompanied by a corresponding isentropic change in density, $d\rho$. Thus, by definition

$$\frac{dp}{d\rho} = \left(\frac{\partial p}{\partial \rho}\right)_s \tag{11.7}$$

The right-hand side of Eq. (11.7) is simply the square of the speed of sound. Thus, Eq. (11.7) yields

$$dp = a^2\,d\rho \tag{11.8}$$

Substituting Eq. (11.8) for the left side of Eq. (11.6), we have

$$dp = -\frac{\rho}{2a^2} d\left[\left(\frac{\partial\phi}{\partial x}\right)^2 + \left(\frac{\partial\phi}{\partial y}\right)^2\right] \tag{11.9}$$

Considering changes in the x direction, Eq. (11.9) directly yields

$$\frac{\partial\rho}{\partial x} = -\frac{\rho}{2a^2}\frac{\partial}{\partial x}\left[\left(\frac{\partial\phi}{\partial x}\right)^2 + \left(\frac{\partial\phi}{\partial y}\right)^2\right]$$

or

$$\frac{\partial\rho}{\partial x} = -\frac{\rho}{a^2}\left(\frac{\partial\phi}{\partial x}\frac{\partial^2\phi}{\partial x^2} + \frac{\partial\phi}{\partial y}\frac{\partial^2\phi}{\partial x\,\partial y}\right) \tag{11.10}$$

Similarly, for changes in the y direction, Eq. (11.9) gives

$$\frac{\partial\rho}{\partial y} = -\frac{\rho}{a^2}\left(\frac{\partial\phi}{\partial x}\frac{\partial^2\phi}{\partial x\,\partial y} + \frac{\partial\phi}{\partial y}\frac{\partial^2\phi}{\partial y^2}\right) \tag{11.11}$$

Substituting Eqs. (11.10) and (11.11) into (11.5), canceling the ρ which appears in each term, and factoring out the second derivatives of ϕ, we obtain

$$\boxed{\left[1 - \frac{1}{a^2}\left(\frac{\partial\phi}{\partial x}\right)^2\right]\frac{\partial^2\phi}{\partial x^2} + \left[1 - \frac{1}{a^2}\left(\frac{\partial\phi}{\partial y}\right)^2\right]\frac{\partial^2\phi}{\partial y^2} - \frac{2}{a^2}\left(\frac{\partial\phi}{\partial x}\right)\left(\frac{\partial\phi}{\partial y}\right)\frac{\partial^2\phi}{\partial x\,\partial y} = 0}$$

$$\tag{11.12}$$

which is called the *velocity potential equation*. It is almost completely in terms of ϕ; only the speed of sound appears in addition to ϕ. However, a can be readily expressed in terms of ϕ as follows. From Eq. (8.33), we have

$$a^2 = a_0^2 - \frac{\gamma-1}{2}V^2 = a_0^2 - \frac{\gamma-1}{2}(u^2 + v^2)$$

$$= a_0^2 - \frac{\gamma-1}{2}\left[\left(\frac{\partial\phi}{\partial x}\right)^2 + \left(\frac{\partial\phi}{\partial y}\right)^2\right] \tag{11.13}$$

Since a_0 is a known constant of the flow, Eq. (11.13) gives the speed of sound, a, as a function of ϕ. Hence, substitution of Eq. (11.13) into (11.12) yields a single partial differential equation in terms of the unknown ϕ. This equation represents a combination of the continuity, momentum, and energy equations. In principle, it can be solved to obtain ϕ for the flow field around any two-dimensional shape, subject of course to the usual boundary conditions at infinity and along the body surface. These boundary conditions on ϕ are detailed in Sec. 3.7, and are given by Eqs. (3.47a and b) and (3.48b).

Because Eq. (11.12) [along with Eq. (11.13)] is a single equation in terms of one dependent variable, ϕ, the analysis of isentropic, irrotational, steady, compressible flow is greatly simplified—we only have to solve one equation instead of three or more. Once ϕ is known, all the other flow variables are directly obtained as follows:

1. Calculate u and v from Eqs. (11.2a and b).
2. Calculate a from Eq. (11.13).
3. Calculate $M = V/a = \sqrt{u^2 + v^2}/a$.
4. Calculate T, p, and ρ from Eqs. (8.40), (8.42), and (8.43), respectively. In these equations, the total conditions T_0, p_0, and ρ_0 are known quantities; they are constant throughout the flow field and hence are obtained from the given freestream conditions.

Although Eq. (11.12) has the advantage of being one equation with one unknown, it also has the distinct disadvantage of being a *nonlinear* partial differential equation. Such nonlinear equations are very difficult to solve analytically, and in modern aerodynamics, solutions of Eq. (11.12) are usually sought by means of sophisticated finite-difference numerical techniques. Indeed, no general analytical solution of Eq. (11.12) has been found to this day. Contrast this situation with that for incompressible flow, which is governed by Laplace's equation—a *linear* partial differential equation for which numerous analytical solutions are well known.

Given this situation, aerodynamicists over the years have made assumptions regarding the physical nature of the flow field which are designed to simplify Eq. (11.12). These assumptions limit our considerations to the flow over slender bodies at small angles of attack. For subsonic and supersonic flows, these assumptions lead to an *approximate* form of Eq. (11.12) which is linear, and hence can be solved analytically. These matters are the subject of the next section.

Keep in mind that, within the framework of steady, irrotational, isentropic flow, Eq. (11.12) is *exact* and holds for all Mach numbers, from subsonic to hypersonic, and for all two-dimensional body shapes, thin and thick.

11.3 THE LINEARIZED VELOCITY POTENTIAL EQUATION

Consider the two-dimensional, irrotational, isentropic flow over the body shown in Fig. 11.2. The body is immersed in a uniform flow with velocity V_∞ oriented in the x direction. At an arbitrary point P in the flow field, the velocity is $\mathbf{V}$ with the x and y components given by u and v, respectively. Let us now visualize the velocity $\mathbf{V}$ as the sum of the uniform flow velocity plus some extra increments in velocity. For example, the x component of velocity, u, in Fig. 11.2 can be visualized as V_∞ plus an increment in velocity (positive or negative). Similarly, the y component of velocity, v, can be visualized as a simple increment itself, because the uniform flow has a zero component in the y direction. These increments are called *perturbations*, and

$$u = V_\infty + \hat{u} \qquad v = \hat{v}$$

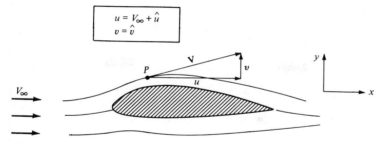

FIGURE 11.2
Uniform flow and perturbed flow.

where $\hat{u}$ and $\hat{v}$ are called the *perturbation velocities*. These perturbation velocities are not necessarily small; indeed, they can be quite large in the stagnation region in front of the blunt nose of the body shown in Fig. 11.2. In the same vein, because $\mathbf{V} = \nabla \phi$, we can define a perturbation velocity potential, $\hat{\phi}$, such that

$$\phi = V_\infty x + \hat{\phi}$$

where

$$\frac{\partial \hat{\phi}}{\partial x} = \hat{u}$$

$$\frac{\partial \hat{\phi}}{\partial y} = \hat{v}$$

Hence,

$$\frac{\partial \phi}{\partial x} = V_\infty + \frac{\partial \hat{\phi}}{\partial x} \qquad \frac{\partial \phi}{\partial y} = \frac{\partial \hat{\phi}}{\partial y}$$

$$\frac{\partial^2 \phi}{\partial x^2} = \frac{\partial^2 \hat{\phi}}{\partial x^2} \qquad \frac{\partial^2 \phi}{\partial y^2} = \frac{\partial^2 \hat{\phi}}{\partial y^2} \qquad \frac{\partial^2 \phi}{\partial x\,\partial y} = \frac{\partial^2 \hat{\phi}}{\partial x\,\partial y}$$

Substituting the above definitions into Eq. (11.12), and multiplying by a^2, we obtain

$$\left[a^2 - \left(V_\infty + \frac{\partial \hat{\phi}}{\partial x} \right)^2 \right] \frac{\partial^2 \hat{\phi}}{\partial x^2} + \left[a^2 - \left(\frac{\partial \hat{\phi}}{\partial y} \right)^2 \right] \frac{\partial^2 \hat{\phi}}{\partial y^2} - 2 \left(V_\infty + \frac{\partial \hat{\phi}}{\partial x} \right) \left(\frac{\partial \hat{\phi}}{\partial y} \right) \frac{\partial^2 \hat{\phi}}{\partial x\,\partial y} = 0$$

(11.14)

Equation (11.14) is called the *perturbation velocity potential equation*. It is precisely the same equation as Eq. (11.12) except that it is expressed in terms of $\hat{\phi}$ instead of ϕ. It is still a nonlinear equation.

To obtain better physical insight in some of our subsequent discussion, let us recast Eq. (11.14) in terms of the perturbation velocities. From the definition of $\hat{\phi}$ given earlier, Eq. (11.14) can be written as

$$[a^2 - (V_\infty + \hat{u})^2] \frac{\partial \hat{u}}{\partial x} + (a^2 - \hat{v}^2) \frac{\partial \hat{v}}{\partial y} - 2(V_\infty + \hat{u})\hat{v} \frac{\partial \hat{u}}{\partial y} = 0 \qquad (11.14a)$$

From the energy equation in the form of Eq. (8.32), we have

$$\frac{a_\infty^2}{\gamma-1}+\frac{V_\infty^2}{2}=\frac{a^2}{\gamma-1}+\frac{(V_\infty+\hat{u})^2+\hat{v}^2}{2} \tag{11.15}$$

Substituting Eq. (11.15) into (11.14a), and algebraically rearranging, we obtain

$$(1-M_\infty^2)\frac{\partial\hat{u}}{\partial x}+\frac{\partial\hat{v}}{\partial y}=M_\infty^2\left[(\gamma+1)\frac{\hat{u}}{V_\infty}+\frac{\gamma+1}{2}\frac{\hat{u}^2}{V_\infty^2}+\frac{\gamma-1}{2}\frac{\hat{v}^2}{V_\infty^2}\right]\frac{\partial\hat{u}}{\partial x}$$

$$+M_\infty^2\left[(\gamma-1)\frac{\hat{u}}{V_\infty}+\frac{\gamma+1}{2}\frac{\hat{v}^2}{V_\infty^2}+\frac{\gamma-1}{2}\frac{\hat{u}^2}{V_\infty^2}\right]\frac{\partial\hat{v}}{\partial y}$$

$$+M_\infty^2\left[\frac{\hat{v}}{V_\infty}\left(1+\frac{\hat{u}}{V_\infty}\right)\left(\frac{\partial\hat{u}}{\partial y}+\frac{\partial\hat{v}}{\partial x}\right)\right] \tag{11.16}$$

Equation (11.16) is still exact for irrotational, isentropic flow. Note that the left-hand side of Eq. (11.16) is linear but the right-hand side is nonlinear. Also, keep in mind that the size of the perturbations $\hat{u}$ and $\hat{v}$ can be large or small; Eq. (11.16) holds for both cases.

Let us now limit our considerations to *small* perturbations; i.e., assume that the body in Fig. 11.2 is a *slender* body at *small* angle of attack. In such a case, $\hat{u}$ and $\hat{v}$ will be small in comparison with V_∞. Therefore, we have

$$\frac{\hat{u}}{V_\infty},\frac{\hat{v}}{V_\infty}\ll 1 \qquad \frac{\hat{u}^2}{V_\infty^2},\frac{\hat{v}^2}{V_\infty^2}\lll 1$$

Keep in mind that products of $\hat{u}$ and $\hat{v}$ with their derivatives are also very small. With this in mind, examine Eq. (11.16). Compare like terms (coefficients of like derivatives) on the left- and right-hand sides of Eq. (11.16). We find

1. For $0\le M_\infty\le 0.8$ or $M_\infty\ge 1.2$, the magnitude of

$$M_\infty^2\left[(\gamma+1)\frac{\hat{u}}{V_\infty}+\cdots\right]\frac{\partial\hat{u}}{\partial x}$$

is small in comparison with the magnitude of

$$(1-M_\infty^2)\frac{\partial\hat{u}}{\partial x}$$

Thus, *ignore* the former term.
2. For $M_\infty<5$ (approximately),

$$M_\infty^2\left[(\gamma-1)\frac{\hat{u}}{V_\infty}+\cdots\right]\frac{\partial\hat{v}}{\partial y}$$

is small in comparison with $\partial\hat{v}/\partial y$. So ignore the former term. Also,

$$M_\infty^2\left[\frac{\hat{v}}{V_\infty}\left(1+\frac{\hat{u}}{V_\infty}\right)\left(\frac{\partial\hat{u}}{\partial y}+\frac{\partial\hat{v}}{\partial x}\right)\right]\approx 0$$

With the above order-of-magnitude comparisons, Eq. (11.16) reduces to

$$(1 - M_\infty^2)\frac{\partial \hat{u}}{\partial x} + \frac{\partial \hat{v}}{\partial y} = 0 \tag{11.17}$$

or in terms of the perturbation velocity potential,

$$\boxed{(1 - M_\infty^2)\frac{\partial^2 \hat{\phi}}{\partial x^2} + \frac{\partial^2 \hat{\phi}}{\partial y^2} = 0} \tag{11.18}$$

Examine Eq. (11.18). It is a *linear* partial differential equation, and therefore is inherently simpler to solve than its parent equation, Eq. (11.16). However, we have paid a price for this simplicity. Equation (11.18) is no longer exact. It is only an approximation to the physics of the flow. Due to the assumptions made in obtaining Eq. (11.18), it is reasonably valid (but not exact) for the following combined situations:

1. *Small* perturbation, i.e., thin bodies at small angles of attack
2. *Subsonic* and *supersonic* Mach numbers

In contrast, Eq. (11.18) is not valid for thick bodies and for large angles of attack. Moreover, it cannot be used for transonic flow, where $0.8 < M_\infty < 1.2$, or for hypersonic flow, where $M_\infty > 5$.

We are interested in solving Eq. (11.18) in order to obtain the pressure distribution along the surface of a slender body. Since we are now dealing with approximate equations, it is consistent to obtain a linearized expression for the pressure coefficient—an expression which is approximate to the same degree as Eq. (11.18), but which is extremely simple and convenient to use. The linearized pressure coefficient can be derived as follows.

First, recall the definition of the pressure coefficient C_p given in Sec. 1.5:

$$C_p \equiv \frac{p - p_\infty}{q_\infty} \tag{11.19}$$

where $q_\infty = \frac{1}{2}\rho_\infty V_\infty^2 = $ dynamic pressure. The dynamic pressure can be expressed in terms of M_∞ as follows:

$$q_\infty = \frac{1}{2}\rho_\infty V_\infty^2 = \frac{1}{2}\frac{\gamma p_\infty}{\gamma p_\infty}\rho_\infty V_\infty^2 = \frac{\gamma}{2}p_\infty\left(\frac{\rho_\infty}{\gamma p_\infty}\right)V_\infty^2 \tag{11.20}$$

From Eq. (8.23), we have $a_\infty^2 = \gamma p_\infty/\rho_\infty$. Hence, Eq. (11.20) becomes

$$q_\infty = \frac{\gamma}{2}p_\infty\frac{V_\infty^2}{a_\infty^2} = \frac{\gamma}{2}p_\infty M_\infty^2 \tag{11.21}$$

Substituting Eq. (11.21) into (11.19), we have

$$C_p = \frac{2}{\gamma M_\infty^2}\left(\frac{p}{p_\infty}-1\right)$$

(11.22)

Equation (11.22) is simply an alternate form of the pressure coefficient expressed in terms of M_∞. It is still an exact representation of the definition of C_p.

To obtain a linearized form of the pressure coefficient, recall that we are dealing with an adiabatic flow of a calorically perfect gas; hence, from Eq. (8.39),

$$T+\frac{V^2}{2c_p} = T_\infty+\frac{V_\infty^2}{2c_p}$$

(11.23)

Recalling from Eq. (7.9) that $c_p = \gamma R/(\gamma-1)$, Eq. (11.23) can be written as

$$T-T_\infty = \frac{V_\infty^2 - V^2}{2\gamma R/(\gamma-1)}$$

(11.24)

Also, recalling that $a_\infty = \sqrt{\gamma R T_\infty}$, Eq. (11.24) becomes

$$\frac{T}{T_\infty}-1 = \frac{\gamma-1}{2}\frac{V_\infty^2 - V^2}{\gamma R T_\infty} = \frac{\gamma-1}{2}\frac{V_\infty^2 - V^2}{a_\infty^2}$$

(11.25)

In terms of the perturbation velocities

$$V^2 = (V_\infty+\hat{u})^2 + \hat{v}^2$$

Eq. (11.25) can be written as

$$\frac{T}{T_\infty} = 1-\frac{\gamma-1}{2a_\infty^2}(2\hat{u}V_\infty+\hat{u}^2+\hat{v}^2)$$

(11.26)

Since the flow is isentropic, $p/p_\infty = (T/T_\infty)^{\gamma/(\gamma-1)}$, and Eq. (11.26) becomes

$$\frac{p}{p_\infty} = \left[1-\frac{\gamma-1}{2a_\infty^2}(2\hat{u}V_\infty+\hat{u}^2+\hat{v}^2)\right]^{\gamma/(\gamma-1)}$$

or

$$\frac{p}{p_\infty} = \left[1-\frac{\gamma-1}{2}M_\infty^2\left(\frac{2\hat{u}}{V_\infty}+\frac{\hat{u}^2+\hat{v}^2}{V_\infty^2}\right)\right]^{\gamma/(\gamma-1)}$$

(11.27)

Equation (11.27) is still an exact expression. However, let us now make the assumption that the perturbations are small, i.e., $\hat{u}/V_\infty \ll 1$, $\hat{u}^2/V_\infty^2 \ll 1$, and $\hat{v}^2/V_\infty^2 \ll 1$. In this case, Eq. (11.27) is of the form

$$\frac{p}{p_\infty} = (1-\varepsilon)^{\gamma/(\gamma-1)}$$

(11.28)

where ε is small. From the binomial expansion, neglecting higher-order terms, Eq. (11.28) becomes

$$\frac{p}{p_\infty} = 1-\frac{\gamma}{\gamma-1}\varepsilon+\cdots$$

(11.29)

Comparing Eq. (11.27) to (11.29), we can express Eq. (11.27) as

$$\frac{p}{p_\infty} = 1 - \frac{\gamma}{2} M_\infty^2 \left(\frac{2\hat{u}}{V_\infty} + \frac{\hat{u}^2 + \hat{v}^2}{V_\infty^2} \right) + \cdots \tag{11.30}$$

Substituting Eq. (11.30) into the expression for the pressure coefficient, Eq. (11.22), we obtain

$$C_p = \frac{2}{\gamma M_\infty^2} \left[1 - \frac{\gamma}{2} M_\infty^2 \left(\frac{2\hat{u}}{V_\infty} + \frac{\hat{u}^2 + \hat{v}^2}{V_\infty^2} \right) + \cdots - 1 \right]$$

or

$$C_p = -\frac{2\hat{u}}{V_\infty} + \frac{\hat{u}^2 + \hat{v}^2}{V_\infty^2} \tag{11.31}$$

Since $\hat{u}^2/V_\infty^2$ and $\hat{v}^2/V_\infty^2 \lll 1$, Eq. (11.31) becomes

$$\boxed{C_p = -\frac{2\hat{u}}{V_\infty}} \tag{11.32}$$

Equation (11.32) is the linearized form for the pressure coefficient; it is valid only for *small* perturbations. Equation (11.32) is consistent with the linearized perturbation velocity potential equation, Eq. (11.18). Note the simplicity of Eq. (11.32); it depends only on the x component of the velocity perturbation, namely, $\hat{u}$.

    To round out our discussion on the basics of the linearized equations, we note that any solution to Eq. (11.18) must satisfy the usual boundary conditions at infinity and at the body surface. At infinity, clearly $\hat{\phi} = $ constant; i.e., $\hat{u} = \hat{v} = 0$. At the body, the flow-tangency condition holds. Let θ be the angle between the tangent to the surface and the freestream. Then, at the surface, the boundary condition is obtained from Eq. (3.48e):[17]

$$\tan \theta = \frac{v}{u} = \frac{\hat{v}}{V_\infty + \hat{u}} \tag{11.33}$$

which is an exact expression for the flow-tangency condition at the body surface. A simpler, approximate expression for Eq. (11.33), consistent with linearized theory, can be obtained by noting that for small perturbations, $\hat{u} \ll V_\infty$. Hence, Eq. (11.33) becomes

$$\hat{v} = V_\infty \tan \theta$$

or

$$\boxed{\frac{\partial \hat{\phi}}{\partial y} = V_\infty \tan \theta} \tag{11.34}$$

Equation (11.34) is an *approximate* expression for the flow-tangency condition at the body surface, with accuracy of the same order as Eqs. (11.18) and (11.32).

11.4 PRANDTL-GLAUERT COMPRESSIBILITY CORRECTION

The aerodynamic theory for incompressible flow over thin airfoils at small angles of attack was presented in Chap. 4. For aircraft of the period 1903–1940, such theory was adequate for predicting airfoil properties. However, with the rapid evolution of high-power reciprocating engines spurred by World War II, the velocities of military fighter planes began to push close to 450 mi/h. Then, with the advent of the first operational jet-propelled airplanes in 1944 (the German Me 262), flight velocities took a sudden spurt into the 550 mi/h range and faster. As a result, the incompressible flow theory of Chap. 4 was no longer applicable to such aircraft; rather, high-speed airfoil theory had to deal with compressible flow. Because a vast bulk of data and experience had been collected over the years in low-speed aerodynamics, and because there was no desire to totally discard such data, the natural approach to high-speed subsonic aerodynamics was to search for methods that would allow relatively simple *corrections* to existing incompressible flow results which would approximately take into account the effects of compressibility. Such methods are called *compressibility corrections.* The first, and most widely known of these corrections is the Prandtl-Glauert compressibility correction, to be derived in this section. The Prandtl-Glauert method is based on the linearized perturbation velocity potential equation given by Eq. (11.18). Therefore, it is limited to thin airfoils at small angles of attack. Moreover, it is purely a subsonic theory and begins to give inappropriate results at values of $M_\infty = 0.7$ and above.

Consider the subsonic, compressible, inviscid flow over the airfoil sketched in Fig. 11.3. The shape of the airfoil is given by $y = f(x)$. Assume that the airfoil is thin and that the angle of attack is small; in such a case, the flow is reasonably approximated by Eq. (11.18). Define

$$\beta^2 \equiv 1 - M_\infty^2$$

so that Eq. (11.18) can be written as

$$\beta^2 \frac{\partial^2 \hat{\phi}}{\partial x^2} + \frac{\partial^2 \hat{\phi}}{\partial y^2} = 0 \qquad (11.35)$$

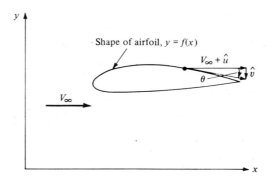

FIGURE 11.3
Airfoil in physical space.

Let us transform the independent variables x and y into a new space, ξ and η, such that

$$\xi = x \qquad (11.36a)$$

$$\eta = \beta y \qquad (11.36b)$$

Moreover, in this transformed space, consider a new velocity potential, $\bar{\phi}$, such that

$$\bar{\phi}(\xi, \eta) = \beta \hat{\phi}(x, y) \qquad (11.36c)$$

To recast Eq. (11.35) in terms of the transformed variables, recall the chain rule of partial differentiation; i.e.,

$$\frac{\partial \hat{\phi}}{\partial x} = \frac{\partial \hat{\phi}}{\partial \xi}\frac{\partial \xi}{\partial x} + \frac{\partial \hat{\phi}}{\partial \eta}\frac{\partial \eta}{\partial x} \qquad (11.37)$$

and

$$\frac{\partial \hat{\phi}}{\partial y} = \frac{\partial \hat{\phi}}{\partial \xi}\frac{\partial \xi}{\partial y} + \frac{\partial \hat{\phi}}{\partial \eta}\frac{\partial \eta}{\partial y} \qquad (11.38)$$

From Eqs. (11.36a and b), we have

$$\frac{\partial \xi}{\partial x} = 1 \qquad \frac{\partial \xi}{\partial y} = 0 \qquad \frac{\partial \eta}{\partial x} = 0 \qquad \frac{\partial \eta}{\partial y} = \beta$$

Hence, Eqs. (11.37) and (11.38) become

$$\frac{\partial \hat{\phi}}{\partial x} = \frac{\partial \hat{\phi}}{\partial \xi} \qquad (11.39)$$

$$\frac{\partial \hat{\phi}}{\partial y} = \beta \frac{\partial \hat{\phi}}{\partial \eta} \qquad (11.40)$$

Recalling Eq. (11.36c), Eqs. (11.39) and (11.40) become

$$\frac{\partial \hat{\phi}}{\partial x} = \frac{1}{\beta}\frac{\partial \bar{\phi}}{\partial \xi} \qquad (11.41)$$

and

$$\frac{\partial \hat{\phi}}{\partial y} = \frac{\partial \bar{\phi}}{\partial \eta} \qquad (11.42)$$

Differentiating Eq. (11.41) with respect to x (again using the chain rule), we obtain

$$\frac{\partial^2 \hat{\phi}}{\partial x^2} = \frac{1}{\beta}\frac{\partial^2 \bar{\phi}}{\partial \xi^2} \qquad (11.43)$$

Differentiating Eq. (11.42) with respect to y, we find that the result is

$$\frac{\partial^2 \hat{\phi}}{\partial y^2} = \beta \frac{\partial^2 \bar{\phi}}{\partial \eta^2} \qquad (11.44)$$

Substitute Eqs. (11.43) and (11.44) into (11.35):

$$\beta^2 \frac{1}{\beta} \frac{\partial^2 \bar{\phi}}{\partial \xi^2} + \beta \frac{\partial^2 \bar{\phi}}{\partial \eta^2} = 0$$

or

$$\frac{\partial^2 \bar{\phi}}{\partial \xi^2} + \frac{\partial^2 \bar{\phi}}{\partial \eta^2} = 0 \qquad (11.45)$$

Examine Eq. (11.45)—it should look familiar. Indeed, Eq. (11.45) is Laplace's equation. Recall from Chap. 3 that Laplace's equation is the governing relation for *incompressible flow*. Hence, starting with a subsonic compressible flow in physical (x, y) space where the flow is represented by $\hat{\phi}(x, y)$ obtained from Eq. (11.35), we have related this flow to an incompressible flow in transformed (ξ, η) space, where the flow is represented by $\bar{\phi}(\xi, \eta)$ obtained from Eq. (11.45). The relation between $\bar{\phi}$ and $\hat{\phi}$ is given by Eq. (11.36c).

Consider again the shape of the airfoil given in physical space by $y = f(x)$. The shape of the airfoil in the transformed space is expressed as $\eta = q(\xi)$. Let us compare the two shapes. First, apply the approximate boundary condition, Eq. (11.34), in physical space, noting that $df/dx = \tan \theta$. We obtain

$$V_\infty \frac{df}{dx} = \frac{\partial \hat{\phi}}{\partial y} = \frac{1}{\beta} \frac{\partial \bar{\phi}}{\partial y} = \frac{\partial \bar{\phi}}{\partial \eta} \qquad (11.46)$$

Similarly, apply the flow-tangency condition in transformed space, which from Eq. (11.34) is

$$V_\infty \frac{dq}{d\xi} = \frac{\partial \bar{\phi}}{\partial \eta} \qquad (11.47)$$

Examine Eqs. (11.46) and (11.47) closely. Note that the right-hand sides of these two equations are identical. Thus, from the left-hand sides, we obtain

$$\frac{df}{dx} = \frac{dq}{d\xi} \qquad (11.48)$$

Equation (11.48) implies that the shape of the airfoil in the transformed space is the same as in the physical space. Hence, the above transformation relates the compressible flow over an airfoil in (x, y) space to the incompressible flow in (ξ, η) space over the *same* airfoil.

The above theory leads to an immensely practical result, as follows. Recall Eq. (11.32) for the linearized pressure coefficient. Inserting the above transformation into Eq. (11.32), we obtain

$$C_p = -\frac{2\hat{u}}{V_\infty} = -\frac{2}{V_\infty} \frac{\partial \hat{\phi}}{\partial x} = -\frac{2}{V_\infty} \frac{1}{\beta} \frac{\partial \bar{\phi}}{\partial x} = -\frac{2}{V_\infty} \frac{1}{\beta} \frac{\partial \bar{\phi}}{\partial \xi} \qquad (11.49)$$

Question: What is the significance of $\partial \bar{\phi}/\partial \xi$ in Eq. (11.49)? Recall that $\bar{\phi}$ is the perturbation velocity potential for an incompressible flow in transformed space.

Hence, from the definition of velocity potential, $\partial \bar{\phi} / \partial \xi = \bar{u}$, where $\bar{u}$ is a perturbation velocity for the incompressible flow. Hence, Eq. (11.49) can be written as

$$C_p = \frac{1}{\beta}\left(-\frac{2\bar{u}}{V_\infty}\right)$$
(11.50)

From Eq. (11.32), the expression $(-2\bar{u}/V_\infty)$ is simply the linearized pressure coefficient for the incompressible flow. Denote this incompressible pressure coefficient by $C_{p,0}$. Hence, Eq. (11.50) gives

$$C_p = \frac{C_{p,0}}{\beta}$$

or recalling that $\beta \equiv \sqrt{1 - M_\infty^2}$, we have

$$C_p = \frac{C_{p,0}}{\sqrt{1 - M_\infty^2}}$$
(11.51)

Equation (11.51) is called the *Prandtl-Glauert rule*; it states that, if we know the incompressible pressure distribution over an airfoil, then the compressible pressure distribution over the same airfoil can be obtained from Eq. (11.51). Therefore, Eq. (11.51) is truly a *compressibility correction* to incompressible data.

Consider the lift and moment coefficients for the airfoil. For an inviscid flow, the aerodynamic lift and moment on a body are simply integrals of the pressure distribution over the body, as described in Sec. 1.5. (If this is somewhat foggy in your mind, review Sec. 1.5 before progressing further.) In turn, the lift and moment *coefficients* are obtained from the integral of the pressure coefficient via Eqs. (1.15) to (1.19). Since Eq. (11.51) relates the compressible and incompressible pressure coefficients, the same relation must therefore hold for lift and moment coefficients:

$$c_l = \frac{c_{l,0}}{\sqrt{1 - M_\infty^2}}$$
(11.52)

$$c_m = \frac{c_{m,0}}{\sqrt{1 - M_\infty^2}}$$
(11.53)

The Prandtl-Glauert rule, embodied in Eqs. (11.51) to (11.53), was historically the first compressibility correction to be obtained. As early as 1922, Prandtl was using this result in his lectures at Göttingen, although without written proof. The derivation of Eqs. (11.51) to (11.53) was first formally published by the British aerodynamicist, Hermann Glauert, in 1928. Hence, the rule is named after both men. The Prandtl-Glauert rule was used exclusively until 1939, when an improved compressibility correction was developed. Because of their simplicity,

Eqs. (11.51) to (11.53) are still used today for initial estimates of compressibility effects.

Recall that the results of Chaps. 3 and 4 proved that inviscid, incompressible flow over a closed, two-dimensional body theoretically produces zero drag—the well-known d'Alembert's paradox. Does the same paradox hold for inviscid, subsonic, compressible flow? The answer can be obtained by again noting that the only source of drag is the integral of the pressure distribution. If this integral is zero for an incompressible flow, and since the compressible pressure coefficient differs from the incompressible pressure coefficient by only a constant scale factor, β, then the integral must also be zero for a compressible flow. Hence, d'Alembert's paradox also prevails for inviscid, subsonic, compressible flow. However, as soon as the freestream Mach number is high enough to produce locally supersonic flow on the body surface with attendant shock waves, as shown in Fig. 1.30b, then a positive wave drag is produced, and d'Alembert's paradox no longer prevails.

Example 11.1. At a given point on the surface of an airfoil, the pressure coefficient is -0.3 at very low speeds. If the freestream Mach number is 0.6, calculate C_p at this point.

Solution. From Eq. (11.51),

$$C_p = \frac{C_{p,0}}{\sqrt{1-M^2}} = \frac{-0.3}{\sqrt{1-(0.6)^2}} = \boxed{-0.375}$$

Example 11.2. From Chap. 4, the theoretical lift coefficient for a thin, symmetric airfoil in an incompressible flow is $c_l = 2\pi\alpha$. Calculate the lift coefficient for $M_\infty = 0.7$.

Solution. From Eq. (11.52),

$$c_l = \frac{c_{l,0}}{\sqrt{1-M_\infty^2}} = \frac{2\pi\alpha}{\sqrt{1-(0.7)^2}} = \boxed{8.8\alpha}$$

Note: The effect of compressibility at Mach 0.7 is to increase the lift slope by the ratio $8.8/2\pi = 1.4$, or by 40 percent.

11.5 IMPROVED COMPRESSIBILITY CORRECTIONS

The importance of accurate compressibility corrections reached new highs during the rapid increase in airplane speeds spurred by World War II. Efforts were made to improve upon the Prandtl-Glauert rule discussed in Sec. 11.4. Several of the more popular formulas are given below.

The Karman-Tsien rule states

$$C_p = \frac{C_{p,0}}{\sqrt{1-M_\infty^2} + [M_\infty^2/(1+\sqrt{1-M_\infty^2})]C_{p,0}/2} \tag{11.54}$$

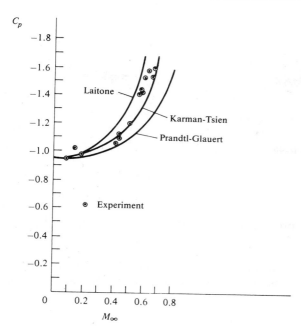

FIGURE 11.4

Several compressibility corrections compared with experimental results for an NACA 4412 airfoil at an angle of attack $\alpha = 1°53'$. The experimental data are chosen for their historical significance; they are from NACA report no. 646, published in 1938 (*Ref. 30*). This was the first major NACA publication to address the compressibility problem in a systematic fashion; it covered work performed in the 2-ft high-speed tunnel at the Langley Aeronautical Laboratory and was carried out during 1935–1936.

This formula, derived in Refs. 27 and 28, has been widely adopted by the aeronautical industry since World War II.

Laitone's rule states

$$C_p = \frac{C_{p,0}}{\sqrt{1-M_\infty^2}+(M_\infty^2\{1+[(\gamma-1)/2]M_\infty^2\}/2\sqrt{1-M_\infty^2})C_{p,0}} \qquad (11.55)$$

This formula is more recent than either the Prandtl-Glauert or the Karman-Tsien rule; it is derived in Ref. 29.

These compressibility corrections are compared in Fig. 11.4, which also shows experimental data for the C_p variation with M_∞ at the 0.3-chord location on an NACA 4412 airfoil. Note that the Prandtl-Glauert rule, although the simplest to apply, underpredicts the experimental data, whereas the improved compressibility corrections are clearly more accurate. Recall that the Prandtl-Glauert rule is based on linear theory. In contrast, both the Laitone and Karman-Tsien rules attempt to account for some of the nonlinear aspects of the flow.

11.6 CRITICAL MACH NUMBER

Return to the road map given in Fig. 11.1. We have now finished our discussion of linearized flow and the associated compressibility corrections. Keep in mind that such linearized theory does *not* apply to the transonic flow regime, $0.8 \le M_\infty \le 1.2$. Transonic flow is highly nonlinear, and theoretical transonic aerodynamics is a challenging and sophisticated subject. For the remainder of this chapter, we deal with several aspects of transonic flow from a qualitative point of view. The theory of transonic aerodynamics is beyond the scope of this book.

Consider an airfoil in a low-speed flow, say, with $M_\infty = 0.3$, as sketched in Fig. 11.5a. In the expansion over the top surface of the airfoil, the local flow Mach number M increases. Let point A represent the location on the airfoil surface where the pressure is a minimum, hence where M is a maximum. In Fig. 11.5a, let us say this maximum is $M_A = 0.435$. Now assume that we gradually increase the freestream Mach number. As M_∞ increases, M_A also increases. For example, if M_∞ is increased to $M = 0.5$, the maximum local value of M will be 0.772, as shown in Fig. 11.5b. Let us continue to increase M_∞ until we achieve just the right value such that the local Mach number at the minimum pressure point equals 1, i.e., such that $M_A = 1.0$, as shown in Fig. 11.5c. When this happens, the freestream Mach number M_∞ is called the *critical Mach number*, denoted by M_{cr}. By definition, the critical Mach number is that *freestream* Mach number at which sonic flow is first achieved on the airfoil surface. In Fig. 11.5c, $M_{cr} = 0.61$.

One of the most important problems in high-speed aerodynamics is the determination of the critical Mach number of a given airfoil, because at values of M_∞ slightly above M_{cr}, the airfoil experiences a dramatic increase in drag

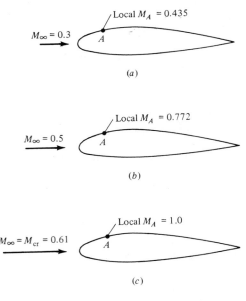

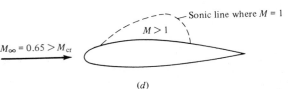

FIGURE 11.5
Definition of critical Mach number. Point A is the location of minimum pressure on the top surface of the airfoil. (See homework problem 11.7 for the calculation of the numbers in this Figure.)

coefficient (discussed in Sec. 11.7). The purpose of the present section is to give a method for estimating M_{cr}.

Let p_∞ and p_A represent the static pressures in the freestream and at point A, respectively, in Fig. 11.5. For isentropic flow, where the total pressure p_0 is constant, these static pressures are related through Eq. (8.42) as follows:

$$\frac{p_A}{p_\infty} = \frac{p_A/p_0}{p_\infty/p_0} = \left(\frac{1+[(\gamma-1)/2]M_\infty^2}{1+[(\gamma-1)/2]M_A^2}\right)^{\gamma/(\gamma-1)} \tag{11.56}$$

The pressure coefficient at point A is given by Eq. (11.22) as

$$C_{p,A} = \frac{2}{\gamma M_\infty^2}\left(\frac{p_A}{p_\infty}-1\right) \tag{11.57}$$

Combining Eqs. (11.56) and (11.57), we have

$$C_{p,A} = \frac{2}{\gamma M_\infty^2}\left[\left(\frac{1+[(\gamma-1)/2]M_\infty^2}{1+[(\gamma-1)/2]M_A^2}\right)^{\gamma/(\gamma-1)}-1\right] \tag{11.58}$$

Equation (11.58) is useful in its own right; for a given freestream Mach number, it relates the local value of C_p to the local Mach number. However, for our purposes here, we ask the question, What is the value of the local C_p when the local Mach number is unity? By definition, this value of the pressure coefficient is called the *critical pressure coefficient*, denoted by $C_{p,cr}$. For a given freestream Mach number M_∞, the value of $C_{p,cr}$ can be obtained by inserting $M_A = 1$ into Eq. (11.58):

$$C_{p,cr} = \frac{2}{\gamma M_\infty^2}\left[\left(\frac{1+[(\gamma-1)/2]M_\infty^2}{1+(\gamma-1)/2}\right)^{\gamma/(\gamma-1)}-1\right] \tag{11.59}$$

Equation (11.59) allows us to calculate the pressure coefficient at any point in the flow where the local Mach number is 1, for a given freestream Mach number M_∞. For example, if M_∞ is slightly greater than M_{cr}, say, $M_\infty = 0.65$ as shown in Fig. 11.5d, then a finite region of supersonic flow will exist above the airfoil; Eq. (11.59) allows us to calculate the pressure coefficient at only those points where $M = 1$, i.e., at only those points that fall on the sonic line in Fig. 11.5d. Now, returning to Fig. 11.5c, when the freestream Mach number is precisely equal to the critical Mach number, there is only one point where $M = 1$, namely, point A. The pressure coefficient at point A will be $C_{p,cr}$, which is obtained from Eq. (11.59). In this case, M_∞ in Eq. (11.59) is precisely M_{cr}. Hence,

$$\boxed{C_{p,cr} = \frac{2}{\gamma M_{cr}^2}\left[\left(\frac{1+[(\gamma-1)/2]M_{cr}^2}{1+(\gamma-1)/2}\right)^{\gamma/(\gamma-1)}-1\right]} \tag{11.60}$$

Equation (11.60) shows that $C_{p,cr}$ is a unique function of M_{cr}; this variation is plotted as curve C in Fig. 11.6. Note that Eq. (11.60) is simply an aerodynamic relation for isentropic flow—it has no connection with the shape of a given airfoil.

In this sense, Eq. (11.60), and hence curve C in Fig. 11.6, is a type of "universal relation" which can be used for all airfoils.

Equation (11.60), in conjunction with any one of the compressibility corrections given by Eqs. (11.51), (11.54), or (11.55), allows us to estimate the critical Mach number for a given airfoil as follows:

1. By some means, either experimental or theoretical, obtain the low-speed incompressible value of the pressure coefficient $C_{p,0}$ at the minimum pressure point on the given airfoil.
2. Using any of the compressibility corrections, Eq. (11.51), (11.54), or (11.55), plot the variation of C_p with M_∞. This is represented by curve B in Fig. 11.6.
3. Somewhere on curve B, there will be a single point where the pressure coefficient corresponds to locally sonic flow. Indeed, this point must coincide with Eq. (11.60), represented by curve C in Fig. 11.6. Hence, the *intersection* of curves B and C represents the point corresponding to sonic flow at the minimum pressure location on the airfoil. In turn, the value of M_∞ at this intersection is, by definition, the critical Mach number, as shown in Fig. 11.6.

The graphical construction in Fig. 11.6 is not an exact determination of M_{cr}. Although curve C is exact, curve B is approximate because it represents the approximate compressibility correction. Hence, Fig. 11.6 gives only an estimation of M_{cr}. However, such an estimation is quite useful for preliminary design, and the results from Fig. 11.6 are accurate enough for most applications.

Consider two airfoils, one thin and the other thick, as sketched in Fig. 11.7. First consider the low-speed incompressible flow over these airfoils. The flow over the thin airfoil is only slightly perturbed from the freestream. Hence, the expansion over the top surface is mild, and $C_{p,0}$ at the minimum pressure point is a negative number of only small absolute magnitude, as shown in Fig. 11.7.

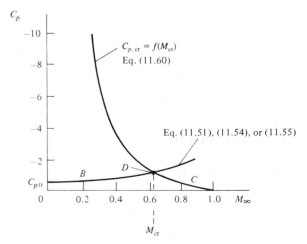

FIGURE 11.6
Estimation of critical Mach number.

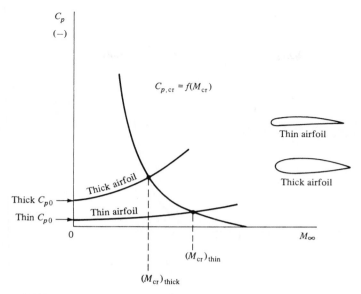

FIGURE 11.7
Effect of airfoil thickness on critical Mach number.

[Recall from Eq. (11.32) that $C_p \propto \hat{u}$; hence, the smaller the perturbation, the smaller is the absolute magnitude of C_p.] In contrast, the flow over the thick airfoil experiences a large perturbation from the freestream. The expansion over the top surface is strong, and $C_{p,0}$ at the minimum pressure point is a negative number of large magnitude, as shown in Fig. 11.7. If we now perform the same construction for each airfoil as given in Fig. 11.6, we see that the thick airfoil will have a lower critical Mach number than the thin airfoil. This is clearly illustrated in Fig. 11.7. For high-speed airplanes, it is desirable to have M_{cr} as high as possible. Hence, modern high-speed subsonic airplanes are usually designed with relatively thin airfoils. (The development of the supercritical airfoil has somewhat loosened this criterion, as discussed in Sec. 11.8.) For example, the Gates Lear jet high-speed jet executive transport utilizes a 9 percent thick airfoil; contrast this with the low-speed Piper Aztec, a twin-engine propeller-driven general aviation aircraft designed with a 14 percent thick airfoil.

11.7 DRAG-DIVERGENCE MACH NUMBER: THE SOUND BARRIER

Imagine that we have a given airfoil at a fixed angle of attack in a wind tunnel, and we wish to measure its drag coefficient c_d as a function of M_∞. To begin with, we measure the drag coefficient at low subsonic speed to be $c_{d,0}$, shown in Fig. 11.8. Now, as we gradually increase the freestream Mach number, we observe

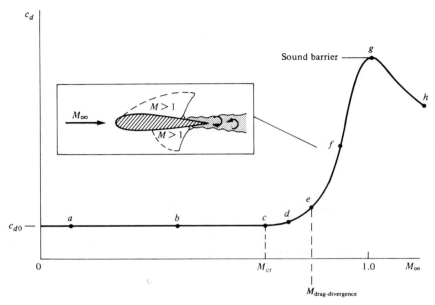

FIGURE 11.8
Sketch of the variation of profile drag coefficient with freestream Mach number, illustrating the critical and drag-divergence Mach numbers and showing the large drag rise near Mach 1.

that c_d remains relatively constant all the way to the critical Mach number, as illustrated in Fig. 11.8. The flow fields associated with points a, b, and c in Fig. 11.8 are represented by Fig. 11.5a, b, and c, respectively. As we very carefully increase M_∞ slightly above M_{cr}, say, to point d in Fig. 11.8, a finite region of supersonic flow appears on the airfoil, as shown in Fig. 11.5d. The Mach number in this bubble of supersonic flow is only slightly above Mach 1, typically 1.02 to 1.05. However, as we continue to nudge M_∞ higher, we encounter a point where the drag coefficient suddenly starts to increase. This is given as point e in Fig. 11.8. The value of M_∞ at which this sudden increase in drag starts is defined as the *drag-divergence Mach number*. Beyond the drag-divergence Mach number, the drag coefficient can become very large, typically increasing by a factor of 10 or more. This large increase in drag is associated with an extensive region of supersonic flow over the airfoil, terminating in a shock wave, as sketched in the insert in Fig. 11.8. Corresponding to point f on the drag curve, this insert shows that as M_∞ approaches unity, the flow on both the top and bottom surfaces can be supersonic, both terminated by shock waves. For example, consider the case of a reasonably thick airfoil, designed originally for low-speed applications, when M_∞ is beyond drag-divergence; in such a case, the local Mach number can reach 1.2 or higher. As a result, the terminating shock waves can be relatively strong. These shocks generally cause severe flow separation downstream of the shocks, with an attendant large increase in drag.

Now, put yourself in the place of an aeronautical engineer in 1936. You are familiar with the Prandtl-Glauert rule, given by Eq. (11.51). You recognize that as $M_\infty \to 1$, this equation shows the absolute magnitude of C_p approaching infinity. This hints at some real problems near Mach 1. Furthermore, you know of some initial high-speed subsonic wind-tunnel tests that have generated drag curves which resemble the portion of Fig. 11.8 from points a to f. How far will the drag coefficient increase as we get closer to $M_\infty = 1$? Will c_d go to infinity? At this stage, you might be pessimistic. You might visualize the drag increase to be so large that no airplane with the power plants existing in 1936, or even envisaged for the future, could ever overcome this "barrier." It was this type of thought that led to the popular concept of a sound barrier and that prompted many people to claim that humans would never fly faster than the speed of sound.

Of course, today we know the sound barrier was a myth. We cannot use the Prandtl-Glauert rule to argue that c_d will become infinite at $M_\infty = 1$, because the Prandtl-Glauert rule is invalid at $M_\infty = 1$ (see Secs. 11.3 and 11.4). Moreover, early transonic wind-tunnel tests carried out in the late 1940s clearly indicated that c_d peaks at or around Mach 1 and then actually decreases as we enter the

FIGURE 11.9
The Bell XS-1—the first manned airplane to fly faster than sound, October 14, 1947. (*Courtesy of the National Air and Space Museum.*)

supersonic regime, as shown by points *g* and *h* in Fig. 11.8. All we need is an aircraft with an engine powerful enough to overcome the large drag rise at Mach 1. The myth of the sound barrier was finally put to rest on October 14, 1947, when Captain Charles (Chuck) Yeager became the first human being to fly faster than sound in the sleek, bullet-shaped Bell XS-1. This rocket-propelled research aircraft is shown in Fig. 11.9. Of course, today supersonic flight is a common reality; we have jet engines powerful enough to accelerate military fighters through Mach 1 flying straight up! Such airplanes can fly at Mach 3 and beyond. Indeed, we are limited only by aerodynamic heating at high speeds (and the consequent structural problems). Right now, NASA is conducting research on supersonic combustion ramjet engines for flight in the Mach 4 to 7 range. Keep in mind, however, that because of the large power requirements for very high-speed flight, the fuel consumption becomes large. In today's energy-conscious world, this constraint can be as much a barrier to high-speed flight as the sound barrier was once envisaged.

Since 1945, research in transonic aerodynamics has focused on reducing the large drag rise shown in Fig. 11.8. Instead of living with a factor of 10 increase in drag at Mach 1, can we reduce it to a factor of 2 or 3? This is the subject of the remaining sections of this chapter.

11.8 THE AREA RULE

For a moment, let us expand our discussion from two-dimensional airfoils to a consideration of a complete airplane. In this section, we introduce a design concept which has effectively reduced the drag rise near Mach 1 for a complete airplane.

As stated before, the first practical jet-powered aircraft appeared at the end of World War II in the form of the German Me 262. This was a subsonic fighter plane with a top speed near 550 mi/h. The next decade saw the design and production of many types of jet aircraft—all limited to subsonic flight by the large drag near Mach 1. Even the "century" series of fighter aircraft designed to give the U.S. Air Force supersonic capability in the early 1950s, such as the Convair F-102 delta-wing airplane, ran into difficulty and could not at first readily penetrate the sound barrier in level flight. The thrust of jet engines at that time simply could not overcome the large peak drag near Mach 1.

A planview, cross section, and area distribution (cross-sectional area versus distance along the axis of the airplane) for a typical airplane of that decade are sketched in Fig. 11.10. Let A denote the total cross-sectional area at any given station. Note that the cross-sectional area distribution experiences some abrupt changes along the axis, with discontinuities in both A and dA/dx in the regions of the wing.

In contrast, for almost a century, it was well known by ballisticians that the speed of a supersonic bullet or artillery shell with a *smooth* variation of cross-sectional area was higher than projectiles with abrupt or discontinuous area distributions. In the mid-1950s, an aeronautical engineer at the NACA Langley

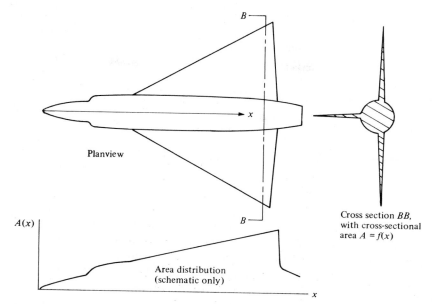

Cross section *BB*,
with cross-sectional
area $A = f(x)$

FIGURE 11.10
A schematic of a non-area-ruled aircraft.

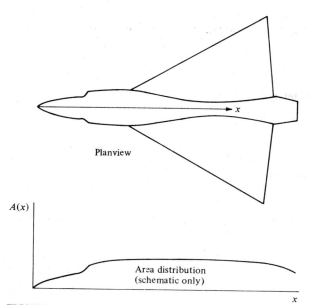

FIGURE 11.11
A schematic of an area-ruled aircraft.

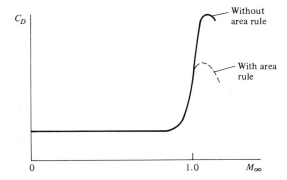

FIGURE 11.12
The drag-rise properties of area-ruled and non-area-ruled aircraft (schematic only).

Aeronautical Laboratory, Richard T. Whitcomb, put this knowledge to work on the problem of transonic flight of airplanes. Whitcomb reasoned that the variation of cross-sectional area for an airplane should be smooth, with no discontinuities. This meant that, in the region of the wings and tail, the fuselage cross-sectional area should decrease to compensate for the addition of the wing and tail cross-sectional area. This led to a "coke bottle" fuselage shape, as shown in Fig. 11.11. Here, the planview and area distribution are shown for an aircraft with a relatively smooth variation of $A(x)$. This design philosophy is called the *area rule*, and it successfully reduced the peak drag near Mach 1 such that practical airplanes could fly supersonically by the mid-1950s. The variations of drag coefficient with M_∞ for an area-ruled and non-area-ruled airplane are schematically compared in Fig. 11.12; typically, the area rule leads to a factor-of-2 reduction in the peak drag near Mach 1.

The development of the area rule was a dramatic breakthrough in high-speed flight, and it earned a substantial reputation for Richard Whitcomb—a reputation which was to be later garnished by a similar breakthrough in transonic airfoil design, to be discussed in Sec. 11.9. The original work on the area rule was presented by Whitcomb in Ref. 31, which should be consulted for more details.

11.9 THE SUPERCRITICAL AIRFOIL

Let us return to a consideration of two-dimensional airfoils. A natural conclusion from the material in Sec. 11.6, and especially from Fig. 11.8, is that an airfoil with a high critical Mach number is very desirable, indeed necessary, for high-speed subsonic aircraft. If we can increase M_{cr}, then we can increase $M_{\text{drag-divergence}}$, which follows closely after M_{cr}. This was the philosophy employed in aircraft design from 1945 to approximately 1965. Almost by accident, the NACA 64-series airfoils (see Sec. 4.2), although originally designed to encourage laminar flow, turned out to have relative high values of M_{cr} in comparison with other NACA shapes. Hence, the NACA 64 series has seen wide application on high-speed airplanes. Also, we know that thinner airfoils have higher values of

M_{cr} (see Fig. 11.7); hence, aircraft designers have used relatively thin airfoils on high-speed airplanes.

However, there is a limit to how thin a practical airfoil can be. For example, considerations other than aerodynamic influence the airfoil thickness; the airfoil requires a certain thickness for structural strength, and there must be room for the storage of fuel. This prompts the following question: For an airfoil of given thickness, how can we delay the large drag rise to higher Mach numbers? To increase M_{cr} is one obvious tack, as described above, but there is another approach. Rather than increasing M_{cr}, let us strive to increase the Mach number *increment* between M_{cr} and $M_{\text{drag-divergence}}$. That is, referring to Fig. 11.8, let us increase the distance between points *e* and *c*. This philosophy has been pursued since 1965, leading to the design of a new family of airfoils called *supercritical airfoils*, which are the subject of this section.

The purpose of a supercritical airfoil is to increase the value of $M_{\text{drag-divergence}}$, although M_{cr} may change very little. The shape of a supercritical airfoil is compared with an NACA 64-series airfoil in Fig. 11.13. Here, an NACA 64_2-A215 airfoil is sketched in Fig. 11.13a, and a 13-percent thick supercritical airfoil is shown in Fig. 11.13c. (Note the similarity between the supercritical profile and

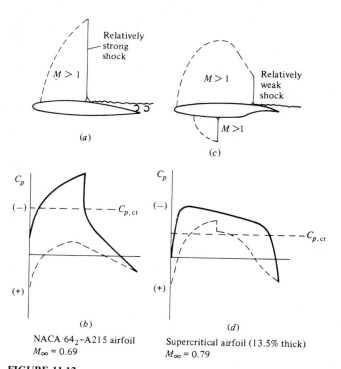

(a)

(c)

(b)

(d)

NACA 64_2–A215 airfoil
$M_\infty = 0.69$

Supercritical airfoil (13.5% thick)
$M_\infty = 0.79$

FIGURE 11.13
Standard NACA 64-series airfoil compared with a supercritical airfoil at cruise lift conditions. (*From Ref. 32.*)

the modern low-speed airfoils discussed in Sec. 4.10.) The supercritical airfoil has a relatively flap top, thus encouraging a region of supersonic flow with lower local values of M than the NACA 64 series. In turn, the terminating shock is weaker, thus creating less drag. Similar trends can be seen by comparing the C_p distributions for the NACA 64 series (Fig. 11.13b) and the supercritical airfoil (Fig. 11.13d). Indeed, Fig. 11.13a and b for the NACA 64-series airfoil pertain to a lower freestream Mach number, $M_\infty = 0.69$, than Fig. 11.13c and d, which pertain to the supercritical airfoil at a higher freestream Mach number, $M_\infty = 0.79$. In spite of the fact that the 64-series airfoil is at a lower M_∞, the extent of the supersonic flow reaches farther above the airfoil, the local supersonic Mach numbers are higher, and the terminating shock wave is stronger. Clearly, the supercritical airfoil shows more desirable flow-field characteristics; namely, the extent of the supersonic flow is closer to the surface, the local supersonic Mach numbers are lower, and the terminating shock wave is weaker. As a result, the value of $M_{\text{drag-divergence}}$ will be higher for the supercritical airfoil. This is verified by the experimental data given in Fig. 11.14, taken from Ref. 32. Here, the value of $M_{\text{drag-divergence}}$ is 0.79 for the supercritical airfoil in comparison with 0.67 for the NACA 64 series.

Because the top of the supercritical airfoil is relatively flat, the forward 60 percent of the airfoil has negative camber, which lowers the lift. To compensate, the lift is increased by having extreme positive camber on the rearward 30 percent of the airfoil. This is the reason for the cusplike shape of the bottom surface near the trailing edge.

The supercritical airfoil was developed by Richard Whitcomb in 1965 at the NASA Langley Research Center. A detailed description of the rationale as well as some early experimental data for supercritical airfoils is given by Whitcomb in Ref. 32, which should be consulted for more details. The supercritical airfoil, and many variations of such, are now used by the aircraft industry on modern high-speed airplane designs. Examples are the Boeing 757 and 767, and the latest model Lear jets. The supercritical airfoil is one of two major breakthroughs made

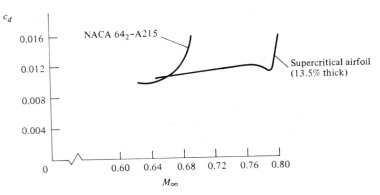

FIGURE 11.14
The drag-divergence properties of a standard NACA 64-series airfoil and a supercritical airfoil.

in transonic airplane aerodynamics since 1945, the other being the area rule discussed in Sec. 11.8. It is a testimonial to the man that Richard Whitcomb was mainly responsible for both.

11.10 HISTORICAL NOTE: HIGH-SPEED AIRFOILS—EARLY RESEARCH AND DEVELOPMENT

Twentieth-century aerodynamics does not have the exclusive rights to the observation of the large drag rise on bodies flying at near the speed of sound; rather, in the eighteenth century the Englishman Benjamin Robins, inventor of the ballistic pendulum, reported that "the velocity at which the body shifts its resistance (from a V^2 to a V^3 relation) is nearly the same with which sound is propagated through air." His statement was based on a large number of experiments during which projectiles were fired into his ballistic pendulum. However, these results had little relevance to the early aerodynamicists of this century, who were struggling to push aircraft speeds to 150 mi/h during and just after World War I. To these people, flight near the speed of sound was just fantasy.

With one exception! World War I airplanes such as the Spad and Nieuport had propeller blades where the tips were moving at near the speed of sound. By 1919, British researchers had already observed the loss in thrust and large increase in blade drag for a propeller with tip speeds up to 1180 ft/s—slightly above the speed of sound. To examine this effect further, F. W. Caldwell and E. N. Fales, both engineers at the U.S. Army's Engineering Division at McCook Field near Dayton, Ohio (the forerunner of the massive Air Force research and development facilities at Wright-Patterson Air Force Base today), conducted a series of high-speed airfoil tests. They designed and built the first high-speed wind tunnel—a facility with a 14-in-diameter test section capable of velocities up to 675 ft/s. In 1918, they conducted the first wind-tunnel tests involving the high-speed flow over a stationary airfoil. Their results showed large decreases in lift coefficient and major increases in drag coefficient for the thicker airfoils at angle of attack. These were the first measured "compressibility effects" on an airfoil in history. Caldwell and Fales noted that such changes occurred at a certain air velocity, which they denoted as the "critical speed"—a term that was to evolve into the critical Mach number at a later date. It is interesting to note that Orville Wright was a consultant to the Army at this time (Wilbur had died prematurely in 1912 of typhoid fever) and observed some of the Caldwell and Fales tests. However, a fundamental understanding and explanation of this critical-speed phenomenon was completely lacking. Nobody at that time had even the remotest idea of what was really happening in this high-speed flow over the airfoil.

Members of the National Advisory Committee for Aeronautics were well aware of the Caldwell-Fales results. Rather than let the matter die, in 1922 the NACA contracted with the National Bureau of Standards (NBS) for a study of high-speed flows over airfoils, with an eye toward improved propeller sections.

The work at NBS included the building of a high-speed wind tunnel with a 12-in-diameter test section, capable of producing a Mach number of 0.95. The aerodynamic testing was performed by Lyman J. Briggs (soon to become director of NBS) and Hugh Dryden (soon to become one of the leading aerodynamicists of the twentieth century). In addition to the usual force data, Briggs and Dryden also measured pressure distributions over the airfoil surface. These pressure distributions allowed more insight into the nature of the flow and definitely indicated flow separation on the top surface of the airfoil. We now know that such flow separation is induced by a shock wave, but these early researchers did not at that time know about the presence of such shocks.

During the same period, the only meaningful theoretical work on high-speed airfoil properties was carried out by Ludwig Prandtl in Germany and Hermann Glauert in England—work which led to the Prandtl-Glauert compressibility correction, given by Eq. (11.51). As early as 1922, Prandtl is quoted as stating that the lift coefficient increased according to $(1 - M_\infty^2)^{-1/2}$; he mentioned this conclusion in his lectures at Göttingen, but without written proof. This result was mentioned again 6 years later by Jacob Ackeret, a colleague of Prandtl, in the famous German series *Handbuch der Physik*, again without proof. Subsequently, in 1928 the concept was formally established by Hermann Glauert, a British aerodynamicist working for the Royal Aircraft Establishment. (See chap. 9 of Ref. 21 for a biographical sketch of Glauert.) Using only six pages in the *Proceedings of the Royal Society*, vol. 118, p. 113, Glauert presented a derivation based on linearized small-perturbation theory (similar to that described in Sec. 11.4) which confirmed the $(1 - M_\infty^2)^{-1/2}$ variation. In this paper, entitled "The Effect of Compressibility on the Lift of an Airfoil," Glauert derived the famous Prandtl-Glauert compressibility correction, given here as Eqs. (11.51) to (11.53). This result was to stand alone, unaltered, for the next 10 years.

Hence, in 1930 the state of the art of high-speed subsonic airfoil research was characterized by experimental proof of the existence of the drag-divergence phenomenon, some idea that it was caused by flow separation, but no fundamental understanding of the basic flow field. In turn, there was virtually no theoretical background outside of the Prandtl-Glauert rule. Also, keep in mind that all the above work was paced by the need to understand propeller performance, because in that day the only component of airplanes to encounter compressibility effects was the propeller tips.

All this changed in the 1930s. In 1928, the NACA had constructed its first rudimentary high-speed subsonic wind tunnel at the Langley Aeronautical Laboratory, utilizing a 1-ft-diameter test section. With Eastman Jacobs as tunnel director and John Stack as the chief researcher, a series of tests was run on various standard airfoil shapes. Frustrated by their continual lack of understanding about the flow field, they turned to optical techniques, following in the footsteps of Ernst Mach (see Sec. 9.8). In 1933, they assembled a crude schlieren optical system consisting of 3-in-diameter reading-glass-quality lenses and a short-duration-spark light source. In their first test using the schlieren system, dealing with flow over a cylinder, the results were spectacular. Shock waves were seen, along with the resulting flow separation. Visitors flocked to the wind tunnel to observe the results, including Theodore Theodorsen, one of the ranking theoretical aerodynamicists of that period. An indicator of the psychology at that time is given by Theodorsen's comment that since the freestream flow was subsonic, what appeared as shock waves must be an "optical illusion." However, Eastman Jacobs and John Stack knew differently. They proceeded with a major series of airfoil testing, using standard NACA sections. Their schlieren pictures revealed the secrets of flow over the airfoils above the critical Mach number. (See Fig. 1.30*b* and its attendant discussion of such supercritical flow.) In 1935, Jacobs traveled to Italy, where he presented results of the NACA high-speed airfoil research at the fifth Volta Conference (see Sec. 7.1). This is the first time in history that photographs of the transonic flow field over standard-shaped airfoils were presented in a large public forum.

During the course of such work in the 1930s, the incentive for high-speed aerodynamic research shifted from propeller applications to concern about the airframe of the airplane itself. By the mid-1930s, the possibility of the 550 mi/h airplane was more than a dream—reciprocating engines were becoming powerful enough to consider such a speed regime for propeller-driven aircraft. In turn, the entire airplane itself (wings, cowling, tail, etc.) would encounter compressibility effects. This led to the design of a large 8-ft high-speed tunnel at Langley, capable of test-section velocities above 500 mi/h. This tunnel, along with the earlier 1-ft tunnel, established the NACA's dominance in high-speed subsonic research in the late 1930s.

In the decade following 1930, the picture had changed completely. By 1940, the high-speed flow over airfoils was relatively well understood. During this period, Stack and Jacobs had not only highlighted the experimental aspects of such high-speed flow, but they also derived the expression for $C_{p,cr}$ as a function of M_{cr} given by Eq. (11.60), and had shown how to estimate the critical Mach number for a given airfoil as discussed in Sec. 11.6. Figure 11.15 shows some representative schlieren photographs taken by the NACA of the flow over standard NACA airfoils. Although these photographs were taken in 1949, they are similar to the results obtained by Stack and Jacobs in the 1930s. Superimposed on these photographs are the measured pressure distributions over the top (solid curve) and bottom (dashed curve) surfaces of the airfoil. Study these pictures carefully. Moving from bottom to top, you can see the influence of increasing freestream Mach number, and going from left to right, you can observe the effect of increasing airfoil thickness. Note how the shock wave moves downstream as M_∞ is increased, finally reaching the trailing edge at $M_\infty = 1.0$. For this case, the top row of pictures shows almost completely supersonic flow over the airfoil. Note also the large regions of separated flow downstream of the shock waves for the Mach numbers of 0.79, 0.87, and 0.94—this separated flow is the primary reason for the large increase in drag near Mach 1. By 1940, it was well understood that the almost discontinuous pressure increase across the shock wave creates a strong adverse pressure gradient on the airfoil surface, and this adverse pressure gradient is responsible for separating the flow.

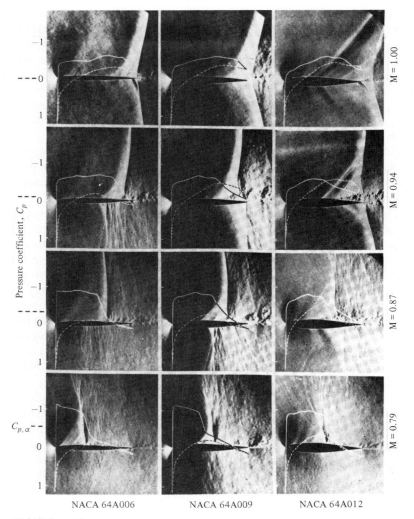

FIGURE 11.15
Schlieren pictures and pressure distributions for transonic flows over several NACA airfoils. These pictures were taken by the NACA in 1949. (*From John V. Becker, "The High-Speed Frontier," NASA SP-445, 1980.*)

The high-speed airfoil research program continues today within NASA. It led to the supercritical airfoils in the 1960s (see Secs. 11.9 and 11.11). It has produced a massive effort in modern times to use computational techniques for theoretically solving the transonic flow over airfoils. Such efforts are beginning to be successful, and in many respects, today we have the capability to design transonic airfoils on the computer. However, such abilities today have roots which reach all the way back to Caldwell and Fales in 1918.

For a more detailed account of the history of high-speed airfoil research, you are encouraged to read the entertaining story portrayed by John V. Becker in *The High-Speed Frontier*, NASA SP-445, 1980.

11.11 HISTORICAL NOTE: RICHARD T. WHITCOMB— ARCHITECT OF THE AREA RULE AND THE SUPERCRITICAL WING

The developments of the area rule (Sec. 11.8) and the supercritical airfoil (Sec. 11.9) are two of the most important advancements in aerodynamics since 1950. That both developments were made by the same man—Richard T. Whitcomb—is remarkable. Who is this man? What qualities lead to such accomplishments? Let us pursue these matters further.

Richard Whitcomb was born on February 21, 1921, in Evanston, Illinois. At an early age, he was influenced by his grandfather, who had known Thomas A. Edison. In an interview with *The Washington Post* on August 31, 1969, Whitcomb is quoted as saying: "I used to sit around and hear stories about Edison. He sort of developed into my idol." Whitcomb entered the Worcester Polytechnic Institute in 1939. (This is the same school from which the rocket pioneer, Robert H. Goddard, had graduated 31 years earlier.) Whitcomb distinguished himself in college and graduated with a mechanical engineering degree with honors in 1943. Informed by a *Fortune* magazine article on the research facilities at the NACA Langley Memorial Laboratory, Whitcomb immediately joined the NACA. He became a wind-tunnel engineer, and as an early assignment he worked on design problems associated with the Boeing B-29 Superfortress. He remained with the NACA and later its successor, NASA, until his retirement in 1980—spending his entire career with the wind tunnels at the Langley Research Center. In the process, he rose to become head of the Eight-foot Tunnel Branch at Langley.

Whitcomb conceived the idea of the area rule as early as 1951. He tested his idea in the transonic wind tunnel at Langley. The results were so promising that the aeronautical industry changed designs in midstream. For example, the Convair F-102 delta-wing fighter had been designed for supersonic flight, but was having major difficulty even exceeding the speed of sound—the increase in drag near Mach 1 was simply too large. The F-102 was redesigned to incorporate Whitcomb's area rule and afterward was able to achieve its originally intended supersonic Mach number. The area rule was such an important aerodynamic breakthrough that it was classified "secret" from 1952 to 1954, when airplanes incorporating the area rule began to roll off the production line. In 1954, Whitcomb was given the Collier Trophy—an annual award for the "greatest achievement in aviation in America."

In the early 1960s, Whitcomb turned his attention to airfoil design, with the objective again of decreasing the large drag rise near Mach 1. Using the existing knowledge about airfoil properties, a great deal of wind-tunnel testing, and intuition honed by years of experience, Whitcomb produced the supercritical airfoil. Again, this development had a major impact on the aeronautical industry, and today virtually all new commercial transport and executive aircraft designs are incorporating a supercritical wing. Because of his development of the super-critical airfoil, in 1974 NASA gave Whitcomb a cash award of $25,000—the largest cash award ever given by NASA to a single individual.

There are certain parallels between the personalities of the Wright brothers and Richard Whitcomb: (1) they all had powerful intuitive abilities which they brought to bear on the problem of flight, (2) they were totally dedicated to their work (none of them ever married), and (3) they did a great deal of their work themselves, trusting only their own results. For example, here is a quote from Whitcomb which appears in the same *Washington Post* interview mentioned above. Concerning the detailed work on the development of the supercritical airfoil, Whitcomb says:

> I modified the shape of the wing myself as we tested it. It's just plain easier this way. In fact my reputation for filing the wing's shape has become so notorious that the people at North American have threatened to provide me with a 10-foot file to work on the real airplane, also.

Perhaps the real ingredient for Whitcomb's success is his personal philosophy, as well as his long hours at work daily. In his own words:

> There's been a continual drive in me ever since I was a teenager to find a better way to do everything. A lot of very intelligent people are willing to adapt, but only to a certain extent. If a human mind can figure out a better way to do something, let's do it. I can't just sit around. I have to think.

Students take note!

11.12 SUMMARY

Review the road map in Fig. 11.1, and make certain that you have all the concepts listed on this map well in mind. Some of the highlights of this chapter are as follows:

For two-dimensional, irrotational, isentropic, steady flow of a compressible fluid, the exact velocity potential equation is

$$\left[1 - \frac{1}{a^2}\left(\frac{\partial \phi}{\partial x}\right)^2\right]\frac{\partial^2 \phi}{\partial x^2} + \left[1 - \frac{1}{a^2}\left(\frac{\partial \phi}{\partial y}\right)^2\right]\frac{\partial^2 \phi}{\partial y^2} - \frac{2}{a^2}\left(\frac{\partial \phi}{\partial x}\right)\left(\frac{\partial \phi}{\partial y}\right)\frac{\partial^2 \phi}{\partial x\,\partial y} = 0$$

$$(11.12)$$

where

$$a^2 = a_0^2 - \frac{\gamma - 1}{2}\left[\left(\frac{\partial \phi}{\partial x}\right)^2 + \left(\frac{\partial \phi}{\partial y}\right)^2\right]$$

$$(11.13)$$

This equation is exact, but it is nonlinear and hence difficult to solve. At present, no general analytical solution to this equation exists.

For the case of small perturbations (slender bodies at low angles of attack), the exact velocity potential equation can be approximated by

$$(1 - M_\infty^2)\frac{\partial^2 \hat{\phi}}{\partial x^2} + \frac{\partial^2 \hat{\phi}}{\partial y^2} = 0 \tag{11.18}$$

This equation is approximate, but linear, and hence more readily solved. This equation holds for subsonic ($0 \le M_\infty \le 0.8$) and supersonic ($1.2 \le M_\infty \le 5$) flows; it does not hold for transonic ($0.8 \le M_\infty \le 1.2$) or hypersonic ($M_\infty > 5$) flows.

The Prandtl-Glauert rule is a compressibility correction that allows the modification of existing incompressible flow data to take into account compressibility effects:

$$C_p = \frac{C_{p,0}}{\sqrt{1 - M_\infty^2}} \tag{11.51}$$

Also,

$$c_l = \frac{c_{l,0}}{\sqrt{1 - M_\infty^2}} \tag{11.52}$$

and

$$c_m = \frac{c_{m,0}}{\sqrt{1 - M_\infty^2}} \tag{11.53}$$

The critical Mach number is that freestream Mach number at which sonic flow is first obtained at some point on the surface of a body. For thin airfoils, the critical Mach number can be estimated as shown in Fig. 11.6.

The drag-divergence Mach number is that freestream Mach number at which a large rise in the drag coefficient occurs, as shown in Fig. 11.8.

The area rule for transonic flow states that the cross-sectional area distribution of an airplane, including fuselage, wing, and tail, should have a smooth distribution along the axis of the airplane.

Supercritical airfoils are specially designed profiles to increase the drag-divergence Mach number.

PROBLEMS

11.1. Consider a subsonic compressible flow in cartesian coordinates where the velocity potential is given by

$$\phi(x, y) = V_\infty x + \frac{70}{\sqrt{1 - M_\infty^2}} e^{-2\pi\sqrt{1 - M_\infty^2}\, y} \sin 2\pi x$$

If the freestream properties are given by $V_\infty = 700$ ft/s, $p_\infty = 1$ atm, and $T_\infty = 519°$R, calculate the following properties at the location $(x, y) = (0.2$ ft, 0.2 ft): M, p, and T.

11.2. Using the Prandtl-Glauert rule, calculate the lift coefficient for an NACA 2412 airfoil at 5° angle of attack in a Mach 0.6 freestream. (Refer to Fig. 4.5 for the original airfoil data.)

11.3. Under low-speed incompressible flow conditions, the pressure coefficient at a given point on an airfoil is -0.54. Calculate C_p at this point when the freestream Mach number is 0.58, using
  (a) The Prandtl-Glauert rule
  (b) The Karman-Tsien rule
  (c) Laitone's rule

11.4. In low-speed incompressible flow, the peak pressure coefficient (at the minimum pressure point) on an NACA 0012 airfoil is -0.41. Estimate the critical Mach number for this airfoil, using the Prandtl-Glauert rule.

11.5. For a given airfoil, the critical Mach number is 0.8. Calculate the value of p/p_∞ at the minimum pressure point when $M_\infty = 0.8$.

11.6. Consider an airfoil in a Mach 0.5 freestream. At a given point on the airfoil, the local Mach number is 0.86. Using the compressible flow tables at the back of this book, calculate the pressure coefficient at that point. Check your answer using the appropriate analytical equation from this chapter. [Note: This problem is analogous to an incompressible problem where the freestream velocity and the velocity at a point are given, and the pressure coefficient is calculated from Eq. (3.38). In an incompressible flow, the pressure coefficient at any point in the flow is a unique function of the local velocity at that point and the freestream velocity. In the present problem, we see that Mach number is the relevant property for a compressible flow—not velocity. The pressure coefficient for an inviscid compressible flow is a unique function of the local Mach number and the freestream Mach number.]

11.7. Figure 11.5 shows four cases for the flow over the same airfoil wherein M_∞ is progressively increased from 0.3 to $M_{cr} = 0.61$. Have you wondered where the numbers on Fig. 11.5 came from? Here is your chance to find out. Point A on the airfoil is the point of minimum pressure (hence maximum M) on the airfoil. Assume that the minimum pressure (maximum Mach number) continues to occur at this same point as M_∞ is increased. In part (a) of Fig. 11.5, for $M_\infty = 0.3$, the local Mach number at point A was arbitrarily chosen as $M_A = 0.435$, this arbitrariness is legitimate because we have not specified the airfoil shape, but rather are stating that, whatever the shape is, a maximum Mach number of 0.435 occurs at point A on the airfoil surface. However, once the numbers are given for part (a), then the numbers for parts (b), (c), and (d) are *not* arbitrary. Rather, M_A is a unique function of M_∞ for the remaining pictures. With all this as background information, starting with the data shown in Fig. 11.5(a), *calculate M_A when $M_\infty = 0.61$.* Obviously, from Fig. 11.5(d), your result should turn out to be $M_A = 1.0$ because $M_\infty = 0.61$ is said to be the critical Mach number. Said in another way, you are being asked to prove that the critical Mach number for this airfoil is 0.61. *Hint:* For simplicity, assume that the Prandtl-Glauert rule holds for the conditions of this problem.

11.8. Consider the flow over a circular cylinder; the incompressible flow over such a cylinder is discussed in Sec. 3.13. Consider also the flow over a sphere; the incompressible flow over a sphere is described in Sec. 6.4. The subsonic compressible flow over both the cylinder and the sphere is qualitatively similar but quantitatively different from their incompressible counterparts. Indeed, because of the "bluntness" of these bodies, their critical Mach numbers are relatively low. In particular:

$$\text{For a cylinder:} \qquad M_{cr} = 0.404$$

$$\text{For a sphere:} \qquad M_{cr} = 0.57$$

Explain on a physical basis why the sphere has a higher M_{cr} than the cylinder.

LINEARIZED SUPERSONIC FLOW

With the stabilizer setting at 2° the speed was allowed to increase to approximately 0.98 to 0.99 Mach number where elevator and rudder effectiveness were regained and the airplane seemed to smooth out to normal flying characteristics. This development lent added confidence and the airplane was allowed to continue until an indication of 1.02 on the cockpit Mach meter was obtained. At this indication the meter momentarily stopped and then jumped at 1.06, and this hesitation was assumed to be caused by the effect of shock waves on the static source. At this time the power units were cut and the airplane allowed to decelerate back to the subsonic flight condition.

Captain Charles Yeager, describing his flight on October 14, 1947—the first manned flight to exceed the speed of sound.

12.1 INTRODUCTION

The linearized perturbation velocity potential equation derived in Chap. 11, Eq. (11.18), is

$$(1 - M_\infty^2)\frac{\partial^2 \hat{\phi}}{\partial x^2} + \frac{\partial^2 \hat{\phi}}{\partial y^2} = 0 \qquad (11.18)$$

and holds for both subsonic and supersonic flow. In Chap. 11, we treated the case of subsonic flow, where $1 - M_\infty^2 > 0$ in Eq. (11.18). However, for supersonic flow, $1 - M_\infty^2 < 0$. This seemingly innocent change in sign on the first term of Eq. (11.18) is, in reality, a very dramatic change. Mathematically, when $1 - M_\infty^2 > 0$ for subsonic flow, Eq. (11.18) is an *elliptic* partial differential equation, whereas when $1 - M_\infty^2 < 0$ for supersonic flow, Eq. (11.18) becomes a *hyperbolic* differential equation. The details of this mathematical difference are beyond the scope of this book; however, the important point is that there *is* a difference. Moreover, this portends a fundamental difference in the physical aspects of subsonic and supersonic flow—something we have already demonstrated in previous chapters.

The purpose of this chapter is to obtain a solution of Eq. (11.18) for supersonic flow and to apply this solution to the calculation of supersonic airfoil properties. Since our purpose is straightforward, and since this chapter is relatively short, there is no need for a chapter road map to provide guidance on the flow of our ideas.

12.2 DERIVATION OF THE LINEARIZED SUPERSONIC PRESSURE COEFFICIENT FORMULA

For the case of supersonic flow, let us write Eq. (11.18) as

$$\lambda^2 \frac{\partial^2 \hat{\phi}}{\partial x^2} - \frac{\partial^2 \hat{\phi}}{\partial y^2} = 0 \tag{12.1}$$

where $\lambda = \sqrt{M_\infty^2 - 1}$. A solution to this equation is the functional relation

$$\hat{\phi} = f(x - \lambda y) \tag{12.2}$$

We can demonstrate this by substituting Eq. (12.2) into Eq. (12.1) as follows. The partial derivative of Eq. (12.2) with respect to x can be written as

$$\frac{\partial \hat{\phi}}{\partial x} = f'(x - \lambda y) \frac{\partial (x - \lambda y)}{\partial x}$$

or

$$\frac{\partial \hat{\phi}}{\partial x} = f' \tag{12.3}$$

In Eq. (12.3), the prime denotes differentiation of f with respect to its argument, $x - \lambda y$. Differentiating Eq. (12.3) again with respect to x, we obtain

$$\frac{\partial^2 \hat{\phi}}{\partial x^2} = f'' \tag{12.4}$$

Similarly,

$$\frac{\partial \hat{\phi}}{\partial y} = f'(x - \lambda y) \frac{\partial (x - \lambda y)}{\partial y}$$

or

$$\frac{\partial \hat{\phi}}{\partial y} = f'(-\lambda) \tag{12.5}$$

Differentiating Eq. (12.5) again with respect to y, we have

$$\frac{\partial^2 \hat{\phi}}{\partial y^2} = \lambda^2 f'' \tag{12.6}$$

Substituting Eqs. (12.4) and (12.6) into (12.1), we obtain the identity

$$\lambda^2 f'' - \lambda^2 f'' = 0$$

Hence, Eq. (12.2) is indeed a solution of Eq. (12.1).

Examine Eq. (12.2) closely. This solution is not very specific, because f can be *any* function of $x - \lambda y$. However, Eq. (12.2) tells us something specific about the flow, namely, that $\hat{\phi}$ is *constant* along lines of $x - \lambda y = $ constant. The slope of these lines is obtained from

$$x - \lambda y = \text{const}$$

Hence,

$$\frac{dy}{dx} = \frac{1}{\lambda} = \frac{1}{\sqrt{M_\infty^2 - 1}} \tag{12.7}$$

From Eq. (9.31) and the accompanying Fig. 9.22, we know that

$$\tan \mu = \frac{1}{\sqrt{M_\infty^2 - 1}} \tag{12.8}$$

where μ is the Mach angle. Therefore, comparing Eqs. (12.7) and (12.8), we see that a line along which $\hat{\phi}$ is constant is a *Mach line*. This result is sketched in Fig. 12.1, which shows supersonic flow over a surface with a small hump in the middle, where θ is the angle of the surface relative to the horizontal. According to Eqs. (12.1) to (12.8), all disturbances created at the wall (represented by the perturbation potential $\hat{\phi}$) propagate unchanged away from the wall along Mach waves. All the Mach waves have the same slope, namely, $dy/dx = (M_\infty^2 - 1)^{-1/2}$. Note that the Mach waves slope *downstream* above the wall. Hence, *any disturbance at the wall cannot propagate upstream*; its effect is limited to the region of the flow downstream of the Mach wave emanating from the point of the disturbance. This is a further substantiation of the major difference between subsonic and supersonic flows mentioned in previous chapters, namely, disturbances propagate *everywhere* throughout a subsonic flow, whereas they cannot propagate upstream in a steady supersonic flow.

Keep in mind that the above results, as well as the picture in Fig. 12.1, pertain to *linearized* supersonic flow [because Eq. (12.1) is a linear equation]. Hence, these results assume *small perturbations*; i.e., the hump in Fig. 12.1 is small, and thus θ is small. Of course, we know from Chap. 9 that in reality a

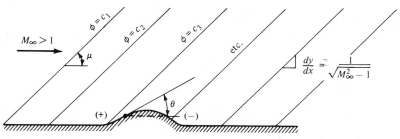

FIGURE 12.1
Linearized supersonic flow.

shock wave will be induced by the forward part of the hump, and an expansion wave will emanate from the rearward part of the hump. These are waves of finite strength and are not a part of linearized theory. Linearized theory is approximate; one of the consequences of this approximation is that waves of finite strength (shock and expansion waves) are not admitted.

The above results allow us to obtain a simple expression for the pressure coefficient in supersonic flow, as follows. From Eq. (12.3),

$$\hat{u} = \frac{\partial \hat{\phi}}{\partial x} = f'$$

(12.9)

and from Eq. (12.5),

$$\hat{v} = \frac{\partial \hat{\phi}}{\partial y} = -\lambda f'$$

(12.10)

Eliminating f' from Eqs. (12.9) and (12.10), we obtain

$$\hat{u} = -\frac{\hat{v}}{\lambda}$$

(12.11)

Recall the linearized boundary condition given by Eq. (11.34), repeated below:

$$\hat{v} = \frac{\partial \hat{\phi}}{\partial y} = V_\infty \tan \theta$$

(12.12)

We can further reduce Eq. (12.12) by noting that, for small perturbations, θ is small. Hence, $\tan \theta \approx \theta$, and Eq. (12.12) becomes

$$\hat{v} = V_\infty \theta$$

(12.13)

Substituting Eq. (12.13) into (12.11), we obtain

$$\hat{u} = -\frac{V_\infty \theta}{\lambda}$$

(12.14)

Recall the linearized pressure coefficient given by Eq. (11.32):

$$C_p = -\frac{2\hat{u}}{V_\infty}$$

(11.32)

Substituting Eq. (12.14) into (11.32), and recalling that $\lambda \equiv \sqrt{M_\infty^2 - 1}$, we have

$$\boxed{C_p = \frac{2\theta}{\sqrt{M_\infty^2 - 1}}}$$

(12.15)

Equation (12.15) is important. It is the linearized supersonic pressure coefficient, and it states that C_p is directly proportional to the local surface inclination with respect to the freestream. It holds for any slender two-dimensional body where θ is small.

Return again to Fig. 12.1. Note that θ is positive when measured above the horizontal, and negative when measured below the horizontal. Hence, from Eq. (12.15), C_p is positive on the forward portion of the hump, and negative on the rear portion. This is denoted by the $(+)$ and $(-)$ signs in front of and behind the hump shown in Fig. 12.1. This is also somewhat consistent with our discussions in Chap. 9; in the real flow over the hump, a shock wave forms above the front portion where the flow is being turned into itself, and hence $p > p_\infty$, whereas an expansion wave occurs over the remainder of the hump, and the pressure decreases. Think about the picture shown in Fig. 12.1; the pressure is higher on the front section of the hump, and lower on the rear section. As a result, a drag force exists on the hump. This drag is called *wave drag* and is a characteristic of supersonic flows. Wave drag was discussed in Sec. 9.7 in conjunction with shock-expansion theory applied to supersonic airfoils. It is interesting that linearized supersonic theory also predicts a finite wave drag, although shock waves themselves are not treated in such linearized theory.

Examining Eq. (12.15), we note that $C_p \propto (M_\infty^2 - 1)^{-1/2}$; hence, for supersonic flow, C_p decreases as M_∞ increases. This is in direct contrast with subsonic flow, where Eq. (11.51) shows that $C_p \propto (1 - M_\infty^2)^{-1/2}$; hence, for subsonic flow, C_p increases as M_∞ increases. These trends are illustrated in Fig. 12.2. Note that both results predict $C_p \to \infty$ as $M_\infty \to 1$ from either side. However, keep in mind that neither Eq. (12.15) nor (11.51) is valid in the transonic range around Mach 1.

12.3 APPLICATION TO SUPERSONIC AIRFOILS

Equation (12.15) is very handy for estimating the lift and wave drag for thin supersonic airfoils, such as sketched in Fig. 12.3. When applying Eq. (12.15) to any surface, one can follow a formal sign convention for θ, which is different for regions of left-running waves (such as above the airfoil in Fig. 12.3) than for regions of right-running waves (such as below the airfoil in Fig. 12.3). This sign

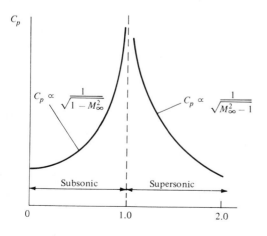

FIGURE 12.2
Variation of the linearized pressure coefficient with Mach number (schematic).

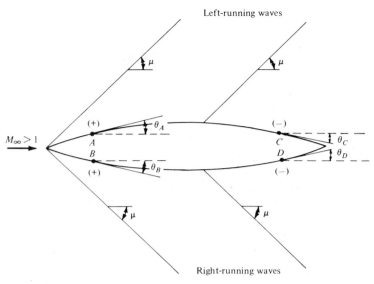

FIGURE 12.3
Linearized supersonic flow over an airfoil.

convention is developed in detail in Ref. 21. However, for our purpose here, there is no need to be concerned about the sign associated with θ in Eq. (12.15). Rather, keep in mind that when the surface is inclined *into* the freestream direction, linearized theory predicts a positive C_p. For example, points A and B in Fig. 12.3 are on surfaces inclined into the freestream, and hence $C_{p,A}$ and $C_{p,B}$ are positive values given by

$$C_{p,A} = \frac{2\theta_A}{\sqrt{M_\infty^2 - 1}} \quad \text{and} \quad C_{p,B} = \frac{2\theta_B}{\sqrt{M_\infty^2 - 1}}$$

In contrast, when the surface is inclined *away from* the freestream direction, linearized theory predicts a negative C_p. For example, points C and D in Fig. 12.3 are on surfaces inclined away from the freestream, and hence $C_{p,C}$ and $C_{p,D}$ are negative values, given by

$$C_{p,C} = -\frac{2\theta_C}{\sqrt{M_\infty^2 - 1}} \quad \text{and} \quad C_{p,D} = -\frac{2\theta_D}{\sqrt{M_\infty^2 - 1}}$$

In the above expressions, θ is always treated as a positive quantity, and the sign of C_p is determined simply by looking at the body and noting whether the surface is inclined into or away from the freestream.

With the distribution of C_p over the airfoil surface given by Eq. (12.15), the lift and drag coefficients, c_l and c_d, respectively, can be obtained from the integrals given by Eqs. (1.15) to (1.19).

Let us consider the simplest possible airfoil, namely, a flat plate at a small angle of attack, α, as shown in Fig. 12.4. Looking at this picture, the lower surface of the plate is a compression surface inclined at the angle α into the freestream, and from Eq. (12.15),

$$C_{p,l} = \frac{2\alpha}{\sqrt{M_\infty^2 - 1}} \tag{12.16}$$

Since the surface inclination angle is constant along the entire lower surface, $C_{p,l}$ is a constant value over the lower surface. Similarly, the top surface is an expansion surface inclined at the angle α away from the freestream, and from Eq. (12.15),

$$C_{p,u} = -\frac{2\alpha}{\sqrt{M_\infty^2 - 1}} \tag{12.17}$$

$C_{p,u}$ is constant over the upper surface. The normal force coefficient for the flat plate can be obtained from Eq. (1.15):

$$c_n = \frac{1}{c} \int_0^c (C_{p,l} - C_{p,u}) \, dx \tag{12.18}$$

Substituting Eqs. (12.16) and (12.17) into (12.18), we obtain

$$c_n = \frac{4\alpha}{\sqrt{M_\infty^2 - 1}} \frac{1}{c} \int_0^c dx = \frac{4\alpha}{\sqrt{M_\infty^2 - 1}} \tag{12.19}$$

The axial force coefficient is given by Eq. (1.16):

$$c_a = \frac{1}{c} \int_{LE}^{TE} (C_{p,u} - C_{p,l}) \, dy \tag{12.20}$$

However, the flat plate has (theoretically) zero thickness. Hence, in Eq. (12.20), $dy = 0$, and as a result, $c_a = 0$. This is also clearly seen in Fig. 12.4; the pressures act normal to the surface, and hence there is no component of the pressure force in the x direction. From Eqs. (1.18) and (1.19),

$$c_l = c_n \cos \alpha - c_a \sin \alpha \tag{1.18}$$

$$c_d = c_n \sin \alpha + c_a \cos \alpha \tag{1.19}$$

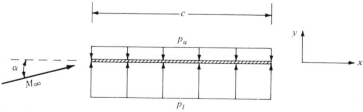

FIGURE 12.4
A flat plate at angle of attack in a supersonic flow.

and, along with the assumption that α is small and hence $\cos \alpha \approx 1$ and $\sin \alpha \approx \alpha$, we have

$$c_l = c_n - c_a \alpha \tag{12.21}$$

$$c_d = c_n \alpha + c_a \tag{12.22}$$

Substituting Eq. (12.19) and the fact that $c_a = 0$ into Eqs. (12.21) and (12.22), we obtain

$$c_l = \frac{4\alpha}{\sqrt{M_\infty^2 - 1}} \tag{12.23}$$

$$c_d = \frac{4\alpha^2}{\sqrt{M_\infty^2 - 1}} \tag{12.24}$$

Equations (12.23) and (12.24) give the lift and wave-drag coefficients, respectively, for the supersonic flow over a flat plate. Keep in mind that they are results from linearized theory and therefore are valid only for small α.

For a thin airfoil of arbitrary shape at small angle of attack, linearized theory gives an expression for c_l identical to Eq. (12.23); i.e.,

$$c_l = \frac{4\alpha}{\sqrt{M_\infty^2 - 1}}$$

Within the approximation of linearized theory, c_l depends only on α and is independent of the airfoil shape and thickness. However, the same linearized theory gives a wave-drag coefficient in the form of

$$c_d = \frac{4}{\sqrt{M_\infty^2 - 1}} (\alpha^2 + g_c^2 + g_t^2)$$

where g_c and g_t are functions of the airfoil camber and thickness, respectively. For more details, see Refs. 25 and 26.

Example 12.1. Using linearized theory, calculate the lift and drag coefficients for a flat plate at a 5° angle of attack in a Mach 3 flow. Compare with the exact results obtained in Example 9.8.

Solution

$$\alpha = 5° = 0.087 \text{ rad}$$

From Eq. (12.23),

$$c_l = \frac{4\alpha}{\sqrt{M_\infty^2 - 1}} = \frac{(4)(0.087)}{\sqrt{(3)^2 - 1}} = \boxed{0.123}$$

From Eq. (12.24),

$$c_d = \frac{4\alpha^2}{\sqrt{M_\infty^2 - 1}} = \frac{4(0.087)}{\sqrt{(3)^2 - 1}} = \boxed{0.011}$$

The results calculated in Example 9.8 for the same problem are *exact results*, utilizing the exact oblique shock theory and the exact Prandtl-Meyer expansion-wave analysis. These results were

$$\left.\begin{array}{r} c_l = 0.125 \\ c_d = 0.011 \end{array}\right\} \quad \text{exact results from Example 9.8}$$

Note that, for the relatively small angle of attack of 5°, the linearized theory results are quite accurate—to within 1.6 percent.

12.4 SUMMARY

In linearized supersonic flow, information is propagated along Mach lines where the Mach angle $\mu = \sin^{-1}(1/M_\infty)$. Since these Mach lines are all based on M_∞, they are straight, parallel lines which propagate away from and downstream of a body. For this reason, disturbances cannot propagate upstream in a steady supersonic flow.

The pressure coefficient, based on linearized theory, on a surface inclined at a small angle θ to the freestream is

$$C_p = \frac{2\theta}{\sqrt{M_\infty^2 - 1}} \tag{12.15}$$

If the surface is inclined into the freestream, C_p is positive; if the surface is inclined away from the freestream, C_p is negative.

Based on linearized supersonic theory, the lift and wave-drag coefficients for a flat plate at an angle of attack are

$$c_l = \frac{4\alpha}{\sqrt{M_\infty^2 - 1}} \tag{12.23}$$

and

$$c_d = \frac{4\alpha^2}{\sqrt{M_\infty^2 - 1}} \tag{12.24}$$

Equation (12.23) also holds for a thin airfoil of arbitrary shape. However, for such an airfoil, the wave-drag coefficient depends on both the shape of the mean camber line and the airfoil thickness.

PROBLEMS

12.1. Using the results of linearized theory, calculate the lift and wave-drag coefficients for an infinitely thin flat plate in a Mach 2.6 freestream at angles of attack of
   (a) $\alpha = 5°$ (b) $\alpha = 15°$ (c) $\alpha = 30°$

Compare these approximate results with those from the exact shock-expansion theory obtained in Prob. 9.13. What can you conclude about the accuracy of linearized theory in this case?

12.2. For the conditions of Prob. 12.1, calculate the pressures (in the form of p/p_∞) on the top and bottom surfaces of the flat plate, using linearized theory. Compare these approximate results with those obtained from exact shock-expansion theory in Prob. 9.13. Make some appropriate conclusions regarding the accuracy of linearized theory for the calculation of pressures.

12.3. Consider a diamond-wedge airfoil such as shown in Fig. 9.24, with a half-angle $\varepsilon = 10°$. The airfoil is at an angle of attack $\alpha = 15°$ to a Mach 3 freestream. Using linear theory, calculate the lift and wave-drag coefficients for the airfoil. Compare these approximate results with those from the exact shock-expansion theory obtained in Prob. 9.14.

CHAPTER 13

INTRODUCTION TO NUMERICAL TECHNIQUES FOR NONLINEAR SUPERSONIC FLOW

Regarding computing as a straightforward routine, some theoreticians still tend to underestimate its intellectual value and challenge, while practitioners often ignore its accuracy and overrate its validity.

C. K. Chu, 1978
Columbia University

13.1 INTRODUCTION: PHILOSOPHY OF COMPUTATIONAL FLUID DYNAMICS

The above quotation underscores the phenomenally rapid increase in computer power available to engineers and scientists during the two decades between 1960 and 1980. This explosion in computer capability is still going on, with no specific limits in sight. As a result, an entirely new discipline in aerodynamics has evolved over the past two decades, namely, computational fluid dynamics (CFD). CFD is a new "third dimension" in aerodynamics, complementing the previous dimensions of both pure experiment and pure theory. It allows us to obtain answers to fluid dynamic problems which heretofore were intractable by classical analytical methods. Consequently, CFD is revolutionizing the airplane design process, and in many ways is modifying the way we conduct modern aeronautical research and development. For these reasons, every modern student of aerodynamics should be aware of the overall philosophy of CFD, because you are bound to be affected by it to some greater or lesser degree in your education and professional life.

Computational fluid dynamics is the art of replacing the governing partial differential equations of fluid flow with numbers, and advancing these numbers

579

in space and/or time to obtain a final numerical description of the complete flow field of interest. The end product of CFD is indeed a collection of numbers, in contrast to a closed-form analytical solution. However, in the long run the objective of most engineering analyses, closed form or otherwise, is a quantitative description of the problem, i.e., numbers (see, e.g., Ref. 33).

The purpose of this chapter is to provide an introduction to some of the basic ideas of CFD as applied to inviscid supersonic flows. More details are given in Ref. 21. Because CFD has developed so rapidly in recent years, we can only scratch the surface here. Indeed, the present chapter is intended to give you only some basic background as well as the incentive to pursue the subject further in the modern literature.

The road map for this chapter is given in Fig. 13.1. We begin by introducing the classical method of characteristics—a numerical technique that has been available in aerodynamics since 1929, but which had to wait on the modern computer for practical, everyday implementation. For this reason, the author classifies the method of characteristics under the general heading of numerical techniques, although others may prefer to list it under a more classical heading. We also show how the method of characteristics is applied to design the divergent contour of a supersonic nozzle. Then we move to a discussion of the finite-difference approach. Finite-difference methods are the "bread and butter" of modern CFD. A special, but very powerful, finite-difference technique called the *time-dependent technique* is used to illustrate the application of CFD to nozzle flows and the flow over a supersonic blunt body.

In contrast to the linearized solutions discussed in Chaps. 11 and 12, CFD represents numerical solutions to the *exact* nonlinear governing equations, i.e., the equations without simplifying assumptions such as small perturbations, and

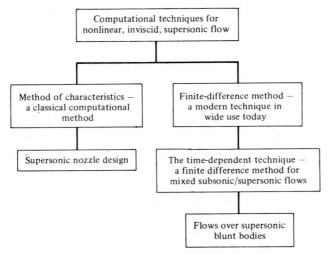

FIGURE 13.1
Road map for Chap. 13.

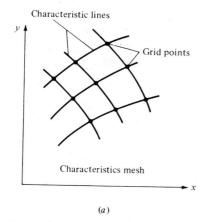

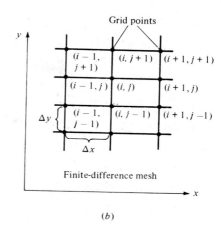

FIGURE 13.2
Grid points.

which apply to all speed regimes, transonic and hypersonic as well as subsonic and supersonic. Although numerical roundoff and truncation errors are always present in any numerical representation of the governing equations, we still think of CFD solutions as being "exact solutions."

Both the method of characteristics and finite-difference methods have one thing in common: They represent a continuous flow field by a series of distinct grid points in space, as shown in Fig. 13.2. The flow-field properties (u, v, p, T, etc.) are calculated at each one of these grid points. The mesh generated by these grid points is generally skewed for the method of characteristics, as shown in Fig. 13.2a, but is usually rectangular for finite-difference solutions, as shown in Fig. 13.2b. We will soon appreciate why these different meshes occur.

13.2 ELEMENTS OF THE METHOD OF CHARACTERISTICS

In this section, we only introduce the basic elements of the method of characteristics. A full discussion is beyond the scope of this book; see Refs. 21, 25, and 34 for more details.

Consider a two-dimensional, steady, inviscid, supersonic flow in xy space, as given in Fig. 13.2a. The flow variables (p, u, T, etc.) are continuous throughout this space. However, there are certain lines in xy space along which the *derivatives* of the flow-field variables ($\partial p/\partial x$, $\partial u/\partial y$, etc.) are *indeterminate* and across which may even be discontinuous. Such lines are called *characteristic lines*. This may sound strange at first; however, let us prove that such lines exist, and let us find their precise directions in the xy plane.

In addition to the flow being supersonic, steady, inviscid, and two-dimensional, assume that it is also irrotational. The exact governing equation for such a flow is given by Eq. (11.12):

$$\left[1-\frac{1}{a^2}\left(\frac{\partial\phi}{\partial x}\right)^2\right]\frac{\partial^2\phi}{\partial x^2}+\left[1-\frac{1}{a^2}\left(\frac{\partial\phi}{\partial y}\right)^2\right]\frac{\partial^2\phi}{\partial y^2}-\frac{2}{a^2}\frac{\partial\phi}{\partial x}\frac{\partial\phi}{\partial y}\frac{\partial^2\phi}{\partial x\,\partial y}=0$$

(11.12)

[Keep in mind that we are dealing with the full velocity potential ϕ in Eq. (11.12), not the perturbation potential.] Since $\partial\phi/\partial x = u$ and $\partial\phi/\partial y = v$, Eq. (11.12) can be written as

$$\left(1-\frac{u^2}{a^2}\right)\frac{\partial^2\phi}{\partial x^2}+\left(1-\frac{v^2}{a^2}\right)\frac{\partial^2\phi}{\partial y^2}-\frac{2uv}{a^2}\frac{\partial^2\phi}{\partial x\,\partial y}=0$$

(13.1)

The velocity potential and its derivatives are functions of x and y, e.g.,

$$\frac{\partial\phi}{\partial x}=f(x,y)$$

Hence, from the relation for an exact differential,

$$df=\frac{\partial f}{\partial x}dx+\frac{\partial f}{\partial y}dy$$

we have

$$d\left(\frac{\partial\phi}{\partial x}\right)=du=\frac{\partial^2\phi}{\partial x^2}dx+\frac{\partial^2\phi}{\partial x\,\partial y}dy$$

(13.2)

Similarly,

$$d\left(\frac{\partial\phi}{\partial y}\right)=dv=\frac{\partial^2\phi}{\partial x\,\partial y}dx+\frac{\partial^2\phi}{\partial y^2}dy$$

(13.3)

Examine Eqs. (13.1) to (13.3) closely. Note that they contain the second derivatives $\partial^2\phi/\partial x^2$, $\partial^2\phi/\partial y^2$, and $\partial^2\phi/\partial x\,\partial y$. If we imagine these derivatives as "unknowns," then Eqs. (13.1), (13.2), and (13.3) represent three equations with three unknowns. For example, to solve for $\partial^2\phi/\partial x\,\partial y$, use Cramer's rule as follows:

$$\frac{\partial^2\phi}{\partial x\,\partial y}=\frac{\begin{vmatrix}1-\dfrac{u^2}{a^2} & 0 & 1-\dfrac{v^2}{a^2}\\[2mm] dx & du & 0\\[2mm] 0 & dv & dy\end{vmatrix}}{\begin{vmatrix}1-\dfrac{u^2}{a^2} & -\dfrac{2uv}{a^2} & 1-\dfrac{v^2}{a^2}\\[2mm] dx & dy & 0\\[2mm] 0 & dx & dy\end{vmatrix}}=\frac{N}{D}$$

(13.4)

where N and D represent the numerator and denominator determinants, respectively. The physical meaning of Eq. (13.4) can be seen by considering point A and its surrounding neighborhood in the flow, as sketched in Fig. 13.3. The derivative $\partial^2\phi/\partial x \partial y$ has a specific value at point A. Equation (13.4) gives the solution for $\partial^2\phi/\partial x \partial y$ for an *arbitrary* choice of dx and dy. The combination of dx and dy defines an arbitrary direction ds away from point A as shown in Fig. 13.3. In general, this direction is different from the streamline direction going through point A. In Eq. (13.4), the differentials du and dv represent the changes in velocity that take place over the increments dx and dy. Hence, although the choice of dx and dy is arbitrary, the values of du and dv in Eq. (13.4) must correspond to this choice. No matter what values of dx and dy are arbitrarily chosen, the corresponding values of du and dv will always ensure obtaining the same value of $\partial^2\phi/\partial x \partial y$ at point A from Eq. (13.4).

The single exception to the above comments occurs when dx and dy are chosen so that $D = 0$ in Eq. (13.4). In this case, $\partial^2\phi/\partial x \partial y$ is not defined. This situation will occur for a specific direction ds away from point A in Fig. 13.3, defined for that specific combination of dx and dy for which $D = 0$. However, we know that $\partial^2\phi/\partial x \partial y$ has a specific defined value at point A. Therefore, the only consistent result associated with $D = 0$ is that $N = 0$, also; i.e.,

$$\frac{\partial^2\phi}{\partial x \partial y} = \frac{N}{D} = \frac{0}{0} \tag{13.5}$$

Here, $\partial^2\phi/\partial x \partial y$ is an indeterminate form, which is allowed to be a finite value, i.e., that value of $\partial^2\phi/\partial x \partial y$ which we know exists at point A. *The important conclusion here is that there is some direction (or directions) through point A along which $\partial^2\phi/\partial x \partial y$ is indeterminate.* Since $\partial^2\phi/\partial x \partial y = \partial u/\partial y = \partial v/\partial x$, this implies that the derivatives of the flow variables are indeterminate along these lines. Hence, we have proven that lines do exist in the flow field along which derivatives of the flow variables are indeterminate; earlier, we defined such lines as *characteristic lines.*

Consider again point A in Fig. 13.3. From our previous discussion, there are one or more characteristic lines through point A. *Question:* How can we calculate the precise direction of these characteristic lines? The answer can be obtained by setting $D = 0$ in Eq. (13.4). Expanding the denominator determinant in Eq. (13.4), and setting it equal to zero, we have

$$\left(1 - \frac{u^2}{a^2}\right)(dy)^2 + \frac{2uv}{a^2}\, dx\, dy + \left(1 - \frac{v^2}{a^2}\right)(dx)^2 = 0$$

or

$$\left(1 - \frac{u^2}{a^2}\right)\left(\frac{dy}{dx}\right)^2_{\text{char}} + \frac{2uv}{a^2}\left(\frac{dy}{dx}\right)_{\text{char}} + \left(1 - \frac{v^2}{a^2}\right) = 0 \tag{13.6}$$

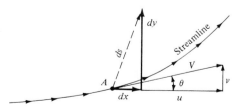

FIGURE 13.3
An arbitrary direction, ds, away from point A.

In Eq. (13.6), dy/dx is the slope of the characteristic lines; hence, the subscript "char" has been added to emphasize this fact. Solving Eq. (13.6) for $(dy/dx)_{\text{char}}$ by means of the quadratic formula, we obtain

$$\left(\frac{dy}{dx}\right)_{\text{char}} = \frac{-2uv/a^2 \pm \sqrt{(2uv/a^2)^2 - 4(1-u^2/a^2)(1-v^2/a^2)}}{2(1-u^2/a^2)}$$

or

$$\left(\frac{dy}{dx}\right)_{\text{char}} = \frac{-uv/a^2 \pm \sqrt{(u^2+v^2)/a^2 - 1}}{1-u^2/a^2} \tag{13.7}$$

From Fig. 13.3, we see that $u = V \cos \theta$ and $v = V \sin \theta$. Hence, Eq. (13.7) becomes

$$\left(\frac{dy}{dx}\right)_{\text{char}} = \frac{(-V^2 \cos \theta \sin \theta)/a^2 \pm \sqrt{(V^2/a^2)(\cos^2 \theta + \sin^2 \theta) - 1}}{1 - [(V^2/a^2) \cos^2 \theta]} \tag{13.8}$$

Recall that the local Mach angle μ is given by $\mu = \sin^{-1}(1/M)$, or $\sin \mu = 1/M$. Thus, $V^2/a^2 = M^2 = 1/\sin^2 \mu$, and Eq. (13.8) becomes

$$\left(\frac{dy}{dx}\right)_{\text{char}} = \frac{(-\cos \theta \sin \theta)/\sin^2 \mu \pm \sqrt{(\cos^2 \theta + \sin^2 \theta)/\sin^2 \mu - 1}}{1 - (\cos^2 \theta)/\sin^2 \mu} \tag{13.9}$$

After considerable algebraic and trigonometric manipulation, Eq. (13.9) reduces to

$$\boxed{\left(\frac{dy}{dx}\right)_{\text{char}} = \tan(\theta \mp \mu)} \tag{13.10}$$

Equation (13.10) is an important result; it states that *two* characteristic lines run through point A in Fig. 13.3, namely, one line with a slope equal to $\tan(\theta - \mu)$ and the other with a slope equal to $\tan(\theta + \mu)$. The physical significance of this result is illustrated in Fig. 13.4. Here, a streamline through point A is inclined

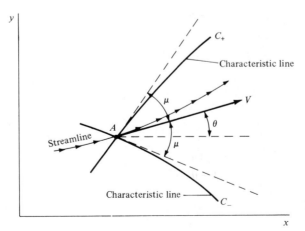

FIGURE 13.4
Left- and right-running characteristic lines through point A.

at the angle θ with respect to the horizontal. The velocity at point A is V, which also makes the angle θ with respect to the horizontal. Equation (13.10) states that one characteristic line at point A is inclined *below* the streamline direction by the angle μ; this characteristic line is labeled as C_- in Fig. 13.4. Equation (13.10) also states that the other characteristic line at point A is inclined *above* the streamline direction by the angle μ; this characteristic line is labeled as C_+ in Fig. 13.4. Examining Fig. 13.4, we see that the characteristic lines through point A are simply the left- and right-running *Mach waves* through point A. Hence, *the characteristic lines are Mach lines.* In Fig. 13.4, the left-running Mach wave is denoted by C_+, and the right-running Mach wave is denoted by C_-. Hence, returning to Fig. 13.2a, the characteristics mesh consists of left- and right-running Mach waves which crisscross the flow field. There are an infinite number of these waves; however, for practical calculations we deal with a finite number of waves, the intersections of which define the grid points shown in Fig. 13.2a. Note that the characteristic lines are curved in space because (1) the local Mach angle depends on the local Mach number, which is a function of x and y, and (2) the local streamline direction θ varies throughout the flow.

The characteristic lines in Fig. 13.2a are of no use to us by themselves. The practical consequence of these lines is that *the governing partial differential equations which describe the flow reduce to ordinary differential equations along the characteristic lines.* These equations are called the *compatibility equations,* which can be found by setting $N = 0$ in Eq. (13.4), as follows. When $N = 0$, the numerator determinant yields

$$\left(1 - \frac{u^2}{a^2}\right) du\, dy + \left(1 - \frac{v^2}{a^2}\right) dx\, dv = 0$$

or

$$\frac{dv}{du} = \frac{-(1 - u^2/a^2)}{1 - v^2/a^2} \frac{dy}{dx} \qquad (13.11)$$

Keep in mind that N is set to zero only when $D = 0$ in order to keep the flow-field derivatives finite, albeit of the indeterminate form $0/0$. When $D = 0$, we are restricted to considering directions *only* along the characteristic lines, as explained earlier. Hence, when $N = 0$, we are held to the same restriction. Therefore, Eq. (13.11) holds only along the characteristic lines. Therefore, in Eq. (13.11),

$$\frac{dy}{dx} \equiv \left(\frac{dy}{dx}\right)_{\text{char}} \qquad (13.12)$$

Substituting Eqs. (13.12) and (13.7) into (13.11), we obtain

$$\frac{dv}{du} = -\frac{1 - u^2/a^2}{1 - v^2/a^2} \frac{-uv/a^2 \pm \sqrt{(u^2 + v^2)/a^2 - 1}}{1 - u^2/a^2}$$

or

$$\frac{dv}{du} = \frac{uv/a^2 \mp \sqrt{(u^2 + v^2)/a^2 - 1}}{1 - v^2/a^2} \qquad (13.13)$$

Recall from Fig. 13.3 that $u = V \cos \theta$ and $v = V \sin \theta$. Also, $(u^2 + v^2)/a^2 = V^2/a^2 = M^2$. Hence, Eq. (13.13) becomes

$$\frac{d(V \sin \theta)}{d(V \cos \theta)} = \frac{M^2 \cos \theta \sin \theta \mp \sqrt{M^2 - 1}}{1 - M^2 \sin^2 \theta}$$

which, after some algebraic manipulations, reduces to

$$d\theta = \mp \sqrt{M^2 - 1}\, \frac{dV}{V} \qquad (13.14)$$

Examine Eq. (13.14). It is an *ordinary differential equation* obtained from the original governing partial differential equation, Eq. (13.1). However, Eq. (13.14) contains the restriction given by Eq. (13.12); i.e., Eq. (13.14) holds *only* along the characteristic lines. Hence, Eq. (13.14) gives the *compatibility relations* along the characteristic lines. In particular, comparing Eq. (13.14) with Eq. (13.10), we see that

$$d\theta = -\sqrt{M^2 - 1}\, \frac{dV}{V} \qquad \text{(applies along the } C_- \text{ characteristic)} \qquad (13.15)$$

$$d\theta = \sqrt{M^2 - 1}\, \frac{dV}{V} \qquad \text{(applies along the } C_+ \text{ characteristic)} \qquad (13.16)$$

Examine Eq. (13.14) further. It should look familiar; indeed, Eq. (13.14) is identical to the expression obtained for Prandtl-Meyer flow in Sec. 9.6, namely, Eq. (9.32). Hence, Eq. (13.14) can be integrated to obtain a result in terms of the Prandtl-Meyer function, given by Eq. (9.42). In particular, the integration of Eqs. (13.15) and (13.16) yields

$$\theta + \nu(M) = \text{const} = K_- \qquad \text{(along the } C_- \text{ characteristic)} \qquad (13.17)$$

$$\theta - \nu(M) = \text{const} = K_+ \qquad \text{(along the } C_+ \text{ characteristic)} \qquad (13.18)$$

In Eq. (13.17), K_- is a constant along a given C_- characteristic; it has different values for different C_- characteristics. In Eq. (13.18), K_+ is a constant along a given C_+ characteristic; it has different values for different C_+ characteristics. Note that our compatibility relations are now given by Eqs. (13.17) and (13.18), which are *algebraic* equations which hold only along the characteristic lines. In a general inviscid, supersonic, steady flow, the compatibility equations are ordinary differential equations; only in the case of two-dimensional irrotational flow do they further reduce to algebraic equations.

What is the advantage of the characteristic lines and their associated compatibility equations discussed above? Simply this—to solve the nonlinear supersonic flow, we need deal only with ordinary differential equations (or in the present case, algebraic equations) instead of the original partial differential equations. Finding the solution of such ordinary differential equations is usually much simpler than dealing with partial differential equations.

How do we use the above results to solve a practical problem? The purpose of the next section is to give such an example, namely, the calculation of the supersonic flow inside a nozzle and the determination of a proper wall contour so that shock waves do not appear inside the nozzle. To carry out this calculation, we deal with two types of grid points: (1) internal points, away from the wall, and (2) wall points. Characteristics calculations at these two sets of points are carried out as follows.

13.2.1 Internal Points

Consider the internal grid points 1, 2, and 3 as shown in Fig. 13.5. Assume that we know the location of points 1 and 2, as well as the flow properties at these points. Define point 3 as the intersection of the C_- characteristic through point 1 and the C_+ characteristic through point 2. From our previous discussion, $(K_-)_1 = (K_-)_3$ because K_- is constant along a given C_- characteristic. The value of $(K_-)_1 = (K_-)_3$ is obtained from Eq. (13.17) evaluated at point 1:

$$(K_-)_3 = (K_-)_1 = \theta_1 + \nu_1 \tag{13.19}$$

Similarly, $(K_+)_2 = (K_+)_3$ because K_+ is constant along a given C_+ characteristic. The value of $(K_+)_2 = (K_+)_3$ is obtained from Eq. (13.18) evaluated at point 2:

$$(K_+)_3 = (K_+)_2 = \theta_2 - \nu_2 \tag{13.20}$$

Now evaluate Eqs. (13.17) and (13.18) at point 3:

$$\theta_3 + \nu_3 = (K_-)_3 \tag{13.21}$$

and
$$\theta_3 - \nu_3 = (K_+)_3 \tag{13.22}$$

In Eqs. (13.21) and (13.22), $(K_-)_3$ and $(K_+)_3$ are known values, obtained from Eqs. (13.19) and (13.20). Hence, Eqs. (13.21) and (13.22) are two algebraic equations for the two unknowns, θ_3 and ν_3. Solving these equations, we obtain

$$\theta_3 = \tfrac{1}{2}[(K_-)_1 + (K_+)_2] \tag{13.23}$$

$$\nu_3 = \tfrac{1}{2}[(K_-)_1 - (K_+)_2] \tag{13.24}$$

Knowing θ_3 and ν_3, all other flow properties at point 3 can be obtained as follows:

1. From ν_3, obtain the associated M_3 from App. C.

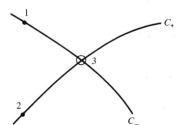

FIGURE 13.5
Characteristic mesh used for the location of point 3 and the calculation of flow conditions at point 3, knowing the locations and flow properties at points 1 and 2.

2. From M_3 and the known p_0 and T_0 for the flow (recall that for inviscid, adiabatic flow, the total pressure and total temperature are constants throughout the flow), find p_3 and T_3 from App. A.
3. Knowing T_3, compute $a_3 = \sqrt{\gamma R T_3}$. In turn, $V_3 = M_3 a_3$.

As stated earlier, point 3 is located by the intersection of the C_- and C_+ characteristics through points 1 and 2, respectively. These characteristics are curved lines; however, for purposes of calculation, we assume that the characteristics are straight-line segments between points 1 and 3 and between points 2 and 3. For example, the slope of the C_- characteristic between points 1 and 3 is assumed to be the average value between these two points, i.e., $\frac{1}{2}(\theta_1 + \theta_3) - \frac{1}{2}(\mu_1 + \mu_3)$. Similarly, the slope of the C_+ characteristic between points 2 and 3 is approximated by $\frac{1}{2}(\theta_2 + \theta_3) + \frac{1}{2}(\mu_2 + \mu_3)$.

13.2.2 Wall Points

In Fig. 13.6, point 4 is an internal flow point near a wall. Assume that we know all the flow properties at point 4. The C_- characteristic through point 4 intersects the wall at point 5. At point 5, the slope of the wall, θ_5, is known. The flow properties at the wall point, point 5, can be obtained from the known properties at point 4 as follows. Along the C_- characteristic, K_- is constant. Hence, $(K_-)_4 = (K_-)_5$. Moreover, the value of K_- is known from Eq. (13.17) evaluated at point 4:

$$(K_-)_4 = (K_-)_5 = \theta_4 + \nu_4 \tag{13.25}$$

Evaluating Eq. (13.17) at point 5, we have

$$(K_-)_5 = \theta_5 + \nu_5 \tag{13.26}$$

In Eq. (13.26), $(K_-)_5$ and θ_5 are known; thus ν_5 follows directly. In turn, all other flow variables at point 5 can be obtained from ν_5 as explained earlier. The characteristic line between points 4 and 5 is assumed to be a straight-line segment with average slope given by $\frac{1}{2}(\theta_4 + \theta_5) - \frac{1}{2}(\mu_4 + \mu_5)$.

From the above discussion of both internal and wall points, we see that properties at the grid points are calculated from *known* properties at other grid points. Hence, in order to *start* a calculation using the method of characteristics,

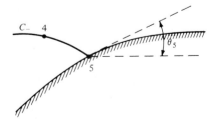

FIGURE 13.6
Wall point.

we have to know the flow properties along some *initial data* line. Then we piece together the characteristics mesh and associated flow properties by "marching downstream" from the initial data line. This is illustrated in the next section.

We emphasize again that the method of characteristics is an exact solution of inviscid, nonlinear supersonic flow. However, in practice, there are numerical errors associated with the finite grid; the approximation of the characteristics mesh by straight-line segments between grid points is one such example. In principle, the method of characteristics is truly exact only in the limit of an infinite number of characteristic lines.

We have discussed the method of characteristics for two-dimensional, irrotational, steady flow. The method of characteristics can also be used for rotational and three-dimensional flows, as well as unsteady flows. See Ref. 21 for more details.

13.3 SUPERSONIC NOZZLE DESIGN

In Chap. 10, we demonstrated that a nozzle designed to expand a gas from rest to supersonic speeds must have a convergent-divergent shape. Moreover, the quasi-one-dimensional analysis of Chap. 10 led to the prediction of flow properties as a function of x through a nozzle of specified shape (see, e.g., Fig. 10.10). The flow properties at any x station obtained from the quasi-one-dimensional analysis represent an *average* of the flow over the given nozzle cross section. The beauty of the quasi-one-dimensional approach is its simplicity. On the other hand, its disadvantages are (1) it cannot predict the details of the actual three-dimensional flow in a convergent-divergent nozzle and (2) it gives no information on the proper wall contour of such nozzles.

The purpose of the present section is to describe how the method of characteristics can supply the above information which is missing from a quasi-one-dimensional analysis. For simplicity, we treat a two-dimensional flow, as sketched in Fig. 13.7. Here, the flow properties are a function of x and y. Such a two-dimensional flow is applicable to supersonic nozzles of rectangular cross section, such as sketched in the insert at the top of Fig. 13.7. Two-dimensional (rectangular) nozzles are used in many supersonic wind tunnels. They are also the heart of gas-dynamic lasers (see Ref. 1). In addition, there is current discussion of employing rectangular exhaust nozzles on advanced military jet airplanes envisaged for the future.

Consider the following problem. We wish to design a convergent-divergent nozzle to expand a gas from rest to a given supersonic Mach number at the exit, M_e. How do we design the proper contour so that we have shock-free, isentropic flow in the nozzle? The answer to this question is discussed in the remainder of this section.

For the convergent, subsonic section, there is no specific contour which is better than any other. There are rules of thumb based on experience and guided by subsonic flow theory; however, we are not concerned with the details here. We simply assume that we have a reasonable contour for the subsonic section.

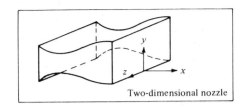

Two-dimensional nozzle

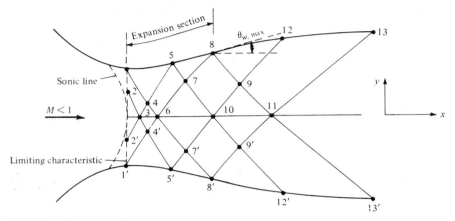

FIGURE 13.7
Schematic of supersonic nozzle design by the method of characteristics.

   Due to the two-dimensional nature of the flow in the throat region, the sonic line is generally curved, as sketched in Fig. 13.7. A line called the *limiting characteristic* is sketched just downstream of the sonic line. The limiting characteristic is defined such that any characteristic line originating downstream of the limiting characteristic does not intersect the sonic line; in contrast, a characteristic line originating in the small region between the sonic line and the limiting characteristic can intersect the sonic line (for more details on the limiting characteristic, see Ref. 21). To begin a method of characteristics solution, we must use an initial data line which is downstream of the limiting characteristic.

   Let us assume that by independent calculation of the subsonic-transonic flow in the throat region, we know the flow properties at all points on the limiting characteristic. That is, we use the limiting characteristic as our initial data line. For example, we know the flow properties at points 1 and 2 on the limiting characteristic in Fig. 13.7. Moreover, consider the nozzle contour just downstream of the throat. Letting θ denote the angle between a tangent to the wall and the horizontal, the section of the divergent nozzle where θ is increasing is called the *expansion section*, as shown in Fig. 13.7. The end of the expansion section occurs where $\theta = \theta_{\max}$ (point 8 in Fig. 13.7). Downstream of this point, θ decreases until it equals zero at the nozzle exit. The portion of the contour where θ decreases is called the *straightening section*. The shape of the expansion section is somewhat arbitrary; typically, a circular arc of large radius is used for the expansion section of many wind-tunnel nozzles. Consequently, in addition to knowing the flow properties along the limiting characteristic, we also have an expansion section of specified shape; i.e., we know θ_1, θ_5, and θ_8 in Fig. 13.7. The purpose of our application of the method of characteristics now becomes the proper design of the contour of the straightening section (from points 8 to 13 in Fig. 13.7).

The characteristics mesh sketched in Fig. 13.7 is very coarse—this is done intentionally to keep our discussion simple. In an actual calculation, the mesh should be much finer. The characteristics mesh and the flow properties at the associated grid points are calculated as follows:

1. Draw a C_- characteristic from point 2, intersecting the centerline at point 3. Evaluating Eq. (13.17) at point 3, we have

$$\theta_3 + \nu_3 = (K_-)_3$$

In the above equation, $\theta_3 = 0$ (the flow is horizontal along the centerline). Also, $(K_-)_3$ is known because $(K_-)_3 = (K_-)_2$. Hence, the above equation can be solved for ν_3.

2. Point 4 is located by the intersection of the C_- characteristic from point 1 and the C_+ characteristic from point 3. In turn, the flow properties at the internal point 4 are determined as discussed in the last part of Sec. 13.2.

3. Point 5 is located by the intersection of the C_+ characteristic from point 4 with the wall. Since θ_5 is known, the flow properties at point 5 are determined as discussed in Sec. 13.2 for wall points.

4. Points 6 through 11 are located in a manner similar to the above, and the flow properties at these points are determined as discussed before, using the internal point or wall point method as appropriate.

5. Point 12 is a wall point on the straightening section of the contour. The purpose of the straightening section is to cancel the expansion waves generated by the expansion section. Hence, there are no waves which are reflected from the straightening section. In turn, no right-running waves cross the characteristic line between points 9 and 12. As a result, the characteristic line between points 9 and 12 is a straight line, along which θ is constant, that is, $\theta_{12} = \theta_9$. The section of the wall contour between points 8 and 12 is approximated by a straight line with an average slope of $\frac{1}{2}(\theta_8 + \theta_{12})$.

6. Along the centerline, the Mach number continuously increases. Let us assume that at point 11, the design exit Mach number M_e is reached. The characteristic line from points 11 to 13 is the last line of the calculation. Again, $\theta_{13} = \theta_{11}$, and the contour from point 12 to point 13 is approximated by a straight-line segment with an average slope of $\frac{1}{2}(\theta_{12} + \theta_{13})$.

The above description is intended to give you a "feel" for the application of the method of characteristics. If you wish to carry out an actual nozzle design, and/or if you are interested in more details, read the more complete treatments in Refs. 21 and 34.

Note in Fig. 13.7 that the nozzle flow is symmetrical about the centerline. Hence, the points below the centerline $(1', 2', 3',$ etc.) are simply mirror images of the corresponding points above the centerline. In making a calculation of the flow through the nozzle, we need to concern ourselves only with those points in the upper half of Fig. 13.7, above and on the centerline.

13.4 ELEMENTS OF FINITE-DIFFERENCE METHODS

Finite-difference techniques for the solution of partial differential equations have been discussed by mathematicians for at least a century; however, it was not until the advent of the high-speed digital computer that finite-difference solutions became practical. Today, the vast majority of computational fluid dynamic applications are based on finite-difference techniques. Finite-difference solutions of various flow problems are used as design tools by industry and abound in modern aerodynamic research and development. In particular, they have revolutionized the analysis of compressible flow.

The purpose of this section is to give you just the flavor of such finite-difference techniques. The intensive work in this area since 1960 has produced a multitude of different algorithms and philosophies, and it is far beyond the scope of this book to go into the details of such work. See Ref. 21 for an expanded discussion of finite-difference methods. In addition, you are strongly encouraged to read the current literature in this regard, in particular, the *AIAA Journal*, *Computers and Fluids*, and the *Journal of Computational Physics*. Also, Ref. 7 is a modern text on the subject of computational fluid dynamics.

What are finite differences? They are algebraic difference quotients which represent the various partial derivatives that occur in our governing equations. Hence, the solution of the governing partial differential equations involves the manipulation of algebraic quantities—just the sort of operation that digital computers are designed to handle. For example, consider the rectangular grid shown in Fig. 13.2*b*. The various grid points are denoted by the index i in the x direction and by j in the y direction. If we consider point (i, j), then immediately to its right is point $(i+1, j)$, and to its left is point $(i-1, j)$. Similarly, directly above and below point (i, j) are points $(i, j+1)$ and $(i, j-1)$, respectively. The spacing between points in the x direction is Δx, and that in the y direction is Δy. We will deal with a uniform grid, i.e., Δx is the same everywhere, and Δy is also the same everywhere; however, in general, $\Delta x \neq \Delta y$. Consider a Taylor series expansion of the velocity component u about point (ij) taken in the positive x direction,

$$u_{i+1,j} = u_{i,j} + \left(\frac{\partial u}{\partial x}\right)_{i,j} \Delta x + \left(\frac{\partial^2 u}{\partial x^2}\right)_{i,j} \frac{(\Delta x)^2}{2} + \cdots \tag{13.27}$$

In its present form, Eq. (13.27) is of "second-order accuracy" because terms involving $(\Delta x)^3$, $(\Delta x)^4$, etc., have been assumed small and are neglected. If we are interested in only first-order accuracy, Eq. (13.27) can be written as

$$u_{i+1,j} = u_{i,j} + \left(\frac{\partial u}{\partial x}\right)_{i,j} \Delta x + \cdots \tag{13.28}$$

Solving Eq. (13.28) for $(\partial u/\partial x)_{i,j}$, we obtain

$$\boxed{\left(\frac{\partial u}{\partial x}\right)_{i,j} = \frac{u_{i+1,j} - u_{i,j}}{\Delta x}} \tag{13.29}$$

Examine Eq. (13.29) carefully. It is an algebraic difference quotient which represents (to first-order accuracy) the partial derivative $(\partial u/\partial x)_{i,j}$. The particular form of the finite difference in Eq. (13.29) is called a *forward difference*. Similarly, if Eq. (13.27) is written for a negative Δx, we have

$$u_{i-1,j} = u_{i,j} + \left(\frac{\partial u}{\partial x}\right)_{i,j}(-\Delta x) + \left(\frac{\partial^2 u}{\partial x^2}\right)_{i,j}\frac{(-\Delta x)^2}{2} + \cdots \tag{13.30}$$

which, for first-order accuracy, can be written as

$$u_{i-1,j} = u_{i,j} - \left(\frac{\partial u}{\partial x}\right)_{i,j} \Delta x + \cdots \tag{13.31}$$

Solving Eq. (13.31) for $(\partial u/\partial x)_{i,j}$, we obtain

$$\boxed{\left(\frac{\partial u}{\partial x}\right)_{i,j} = \frac{u_{i,j} - u_{i-1,j}}{\Delta x}} \tag{13.32}$$

Equation (13.32) gives a *rearward difference* for the derivative.

Both Eqs. (13.29) and (13.32) are finite-difference representations of the derivative $\partial u/\partial x$ evaluated at point (i, j) in Fig. 13.2b. They are both of first-order accuracy. A finite difference of second-order accuracy can be obtained by subtracting Eq. (13.30) from (13.27), yielding

$$u_{i+1,j} - u_{i-1,j} = 0 + 2\left(\frac{\partial u}{\partial x}\right)_{i,j} \Delta x + 0 + \cdots \tag{13.33}$$

Solving Eq. (13.33) for $(\partial u/\partial x)_{i,j}$, we have

$$\boxed{\left(\frac{\partial u}{\partial x}\right)_{i,j} = \frac{u_{i+1,j} - u_{i-1,j}}{2\Delta x}} \tag{13.34}$$

Equation (13.34) is called a *central difference*. Because it was obtained from Eqs.

(13.30) and (13.27), which are of second-order accuracy, the central difference defined by Eq. (13.34) is also second-order accurate.

Analogous expressions for the derivatives in the y direction are as follows:

$$\left(\frac{\partial u}{\partial y}\right)_{i,j} = \begin{cases} \dfrac{u_{i,j+1} - u_{i,j}}{\Delta y} & \text{(forward difference)} \\[2ex] \dfrac{u_{i,j} - u_{i,j-1}}{\Delta y} & \text{(rearward difference)} \\[2ex] \dfrac{u_{i,j+1} - u_{i,j-1}}{2\Delta y} & \text{(central difference)} \end{cases}$$

How do we use the finite differences obtained here? Imagine that a flow in xy space is covered by the mesh shown in Fig. 13.2b. Assume there are N grid points. At each one of these grid points, evaluate the continuity, momentum, and energy equations with their partial derivatives replaced by the finite-difference expressions derived above. For example, replacing the derivatives in Eqs. (7.40), (7.42a and b), and (7.44) with finite differences, along with Eqs. (7.1) and (7.6a), we obtain a system (over all N grid points) of $6N$ simultaneous nonlinear algebraic equations in terms of the $6N$ unknowns, namely, ρ, u, v, p, T, and e, at each of the N grid points. In principle, we could solve this system for the unknown flow variables at all the grid points. In practice, this is easier said than done. There are severe problems in solving such a large number of simultaneous nonlinear equations. Moreover, we have to deal with problems associated with numerical instabilities that sometimes cause such attempted solutions to "blow up" on the computer. Finally, and most importantly, we must properly account for the boundary conditions. These considerations make all finite-difference solutions a nontrivial endeavor. As a result, a number of specialized finite-difference techniques have evolved, directed at solving different types of flow problems and attempting to increase computational efficiency and accuracy. It is beyond the scope of this book to describe these difference techniques in detail. However, one technique in particular was widely used during the 1970s. This is an approach developed in 1969 by Robert MacCormack at the NASA Ames Research Center. Because of its widespread use and acceptance, as well as its relative simplicity, we will describe MacCormack's technique in enough detail to give you a reasonable understanding of the method. This description will be carried out in the context of the following example.

Consider the two-dimensional supersonic flow through the divergent duct shown in Fig. 13.8a. Assume the flow is supersonic at the inlet, and that all properties are known at the inlet. That is, the flow-field variables at grid points $(1, 1)$, $(1, 2)$, $(1, 3)$, $(1, 4)$, and $(1, 5)$ are known. The duct is formed by a flat surface at the bottom and a specified contour, $y_s = f(x)$, at the top. In addition, assume that the flow is inviscid, adiabatic, and steady, with no body forces. It can be rotational or irrotational—the method of solution is the same. The governing equations are obtained from Eqs. (7.40), (7.42a and b), (7.44), (7.1),

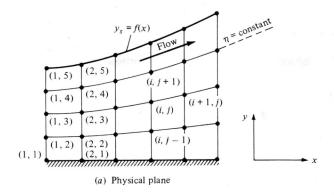

(a) Physical plane

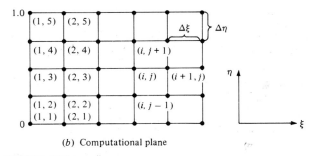

(b) Computational plane

FIGURE 13.8
Finite-difference meshes in both the physical and computational planes.

and $(7.6a)$, which yield

$$\frac{\partial(\rho u)}{\partial x} + \frac{\partial(\rho v)}{\partial y} = 0 \tag{13.35}$$

$$\rho u \frac{\partial u}{\partial x} + \rho v \frac{\partial u}{\partial y} = -\frac{\partial p}{\partial x} \tag{13.36}$$

$$\rho u \frac{\partial v}{\partial x} + \rho v \frac{\partial v}{\partial y} = -\frac{\partial p}{\partial y} \tag{13.37}$$

$$\rho u \frac{\partial(e + V^2/2)}{\partial x} + \rho v \frac{\partial(e + V^2/2)}{\partial y} = -\frac{\partial(pu)}{\partial x} - \frac{\partial(pv)}{\partial y} \tag{13.38}$$

$$p = \rho RT \tag{13.39}$$

$$e = c_v T \tag{13.40}$$

Let us express these equations in slightly different form, as follows. Multiplying

Eq. (13.35) by u, and adding the result to Eq. (13.36), we have

$$u\frac{\partial(\rho u)}{\partial x}+\rho u\frac{\partial u}{\partial x}+u\frac{\partial(\rho v)}{\partial y}+\rho v\frac{\partial u}{\partial y}=-\frac{\partial p}{\partial x}$$

or

$$\frac{\partial(\rho u^2)}{\partial x}+\frac{\partial(\rho uv)}{\partial y}=-\frac{\partial p}{\partial x}$$

or

$$\frac{\partial}{\partial x}(\rho u^2+p)=-\frac{\partial(\rho uv)}{\partial y} \tag{13.41}$$

Similarly, multiplying Eq. (13.35) by v, and adding the result to Eq. (13.37), we obtain

$$\frac{\partial(\rho uv)}{\partial x}=-\frac{\partial(\rho v^2+p)}{\partial y} \tag{13.42}$$

Multiplying Eq. (13.35) by $e+V^2/2$, and adding the result to Eq. (13.38), we obtain

$$\frac{\partial}{\partial x}\left[\rho u\left(e+\frac{V^2}{2}\right)+pu\right]=-\frac{\partial}{\partial y}\left[\rho v\left(e+\frac{V^2}{2}\right)+pv\right] \tag{13.43}$$

Define the following symbols:

$$F=\rho u \tag{13.44a}$$

$$G=\rho u^2+p \tag{13.44b}$$

$$H=\rho uv \tag{13.44c}$$

$$K=\rho u\left(e+\frac{V^2}{2}\right)+pu \tag{13.44d}$$

Then, Eqs. (13.35) and (13.41) to (13.43) become

$$\frac{\partial F}{\partial x}=-\frac{\partial(\rho v)}{\partial y} \tag{13.45}$$

$$\frac{\partial G}{\partial x}=-\frac{\partial(\rho uv)}{\partial y} \tag{13.46}$$

$$\frac{\partial H}{\partial x}=-\frac{\partial(\rho v^2+p)}{\partial y} \tag{13.47}$$

$$\frac{\partial K}{\partial x}=-\frac{\partial}{\partial y}\left[\rho v\left(e+\frac{V^2}{2}\right)+pv\right] \tag{13.48}$$

Equations (13.45) to (13.48) are the continuity, x and y momentum, and energy equations, respectively—but in a slightly different form from those we are used to seeing. The above form of these equations is frequently called the *conservation form.* Let us now treat F, G, H, and K as our primary dependent variables; these quantities are called *flux variables*, in contrast to the usual p, ρ, T, u, v, e, etc., which are called *primitive variables*. It is important to note that once the values of F, G, H, and K are known at a given grid point, the primitive variables at that point can be found from Eqs. (13.44*a* to *d*) and

$$p = \rho R T \tag{13.49}$$

$$e = c_v T \tag{13.50}$$

$$V^2 = u^2 + v^2 \tag{13.51}$$

That is, Eqs. (13.44*a* to *d*) and (13.49) to (13.51) constitute seven algebraic equations for the seven primitive variables, ρ, u, v, p, e, T, and V.

Let us return to the physical problem given in Fig. 13.8*a*. Because the duct diverges, it is difficult to deal with an orthogonal, rectangular mesh; rather, a mesh which conforms to the boundary of the system will be curved, as shown in Fig. 13.8*a*. On the other hand, to use our finite-difference quotients as given in Eq. (13.29), (13.32), or (13.34), we desire a rectangular computational mesh. Therefore, we must *transform* the curved mesh shown in Fig. 13.8*a*, known as the *physical plane*, to a rectangular mesh shown in Fig. 13.8*b*, known as the *computational plane*. This transformation can be carried out as follows. Define

$$\xi = x \tag{13.52a}$$

$$\eta = \frac{y}{y_s}$$

where

$$y_s = f(x) \tag{13.52b}$$

In the above transformation, η ranges from 0 at the bottom wall to 1.0 at the top wall. In the computational plane (Fig. 13.8*b*), $\eta = \text{constant}$ is a straight horizontal line, whereas in the physical plane, $\eta = \text{constant}$ corresponds to the curved line shown in Fig. 13.8. Because we wish to apply our finite differences in the computational plane, we need the governing equations in terms of ξ and η rather than x and y. To accomplish this transformation, apply the chain rule of differentiation, using Eqs. (13.52*a* and *b*) as follows:

$$\frac{\partial}{\partial x} = \frac{\partial}{\partial \xi}\frac{\partial \xi}{\partial x} + \frac{\partial}{\partial \eta}\frac{\partial \eta}{\partial x} = \frac{\partial}{\partial \xi} - \frac{y}{y_s^2}\frac{dy_s}{dx}\frac{\partial}{\partial \eta}$$

or

$$\frac{\partial}{\partial x} = \frac{\partial}{\partial \xi} - \left(\frac{\eta}{y_s}\frac{dy_s}{dx}\right)\frac{\partial}{\partial \eta} \tag{13.53}$$

and

$$\frac{\partial}{\partial y} = \frac{\partial}{\partial \xi}\frac{\partial \xi}{\partial y} + \frac{\partial}{\partial \eta}\frac{\partial \eta}{\partial y} = \frac{1}{y_s}\frac{\partial}{\partial \eta} \tag{13.54}$$

Using Eqs. (13.53) and (13.54), we see that Eqs. (13.45) to (13.48) become

$$\frac{\partial F}{\partial \xi} = \left(\frac{\eta}{y_s}\frac{dy_s}{dx}\right)\left(\frac{\partial F}{\partial \eta}\right) - \frac{1}{y_s}\frac{\partial(\rho v)}{\partial \eta} \tag{13.55}$$

$$\frac{\partial G}{\partial \xi} = \left(\frac{\eta}{y_s}\frac{dy_s}{dx}\right)\frac{\partial G}{\partial \eta} - \frac{1}{y_s}\frac{\partial(\rho u v)}{\partial \eta} \tag{13.56}$$

$$\frac{\partial H}{\partial \xi} = \left(\frac{\eta}{y_s}\frac{dy_s}{dx}\right)\frac{\partial H}{\partial \eta} - \frac{1}{y_s}\frac{\partial(\rho v^2 + p)}{\partial \eta} \tag{13.57}$$

$$\frac{\partial K}{\partial \xi} = \left(\frac{\eta}{y_s}\frac{dy_s}{dx}\right)\frac{\partial K}{\partial \eta} - \frac{1}{y_s}\frac{\partial}{\partial \eta}\left[\rho v\left(e + \frac{V^2}{2}\right) + pv\right] \tag{13.58}$$

Note in the above equations that the ξ derivatives are on the left and the η derivatives are all grouped on the right.

Let us now concentrate on obtaining a numerical, finite-difference solution of the problem shown in Fig. 13.8. We will deal exclusively with the computational plane, Fig. 13.8b, where the governing continuity, x and y momentum, and energy equations are given by Eqs. (13.55) to (13.58), respectively. Grid points (1, 1), (2, 1), (1, 2), (2, 2), etc., in the computational plane are the same as grid points (1, 1), (2, 1), (1, 2), (2, 2), etc., in the physical plane. All the flow variables are known at the inlet, including F, G, H, K. The solution for the flow variables downstream of the inlet can be found by using MacCormack's method, which is based on Taylor's series expansions for F, G, H, and K as follows:

$$F_{i+1,j} = F_{i,j} + \left(\frac{\partial F}{\partial \xi}\right)_{ave}\Delta\xi \tag{13.59a}$$

$$G_{i+1,j} = G_{i,j} + \left(\frac{\partial G}{\partial \xi}\right)_{ave}\Delta\xi \tag{13.59b}$$

$$H_{i+1,j} = H_{i,j} + \left(\frac{\partial H}{\partial \xi}\right)_{ave}\Delta\xi \tag{13.59c}$$

$$K_{i+1,j} = K_{i,j} + \left(\frac{\partial K}{\partial \xi}\right)_{ave}\Delta\xi \tag{13.59d}$$

In Eqs. (13.59a to d), F, G, H, and K at point (i, j) are considered known, and these equations are used to find F, G, H, and K at point $(i+1, j)$ assuming that we can calculate the values of $(\partial F/\partial \xi)_{ave}$, $(\partial G/\partial \xi)_{ave}$, etc. The main thrust of MacCormack's method is the calculation of these average derivatives. Examining Eqs. (13.59a to d), we find that this finite-difference method is clearly a "downstream marching" method; given the flow at point (i, j), we use Eqs. (13.59a to d) to find the flow at point $(i+1, j)$. Then the process is repeated to find the flow at point $(i+2, j)$, etc. This downstream marching is similar to that performed with the method of characteristics.

The average derivatives in Eqs. (13.59a to d) are found by means of a "predictor-corrector" approach, outlined below. In carrying out this approach, we assume that the flow properties are known at grid point (i, j), as well as at all points directly above and below (i, j), namely, at $(i, j+1)$, $(i, j+2)$, $(i, j-1)$, $(i, j-2)$, etc.

13.4.1 Predictor Step

First, predict the value of $F_{i+1,j}$ by using a Taylor series where $\partial F/\partial \xi$ is evaluated at point (i, j). Denote this predicted value by $\bar{F}_{i+1,j}$:

$$\bar{F}_{i+1,j} = F_{i,j} + \left(\frac{\partial F}{\partial \xi}\right)_{i,j} \Delta \xi \tag{13.60}$$

In Eq. (13.60), $(\partial F/\partial \xi)_{i,j}$ is obtained from the continuity equation, Eq. (13.55), using *forward differences* for the η derivatives; i.e.,

$$\left(\frac{\partial F}{\partial \xi}\right)_{i,j} = \left(\frac{\eta}{y_s}\frac{dy_s}{dx}\right)_{i,j}\left(\frac{F_{i,j+1} - F_{i,j}}{\Delta \eta}\right) - \frac{1}{y_s}\left[\frac{(\rho v)_{i,j+1} - (\rho v)_{i,j}}{\Delta \eta}\right] \tag{13.61}$$

In Eq. (13.61), all quantities on the right-hand side are known and allow the calculation of $(\partial F/\partial \xi)_{i,j}$ which is, in turn, inserted into Eq. (13.60). A similar procedure is used to find predicted values of G, H, and K, namely, $\bar{G}_{i+1,j}$, $\bar{H}_{i+1,j}$, and $\bar{K}_{i+1,j}$, using forward differences in Eqs. (13.56) to (13.58). In turn, predicted values of the primitive variables, $\bar{p}_{i+1,j}$, $\bar{\rho}_{i+1,j}$, etc., can be obtained from Eqs. (13.44a to d) and (13.49) to (13.51).

13.4.2 Corrector Step

The predicted values obtained above are used to obtain predicted values of the derivative $(\overline{\partial F/\partial \xi})_{i+1,j}$, using *rearward differences* in Eq. (13.55):

$$\left(\overline{\frac{\partial F}{\partial \xi}}\right)_{i+1,j} = \left(\frac{\eta}{y_s}\frac{dy_s}{dx}\right)_{i+1,j} \frac{\bar{F}_{i+1,j} - \bar{F}_{i+1,j-1}}{\Delta \eta} - \frac{1}{y_s}\frac{\overline{(\rho v)}_{i+1,j} - \overline{(\rho v)}_{i+1,j-1}}{\Delta \eta} \tag{13.62}$$

In turn, the results from Eqs. (13.61) and (13.62) allow the calculation of the average derivative

$$\left(\frac{\partial F}{\partial \xi}\right)_{\text{ave}} = \frac{1}{2}\left[\left(\frac{\partial F}{\partial \xi}\right)_{i,j} + \left(\overline{\frac{\partial F}{\partial \xi}}\right)_{i+1,j}\right] \tag{13.63}$$

Finally, this average derivative is used in Eq. (13.59a) to obtain the corrected value of $F_{i+1,j}$. The same process is followed to find the corrected values of $G_{i+1,j}$, $H_{i+1,j}$, and $K_{i+1,j}$ using rearward differences in Eqs. (13.56) to (13.58) and calculating the average derivatives $(\partial G/\partial \xi)_{\text{ave}}$, etc., in the same manner as Eq. (13.63).

The above finite-difference procedure allows the step-by-step calculation of the flow field, marching downstream from some initial data line. In the flow given in Fig. 13.8, the initial data line is the inlet, where properties are considered known. Although all the calculations are carried out in the transformed, computational plane, the flow-field results obtained at points $(2, 1)$, $(2, 2)$, etc., in the computational plane are the same values at points $(2, 1)$, $(2, 2)$, etc., in the physical plane.

There are other aspects of the finite-difference solution which have not been described above. For example, what values of $\Delta\eta$ and $\Delta\xi$ in Eqs. $(13.59a$ to $d)$, (13.60), (13.61), and (13.62) are allowed in order to maintain numerical stability? How is the flow-tangency condition at the walls imposed on the finite-difference calculations? These are important matters, but we do not take the additional space to discuss them here. See chap. 11 of Ref. 21 for details on these questions. Our purpose here has been to give you only a feeling for the nature of the finite-difference method.

13.5 THE TIME-DEPENDENT TECHNIQUE: APPLICATION TO SUPERSONIC BLUNT BODIES

The method of characteristics described in Sec. 13.2 is applicable only to supersonic flows; the characteristic lines are not defined in a practical fashion for steady, subsonic flow. Also, the particular finite-difference method outlined in Sec. 13.4 applies only to supersonic flows; if it were to be used in a locally subsonic region, the calculation would blow up. The reason for both of the above comments is that the method of characteristics and the steady flow, forward-marching finite-difference technique depend on the governing equations being mathematically "hyperbolic." In contrast, the equations for steady subsonic flow are "elliptic." (See Ref. 21 for a description of these mathematical classifications.) The fact that the governing equations change their mathematical nature in going from locally supersonic to locally subsonic flow has historically caused theoretical aerodynamicists much grief. One problem in particular, namely, the mixed subsonic-supersonic flow over a supersonic blunt body as described in Sec. 9.5, was a major research area until a breakthrough was made in the late 1960s for its proper numerical solution. The purpose of this section is to describe a numerical finite-difference solution which readily allows the calculation of mixed subsonic-supersonic flows—the *time-dependent method*—and to show how it is used to solve supersonic blunt-body flows. Time-dependent techniques are very common in modern computational fluid dynamics, and as a student of aerodynamics, you should be familiar with their philosophy.

Consider a blunt body in a supersonic stream, as sketched in Fig. 13.9a. The shape of the body is known and is given by $b = b(y)$. For a given freestream Mach number, M_∞, we wish to calculate the shape and location of the detached shock wave, as well as the flow-field properties between the shock and the body. The physical aspects of this flow field were described in Sec. 9.5, which you should review before progressing further.

The flow around a blunt body in a supersonic stream is rotational. Why? Examine Fig. 13.10, which illustrates several streamlines around the blunt body.

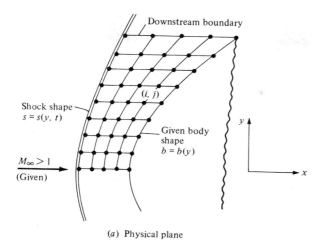

(a) Physical plane

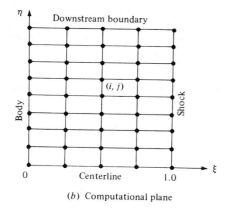

(b) Computational plane

FIGURE 13.9
Blunt-body flow field in both the physical and computational planes.

The flow is inviscid and adiabatic. In the uniform freestream ahead of the shock wave, the entropy is the same for each streamline. However, in crossing the shock wave, each streamline traverses a different part of the wave, and hence experiences a different increase in entropy. That is, the streamline at point a in Fig. 13.10 crosses a normal shock, and hence experiences a large increase in entropy, whereas the streamline at point b crosses a weaker, oblique shock, and therefore experiences a smaller increase in entropy, $s_b < s_a$. The streamline at point c experiences an even weaker portion of the shock, and hence $s_c < s_b < s_a$. The net result is that in the flow between the shock and the body, the entropy *along* a given streamline is constant, whereas the entropy changes from one streamline to the next; i.e., an *entropy gradient* exists normal to the streamlines. It can readily be shown (see chap. 6 of Ref. 21) that an adiabatic flow with entropy gradients is *rotational*. Hence, the flow field over a supersonic blunt body is rotational.

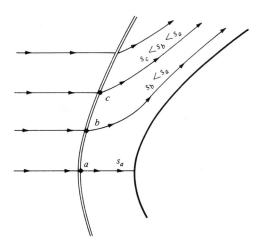

FIGURE 13.10
In a supersonic blunt-body flow field, the entropy is different for different streamlines.

In light of the above, we cannot use the velocity potential equation to analyze the blunt-body flow. Rather, the basic continuity, momentum, and energy equations must be employed in their fundamental form, given by Eqs. (7.40), (7.42*a* and *b*), and (7.44). With no body forces, these equations are

Continuity:
$$\frac{\partial \rho}{\partial t} = -\left(\frac{\partial(\rho u)}{\partial x} + \frac{\partial(\rho v)}{\partial y}\right) \tag{13.64}$$

x momentum:
$$\frac{\partial u}{\partial t} = -\left(u\frac{\partial u}{\partial x} + v\frac{\partial u}{\partial y} + \frac{1}{\rho}\frac{\partial p}{\partial x}\right) \tag{13.65}$$

y momentum:
$$\frac{\partial v}{\partial t} = -\left(u\frac{\partial v}{\partial x} + v\frac{\partial v}{\partial y} + \frac{1}{\rho}\frac{\partial p}{\partial y}\right) \tag{13.66}$$

Energy:
$$\frac{\partial(e + V^2/2)}{\partial t} = -\left(u\frac{\partial(e + V^2/2)}{\partial x} + v\frac{\partial(e + V^2/2)}{\partial y}\right.$$
$$\left. + \frac{1}{\rho}\frac{\partial(pu)}{\partial x} + \frac{1}{\rho}\frac{\partial(pv)}{\partial y}\right) \tag{13.67}$$

Notice the form of the above equations; the time derivatives are on the left, and all spatial derivatives are on the right. These equations are in the form necessary for a time-dependent finite-difference solution, as described below.

Return to Fig. 13.9*a*. Recall that the body shape and freestream conditions are given, and we wish to calculate the shape and location of the shock wave as well as the flow field between the shock and body. We are interested in the *steady flow* over the blunt body; however, we use a time-dependent method to obtain the steady flow. The basic philosophy of this method is as follows. First, *assume* a shock-wave shape and location. Also, cover the flow field between the shock and body with a series of grid points, as sketched in Fig. 13.9a. At each of these grid points, *assume* values of all the flow variables, ρ, u, v, etc. These assumed values are identified as *initial conditions* at time $t = 0$. With these assumed values, the spatial derivatives on the right sides of Eqs. (13.64) to (13.67) are known values (obtained from finite differences). Hence, Eqs. (13.64) to (13.67) allow the calculation of the time derivatives $\partial\rho/\partial t$, $\partial u/\partial t$, etc. In turn, these time derivatives allow us to calculate the flow properties at each grid point at a later instant in time, say, Δt. The flow properties at time $t = \Delta t$ are different from at $t = 0$. A repetition of this cycle gives the flow-field variables at all grid points at time $t = 2\Delta t$. As this cycle is repeated many hundreds of times, the flow-field properties at each grid point are calculated as a function of time. For example, the time variation of $u_{i,j}$ is sketched in Fig. 13.11. At each time step, the value of $u_{i,j}$ is different; however, at large times the changes in $u_{i,j}$ from one time step to another become small, and $u_{i,j}$ approaches a steady-state value, as shown in Fig. 13.11. *It is this steady-state value that we want*; *the time-dependent approach is simply a means to that end.* Moreover, the shock-wave shape and location will change with time; the new shock location and shape at each time step are calculated so as to satisfy the shock relations across the wave at each of the grid points immediately behind the wave. At large times, as the flow-field variables approach a steady state, the shock shape and location also approach a steady state. Because of the time-dependent motion of the shock wave, the wave shape is a function of both t and y as shown in Fig. 13.9a, $s = s(y, t)$.

Given the above philosophy, let us examine a few details of the method. First, note that the finite-difference grid in Fig. 13.9a is curved. We would like to apply our finite differences in a rectangular grid; hence, in Eqs. (13.64) to (13.67) the independent variables can be transformed as

$$\xi = \frac{x - b}{s - b} \quad \text{and} \quad \eta = y$$

where $b = b(y)$ gives the abscissa of the body and $s = s(y, t)$ gives the abscissa of the shock. The above transformation produces a rectangular grid in the

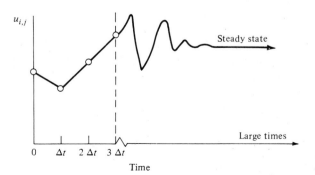

FIGURE 13.11
Schematic of the time variation of a typical flow variable—the time-dependent method.

computational plane, shown in Fig. 13.9b, where the body corresponds to $\xi = 0$ and the shock corresponds to $\xi = 1$. All calculations are made in this transformed, computational plane.

The finite-difference calculations themselves can be carried out using Mac-Cormack's method (see Sec. 13.4) applied as follows. The flow-field variables can be advanced in time using a Taylor series in time; e.g.,

$$\rho_{i,j}(t + \Delta t) = \rho_{i,j}(t) + \left[\left(\frac{\partial \rho}{\partial t} \right)_{i,j} \right]_{\text{ave}} \Delta t \tag{13.68}$$

In Eq. (13.68), we know the density at grid point (i, j) at time t; i.e., we know $\rho_{i,j}(t)$. Then Eq. (13.68) allows us to calculate the density at the same grid point at time $t + \Delta t$, i.e., $\rho_{i,j}(t + \Delta t)$, if we know a value of the average time derivative $[(\partial \rho / \partial t)_{i,j}]_{\text{ave}}$. This time derivative is an average between times t and $t + \Delta t$ and is obtained from a predictor-corrector process as follows.

13.5.1 Predictor Step

All the flow variables are known at time t at all the grid points. This allows us to replace the spatial derivatives on the right of Eqs. (13.64) to (13.67) (suitably transformed into $\xi \eta$ space) with known *forward differences*. These equations then give values of the time derivatives at time t, which are used to obtain *predicted* values of the flow-field variables at time $t + \Delta t$; e.g.,

$$\bar{\rho}_{i,j}(t + \Delta t) = \rho_{i,j}(t) + \left[\left(\frac{\partial \rho}{\partial t} \right)_{i,j} \right]_t \Delta t$$

where $\rho_{i,j}(t)$ is known, $[(\partial \rho / \partial t)_{i,j}]_t$ is obtained from the governing equation, Eq. (13.64) (suitably transformed), using *forward differences* for the spatial derivatives, and $\bar{\rho}_{i,j}(t + \Delta t)$ is the predicted density at time $t + \Delta t$. Predicted values of all other flow variables $\bar{u}_{i,j}(t + \Delta t)$, etc., are obtained at all the grid points in a likewise fashion.

13.5.2 Corrector Step

Inserting the flow variables obtained above into the governing equations, Eqs. (13.64) to (13.67), using *rearward differences* for the spatial derivatives, predicted values of the time derivatives at $t + \Delta t$ are obtained, e.g., $[(\overline{\partial \rho / \partial t})_{i,j}]_{(t+\Delta t)}$. In turn, these are averaged with the time derivatives from the predictor step to obtain; e.g.,

$$\left[\left(\frac{\partial \rho}{\partial t} \right)_{i,j} \right]_{\text{ave}} = \frac{1}{2} \left\{ \left[\left(\frac{\partial \rho}{\partial t} \right)_{i,j} \right]_t + \left[\left(\frac{\overline{\partial \rho}}{\partial t} \right)_{i,j} \right]_{(t+\Delta t)} \right\} \tag{13.69}$$

Finally, the average time derivative obtained from Eq. (13.69) is inserted into

Eq. (13.68) to yield the corrected value of density at time $t + \Delta t$. The same procedure is used for all the dependent variables, u, v, etc.

Starting from the assumed initial conditions at $t = 0$, the repeated application of Eq. (13.68) along with the above predictor-corrector algorithm at each time step allows the calculation of the flow-field variables and shock shape and location as a function of time. As stated above, after a large number of time steps, the calculated flow-field variables approach a steady state, where $[(\partial\rho/\partial t)_{i,j}]_{\text{ave}} \to 0$ in Eq. (13.68). Once again, we emphasize that we are interested in the steady-state answer, and the time-dependent technique is simply a means to that end.

Note that the applications of MacCormack's technique to both the steady flow calculations described in Sec. 13.4 and the time-dependent calculations described in the present section are analogous; in the former, we march forward in the spatial coordinate x, starting with known values along with a constant y line, whereas, in the latter, we march forward in time starting with a known flow field at $t = 0$.

Why do we bother with a time-dependent solution? Is it not an added complication to deal with an extra independent variable, t, in addition to the spatial variables x and y? The answers to these questions are as follows. The governing unsteady flow equations given by Eqs. (13.64) to (13.67) are hyperbolic with respect to time, independent of whether the flow is locally subsonic or supersonic. In Fig. 13.9a, some of the grid points are in the subsonic region and others are in the supersonic region. However, the time-dependent solution progresses in the same manner at all these points, independent of the local Mach number. Hence, the time-dependent technique is the only approach known today which allows the uniform calculation of a mixed subsonic-supersonic flow field of arbitrary extent. For this reason, the application of the time-dependent technique, although it adds one additional independent variable, allows the straightforward solution of a flow field which is extremely difficult to solve by a purely steady-state approach.

A much more detailed description of the time-dependent technique is given in chap. 12 of Ref. 21, which you should study before attempting to apply this technique to a specific problem. The intent of our description here has been to give you simply a "feeling" for the philosophy and general approach of the technique.

Some typical results for supersonic blunt-body flow fields are given in Figs. 13.12 to 13.15. These results were obtained with a time-dependent solution described in Ref. 35. Figures 13.12 and 13.13 illustrate the behavior of a time-dependent solution during its approach to the steady state. In Fig. 13.12, the time-dependent motion of the shock wave is shown for a parabolic cylinder in a Mach 4 freestream. The shock labeled $0\,\Delta t$ is the initially assumed shock wave at $t = 0$. At early times, the shock wave rapidly moves away from the body; however, after about 300 time steps, it has slowed considerably, and between 300 and 500 time steps, the shock wave is virtually motionless—it has reached its steady-state shape and location. The time variation of the stagnation point pressure is given in Fig. 13.13. Note that the pressure shows strong timewise

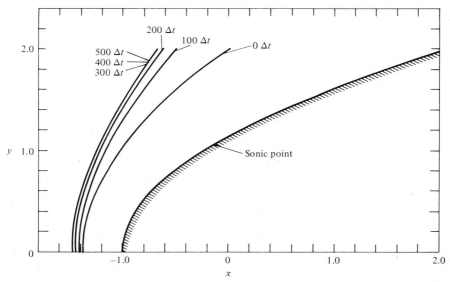

FIGURE 13.12
Time-dependent shock-wave motion, parabolic cylinder, $M_\infty = 4$.

oscillations at early times, but then it asymptotically approaches a steady value at large times. Again, it is this asymptotic steady state that we want, and the intermediate transient results are just a means to that end. Concentrating on just the steady-state results, Fig. 13.14 gives the pressure distribution (nondimensionalized by stagnation point pressure) over the body surface for the cases of both $M_\infty = 4$ and 8. The time-dependent numerical results are shown as the solid

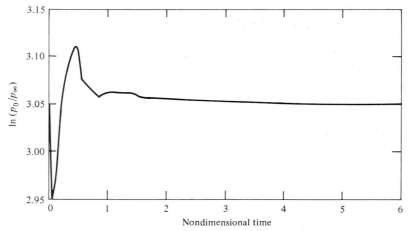

FIGURE 13.13
Time variation of stagnation point pressure, parabolic cylinder, $M_\infty = 4$.

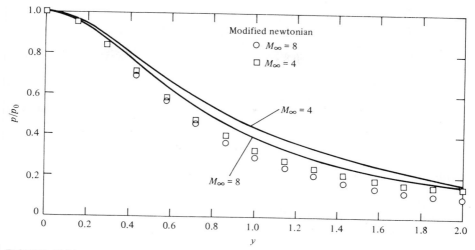

FIGURE 13.14
Surface pressure distributions, parabolic cylinder.

curves, whereas the open symbols are from newtonian theory, to be discussed in Chap. 14. Note that the pressure is a maximum at the stagnation point and decreases as a function of distance away from the stagnation point—a variation that we most certainly would expect based on our previous aerodynamic experience. The steady shock shapes and sonic lines are shown in Fig. 13.15 for the cases of $M_\infty = 4$ and 8. Note that as the Mach number increases, the shock wave moves closer to the body.

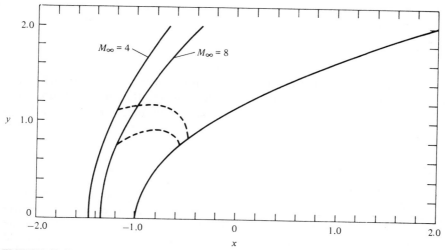

FIGURE 13.15
Shock shapes and sonic lines, parabolic cylinder.

13.6 SUMMARY

We have now completed both branches of our road map shown in Fig. 13.1. Make certain that you feel comfortable with all the material represented by this road map. A short summary of the highlights is given below:

For a steady, two-dimensional, irrotational, supersonic flow, the characteristic lines are Mach lines, and the compatibility equations which hold along these characteristic lines are

$$\theta + \nu = K_- \quad \text{(along a } C_- \text{ characteristic)}$$

and

$$\theta - \nu = K_+ \quad \text{(along a } C_+ \text{ characteristic)}$$

The numerical solution of such a flow can be carried out by solving the compatibility equations along the characteristic lines in a step-by-step fashion, starting from an appropriate initial data line.

The contour of a supersonic nozzle can be obtained by applying the method of characteristics downstream of the limiting characteristic (which is usually downstream of the geometric throat).

The essence of finite-difference methods is to replace the partial derivatives in the governing flow equations with finite-difference quotients. For supersonic steady flows, this allows us to march downstream, starting from known data along an initial data line in the supersonic flow. For the solution of mixed subsonic-supersonic flows, a time-dependent technique can be used which allows us to march forward in time, starting with assumed initial conditions at time $t = 0$ and achieving a steady-state result in the limit of large times.

A popular technique for carrying out finite-difference solutions, whether for supersonic steady flow or for a time-dependent solution of mixed subsonic and supersonic flow, is the predictor-corrector technique by MacCormack.

PROBLEM

Note: The purpose of the following problem is to provide an exercise in carrying out a unit process for the method of characteristics. A more extensive application to a complete flow field is left to your specific desires. Also, an extensive practical problem utilizing the finite-difference method requires a large number of arithmetic operations and is practical only on a digital computer. You are encouraged to set up such a problem at your leisure. The main purpose of the present chapter is to present the essence of several numerical methods, not to burden the reader with a lot of calculations or the requirement to write an extensive computer program.

13.1. Consider two points in a supersonic flow. These points are located in a cartesian coordinate system at $(x_1, y_1) = (0, 0.0684)$ and $(x_2, y_2) = (0.0121, 0)$, where the units are meters. At point (x_1, y_1): $u_1 = 639$ m/s, $v_1 = 232.6$ m/s, $p_1 = 1$ atm, $T_1 = 288$ K. At point (x_2, y_2): $u_2 = 680$ m/s, $v_2 = 0$, $p_2 = 1$ atm, $T_2 = 288$ K. Consider point 3 downstream of points 1 and 2 located by the intersection of the C_+ characteristic through point 2 and the C_- characteristic through point 1. At point 3, calculate: $u_3, v_3, p_3,$ and T_3. Also, calculate the location of point 3, assuming the characteristics between these points are straight lines.

CHAPTER 14

ELEMENTS OF HYPERSONIC FLOW

Almost everyone has their own definition of the term hypersonic. If we were to conduct something like a public opinion poll among those present, and asked everyone to name a Mach number above which the flow of a gas should properly be described as hypersonic there would be a majority of answers round about five or six, but it would be quite possible for someone to advocate, and defend, numbers as small as three, or as high as 12.

P. L. Roe,
comment made in a lecture
at the von Karman Institute, Belgium,
January 1970

14.1 INTRODUCTION

The history of aviation has always been driven by the philosophy of "faster and higher," starting with the Wright brothers' sea level flights at 35 mi/h in 1903, and progressing exponentially to the manned space flight missions of the 1960s and 1970s. The current altitude and speed records for manned flight are the moon and 36,000 ft/s—more than 36 times the speed of sound—set by the Apollo lunar capsule in 1969. Although most of the flight of the Apollo took place in space, outside the earth's atmosphere, one of its most critical aspects was reentry into the atmosphere after completion of the lunar mission. The aerodynamic phenomena associated with very high-speed flight, such as encountered during atmospheric reentry, are classified as *hypersonic aerodynamics*—the subject of this chapter. In addition to reentry vehicles, both manned and unmanned, there are other hypersonic applications on the horizon, such as ramjet-powered hypersonic missiles now under consideration by the military and the concept of a hypersonic transport, the basic technology of which is now being studied by NASA. Therefore, although hypersonic aerodynamics is at one extreme end of the whole flight spectrum (see Sec. 1.10), it is important enough to justify one small chapter in our presentation of the fundamentals of aerodynamics.

This chapter is short; its purpose is simply to introduce some basic considerations of hypersonic flow. Therefore, we have no need for a chapter road map or a summary at the end. Also, before progressing further, return to Chap. 1 and review the short discussion on hypersonic flow given in Sec. 1.10. For an in-depth study of hypersonic flow, see the author's book listed as Ref. 55.

14.2 QUALITATIVE ASPECTS OF HYPERSONIC FLOW

Consider a 15° half-angle wedge flying at $M_\infty = 36$. From Fig. 9.7, we see that the wave angle of the oblique shock is only 18°; i.e., the oblique shock wave is very close to the surface of the body. This situation is sketched in Fig. 14.1. Clearly, the shock layer between the shock wave and the body is very thin. Such thin shock layers are one characteristic of hypersonic flow. A practical consequence of a thin shock layer is that a major interaction frequently occurs between the inviscid flow behind the shock and the viscous boundary layer on the surface. Indeed, hypersonic vehicles generally fly at high altitudes where the density, hence Reynolds number, is low, and therefore the boundary layers are thick. Moreover, at hypersonic speeds, the boundary-layer thickness on slender bodies is approximately proportional to M_∞^2; hence, the high Mach numbers further contribute to a thickening of the boundary layer. In many cases, the boundary-layer thickness is of the same magnitude as the shock-layer thickness, such as sketched in the insert at the top of Fig. 14.1. Here, the shock layer is fully viscous, and the shock-wave shape and surface pressure distribution are affected by such viscous effects. These phenomena are called *viscous interaction phenomena*—where the viscous flow greatly affects the external inviscid flow, and, of course, the external inviscid flow affects the boundary layer. A graphic example of such viscous interaction occurs on a flat plate at hypersonic speeds, as sketched in Fig. 14.2. If the flow were completely inviscid, then we would have the case shown in Fig. 14.2a, where a Mach wave trails downstream from the leading edge. Since there is no deflection of the flow, the pressure distribution

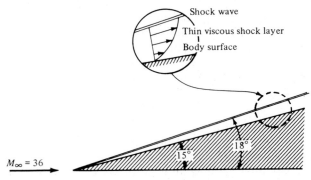

FIGURE 14.1
For hypersonic flow, the shock layers are thin and viscous.

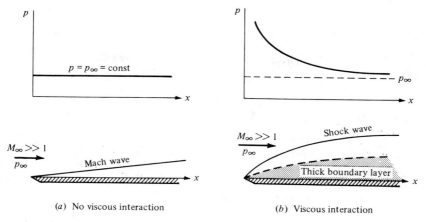

(a) No viscous interaction

(b) Viscous interaction

FIGURE 14.2
The viscous interaction on a flat plate at hypersonic speeds.

over the surface of the plate is constant and equal to p_∞. In contrast, in real life there is a boundary layer over the flat plate, and at hypersonic conditions this boundary layer can be thick, as sketched in Fig. 14.2b. The thick boundary layer deflects the external, inviscid flow, creating a comparably strong, curved shock wave which trails downstream from the leading edge. In turn, the surface pressure from the leading edge is considerably higher than p_∞, and only approaches p_∞ far downstream of the leading edge, as shown in Fig. 14.2b. In addition to influencing the aerodynamic force, such high pressures increase the aerodynamic heating at the leading edge. Therefore, hypersonic viscous interaction can be important, and this has been one of the major areas of modern hypersonic aerodynamic research.

There is a second and frequently more dominant aspect of hypersonic flow, namely, high temperatures in the shock layer, along with large aerodynamic heating of the vehicle. For example, consider a blunt body reentering the atmosphere at Mach 36, as sketched in Fig. 14.3. Let us calculate the temperature in

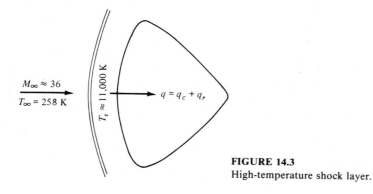

FIGURE 14.3
High-temperature shock layer.

the shock layer immediately behind the normal portion of the bow shock wave. From App. B, we find that the static temperature ratio across a normal shock wave with $M_\infty = 36$ is 252.9; this is denoted by T_s/T_∞ in Fig. 14.3. Moreover, at a standard altitude of 59 km, $T_\infty = 258$ K. Hence, we obtain $T_s = 65,248$ K—an incredibly high temperature, which is more than six times hotter than the surface of the sun! This is, in reality, an incorrect value, because we have used App. B which is good only for a calorically perfect gas with $\gamma = 1.4$. However, at high temperatures, the gas will become chemically reacting; γ will no longer equal 1.4 and will no longer be constant. Nevertheless, we get the impression from this calculation that the temperature in the shock layer will be very high, albeit something less than 65,248 K. Indeed, if a proper calculation of T_s is made taking into account the chemically reacting gas, we would find that $T_s \approx 11,000$ K—still a very high value. Clearly, high-temperature effects are very important in hypersonic flow.

Let us examine these high-temperature effects in more detail. If we consider air at $p = 1$ atm and $T = 288$ K (standard sea level), the chemical composition is essentially 20 percent O_2 and 80 percent N_2 by volume. The temperature is too low for any significant chemical reaction to take place. However, if we were to increase T to 2000 K, we would observe that the O_2 begins to dissociate; i.e.,

$$O_2 \to 2O \qquad 2000\ K < T < 4000\ K$$

If the temperature were increased to 4000 K, most of the O_2 would be dissociated, and N_2 dissociation would commence:

$$N_2 \to 2N \qquad 4000\ K < T < 9000\ K$$

If the temperature were increased to 9000 K, most of the N_2 would be dissociated, and ionization would commence:

$$\begin{matrix} N \to N^+ + e^- \\ O \to O^+ + e^- \end{matrix} \qquad T > 9000\ K$$

Hence, returning to Fig. 14.3, the shock layer in the nose region of the body is a partially ionized plasma, consisting of the atoms N and O, the ions N^+ and O^+, and electrons, e^-. Indeed, the presence of these free electrons in the shock layer is responsible for the "communications blackout" experienced over portions of the trajectory of a reentry vehicle.

One consequence of these high-temperature effects is that all our equations and tables obtained in Chaps. 7 to 13 which depended on a constant $\gamma = 1.4$ are no longer valid. Indeed, the governing equations for the high-temperature, chemically reacting shock layer in Fig. 14.3 must be solved numerically, taking into account the proper physics and chemistry of the gas itself. The analysis of aerodynamic flows with such real physical effects is discussed in detail in chaps. 16 and 17 of Ref. 21; such matters are beyond the scope of this book.

Associated with the high-temperature shock layers is a large amount of heat transfer to the surface of a hypersonic vehicle. Indeed, for reentry velocities, aerodynamic heating dominates the design of the vehicle, as explained at the end of Sec. 1.1. (Recall that the third historical example discussed in Sec. 1.1 was the evolution of the blunt-body concept to reduce aerodynamic heating; review this material before progressing further.) The usual mode of aerodynamic heating is the transfer of energy from the hot shock layer to the surface by means of thermal conduction at the surface; i.e., if $\partial T/\partial n$ represents the temperature gradient in the gas normal to the surface, then $q_c = -k(\partial T/\partial n)$ is the heat transfer into the surface. Because $\partial T/\partial n$ is a flow-field property generated by the flow of the gas over the body, q_c is called *convective heating*. For reentry velocities associated with ICBMs (about 28,000 ft/s), this is the only meaningful mode of heat transfer to the body. However, at higher velocities, the shock-layer temperature becomes even hotter. From experience, we know that all bodies emit thermal radiation, and from physics you know that blackbody radiation varies as T^4; hence, radiation becomes a dominant mode of heat transfer at high temperatures. (For example, the heat you feel by standing beside a fire in a fireplace is radiative heating from the flames and the hot walls.) When the shock layer reaches temperatures on the order of 11,000 K, as for the case given in Fig. 14.3, thermal radiation from the hot gas becomes a substantial portion of the total heat transfer to the body surface. Denoting radiative heating by q_r, we can express the total aerodynamic heating q as the sum of convective and radiative heating; $q = q_c + q_r$. For Apollo reentry, $q_r/q \approx 0.3$, and hence radiative heating was an important consideration in the design of the Apollo heat shield. For the entry of a space probe into the atmosphere of Jupiter, the velocities will be so high and the shock-layer temperatures so large that the convective heating is negligible, and in this case, $q \approx q_r$. For such a vehicle, radiative heating becomes the dominant aspect in its design. Figure 14.4 illustrates the relative importance of q_c and q_r for a typical manned reentry vehicle in the earth's atmosphere; note how rapidly q_r dominates the aerodynamic heating of the body as velocities increase above 36,000 ft/s. The details of shock-layer radiative heating are interesting and important; however, they are beyond the scope of this book. For a

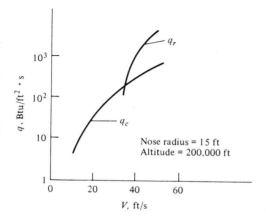

FIGURE 14.4
Convective and radiative heating rates of a blunt reentry vehicle as a function of flight velocity. (*From Ref. 36.*)

thorough survey of the engineering aspects of shock-layer radiative heat transfer, see Ref. 36.

In summary, the aspects of thin shock-layer viscous interaction and high-temperature, chemically reacting and radiative effects distinguish hypersonic flow from the more moderate supersonic regime. Hypersonic flow has been the subject of several complete books; see, e.g., Refs. 37 to 41. In particular, see Ref. 55 for a modern textbook on the subject.

14.3 NEWTONIAN THEORY

Return to Fig. 14.1; note how close the shock wave lies to the body surface. This figure is redrawn in Fig. 14.5 with the streamlines added to the sketch. When viewed from afar, the straight, horizontal streamlines in the freestream appear to almost impact the body, and then move tangentially along the body. Return to Fig. 1.1, which illustrates Isaac Newton's model for fluid flow, and compare it with the hypersonic flow field shown in Fig. 14.5; they have certain distinct similarities. (Also, review the discussion surrounding Fig. 1.1 before progressing further.) Indeed, the thin shock layers around hypersonic bodies are the closest example in fluid mechanics to Newton's model. Therefore, we might expect that results based on Newton's model would have some applicability in hypersonic flows. This is indeed the case; newtonian theory is used frequently to estimate the pressure distribution over the surface of a hypersonic body. The purpose of this section is to derive the famous newtonian sine-squared law first mentioned in Sec. 1.1 and to show how it is applied to hypersonic flows.

Consider a surface inclined at the angle θ to the freestream, as sketched in Fig. 14.6. According to the newtonian model, the flow consists of a large number of individual particles which impact the surface and then move tangentially to the surface. During collision with the surface, the particles lose their component of momentum normal to the surface, but the tangential component is preserved. The time rate of change of the normal component of momentum equals the force exerted on the surface by the particle impacts. To quantify this model, examine Fig. 14.6. The component of the freestream velocity normal to the surface is $V_\infty \sin \theta$. If the area of the surface is A, the mass flow incident on the surface is $\rho_\infty (A \sin \theta) V_\infty$. Hence, the time rate of change of momentum is

Mass flow × change in normal component of velocity

or
$$(\rho_\infty V_\infty A \sin \theta)(V_\infty \sin \theta) = \rho_\infty V_\infty^2 A \sin^2 \theta$$

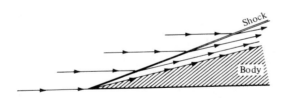

FIGURE 14.5
Streamlines in a hypersonic flow.

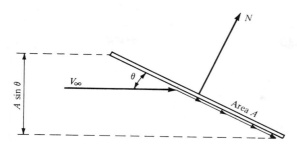

FIGURE 14.6
Schematic for newtonian impact theory.

In turn, from Newton's second law, the force on the surface is

$$N = \rho_\infty V_\infty^2 A \sin^2 \theta \qquad (14.1)$$

This force acts along the same line as the time rate of change of momentum, i.e., normal to the surface, as sketched in Fig. 14.6. From Eq. (14.1), the normal force per unit area is

$$\frac{N}{A} = \rho_\infty V_\infty^2 \sin^2 \theta \qquad (14.2)$$

Let us now interpret the physical meaning of the normal force per unit area in Eq. (14.2), N/A, in terms of our modern knowledge of aerodynamics. Newton's model assumes a stream of individual particles all moving in straight, parallel paths toward the surface; i.e., the particles have a completely directed, rectilinear motion. There is no random motion of the particles—it is simply a stream of particles such as pellets from a shotgun. In terms of our modern concepts, we know that a moving gas has molecular motion that is a composite of random motion of the molecules as well as a directed motion. Moreover, we know that the freestream static pressure p_∞ is simply a measure of the purely *random motion* of the molecules. Therefore, when the purely *directed motion* of the particles in Newton's model results in the normal force per unit area, N/A in Eq. (14.2), this normal force per unit area must be construed as the *pressure difference* above p_∞, namely, $p - p_\infty$ on the surface. Hence, Eq. (14.2) becomes

$$p - p_\infty = \rho_\infty V_\infty^2 \sin^2 \theta \qquad (14.3)$$

Equation (14.3) can be written in terms of the pressure coefficient $C_p = (p - p_\infty)/\frac{1}{2}\rho_\infty V_\infty^2$, as follows

$$\frac{p - p_\infty}{\frac{1}{2}\rho_\infty V_\infty^2} = 2 \sin^2 \theta$$

or

$$\boxed{C_p = 2 \sin^2 \theta} \qquad (14.4)$$

Equation (14.4) is Newton's sine-squared law; it states that the pressure coefficient is proportional to the sine square of the angle between a tangent to the surface

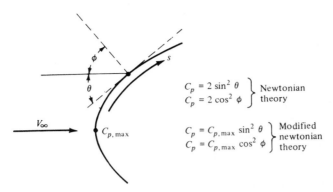

$$C_p = 2 \sin^2 \theta$$
$$C_p = 2 \cos^2 \phi$$
Newtonian theory

$$C_p = C_{p,\max} \sin^2 \theta$$
$$C_p = C_{p,\max} \cos^2 \phi$$
Modified newtonian theory

FIGURE 14.7
Definition of angles for newtonian theory.

and the direction of the freestream. This angle, θ, is illustrated in Fig. 14.7. Frequently, the results of newtonian theory are expressed in terms of the angle between a normal to the surface and the freestream direction, denoted by ϕ as shown in Fig. 14.7. In terms of ϕ, Eq. (14.4) becomes

$$C_p = 2 \cos^2 \phi \qquad (14.5)$$

which is an equally valid expression of newtonian theory.

Consider the blunt body sketched in Fig. 14.7. Clearly, the maximum pressure, hence the maximum value of C_p, occurs at the stagnation point, where $\theta = \pi/2$ and $\phi = 0$. Equation (14.4) predicts $C_p = 2$ at the stagnation point. Contrast this hypersonic result with the result obtained for incompressible flow theory in Chap. 3, where $C_p = 1$ at a stagnation point. Indeed, the stagnation pressure coefficient increases continuously from 1.0 at $M_\infty = 0$ to 1.28 at $M_\infty = 1.0$ to 1.86 for $\gamma = 1.4$ as $M_\infty \to \infty$. (Prove this to yourself.)

The result that the maximum pressure coefficient approaches 2 at $M_\infty \to \infty$ can be obtained independently from the one-dimensional momentum equation, namely, Eq. (8.6). Consider a normal shock wave at hypersonic speeds, as sketched in Fig. 14.8. For this flow, Eq. (8.6) gives

$$p_\infty + \rho_\infty V_\infty^2 = p_2 + \rho_2 V_2^2 \qquad (14.6)$$

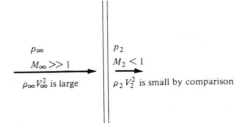

p_∞

$M_\infty \gg 1$

$\rho_\infty V_\infty^2$ is large

p_2

$M_2 < 1$

$\rho_2 V_2^2$ is small by comparison

FIGURE 14.8
Hypersonic flow across a normal shock wave.

Recall that across a normal shock wave the flow velocity decreases, $V_2 < V_\infty$; indeed, the flow behind the normal shock is subsonic. This change becomes more severe as M_∞ increases. Hence, at hypersonic speeds, we can assume that $(\rho_\infty V_\infty^2) \gg (\rho_2 V_2^2)$, and we can neglect the latter term in Eq. (14.6). As a result, Eq. (14.6) becomes, at hypersonic speeds in the limiting case as $M_\infty \to \infty$,

$$p_2 - p_\infty = \rho_\infty V_\infty^2$$

or

$$C_p = \frac{p_2 - p_\infty}{\frac{1}{2}\rho_\infty V_\infty^2} = 2$$

thus confirming the newtonian results from Eq. (14.4).

As stated above, the result that $C_p = 2$ at a stagnation point is a limiting value as $M_\infty \to \infty$. For large but finite Mach numbers, the value of C_p at a stagnation point is less than 2. Return again to the blunt body shown in Fig. 14.7. Considering the distribution of C_p as a function of distance s along the surface, the largest value of C_p will occur at the stagnation point. Denote the stagnation point value of C_p by $C_{p,\text{max}}$, as shown in Fig. 14.7. $C_{p,\text{max}}$ for a given M_∞ can be readily calculated from normal shock-wave theory. [If $\gamma = 1.4$, then $C_{p,\text{max}}$ can be obtained from $p_{0,2}/p_1 = p_{0,2}/p_\infty$, tabulated in App. B. Recall from Eq. (11.22) that $C_{p,\text{max}} = (2/\gamma M_\infty^2)(p_{0,2}/p_\infty - 1)$.] Downstream of the stagnation point, C_p can be assumed to follow the sine-squared variation predicted by newtonian theory; i.e.,

$$C_p = C_{p,\text{max}} \sin^2 \theta \qquad (14.7)$$

Equation (14.7) is called the *modified* newtonian law. For the calculation of the C_p distribution around blunt bodies, Eq. (14.7) is more accurate than Eq. (14.4).

Return to Fig. 13.14, which gives the numerical results for the pressure distributions around a blunt, parabolic cylinder at $M_\infty = 4$ and 8. The open symbols in this figure represent the results of modified newtonian theory, namely, Eq. (14.7). For this two-dimensional body, modified newtonian theory is reasonably accurate only in the nose region, although the comparison improves at the higher Mach numbers. It is generally true that newtonian theory is more accurate at larger values of both M_∞ and θ. The case for an axisymmetric body, a paraboloid at $M_\infty = 4$, is given in Fig. 14.9. Here, although M_∞ is relatively low, the agreement between the time-dependent numerical solution (see Chap. 13) and newtonian theory is much better. It is generally true that newtonian theory works better for three-dimensional bodies. In general, the modified newtonian law, Eq. (14.7), is sufficiently accurate that it is used very frequently in the preliminary design of hypersonic vehicles. Indeed, extensive computer codes have been developed to apply Eq. (14.7) to three-dimensional hypersonic bodies of general shape. Therefore, we can be thankful to Isaac Newton for supplying us with a law which holds reasonably well at hypersonic speeds, although such an application most likely never crossed his mind. Nevertheless, it is fitting that three centuries later, Newton's fluid mechanics has finally found a reasonable application.

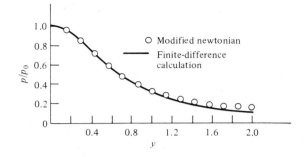

FIGURE 14.9
Surface pressure distribution, paraboloid, $M_\infty = 4$. Comparison of modified newtonian theory and time-dependent finite-difference calculations.

14.4 THE LIFT AND DRAG OF WINGS AT HYPERSONIC SPEEDS: NEWTONIAN RESULTS FOR A FLAT PLATE AT ANGLE OF ATTACK

Question: At *subsonic* speeds, how do the lift coefficient, C_L, and drag coefficient, C_D, for a wing vary with angle of attack α?

Answer: As shown in Chap. 5, we know that:

1. The lift coefficient varies *linearly* with angle of attack, at least up to the stall; see, e.g., Fig. 5.22.

2. The drag coefficient is given by the drag polar, as expressed in Eq. (5.63), repeated below:

$$C_D = c_d + \frac{C_L^2}{\pi e \mathrm{AR}} \tag{5.63}$$

Since C_L is proportional to α, then C_D varies as the *square* of α.

Question: At *supersonic* speeds, how do C_L and C_D for a wing vary with α?

Answer: In Chap. 12, we demonstrated for an airfoil at supersonic speeds that:

1. Lift coefficient varies *linearly* with α, as seen from Eq. (12.23), repeated below:

$$c_l = \frac{4\alpha}{\sqrt{M_\infty^2 - 1}} \tag{12.23}$$

2. Drag coefficient varies as the *square* of α, as seen from Eq. (12.24) for the flat plate, repeated below:

$$c_d = \frac{4\alpha^2}{\sqrt{M_\infty^2 - 1}} \tag{12.24}$$

The characteristics of a finite wing at supersonic speeds follow essentially the same functional variation with the angle of attack, namely, C_L is proportional to α and C_D is proportional to α^2.

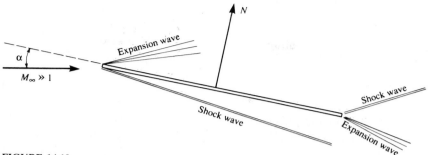

FIGURE 14.10
Wave system on a flat plate in hypersonic flow.

Question: At *hypersonic* speeds, how do C_L and C_D for a wing vary with α? We have shown that C_L is proportional to α for both subsonic and supersonic speeds—does the same proportionality hold for hypersonic speeds? We have shown that C_D is proportional to α^2 for both subsonic and supersonic speeds—does the same proportionality hold for hypersonic speeds? The purpose of the present section is to address these questions.

In an approximate fashion, the lift and drag characteristics of a wing in hypersonic flow can be modeled by a flat plate at an angle of attack, as sketched in Fig. 14.10. The exact flow field over the flat plate involves a series of expansion and shock waves as shown in Fig. 14.10; the exact lift- and wave-drag coefficients can be obtained from the shock-expansion method as described in Sec. 9.7. However, for hypersonic speeds, the lift- and wave-drag coefficients can be further approximated by the use of newtonian theory, as described in this section.

Consider Fig. 14.11. Here, a two-dimensional flat plate with chord length c is at an angle of attack α to the freestream. Since we are not including friction, and because surface pressure always acts normal to the surface, the resultant

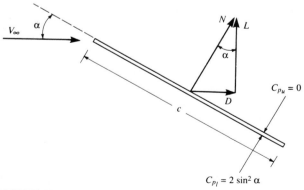

FIGURE 14.11
Flat plate at angle of attack. Illustration of aerodynamic forces.

aerodynamic force is perpendicular to the plate; i.e., in this case, the normal force N is the resultant aerodynamic force. (For an infinitely thin flat plate, this is a general result which is not limited to newtonian theory, or even to hypersonic flow.) In turn, N is resolved into lift and drag, denoted by L and D, respectively, as shown in Fig. 14.11. According to newtonian theory, the pressure coefficient on the lower surface is

$$C_{p,l} = 2 \sin^2 \alpha \qquad (14.8)$$

The upper surface of the flat plate shown in Fig. 14.11, in the spirit of newtonian theory, receives no direct "impact" of the freestream particles; the upper surface is said to be in the "shadow" of the flow. Hence, consistent with the basic model of newtonian flow, only freestream pressure acts on the upper surface, and we have

$$C_{p,u} = 0 \qquad (14.9)$$

Returning to the discussion of aerodynamic force coefficients in Sec. 1.5, we note that the normal force coefficient is given by Eq. (1.15). Neglecting friction, this becomes

$$c_n = \frac{1}{c} \int_0^c (C_{p,l} - C_{p,u}) \, dx \qquad (14.10)$$

where x is the distance along the chord from the leading edge. (*Please note*: In this section, we treat a flat plate as an airfoil section; hence, we will use lowercase letters to denote the force coefficients, as first described in Chap. 1.) Substituting Eqs. (14.8) and (14.9) into (14.10), we obtain

$$c_n = \frac{1}{c} (2 \sin^2 \alpha) c$$

or

$$= 2 \sin^2 \alpha \qquad (14.11)$$

From the geometry of Fig. 14.11, we see that the lift and drag coefficients, defined as $c_l = L/q_\infty S$ and $c_d = D/q_\infty S$, respectively, where $S = (c)(l)$, are given by

$$c_l = c_n \cos \alpha \qquad (14.12)$$

and

$$c_d = c_n \sin \alpha \qquad (14.13)$$

Substituting Eq. (14.11) into Eqs. (14.12) and (14.13), we obtain

$$c_l = 2 \sin^2 \alpha \cos \alpha \qquad (14.14)$$

$$c_d = 2 \sin^3 \alpha \qquad (14.15)$$

Finally, from the geometry of Fig. 14.11, the lift-to-drag ratio is given by

$$\frac{L}{D} = \cot \alpha \qquad (14.16)$$

[Note: Equation (14.16) is a general result for inviscid supersonic or hypersonic flow over a flat plate. For such flows, the resultant aerodynamic force is the normal force N. From the geometry shown in Fig. 14.11, the resultant aerodynamic force makes the angle α with respect to lift, and clearly, from the right triangle between L, D, and N, we have $L/D = \cot \alpha$. Hence, Eq. (14.16) is not limited to newtonian theory.]

The aerodynamic characteristics of a flat plate based on newtonian theory are shown in Fig. 14.12. Although an infinitely thin flat plate, by itself, is not a practical aerodynamic configuration, its aerodynamic behavior at hypersonic speeds is consistent with some of the basic characteristics of other hypersonic shapes. For example, consider the variation of c_l shown in Fig. 14.12. First, note that, at a small angle of attack, say, in the range of α from 0 to 15°, c_l varies in a *nonlinear* fashion; i.e., the slope of the lift curve is *not* constant. This is in direct contrast to the subsonic case we studied in Chaps. 4 and 5, where the lift coefficient for an airfoil or a finite wing was shown to vary linearly with α at small angles of attack, up to the stalling angle. This is also in contrast with the results from linearized supersonic theory as itemized in Sec. 12.3, leading to Eq. (12.23) where a linear variation of c_l with α for a flat plate is indicated. However, the nonlinear lift curve shown in Fig. 14.12 is *totally consistent* with the results discussed in Sec. 11.3, where hypersonic flow was shown to be governed by the

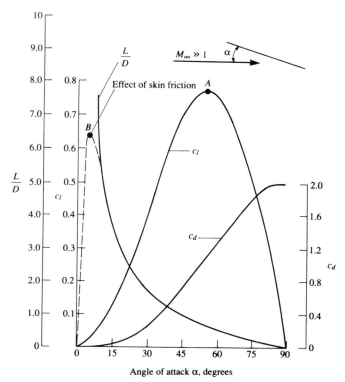

FIGURE 14.12
Aerodynamic properties of a flat plate based on newtonian theory.

nonlinear velocity potential equation, *not* by the linear equation expressed by Eq. (11.8). In that section, we noted that both transonic and hypersonic flow cannot be described by a linear theory—both these flows are inherently nonlinear regimes, even for low angles of attack. Once again, the flat-plate lift curve shown in Fig. 14.12 certainly demonstrates the nonlinearity of hypersonic flow.

Also, note from the lift curve in Fig. 14.12 that c_l first increases as α increases, reaches a maximum value at an angle of attack of about 55° (54.7° to be exact), and then decreases, reaching zero at $\alpha = 90°$. However, the attainment of $c_{l,max}$ (point A) in Fig. 14.12 is *not* due to any viscous, separated flow phenomenon analogous to that which occurs in subsonic flow. Rather, in Fig. 14.12, the attainment of a maximum c_l is purely a geometric effect. To understand this better, return to Fig. 14.11. Note that, as α increases, C_p continues to increase via the newtonian expression

$$C_p = 2 \sin^2 \alpha$$

That is, C_p reaches a maximum value at $\alpha = 90°$. In turn, the *normal force, N,* shown in Fig. 14.11 continues to increase as α increases, also reaching a maximum value at $\alpha = 90°$. However, recall from Eq. (14.12) that the vertical component of the aerodynamic force, namely, the lift, is given by

$$L = N \cos \alpha \tag{14.17}$$

Hence, as α increases to 90°, although N continues to increase monotonically, the value of L reaches a maximum value around $\alpha = 55°$, and then begins to decrease at higher α due to the effect of the cosine variation shown in Eq. (14.17)—strictly a geometric effect. In other words, in Fig. 14.11, although N is increasing with α, it eventually becomes inclined enough relative to the vertical that its vertical component (lift) begins to decrease gradually. It is interesting to note that a large number of practical hypersonic configurations achieve a maximum C_L at an angle of attack in the neighborhood of that shown in Fig. 14.12, namely, around 55°.

The maximum lift coefficient for a hypersonic flat plate, and the angle at which it occurs, is easily quantified using newtonian theory. Differentiating Eq. (14.14) with respect to α, and setting the derivative equal to zero (for the condition of maximum c_l), we have

$$\frac{dc_l}{d\alpha} = (2 \sin^2 2)(-\sin \alpha) + 4 \cos^2 \alpha \sin \alpha = 0$$

or

$$\sin^2 \alpha = 2 \cos^2 \alpha = 2(1 - \sin^2 \alpha)$$

or

$$\sin^2 \alpha = \frac{2}{3}$$

Hence,

$$\alpha = 54.7°$$

This is the angle of attack at which c_l is a maximum. The maximum value of c_l is obtained by substituting the above result for α into Eq. (14.14):

$$c_{l,\max} = 2 \sin^2 (54.7°) \cos (54.7°) = 0.77$$

Note, although c_l increases over a wide latitude in the angle of attack (c_l increases in the range from $\alpha = 0$ to $\alpha = 54.7°$), its rate of increase is small (i.e., the effective lift slope is small). In turn, the resulting value for the maximum lift coefficient is relatively small—at least in comparison to the much higher $c_{l,\max}$ values associated with low-speed flows (see, e.g., Figs. 4.10 and 4.28). Returning to Fig. 14.12, we now note the *precise* values associated with the peak of the lift curve (point A), namely, the peak value of c_l is 0.77, and it occurs at an angle of attack of 54.7°.

Examining the variation of drag coefficient, c_d, in Fig. 14.12, we note that it monotonically increases from zero at $\alpha = 0$ to a maximum of 2 at $\alpha = 90°$. The newtonian result for drag is essentially *wave drag* at hypersonic speeds because we are dealing with an inviscid flow, hence no friction drag. The variation of c_d with α for the low angle of attack in Fig. 14.12 is essentially a *cubic* variation, in contrast to the result from linearized supersonic flow, namely, Eq. (12.24), which shows that c_d varies as the square angle of attack. The hypersonic result that c_d varies as α^3 is easily obtained from Eq. (14.15), which for small α becomes

$$c_d = 2\alpha^3 \tag{14.18}$$

The variation of the lift-to-drag ratio as predicted by newtonian theory is also shown in Fig. 14.12. The solid curve is the pure newtonian result; it shows that L/D is infinitely large at $\alpha = 0$ and monotonically decreases to zero at $\alpha = 90°$. The infinite value of L/D at $\alpha = 0$ is purely fictional—it is due to the neglect of skin friction. When skin friction is added to the picture, denoted by the dashed curve in Fig. 14.12, L/D reaches a maximum value at a small angle of attack (point B in Fig. 14.12) and is equal to zero at $\alpha = 0$. (At $\alpha = 0$, no lift is produced, but there is a finite drag due to friction; hence, $L/D = 0$ at $\alpha = 0$.)

Let us examine the conditions associated with $(L/D)_{\max}$ more closely. The value of $(L/D)_{\max}$ and the angle of attack at which it occurs (i.e., the coordinates of point B in Fig. 14.12) are strictly a function of the zero-lift drag coefficient, denoted by $c_{d,0}$. The zero-lift drag coefficient is simply due to the integrated effect of skin friction over the plate surface at zero angle of attack. At small angles of attack, the skin friction exerted on the plate should be essentially that at zero angle of attack; hence, we can write the total drag coefficient [referring to Eq. (14.15)] as

$$c_d = 2 \sin^3 \alpha + c_{d,0} \tag{14.19}$$

Furthermore, when α is small, we can write Eqs. (14.14) and (14.19) as

$$c_l = 2\alpha^2 \tag{14.20}$$

and

$$c_d = 2\alpha^3 + c_{d,0} \tag{14.21}$$

Dividing Eq. (14.20) by (14.21), we have

$$\frac{c_l}{c_d} = \frac{2\alpha^2}{2\alpha^3 + c_{d,0}} \tag{14.22}$$

The conditions associated with maximum lift-to-drag ratio can be found by differentiating Eq. (14.22) and setting the result equal to zero:

$$\frac{d(c_l/c_d)}{d\alpha} = \frac{(2\alpha^3 + c_{d,0})4\alpha - 2\alpha^2(6\alpha^2)}{(2\alpha^3 + c_{d,0})} = 0$$

or

$$8\alpha^4 + 4\alpha c_{d,0} - 12\alpha^4 = 0$$

$$4\alpha^3 = 4c_{d,0}$$

Hence,

$$\boxed{\alpha = (c_{d,0})^{1/3}} \tag{14.23}$$

Substituting Eq. (14.23) into Eq. (14.21), we obtain

$$\left(\frac{c_l}{c_d}\right)_{max} = \frac{2(c_{d,0})^{2/3}}{2c_{d,0} + c_{d,0}} = \frac{2/3}{(c_{d,0})^{1/3}}$$

or

$$\left(\frac{L}{D}\right)_{max} = \left(\frac{c_l}{c_d}\right)_{max} = \boxed{0.67/(c_{d,0})^{1/3}} \tag{14.24}$$

Equations (14.23) and (14.24) are important results. They clearly state that the coordinates of the maximum L/D point in Fig. 14.12, when friction is included (point B in Fig. 14.12), are strictly a function of $c_{d,0}$. In particular, note the expected trend that $(L/D)_{max}$ decreases as $c_{d,0}$ increases—the higher the friction drag, the lower is L/D. Also, the angle of attack at which maximum L/D occurs increases as $c_{d,0}$ increases. There is yet another interesting aerodynamic condition that holds at $(L/D)_{max}$, derived as follows. Substituting Eq. (14.23) into (14.21), we have

$$c_d = 2c_{d,0} + c_{d,0} = 3c_{d,0} \tag{14.25}$$

Since the total drag coefficient is the sum of the wave-drag coefficient, $c_{d,w}$, and the friction drag coefficient, $c_{d,0}$, we can write

$$c_d = c_{d,w} + c_{d,0} \tag{14.26}$$

However, at the point of maximum L/D (point B in Fig. 14.12), we know from Eq. (14.25) that $c_d = 3c_{d,0}$. Substituting this result into Eq. (14.26), we obtain

$$3c_{d,0} = c_{d,w} + c_{d,0}$$

or

$$\boxed{c_{d,w} = 2c_{d,0}} \tag{14.27}$$

This clearly shows that, for the hypersonic flat plate using newtonian theory, at the flight condition associated with maximum lift-to-drag ratio, wave drag is twice the friction drag.

This brings to an end our short discussion of the lift and drag of wings at hypersonic speeds as modeled by the newtonian flat-plate problem. The quantitative and qualitative results presented here are reasonable representations of the hypersonic aerodynamic characteristics of a number of practical hypersonic vehicles; the flat-plate problem is simply a straightforward way of demonstrating these characteristics.

14.5 HYPERSONIC SHOCK-WAVE RELATIONS AND ANOTHER LOOK AT NEWTONIAN THEORY

The basic oblique shock relations are derived and discussed in Chap. 9. These are exact shock relations and hold for all Mach numbers greater than unity, supersonic or hypersonic (assuming a calorically perfect gas). However, some interesting approximate and simplified forms of these shock relations are obtained in the limit of a high Mach number. These limiting forms are called the hypersonic shock relations; they are obtained below.

Consider the flow through a straight oblique shock wave. See, e.g., Fig. 9.1. Upstream and downstream conditions are denoted by subscripts 1 and 2, respectively. For a calorically perfect gas, the classical results for changes across the shock are given in Chap. 9. To begin with, the exact oblique shock relation for pressure ratio across the wave is given by Eq. (9.16). Since $M_{n,1} = M_1 \sin \beta$, this equation becomes

Exact:
$$\frac{p_2}{p_1} = 1 + \frac{2\gamma}{\gamma+1}(M_1^2 \sin^2 \beta - 1) \tag{14.28}$$

where β is the wave angle. In the limit as M_1 goes to infinity, the term $M_1^2 \sin^2 \beta \gg 1$, and hence Eq. (14.28) becomes

as $M_1 \to \infty$:
$$\boxed{\frac{p_2}{p_1} = \frac{2\gamma}{\gamma+1} M_1^2 \sin^2 \beta} \tag{14.29}$$

In a similar vein, the density and temperature ratios are given by Eqs. (9.15) and (9.17), respectively. These can be written as follows:

Exact:
$$\frac{\rho_2}{\rho_1} = \frac{(\gamma+1)M_1^2 \sin^2 \beta}{(\gamma-1)M_1^2 \sin^2 \beta + 2} \tag{14.30}$$

as $M_1 \to \infty$:
$$\boxed{\frac{\rho_2}{\rho_1} = \frac{\gamma+1}{\gamma-1}} \tag{14.31}$$

$$\frac{T_2}{T_1} = \frac{(p_2/p_1)}{(\rho_2/\rho_1)} \qquad \text{(from the equation of state: } p = \rho RT\text{)}$$

as $M_1 \to \infty$:

$$\frac{T_2}{T_1} = \frac{2\gamma(\gamma-1)}{(\gamma+1)^2} M_1^2 \sin^2 \beta \qquad (14.32)$$

The relationship among Mach number M_1, shock angle β, and deflection angle θ is expressed by the so-called θ-β-M relation given by Eq. (9.23), repeated below:

Exact:
$$\tan \theta = 2 \cot \beta \left[\frac{M_1^2 \sin^2 \beta - 1}{M_1^2(\gamma + \cos 2\beta) + 2} \right] \qquad (9.23)$$

This relation is plotted in Fig. 9.7, which is a standard plot of the wave angle versus the deflection angle, with the Mach number as a parameter. Returning to Fig. 9.7, we note that, in the hypersonic limit, where θ is small, β is also small. Hence, in this limit, we can insert the usual small-angle approximation into Eq. (9.23):

$$\sin \beta \approx \beta$$

$$\cos 2\beta \approx 1$$

$$\tan \theta \approx \sin \theta \approx \theta$$

resulting in

$$\theta = \frac{2}{\beta} \left[\frac{M_1^2 \beta^2 - 1}{M_1^2(\gamma + 1) + 2} \right] \qquad (14.33)$$

Applying the high Mach number limit to Eq. (14.33), we have

$$\theta = \frac{2}{\beta} \left[\frac{M_1^2 \beta^2}{M_1^2(\gamma + 1)} \right] \qquad (14.34)$$

In Eq. (14.34), M_1 cancels, and we finally obtain in both the small-angle and hypersonic limits,
as $M_1 \to \infty$ and θ, hence β is small:

$$\frac{\beta}{\theta} = \frac{\gamma + 1}{2} \qquad (14.35)$$

Note that, for $\gamma = 1.4$,

$$\beta = 1.2\theta \qquad (14.36)$$

It is interesting to observe that, in the hypersonic limit for a slender wedge, the wave angle is only 20 percent larger than the wedge angle—a graphic demonstration of a thin shock layer in hypersonic flow.

In aerodynamics, pressure distributions are usually quoted in terms of the nondimensional pressure coefficient C_p, rather than the pressure itself. The pressure coefficient is defined as

$$C_p = \frac{p_2 - p_1}{q_1} \tag{14.37}$$

where p_1 and q_1 are the upstream (freestream) static pressure and dynamic pressure, respectively. Recall from Sec. 11.3 that Eq. (14.37) can also be written as Eq. (11.22), repeated below:

$$C_p = \frac{2}{\gamma M_1^2} \left(\frac{p_2}{p_1} - 1 \right) \tag{11.22}$$

Combining Eqs. (11.22) and (14.28), we obtain an exact relation for C_p behind an oblique shock wave as follows:

Exact:
$$C_p = \frac{4}{\gamma + 1} \left(\sin^2 \beta - \frac{1}{M_1^2} \right) \tag{14.38}$$

In the hypersonic limit,

as $M_1 \to \infty$:
$$\boxed{C_p = \left(\frac{4}{\gamma + 1} \right) \sin^2 \beta} \tag{14.39}$$

Pause for a moment, and review our results. We have obtained limiting forms of the oblique shock equations, valid for the case when the upstream Mach number becomes very large. These limiting forms, called the hypersonic shock-wave relations, are given by Eqs. (14.29), (14.31), and (14.32), which yield the pressure ratio, density ratio, and temperature ratio across the shock when $M_1 \to \infty$. Furthermore, in the limit of both $M_1 \to \infty$ and small θ (such as the hypersonic flow over a slender airfoil shape), the limiting relation for the wave angle as a function of the deflection angle is given by Eq. (14.35). Finally, the form of the pressure coefficient behind an oblique shock is given in the limit of hypersonic Mach numbers by Eq. (14.39). Note that the limiting forms of the equations are always simpler than their corresponding exact counterparts.

In terms of actual *quantitative* results, it is always recommended that the *exact* oblique shock equations be used, even for hypersonic flow. This is particularly convenient because the exact results are tabulated in App. B. The value of the relations obtained in the hypersonic limit (as described above) is more for theoretical analysis rather than for the calculation of actual numbers. For example, in this section, we use the hypersonic shock relations to shed additional understanding of the significance of newtonian theory. In the next section, we will examine the same hypersonic shock relations to demonstrate the principle of Mach number independence.

Newtonian theory was discussed at length in Secs. 14.3 and 14.4. For our purposes here, temporarily discard any thoughts of newtonian theory, and simply recall the exact oblique shock relation for C_p as given by Eq. (14.38), repeated below (with freestream conditions now denoted by a subscript ∞ rather than a subscript 1, as used earlier):

$$C_p = \frac{4}{\gamma+1}\left[\sin^2\beta - \frac{1}{M_\infty^2}\right] \tag{14.38}$$

Equation (14.39) gave the limiting value of C_p as $M_\infty \to \infty$, repeated below:

as $M_\infty \to \infty$:
$$C_p \to \frac{4}{\gamma+1}\sin^2\beta \tag{14.39}$$

Now take the additional limit of $\gamma \to 1.0$. From Eq. (14.39), in both limits as $M_\infty \to \infty$ and $\gamma \to 1.0$, we have

$$C_p \to 2\sin^2\beta \tag{14.40}$$

Equation (14.40) is a result from exact oblique shock theory; it has nothing to do with newtonian theory (as yet). Keep in mind that β in Eq. (14.40) is the wave angle, not the deflection angle.

Let us go further. Consider the exact oblique shock relation for ρ/ρ_∞, given by Eq. (14.30), repeated below (again with a subscript ∞ replacing the subscript 1):

$$\frac{\rho_2}{\rho_\infty} = \frac{(\gamma+1)M_\infty^2\sin^2\beta}{(\gamma-1)M_\infty^2\sin^2\beta + 2} \tag{14.41}$$

Equation (14.31) was obtained as the limit where $M_\infty \to \infty$, namely,

as $M_\infty \to \infty$:
$$\frac{\rho_2}{\rho_\infty} \to \frac{\gamma+1}{\gamma-1} \tag{14.42}$$

In the additional limit as $\gamma \to 1$, we find

as $\gamma \to 1$ and $M_\infty \to \infty$:
$$\boxed{\frac{\rho_2}{\rho_\infty} \to \infty} \tag{14.43}$$

i.e., the density behind the shock is infinitely large. In turn, mass flow considerations then dictate that the shock wave is coincident with the body surface. This is further substituted by Eq. (14.35), which is good for $M_\infty \to \infty$ and small deflection angles:

$$\frac{\beta}{\theta} \to \frac{\gamma+1}{2} \tag{14.35}$$

In the additional limit as $\gamma \to 1$, we have

as $\gamma \to 1$ and $M_\infty \to \infty$ and
θ and β are small:

$$\boxed{\beta = \theta}$$

i.e., the shock wave lies on the body. In light of this result, Eq. (14.40) is written as

$$\boxed{C_p = 2 \sin^2 \theta} \tag{14.44}$$

Examine Eq. (14.44). It is a result from exact oblique shock theory, taken in the combined limit of $M_\infty \to \infty$ and $\gamma \to 1$. However, it is also precisely the newtonian results given by Eq. (14.4). Therefore, we make the following conclusion. The closer the actual hypersonic flow problem is to the limits $M_\infty \to \infty$ and $\gamma \to 1$, the closer it should be physically described by newtonian flow. In this regard, we gain a better appreciation of the true significance of newtonian theory. We can also state that the application of newtonian theory to practical hypersonic flow problems, where γ is always greater than unity, is theoretically not proper, and the agreement that is frequently obtained with experimental data has to be viewed as somewhat fortuitous. Nevertheless, the simplicity of newtonian theory along with its (sometimes) reasonable results (no matter how fortuitous) has made it a widely used and popular engineering method for the estimation of surface pressure distributions, hence lift- and wave-drag coefficients, for hypersonic bodies.

14.6 MACH NUMBER INDEPENDENCE

Examine again the hypersonic shock-wave relation for pressure ratio as given by Eq. (14.29); note that, as the freestream Mach number approaches infinity, the pressure ratio itself also becomes infinitely large. On the other hand, the *pressure coefficient* behind the shock, given in the hypersonic limit by Eq. (14.39), *is a constant value at high values of the Mach number*. This hints strongly of a situation where certain aspects of a hypersonic flow do not depend on Mach number, as long as the Mach number is sufficiently high. This is a type of "independence" from the Mach number, formally called the *hypersonic Mach number independence principle*. From the above argument, C_p clearly demonstrates Mach number independence. In turn, recall that the lift- and wave-drag coefficients for a body shape are obtained by integrating the local C_p, as shown by Eqs. (1.15), (1.16), (1.18), and (1.19). These equations demonstrate that, since C_p is independent of the Mach number at high values of M_∞, the lift and drag coefficients are also Mach number independent. Keep in mind that these conclusions are theoretical, based on the limiting form of the hypersonic shock relations.

Let us examine an example that clearly illustrates the Mach number independence principle. In Fig. 14.13, the pressure coefficients for a 15° half-angle wedge

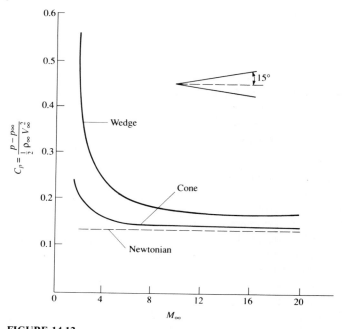

FIGURE 14.13
Comparison between newtonian and exact results for the pressure coefficient on a sharp wedge and a sharp cone. Also, an illustration of Mach number independence at high Mach numbers.

and a 15° half-angle cone are plotted versus freestream Mach number for $\gamma = 1.4$. The exact wedge results are obtained from Eq. (14.38), and the exact cone results are obtained from the solution of the classical Taylor-Maccoll equation. (See Ref. 21 for a detailed discussion of the solution of the supersonic flow over a cone. There, you will find that the governing continuity, momentum, and energy equations for a conical flow cascade into a single differential equation called the Taylor-Maccoll equation. In turn, this equation allows the *exact* solution of this conical flow field.) Both sets of results are compared with newtonian theory, $C_p = 2 \sin^2 \theta$, shown as the dashed line in Fig. 14.13. This comparison demonstrates two general aspects of newtonian results:

1. The accuracy of the newtonian results improves as M_∞ increases. This is to be expected from our discussion in Sec. 14.5. Note from Fig. 14.13 that below $M_\infty = 5$ the newtonian results are not even close, but the comparison becomes much closer as M_∞ increases above 5.
2. Newtonian theory is usually more accurate for three-dimensional bodies (e.g., the cone) than for two-dimensional bodies (e.g., the wedge). This is clearly evident in Fig. 14.13 where the newtonian result is much closer to the cone results than to the wedge results.

However, more to the point of Mach number independence, Fig. 14.13 also shows the following trends. For both the wedge and the cone, the exact results show that, at low supersonic Mach numbers, C_p decreases rapidly as M_∞ is increased. However, at hypersonic speeds, the rate of decrease diminishes considerably, and C_p appears to reach a plateau as M_∞ becomes large; i.e., C_p becomes relatively independent of M_∞ at high values of the Mach number. This is the essence of the Mach number independence principle; at high Mach numbers, certain aerodynamic quantities such as pressure coefficient, lift- and wave-drag coefficients, and flow-field structure (such as shock-wave shapes and Mach wave patterns) become essentially independent of the Mach number. Indeed, newtonian theory gives results that are totally independent of the Mach number, as clearly demonstrated by Eq. (14.4).

Another example of Mach number independence is shown in Fig. 14.14. Here, the measured drag coefficients for spheres and for a large-angle cone cylinder are plotted versus the Mach number, cutting across the subsonic, supersonic, and hypersonic regimes. Note the large drag rise in the subsonic regime associated with the drag-divergence phenomenon near Mach 1 and the decrease in C_D in the supersonic regime beyond Mach 1. Both of these variations are expected and well understood. For our purposes in the present section, note, in particular, the variation of C_D in the hypersonic regime; for both the sphere and cone cylinder, C_D approaches a plateau and becomes relatively independent of the Mach number as M_∞ becomes large. Note also that the sphere data appear to achieve "Mach number independence" at lower Mach numbers than the cone cylinder.

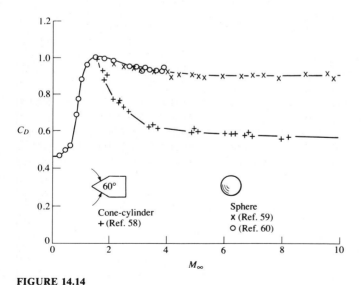

FIGURE 14.14
Drag coefficient for a sphere and a cone cylinder from ballistic range measurements; an example of Mach number independence at hypersonic speeds. (*From Ref. 61.*)

Keep in mind from the above analysis that it is the nondimensional variables that become Mach number independent. Some of the dimensional variables, such as p, are not Mach number independent; indeed, $p \to \infty$ and $M_\infty \to \infty$.

Finally, the Mach number independence principle is well grounded mathematically. The governing inviscid flow equations (the Euler equations) expressed in terms of suitable nondimensional quantities, along with the boundary conditions for the limiting hypersonic case, do not have the Mach number appearing in them—hence, by definition, the solution to these equations is independent of the Mach number. See Refs. 21 and 55 for more details.

14.7 SUMMARY

Only a few of the basic elements of hypersonic flow are presented here, with special emphasis on newtonian flow results. Useful information on hypersonic flows can be extracted from such results. We have derived the basic newtonian sine-squared law:

$$C_p = 2 \sin^2 \theta \tag{14.4}$$

and used this result to treat the case of a hypersonic flat plate in Sec. 14.4. We also obtained the limiting form of the oblique shock relations as $M_\infty \to \infty$, i.e., the hypersonic shock relations. From these relations, we were able to examine the significance of newtonian theory more thoroughly, namely, Eq. (14.4) becomes an exact relation for a hypersonic flow in the combined limit of $M_\infty \to \infty$ and $\gamma \to 1$. Finally, these hypersonic shock relations illustrate the existence of the Mach number independence principle.

PROBLEMS

14.1. Repeat Prob. 9.13 using
   (a) Newtonian theory
   (b) Modified newtonian theory
   Compare these results with those obtained from exact shock-expansion theory (Prob. 9.13). From this comparison, what comments can you make about the accuracy of newtonian and modified newtonian theories at low supersonic Mach numbers?

14.2. Consider a flat plate at $\alpha = 20°$ in a Mach 20 freestream. Using straight newtonian theory, calculate the lift and wave-drag coefficients. Compare these results with exact shock-expansion theory.

PART
IV

VISCOUS
FLOW

In Part IV, we deal with flows that are dominated by viscosity and thermal conduction—viscous flows. We will treat both incompressible and compressible viscous flows.

INTRODUCTION TO THE FUNDAMENTAL PRINCIPLES AND EQUATIONS OF VISCOUS FLOW

I do not see then, I admit, how one can explain the resistance of fluids by the theory in a satisfactory manner. It seems to me on the contrary that this theory, dealt with and studied with profound attention gives, at least in most cases, resistance absolutely zero: a singular paradox which I leave to geometricians to explain.

Jean LeRond d'Alembert, 1768

15.1 INTRODUCTION

In the above quotation, the "theory" referred to by d'Alembert is inviscid, incompressible flow theory; we have seen in Chap. 3 that such theory leads to a prediction of zero drag on a closed two-dimensional body—this is d'Alembert's paradox. In reality, there is always a finite drag on any body immersed in a moving fluid. Our earlier predictions of zero drag are a result of the inadequacy of the theory rather than some fluke of nature. With the exception of induced drag and supersonic wave drag, which can be obtained from inviscid theory, the calculation of all other forms of drag must explicitly take into account the presence of viscosity, which has not been included in our previous inviscid analyses. The purpose of the remaining chapters in this book is to discuss the basic aspects of viscous flows, thus "rounding out" our overall presentation of the fundamentals of aerodynamics. In so doing, we address the predictions of aerodynamic drag and aerodynamic heating. To help put our current discussion in perspective, return to the block diagram of flow categories given in Fig. 1.31. All of our previous discussions have focused on blocks D, E, and F—inviscid, incompressible and compressible flows. Now, for the remaining four chapters, we move to the left branch in Fig. 1.31, and deal with blocks C, E, and F—*viscous*, incompressible and compressible flows.

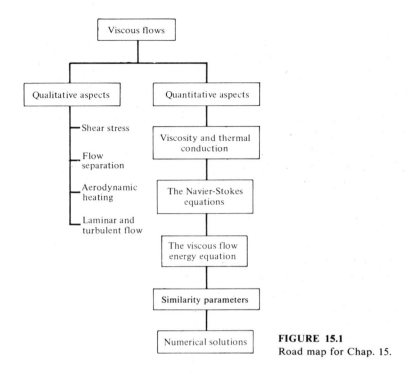

FIGURE 15.1
Road map for Chap. 15.

Our treatment of viscous flows will be intentionally brief—our purpose is to present enough of the fundamental concepts and equations to give you the flavor of viscous flows. A thorough presentation of viscous flow theory would double the size of this book (at the very least) and is clearly beyond our scope. A study of viscous flow is an essential part of any serious study of aerodynamics. Many books have been exclusively devoted to the presentation of viscous flows; Refs. 42 and 43 are two good examples. You are encouraged to examine these references closely.

The road map for the present chapter is given in Fig. 15.1. Our course is to first examine some qualitative aspects of viscous flow as shown on the left branch of Fig. 15.1. Then we quantify some of these aspects as given on the right branch. In the process, we obtain the governing equations for a general viscous flow—in particular, the Navier-Stokes equations (the momentum equations) and the viscous flow energy equation. Finally, we examine a numerical solution to these equations.

15.2 QUALITATIVE ASPECTS OF VISCOUS FLOW

What is a viscous flow? *Answer*: A flow where the effects of viscosity, thermal conduction, and mass diffusion are important. The phenomenon of mass diffusion is important in a gas with gradients in its chemical species, e.g., the flow of air over a surface through which helium is being injected or the chemically reacting flow through a jet engine or over a high-speed reentry body. In this book, we are not concerned with the effects of diffusion, and therefore we treat a viscous flow as one where only viscosity and thermal conduction are important.

First, consider the influence of viscosity. Imagine two solid surfaces slipping over each other, such as this book being pushed across a table. Clearly, there will be a frictional force between these objects which will retard their relative motion. The same is true for the flow of a fluid over a solid surface; the influence of friction between the surface and the fluid adjacent to the surface acts to create a frictional force which retards the relative motion. This has an effect on both the surface and the fluid. The surface feels a "tugging" force in the direction of the flow, tangential to the surface. This tangential force per unit area is defined as the *shear stress* τ, first introduced in Sec. 1.5 and illustrated in Fig. 15.2. As an equal and opposite reaction, the fluid adjacent to the surface feels a retarding force which decreases its local flow velocity, as shown in insert *a* of Fig. 15.2. Indeed, the influence of friction is to create $V = 0$ right at the body surface—this is called the *no-slip* condition which dominates viscous flow. In any real continuum fluid flow over a solid surface, the flow velocity is zero at the surface. Just above the surface, the flow velocity is finite, but retarded, as shown in insert *a*. If n represents the coordinate normal to the surface, then in the region near the surface, $V = V(n)$, where $V = 0$ at $n = 0$, and V increases as n increases. The plot of V versus n as shown in insert *a* is called a *velocity profile*. Clearly, the region of flow near the surface has velocity gradients, $\partial V/\partial n$, which are due to the frictional force between the surface and the fluid.

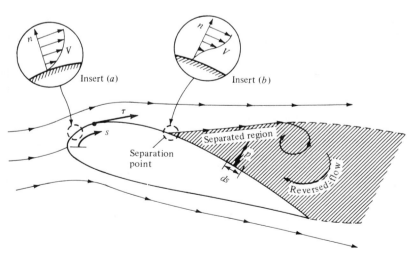

FIGURE 15.2
Effect of viscosity on a body in a moving fluid: shear stress and separated flow.

In addition to the generation of shear stress, friction also plays another (but related) role in dictating the flow over the body in Fig. 15.2. Consider a fluid element moving in the viscous flow near a surface, as sketched in Fig. 15.3. Assume that the flow is in its earliest moments of being started. At the station s_1, the velocity of the fluid element is V_1. Assume that the flow over the surface produces an increasing pressure distribution in the flow direction; i.e., assume $p_3 > p_2 > p_1$. Such a region of increasing pressure is called an *adverse pressure gradient*. Now follow the fluid element as it moves downstream. The motion of the element is already retarded by the effect of friction; in addition, it must work its way along the flow against an increasing pressure, which tends to further reduce its velocity. Consequently, at station 2 along the surface, its velocity V_2 is less than V_1. As the fluid element continues to move downstream, it may completely "run out of steam," come to a stop, and then, under the action of the adverse pressure gradient, actually reverse its direction and start moving back upstream. This "reversed flow" is illustrated at station s_3 in Fig. 15.3, where the fluid element is now moving upstream at the velocity V_3. The picture shown in Fig. 15.3 is meant to show the flow details very near the surface at the very initiation of the flow. In the bigger picture of this flow at later times shown in Fig. 15.2, the consequence of such reversed-flow phenomena is to cause the flow to *separate from the surface* and create a large wake of recirculating flow downstream of the surface. The point of separation on the surface in Fig. 15.2 occurs where $\partial V/\partial n = 0$ at the surface, as sketched in insert b of Fig. 15.2. Beyond this point, reversed flow occurs. Therefore, in addition to the generation of shear stress, the influence of friction can cause the flow over a body to separate from

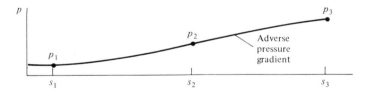

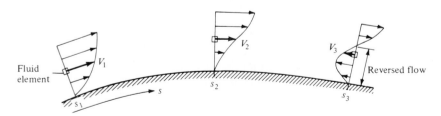

FIGURE 15.3
Separated flow induced by an adverse pressure gradient. This picture corresponds to the early evolution of the flow; once the flow separates from the surface between points 2 and 3, the fluid element shown at s_3 is in reality different from that shown at s_1 and s_2 because the primary flow moves away from the surface, as shown in Fig. 15.2.

the surface. When such separated flow occurs, the pressure distribution over the surface is greatly altered. The primary flow over the body in Fig. 15.2 no longer sees the complete body shape; rather, it sees the body shape upstream of the separation point, but downstream of the separation point it sees a greatly deformed "effective body" due to the large separated region. The net effect is to create a pressure distribution over the actual body surface which results in an integrated force in the flow direction, i.e., a drag. To see this more clearly, consider the pressure distribution over the upper surface of the body as sketched in Fig. 15.4. If the flow were attached, the pressure over the downstream portion of the body would be given by the dashed curve. However, for separated flow, the pressure over the downstream portion of the body is smaller, given by the solid curve in Fig. 15.4. Now return to Fig. 15.2. Note that the pressure over the upper rearward surface contributes a force in the negative drag direction; i.e., p acting over the element of surface ds shown in Fig. 15.2 has a horizontal component in the upstream direction. If the flow were inviscid, subsonic, and attached and the body were two-dimensional, the forward-acting components of the pressure distribution shown in Fig. 15.2 would exactly cancel the rearward-acting components due to the pressure distribution over other parts of the body such that the net, integrated pressure distribution would give zero drag. This would be d'Alembert's paradox discussed in Chap. 3. However, for the viscous, separated flow, we see that p is reduced in the separated region; hence, it can no longer fully cancel the pressure distribution over the remainder of the body. The net result is the production of drag; this is called the *pressure drag due to flow separation* and is denoted by D_p.

In summary, we see that the effects of viscosity are to produce two types of drag as follows:

D_f is the skin friction drag, i.e., the component in the drag direction of the integral of the shear stress τ over the body.

D_p is the pressure drag due to separation, i.e., the component in the drag direction of the integral of the pressure distribution over the body.

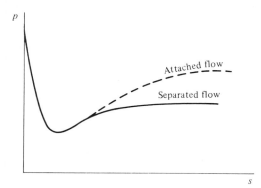

FIGURE 15.4
Schematic of the pressure distributions for attached and separated flow over the upper surface of the body illustrated in Fig. 15.2.

D_p is sometimes called *form drag.* The sum $D_f + D_p$ is called the *profile drag* of a two-dimensional body. For a three-dimensional body such as a complete airplane, the sum $D_f + D_p$ is frequently called *parasite drag.* (See Ref. 2 for a more extensive discussion of the classification of different drag contributions.)

The occurrence of separated flow over an aerodynamic body not only increases the drag but also results in a substantial loss of lift. Such separated flow is the cause of airfoil stall as discussed in Sec. 4.3. For these reasons, the study, understanding, and prediction of separated flow is an important aspect of viscous flow.

Let us turn our attention to the influence of thermal conduction—another overall physical characteristic of viscous flow in addition to friction. Again, let us draw an analogy from two solid bodies slipping over each other, such as the motion of this book over the top of a table. If we would press hard on the book, and vigorously rub it back and forth over the table, the cover of the book as well as the table top would soon become warm. Some of the energy we expend in pushing the book over the table will be dissipated by friction, and this shows up as a form of heating of the bodies. The same phenomenon occurs in the flow of a fluid over a body. The moving fluid has a certain amount of kinetic energy; in the process of flowing over a surface, the flow velocity is decreased by the influence of friction, as discussed earlier, and hence the kinetic energy is decreased. This lost kinetic energy reappears in the form of internal energy of the fluid, hence causing the temperature to rise. This phenomenon is called *viscous dissipation* within the fluid. In turn, when the fluid temperature increases, there is an overall temperature difference between the warmer fluid and the cooler body. We know from experience that heat is transferred from a warmer body to a cooler body; therefore, heat will be transferred from the warmer fluid to the cooler surface. This is the mechanism of *aerodynamic heating* of a body. Aerodynamic heating becomes more severe as the flow velocity increases, because more kinetic energy is dissipated by friction, and hence the overall temperature difference between the warm fluid and the cool surface increases. As discussed in Chap. 14, at hypersonic speeds, aerodynamic heating becomes a dominant aspect of the flow.

All the aspects discussed above—shear stress, flow separation, aerodynamic heating, etc.—are dominated by a single major question in viscous flow, namely, Is the flow laminar or turbulent? Consider the viscous flow over a surface as sketched in Fig. 15.5. If the path lines of various fluid elements are smooth and

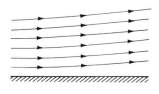

(a) Laminar flow

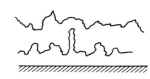

(b) Turbulent flow

FIGURE 15.5
Path lines for laminar and turbulent flows.

regular, as shown in Fig. 15.5a, the flow is called *laminar flow*. In contrast, if the motion of a fluid element is very irregular and tortuous, as shown in Fig. 15.5b, the flow is called *turbulent flow*. Because of the agitated motion in a turbulent flow, the higher-energy fluid elements from the outer regions of the flow are pumped close to the surface. Hence, the average flow velocity near a solid surface is larger for a turbulent flow in comparison with laminar flow. This comparison is shown in Fig. 15.6, which gives velocity profiles for laminar and turbulent flow. Note that immediately above the surface, the turbulent flow velocities are much larger than the laminar values. If $(\partial V/\partial n)_{n=0}$ denotes the velocity gradient at the surface, we have

$$\left[\left(\frac{\partial V}{\partial n}\right)_{n=0}\right]_{\text{turbulent}} > \left[\left(\frac{\partial V}{\partial n}\right)_{n=0}\right]_{\text{laminar}}$$

Because of this difference, the frictional effects are more severe for a turbulent flow; both the shear stress and aerodynamic heating are larger for the turbulent flow in comparison with laminar flow. However, turbulent flow has a major redeeming value; because the energy of the fluid elements close to the surface is larger in a turbulent flow, a turbulent flow does not separate from the surface as readily as a laminar flow. If the flow over a body is turbulent, it is less likely to separate from the body surface, and if flow separation does occur, the separated region will be smaller. As a result, the pressure drag due to flow separation, D_p, will be smaller for turbulent flow.

The above discussion points out one of the great compromises in aerodynamics. For the flow over a body, is laminar or turbulent flow preferable? There is no pat answer; it depends on the shape of the body. In general, if the body is slender, as sketched in Fig. 15.7a, the friction drag D_f is much greater than D_p. For this case, because D_f is smaller for laminar than for turbulent flow, laminar flow is desirable for slender bodies. In contrast, if the body is blunt, as sketched

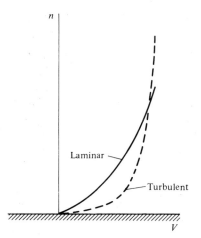

FIGURE 15.6
Schematic of velocity profiles for laminar and turbulent flows.

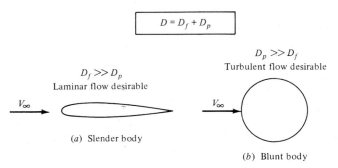

FIGURE 15.7
Drag on slender and blunt bodies.

in Fig. 15.7b, D_p is much greater than D_f. For this case, because D_p is smaller for turbulent than for laminar flow, turbulent flow is desirable for blunt bodies. The above comments are not all-inclusive; they simply state general trends, and for any given body, the aerodynamic virtues of laminar versus turbulent flow must always be assessed.

Although, from the above discussion, laminar flow is preferable for some cases, and turbulent flow for other cases, in reality we have little control over what actually happens. Nature makes the ultimate decision as to whether a flow will be laminar or turbulent. There is a general principle in nature that a system, when left to itself, will always move toward its state of maximum disorder. To bring order to the system, we generally have to exert some work on the system or expend energy in some manner. (This analogy can be carried over to daily life; a room will soon become cluttered and disordered unless we exert some effort to keep it clean.) Since turbulent flow is much more "disordered" than laminar flow, nature will always favor the occurrence of turbulent flow. Indeed, in the vast majority of practical aerodynamic problems, turbulent flow is usually present.

Let us examine this phenomenon in more detail. Consider the viscous flow over a flat plate, as sketched in Fig. 15.8. The flow immediately upstream of the

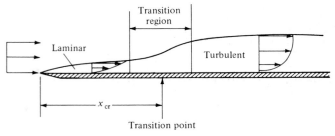

FIGURE 15.8
Transition from laminar to turbulent flow.

leading edge is uniform at the freestream velocity. However, downstream of the leading edge, the influence of friction will begin to retard the flow adjacent to the surface, and the extent of this retarded flow will grow higher above the plate as we move downstream, as shown in Fig. 15.8. To begin with, the flow just downstream of the leading edge will be laminar. However, after a certain distance, instabilities will appear in the laminar flow; these instabilities rapidly grow, causing transition to turbulent flow. The transition from laminar to turbulent flow takes place over a finite region, as sketched in Fig. 15.8. However, for purposes of analysis, we frequently model the transition region as a single point, called the *transition point*, upstream of which the flow is laminar and downstream of which the flow is turbulent. The distance from the leading edge to the transition point is denoted by x_{cr}. The value of x_{cr} depends on a whole host of phenomena. For example, some characteristics which encourage transition from laminar to turbulent flow, and hence reduce x_{cr}, are:

1. *Increased surface roughness.* Indeed, to promote turbulent flow over a body, rough grit can be placed on the surface near the leading edge to "trip" the laminar flow into turbulent flow. This is a frequently used technique in wind-tunnel testing. Also, the dimples on the surface of a golf ball are designed to encourage turbulent flow, thus reducing D_p. In contrast, in situations where we desire large regions of laminar flow, such as the flow over the NACA six-series laminar-flow airfoils, the surface should be as smooth as possible. The main reason why such airfoils do not produce in actual flight the large regions of laminar flow observed in the laboratory is that manufacturing irregularities and bug spots (believe it or not) roughen the surface and promote early transition to turbulent flow.

2. *Increased turbulence in the freestream.* This is particularly a problem in wind-tunnel testing; if two wind tunnels have different levels of freestream turbulence, then data generated in one tunnel are not repeatable in the other.

3. *Adverse pressure gradients.* In addition to causing flow-field separation as discussed earlier, an adverse pressure gradient strongly favors transition to turbulent flow. In contrast, strong favorable pressure gradients (where p decreases in the downstream direction) tend to preserve initially laminar flow.

4. *Heating of the fluid by the surface.* If the surface temperature is warmer than the adjacent fluid, such that heat is transferred to the fluid from the surface, the instabilities in the laminar flow will be amplified, thus favoring early transition. In contrast, a cold wall will tend to encourage laminar flow.

There are many other parameters which influence transition; see Ref. 42 for a more extensive discussion. Among these are the similarity parameters of the flow, principally Mach number and Reynolds number. High values of M_∞ and low values of Re tend to encourage laminar flow; hence, for high-altitude hypersonic flight, laminar flow can be quite extensive. The Reynolds number itself is a dominant factor in transition to turbulent flow. Referring to Fig. 15.8, we define a *critical* Reynolds number, Re_{cr}, as

$$\text{Re}_{cr} \equiv \frac{\rho_\infty V_\infty x_{cr}}{\mu_\infty}$$

The value of Re_{cr} for a given body under specified conditions is difficult to predict; indeed, the analysis of transition is still a very active area of modern aerodynamic research. As a rule of thumb in practical applications, we frequently take $\text{Re}_{cr} \approx$ 500,000; if the flow at a given x station is such that $\text{Re} = \rho_\infty V_\infty x/\mu_\infty$ is considerably below 500,000, then the flow at that station is most likely laminar, and if the value of Re is much larger than 500,000, then the flow is most likely turbulent.

To obtain a better feeling for Re_{cr}, let us imagine that the flat plate in Fig. 15.8 is a wind-tunnel model. Assume that we carry out an experiment under standard sea level conditions [$\rho_\infty = 1.23 \text{ kg/m}^3$ and $\mu_\infty = 1.79 \times 10^{-5} \text{ kg/(m} \cdot \text{s)}$] and *measure* x_{cr} for a certain freestream velocity; e.g., say that $x_{cr} = 0.05$ m when $V_\infty = 120$ m/s. In turn, this measured value of x_{cr} determines the measured Re_{cr} as

$$\text{Re}_{cr} = \frac{\rho_\infty V_\infty x_{cr}}{\mu_\infty} = \frac{1.23(120)(0.05)}{1.79 \times 10^{-5}} = 412,000$$

Hence, for the given flow conditions and the surface characteristics of the flat plate, transition will occur whenever the local Re exceeds 412,000. For example, if we double V_∞, that is, $V_\infty = 240$ m/s, then we will observe transition to occur at $x_{cr} = 0.05/2 = 0.025$ m, such that Re_{cr} remains the same value of 412,000.

This brings to an end our introductory qualitative discussion of viscous flow. The physical principles and trends discussed in this section are very important, and you should study them carefully and feel comfortable with them before progressing further.

15.3 VISCOSITY AND THERMAL CONDUCTION

The basic physical phenomena of viscosity and thermal conduction in a fluid are due to the transport of momentum and energy via random molecular motion. Each molecule in a fluid has momentum and energy, which it carries with it when it moves from one location to another in space before colliding with another molecule. The transport of molecular momentum gives rise to the macroscopic effect we call viscosity, and the transport of molecular energy gives rise to the macroscopic effect we call thermal conduction. This is why viscosity and thermal conduction are labeled as *transport phenomena*. A study of these transport phenomena at the molecular level is part of kinetic theory, which is beyond the scope of this book. Instead, in this section we simply state the macroscopic results of such molecular motion.

Consider the flow sketched in Fig. 15.9. For simplicity, we consider a one-dimensional shear flow, i.e., a flow with horizontal streamlines in the x direction but with gradients in the y direction of velocity, $\partial u/\partial y$, and temperature, $\partial T/\partial y$. Consider a plane ab perpendicular to the y axis, as shown in Fig. 15.9.

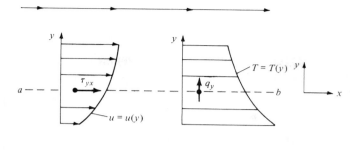

FIGURE 15.9
Relationship of shear stress and thermal conduction to velocity and temperature gradients, respectively.

The shear stress exerted on plane ab by the flow is denoted by τ_{yx} and is proportional to the velocity gradient in the y direction, $\tau_{yx} \propto \partial u / \partial y$. The constant of proportionality is defined as the *viscosity coefficient* μ. Hence,

$$\tau_{yx} = \mu \frac{\partial u}{\partial y} \tag{15.1}$$

The subscripts on τ_{yx} denote that the shear stress is acting in the x direction and is being exerted on a plane perpendicular to the y axis. The velocity gradient $\partial u / \partial y$ is also taken perpendicular to this plane, i.e., in the y direction. The dimensions of μ are mass/length × time, as originally stated in Sec. 1.7 and as can be seen from Eq. (15.1). In addition, the time rate of heat conducted per unit area across plane ab in Fig. 15.9 is denoted by $\dot{q}_y$ and is proportional to the temperature gradient in the y direction, $\dot{q}_y \propto \partial T / \partial y$. The constant of proportionality is defined as the *thermal conductivity* k. Hence,

$$\dot{q}_y = -k \frac{\partial T}{\partial y} \tag{15.2}$$

where the minus sign accounts for the fact that the heat is transferred from a region of high temperature to a region of lower temperature; i.e., $\dot{q}_y$ is in the opposite direction of the temperature gradient. The dimensions of k are mass × length/$(s^2 \cdot K)$, which can be obtained from Eq. (15.2) keeping in mind that $\dot{q}_y$ is energy per second per unit area.

Both μ and k are physical properties of the fluid and, for most normal situations, are functions of temperature only. A conventional relation for the temperature variation of μ for air is given by Sutherland's law,

$$\frac{\mu}{\mu_0} = \left(\frac{T}{T_0} \right)^{3/2} \frac{T_0 + 110}{T + 110} \tag{15.3}$$

where T is in kelvin and μ_0 is a reference viscosity at a reference temperature, T_0. For example, if we choose reference conditions to be standard sea level values, then $\mu_0 = 1.7894 \times 10^{-5}$ kg/(m · s) and $T_0 = 288.16$ K. The temperature variation of k is analogous to Eq. (15.3) because the results of elementary kinetic theory show that $k \propto \mu c_p$; for air at standard conditions,

$$k = 1.45 \mu c_p \tag{15.4}$$

where $c_p = 1000$ J/(kg · K).

Equations (15.3) and (15.4) are only approximate and do not hold at high temperatures. They are given here as representative expressions which are handy to use. For any detailed viscous flow calculation, you should consult the published literature for more precise values of μ and k.

In order to simplify our introduction of the relation between shear stress and viscosity, we considered the case of a one-dimensional shear flow in Fig. 15.9. In this picture, the y and z components of velocity, v and w, respectively, are zero. However, in a general three-dimensional flow, u, v, and w are finite, and this requires a generalization of our treatment of stress in the fluid. Consider the fluid element sketched in Fig. 15.10. In a three-dimensional flow, each face of the fluid element experiences both tangential and normal stresses. For example, on face $abcd$, τ_{xy} and τ_{xz} are the tangential stresses, and τ_{xx} is the normal stress. As before, the nomenclature τ_{ij} denotes a stress in the j direction exerted on a plane perpendicular to the i axis. Similarly, on face $abfe$, we have the tangential stresses τ_{yx} and τ_{yz}, and the normal stress τ_{yy}. On face $adge$, we have the tangential stresses τ_{zx} and τ_{zy}, and the normal stress τ_{zz}. Now recall the discussion in the last part of Sec. 2.12 concerning the *strain* of a fluid element, i.e., the change in the angle κ shown in Fig. 2.28. What is the force which causes this deformation shown in Fig. 2.28? Returning to Fig. 15.10, we have to say that the strain is caused by the tangential shear stress. However, in contrast to solid mechanics where stress is proportional to strain, in fluid mechanics the stress is proportional

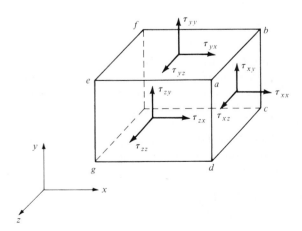

FIGURE 15.10
Shear and normal stresses caused by viscous action on a fluid element.

to the *time rate of strain*. The time rate of strain in the xy plane was given in Sec. 2.12 as Eq. (2.126a):

$$\varepsilon_{xy} = \frac{\partial v}{\partial x} + \frac{\partial u}{\partial y} \tag{2.126a}$$

Examining Fig. 15.10, the strain in the xy plane must be carried out by τ_{xy} and τ_{yx}. Moreover, we assume that moments on the fluid element in Fig. 15.10 are zero; hence, $\tau_{xy} = \tau_{yx}$. Finally, from the above, we know that $\tau_{xy} = \tau_{yx} \propto \varepsilon_{xy}$. The proportionality constant is the viscosity coefficient μ. Hence, from Eq. (2.126a), we have

$$\tau_{xy} = \tau_{yx} = \mu \left(\frac{\partial v}{\partial x} + \frac{\partial u}{\partial y} \right) \tag{15.5}$$

which is a generalization of Eq. (15.1), extended to the case of multidimensional flow. For the shear stresses in the other planes, Eqs. (2.126b and c) yield

$$\tau_{yz} = \tau_{zy} = \mu \left(\frac{\partial w}{\partial y} + \frac{\partial v}{\partial z} \right) \tag{15.6}$$

and

$$\tau_{zx} = \tau_{xz} = \mu \left(\frac{\partial u}{\partial z} + \frac{\partial w}{\partial x} \right) \tag{15.7}$$

The normal stresses τ_{xx}, τ_{yy}, and τ_{zz} shown in Fig. 15.10 may at first seem strange. In our previous treatments of inviscid flow, the only force normal to a surface in a fluid is the pressure force. However, if the gradients in velocity $\partial u / \partial x$, $\partial v / \partial y$, and $\partial w / \partial z$ are *extremely large* on the faces of the fluid element, there can be a meaningful viscous-induced normal force on each face which acts *in addition to* the pressure. These normal stresses act to compress or expand the fluid element, hence changing its volume. Recall from Sec. 2.12 that the derivatives $\partial u / \partial x$, $\partial v / \partial y$, and $\partial w / \partial z$ are related to the dilatation of a fluid element, i.e., to $\nabla \cdot \mathbf{V}$. Hence, the normal stresses should in turn be related to these derivatives. Indeed, it can be shown that

$$\tau_{xx} = \lambda (\nabla \cdot \mathbf{V}) + 2\mu \frac{\partial u}{\partial x} \tag{15.8}$$

$$\tau_{yy} = \lambda (\nabla \cdot \mathbf{V}) + 2\mu \frac{\partial v}{\partial y} \tag{15.9}$$

$$\tau_{zz} = \lambda (\nabla \cdot \mathbf{V}) + 2\mu \frac{\partial w}{\partial z} \tag{15.10}$$

In Eqs. (15.8) to (15.10), λ is called the *bulk viscosity coefficient*. In 1845, the Englishman George Stokes hypothesized that

$$\lambda = -\tfrac{2}{3}\mu \tag{15.11}$$

To this day, the correct expression for the bulk viscosity is still somewhat controversial, and so we continue to use the above expression given by Stokes.

Once again, the normal stresses are important only where the derivatives $\partial u/\partial x$, $\partial v/\partial y$, and $\partial w/\partial z$ are very large. For most practical flow problems, τ_{xx}, τ_{yy}, and τ_{zz} are small, and hence the uncertainty regarding λ is essentially an academic question. An example where the normal stress is important is inside the internal structure of a shock wave. Recall that, in real life, shock waves have a finite but small thickness. If we consider a normal shock wave across which large changes in velocity occur over a small distance (typically 10^{-5} cm), then clearly $\partial u/\partial x$ will be very large, and τ_{xx} becomes important inside the shock wave.

To this point in our discussion, the transport coefficients μ and k have been considered molecular phenomena, involving the transport of momentum and energy by random molecular motion. This molecular picture prevails in a laminar flow. The values of μ and k are physical properties of the fluid; i.e., their values for different gases can be found in standard reference sources, such as the *Handbook of Chemistry and Physics* (The Chemical Rubber Co.). In contrast, for a turbulent flow the transport of momentum and energy can also take place by random motion of large turbulent eddies, or globs of fluid. This turbulent transport gives rise to effective values of viscosity and thermal conductivity defined as *eddy viscosity*, ε, and *eddy thermal conductivity*, κ, respectively. (Please do not confuse this use of the symbols ε and κ with the time rate of strain and strain itself, as used earlier.) These turbulent transport coefficients ε and κ can be much larger (typically 10 to 100 times larger) than the respective molecular values μ and k. Moreover, ε and κ predominantly depend on characteristics of the flow field, such as velocity gradients; they are not just a molecular property of the fluid such as μ and k. The proper calculation of ε and κ for a given flow has remained a state-of-the-art research question for the past 80 years; indeed, the attempt to model the complexities of turbulence by defining an eddy viscosity and thermal conductivity is even questionable. The details and basic understanding of turbulence remain one of the greatest unsolved problems in physics today. For our purpose here, we simply adopt the ideas of eddy viscosity and eddy thermal conductivity, and for the transport of momentum and energy in a turbulent flow, we replace μ and k in Eqs. (15.1) to (15.10) by the combination $\mu + \varepsilon$ and $k + \kappa$; i.e.,

$$\tau_{yx} = (\mu + \varepsilon)\left(\frac{\partial v}{\partial x} + \frac{\partial u}{\partial y}\right)$$

$$q_y = -(k + \kappa)\frac{\partial T}{\partial y}$$

An example of the calculation of ε and κ is as follows. In 1925, Prandtl suggested that

$$\varepsilon = \rho l^2 \left|\frac{\partial u}{\partial y}\right| \tag{15.12}$$

for a flow where the dominant velocity gradient is in the y direction. In Eq.

(15.12), l is called the *mixing length,* which is different for different applications; it is an empirical constant which must be obtained from experiment. Indeed, *all* turbulence models require the input of empirical data; no self-contained purely theoretical turbulence model exists today. Prandtl's mixing length theory, embodied in Eq. (15.12), is a simple relation which appears to be adequate for a number of engineering problems. For these reasons, the mixing length model for ε has been used extensively since 1925. In regard to κ, a relation similar to Eq. (15.4) can be assumed (using 1.0 for the constant); i.e.,

$$\kappa = \varepsilon c_p \qquad (15.13)$$

The above comments on eddy viscosity and thermal conductivity are purely introductory. The modern aerodynamicist has a whole stable of turbulence models to choose from, and before tackling the analysis of a turbulent flow, you should be familiar with the modern approaches described in such books as Refs. 42 to 45.

15.4 THE NAVIER-STOKES EQUATIONS

In Chap. 2, Newton's second law was applied to obtain the fluid-flow momentum equation in both integral and differential forms. In particular, recall Eqs. (2.104a to c), where the influence of viscous forces was expressed simply by the generic terms $(\mathscr{F}_x)_{\text{viscous}}$, $(\mathscr{F}_y)_{\text{viscous}}$, and $(\mathscr{F}_z)_{\text{viscous}}$. The purpose of this section is to obtain the analogous forms of Eqs. (2.104a to c) where the viscous forces are expressed explicitly in terms of the appropriate flow-field variables. The resulting equations are called the *Navier-Stokes equations*—probably the most pivotal equations in all of theoretical fluid dynamics.

In Sec. 2.3, we discussed the philosophy behind the derivation of the governing equations, namely, certain physical principles are applied to a suitable *model* of the fluid flow. Moreover, we saw that such a model could be either a finite control volume (moving or fixed in space) or an infinitesimally small element (moving or fixed in space). In Chap. 2, we chose the fixed, finite control volume for our model and obtained integral forms of the continuity, momentum, and energy equations directly from this model. Then, indirectly, we went on to extract partial differential equations from the integral forms. Before progressing further, it would be wise for you to review these matters from Chap. 2.

For the sake of variety, let us not use the fixed, finite control volume employed in Chap. 2; rather, in this section, let us adopt an infinitesimally small moving fluid element of fixed mass as our model of the flow, as sketched in Fig. 15.11. To this model let us apply Newton's second law in the form $\mathbf{F} = m\mathbf{a}$. Moreover, for the time being consider only the x component of Newton's second law:

$$F_x = ma_x \qquad (15.14)$$

In Eq. (15.14), F_x is the sum of all the body and surface forces acting on the fluid element in the x direction. Let us ignore body forces; hence, the net force acting on the element in Fig. 15.11 is simply due to the pressure and viscous stress distributions over the surface of the element. For example, on face *abcd*, the only force in the x direction is that due to shear stress, $\tau_{yx} \, dx \, dz$. Face *efgh* is a distance dy above face *abcd*; hence, the shear force in the x direction on face *efgh* is $[\tau_{yx} + (\partial \tau_{yx}/\partial y) \, dy] \, dx \, dz$. Note the directions of the shear stress on faces *abcd* and *efgh*; on the bottom face, τ_{yx} is to the left (the negative x direction), whereas on the top face, $\tau_{yx} + (\partial \tau_{yx}/\partial y) \, dy$ is to the right (the positive x direction). These directions are due to the convention that positive increases in all three components of velocity, u, v, and w, occur in the positive directions of the axes. For example, in Fig. 15.11, u increases in the positive y direction. Therefore, concentrating on face *efgh*, u is higher just above the face than on the face; this causes a "tugging" action which tries to pull the fluid element in the positive x direction (to the right) as shown in Fig. 15.11. In turn, concentrating on face *abcd*, u is lower just beneath the face than on the face; this causes a retarding or dragging action on the fluid element, which acts in the negative x direction (to the left), as shown in Fig. 15.11. The directions of all the other viscous stresses shown in Fig. 15.11, including τ_{xx}, can be justified in a like fashion. Specifically, on face *dcgh*, τ_{zx} acts in the negative x direction, whereas on face *abfe*, $\tau_{zx} + (\partial \tau_{zx}/\partial z) \, dz$ acts in the positive x direction. On face *adhe*, which is perpendicular to the x axis, the only forces in the x direction are the pressure force $p \, dy \, dz$, which always acts in the direction *into* the fluid element, and $\tau_{xx} \, dy \, dz$, which is

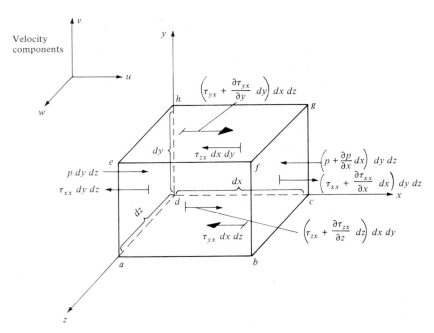

FIGURE 15.11
Infinitesimally small, moving fluid element. Only the forces in the x direction are shown.

in the negative x direction. In Fig. 15.11, the reason why τ_{xx} on face *adhe* is to the left hinges on the convention mentioned earlier for the direction of increasing velocity. Here, by convention, a positive increase in u takes place in the positive x direction. Hence, the value of u just to the left of face *adhe* is smaller than the value of u on the face itself. As a result, the viscous action of the normal stress acts as a "suction" on face *adhe*; i.e., there is a dragging action toward the left that wants to retard the motion of the fluid element. In contrast, on face *bcgf*, the pressure force $[\,p+(\partial p/\partial x)\,dx\,]\,dy\,dz$ presses inward on the fluid element (in the negative x direction), and because the value of u just to the right of face *bcgf* is larger than the value of u on the face, there is a "suction" due to the viscous normal stress which tries to pull the element to the right (in the positive x direction) with a force equal to $[\tau_{xx}+(\partial\tau_{xx}/\partial x)\,dx\,]\,dy\,dz$.

Return to Eq. (15.14). Examining Fig. 15.11 in light of our previous discussion, we can write for the net force in the x direction acting on the fluid element:

$$F_x=\left[\,p-\left(p+\frac{\partial p}{\partial x}\,dx\right)\right]dy\,dz+\left[\left(\tau_{xx}+\frac{\partial\tau_{xx}}{\partial x}\,dx\right)-\tau_{xx}\right]dy\,dz$$

$$+\left[\left(\tau_{yx}+\frac{\partial\tau_{yx}}{\partial y}\,dy\right)-\tau_{yx}\right]dx\,dz+\left[\left(\tau_{zx}+\frac{\partial\tau_{zx}}{\partial z}\,dz\right)-\tau_{zx}\right]dx\,dy$$

or $\qquad F_x=\left(-\frac{\partial p}{\partial x}+\frac{\partial\tau_{xx}}{\partial x}+\frac{\partial\tau_{yx}}{\partial y}+\frac{\partial\tau_{zx}}{\partial z}\right)dx\,dy\,dz$ $\qquad\qquad$ (15.15)

Equation (15.15) represents the left-hand side of Eq. (15.14). Considering the right-hand side of Eq. (15.14), recall that the mass of the fluid element is fixed and is equal to

$$m=\rho\,dx\,dy\,dz \qquad\qquad (15.16)$$

Also, recall that the acceleration of the fluid element is the time rate of change of its velocity. Hence, the component of acceleration in the x direction, denoted by a_x, is simply the time rate of change of u; since we are following a moving fluid element, this time rate of change is given by the *substantial derivative* (see Sec. 2.9 for a review of the meaning of substantial derivative). Thus,

$$a_x=\frac{Du}{Dt} \qquad\qquad (15.17)$$

Combining Eqs. (15.14) to (15.17), we obtain

$$\boxed{\rho\frac{Du}{Dt}=-\frac{\partial p}{\partial x}+\frac{\partial\tau_{xx}}{\partial x}+\frac{\partial\tau_{yx}}{\partial y}+\frac{\partial\tau_{zx}}{\partial z}} \qquad\qquad (15.18a)$$

which is the x component of the momentum equation for a viscous flow. In a similar fashion, the y and z components can be obtained as

$$\rho \frac{Dv}{Dt} = -\frac{\partial p}{\partial y} + \frac{\partial \tau_{xy}}{\partial x} + \frac{\partial \tau_{yy}}{\partial y} + \frac{\partial \tau_{zy}}{\partial z} \tag{15.18b}$$

$$\rho \frac{Dw}{Dt} = -\frac{\partial p}{\partial z} + \frac{\partial \tau_{xz}}{\partial x} + \frac{\partial \tau_{yz}}{\partial y} + \frac{\partial \tau_{zz}}{\partial z} \tag{15.18c}$$

Equations (15.18a to c) are the momentum equations in the x, y, and z directions, respectively. They are scalar equations and are called the *Navier-Stokes* equations in honor of two men—the Frenchman M. Navier and the Englishman G. Stokes—who independently obtained the equations in the first half of the nineteenth century.

With the expressions for $\tau_{xy} = \tau_{yx}$, $\tau_{yz} = \tau_{zy}$, $\tau_{zx} = \tau_{xz}$, τ_{xx}, τ_{yy}, and τ_{zz} from Eqs. (15.5) to (15.10), the Navier-Stokes equations, Eqs. (15.18a to c), can be written as

$$\rho \frac{\partial u}{\partial t} + \rho u \frac{\partial u}{\partial x} + \rho v \frac{\partial u}{\partial y} + \rho w \frac{\partial u}{\partial z} = -\frac{\partial p}{\partial x} + \frac{\partial}{\partial x}\left(\lambda \nabla \cdot \mathbf{V} + 2\mu \frac{\partial u}{\partial x}\right)$$
$$+ \frac{\partial}{\partial y}\left[\mu\left(\frac{\partial v}{\partial x} + \frac{\partial u}{\partial y}\right)\right] + \frac{\partial}{\partial z}\left[\mu\left(\frac{\partial u}{\partial z} + \frac{\partial w}{\partial x}\right)\right] \tag{15.19a}$$

$$\rho \frac{\partial v}{\partial t} + \rho u \frac{\partial v}{\partial x} + \rho v \frac{\partial v}{\partial y} + \rho w \frac{\partial v}{\partial z} = -\frac{\partial p}{\partial y} + \frac{\partial}{\partial x}\left[\mu\left(\frac{\partial v}{\partial x} + \frac{\partial u}{\partial y}\right)\right]$$
$$+ \frac{\partial}{\partial y}\left(\lambda \nabla \cdot \mathbf{V} + 2\mu \frac{\partial v}{\partial y}\right) + \frac{\partial}{\partial z}\left[\mu\left(\frac{\partial w}{\partial y} + \frac{\partial v}{\partial z}\right)\right] \tag{15.19b}$$

$$\rho \frac{\partial w}{\partial t} + \rho u \frac{\partial w}{\partial x} + \rho v \frac{\partial w}{\partial y} + \rho w \frac{\partial w}{\partial z} = -\frac{\partial p}{\partial z} + \frac{\partial}{\partial x}\left[\mu\left(\frac{\partial u}{\partial z} + \frac{\partial w}{\partial x}\right)\right]$$
$$+ \frac{\partial}{\partial y}\left[\mu\left(\frac{\partial w}{\partial y} + \frac{\partial v}{\partial z}\right)\right] + \frac{\partial}{\partial z}\left(\lambda \nabla \cdot \mathbf{V} + 2\mu \frac{\partial w}{\partial z}\right) \tag{15.19c}$$

Equations (15.19a to c) represent the complete Navier-Stokes equations for an unsteady, compressible, three-dimensional viscous flow. To analyze incompressible viscous flow, Eqs. (15.19a to c) and the continuity equation [say, Eq. (2.43)] are sufficient. However, for a compressible flow, we need an additional equation, namely, the energy equation to be discussed in the next section.

In the above form, the Navier-Stokes equations are suitable for the analysis of laminar flow. For a turbulent flow, the flow variables in Eqs. (15.19*a* to *c*) can be assumed to be time-mean values over the turbulent fluctuations, and μ can be replaced by $\mu + \varepsilon$, as discussed in Sec. 15.3. For more details, see Refs. 42 and 43.

15.5 THE VISCOUS FLOW ENERGY EQUATION

The energy equation was derived in Sec. 2.7, where the first law of thermodynamics was applied to a finite control volume fixed in space. The resulting integral form of the energy equation was given by Eq. (2.86), and differential forms were obtained in Eqs. (2.87) and (2.105). In these equations, the influence of viscous effects was expressed generically by such terms as $\dot{Q}'_{\text{viscous}}$ and $\dot{W}'_{\text{viscous}}$. It is recommended that you review Sec. 2.7 before progressing further.

In the present section, we derive the energy equation for a viscous flow using as our model an infinitesimal moving fluid element. This will be in keeping with our derivation of the Navier-Stokes equation in Sec. 15.4, where the infinitesimal element was shown in Fig. 15.11. In the process, we obtain explicit expressions for $\dot{Q}'_{\text{viscous}}$ and $\dot{W}'_{\text{viscous}}$ in terms of the flow-field variables. That is, we once again derive Eq. (2.105), except the viscous terms are now displayed in detail.

Consider again the moving fluid element shown in Fig. 15.11. To this element, apply the first law of thermodynamics, which states

$$
\begin{array}{ccccc}
\text{Rate of change} & \text{net flux of} & \text{rate of work} & & \\
\text{of energy inside} & = \text{heat into} & + \text{ done on element} & & \\
\text{fluid element} & \text{element} & \text{due to pressure and} & & \\
& & \text{stress forces on surface} & &
\end{array}
$$

or
$$
A \quad = \quad B \quad + \quad C \tag{15.20}
$$

where A, B, and C denote the respective terms above.

Let us first evaluate C; i.e., let us obtain an expression for the rate of work done on the moving fluid element due to the pressure and stress forces on the surface of the element. (Note that we are neglecting body forces in this derivation.) These surface forces are illustrated in Fig. 15.11, which for simplicity shows only the forces in the x direction. Recall from Sec. 2.7 that the rate of doing work by a force exerted on a moving body is equal to the product of the force and the component of velocity in the direction of the force. Hence, the rate of work done on the moving fluid element by the forces in the x direction shown in Fig. 15.11 is simply the x component of velocity, u, multiplied by the forces; e.g., on face *abcd* the rate of work done by $\tau_{yx}\, dx\, dz$ is $u\tau_{yx}\, dx\, dz$, with similar expressions for the other faces. To emphasize these energy considerations, the moving fluid element is redrawn in Fig. 15.12, where the rate of work done on each face by

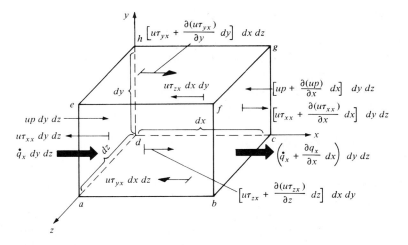

FIGURE 15.12
Energy fluxes associated with an infinitesimally small, moving fluid element. For simplicity, only the fluxes in the x direction are shown.

forces in the x direction is shown explicitly. Study this figure carefully, referring frequently to its companion in Fig. 15.11, until you feel comfortable with the work terms given in each face. To obtain the *net* rate of work done on the fluid element by the forces in the x direction, note that forces in the positive x direction do positive work and that forces in the negative x direction do negative work. Hence, comparing the pressure forces on faces *adhe* and *bcgf* in Fig. 15.12, the net rate of work done by pressure in the x direction is

$$\left[up - \left(up + \frac{\partial(up)}{\partial x}\, dx \right) \right] dy\, dz = -\frac{\partial(up)}{\partial x}\, dx\, dy\, dz$$

Similarly, the net rate of work done by the shear stresses in the x direction on faces *abcd* and *efgh* is

$$\left[\left(u\tau_{yx} + \frac{\partial(u\tau_{yx})}{\partial y}\, dy \right) - u\tau_{yx} \right] dx\, dz = \frac{\partial(u\tau_{yx})}{\partial y}\, dx\, dy\, dz$$

Considering all the forces shown in Fig. 15.12, the net rate of work done on the moving fluid element is simply

$$\left[-\frac{\partial(up)}{\partial x} + \frac{\partial(u\tau_{xx})}{\partial x} + \frac{\partial(u\tau_{yx})}{\partial y} + \frac{\partial(u\tau_{zx})}{\partial z} \right] dx\, dy\, dz$$

The above expression considers only forces in the x direction. When the forces in the y and z directions are also included, similar expressions are obtained (draw some pictures and obtain these expressions yourself). In total, the net rate of work done on the moving fluid element is the sum of all contributions in the

x, y, and z directions; this is denoted by C is Eq. (15.20) and is given by

$$C = \left[-\left(\frac{\partial(up)}{\partial x} + \frac{\partial(vp)}{\partial y} + \frac{\partial(wp)}{\partial z} \right) + \frac{\partial(u\tau_{xx})}{\partial x} + \frac{\partial(u\tau_{yx})}{\partial y} \right.$$

$$+ \frac{\partial(u\tau_{zx})}{\partial z} + \frac{\partial(v\tau_{xy})}{\partial x} + \frac{\partial(v\tau_{yy})}{\partial y} + \frac{\partial(v\tau_{zy})}{\partial z} + \frac{\partial(w\tau_{xz})}{\partial x}$$

$$\left. + \frac{\partial(w\tau_{yz})}{\partial y} + \frac{\partial(w\tau_{zz})}{\partial z} \right] dx \, dy \, dz \tag{15.21}$$

Note in Eq. (15.21) that the term in large parentheses is simply $\nabla \cdot p\mathbf{V}$.

Let us turn our attention to B in Eq. (15.20), i.e., the net flux of heat into the element. This heat flux is due to (1) volumetric heating such as absorption or emission of radiation and (2) heat transfer across the surface due to temperature gradients (i.e., thermal conduction). Let us treat the volumetric heating the same as was done in Sec. 2.7; i.e., define $\dot{q}$ as the rate of volumetric heat addition per unit mass. Noting that the mass of the moving fluid element in Fig. 15.12 is $\rho \, dx \, dy \, dz$, we obtain

$$\text{Volumetric heating of element} = \rho\dot{q} \, dx \, dy \, dz \tag{15.22}$$

Thermal conduction was discussed in Sec. 15.3. In Fig. 15.12, the heat transferred by thermal conduction into the moving fluid element across face *adhe* is $\dot{q}_x \, dy \, dz$, and the heat transferred out of the element across face *bcgf* is $[\dot{q}_x + (\partial\dot{q}_x/\partial x) \, dx] \, dy \, dz$. Thus, the net heat transferred in the x direction into the fluid element by thermal conduction is

$$\left[\dot{q}_x - \left(\dot{q}_x + \frac{\partial\dot{q}_x}{\partial x} dx \right) \right] dy \, dz = -\frac{\partial\dot{q}_x}{\partial x} dx \, dy \, dz$$

Taking into account heat transfer in the y and z directions across the other faces in Fig. 15.12, we obtain

$$\begin{array}{l} \text{Heating of fluid element} \\ \text{by thermal conduction} \end{array} = -\left(\frac{\partial\dot{q}_x}{\partial x} + \frac{\partial\dot{q}_y}{\partial y} + \frac{\partial\dot{q}_z}{\partial z} \right) dx \, dy \, dz \tag{15.23}$$

The term B in Eq. (15.20) is the sum of Eqs. (15.22) and (15.23). Also, recalling that thermal conduction is proportional to temperature gradient, as exemplified by Eq. (15.2), we have

$$B = \left[\rho\dot{q} + \frac{\partial}{\partial x}\left(k\frac{\partial T}{\partial x} \right) + \frac{\partial}{\partial y}\left(k\frac{\partial T}{\partial y} \right) + \frac{\partial}{\partial z}\left(k\frac{\partial T}{\partial z} \right) \right] dx \, dy \, dz \tag{15.24}$$

Finally, the term A in Eq. (15.20) denotes the time rate of change of energy of the fluid element. In Sec. 2.7, we stated that the energy of a moving fluid per unit mass is the sum of the internal and kinetic energies, e.g., $e + V^2/2$. Since we are following a moving fluid element, the time rate of change of energy per unit mass is given by the substantial derivative (see Sec. 2.9). Since the mass of the fluid element is $\rho\, dx\, dy\, dz$, we have

$$A = \rho \frac{D}{Dt}\left(e + \frac{V^2}{2}\right) dx\, dy\, dz \tag{15.25}$$

The final form of the energy equation for a viscous flow is obtained by substituting Eqs. (15.21), (15.24), and (15.25) into Eq. (15.20), obtaining

$$
\begin{aligned}
\rho \frac{D(e + V^2/2)}{Dt} = {}& \rho\dot{q} + \frac{\partial}{\partial x}\left(k\frac{\partial T}{\partial x}\right) + \frac{\partial}{\partial y}\left(k\frac{\partial T}{\partial y}\right) \\
& + \frac{\partial}{\partial z}\left(k\frac{\partial T}{\partial z}\right) - \nabla \cdot p\mathbf{V} + \frac{\partial(u\tau_{xx})}{\partial x} + \frac{\partial(u\tau_{yx})}{\partial y} \\
& + \frac{\partial(u\tau_{zx})}{\partial z} + \frac{\partial(v\tau_{xy})}{\partial x} + \frac{\partial(v\tau_{yy})}{\partial y} + \frac{\partial(v\tau_{zy})}{\partial z} \\
& + \frac{\partial(w\tau_{xz})}{\partial x} + \frac{\partial(w\tau_{yz})}{\partial y} + \frac{\partial(w\tau_{zz})}{\partial z}
\end{aligned}
\tag{15.26}
$$

Equation (15.26) is the general energy equation for unsteady, compressible, three-dimensional, viscous flow. Compare Eq. (15.26) with Eq. (2.105); the viscous terms are now explicitly spelled out in Eq. (15.26). [Note that the body force term in Eq. (15.26) has been neglected.] Moreover, the normal and shear stresses that appear in Eq. (15.26) can be expressed in terms of the velocity field via Eqs. (15.5) to (15.10). This substitution will not be made here because the resulting equation would simply occupy too much space.

Reflect on the viscous flow equations obtained in this chapter—the Navier-Stokes equations given by Eqs. (15.19a to c) and the energy equation given by Eq. (15.26). These equations are obviously more complex than the inviscid flow equations dealt with in previous chapters. This underscores the fact that viscous flows are inherently more difficult to analyze than inviscid flows. This is why, in the study of aerodynamics, the student is first introduced to the concepts associated with inviscid flow. Moreover, this is why we attempt to model a number of practical aerodynamic problems in real life as inviscid flows—simply to allow a reasonable analysis of such flows. However, there are many aerodynamic problems, especially those involving the prediction of drag and flow separation, which must take into account viscous effects. For the analysis of such problems, the basic equations derived in this chapter form a starting point.

Question: What is the form of the continuity equation for a viscous flow? To answer this question, review the derivation of the continuity equation in Sec. 2.4. You will note that the consideration of the viscous or inviscid nature of the flow never enters the derivation—the continuity equation is simply a statement that mass is conserved, which is independent of whether the flow is viscous or inviscid. Hence, Eq. (2.43) holds in general.

15.6 SIMILARITY PARAMETERS

In Sec. 1.7, we introduced the concept of dimensional analysis, from which sprung the similarity parameters necessary to ensure the dynamic similarity between two or more different flows (see Sec. 1.8). In the present section, we revisit the governing similarity parameters, but cast them in a slightly different light.

Consider a steady two-dimensional, viscous, compressible flow. The x-momentum equation for such a flow is given by Eq. (15.19a), which for the present case reduces to

$$\rho u \frac{\partial u}{\partial x} + \rho v \frac{\partial u}{\partial y} = -\frac{\partial p}{\partial x} + \frac{\partial}{\partial y}\left[\mu\left(\frac{\partial v}{\partial x} + \frac{\partial u}{\partial y}\right)\right] \tag{15.27}$$

In Eq. (15.27), ρ, u, p, etc., are the actual dimensional variables, say, $[\rho] = \text{kg/m}^3$, etc. Let us introduce the following dimensionless variables:

$$\rho' = \frac{\rho}{\rho_\infty} \qquad u' = \frac{u}{V_\infty} \qquad v' = \frac{v}{V_\infty} \qquad p' = \frac{p}{p_\infty}$$

$$\mu' = \frac{\mu}{\mu_\infty} \qquad x' = \frac{x}{c} \qquad y' = \frac{y}{c}$$

where ρ_∞, V_∞, p_∞, and μ_∞ are reference values (say, e.g., freestream values) and c is a reference length (say, the chord of an airfoil). In terms of these dimensionless variables, Eq. (15.27) becomes

$$\rho'u'\frac{\partial u'}{\partial x'} + \rho'v'\frac{\partial u'}{\partial y'} = -\left(\frac{p_\infty}{\rho_\infty V_\infty^2}\right)\frac{\partial p'}{\partial x'} + \left(\frac{\mu_\infty}{\rho_\infty V_\infty c}\right)\frac{\partial}{\partial y'}\left[\mu'\left(\frac{\partial v'}{\partial x'} + \frac{\partial u'}{\partial y'}\right)\right] \tag{15.28}$$

Noting that

$$\frac{p_\infty}{\rho_\infty V_\infty^2} = \frac{\gamma p_\infty}{\gamma \rho_\infty V_\infty^2} = \frac{a_\infty^2}{\gamma V_\infty^2} = \frac{1}{\gamma M_\infty^2}$$

and

$$\frac{\mu_\infty}{\rho_\infty V_\infty c} = \frac{1}{\text{Re}_\infty}$$

where M_∞ and Re_∞ are the freestream Mach and Reynolds numbers, respectively, Eq. (15.28) becomes

$$\rho'u'\frac{\partial u'}{\partial x'} + \rho'v'\frac{\partial u'}{\partial y'} = -\frac{1}{\gamma M_\infty^2}\frac{\partial p'}{\partial x'} + \frac{1}{\text{Re}_\infty}\frac{\partial}{\partial y'}\left[\mu'\left(\frac{\partial v'}{\partial x'} + \frac{\partial u'}{\partial y'}\right)\right] \tag{15.29}$$

Equation (15.29) tells us something important. Consider two different flows over two bodies of different shapes. In one flow, the ratio of specific heats, Mach number, and Reynolds number are γ_1, $M_{\infty 1}$, and $\text{Re}_{\infty 1}$, respectively; in the other flow, these parameters have different values, γ_2, $M_{\infty 2}$, and $\text{Re}_{\infty 2}$. Equation (15.29) is valid for both flows. It can, in principle, be solved to obtain u' as a function of x' and y'. However, since γ, M_∞, and Re_∞ are different for the two cases, the coefficients of the derivatives in Eq. (15.29) will be different. This will ensure, if

$$u' = f_1(x', y')$$

represents the solution for one flow and

$$u' = f_2(x', y')$$

represents the solution for the other flow, that

$$f_1 \neq f_2$$

However, consider now the case where the two different flows have the *same* values of γ, M_∞, and Re_∞. Now the coefficients of the derivatives in Eq. (15.29) will be the same for both flows; i.e., Eq. (15.29) is *numerically identical* for the two flows. In addition, assume the two bodies are geometrically similar, so that the boundary conditions in terms of the nondimensional variables are the same. Then, the solutions of Eq. (15.29) for the two flows in terms of $u' = f_1(x', y')$ and $u' = f_2(x', y')$ must be identical; i.e.,

$$f_1(x', y') \equiv f_2(x', y') \tag{15.30}$$

Recall the definition of dynamically similar flows given in Sec. 1.8. There, we stated in part that two flows are dynamically similar if the distributions of V/V_∞, p/p_∞, etc., are the same throughout the flow field when plotted against common nondimensional coordinates. This is precisely what Eq. (15.30) is saying—that u' as a function of x' and y' is the same for the two flows. That is, the variation of the *nondimensional* velocity as a function of the *nondimensional* coordinates is the same for the two flows. How did we obtain Eq. (15.30)? Simply by saying that γ, M_∞, and Re_∞ are the same for the two flows and that the two bodies are geometrically similar. *These are precisely the criteria for two flows to be dynamically similar*, as originally stated in Sec. 1.8.

What we have seen in the above derivation is a formal mechanism to identify governing similarity parameters for a flow. By couching the governing flow equations in terms of nondimensional variables, we find that the coefficients of the derivatives in these equations are dimensionless similarity parameters or combinations thereof.

To see this more clearly, and to extend our analysis further, consider the energy equation for a steady, two-dimensional, viscous, compressible flow, which from Eq. (15.26) can be written as (assuming no volumetric heating and neglecting the normal stresses)

$$\rho u \frac{\partial(e + V^2/2)}{\partial x} + \rho v \frac{\partial(e + V^2/2)}{\partial y} = \frac{\partial}{\partial x}\left(k\frac{\partial T}{\partial x}\right) + \frac{\partial}{\partial y}\left(k\frac{\partial T}{\partial y}\right) - \frac{\partial(up)}{\partial x}$$

$$- \frac{\partial(vp)}{\partial y} + \frac{\partial(v\tau_{xy})}{\partial x} + \frac{\partial(u\tau_{yx})}{\partial y} \tag{15.31}$$

Substituting Eq. (15.5) into (15.31), we have

$$\rho u \frac{\partial(e+V^2/2)}{\partial x} + \rho v \frac{\partial(e+V^2/2)}{\partial y} = \frac{\partial}{\partial x}\left(k\frac{\partial T}{\partial x}\right) + \frac{\partial}{\partial y}\left(k\frac{\partial T}{\partial y}\right)$$

$$-\frac{\partial(up)}{\partial x} - \frac{\partial(vp)}{\partial y}$$

$$+\frac{\partial}{\partial x}\left[\mu v\left(\frac{\partial v}{\partial x}+\frac{\partial u}{\partial y}\right)\right]$$

$$+\frac{\partial}{\partial y}\left[\mu u\left(\frac{\partial v}{\partial x}+\frac{\partial u}{\partial y}\right)\right] \qquad (15.32)$$

Using the same nondimensional variables as before, and introducing

$$e' = \frac{e}{c_v T_\infty} \qquad k' = \frac{k}{k_\infty} \qquad V'^2 = \frac{V^2}{V_\infty^2} = \frac{u^2+v^2}{V_\infty^2} = (u')^2 + (v')^2$$

Eq. (15.32) can be written as

$$\frac{\rho_\infty V_\infty c_v T_\infty}{c}\left(\rho'u'\frac{\partial e'}{\partial x'} + \rho'v'\frac{\partial e'}{\partial y'}\right)$$

$$= -\frac{\rho_\infty V_\infty^3}{2c}\left[\rho'u'\frac{\partial}{\partial x'}(u'^2+v'^2) + \rho'v'\frac{\partial}{\partial y'}(u'^2+v'^2)\right]$$

$$+\frac{k_\infty T_\infty}{c^2}\left[\frac{\partial}{\partial x'}\left(k'\frac{\partial T'}{\partial x'}\right) + \frac{\partial}{\partial y'}\left(k'\frac{\partial T'}{\partial y'}\right)\right] - \frac{V_\infty p_\infty}{c}\left(\frac{\partial(u'p')}{\partial x'} + \frac{\partial(v'p')}{\partial y'}\right)$$

$$+\frac{\mu_\infty V_\infty^2}{c^2}\left\{\frac{\partial}{\partial x'}\left[\mu'v'\left(\frac{\partial v'}{\partial x'}+\frac{\partial u'}{\partial y'}\right)\right] + \frac{\partial}{\partial y'}\left[\mu'u'\left(\frac{\partial v'}{\partial x'}+\frac{\partial u'}{\partial y'}\right)\right]\right\}$$

or $\quad\rho'u'\dfrac{\partial e'}{\partial x'} + \rho'v'\dfrac{\partial e'}{\partial y'} = \dfrac{V_\infty^2}{2c_v T_\infty}\left[\rho'u'\dfrac{\partial}{\partial x'}(u'^2+v'^2) + \rho'v'\dfrac{\partial}{\partial y'}(u'^2+v'^2)\right]$

$$+\frac{k_\infty}{c\rho_\infty V_\infty c_v}\left[\frac{\partial}{\partial x'}\left(k'\frac{\partial T'}{\partial x'}\right) + \frac{\partial}{\partial y'}\left(k'\frac{\partial T'}{\partial y'}\right)\right]$$

$$-\frac{p_\infty}{\rho_\infty c_v T_\infty}\left(\frac{\partial(u'p')}{\partial x'} + \frac{\partial(v'p')}{\partial y'}\right)$$

$$+\frac{\mu_\infty V_\infty}{c\rho_\infty c_v T_\infty}\left\{\frac{\partial}{\partial x'}\left[\mu'v'\left(\frac{\partial v'}{\partial x'}+\frac{\partial u'}{\partial y'}\right)\right]\right.$$

$$\left.+\frac{\partial}{\partial y'}\left[\mu'u'\left(\frac{\partial v'}{\partial x'}+\frac{\partial u'}{\partial y'}\right)\right]\right\} \qquad (15.32a)$$

Examining the coefficients of each term on the right-hand side of Eq. (15.32a), we find, consecutively,

$$\frac{V_\infty^2}{c_v T_\infty} = \frac{(\gamma-1)V_\infty^2}{RT_\infty} = \frac{\gamma(\gamma-1)V_\infty^2}{\gamma RT_\infty} = \frac{\gamma(\gamma-1)V_\infty^2}{a_\infty^2} = \gamma(\gamma-1)M_\infty^2$$

$$\frac{k_\infty}{c p_\infty V_\infty c_v} = \frac{k_\infty \gamma \mu_\infty}{c p_\infty V_\infty c_p \mu_\infty} = \frac{\gamma}{\mathrm{Pr}_\infty \, \mathrm{Re}_\infty}$$

Note: In the above, we have introduced a new dimensionless parameter, the *Prandtl number*, $\mathrm{Pr}_\infty \equiv \mu_\infty c_p / k_\infty$, the significance of which will be discussed later.

$$\frac{p_\infty}{\rho_\infty c_v T_\infty} = \frac{(\gamma-1)p_\infty}{\rho_\infty RT_\infty} = \frac{(\gamma-1)p_\infty}{p_\infty} = \gamma-1$$

$$\frac{\mu_\infty V_\infty}{c p_\infty c_v T_\infty} = \frac{\mu_\infty}{\rho_\infty V_\infty c}\left(\frac{V_\infty^2}{c_v T_\infty}\right) = \frac{1}{\mathrm{Re}_\infty}(\gamma-1)\frac{V_\infty^2}{RT_\infty} = \gamma(\gamma-1)\frac{M_\infty^2}{\mathrm{Re}_\infty}$$

Hence, Eq. (15.32) can be written as

$$\rho'u'\frac{\partial e'}{\partial x'} + \rho'v'\frac{\partial e'}{\partial y'} = \frac{\gamma(\gamma-1)}{2}M_\infty^2\left[\rho'u'\frac{\partial}{\partial x'}(u'^2+v'^2) + \rho'v'\frac{\partial}{\partial y'}(u'^2+v'^2)\right]$$

$$+\frac{\gamma}{\mathrm{Pr}_\infty \mathrm{Re}_\infty}\left[\frac{\partial}{\partial x'}\left(k'\frac{\partial T'}{\partial x'}\right) + \frac{\partial}{\partial y'}\left(k'\frac{\partial T'}{\partial y'}\right)\right]$$

$$-(\gamma-1)\left(\frac{\partial(u'p')}{\partial x'} + \frac{\partial(v'p')}{\partial y'}\right)$$

$$+\gamma(\gamma-1)\frac{M_\infty^2}{\mathrm{Re}_\infty}\left\{\frac{\partial}{\partial x'}\left[\mu'v'\left(\frac{\partial v'}{\partial x'} + \frac{\partial u'}{\partial y'}\right)\right]\right.$$

$$\left.+\frac{\partial}{\partial y'}\left[\mu'u'\left(\frac{\partial v'}{\partial x'} + \frac{\partial u'}{\partial y'}\right)\right]\right\} \tag{15.33}$$

Examine Eq. (15.33). It is a nondimensional equation which, in principle, can be solved for $e' = f(x', y')$. If we have two different flows, but with the same values γ, M_∞, Re_∞, and Pr_∞, Eq. (15.33) will be numerically identical for the two flows, and if we are considering geometrically similar bodies, then the solution $e' = f(x', y')$ will be identical for the two flows.

Reflecting upon Eqs. (15.29) and (15.33), which are the nondimensional x-momentum and energy equations, respectively, we clearly see that the governing similarity parameters for a viscous, compressible flow are γ, M_∞, Re_∞, and Pr_∞. If the above parameters are the same for two different flows with geometrically similar bodies, then the flows are dynamically similar. We obtained these results by considering the x-momentum equation and the energy equation, both in two dimensions. The same results would have occurred if we had considered three-dimensional flow and the y- and z-momentum equations.

Note that the similarity parameters γ, M_∞, and Re_∞ were obtained from the momentum equation. When the energy equation is considered, an additional similarity parameter is introduced, namely, the Prandtl number. On a physical basis, the Prandtl number is an index which is proportional to the ratio of energy dissipated by friction to the energy transported by thermal conduction; i.e.,

$$\mathrm{Pr} = \frac{\mu_\infty c_p}{k} \propto \frac{\text{frictional dissipation}}{\text{thermal conduction}}$$

In the study of compressible, viscous flow, Prandtl number is just as important as γ, Re_∞, or M_∞. For air at standard conditions, $Pr_\infty = 0.71$. Note that Pr_∞ is a property of the gas. For different gases, Pr_∞ is different. Also, like μ and k, Pr_∞ is, in general, a function of temperature; however, for air over a reasonable temperature range (up to $T_\infty = 600$ K), it is safe to assume $Pr_\infty = \text{constant} = 0.71$.

15.7 SOLUTIONS OF VISCOUS FLOWS: A PRELIMINARY DISCUSSION

The governing continuity, momentum, and energy equations for a general unsteady, compressible, viscous, three-dimensional flow are given by Eqs. (2.43), (15.19a to c), and (15.26), respectively. Examine these equations closely. They are nonlinear, coupled, partial differential equations. Moreover, they have additional terms—namely, the viscous terms—in comparison to the analogous equations for an inviscid flow treated in Part III. Since we have already seen that the nonlinear inviscid flow equations do not lend themselves to a general analytical solution, we can certainly expect the viscous flow equations also not to have any general solutions (at least, at the time of this writing, no general analytical solutions have been found). This leads to the following question: How, then, can we make use of the viscous flow equations in order to obtain some practical results? The answer is much like our approach to the solution of inviscid flows. We have the following options:

1. There are a few viscous flow problems which, by their physical and geometrical nature, allow many terms in the Navier-Stokes solutions to be precisely zero, with the resulting equations being simple enough to solve, either analytically or by simple numerical methods. Sometimes this class of solutions is called "exact solutions" of the Navier-Stokes equations, because no simplifying *approximations* are made to reduce the equations—just *precise* conditions are applied to reduce the equations. Chapter 16 is devoted to this class of solutions; examples are Couette flow and Poiseuille flow (to be defined later).

2. We can simplify the equations by treating certain classes of physical problems for which some terms in the viscous flow equations are small and can be neglected. This is an approximation, not a precise condition. The boundary-layer equations developed and discussed in Chap. 17 are a case in point. However, as we will see, the boundary-layer equations may be simpler than the full viscous flow equations, but they are still nonlinear.

3. We can tackle the solution of the full viscous flow equations by modern numerical techniques. For example, some of the computational fluid dynamic algorithms discussed in Chap. 13 in conjunction with "exact" solutions for the inviscid flow equations carry over to exact solutions for the viscous flow equations. These matters will be discussed in Chap. 18.

There are some inherent very important differences between the analysis of viscous flows and the study of inviscid flows that were presented in Parts II and III. The remainder of this section highlights these differences.

First, we have already demonstrated in Example 2.4 that viscous flows are *rotational flows*. Therefore, a velocity potential cannot be defined for a viscous flow, thus losing the attendant advantages that were discussed in Secs. 2.15 and 11.2. On the other hand, a stream function can be defined, because the stream function satisfies the continuity equation and has nothing to do with the flow being rotational or irrotational (see Sec. 2.14).

Secondly, the boundary condition at a solid surface for a viscous flow is the *no-slip* condition. Due to the presence of friction between the surface material and the adjacent layer of fluid, the fluid velocity right at the surface is zero. This no-slip condition was discussed in Sec. 15.2. For example, if the surface is located at $y = 0$ in a cartesian coordinate system, then the no-slip boundary condition on velocity is

$$At\ y = 0: \qquad\qquad u = 0, \qquad v = 0, \qquad w = 0$$

This is in contrast to the analogous boundary condition for an inviscid flow, namely, the flow-tangency condition at a surface as discussed in Sec. 3.7, where only the component of the velocity normal to the surface is zero. Also, recall that for an inviscid flow, there is no boundary condition on the temperature; the temperature of the gas adjacent to a solid surface in an inviscid flow is governed by the physics of the flow field and has no connection whatsoever with the actual wall temperature. However, for a viscous flow, the mechanism of thermal conduction ensures that the temperature of the fluid immediately adjacent to the surface is the same as the temperature of the material surface. In this respect, the no-slip condition is more general than that applied to the velocity; in addition to $u = v = 0$ at the wall, we also have $T = T_w$ at the wall, where T is the gas temperature immediately adjacent to the wall and T_w is the temperature of the surface material. Thus,

$$At\ y = 0: \qquad\qquad T = T_w \qquad\qquad (15.34)$$

In many problems, T_w is specified and held constant; this boundary condition is easily applied. However, consider the following, more general case. Imagine a viscous flow over a surface where heat is being transferred from the gas to the surface, or vice versa. Also, assume that the surface is at a certain temperature, T_w, when the flow first starts, but that T_w changes as a function of time as the surface is either heated or cooled by the flow; i.e., $T_w = T_w(t)$. Because this time-wise variation is dictated in part by the flow which is being calculated, T_w becomes an unknown in the problem and must be calculated along with the solution of the viscous flow. For this general case, the boundary condition at the surface is obtained from Eq. (15.2) applied at the wall; i.e.,

$$At \ y = 0: \qquad \dot{q}_w = -\left(k\frac{\partial T}{\partial y}\right)_w \qquad (15.35)$$

Here, the surface material is responding to the heat transfer to the wall, $\dot{q}_w$, hence changing T_w, which in turn affects $\dot{q}_w$. This general, unsteady heat transfer problem must be solved by treating the viscous flow and the thermal response of the material simultaneously. This problem is beyond the scope of the present book.

Finally, let us imagine the above, unsteady case carried out to the limit of large times. That is, imagine a wind-tunnel model which is at room temperature suddenly inserted in a supersonic or hypersonic stream. At early times, say, for the first few seconds, the surface temperature remains relatively cool, and the assumption of constant wall temperature, T_w, is reasonable [Eq. (15.34)]. However, due to the heat transfer to the model [Eq. (15.35)], the surface temperature soon starts to increase and becomes a function of time, as discussed in the previous paragraph. However, as T_w increases, the heating rate decreases. Finally, at large times, T_w increases to a high enough value that the net heat transfer rate to the surface becomes zero, i.e., from Eq. (15.35),

$$\dot{q}_w = -\left(k\frac{\partial T}{\partial y}\right)_w = 0$$

or

$$\left(\frac{\partial T}{\partial y}\right)_w = 0 \qquad (15.36)$$

When the situation of zero heat transfer is achieved, a state of equilibrium exists; the wall temperature at which this occurs is, by definition, the equilibrium wall temperature, or, as it is more commonly denoted, the *adiabatic wall temperature*, T_{aw}. Hence, for the case of an *adiabatic wall* (no heat transfer), the wall boundary condition is given by Eq. (15.36).

In summary, for the wall boundary condition associated with the solution of the energy equation [Eq. (15.26)], we have three possible cases:

1. *Constant temperature wall, where T_w is a specified constant* [Eq. (15.34)]. For this given wall temperature, the temperature gradient at the wall, $(\partial T/\partial y)_w$, is obtained as part of the flow-field solution and allows the direct calculation of the aerodynamic heating to the wall via Eq. (15.35).

2. *The general, unsteady case, where the heat transfer to the wall, $\dot{q}_w$, causes the wall temperature, T_w, to change, which in turn causes $\dot{q}_w$ to change.* Here, both T_w and $(\partial T/\partial y)_w$ change as a function of time, and the problem must be solved by treating jointly the viscous flow as well as the thermal response of the wall material (which usually implies a separate thermal conduction heat transfer analysis).

3. *The adiabatic wall case (zero heat transfer), where $(\partial T/\partial y)_w = 0$* [Eq. (15.36)]. Here, the boundary condition is applied to the temperature *gradient* at the

wall, not to the wall temperature itself. Indeed, the wall temperature for this case is defined as the adiabatic wall temperature, T_{aw}, and is obtained as part of the flow-field solution.

Finally, we emphasize again that, from the point of view of applied aerodynamics, the practical results obtained from a viscous flow analysis are the skin friction and heat transfer at the surface. However, to obtain these quantities, we usually need a complete solution of the viscous flow field; among the data obtained from such a solution are the velocity and temperature gradients at the wall. These, in turn, allow the direct calculation of τ_w and $\dot{q}_w$ from

$$\tau_w = \mu \left(\frac{\partial u}{\partial y}\right)_w$$

and

$$\dot{q}_w = -k \left(\frac{\partial T}{\partial y}\right)_w$$

Another practical result provided by a viscous flow analysis is the prediction and calculation of flow separation; we have discussed numerous cases in the preceding chapters where the pressure field around an aerodynamic body can be greatly changed by flow separation; the flows over cylinders and spheres (see Secs. 3.18 and 6.7) are cases in point.

Clearly, the study of viscous flow is important within the entire scope of aerodynamics. The purpose of the following chapters is to provide an introduction to such flows. We will organize our study following the three options itemized at the beginning of this section; i.e., we will treat, in turn, certain specialized "exact" solutions of the Navier-Stokes equations, boundary-layer solutions, and then "exact" numerical solutions of Navier-Stokes equations. In so doing, we hope that the reader will gain an overall, introductory picture of the whole area of viscous flow. Entire books have been written on this subject, see, e.g., Refs. 42 and 43. We cannot possibly present such detail here; rather, our objective is simply to provide a "feel" for and a basic understanding of the material. Let us proceed.

15.8 SUMMARY

We have now completed the road map given in Fig. 15.1. The main results of this chapter are summarized below:

Shear stress and flow separation are two major ramifications of viscous flow. Shear stress is the cause of skin friction drag D_f, and flow separation is the source of pressure drag D_p, sometimes called form drag. Transition from laminar to turbulent flow causes D_f to increase and D_p to decrease.

Shear stress in a flow is due to velocity gradients; e.g., $\tau_{yx} = \mu\, \partial u/\partial y$ for a flow with gradients in the y direction. Similarly, heat conduction is due to temperature gradients; e.g., $\dot{q}_y = -k\, \partial T/\partial y$, etc. Both μ and k are physical properties of the gas and are functions of temperature.

The general equations of viscous flow are

x momentum:
$$\rho \frac{Du}{Dt} = -\frac{\partial p}{\partial x} + \frac{\partial \tau_{xx}}{\partial x} + \frac{\partial \tau_{yx}}{\partial y} + \frac{\partial \tau_{zx}}{\partial z}$$
(15.18a)

y momentum:
$$\rho \frac{Dv}{Dt} = -\frac{\partial p}{\partial y} + \frac{\partial \tau_{xy}}{\partial x} + \frac{\partial \tau_{yy}}{\partial y} + \frac{\partial \tau_{zy}}{\partial z}$$
(15.18b)

z momentum:
$$\rho \frac{Dw}{Dt} = -\frac{\partial p}{\partial z} + \frac{\partial \tau_{xz}}{\partial x} + \frac{\partial \tau_{yz}}{\partial y} + \frac{\partial \tau_{zz}}{\partial z}$$
(15.18c)

Energy:

$$\rho \frac{D(e + V^2/2)}{Dt} = \rho\dot{q} + \frac{\partial}{\partial x}\left(k\frac{\partial T}{\partial x}\right) + \frac{\partial}{\partial y}\left(k\frac{\partial T}{\partial y}\right) + \frac{\partial}{\partial z}\left(k\frac{\partial T}{\partial z}\right) - \nabla \cdot p\mathbf{V}$$

$$+ \frac{\partial(u\tau_{xx})}{\partial x} + \frac{\partial(u\tau_{yx})}{\partial y} + \frac{\partial(u\tau_{zx})}{\partial z} + \frac{\partial(v\tau_{xy})}{\partial x} + \frac{\partial(v\tau_{yy})}{\partial y}$$

$$+ \frac{\partial(v\tau_{zy})}{\partial z} + \frac{\partial(w\tau_{xz})}{\partial x} + \frac{\partial(w\tau_{yz})}{\partial y} + \frac{\partial(w\tau_{zz})}{\partial z}$$
(15.26)

where

$$\tau_{xy} = \tau_{yx} = \mu\left(\frac{\partial v}{\partial x} + \frac{\partial u}{\partial y}\right)$$

$$\tau_{yz} = \tau_{zy} = \mu\left(\frac{\partial w}{\partial y} + \frac{\partial v}{\partial z}\right)$$

$$\tau_{zx} = \tau_{xz} = \mu\left(\frac{\partial u}{\partial z} + \frac{\partial w}{\partial x}\right)$$

$$\tau_{xx} = \lambda(\nabla \cdot \mathbf{V}) + 2\mu\frac{\partial u}{\partial x}$$

$$\tau_{yy} = \lambda(\nabla \cdot \mathbf{V}) + 2\mu\frac{\partial v}{\partial y}$$

$$\tau_{zz} = \lambda(\nabla \cdot \mathbf{V}) + 2\mu\frac{\partial w}{\partial z}$$

The similarity parameters for a flow can be obtained by nondimensionalizing the governing equations; the coefficients in front of the nondimensionalized derivatives give the similarity parameters or combinations thereof. For a viscous, compressible flow, the main similarity parameters are γ, M_∞, Re_∞, and Pr_∞.

Exact analytical solutions of the complete Navier-Stokes equations exist for only a few very specialized cases. Instead, the equations are frequently simplified by making appropriate approximations about the flow. In modern times, exact solutions of the complete Navier-Stokes equations for many practical problems can be obtained numerically, using various techniques of computational fluid dynamics.

PROBLEMS

15.1. Consider the incompressible viscous flow of air between two infinitely long parallel plates separated by a distance h. The bottom plate is stationary, and the top plate is moving at the constant velocity u_e in the direction of the plate. Assume that no pressure gradient exists in the flow direction.

(*a*) Obtain an expression for the variation of velocity between the plates.

(*b*) If $T = \text{constant} = 320$ K, $u_e = 30$ m/s, and $h = 0.01$ m, calculate the shear stress on the top and bottom plates.

15.2. Assume that the two parallel plates in Prob. 15.1 are both stationary but that a constant pressure gradient exists in the flow direction; i.e., $dp/dx = \text{constant}$.

(*a*) Obtain an expression for the variation of velocity between the plates.

(*b*) Obtain an expression for the shear stress on the plates in terms of dp/dx.

CHAPTER
16

SOME
SPECIAL
CASES;
COUETTE AND
POISEUILLE
FLOWS

The resistance arising from the want of lubricity in the parts of a fluid is, other things being equal, proportional to the velocity with which the parts of the fluid are separated from one another.

Isaac Newton, 1687,
from Section IX of Book II
of his Principia

16.1 INTRODUCTION

The general equations of viscous flow were derived and discussed in Chap. 15. In particular, the viscous flow momentum equations were treated in Sec. 15.4 and are given in partial differential equation form by Eqs. (15.19a to c)—the Navier-Stokes equations. These, along with the viscous flow energy equation, Eq. (15.26), derived in Sec. 15.5, are the theoretical tools for the study of viscous flows. However, examine these equations closely; as discussed in Sec. 15.7, they are a system of coupled, nonlinear partial differential equations—equations which contain more terms and which are inherently more elaborate than the inviscid flow equations treated in Parts II and III of this book. Three classes of solutions of these equations were itemized in Sec. 15.5. The first itemized class was that of "exact" solutions of the Navier-Stokes equations for a few specific physical problems which, by their physical and geometrical nature, allow many terms in the governing equations to be precisely zero, resulting in a system of equations simple enough to solve, either analytically or by simple numerical methods. Such exact problems are the subject of this chapter.

669

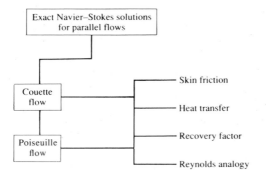

FIGURE 16.1
Road map for Chap. 16.

The road map for this chapter is given in Fig. 16.1. The types of flows considered here are generally labeled as *parallel flows* because the streamlines are straight and parallel to each other. We will consider two of these flows, Couette and Poiseuille, which will be defined in due course. In addition to representing exact solutions of the Navier-Stokes equations, these flows illustrate some of the important practical facets of any viscous flow, as itemized on the right side of the road map. In a clear, uncomplicated fashion, we will be able to calculate and study the surface skin friction and heat transfer. We will also use the results to define the recovery factor and Reynolds analogy—two practical engineering tools that are frequently used in the analysis of skin friction and heat transfer.

16.2 COUETTE FLOW: GENERAL DISCUSSION

Consider the flow model shown in Fig. 16.2. Here we see a viscous fluid contained between two parallel plates separated by a distance D. The upper plate is moving to the right at velocity u_e. Due to the no-slip condition, there can be no relative motion between the plate and the fluid; hence, at $y = D$ the flow velocity is $u = u_e$ and is directed toward the right. Similarly, the flow velocity at $y = 0$, which is the surface of the stationary lower plate, is $u = 0$. In addition, the two plates may be at different temperatures; the upper plate is at temperature T_e and the lower plate is at temperature T_w. Again, due to the no-slip condition as discussed in Sec. 15.7, the fluid temperature at $y = D$ is $T = T_e$ and that at $y = 0$ is $T = T_w$.

Clearly, there is a flow field between the two plates; the driving force for this flow is the motion of the upper plate, dragging the flow along with it through the mechanism of friction. The upper plate is exerting a shear stress, τ_e, acting toward the right on the fluid at $y = D$, thus causing the fluid to move toward the right. By an equal and opposite reaction, the fluid is exerting a shear stress, τ_e, on the upper plate acting toward the left, tending to retard its motion. We assume that the upper plate is being driven by some external force which is sufficient to

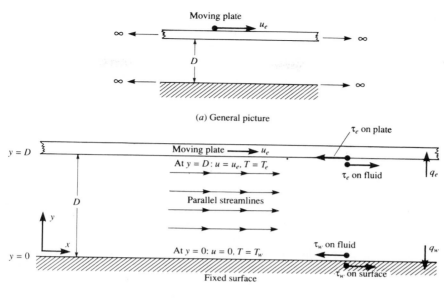

FIGURE 16.2
Model for Couette flow.

overcome the retarding shear stress and to allow the plate to move at the constant velocity, u_e. Similarly, the lower plate is exerting a shear stress, τ_w, acting toward the left on the fluid at $y = 0$. By an equal and opposite reaction, the fluid is exerting a shear stress, τ_w, acting toward the right on the lower plate. (In all subsequent diagrams dealing with viscous flow, the only shear stresses shown will be those due to the fluid acting on the surface, unless otherwise noted.)

In addition to the velocity field induced by the relative motion of the two plates, there will also be a temperature field induced by the following two mechanisms:

1. The plates in general will be at different temperatures, thus causing temperature gradients in the flow.
2. The kinetic energy of the flow will be partially dissipated by the influence of friction and will be transformed into internal energy within the fluid. These changes in internal energy will be reflected by changes in temperature. This phenomenon is called *viscous dissipation*.

Consequently, temperature gradients will exist within the flow; in turn, these temperature gradients result in the transfer of heat through the fluid. Of particular interest is the heat transfer at the upper and lower surfaces, denoted by $\dot{q}_e$ and

$\dot{q}_w$, respectively. These heat transfers are shown in Fig. 16.2; the directions for $\dot{q}_e$ and $\dot{q}_w$ show heat being transferred from the fluid to the wall in both cases. When heat flows from the fluid to the wall, this is called a *cold wall* case, such as sketched in Fig. 16.2. When heat flows from the wall into the fluid, this is called a *hot wall* case. Keep in mind that the heat flux through the fluid at any point is given by the Fourier law expressed by Eq. (15.2); i.e., the heat flux in the y direction is expressed as

$$\dot{q}_y = -k \frac{\partial T}{\partial y} \qquad (15.2)$$

where the minus sign accounts for the fact that heat is transferred from a region of high temperature to a region of lower temperature; i.e., $\dot{q}_y$ is in the opposite direction of the temperature gradient.

Let us examine the geometry of Couette flow as illustrated in Fig. 16.2. An x-y cartesian coordinate system is oriented with the x axis in the direction of the flow and the y axis perpendicular to the flow. Since the two plates are parallel, the only possible flow pattern consistent with this picture is that of straight, parallel streamlines. Moreover, since the plates are infinitely long (i.e., stretching to plus and minus infinity in the x direction), then the flow properties cannot change with x. (If the properties did change with x, then the flow-field properties would become infinitely large or infinitesimally small at large values of x—a physical inconsistency.) Thus, all partial derivatives with respect to x are zero. The only changes in the flow-field variables take place in the y direction. Moreover, the flow is steady, so that all time derivatives are zero. With this geometry in mind, return to the governing Navier-Stokes equations given by Eqs. (15.19a to c) and Eq. (15.26). In these equations, for Couette flow,

$$v = w = 0 \qquad \frac{\partial u}{\partial x} = \frac{\partial T}{\partial x} = \frac{\partial p}{\partial x} = 0$$

Hence, from Eqs. (15.19a to c) and Eq. (15.26), we have

x-momentum equation:
$$\frac{\partial}{\partial y}\left(\mu \frac{\partial u}{\partial y}\right) = 0 \qquad (16.1)$$

y-momentum equation:
$$\frac{\partial p}{\partial y} = 0 \qquad (16.2)$$

Energy equation:
$$\frac{\partial}{\partial y}\left(k \frac{\partial T}{\partial y}\right) + \frac{\partial}{\partial y}\left(\mu u \frac{\partial u}{\partial y}\right) = 0 \qquad (16.3)$$

Equations (16.1) to (16.3) are the governing equations for Couette flow. Note that these equations are exact forms of the Navier-Stokes equations applied to the geometry of Couette flow—no approximations have been made. Also, note from Eq. (16.2) that the variation of pressure in the y direction is zero; this in combination with the earlier result that $\partial p/\partial x = 0$ implies that the *pressure is constant* throughout the entire flow field. Couette flow is a constant pressure flow. It is interesting to note that all the previous flow problems discussed in Parts II and III, being *inviscid* flows, were established and maintained by the existence of *pressure gradients* in the flow. In these problems, the pressure gradient was nature's mechanism of grabbing hold of the flow and making it move. However, in the problem we are discussing now—being a viscous flow—shear stress is another mechanism by which nature can exert a force on a flow. For Couette flow, the shear stress exerted by the moving plate on the fluid is the exclusive driving mechanism that maintains the flow; clearly, no pressure gradient is present, nor is it needed.

This section has presented the general nature of Couette flow. Note that we have made no distinction between incompressible and compressible flow; all aspects discussed here apply to both cases. Also, we note that, although Couette flow appears to be a rather academic problem, the following sections illustrate, in a simple fashion, many of the important characteristics of practical viscous flows in real engineering applications.

The next two sections will treat the separate cases of incompressible and compressible Couette flow. Incompressible flow will be discussed first because of its relative simplicity; this is the subject of Sec. 16.3. Then compressible Couette flow, and how it differs from the incompressible case, will be examined in Sec. 16.4.

As a final note in this section, it is obvious from our general discussion of Couette flow that the flow-field properties vary only in the y direction; all derivatives in the x direction are zero. Therefore, as a matter of mathematical preciseness, all the partial derivatives in Eqs. (16.1) to (16.3) can be written as ordinary derivatives. For example, Eq. (16.1) can be written as

$$\frac{d}{dy}\left(\mu\,\frac{du}{dy}\right)=0$$

However, our discussion of Couette flow is intended to serve as a straightforward example of a viscous flow problem, "breaking the ice" so-to-speak for the more practical but more complex problems to come—problems which involve changes in both the x and y directions, and which are described by *partial* differential equations. Therefore, on pedagogical grounds, we choose to continue the partial differential notation here, simply to make the reader feel more comfortable when we extend these concepts to the boundary layer and full Navier-Stokes solutions in Chaps. 17 and 18, respectively.

16.3   INCOMPRESSIBLE (CONSTANT PROPERTY) COUETTE FLOW

In the study of viscous flows, a flow field in which ρ, μ, and k are treated as constants is sometimes labeled as "constant property" flow. This assumption is made in the present section. On a physical basis, this means that we are dealing with an incompressible flow, where ρ is constant. Also, since μ and k are functions of temperature (see Sec. 15.3), constant property flow implies that T is constant also. (We will relax this assumption slightly at the end of this section.)

The governing equations for Couette flow were derived in Sec. 16.2. In particular, the y-momentum equation, Eq. (16.2), along with the geometrical property that $\partial p/\partial x = 0$, states that the pressure is constant throughout the flow. Consequently, all the information about the velocity field comes from the x-momentum equation, Eq. (16.1), repeated below:

$$\frac{\partial}{\partial y}\left(\mu \frac{\partial u}{\partial y}\right) = 0 \tag{16.1}$$

For constant μ, this becomes

$$\frac{\partial^2 u}{\partial y^2} = 0 \tag{16.4}$$

Integrating with respect to y twice, we obtain

$$u = ay + b \tag{16.5}$$

where a and b are constants of integration. These constants can be obtained from the boundary conditions illustrated in Fig. 16.2, as follows:

At $y = 0$, $u = 0$; hence, $b = 0$.
At $y = D$, $u = u_e$; hence, $a = u_e/D$.

Thus, the variation of velocity for incompressible Couette flow is given by Eq. (16.5) as

$$\boxed{u = u_e\left(\frac{y}{d}\right)} \tag{16.6}$$

Note the important result that the velocity varies *linearly* across the flow. This result is sketched in Fig. 16.3.

Once the velocity profile is obtained, we can obtain the shear stress at any point in the flow from Eq. (15.1), repeated below (the subscript yx is dropped here because we know the only shear stress acting in this problem is that in the x direction).

$$\tau = \mu \frac{\partial u}{\partial y} \tag{16.7}$$

From Eq. (16.6),

$$\frac{\partial u}{\partial y} = \frac{u_e}{D} \tag{16.8}$$

Hence, from Eqs. (16.7) and (16.8), we have

$$\tau = \mu \left(\frac{u_e}{D} \right) \qquad (16.9)$$

Note that the shear stress is *constant* throughout the flow. Moreover, the straightforward result given by Eq. (16.9) illustrates two important physical trends—trends that we will find to be almost universally present in all viscous flows:

1. As u_e increases, the shear stress increases. From Eq. (16.9), τ increases linearly with u_e; this is a specific result germane to Couette flow. For other problems, the increase is not necessarily linear.

2. As D increases, the shear stress decreases; i.e., as the thickness of the viscous shear layer increases, all other things being equal, the shear stress becomes smaller. From Eq. (16.9), τ is inversely proportional to D—again a result germane to Couette flow. For other problems, the decrease in τ is not necessarily in *direct* inverse proportion to the shear-layer thickness.

With the above results in mind, reflect for a moment on the quotation from Isaac Newton's *Principia* given at the beginning of this chapter. Here, the "want of lubricity" is, in modern terms, interpreted as the shear stress. This want of lubricity is, according to Newton, "proportional to the velocity with which the parts of the fluid are separated from one another," i.e., in the context of the present problem proportional to u_e/D. This is precisely the statement contained in Eq. (16.9). In more recent times, Newton's statement is generalized to the form given by Eq. (16.7), and even more generalized by Eq. (15.1). For this reason, Eqs. (15.1) and (16.7) are frequently called the newtonian shear stress law, and fluids which obey this law are called *newtonian fluids*. [There are some specialized fluids which do not obey Eq. (15.1) or (16.7); they are called non-newtonian fluids—some polymers and blood are two such examples.] By far, the vast majority of aeronautical applications deal with air or other gases, which are newtonian fluids. In hydrodynamics, water is the primary medium, and it is a newtonian fluid. Therefore, we will deal exclusively with newtonian fluids in this book. Consequently, the quote given at the beginning of this chapter is one of Newton's most important contributions to fluid mechanics—it represents the first time in history where shear stress is recognized as being proportional to a velocity gradient.

Let us now turn our attention to heat transfer in a Couette flow. Here, we continue our assumptions of constant ρ, μ, and k, but at the same time, we will allow a temperature gradient to exist in the flow. In an exact sense, this is inconsistent; if T varies throughout the flow, then ρ, μ, and k also vary. However, for this application, we assume that the temperature variations are small—indeed, small enough such that ρ, μ, and k are *approximately* constant—and treat them so in the equations. On the other hand, the temperature changes, although small on an absolute basis, are sufficient to result in meaningful heat flux through the fluid. The results obtained will reflect some of the important trends in aerodynamic heating associated with high-speed flows, to be discussed in subsequent chapters.

For Couette flow with heat transfer, return to Fig. 16.2. Here, the temperature of the upper plate is T_e and that of the lower plate is T_w. Hence, we have as boundary conditions for the temperature of the fluid:

At $y = 0$: $\qquad\qquad\qquad\qquad\qquad T = T_w$

At $y = D$: $\qquad\qquad\qquad\qquad\qquad T = T_e$

The temperature profile in the flow is governed by the energy equation, Eq. (16.3). For constant μ and k, this equation is written as

$$\frac{k}{\mu}\left(\frac{\partial^2 T}{\partial y^2}\right) + \frac{\partial}{\partial y}\left(u\frac{\partial u}{\partial y}\right) = 0 \qquad (16.10)$$

Also, since μ is assumed to be constant, Eqs. (16.10) and (16.1) are *totally uncoupled.* That is, for the constant property flow considered here, the solution of the momentum equation [Eq. (16.1)] is totally separate from the solution of the energy equation [Eq. (16.10)]. Therefore, in this problem, although the temperature is allowed to vary, the velocity field is still given by Eq. (16.6), as sketched in Fig. 16.3.

In dealing with flows where energy concepts are important, the enthalpy, h, is frequently a more fundamental variable than temperature; we have seen much evidence of this in Part III, where energy changes were a vital consideration. In the present problem, where the temperature changes are small enough to justify the assumptions of constant ρ, μ, and k, this is not quite the same situation. However, because we will need to solve Eq. (16.10), which is an energy equation for a flow (no matter how small the energy changes), and because we are using Couette flow as an example to set the stage for more complex problems, it is instructional (but by no means necessary) to couch Eq. (16.10) in terms of enthalpy. Assuming constant specific heat, we have

$$h = c_p T \qquad (16.11)$$

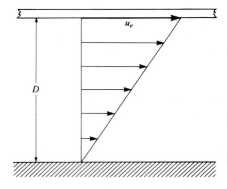

FIGURE 16.3
Velocity profile for incompressible Couette flow.

Equation (16.11) is valid for the Couette flow of any fluid with constant heat capacity; here, the germane specific heat is that at constant pressure, c_p, because the entire flow field is at constant pressure. In this sense, Eq. (16.11) is a result of applying the first law of thermodynamics to a constant pressure process and recalling the fundamental definition of heat capacity as the heat added per unit change in temperature, $\delta q/dT$. Of course, if the fluid is a calorically perfect gas, then Eq. (16.11) is a basic thermodynamic property of such a gas quite independent of what the process may be [see Sec. 7.2 and Eq. (7.6b)]. Inserting Eq. (16.11) into Eq. (16.10), we have

$$\frac{k}{\mu c_p}\frac{\partial^2 h}{\partial y^2}+\frac{\partial}{\partial y}\left(u\frac{\partial u}{\partial y}\right)=0 \tag{16.12}$$

Recall the definition of the Prandtl number from Sec. 15.6, namely,

$$\Pr=\frac{\mu c_p}{k}$$

Equation (16.12) can be written in terms of the Prandtl number as

$$\frac{1}{\Pr}\frac{\partial^2 h}{\partial y^2}+\frac{\partial}{\partial y}\left(u\frac{\partial u}{\partial y}\right)=0$$

or

$$\frac{\partial^2 h}{\partial y^2}+\frac{\Pr}{2}\frac{\partial}{\partial y}\left(\frac{\partial u^2}{\partial y}\right)=0 \tag{16.13}$$

Integrating twice in the y direction, we find that Eq. (16.13) yields

$$h+\left(\frac{\Pr}{2}\right)u^2=ay+b \tag{16.14}$$

where a and b are constants of integration [different from the a and b in Eq. (16.5)]. Expressions for a and b are found by applying Eq. (16.14) at the boundaries, as follows:

At $y=0$: $\qquad h=h_w \qquad$ and $\qquad u=0$

At $y=D$: $\qquad h=h_e \qquad$ and $\qquad u=u_e$

Hence, from Eq. (16.14) at the boundaries,

$$b=h_w$$

and

$$a=\frac{h_e-h_w+(\Pr/2)\,u_e^2}{D}$$

Inserting these values into Eq. (16.14) and rearranging, we have

$$h=h_w+\left[h_e-h_w+\left(\frac{\Pr}{2}\right)u_e^2\right]\frac{y}{D}-\left(\frac{\Pr}{2}\right)u^2 \tag{16.15}$$

Inserting Eq. (16.6) for the velocity profile in Eq. (16.15) yields

$$h = h_w + \left[h_e - h_w + \left(\frac{\mathrm{Pr}}{2} \right) u_e^2 \right] \frac{y}{D} - \left(\frac{\mathrm{Pr}}{2} \right) u_e^2 \left(\frac{y}{D} \right)^2 \qquad (16.16)$$

Note that h varies *parabolically* with y/D across the flow. Since $T = h/c_p$, then the temperature profile across the flow is also parabolic. The precise shape of the parabolic curve depends on h_w (or T_w), h_e (or T_e), and Pr. Also note that, as expected from our discussion of the viscous flow similarity parameters in Sec. 15.6, the Prandtl number is clearly a strong player in the results; Eq. (16.16) is one such example.

Once the enthalpy (or temperature) profile is obtained, we can obtain the heat flux at any point in the flow from Eq. (15.2), repeated below (the subscript y is dropped here because we know the only direction of heat transfer is in the y direction for this problem):

$$\dot{q} = -k \frac{\partial T}{\partial y} \qquad (16.17)$$

Equation (16.17) can be written as

$$\dot{q} = -\frac{k}{c_p} \frac{\partial h}{\partial y} \qquad (16.18)$$

In Eq. (16.18), the enthalpy gradient is obtained by differentiating Eq. (16.16) as follows:

$$\frac{\partial h}{\partial y} = \left[h_e - h_w + \left(\frac{\mathrm{Pr}}{2} \right) u_e^2 \right] \frac{1}{D} - \mathrm{Pr}\, u_e^2 \frac{y}{D^2} \qquad (16.19)$$

Inserting Eq. (16.19) into Eq. (16.18), and writing k/c_p as μ/Pr, we have

$$\dot{q} = -\mu \left(\frac{h_e - h_w}{\mathrm{Pr}} + \frac{u_e^2}{2} \right) \frac{1}{D} + \mu u_e^2 \frac{y}{D^2} \qquad (16.20)$$

From Eq. (16.20), note that $\dot{q}$ is *not* constant across the flow, unlike the shear stress discussed earlier. Rather, $\dot{q}$ varies linearly with y. The physical reason for the variation of $\dot{q}$ is *viscous dissipation* which takes place within the flow, and which is associated with the shear stress in the flow. Indeed, the last term in Eq. (16.20), in light of Eqs. (16.6) and (16.9), can be written as

$$\mu u_e^2 \frac{y}{D^2} = \tau u_e \left(\frac{y}{D} \right) = \tau u$$

Hence, Eq. (16.20) becomes

$$\dot{q} = -\mu \left(\frac{h_e - h_w}{\mathrm{Pr}} + \frac{u_e^2}{2} \right) \frac{1}{D} + \tau u \qquad (16.21)$$

The variation of $\dot{q}$ across the flow is due to the last term in Eq. (16.21), and this term involves shear stress multiplied by velocity. The term τu is *viscous dissipation*; it is the time rate of heat generated at a point in the flow by one streamline at a given velocity "rubbing" against an adjacent streamline at a slightly different velocity—analogous to the heat you feel when rubbing your hands together vigorously. Note that, if u_e is negligibly small, then the viscous dissipation is small and can be neglected; i.e., in Eq. (16.20) the last term can be neglected (u_e is small), and in Eq. (16.21) the last term can be neglected (τ is small if u_e is small). In this case, the heat flux becomes constant across the flow, simply equal to

$$\dot{q} \approx -\frac{\mu}{\text{Pr}}\left(\frac{h_e - h_w}{D}\right) \tag{16.22}$$

In this case, the "driving potential" for heat transfer across the flow is simply the enthalpy difference $(h_e - h_w)$ or, in other words, the temperature difference $(T_e - T_w)$ across the flow. However, as we have emphasized, if u_e is not negligible, then viscous dissipation becomes another factor that drives the heat transfer across the flow.

Of particular practical interest is the heat flux at the walls—the *aerodynamic heating* as we label it here. We denote the heat transfer at a wall as $\dot{q}_w$. Moreover, it is conventional to quote aerodynamic heating at a wall without any sign convention. For example, if the heat transfer from the fluid to the wall is 10 W/cm^2, or, if in reverse the heat transfer from the wall to the fluid is 10 W/cm^2, it is simply quoted as such; in both cases, $\dot{q}_w$ is given as 10 W/cm^2 without any sign convention. In this sense, we write Eq. (16.18) as

$$\dot{q}_w = \frac{k}{c_p}\left|\frac{\partial h}{\partial y}\right|_w = \frac{\mu}{\text{Pr}}\left|\frac{\partial h}{\partial y}\right|_w \tag{16.23}$$

where the subscript w implies conditions at the wall. The direction of the net heat transfer at the wall, whether it is from the fluid to the wall or from the wall to the fluid, is easily seen from the temperature gradient at the wall; if the wall is cooler than the adjacent fluid, heat is transferred into the wall, and if the wall is hotter than the adjacent fluid, heat is transferred into the fluid. Another criterion is to compare the wall temperature with the adiabatic wall temperature, to be defined shortly.

Return to the picture of Couette flow in Fig. 16.2. To calculate the heat transfer at the lower wall, use Eq. (16.23) with the enthalpy gradient given by Eq. (16.19) evaluated at $y = 0$.

$$\text{At } y = 0: \qquad \dot{q}_w = \frac{\mu}{\text{Pr}}\left|\frac{h_e - h_w + \frac{1}{2}\text{Pr}\, u_e^2}{D}\right| \tag{16.24}$$

To calculate the heat transfer at the upper wall, use Eq. (16.23) with the enthalpy gradient given by Eq. (16.19) evaluated at $y = D$. In this case, Eq. (16.19) yields

$$\frac{\partial h}{\partial y} = \frac{h_e - h_w + \frac{1}{2} \Pr u_e^2}{D} - \frac{\Pr u_e^2}{D} = \frac{h_e - h_w - \frac{1}{2} \Pr u_e^2}{D}$$

In turn, from Eq. (16.23),

$$\boxed{\text{At } y = D: \qquad \dot{q}_w = \frac{\mu}{\Pr} \left| \frac{h_e - h_w - \frac{1}{2} \Pr u_e^2}{D} \right|} \qquad (16.25)$$

Let us examine the above results for three different scenarios, namely, (1) *negligible viscous dissipation*, (2) *equal wall temperature*, and (3) *adiabatic wall conditions* (no heat transfer to the wall). In the process, we define three important concepts in the analysis of aerodynamic heating: (1) *adiabatic wall temperature*, (2) *recovery factor*, and (3) *Reynolds analogy*.

16.3.1 Negligible Viscous Dissipation

To some extent, we have already discussed this case in regard to the local heat flux at any point within the flow. If u_e is very small, hence τ is very small, then the amount of viscous dissipation is negligibly small, and Eq. (16.21) becomes

$$\dot{q} = -\frac{\mu}{\Pr} \left(\frac{h_e - h_w}{D} \right) \qquad (16.26)$$

Clearly, for this case, the heat flux is constant across the flow. Moreover, the enthalpy profile given by Eq. (16.16) becomes

$$h = h_w + (h_e - h_w) \frac{y}{D} \qquad (16.27)$$

Since $h = c_p T$, the temperature profile is identical to the enthalpy profile:

$$T = T_w + (T_e - T_w) \frac{y}{D} \qquad (16.28)$$

Note that the temperature varies *linearly* across the flow, as sketched in Fig. 16.4.

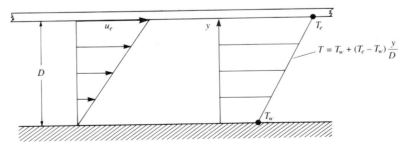

FIGURE 16.4
Couette flow temperature profile for negligible viscous dissipation.

The case shown here is for the upper wall at a higher temperature than the lower wall. The heat transfer at the lower wall is obtained from Eq. (16.24) with a negligible u_e:

At $y = 0$:
$$\dot{q}_w = \frac{\mu}{\text{Pr}} \left| \frac{h_e - h_w}{D} \right| \tag{16.29}$$

The heat transfer at the upper wall is similarly obtained as

At $y = D$:
$$\dot{q}_w = \frac{\mu}{\text{Pr}} \left| \frac{h_e - h_w}{D} \right| \tag{16.30}$$

Equations (16.29) and (16.30) are identical; this is no surprise, since we have already shown that the heat flux is constant across the flow, as shown by Eq. (16.26), and therefore the heat transfer at both walls should be the same. Equations (16.29) and (16.30) can also be written in terms of temperature as

$$\dot{q}_w = k \left| \frac{T_e - T_w}{D} \right| \tag{16.31}$$

Examining Eqs. (16.29) to (16.31), we can make some conclusions which can be generalized to most viscous flow problems, as follows:

1. Everything else being equal, the larger the temperature difference across the viscous layer, the greater the heat transfer at the wall. The temperature difference $(T_e - T_w)$ or the enthalpy difference $(h_e - h_w)$ takes on the role of a "driving potential" for heat transfer. For the special case treated here, the heat transfer at the wall is directly proportional to this driving potential.
2. Everything else being equal, the thicker the viscous layer (the larger D is) the smaller the heat transfer is at the wall. For the special case treated here, $\dot{q}_w$ is inversely proportional to D.
3. Heat flows from a region of high temperature to low temperature. For negligible viscous dissipation, if the temperature at the top of the viscous layer is higher than that at the bottom, heat flows from the top to the bottom. In the case sketched in Fig. 16.4, heat is transferred from the upper plate into the fluid, and then is transferred from the fluid to the lower plate.

16.3.2 Equal Wall Temperatures

Here we assume that $T_e = T_w$; i.e., $h_e = h_w$. The enthalpy profile for this case, from Eq. (16.16), is

$$h = h_w + \frac{1}{2} \text{Pr}\, u_e^2 \left(\frac{y}{D} \right) - \frac{1}{2} \text{Pr}\, u_e^2 \left(\frac{y}{D} \right)^2$$

or
$$= h_w + \frac{1}{2} \text{Pr}\, u_e^2 \left[\frac{y}{D} - \left(\frac{y}{D} \right)^2 \right] \tag{16.32}$$

In terms of temperature, this becomes

$$T = T_w + \frac{\text{Pr } u_e^2}{2c_p}\left[\frac{y}{D} - \left(\frac{y}{D}\right)^2\right] \tag{16.33}$$

Note that the temperature varies parabolically with y, as sketched in Fig. 16.5. The maximum value of temperature occurs at the midpoint, $y = D/2$. This maximum value is obtained by evaluating Eq. (16.33) at $y = D/2$.

$$T_{\max} = T_w + \frac{\text{Pr } u_e^2}{8c_p} \tag{16.34}$$

The heat transfer at the walls is obtained from Eqs. (16.24) and (16.25) as

At $y = 0$: $$\dot{q}_w = \frac{\mu}{2}\frac{u_e^2}{D} \tag{16.35}$$

At $y = D$: $$\dot{q}_w = \frac{\mu}{2}\frac{u_e^2}{D} \tag{16.36}$$

Equations (16.35) and (16.36) are identical; the heat transfers at the upper and lower walls are equal. In this case, as can be seen by inspecting the temperature distribution shown in Fig. 16.5, the upper and lower walls are both cooler than the adjacent fluid. Hence, at both the upper and lower walls, heat is transferred from the fluid to the wall.

Question: Since the walls are at equal temperature, where is the heat transfer coming from? Answer: Viscous dissipation. The local temperature increase in the flow as sketched in Fig. 16.5 is due solely to viscous dissipation within the fluid. In turn, both walls experience an aerodynamic heating effect due to this viscous dissipation. This is clearly evident in Eqs. (16.35) and (16.36), where $\dot{q}_w$ depends on the velocity, u_e. Indeed, $\dot{q}_w$ is directly proportional to the square of u_e. In light of Eq. (16.9), Eqs. (16.35) and (16.36) can be written as

$$\dot{q}_w = \tau\left(\frac{u_e}{2}\right) \tag{16.37}$$

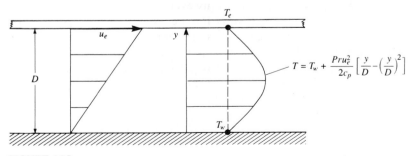

FIGURE 16.5
Couette flow temperature profile for equal wall temperature with viscous dissipation.

which further emphasizes that $\dot{q}_w$ is due entirely to the action of shear stress in the flow. From Eqs. (16.35) to (16.37), we can make the following conclusions that reflect general properties of most viscous flows:

1. Everything else being equal, aerodynamic heating increases as the flow velocity increases. This is why aerodynamic heating becomes an important design factor in high-speed aerodynamics. Indeed, for most hypersonic vehicles, you can begin to appreciate that viscous dissipation generates extreme temperatures within the boundary layer adjacent to the vehicle surface and frequently makes aerodynamic heating the dominant design factor. In the Couette flow case shown here—a far cry from hypersonic flow—we see that $\dot{q}_w$ varies directly as u_e^2.

2. Everything else being equal, aerodynamic heating decreases as the thickness of the viscous layer increases. For the case considered here, $\dot{q}_w$ is inversely proportional to D. This conclusion is the same as that made for the above case of negligible viscous dissipation but with unequal wall temperature.

16.3.3 Adiabatic Wall Conditions (Adiabatic Wall Temperature)

Let us imagine the following situation. Assume that the flow illustrated in Fig. 16.5 is established. We have the parabolic temperature profile established as shown, and we have heat transfer into the walls as discussed above. However, both wall temperatures are considered *fixed*, and both are equal to the same constant value. *Question*: How can the wall temperature remain fixed at the same time that heat is transferred into the wall? *Answer*: There must be some independent mechanism that conducts heat away from the wall at the same rate that the aerodynamic heating is pumping heat into the wall. This is the only way for the wall temperature to remain fixed at some cooler temperature than the adjacent fluid. For example, the wall can be some vast heat sink that can absorb heat without any appreciable change in temperature, or possibly there are cooling coils within the plate that can carry away the heat, much like the water coils that keep the engine of your automobile cool. In any event, to have the picture shown in Fig. 16.5 with a constant wall temperature independent of time, some exterior mechanism must carry away the heat that is transferred from the fluid to the walls. Now imagine that, at the lower wall, this exterior mechanism is suddenly shut off. The lower wall will now begin to grow hotter in response to $\dot{q}_w$, and T_w will begin to increase with time. At any given instant during this transient process, the heat transfer to the lower wall is given by Eq. (16.24), repeated below:

$$\dot{q}_w = \frac{\mu}{\mathrm{Pr}} \left| \frac{h_e - h_w + \frac{1}{2} \mathrm{Pr} \, u_e^2}{D} \right| \tag{16.24}$$

At time $t = 0$, when the exterior cooling mechanism is just shut off, $h_w = h_e$, and

$\dot{q}_w$ is given by Eq. (16.35), namely,

At time $t = 0$:
$$\dot{q}_w = \frac{\mu}{2}\frac{u_e^2}{D}$$

However, as time now progresses, T_w (and therefore h_w) increases. From Eq. (16.24), as h_w increases, the numerator decreases in magnitude, and hence $\dot{q}_w$ decreases. That is,

At $t > 0$:
$$\dot{q}_w < \frac{\mu}{2}\frac{u_e^2}{D}$$

Hence, as time progresses from when the exterior cooling mechanism was first cut off at the lower wall, the wall temperature increases, and the aerodynamic heating to the wall decreases. This in turn slows the rate of increase of T_w as time progresses. The transient variations of both $\dot{q}_w$ and T_w are sketched in Fig. 16.6. In Fig. 16.6a, we see that, as time increases to large values, the heat transfer to the wall approaches zero—*this is defined as the equilibrium, or the adiabatic wall condition.* For an *adiabatic* wall, the heat transfer is, by definition, equal to *zero*. Simultaneously, the wall temperature, T_w, approaches asymptotically a limiting value defined as the *adiabatic wall temperature*, T_{aw}, and the corresponding enthalpy is defined as the *adiabatic wall enthalpy*, h_{aw}.

The purpose of the above discussion is to define an adiabatic wall condition; the example involving a timewise approach to this condition was just for convenience and edification. Let us now assume that the lower wall in our Couette flow is an adiabatic wall. For this case, we already know the value of heat transfer to the wall—by definition, it is zero. The question now becomes, What is the value of the adiabatic wall enthalpy, h_{aw}, and in turn the adiabatic wall temperature, T_{aw}? The answer is given by Eq. (16.23), where $\dot{q}_w = 0$ for an

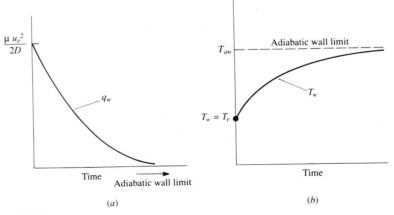

(a)   *(b)*

FIGURE 16.6
Illustration for the definition of an adiabatic wall and the adiabatic wall temperature.

adiabatic wall.

Adiabatic wall: $\qquad\qquad \dot{q}_w = 0 \rightarrow \left(\dfrac{\partial h}{\partial y}\right)_w = \left(\dfrac{\partial T}{\partial y}\right)_w = 0$ $\qquad\qquad$ (16.38)

Therefore, from Eq. (16.19), with $\partial h/\partial y = 0$, $y = 0$, and $h_w = h_{aw}$, by definition,

$$h_e - h_{aw} + \tfrac{1}{2}\, \text{Pr}\, u_e^2 = 0$$

or

$$\boxed{h_{aw} = h_e + \text{Pr}\,\frac{u_e^2}{2}} \qquad\qquad (16.39)$$

In turn, the adiabatic wall temperature is given by

$$\boxed{T_{aw} = T_e + \text{Pr}\,\frac{u_e^2}{2c_p}} \qquad\qquad (16.40)$$

Clearly, the higher the value of u_e, the higher is the adiabatic wall temperature.
The enthalpy profile across the flow for this case is given by a combination of Eqs. (16.16) and (16.40), as follows. Setting $h_w = h_{aw}$ in Eq. (16.16), we obtain

$$h = h_{aw} + \left(h_e + h_{aw} + \text{Pr}\,\frac{u_e^2}{2}\right)\frac{y}{D} - \frac{\text{Pr}}{2}\,u_e^2\left(\frac{y}{D}\right)^2 \qquad\qquad (16.41)$$

From Eq. (16.39),

$$h_e - h_{aw} = -\text{Pr}\,\frac{u_e^2}{2} \qquad\qquad (16.42)$$

Inserting Eq. (16.42) into (16.41), we have

$$h = h_{aw} - \text{Pr}\,\frac{u_e^2}{2}\left(\frac{y}{D}\right)^2 \qquad\qquad (16.43)$$

Equation (16.43) gives the enthalpy profile across the flow. The temperature profile follows from Eq. (16.43) as

$$T = T_{aw} - \text{Pr}\,\frac{u_e^2}{2c_p}\left(\frac{y}{D}\right)^2 \qquad\qquad (16.44)$$

This variation of T is sketched in Fig. 16.7. Note that T_{aw} is the maximum temperature in the flow. Moreover, the temperature curve is perpendicular at the plate for $y = 0$; i.e., the temperature gradient at the lower plate is zero, as expected for an adiabatic wall. This result is also obtained by differentiating Eq. (16.44):

$$\frac{\partial T}{\partial y} = -\text{Pr}\,\frac{u_e^2}{c_p D}\left(\frac{y}{D}\right)$$

which gives $\partial T/\partial y = 0$ at $y = 0$.

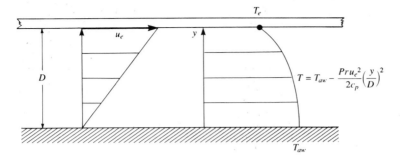

FIGURE 16.7
Couette flow temperature profile for an adiabatic lower wall.

16.3.4 Recovery Factor

As a corollary to the above case for the adiabatic wall, we take this opportunity to define the recovery factor—a useful engineering parameter in the analysis of aerodynamic heating. The total enthalpy of the flow at the upper plate (which represents the upper boundary on a viscous shear layer) is, by definition,

$$h_0 = h_e + \frac{u_e^2}{2} \tag{16.45}$$

(The significance and definition of total enthalpy are discussed in Sec. 7.5.) Compare Eq. (16.45), which is a general definition, with Eq. (16.39), repeated below, which is for the special case of Couette flow:

$$h_{aw} = h_e + \mathrm{Pr}\,\frac{u_e^2}{2} \tag{16.39}$$

Note that h_{aw} is different from h_0, the difference provided by the value of Pr as it appears in Eq. (16.39). We now *generalize* Eq. (16.39) to a form which holds for any viscous flow, as follows:

$$h_{aw} = h_e + r\frac{u_e^2}{2} \tag{16.46a}$$

Similarly, Eq. (16.40) can be generalized to

$$T_{aw} = T_e + r\frac{u_e^2}{2c_p} \tag{16.46b}$$

In Eqs. (16.46a and b), r is defined as the *recovery factor*. It is the factor that tells us how close the adiabatic wall enthalpy is to the total enthalpy at the upper boundary of the viscous flow. If $r = 1$, then $h_{aw} = h_0$. An alternate expression for the recovery factor can be obtained by combining Eqs. (16.46) and (16.45) as follows. From Eq. (16.46),

$$r = \frac{h_{aw} - h_e}{u_e^2/2} \tag{16.47}$$

From Eq. (16.45),

$$\frac{u_e^2}{2} = h_0 - h_e \tag{16.48}$$

Inserting Eq. (16.48) into (16.47), we have

$$\boxed{r = \frac{h_{aw} - h_e}{h_0 - h_e} = \frac{T_{aw} - T_e}{T_0 - T_e}} \tag{16.49}$$

where T_0 is the total temperature. Equation (16.49) is frequently used as an alternate definition of the recovery factor.

In the special case of Couette flow, by comparing Eq. (16.39) or (16.40) with Eq. (16.46a) or (16.46b), we find that

$$r = \text{Pr} \tag{16.50}$$

For Couette flow, the recovery factor is simply the Prandtl number. Note that, if $\text{Pr} < 1$, then $h_{aw} < h_0$; conversely, if $\text{Pr} > 1$, then $h_{aw} > h_0$.

In more general viscous flow cases, the recovery factor is not simply the Prandtl number; however, in general, for incompressible viscous flows, we will find that the recovery *factor* is some *function* of Pr. Hence, the Prandtl number is playing its role as an important viscous flow parameter. As expected from Sec. 15.6, for a compressible viscous flow, the recovery factor is a function of Pr along with the Mach number and the ratio of specific heats.

16.3.5 Reynolds Analogy

Another useful engineering relation for the analysis of aerodynamic heating is Reynolds analogy, which can easily be introduced within the context of our discussion of Couette flow. Reynolds analogy is a relation between the skin friction coefficient and the heat transfer coefficient. The skin friction coefficient, c_f, was first introduced in Sec. 1.5. In our context here, we define the skin friction coefficient as

$$c_f = \frac{\tau_w}{\frac{1}{2}\rho_e u_e^2} \tag{16.51}$$

From Eq. (16.9), we have, for Couette flow,

$$\tau_w = \mu \left(\frac{u_e}{D}\right) \tag{16.52}$$

Combining Eqs. (16.51) and (16.52), we have

$$c_f = \frac{\mu(u_e/D)}{\frac{1}{2}\rho_e u_e^2} = \frac{2\mu}{\rho_e u_e D} \tag{16.53}$$

Let us define the Reynolds number for Couette flow as

$$\mathrm{Re} = \frac{\rho_e u_e D}{\mu}$$

Then, Eq. (16.53) becomes

$$c_f = \frac{2}{\mathrm{Re}} \tag{16.54}$$

Equation (16.54) is interesting in its own right. It demonstrates that the skin friction coefficient is a function of just the Reynolds number—a result which applies in general for other incompressible viscous flows [although the function is not necessarily the same as given in Eq. (16.54)].

Now let us define a *heat transfer coefficient* as

$$C_H = \frac{\dot{q}_w}{\rho_e u_e(h_{aw} - h_e)} \tag{16.55}$$

In Eq. (16.55), C_H is called the *Stanton number*; it is one of several different types of heat transfer coefficient that is used in the analysis of aerodynamic heating. It is a dimensionless quantity, in the same vein as the skin friction coefficient. For Couette flow, from Eq. (16.24), and dropping the absolute value signs for convenience, we have

$$\dot{q}_w = \frac{\mu}{\mathrm{Pr}}\left(\frac{h_e - h_w + \frac{1}{2}\,\mathrm{Pr}\,u_e^2}{D}\right) \tag{16.56}$$

Inserting Eq. (16.39) into (16.56), we have for Couette flow

$$\dot{q}_w = \frac{\mu}{\mathrm{Pr}}\left(\frac{h_{aw} - h_w}{D}\right) \tag{16.57}$$

Inserting Eq. (16.57) into (16.55), we obtain

$$C_H = \frac{(\mu/\mathrm{Pr})[(h_{aw} - h_w)/D]}{\rho_e u_e(h_{aw} - h_w)} = \frac{\mu/\mathrm{Pr}}{\rho_e u_e D} = \frac{1}{\mathrm{Re}\,\mathrm{Pr}} \tag{16.58}$$

Equation (16.58) is interesting in its own right. It demonstrates that the Stanton number is a function of the Reynolds number and Prandtl number—a result that applies generally for other incompressible viscous flows [although the function is not necessarily the same as given in Eq. (16.58)].

We now combine the results for c_f and C_H obtained above. From Eqs. (16.54) and (16.58), we have

$$\frac{C_H}{c_f} = \left(\frac{1}{\text{Re Pr}}\right)\frac{\text{Re}}{2}$$

or

$$\boxed{\frac{C_H}{c_f} = \frac{1}{2}\,\text{Pr}^{-1}}$$

(16.59)

Equation (16.59) is Reynolds analogy as applied to Couette flow. Reynolds analogy is, in general, a relation between the heat transfer coefficient and the skin friction coefficient. For Couette flow, this relation is given by Eq. (16.59). Note that the ratio C_H/c_f is simply a function of the Prandtl number—a result that applies usually for other incompressible viscous flows, although not necessarily the same function as given in Eq. (16.59).

16.3.6 Interim Summary

In this section, we have studied incompressible Couette flow. Although it is a somewhat academic flow, it has all the trappings of many practical viscous flow problems, with the added advantage of lending itself to a simple, straightforward solution. We have taken this advantage, and have discussed incompressible Couette flow in great detail. Our major purpose in this discussion is to make the reader familiar with many concepts used in general in the analysis of viscous flows without clouding the picture with more fluid dynamic complexities. In the context of our study of Couette flow, we have one additional question to address, namely, What is the effect of compressibility? This question is addressed in the next section.

Example 16.1. Consider the geometry sketched in Fig. 16.2. The velocity of the upper plate is 200 ft/s, and the two plates are separated by a distance of 0.01 in. The fluid between the plates is air. Assume incompressible flow. The temperature of both plates is the standard sea level value of 519°R.

    (a) Calculate the velocity in the middle of the flow.
    (b) Calculate the shear stress.
    (c) Calculate the maximum temperature in the flow.
    (d) Calculate the heat transfer to either wall.
    (e) If the lower wall is suddenly made adiabatic, calculate its temperature.

Solution. Assume that μ is constant throughout the flow, and that it is equal to its value of 3.7373×10^{-7} slug/ft/s at the standard sea level temperature of 519°R.
    (a) From Eq. (16.6),

$$u = u_e\left(\frac{y}{D}\right)$$

$$u = (200)\left(\frac{1}{2}\right) = \boxed{100 \text{ ft/s}}$$

(b) From Eq. (16.9),

$$\tau_w = \mu \frac{u_e}{D}, \qquad \text{where } D = 0.01 \text{ in} = 8.33 \times 10^{-4} \text{ ft}$$

$$\tau_w = \frac{(3.7373 \times 10^{-7})(200)}{8.33 \times 10^{-4}} = \boxed{0.09 \text{ lb/ft}^2}$$

Note that the shear stress is relatively small—less than a tenth of a pound acting over a square foot.

(c) From Eq. (16.34), for equal wall temperatures, the maximum temperature, which occurs at $y/D = 0.5$, is

$$T = T_w + \frac{\text{Pr}}{c_p} \frac{u_e^2}{2} \left[\frac{y}{D} - \left(\frac{y}{D} \right)^2 \right] = T_w + \frac{\text{Pr } u_e^2}{8 c_p}$$

For air at standard conditions, $\text{Pr} = 0.71$ and $c_p = 6006 \text{ (ft} \cdot \text{lb)/(slug} \cdot {}^\circ\text{R)}$. Hence,

$$T = 519 + \frac{(0.71)(200)^2}{8(6006)} = 519 + 0.6 = \boxed{519.6{}^\circ\text{R}}$$

Notice that the maximum temperature in the flow is only six-tenths of a degree above the wall temperature—viscous dissipation for this relatively low-speed case is very small. This certainly justifies our assumption of constant ρ, μ, and k in this section, and gives us a feeling for the energy changes associated with an essentially incompressible flow—they are very small.

(d) From Eq. (16.35),

$$\dot{q}_w = \frac{\mu}{2} \left(\frac{u_e^2}{D} \right) = \frac{(3.7373 \times 10^{-7})(200)^2}{(2)(8.33 \times 10^{-4})} = \boxed{8.97 \text{ (ft} \cdot \text{lb)/(ft}^2/\text{s)}}$$

Since there are 778 ft · lb to a Btu (British thermal unit), then

$$\dot{q}_w = 8.97 \text{ (ft} \cdot \text{lb)/(ft}^2/\text{s)} = 0.0115 \text{ Btu/(ft}^2/\text{s)}$$

(e) From Eq. (16.40),

$$T_{aw} = T_e + \frac{\text{Pr}}{c_p} \left(\frac{u_e^2}{2} \right) = 519 + \frac{(0.71)(200)^2}{(2)(6006)}$$

$$= 519 + 2.36 = \boxed{521.36{}^\circ\text{R}}$$

Note in the above example that the adiabatic wall temperature is higher than the maximum flow temperature calculated in part (c) for the cold wall case. In general, for cold wall cases, the viscous dissipation in the flow is not sufficient to heat the gas anywhere in the flow to a temperature as high as the adiabatic wall temperature. Also, we again note the comparatively low temperature increase—T_{aw} is only 2.36° higher than the upper wall temperature. In contrast, for the compressible flow to be treated in the next section, the temperature increases can be substantial—this is one of the major aspects that distinguishes compressible viscous flow from incompressible viscous flow. Note that, in the present problem, the Mach number of the upper plate is

$$M_e = \frac{u_e}{a_e} = \frac{u_e}{\sqrt{\gamma R T_e}} = \frac{200}{\sqrt{(1.4)(1716)(519)}} = 0.18$$

Again, this certainly justifies our assumption of incompressible flow for this problem.

16.4 COMPRESSIBLE COUETTE FLOW

Return to Fig. 6.2, which is our general model of Couette flow. We now assume that u_e is large enough; hence, the changes in temperature within the flow are substantial enough, so that we must treat ρ, μ, and k as variables—this is compressible Couette flow. Since $T = T(y)$, then $\mu = \mu(y)$ and $k = k(y)$. Also, since $\partial p/\partial x = 0$ from the geometry and $\partial p/\partial y = 0$ from Eq. (16.2), then the pressure is constant throughout the compressible Couette flow, just as in the incompressible case discussed in Sec. 16.3. From the equation of state, we have $\rho = p/RT$; because $T = T(y)$, ρ is also a function of y, varying inversely with temperature.

The governing equations for compressible Couette flow are Eqs. (16.1) to (16.3), with μ and k as variables. Let us arrange these equations in a form convenient for solution. From Eq. (16.1),

$$\frac{\partial}{\partial y}\left(\mu\frac{\partial u}{\partial y}\right) = \frac{\partial \tau}{\partial y} = 0 \tag{16.60}$$

or
$$\tau = \text{const} \tag{16.61}$$

Hence, just as in the incompressible case, the shear stress is constant across the flow. However, keep in mind that $\mu = \mu(y)$, and, from $\tau = \mu(\partial u/\partial y)$, clearly the velocity gradient, $\partial u/\partial y$, is *not* constant across the flow—this is an essential difference between compressible and incompressible flows. With all this in mind, Eq. (16.3), repeated below

$$\frac{\partial}{\partial y}\left(k\frac{\partial T}{\partial y}\right) + \frac{\partial}{\partial y}\left(\mu u\frac{\partial u}{\partial y}\right) = 0 \tag{16.3}$$

can be written as

$$\frac{\partial}{\partial y}\left(k\frac{\partial T}{\partial y}\right) + \tau\frac{\partial \mu}{\partial y} = 0 \tag{16.62}$$

The temperature variation of μ is accurately given by Sutherland's law, Eq. (15.3), for the temperature range of interest in this book. Hence, $d\mu/dT$ is easily obtained by differentiating Eq. (15.3) with respect to T. In turn, we can write

$$\frac{d\mu}{dy} = \frac{d\mu}{dT}\left(\frac{dT}{dy}\right) \tag{16.63}$$

Also, the solution for compressible Couette flow requires a numerical solution of Eq. (16.62). Note that, with μ and k as variables, Eq. (16.62) is a *nonlinear* differential equation, and for the conditions stated, it does not have a neat, closed-form, analytic solution. Recognizing the need for a numerical solution, let us write Eq. (16.62) in terms of the ordinary differential equation that it really is. (Recall that we have been using the partial differential notation only as a carry-over from the Navier-Stokes equations and to make the equations for our study of Couette flow look more familiar when treating the two-dimensional and three-dimensional viscous flows discussed in Chaps. 17 and 18—just a pedagogical ploy on our part.) Also, combining Eq. (16.63) with Eq. (16.62) written in terms of ordinary differentials, we have

$$\frac{d}{dy}\left(k\frac{dT}{dy}\right) + \tau\frac{d\mu}{dT}\left(\frac{dT}{dy}\right) = 0 \qquad (16.64)$$

Equation (16.64) must be solved between $y = 0$, where $T = T_w$, and $y = D$, where $T = T_e$. Note that the boundary conditions must be satisfied at two different locations in the flow, namely, at $y = 0$ and $y = D$; this is called a *two-point boundary value problem*. We present two approaches to the numerical solution of this problem. Both approaches are used for the solutions of more complex viscous flows to be discussed in Chaps. 17 and 18, and that is why they are presented here in the context of Couette flow—simply to "break the ice" for our subsequent discussions.

16.4.1 Shooting Method

This method is a classic method for the solution of the boundary-layer equations to be discussed in Chap. 17. For the solution of compressible Couette flow, the same philosophy follows as that to be applied to boundary-layer solutions, and that is why we discuss it now. The method involves a double iteration, i.e., two minor iterations nested within a major iteration. The scheme is as follows:

1. *Assume* a value for τ in Eq. (16.64). A reasonable assumption to start with is the incompressible value, $\tau = \mu(u_e/D)$.
2. Starting at $y = 0$ with the known boundary condition $T = T_w$, integrate Eq. (16.64) across the flow until $y = D$. Use any standard numerical technique for ordinary differential equations, such as the well-known Runge-Kutta method (see, e.g., Ref. 52). However, to start this numerical integration, because Eq. (16.64) is second order, *two* boundary conditions must be specified at $y = 0$. We only have one physical condition, namely, $T = T_w$. Therefore, we have to *assume* a second condition; let us assume a value for the temperature gradient at the wall, i.e., assume a value for $(dT/dy)_w$. A value based on the incompressible flow solution discussed in Sec. 16.3 would be a reasonable assumption. With the *assumed* $(dT/dy)_w$ and the *known* T_w at $y = 0$, then Eq. (16.64) is integrated numerically away from the wall, starting at $y = 0$ and moving in small increments, Δy, in the direction of increasing y. Values of T at each increment in y are produced by the numerical algorithm.
3. Stop the numerical integration when $y = D$ is reached. Check to see if the numerical value of T at $y = D$ equals the specified boundary condition, $T = T_e$. Most likely, it will not because we have had to assume a value for $(dT/dy)_w$ in step 2. Hence, return to step 2, assume another value of $(dT/dy)_w$, and repeat the integration. Continue to repeat steps 2 and 3 until convergence is

obtained, i.e., until a value of $(dT/dy)_w$ is found such that, after the numerical integration, $T = T_e$ at $y = D$. From the converged temperature profile obtained by repetition of steps 2 and 3, we now have numerical values for T as a function of y that satisfy both boundary conditions; i.e., $T = T_w$ at the lower wall and $T = T_e$ at the upper wall. However, do not forget that this converged solution was obtained for the *assumed* value of τ in step 1. Therefore, the converged profile for T is not necessarily the correct profile. We must continue further; this time to find the correct value for τ.

4. From the converged temperature profile obtained by the repetitive iteration in steps 2 and 3, we can obtain $\mu = \mu(y)$ from Eq. (15.3).

5. From the definition of shear stress,

$$\tau = \mu \frac{du}{dy}$$

and we have

$$\frac{du}{dy} = \frac{\tau}{\mu} \tag{16.65}$$

Recall from the solution of the momentum equation, Eq. (16.60), that τ is a constant. Using the assumed value of τ from step 1, and the values of $\mu = \mu(y)$ from step 4, numerically integrate Eq. (16.65) starting at $y = 0$ and using the known boundary condition $u = 0$ at $y = 0$. Since Eq. (16.65) is first order, this single boundary condition is sufficient to initiate the numerical integration. Values of u at each increment in y, Δy, are produced by the numerical algorithm.

6. Stop the numerical integration when $y = D$ is reached. Check to see if the numerical value of u at $y = D$ equals the specified boundary condition, $u = u_e$. Most likely, it will not, because we have had to *assume* a value of τ all the way back in step 1, which has carried through to this point in our iterative solution. Hence, return to step 5, assume another value for τ, and repeat the integration of Eq. (16.65). Continue to repeat steps 5 and 6 [using the same values of $\mu = \mu(y)$ from step 4] until convergence is obtained, i.e., until a value of τ is found that, after the numerical integration of Eq. (16.65), $u = u_e$ at $y = D$. From the converged velocity profile obtained by repetition of steps 5 and 6, we now have numerical values for u as a function of y that satisfy both boundary conditions; i.e., $u = 0$ at $y = 0$ and $u = u_e$ at $y = D$. However, do not forget that this converged solution was obtained using $\mu = \mu(y)$ from step 4, which was obtained using the initially assumed τ from step 1. Therefore, the converged profile for u obtained here is not necessarily the correct profile. We must continue one big step further.

7. Return to step 2, using the value of τ obtained from step 6. Repeat steps 2 through 7 until total convergence is obtained. When this double iteration is completed, then the profile for $T = T(y)$ obtained at the last cycle of step 3, the profile for $u = u(y)$ obtained at the last cycle of step 6, and the value of τ obtained at the last cycle of step 7 are all the correct values for the given boundary conditions. The problem is solved!

Looking over the shooting method as described above, we see two minor iterations nested within a major iteration. Steps 2 and 3 constitute the first minor iteration and provide ultimately the temperature profile. Steps 5 and 6 are the second minor iteration and provide ultimately the velocity profile. Steps 2 to 7 constitute the major iteration and ultimately result in the proper value of τ.

The shooting method described above for the solution of compressible Couette flow is carried over almost directly for the solution of the boundary-layer equations to be described in Chap. 17. In the same vein, there is another completely different approach to the solution of compressible Couette flow which carries over directly for the solution of the Navier-Stokes equations to be described in Chap. 18. This is the time-dependent, finite-difference method, first discussed in Chap. 13 and applied to the inviscid flow over a supersonic blunt body in Sec. 13.5. In order to prepare ourselves for Chap. 18, we briefly discuss the application of this method to the solution of compressible Couette flow.

16.4.2   Time-Dependent Finite-Difference Method

Return to the picture of Couette flow in Fig. 16.2. Imagine, for a moment, that the space between the upper and lower plates is filled with a flow field which is *not* a Couette flow; e.g., imagine some arbitrary flow field with gradients in both the x and y directions, including gradients in pressure. We can imagine such a flow existing at some instant during the start-up process just after the upper plate is set into motion. This would be a transient flow field, where u, T, ρ, etc., would be functions of time, t, as well as of x and y. Finally, after enough time elapses, the flow will approach a steady state, and this steady state will be the Couette flow solution discussed above. Let us track this picture numerically. That is, starting from an assumed initial flow field at time $t = 0$, let us solve the unsteady Navier-Stokes equations in steps of time until a steady flow is obtained at large times. As discussed in Sec. 13.5, the time-asymptotic steady flow is the desired result; the time-dependent approach is just a means to that end. At this stage in our discussion, it would be well for you to review the philosophy (not the details) presented in Sec. 13.5 before progressing further.

The Navier-Stokes equations are given by Eqs. (15.18a to c) and (15.26). For an unsteady, two-dimensional flow, they are

Continuity:

$$\frac{\partial \rho}{\partial t} = -\frac{\partial(\rho u)}{\partial x} - \frac{\partial(\rho v)}{\partial y} \tag{16.66}$$

x momentum:

$$\frac{\partial u}{\partial t} = -u\frac{\partial u}{\partial x} - v\frac{\partial u}{\partial y} - \frac{1}{\rho}\left[\frac{\partial p}{\partial x} - \frac{\partial \tau_{xx}}{\partial x} - \frac{\partial \tau_{yx}}{\partial y}\right] \tag{16.67}$$

y momentum:

$$\frac{\partial v}{\partial t} = -u\frac{\partial v}{\partial x} - v\frac{\partial v}{\partial y} - \frac{1}{\rho}\left[\frac{\partial p}{\partial y} - \frac{\partial \tau_{xy}}{\partial x} - \frac{\partial \tau_{yy}}{\partial y}\right] \tag{16.68}$$

Energy:

$$\frac{\partial(e + V^2/2)}{\partial t} = -u\frac{\partial(e + V^2/2)}{\partial x} - v\frac{\partial(e + V^2/2)}{\partial y}$$

$$+ \frac{1}{\rho}\left\{\frac{\partial}{\partial x}\left(k\frac{\partial T}{\partial x}\right) + \frac{\partial}{\partial y}\left(k\frac{\partial T}{\partial y}\right) - \frac{\partial(pu)}{\partial x} - \frac{\partial(pv)}{\partial y}\right.$$

$$\left. + \frac{\partial(u\tau_{xx})}{\partial x} + \frac{\partial(u\tau_{yx})}{\partial y} + \frac{\partial(v\tau_{xy})}{\partial x} + \frac{\partial(v\tau_{yy})}{\partial y}\right\} \qquad (16.69)$$

Note that Eqs. (16.66) to (16.69) are written with the time derivatives on the left-hand side and spatial derivatives on the right-hand side. These are analogous to the form of the Euler equations given by Eqs. (13.64) to (13.67). In Eqs. (16.67) to (16.69), τ_{xy}, τ_{xx}, and τ_{yy} are given by Eqs. (15.5), (15.8), and (15.9), respectively.

The above equations can be solved by means of MacCormack's method as described in Chap. 13. This is a predictor-corrector approach, and its arrangement for the time-dependent method is described in Sec. 13.5. The application to compressible Couette flow is outlined as follows:

1. Divide the space between the two plates into a finite-difference grid, as sketched in Fig. 16.8a. The length, L, of the grid is somewhat arbitrary, but it must be longer than a certain minimum, to be described shortly.
2. At $x = 0$ (the inflow boundary), specify some inflow conditions for u, v, ρ, and T (hence, e, since $e = c_v T$). The incompressible solution for Couette flow makes reasonable inflow boundary conditions.
3. At all the remaining grid points, arbitrarily assign values for all the flow-field variables, u, v, ρ, and T. This arbitrary flow field, which constitutes the initial conditions at $t = 0$, can have finite values of v, and can include pressure gradients.
4. Starting with the initial flow field established in step 3, solve Eqs. (16.66) to (16.69) in steps of time. For example, consider the x-momentum equation in the form of Eq. (16.67). MacCormack's predictor-corrector method, applied to this equation, is as follows.

Predictor: Assume that we know the complete flow field at time t, and we wish to advance the flow-field variables to time $t + \Delta t$. Replace the spatial derivatives with forward differences:

$$\left(\frac{\partial u}{\partial t}\right)_{i,j} = -u_{i,j}\left(\frac{u_{i+1,j} - u_{i,j}}{\Delta x}\right) - v_{i,j}\left(\frac{u_{i,j+1} - u_{i,j}}{\Delta y}\right)$$

$$-\frac{1}{\rho_{i,j}}\left\{\frac{p_{i+1,j} - p_{i,j}}{\Delta x} - \left[\frac{(\tau_{xx})_{i+1,j} - (\tau_{xx})_{i,j}}{\Delta x}\right] - \left[\frac{(\tau_{yx})_{i,j+1} - (\tau_{yx})_{i,j}}{\Delta y}\right]\right\}$$

$$(16.70)$$

All the quantities on the right-hand side are known at time t; we want to advance the flow-field values to the next time, $t + \Delta t$. That is, the right-hand side of Eq. (16.70) is a known number at time t. Form the *predicted* value of

$u_{i,j}$ at time $t + \Delta t$, denoted by $\bar{u}_{i,j}$ from the first two terms of a Taylor's series as

$$\bar{u}_{i,j} = \underbrace{u_{i,j}}_{\substack{\text{Known at} \\ \text{time } t}} + \underbrace{\left(\frac{\partial u}{\partial t}\right)_{i,j} \Delta t}_{\substack{\text{Calculated} \\ \text{from} \\ \text{Eq. (16.70)}}} \tag{16.71}$$

Calculate predicted values for ρ, v, and e, namely, $\bar{\rho}_{i,j}$, $\bar{v}_{i,j}$, and $\bar{e}_{i,j}$, by the same approach applied to Eqs. (16.66), (16.68), and (16.69), respectively. Do this for all the grid points in Fig. 16.8a.

Corrector: Return to Eq. (16.67), and replace the spatial derivatives with rearward differences using the predicted (barred) quantities obtained from the predictor step.

$$\left(\frac{\partial u}{\partial t}\right)_{i,j} = -\bar{u}_{i,j}\left(\frac{\bar{u}_{i,j} - \bar{u}_{i-1,j}}{\Delta x}\right) - \bar{v}_{i,j}\left(\frac{\bar{u}_{i,j} - \bar{u}_{i,j-1}}{\Delta y}\right)$$
$$-\frac{1}{\bar{\rho}_{i,j}}\left\{\frac{\bar{p}_{i,j} - \bar{p}_{i-1,j}}{\Delta x} - \left[\frac{(\bar{\tau}_{xx})_{i,j} - (\bar{\tau}_{xx})_{i-1,j}}{\Delta x}\right] - \left[\frac{(\bar{\tau}_{yx})_{i,j} - (\bar{\tau}_{yx})_{i,j-1}}{\Delta y}\right]\right\} \tag{16.72}$$

Finally, calculate the corrected value of $u_{i,j}$ at time $t + \Delta t$, denoted by $u_{i,j}^{t+\Delta t}$, from the first two terms of a Taylor's series using an average time derivative obtained from Eqs. (16.70) and (16.72). That is,

$$u_{i,j}^{t+\Delta t} = u_{i,j}^{t} + \frac{1}{2}\left[\underbrace{\left(\frac{\partial u}{\partial t}\right)_{i,j}}_{\substack{\text{From} \\ \text{Eq. (16.70)}}} + \underbrace{\left(\frac{\overline{\partial u}}{\partial t}\right)_{i,j}}_{\substack{\text{From} \\ \text{Eq. (16.72)}}}\right]\Delta t \tag{16.73}$$

Carry out the same process using Eqs. (16.66), (16.68), and (16.69) to obtain $\rho_{i,j}^{t+\Delta t}$, $v_{i,j}^{t+\Delta t}$, and $e_{i,j}^{t+\Delta t}$. The complete flow field at time $t + \Delta t$ is now obtained.

5. Repeat step 4, except starting with the newly calculated flow-field variables at the previous time. The flow-field variables will change from one time step to the next. This transient flow field will not even have parallel streamlines; i.e., there will be finite values of v throughout the flow. This is sketched in Fig. 16.8b. Make the calculations for a large number of time steps; as we go out to large times, the changes in the flow-field variables from one time step to another will become smaller. Finally, if we go out to a large enough time (hundreds, sometimes even thousands, of time steps in some problems), the flow-field variables will not change anymore—a steady flow will be achieved, as sketched in Fig. 16.8c. Moving from left to right in Fig. 16.8c, we see a developing flow near the entrance, influenced by the assumed inflow profile. However, at the right of Fig. 16.8c, the history of the inflow has died out, and the flow-field profiles become independent of distance. Indeed, we have chosen L to be a sufficient length for this to occur. The flow field near the exit is the desired solution to the compressible Couette flow problem.

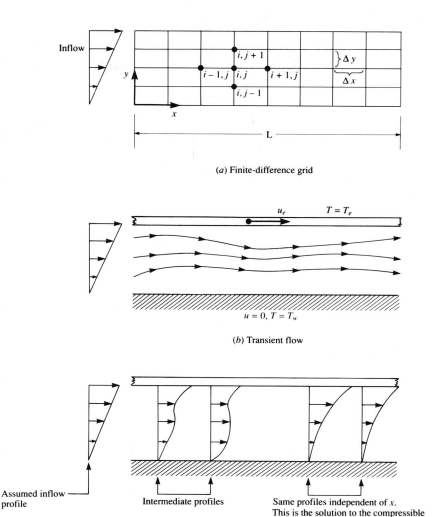

(a) Finite-difference grid

(b) Transient flow

(c) Steady-state flow

FIGURE 16.8
Illustration of the finite-difference grid, and characteristics of the flow during its transient approach
to the steady state.

The value of Δt in Eqs. (16.71) and (16.73) is not arbitrary. The steps outlined
above constitute an *explicit* finite-difference method, and hence there is a stability
bound on Δt. The value of Δt must be less than some prescribed maximum, or
else the numerical solution will become unstable and "blow up" on the computer.
A useful expression for Δt is the Courant-Friedrichs-Lewy (CFL) criterion, which

states that Δt should be the minimum of Δt_x and Δt_y, where

$$\Delta t_x = \frac{\Delta x}{u+a} \qquad \Delta t_y = \frac{\Delta y}{v+a} \tag{16.74}$$

In Eq. (16.74), a is the local speed of sound. Equation (16.74) is evaluated at every grid point, and the minimum value is used to advance the whole flow field.

The time-dependent technique described above is a common approach to the solution of the compressible Navier-Stokes equations, and for that reason, it has been outlined here. Our purpose has been not so much to outline the solution of Couette flow by means of this technique, but rather to present the technique as a precursor to our later discussions on Navier-Stokes solutions.

16.4.3 Results for Compressible Couette Flow

Some typical results for compressible Couette flow are shown in Fig. 16.9 for a cold wall case, and in Fig. 16.10 for an adiabatic lower wall case. These results are obtained from White (Ref. 43); they assume a viscosity-temperature relation of $\mu/\mu_{\max} = (T/T_{\text{ref}})^{2/3}$, which is not quite as accurate for a gas as is Sutherland's law [Eq. (15.3)]. Recall from Sec. 15.6 that a compressible viscous flow is governed by the following similarity parameters: the Mach number, the Prandtl number, and the ratio of specific heats, γ. Therefore, we expect the results for compressible Couette flow to be governed by the same parameters. Such is the case, as illustrated in Figs. 16.9 and 16.10. Here we see the different flow-field profiles for different values of the combined parameter $A = (\gamma - 1)\text{Pr}\, M_e^2$. In particular, examining Fig. 16.9 for the equal temperature, cold wall case, we note that:

1. From Fig. 16.9a, the velocity profiles are not greatly affected by compressibility. The profile labeled $A = 0$ is the familiar linear incompressible case, and that

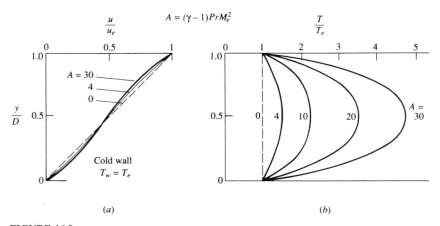

FIGURE 16.9
Velocity and temperature profiles for compressible Couette flow. Cold wall cases. (*From White, Ref. 43.*)

labeled $A = 30$ corresponds to M_e approximately 10. Clearly, the velocity profile (in terms of u/u_e versus y/D) does not change greatly over such a large range of Mach number.

2. In contrast, from Fig. 16.9b, there are huge temperature changes in the flow; these are due exclusively to viscous dissipation, which is a major effect at high Mach numbers. For example, for $A = 30$ ($M_e \approx 10$), the temperature in the middle of the flow is almost five times the wall temperature. Contrast this with the very small temperature increase calculated in Example 16.1 for an incompressible flow. This is why, on the scale in Fig. 16.9b, the incompressible case ($A = 0$) is seen as essentially a vertical line.

For the adiabatic wall case shown in Fig. 16.10, we note the following:

1. From Fig. 16.10a, the velocity profiles show a pronounced curvature due to compressibility.
2. From Fig. 16.10b, the temperature increases are larger than for the cold wall case. Note that, for $A = 30$ ($M_e \approx 10$), the maximum temperature is over 15 times that of the upper wall. Also, note the results, familiar from our discussion in Sec. 16.3, that the temperature is the largest at the adiabatic wall; i.e., T_{aw} is the maximum temperature. As expected, Fig. 16.10b shows that T_{aw} increases markedly as M_e increases.

In summary, in a general comparison between the incompressible flow discussed in Sec. 16.3 and the compressible flow discussed here, there is no tremendous *qualitative* change; i.e., there is no discontinuous change in the flow-field behavior in going from subsonic to supersonic flow as is the case for an inviscid flow, such as discussed in Part III. Qualitatively, a supersonic viscous flow is similar to a subsonic viscous flow. On the other hand, there are tremendous *quantitative*

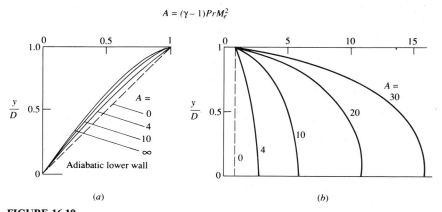

FIGURE 16.10
Velocity and temperature profiles for compressible Couette flow. Adiabatic lower wall. (*From White, Ref. 43.*)

differences, especially in regard to the large temperature changes that occur due to massive viscous dissipation in a high-speed compressible viscous flow. The physical reason for this difference in viscous versus inviscid flow is as follows. In an inviscid flow, information is propagated via the mechanism of pressure waves traveling throughout the flow. This mechanism changes radically when the flow goes from subsonic to supersonic. In contrast, for a viscous flow, information is propagated by the diffusive transport mechanisms of μ and k (a molecular phenomenon), and these mechanisms are not basically changed when the flow goes from subsonic to supersonic. These statements hold in general for any viscous flow, not just for the Couette flow case treated here.

16.4.4 Some Analytical Considerations

For air temperatures up to 1000 K, the specific heats are essentially constant, thus justifying the assumption of a calorically perfect gas for this range. Moreover, the temperature variations of μ and k over this range are virtually identical. As a result, the Prandtl number, $\mu c_p/k$, is essentially *constant* up to temperatures on the order of 1000 K. This is shown in Fig. 16.11, obtained from Schetz (Ref. 53). Note that $\text{Pr} \approx 0.71$ for air; this is the value that was used in Example 16.1.

Question: How high a Mach number can exist before we would expect to encounter temperatures in the flow above 1000 K? *Answer:* An approximate answer is to calculate that Mach number at which the total temperature is 1000 K. Assuming a static temperature $T = 288$ K, from Eq. (8.40),

$$M = \sqrt{\frac{2}{\gamma-1}\left(\frac{T_0}{T}-1\right)} = \sqrt{\frac{2}{0.4}\left(\frac{1000}{288}-1\right)} = 3.5$$

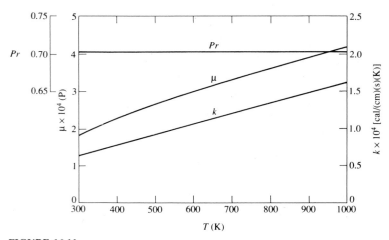

FIGURE 16.11
Variation of viscosity coefficient, thermal conductivity, and Prandtl number for air as a function of temperature. (*From Schetz, Ref. 53.*)

Hence, for most aeronautical applications involving flight at a Mach number of 3.5 or less, the temperature within the viscous portions of the flow field will not exceed 1000 K. A Mach number of 3.5 or less encompasses virtually all operational aircraft today, with the exception of a few hypersonic test vehicles.

In light of the above, many viscous flow solutions are carried out making the justifiable assumption of a *constant Prandtl number*. For the case of compressible Couette flow, the assumption of Pr = constant allows the following analysis. Consider the energy equation, Eq. (16.3), repeated below:

$$\frac{\partial}{\partial y}\left(k\frac{\partial T}{\partial y}\right) + \frac{\partial}{\partial y}\left(\mu u \frac{\partial u}{\partial y}\right) = 0 \qquad (16.3)$$

Since $T = h/c_p$ and $\Pr = \mu c_p / k$, Eq. (16.3) can be written as

$$\frac{\partial}{\partial y}\left(\frac{\mu}{\Pr}\frac{\partial h}{\partial y}\right) + \frac{\partial}{\partial y}\left(\mu u \frac{\partial u}{\partial y}\right) = 0 \qquad (16.75)$$

or

$$\frac{\partial}{\partial y}\left[\mu\left(\frac{1}{\Pr}\frac{\partial h}{\partial y} + u \frac{\partial u}{\partial y}\right)\right] = 0 \qquad (16.76)$$

Integrating Eq. (16.76) with respect to y, we have

$$\frac{1}{\Pr}\frac{\partial h}{\partial y} + u \frac{\partial u}{\partial y} = \frac{a}{\mu} \qquad (16.77)$$

Since $\tau = \mu(\partial u/\partial y)$, we have $\mu = \tau(\partial u/\partial y)^{-1}$. Also, recalling from Eq. (16.61) that τ is constant, we can write the right-hand side of Eq. (16.77) as

$$\frac{a}{\mu} = \frac{a}{\tau}\frac{\partial u}{\partial y} = b \frac{\partial u}{\partial y}$$

where b is a constant. With this, Eq. (16.77) becomes

$$\frac{1}{\Pr}\frac{\partial h}{\partial y} + \frac{\partial(u^2/2)}{\partial y} - b\frac{\partial u}{\partial y} = 0 \qquad (16.78)$$

Integrating Eq. (16.78) with respect to y, remembering that Pr = constant, we have

$$\frac{h}{\Pr} + \frac{u^2}{2} - bu = c \qquad (16.79)$$

where c is another constant of integration. Expressions for b and c can be obtained by evaluating Eq. (16.79) at $y = 0$ and $y = D$. At $y = 0$, $h = h_w$ and $u = 0$; hence,

$$c = \frac{h_w}{\Pr}$$

At $y = D$, $h = h_e$ and $u = u_e$; hence,

$$b = \frac{1}{u_e}\left(\frac{h_e - h_w}{\Pr}\right) + \frac{u_e}{2}$$

Inserting b and c into Eq. (16.79) and simplifying, we obtain

$$h + \Pr\frac{u^2}{2} = h_w + \frac{u}{u_e}(h_e - h_w) + \frac{\Pr}{2}(uu_e) \qquad (16.80)$$

Assume the lower wall is adiabatic; i.e., $(\partial h/\partial y)_w = 0$. Differentiating Eq. (16.80) with respect to y, we have

$$\frac{\partial h}{\partial y} = -\Pr u \frac{\partial u}{\partial y} + \left(\frac{h_e + h_w}{u_e}\right)\frac{\partial u}{\partial y} + \frac{u_e \Pr}{2}\frac{\partial u}{\partial y}$$

or

$$\frac{\partial h}{\partial y} = \left(-u \Pr + \frac{h_e - h_w}{u_e} + \frac{u_e \Pr}{2}\right)\frac{\partial u}{\partial y} \qquad (16.81)$$

Recall that the condition for an adiabatic wall is that $(\partial h/\partial y)_w = 0$. Applying Eq. (16.81) at $y = 0$ for an adiabatic wall, where $u = 0$ and by definition $h_w = h_{aw}$, we have

$$\left(\frac{\partial h}{\partial y}\right)_w = \left(\frac{h_e - h_{aw}}{u_e} + \frac{u_e \Pr}{2}\right)\left(\frac{\partial u}{\partial y}\right)_w = 0$$

Since $(\partial u/\partial y)_w$ is finite, then

$$\frac{h_e - h_{aw}}{u_e} + \frac{u_e \Pr}{2} = 0$$

or

$$\boxed{h_{aw} = h_e + \Pr \frac{u_e^2}{2}} \qquad (16.82)$$

This is identical to Eq. (16.39) obtained for incompressible flow. Hence, we have just proven that the recovery factor for compressible Couette flow, assuming constant Prandtl number, is also

$$r = \Pr \qquad (16.83)$$

Since the recovery factors for the incompressible and compressible cases are the same (as long as $\Pr = $ constant), what can we say about Reynolds analogy? Does Eq. (16.59) hold for the compressible case? Let us examine this question. Return to Eq. (16.3), repeated below:

$$\frac{\partial}{\partial y}\left(k\frac{\partial T}{\partial y}\right) + \frac{\partial}{\partial y}\left(\mu u \frac{\partial u}{\partial y}\right) = 0 \qquad (16.3)$$

Recalling that, from the definitions,

$$\dot{q} = k\frac{\partial T}{\partial y} \qquad (16.84)$$

and

$$\tau = \mu \frac{\partial u}{\partial y} \qquad (16.85)$$

then Eq. (16.3) can be written as

$$\frac{\partial \dot{q}}{\partial y} + \frac{\partial(\tau u)}{\partial y} = 0 \qquad (16.86)$$

Integrating Eq. (16.86) with respect to y, we have

$$\dot{q} + \tau u = a \qquad (16.87)$$

where a is a constant of integration. Evaluating Eq. (16.87) at $y = 0$, where $u = 0$ and $\dot{q} = q_w$, we find that

$$a = \dot{q}_w$$

Hence, Eq. (16.87) is

$$\dot{q} + \tau u = \dot{q}_w \qquad (16.88)$$

Inserting Eqs. (16.84) and (16.85) into (16.88), we have

$$\dot{q}_w = k \frac{\partial T}{\partial y} + \mu u \frac{\partial u}{\partial y} \qquad (16.89)$$

or

$$\frac{\dot{q}_w}{\mu} = \frac{k}{\mu} \frac{\partial T}{\partial y} + u \frac{\partial u}{\partial y} \qquad (16.90)$$

Recall that the shear stress is constant throughout the flow; hence,

$$\tau = \mu \frac{\partial u}{\partial y} = \tau_w$$

or

$$\mu = \frac{\tau_w}{\partial u / \partial y} \qquad (16.91)$$

Also,

$$\frac{k}{\mu} = \frac{c_p}{\text{Pr}} \qquad (16.92)$$

Inserting Eq. (16.91) into the left-hand side of Eq. (16.90), and Eq. (16.92) into the right-hand side of Eq. (16.90), we have

$$\frac{\dot{q}_w}{\tau_w} \frac{\partial u}{\partial y} = \frac{c_p}{\text{Pr}} \frac{\partial T}{\partial y} + \frac{\partial (u^2/2)}{\partial y} \qquad (16.93)$$

Integrate Eq. (16.93) between the two plates, keeping in mind that $\dot{q}_w$, τ_w, c_p, and Pr are all fixed values:

$$\frac{\dot{q}_w}{\tau_w} \int_0^D \frac{\partial u}{\partial y} \, dy = \frac{c_p}{\text{Pr}} \int_0^D \frac{\partial T}{\partial y} \, dy + \int_0^D \frac{\partial (u^2/2)}{\partial y} \, dy$$

or

$$\frac{\dot{q}_w}{\tau_w} \int_0^{u_e} du = \frac{c_p}{\text{Pr}} \int_{T_w}^{T_e} dT + \int_0^{u_e} d \left(\frac{u^2}{2} \right)$$

which yields

$$\frac{\dot{q}_w}{\tau_w} u_e = \frac{c_p}{\text{Pr}} (T_e - T_w) + \frac{u_e^2}{2} \qquad (16.94)$$

Rearranging Eq. (16.94), and recalling that $h = c_p T$, we have

$$\dot{q}_w = \frac{\tau_w}{u_e \, \text{Pr}} \left(h_e - h_w + \text{Pr} \frac{u_e^2}{2} \right) \tag{16.95}$$

Inserting Eq. (16.82) into (16.95), we have

$$\dot{q}_w = \frac{\tau_w}{u_e \, \text{Pr}} (h_{aw} - h_w) \tag{16.96}$$

The skin friction coefficient and Stanton number are defined by Eqs. (16.51) and (16.55), respectively. Thus, their ratio is

$$\frac{C_H}{c_f} = \frac{\dot{q}_w / [\rho_e u_e (h_{aw} - h_w)]}{\tau_w / (\frac{1}{2}\rho_e u_e^2)} = \frac{\dot{q}_w}{\tau_w} \left[\frac{u_e}{2(h_{aw} - h_w)} \right] \tag{16.97}$$

Inserting Eq. (16.96) into (16.97), we have

$$\frac{C_H}{c_f} = \frac{(h_{aw} - h_w)}{u_e \, \text{Pr}} \left[\frac{u_e}{2(h_{aw} - h_w)} \right]$$

or

$$\boxed{\frac{C_H}{c_f} = \frac{1}{2} \text{Pr}^{-1}} \tag{16.98}$$

Equation (16.98) is *Reynolds analogy*—a relation between heat transfer and skin friction coefficients. Moreover, it is precisely the same result as obtained in Eq. (16.59) for incompressible flow. Hence, for a constant Prandtl number, we have shown that Reynolds analogy is precisely the same form for incompressible and compressible flow.

Example 16.2. Consider the geometry given in Fig. 16.2. The two plates are separated by a distance of 0.01 in (the same as in Example 16.1). The temperature of the two plates is equal, at a value of 288 K (standard sea level temperature). The air pressure is constant throughout the flow and equal to 1 atm. The upper plate is moving at Mach 3. The shear stress at the lower wall is 72 N/m². (This is about 1.5 lb/ft²—a much larger value than that associated with the low-speed case treated in Example 16.1.) Calculate the heat transfer to either plate. (Since the shear stress is constant throughout the flow, and the plates are at equal temperature, the heat transfer to the upper and lower plates is the same.)

Solution. The velocity of the upper plate is

$$u_e = M_e a_e = M_e \sqrt{\gamma R T_e} = 3\sqrt{(1.4)(288)(287)} = 1020 \text{ m/s}$$

The air density at both plates is (noting that 1 atm $= 1.01 \times 10^5$ N/m²)

$$\rho_e = \frac{p_e}{R T_e} = \frac{1.01 \times 10^5}{(287)(288)} = 1.22 \text{ kg/m}^3$$

Hence, the skin friction coefficient is

$$c_f = \frac{\tau_w}{\frac{1}{2}\rho_e u_e^2} = \frac{72}{(0.5)(1.22)(1020)^2} = 1.13 \times 10^{-4}$$

From Reynolds analogy, Eq. (16.92), we have

$$C_H = \frac{c_f}{2\,\text{Pr}} = \frac{1.13 \times 10^{-4}}{2(0.71)} = 8 \times 10^{-5}$$

The adiabatic wall enthalpy, from Eq. (16.82), is

$$h_{aw} = h_e + \text{Pr}\,\frac{u_e^2}{2} = c_p T_e + \text{Pr}\,\frac{u_e^2}{2}$$

For air, $c_p = 1004.5$ J/kg · K. Thus,

$$h_{aw} = (1004.5)(288) + (0.71)\frac{(1020)^2}{2} = 6.59 \times 10^5 \text{ J/kg}$$

[*Note*: This gives $T_{aw} = h_{aw}/c_p = (6.59 \times 10^5)/1004.5 = 656$ K. In the adiabatic case, the wall would be quite warm.] Hence, from the definition of the Stanton number [Eq. (16.55)], and noting that $h_w = c_p T_w = (1004.6)(288) = 2.89 \times 10^5$ J/kg,

$$\dot{q}_w = \rho_e u_e (h_{aw} - h_w) C_H = (1.22)(1020)[(6.59 - 2.89) \times 10^5](8 \times 10^{-5})$$
$$= \boxed{3.68 \times 10^4 \text{ W/m}^2}$$

16.5 TWO-DIMENSIONAL POISEUILLE FLOW

Consider the parallel flow between two horizontal plates separated by the distance D as sketched in Fig. 16.12. In this case, unlike Couette flow where one of the plates is in motion, we consider both plates to be stationary. Recall that the driving force which established Couette flow was the shear stress between the moving plate and the fluid. In the present case sketched in Fig. 16.12, what is the driving force; i.e., what makes the fluid move? In Chap. 1, we emphasized that the only way that nature can exert a force on a fluid is by means of shear stress and pressure distributions. In the present problem, since the walls are not moving, there is no shear stress to drive the flow. Hence, the only other possibility is the pressure distribution. Indeed, to establish the flow shown in Fig. 16.12, there must be a *pressure gradient* acting on the gas. Moreover, in Fig. 16.12 the flow extends to infinity in both directions along the x axis. As in the case of

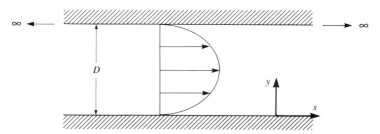

FIGURE 16.12
Schematic for Poiseuille flow.

Couette flow, this implies that the velocity, u, is independent of x; i.e., $u = u(y)$. Since the streamlines are parallel, $v = 0$. This flow is called two-dimensional Poiseuille flow, named after the French physician J. L. M. Poiseuille, who studied similar flows in pipes.

Let us examine the Navier-Stokes equations in light of the problem as outlined above. For simplicity, we will consider only steady, incompressible flow. First, return to the continuity equation for an incompressible flow, given by Eq. (3.39). In cartesian coordinates, this is

$$\frac{\partial u}{\partial x} + \frac{\partial v}{\partial y} = 0 \tag{16.99}$$

Since the flow is parallel, $v = 0$; hence, $\partial v / \partial y = 0$. From Eq. (16.99), then $\partial u / \partial x = 0$; this confirms that u is constant with x; i.e., u is a function of y only. From the y-momentum equation, Eq. (15.19b), we have

$$\frac{\partial p}{\partial y} = 0 \tag{16.100}$$

Hence, p varies only in the x direction; $p = p(x)$. From the x-momentum equation, Eq. (15.19a), we have

$$0 = -\frac{\partial p}{\partial x} + \frac{\partial}{\partial y}\left(\mu \frac{\partial u}{\partial y}\right) \tag{16.101}$$

or
$$\frac{\partial p}{\partial x} = \frac{\partial}{\partial y}\left(\mu \frac{\partial u}{\partial y}\right) = \text{const} \tag{16.102}$$

On the left-hand side of Eq. (16.102), p is a function of x only. On the right-hand side of Eq. (16.102), u is a function of y only. Hence, the left-hand and right-hand sides of Eq. (16.102) must be equal to the same constant. This confirms an important aspect of this flow, namely, the *pressure gradient is constant along the flow direction*. Once again, we emphasize that it is this pressure gradient that drives the flow. The pressure gradient must be provided by an outside mechanism, i.e., some source of high pressure toward the left and low pressure toward the right.

The velocity profile across the flow is obtained by solving Eq. (16.102). For convenience, and to emphasize that $p = p(x)$, we write the pressure derivative as an ordinary derivative, dp/dx. Integrating Eq. (16.102) twice across the flow, we have

$$u = \frac{1}{2\mu}\left(\frac{dp}{dx}\right) y^2 + ay + b \tag{16.103}$$

where a and b are constants of integration. Evaluating Eq. (16.103) at $y = 0$, where $u = 0$, we have

$$b = 0$$

Evaluating Eq. (16.103) at $y = D$, where $u = 0$, we have

$$a = -\frac{1}{2\mu}\left(\frac{dp}{dx}\right) D$$

Hence, Eq. (16.103) becomes

$$u = \frac{1}{2\mu}\left(\frac{dp}{dx}\right)(y^2 - Dy)$$

(16.104)

From Eq. (16.104), the velocity varies *parabolically* across the flow. Moreover, the velocity varies *directly* as the pressure gradient; as the magnitude of dp/dx increases, so does the velocity. The location of the maximum velocity can be found by

$$\frac{\partial u}{\partial y} = 0$$

where $\partial u/\partial y$ is obtained by differentiating Eq. (16.104), as follows:

$$\frac{\partial u}{\partial y} = \frac{1}{2\mu}\left(\frac{dp}{dx}\right)(2y - D) = 0$$

Hence, the maximum velocity in the flow, denoted by u_{max}, occurs at $y = D/2$, i.e., at the midpoint of the flow. The velocity profile is a parabola that is symmetric about the centerline of the flow. The value of u_{max} is found by evaluating Eq. (16.104) at $y = D/2$ as follows:

$$u_{max} = \frac{1}{2\mu}\left(\frac{dp}{dx}\right)\left(\frac{D^2}{4} - \frac{D^2}{2}\right)$$

or

$$u_{max} = -\frac{D^2}{8\mu}\left(\frac{dp}{dx}\right)$$

(16.105)

Note, in order to drive the flow from left to right as shown in Fig. 16.12, the pressure must be *decreasing* in the positive x direction; i.e., dp/dx is itself a negative quantity—a *favorable* pressure gradient. This negative value of dp/dx results in a positive u_{max} in Eq. (16.105). Once again, note that u_{max} varies directly with the magnitude of the pressure gradient; if the pressure gradient doubles, u_{max} doubles.

The shear stress at the walls is obtained from

$$\tau_w = \mu\left(\frac{\partial u}{\partial y}\right)_w$$

where, from Eq. (16.104),

$$\left(\frac{\partial u}{\partial y}\right) = \frac{1}{2\mu}\left(\frac{dp}{dx}\right)(2y - D)$$

(16.106)

At $y = 0$, Eq. (16.106) yields

$$\left(\frac{\partial u}{\partial y}\right)_w = -\frac{D}{2\mu}\left(\frac{dp}{dx}\right)$$

(16.107)

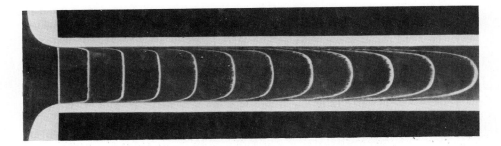

FIGURE 16.13
Photograph of velocity profiles for viscous flow in a duct, showing the approach to a fully developed flow. Flow is from left to right. (*Courtesy of Yasuki Nakayama, Tokai University, Japan.*)

Hence, the wall shear stress is

$$\tau_w = \mu \left(\frac{\partial u}{\partial y} \right)_w = -\frac{D}{2} \left(\frac{dp}{dx} \right) \tag{16.108}$$

Note the interesting fact from Eq. (16.108) that τ_w does not depend on the viscosity coefficient, μ, but rather only on the separation distance of the walls and on the pressure gradient. Clearly, this flow is a force balance between the pressure gradient acting toward the right on the gas and the shear stress at the walls acting toward the left on the gas.

This flow is sometimes called *fully developed flow,* for the following reason. Consider an actual flow in the laboratory wherein a uniform flow enters a channel, such as shown in the photograph in Fig. 16.13. Here, velocity profiles in water flow are made visible by the hydrogen bubble method, where the bubbles are generated by electrolysis on a fine wire used as a cathode at the entrance of the channel. Near the entrance, the flow is uniform over a large portion of the distance across the channel; the viscous effects are limited to a thin boundary layer at the walls. However, as the flow progresses downstream, the viscous effects are felt over a larger portion of the flow. Finally, after the flow has covered a sufficient distance through the channel, the velocity profile is totally dominated by viscosity; a parabolic velocity profile is achieved, and the real flow becomes essentially the Poiseuille flow studied in this section. When this type of real flow is reached in the channel (at the right of Fig. 16.13), it is called "fully developed flow."

16.6 SUMMARY

The parallel flows discussed in this chapter illustrate features common to many more complex viscous flows, with the added advantage of lending themselves to a relatively straightforward solution. The purpose of this discussion has been to introduce many of the basic concepts of viscous flows in a fashion unencumbered by fluid dynamic complexities. In particular, we have studied Couette and two-dimensional Poiseuille flows and found the following.

16.6.1 Couette Flow

1. The driving force is the shear stress between the moving wall and the fluid. Shear stress is constant across the flow for both incompressible and compressible cases.

2. For incompressible Couette flow,

$$u = u_e \left(\frac{y}{D} \right) \tag{16.6}$$

$$\tau = \mu \left(\frac{u_e}{D} \right) \tag{16.9}$$

3. The heat transfer depends on the wall temperatures and the amount of viscous dissipation. For an adiabatic wall, the wall enthalpy is

$$h_{aw} = h_e + r \frac{u_e^2}{2} \tag{16.46a}$$

For incompressible and compressible Couette flow with a constant Prandtl number, the recovery factor is

$$r = \text{Pr}$$

and Reynolds analogy holds in both cases;

$$\frac{C_H}{c_f} = \frac{1}{2} \text{Pr}^{-1} \tag{16.59}$$

16.6.2 Poiseuille Flow

1. The driving force is a pressure gradient in the flow direction, generated by some outside mechanism. This fully developed flow is a force balance between the pressure gradient and shear stress at the walls.

2. For incompressible flow, the velocity profile is parabolic, and is given by

$$u = \frac{1}{2\mu} \left(\frac{dp}{dx} \right) (y^2 - Dy) \tag{16.104}$$

The maximum velocity occurs in the middle of the flow, and is

$$u_{\max} = -\frac{D^2}{8\mu} \left(\frac{dp}{dx} \right) \tag{16.105}$$

The shear stress at the walls is

$$\tau_w = -\frac{D}{2} \left(\frac{dp}{dx} \right) \tag{16.108}$$

A very satisfactory explanation of the physical process in the boundary layer between a fluid and a solid body could be obtained by the hypothesis of an adhesion of the fluid to the walls, that is, by the hypothesis of a zero relative velocity between fluid and wall. If the viscosity was very small and the fluid path along the wall not too long, the fluid velocity ought to resume its normal value at a very short distance from the wall. In the thin transition layer however, the sharp changes of velocity, even with small coefficient of friction, produce marked results.

Ludwig Prandtl, 1904

17.1 INTRODUCTION

The above quotation is taken from an historic paper given by Ludwig Prandtl at the third Congress of Mathematicians at Heidelberg, Germany, in 1904. In this paper, the concept of the boundary layer was first introduced—a concept which eventually revolutionized the analysis of viscous flows in the twentieth century and which allowed the practical calculation of drag and flow separation over aerodynamic bodies. Before Prandtl's 1904 paper, the Navier-Stokes equations discussed in Chap. 15 were well known, but fluid dynamicists were frustrated in their attempts to solve these equations for practical engineering problems. After 1904, the picture changed completely. Using Prandtl's concept of a boundary layer adjacent to an aerodynamic surface, the Navier-Stokes equations can be reduced to a more tractable form called the *boundary-layer equations*. In turn, these boundary-layer equations can be solved to obtain the distributions of shear stress and aerodynamic heat transfer to the surface. Prandtl's boundary-layer concept was an advancement in the science of fluid mechanics of the caliber of a Nobel prize, although he never received that honor. The purpose of this chapter is to present the general concept of the boundary layer and to give a few representative samples of its application. Our purpose here is to provide only an introduction to boundary-layer theory; consult Ref. 42 for a rigorous and thorough discussion of boundary-layer analysis and applications.

What is a boundary layer? We have used this term in several places in our previous chapters, first introducing the idea in Sec. 1.10 and illustrating the concept in Fig. 1.28. The boundary layer is the thin region of flow adjacent to a surface, where the flow is retarded by the influence of friction between a solid surface and the fluid. For example, a photograph of the flow over a supersonic body is shown in Fig. 17.1, where the boundary layer (along with shock and expansion waves and the wake) is made visible by a special optical technique called a *shadowgraph* (see Refs. 25 and 26 for discussions of the shadowgraph method). Note how thin the boundary layer is in comparison with the size of the body; however, although the boundary layer occupies geometrically only a small portion of the flow field, its influence on the drag and heat transfer to the body is immense—in Prandtl's own words as quoted above, it produces "marked results."

The purpose of this chapter is to examine these "marked results." The road map for the present chapter is given in Fig. 17.2. In the next section, we discuss

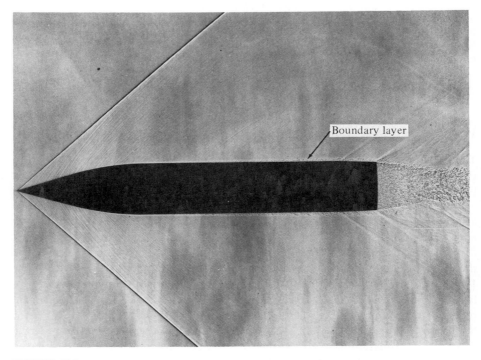

FIGURE 17.1
The boundary layer on an aerodynamic body. (*Courtesy of the U.S. Army Ballistics Laboratory, Aberdeen, Maryland.*)

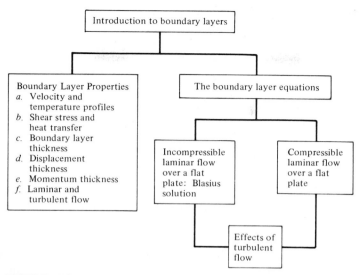

FIGURE 17.2
Road map for Chap. 17.

some fundamental properties of boundary layers. This is followed by a development of the boundary-layer equations, with specific solutions of these equations for the incompressible and compressible laminar flow over a flat plate. Finally, some comments are made on the effects of turbulent flow. As we progress through this chapter, refer to the road map in Fig. 17.2 for orientation on the flow of ideas.

Finally, we note that this chapter represents the second of the three options discussed in Sec. 15.7 for the solution of the viscous flow equations, namely, the simplification of the Navier-Stokes equations by neglecting certain terms that are smaller than other terms. This is an approximation, not a precise condition as in the case of Couette and Poiseuille flows in Chap. 16. In this chapter, we will see that the Navier-Stokes equations, when applied to the thin viscous boundary layer adjacent to a surface, can be reduced to simpler forms, albeit approximate, which lend themselves to simpler solutions. These simpler forms of the equations are called the boundary-layer equations—they are the subject of the present chapter.

17.2 BOUNDARY-LAYER PROPERTIES

Consider the viscous flow over a flat plate as sketched in Fig. 17.3. The viscous effects are contained within a thin layer adjacent to the surface; the thickness is exaggerated in Fig. 17.3 for clarity. Immediately at the surface, the flow velocity is zero; this is the "no-slip" condition discussed in Sec. 15.2. In addition, the temperature of the fluid immediately at the surface is equal to the temperature of the surface; this is called the *wall temperature* T_w, as shown in Fig. 17.3. Above

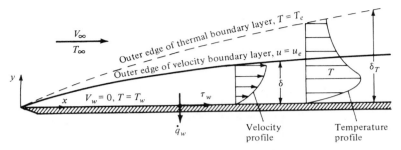

FIGURE 17.3
Boundary-layer properties.

the surface, the flow velocity increases in the y direction until, for all practical purposes, it equals the freestream velocity. This will occur at a height above the wall equal to δ, as shown in Fig. 17.3. More precisely, δ is defined as that distance above the wall where $u = 0.99u_e$; here, u_e is the velocity at the outer edge of the boundary layer. In Fig. 17.3, which illustrates the flow over a flat plate, the velocity at the edge of the boundary layer will be V_∞; i.e., $u_e = V_\infty$. For a body of general shape, u_e is the velocity obtained from an inviscid flow solution evaluated at the body surface (or at the "effective body" surface, as discussed later). The quantity δ is called the *velocity boundary-layer thickness*. At any given x station, the variation of u between $y = 0$ and $y = \delta$, i.e., $u = u(y)$, is defined as the *velocity profile* within the boundary layer, as sketched in Fig. 17.3. This profile is different for different x stations. Similarly, the flow temperature will change above the wall, ranging from $T = T_w$ at $y = 0$ to $T = 0.99T_e$ at $y = \delta_T$. Here, δ_T is defined as the *thermal boundary-layer thickness*. At any given x station, the variation of T between $y = 0$ and $y = \delta_T$, i.e., $T = T(y)$, is called the *temperature profile* within the boundary layer, as sketched in Fig. 17.3. (In the above, T_e is the temperature at the edge of the thermal boundary layer. For the flow over a flat plate, as sketched in Fig. 17.3, $T_e = T_\infty$. For a general body, T_e is obtained from an inviscid flow solution evaluated at the body surface, or at the "effective body" surface, to be discussed later.) Hence, two boundary layers can be defined: a velocity boundary layer with thickness δ and a temperature boundary layer with thickness δ_T. In general, $\delta_T \neq \delta$. The relative thicknesses depend on the Prandtl number: it can be shown that if $\mathrm{Pr} = 1$, then $\delta = \delta_T$; if $\mathrm{Pr} > 1$, then $\delta_T < \delta$; if $\mathrm{Pr} < 1$, then $\delta_T > \delta$. For air at standard conditions, $\mathrm{Pr} = 0.71$; hence, the thermal boundary layer is thicker than the velocity boundary layer, as shown in Fig. 17.3. Note that both boundary-layer thicknesses increase with distance from the leading edge; i.e., $\delta = \delta(x)$ and $\delta_T = \delta_T(x)$.

The consequence of the velocity gradient at the wall is the generation of shear stress at the wall,

$$\tau_w = \mu \left(\frac{\partial u}{\partial y} \right)_w \tag{17.1}$$

where $(\partial u/\partial y)_w$ is the velocity gradient evaluated at $y = 0$, i.e., at the wall. Similarly, the temperature gradient at the wall generates heat transfer at the wall,

$$\dot{q}_w = -k\left(\frac{\partial T}{\partial y}\right)_w \qquad (17.2)$$

where $(\partial T/\partial y)_w$ is the temperature gradient evaluated at $y = 0$, i.e., at the wall. In general, both τ_w and $\dot{q}_w$ are functions of distance from the leading edge; i.e., $\tau_w = \tau_w(x)$ and $\dot{q}_w = \dot{q}_w(x)$. One of the central purposes of boundary-layer theory is to compute τ_w and $\dot{q}_w$.

A frequently used boundary-layer property is the *displacement thickness* δ^*, defined as

$$\boxed{\delta^* \equiv \int_0^{y_1}\left(1 - \frac{\rho u}{\rho_e u_e}\right)dy \qquad \delta \le y_1 \to \infty} \qquad (17.3)$$

The displacement thickness has two physical interpretations:

1. δ^* is an index proportional to the "missing mass flow" due to the presence of the boundary layer. Let us explain. Consider point y_1 above the boundary layer, as shown in Fig. 17.4. Consider also the mass flow (per unit depth perpendicular to the page) across the vertical line connecting $y = 0$ and $y = y_1$. Then

$$A = \text{actual mass flow between 0 and } y_1 = \int_0^{y_1} \rho u \, dy$$

$$B = \begin{array}{l}\text{hypothetical mass flow}\\ \text{between 0 and } y_1 \text{ if boundary} \\ \text{layer were not present}\end{array} = \int_0^{y_1} \rho_e u_e \, dy$$

$$B - A = \begin{array}{l}\text{decrement in mass flow due to}\\ \text{presence of boundary layer, i.e.,} \\ \text{missing mass flow}\end{array} = \int_0^{y_1} (\rho_e u_e - \rho u) \, dy \qquad (17.4)$$

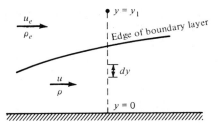

FIGURE 17.4
Construction for the discussion of displacement thickness.

Express this missing mass flow as the product of $\rho_e u_e$ and a height, δ^*; i.e.,

$$\text{Missing mass flow} = \rho_e u_e\, \delta^* \tag{17.5}$$

Equating Eqs. (17.4) and (17.5), we have

$$\rho_e u_e \sigma^* = \int_0^{y_1} (\rho_e u_e - \rho u)\, dy$$

or

$$\delta^* = \int_0^{y_1} \left(1 - \frac{\rho u}{\rho_e u_e}\right) dy \tag{17.6}$$

Equation (17.6) is identical to the definition of δ^* given in Eq. (17.3). Hence, clearly δ^* is a height proportional to the missing mass flow. If this missing mass flow was crammed into a streamtube where the flow properties were constant at ρ_e and u_e, then Eq. (17.5) says that δ^* is the height of this hypothetical streamtube.

2. The second physical interpretation of δ^* is more practical than the one discussed above. Consider the flow over a flat surface as sketched in Fig. 17.5. At the left is a picture of the hypothetical inviscid flow over the surface; a streamline through point y_1 is straight and parallel to the surface. The actual viscous flow is shown at the right of Fig. 17.5; here, the retarded flow inside the boundary layer acts as a partial obstruction to the freestream flow. As a result, the streamline external to the boundary layer passing through point y_1 is deflected upward through a distance δ^*. We now prove that this δ^* is precisely the displacement thickness defined by Eq. (17.3). At station 1 in Fig. 17.5, the mass flow (per unit depth perpendicular to the page) between the surface and the external streamline is

$$\dot{m} = \int_0^{y_1} \rho_e u_e\, dy \tag{17.7}$$

At station 2, the mass flow between the surface and the external streamline is

$$\dot{m} = \int_0^{y_1} \rho u\, dy + \rho_e u_e\, \delta^* \tag{17.8}$$

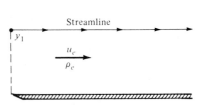

Hypothetical flow with no boundary layer
(inviscid case)

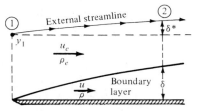

Actual flow with a boundary layer

FIGURE 17.5
Displacement thickness is the distance by which an external flow streamline is displaced by the presence of the boundary layer.

Since the surface and the external streamline form the boundaries of a stream-tube, the mass flows across stations 1 and 2 are equal. Hence, equating Eqs. (17.7) and (17.8), we have

$$\int_0^{y_1} \rho_e u_e \, dy = \int_0^{y_1} \rho u \, dy + \rho_e u_e \, \delta^*$$

or

$$\delta^* = \int_0^{y_1} \left(1 - \frac{\rho u}{\rho_e u_e}\right) dy \qquad (17.9)$$

Hence, the height by which the streamline in Fig. 17.5 is displaced upward by the presence of the boundary layer, namely, δ^*, is given by Eq. (17.9). However, Eq. (17.9) is precisely the definition of the displacement thickness given by Eq. (17.3). Thus, the displacement thickness, first defined by Eq. (17.3), is physically the distance through which the external inviscid flow is displaced by the presence of the boundary layer.

This second interpretation of δ^* gives rise to the concept of an *effective body*. Consider the aerodynamic shape sketched in Fig. 17.6. The actual contour of the body is given by curve *ab*. However, due to the displacement effect of the boundary layer, the shape of the body effectively seen by the freestream is not given by curve *ab*; rather, the freestream sees an effective body given by curve *ac*. In order to obtain the conditions ρ_e, u_e, T_e, etc., at the outer edge of the boundary layer on the actual body *ab*, an inviscid flow solution should be carried out for the effective body, and ρ_e, u_e, T_e, etc., are obtained from this inviscid solution evaluated along curve *ac*.

Note that in order to solve for δ^* from Eq. (17.3), we need the profiles of u and ρ from a solution of the boundary-layer flow. In turn, to solve the boundary-layer flow, we need ρ_e, u_e, T_e, etc. However, ρ_e, u_e, T_e, etc., depend on δ^*. This leads to an iterative solution. To calculate accurately the boundary-layer properties as well as the pressure distribution over the surface of the body in Fig. 17.6, we proceed as follows:

1. Carry out an inviscid solution for the given body shape *ab*. Evaluate ρ_e, u_e, T_e, etc., along curve *ab*.
2. Using these values of ρ_e, u_e, T_e, etc., solve the boundary-layer equations (discussed in Secs. 17.3 to 17.6) for $u = u(y)$, $\rho = \rho(y)$, etc., at various stations along the body.

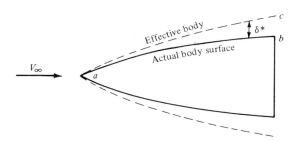

FIGURE 17.6
The "effective body," equal to the actual body shape plus the displacement thickness distribution.

3. Obtain δ^* at these stations from Eq. (17.3). This will not be an accurate δ^* because ρ_e, u_e, T_e, etc., were evaluated on curve ab, not the proper effective body. Using this intermediate δ^*, calculate an effective body, given by a curve ac' (not shown in Fig. 17.6).

4. Carry out an inviscid solution for the flow over the intermediate effective body ac', and evaluate new values of ρ_e, u_e, T_e, etc., along ac'.

5. Repeat steps 2 to 4 above until the solution at one iteration essentially does not deviate from the solution at the previous iteration. At this stage, a converged solution will be obtained, and the final results will pertain to the flow over the proper effective body ac shown in Fig. 17.6.

In some cases, the boundary layers are so thin that the effective body can be ignored, and a boundary-layer solution can proceed directly from ρ_e, u_e, T_e, etc., obtained from an inviscid solution evaluated on the actual body surface (ab in Fig. 17.6). However, for highly accurate solutions, and for cases where the boundary-layer thickness is relatively large (such as for hypersonic flow as discussed in Chap. 14), the iterative procedure described above should be carried out. Also, we note parenthetically that δ^* is usually smaller than δ; typically, $\delta^* \approx 0.3\delta$.

Another boundary-layer property of importance is the *momentum thickness* θ, defined as

$$\theta \equiv \int_0^{y_1} \frac{\rho u}{\rho_e u_e}\left(1 - \frac{u}{u_e}\right) dy \qquad \delta \leq y_1 \to \infty \qquad (17.10)$$

To understand the physical interpretation of θ, return again to Fig. 17.4. Consider the mass flow across a segment dy, given by $dm = \rho u \, dy$. Then

$$A = \text{momentum flow across } dy = dm \, u = \rho u^2 \, dy$$

If this same elemental mass flow were associated with the freestream, where the velocity is u_e, then

$$B = \begin{cases} \text{momentum flow at freestream} \\ \text{velocity associated with mass } dm = dm \, u_e = (\rho u \, dy) \, u_e \end{cases}$$

Hence,

$$B - A = \begin{cases} \text{decrement in momentum flow} \\ \text{(missing momentum flow) associated} = \rho u(u_e - u) \, dy \\ \text{with mass } dm \end{cases} \qquad (17.11)$$

The *total* decrement in momentum flow across the vertical line from $y = 0$ to $y = y_1$ in Fig. 16.4 is the integral of Eq. (17.11),

$$\left.\begin{array}{l} \text{Total decrement in momentum} \\ \text{flow, or missing momentum} \\ \text{flow} \end{array}\right\} = \int_0^{y_1} \rho u(u_e - u) \, dy \qquad (17.12)$$

Assume that the missing momentum flow is the product of $\rho_e u_e^2$ and a height θ. Then,

$$\text{Missing momentum flow} = \rho_e u_e^2 \theta \qquad (17.13)$$

Equating Eqs. (17.12) and (17.13), we obtain

$$\rho_e u_e^2 \theta = \int_0^{y_1} \rho u (u_e - u)\, dy$$

$$\theta = \int_0^{y_1} \frac{\rho u}{\rho_e u_e} \left(1 - \frac{u}{u_e}\right) dy \qquad (17.14)$$

Equation (17.14) is precisely the definition of the momentum thickness given by Eq. (17.10). Therefore, θ is an index that is proportional to the decrement in momentum flow due to the presence of the boundary layer. It is the height of a hypothetical streamtube which is carrying the missing momentum flow at free-stream conditions.

Note that $\theta = \theta(x)$. In more detailed discussions of boundary-layer theory, it can be shown that θ evaluated at a given station $x = x_1$ is proportional to the integrated friction drag coefficient from the leading edge to x_1; i.e.,

$$\theta(x_1) \propto \frac{1}{x_1} \int_0^{x_1} c_f\, dx = C_f$$

where c_f is the local skin friction coefficient defined in Sec. 1.5 and C_f is the total skin friction drag coefficient for the length of surface from $x = 0$ to $x = x_1$. Hence, the concept of momentum thickness is useful in the prediction of drag coefficient.

All the boundary-layer properties discussed above are general concepts; they apply to compressible as well as incompressible flows, and to turbulent as well as laminar flows. The differences between turbulent and laminar flows were introduced in Sec. 15.2. Here, we extend that discussion by noting that the increased momentum and energy exchange that occur within a turbulent flow cause a turbulent boundary layer to be thicker than a laminar boundary layer. That is, for the same edge conditions, ρ_e, u_e, T_e, etc., we have $\delta_{\text{turbulent}} > \delta_{\text{laminar}}$ and $(\delta_T)_{\text{turbulent}} > (\delta_T)_{\text{laminar}}$. When a boundary layer changes from laminar to turbulent flow, as sketched in Fig. 15.8, the boundary-layer thickness markedly increases. Similarly, δ^* and θ are larger for turbulent flows.

17.3 THE BOUNDARY-LAYER EQUATIONS

For the remainder of this chapter, we consider two-dimensional, steady flow. The nondimensionalized form of the x-momentum equation (one of the Navier-Stokes equations) was developed in Sec. 15.6 and was given by Eq. (15.29):

$$\rho' u' \frac{\partial u'}{\partial x'} + \rho' v' \frac{\partial u'}{\partial y'} = -\frac{1}{\gamma M_\infty^2} \frac{\partial p'}{\partial x'} + \frac{1}{\text{Re}_\infty} \frac{\partial}{\partial y'} \left[\mu' \left(\frac{\partial v'}{\partial x'} + \frac{\partial u'}{\partial y'} \right) \right] \qquad (15.29)$$

Let us now reduce Eq. (15.29) to an approximate form which holds reasonably well within a boundary layer.

Consider the boundary layer along a flat plate of length c as sketched in Fig. 17.7. The basic assumption of boundary-layer theory is that a boundary layer is very thin in comparison with the scale of the body; i.e.,

$$\boxed{\delta \ll c} \tag{17.15}$$

Consider the continuity equation for a steady, two-dimensional flow,

$$\frac{\partial(\rho u)}{\partial x} + \frac{\partial(\rho v)}{\partial y} = 0 \tag{17.16}$$

In terms of the nondimensional variables defined in Sec. 15.6, Eq. (17.16) becomes

$$\frac{\partial(\rho' u')}{\partial x'} + \frac{\partial(\rho' v')}{\partial y'} = 0 \tag{17.17}$$

Because u' varies from 0 at the wall to 1 at the edge of the boundary layer, let us say that u' is of the *order of magnitude* equal to 1, symbolized by $O(1)$. Similarly, $\rho' = O(1)$. Also, since x varies from 0 to c, $x' = O(1)$. However, since y varies from 0 to δ, where $\delta \ll c$, then y' is of the *smaller* order of magnitude, denoted by $y' = O(\delta/c)$. Without loss of generality, we can assume that c is a unit length. Therefore, $y' = O(\delta)$. Putting these orders of magnitude in Eq. (17.17), we have

$$\frac{[O(1)][O(1)]}{O(1)} + \frac{[O(1)][v']}{O(\delta)} = 0 \tag{17.18}$$

Hence, from Eq. (17.18), clearly v' must be of an order of magnitude equal to δ; i.e., $v = O(\delta)$. Now examine the order of magnitude of the terms in Eq. (15.29). We have

$$\rho' u' \frac{\partial u'}{\partial x'} = O(1) \qquad \rho' v' \frac{\partial u'}{\partial y'} = O(1) \qquad \frac{\partial p'}{\partial x'} = O(1)$$

$$\frac{\partial}{\partial y'}\left(\mu' \frac{\partial v'}{\partial x'}\right) = O(1) \qquad \frac{\partial}{\partial y'}\left(\mu' \frac{\partial u'}{\partial y'}\right) = O\left(\frac{1}{\delta^2}\right)$$

Hence, the order-of-magnitude equation for Eq. (15.29) can be written as

$$O(1) + O(1) = -\frac{1}{\gamma M_\infty^2} O(1) + \frac{1}{\text{Re}_\infty}\left[O(1) + O\left(\frac{1}{\delta^2}\right)\right] \tag{17.19}$$

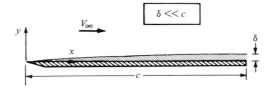

FIGURE 17.7
The basic assumption of boundary-layer theory: A boundary layer is very thin in comparison with the scale of the body.

Let us now introduce another assumption of boundary-layer theory, namely, the *Reynolds number is large*, indeed, large enough such that

$$\frac{1}{\text{Re}_\infty} = O(\delta^2) \tag{17.20}$$

Then, Eq. (17.19) becomes

$$O(1) + O(1) = -\frac{1}{\gamma M_\infty^2} O(1) + O(\delta^2)\left[O(1) + O\left(\frac{1}{\delta^2}\right)\right] \tag{17.21}$$

In Eq. (17.21), there is one term with an order of magnitude that is much smaller than the rest, namely, the product $O(\delta^2)[O(1)] = O(\delta^2)$. This term corresponds to $(1/\text{Re}_\infty)\partial/\partial y'(\mu'\partial v'/\partial x')$ in Eq. (15.29). Hence, *neglect* this term in comparison to the remaining terms in Eq. (15.29). We obtain

$$\rho'u'\frac{\partial u'}{\partial x'} + \rho'v'\frac{\partial u'}{\partial y'} = -\frac{1}{\gamma M_\infty^2}\frac{\partial p'}{\partial x'} + \frac{1}{\text{Re}_\infty}\frac{\partial}{\partial y'}\left(\mu'\frac{\partial u'}{\partial y'}\right) \tag{17.22}$$

In terms of dimensional variables, Eq. (17.22) is

$$\rho u\frac{\partial u}{\partial x} + \rho v\frac{\partial u}{\partial y} = -\frac{\partial p}{\partial x} + \frac{\partial}{\partial y}\left(\mu\frac{\partial u}{\partial y}\right) \tag{17.23}$$

Equation (17.23) is the approximate x-momentum equation which holds for flow in a thin boundary layer at a high Reynolds number.

Consider the y-momentum equation for two-dimensional, steady flow, obtained from Eq. (15.19b) as (neglecting the normal stress τ_{yy})

$$\rho u\frac{\partial v}{\partial x} + \rho v\frac{\partial v}{\partial y} = -\frac{\partial p}{\partial y} + \frac{\partial}{\partial x}\left[\mu\left(\frac{\partial v}{\partial x} + \frac{\partial u}{\partial y}\right)\right] \tag{17.24}$$

In terms of the nondimensional variables, Eq. (17.24) becomes

$$\rho'u'\frac{\partial v'}{\partial x'} + \rho'v'\frac{\partial v'}{\partial y'} = -\frac{1}{\gamma M_\infty^2}\frac{\partial p'}{\partial y'} + \frac{1}{\text{Re}_\infty}\frac{\partial}{\partial x'}\left[\mu'\left(\frac{\partial v'}{\partial x'} + \frac{\partial u'}{\partial y'}\right)\right] \tag{17.25}$$

The order-of-magnitude equation for Eq. (17.25) is

$$O(\delta) + O(\delta) = -\frac{1}{\gamma M_\infty^2}\frac{\partial p'}{\partial y'} + O(\delta^2)\left[O(\delta) + O\left(\frac{1}{\delta}\right)\right] \tag{17.26}$$

From Eq. (17.26), we see that $\partial p'/\partial y' = O(\delta)$ or smaller, assuming that $\gamma M_\infty^2 = O(1)$. Since δ is very small, this implies that $\partial p'/\partial y'$ is very small. Therefore, from the y-momentum equation specialized to a boundary layer, we have

$$\frac{\partial p}{\partial y} = 0 \tag{17.26a}$$

Equation (17.26a) is important; it states that at a given x station, *the pressure is constant through the boundary layer in a direction normal to the surface*. This implies that the pressure distribution at the outer edge of the boundary layer is impressed directly to the surface without change. Hence, throughout the boundary layer, $p = p(x) = p_e(x)$.

It is interesting to note that if M_∞^2 is very large, as in the case of large hypersonic Mach numbers, then from Eq. (17.26) $\partial p'/\partial y'$ does not have to be small. For example, if M_∞ were large enough such that $1/\gamma M_\infty^2 = O(\delta)$, then $\partial p'/\partial y'$ could be as large as $O(1)$, and Eq. (17.26) would still be satisfied. Thus, for very large hypersonic Mach numbers, the assumption that p is constant in the normal direction through a boundary layer is not always valid.

Consider the general energy equation given by Eq. (15.26). The non-dimensional form of this equation for two-dimensional, steady flow is given in Eq. (15.33). Inserting $e = h - p/\rho$ into this equation, subtracting the momentum equation multiplied by velocity, and performing an order-of-magnitude analysis similar to those above, we can obtain the boundary-layer energy equation as

$$\rho u \frac{\partial h}{\partial x} + \rho v \frac{\partial h}{\partial y} = \frac{\partial}{\partial y}\left(k \frac{\partial T}{\partial y}\right) + u \frac{\partial p}{\partial x} + \mu \left(\frac{\partial u}{\partial y}\right)^2 \tag{17.27}$$

The details are left to you.

In summary, by making the combined assumptions of $\delta \ll c$ and $\text{Re} \geq 1/\delta^2$, the complete Navier-Stokes equations derived in Chap. 15 can be reduced to simpler forms which apply to a boundary layer. These boundary-layer equations are

$$\boxed{\quad Continuity: \quad \frac{\partial(\rho u)}{\partial x} + \frac{\partial(\rho v)}{\partial y} = 0 \quad} \tag{17.28}$$

$$\boxed{\quad x\ momentum: \quad \rho u \frac{\partial u}{\partial x} + \rho v \frac{\partial u}{\partial y} = -\frac{dp_e}{dx} + \frac{\partial}{\partial y}\left(\mu \frac{\partial u}{\partial y}\right) \quad} \tag{17.29}$$

$$\boxed{\quad y\ momentum: \quad \frac{\partial p}{\partial y} = 0 \quad} \tag{17.30}$$

$$\boxed{\quad Energy: \quad \rho u \frac{\partial h}{\partial x} + \rho v \frac{\partial h}{\partial y} = \frac{\partial}{\partial y}\left(k \frac{\partial T}{\partial y}\right) + u \frac{dp_e}{dx} + \mu \left(\frac{\partial u}{\partial y}\right)^2 \quad} \tag{17.31}$$

Note that, as in the case of the Navier-Stokes equations, the boundary-layer equations are nonlinear. However, the boundary-layer equations are simpler, and therefore are more readily solved. Also, since $p = p_e(x)$, the pressure gradient expressed as $\partial p/\partial x$ in Eqs. (17.23) and (17.27) is reexpressed as dp_e/dx in Eqs.

(17.29) and (17.31). In the above equations, the unknowns are u, v, ρ, and h; p is known from $p = p_e(x)$, and μ and k are properties of the fluid which vary with temperature. To complete the system, we have

$$p = \rho R T \tag{17.32}$$

and

$$h = c_p T \tag{17.33}$$

Hence, Eqs. (17.28), (17.29), and (17.31) to (17.33) are five equations for the five unknowns, u, v, ρ, T, and h.

The boundary conditions for the above equations are as follows:

At the wall: $y = 0$ $u = 0$ $v = 0$ $T = T_w$

At the boundary-layer edge: $y \to \infty$ $u \to u_e$ $T \to T_e$

Note that since the boundary-layer thickness is not known a priori, the boundary condition at the edge of the boundary layer is given at large y, essentially y approaching infinity.

In general, the boundary-layer equations given by Eqs. (17.28) to (17.31) must be solved numerically. Modern boundary-layer techniques utilize sophisticated finite-difference solutions to obtain the boundary-layer profiles and the resulting distributions of τ_w and $\dot{q}_w$ over bodies of general shapes. Such techniques are beyond the scope of this book. However, in order to give you a feeling for the nature of boundary-layer solutions, the following two sections treat the case of flow over a flat plate, first dealing with incompressible flow in Sec. 17.4, and extending our considerations to compressible flow in Sec. 17.5. The flat plate is a special case among the whole inventory of boundary-layer solutions. It is one of a special class of solutions called *self-similar solutions*, the nature of which is discussed in the remaining sections.

17.4 INCOMPRESSIBLE FLOW OVER A FLAT PLATE: THE BLASIUS SOLUTION

Consider the incompressible, two-dimensional flow over a flat plate at $0°$ angle of attack, such as sketched in Fig. 17.7. For such a flow, $\rho = $ constant, $\mu = $ constant, and $dp_e/dx = 0$ (because the inviscid flow over a flat plate at $\alpha = 0$ yields a constant pressure over the surface). Moreover, recall that the energy equation is not needed to calculate the velocity field for an incompressible flow. Hence, the boundary-layer equations, Eqs. (17.28) to (17.31), reduce to

$$\frac{\partial u}{\partial x} + \frac{\partial v}{\partial y} = 0 \tag{17.34}$$

$$u \frac{\partial u}{\partial x} + v \frac{\partial u}{\partial y} = \nu \frac{\partial^2 u}{\partial y^2} \tag{17.35}$$

$$\frac{\partial p}{\partial y} = 0 \tag{17.36}$$

where ν is the *kinematic viscosity*, defined as $\nu \equiv \mu/\rho$.

We now embark on a procedure which is common to many boundary-layer solutions. Let us transform the independent variables (x, y) to (ξ, η), where

$$\xi = x \quad \text{and} \quad \eta = y\sqrt{\frac{V_\infty}{\nu x}} \tag{17.37}$$

Using the chain rule, we obtain the derivatives

$$\frac{\partial}{\partial x} = \frac{\partial}{\partial \xi}\frac{\partial \xi}{\partial x} + \frac{\partial}{\partial \eta}\frac{\partial \eta}{\partial x} \tag{17.38}$$

$$\frac{\partial}{\partial y} = \frac{\partial}{\partial \xi}\frac{\partial \xi}{\partial y} + \frac{\partial}{\partial \eta}\frac{\partial \eta}{\partial y} \tag{17.39}$$

However, from Eqs. (17.37) we have

$$\frac{\partial \xi}{\partial x} = 1 \qquad \frac{\partial \xi}{\partial y} = 0 \qquad \frac{\partial \eta}{\partial y} = \sqrt{\frac{V_\infty}{\nu x}} \tag{17.40}$$

(We do not have to explicitly obtain $\partial \eta / \partial x$ because these terms will eventually cancel from our equations.) Substituting Eqs. (17.40) into (17.38) and (17.39), we have

$$\frac{\partial}{\partial x} = \frac{\partial}{\partial \xi} + \frac{\partial \eta}{\partial x}\frac{\partial}{\partial \eta} \tag{17.41}$$

$$\frac{\partial}{\partial y} = \sqrt{\frac{V_\infty}{\eta x}}\frac{\partial}{\partial \eta} \tag{17.42}$$

$$\frac{\partial^2}{\partial y^2} = \frac{V_\infty}{\eta x}\frac{\partial^2}{\partial \eta^2} \tag{17.43}$$

Also, let us define a stream function ψ such that

$$\psi = \sqrt{\nu x V_\infty}\, f(\eta) \tag{17.44}$$

where $f(\eta)$ is strictly a function of η only. This expression for ψ identically satisfies the continuity equation, Eq. (17.34); therefore, it is a physically possible stream function. [Show yourself that ψ satisfies Eq. (17.34); to do this, you will have to carry out many of the same manipulations described below.] From the definition of the stream function, and using Eqs. (17.41), (17.42), and (17.44), we have

$$u = \frac{\partial \psi}{\partial y} = \sqrt{\frac{V_\infty}{\nu x}}\frac{\partial \psi}{\partial \eta} = V_\infty f'(\eta) \tag{17.45}$$

$$v = -\frac{\partial \psi}{\partial x} = -\left(\frac{\partial \psi}{\partial \xi} + \frac{\partial \eta}{\partial x}\frac{\partial \psi}{\partial \eta}\right) = -\frac{1}{2}\sqrt{\frac{\nu V_\infty}{x}}f - \sqrt{\nu x V_\infty}\frac{\partial \eta}{\partial x}f' \tag{17.46}$$

Equation (17.45) is of particular note. The function $f(\eta)$ defined in Eq. (17.44) has the property that its derivative, f', gives the x component of velocity as

$$f'(\eta) = \frac{u}{V_\infty}$$

Substitute Eqs. (17.41) to (17.43), (17.45), and (17.46) into the momentum equation, Eq. (17.35). Writing each term explicitly so that you can see what is happening, we have

$$V_\infty f'\left(V_\infty \frac{\partial \eta}{\partial x} f''\right) - \left(\frac{1}{2}\sqrt{\frac{\nu V_\infty}{x}} f + \sqrt{\nu x V_\infty}\frac{\partial \eta}{\partial x} f'\right)V_\infty\sqrt{\frac{V_\infty}{\nu x}} f'' = \nu V_\infty \frac{V_\infty}{\nu x} f'''$$

Simplifying, we obtain

$$V_\infty^2 \frac{\partial \eta}{\partial x} f'f'' - \frac{1}{2}\frac{V_\infty^2}{x} ff'' - V_\infty^2\left(\frac{\partial \eta}{\partial x}\right)f'f'' = \frac{V_\infty^2}{x} f''' \qquad (17.47)$$

The first and third terms cancel, and Eq. (17.47) becomes

$$\boxed{2f''' + ff'' = 0} \qquad (17.48)$$

Equation (17.48) is important; it is called *Blasius' equation*, after H. Blasius, who obtained it in his Ph.D. dissertation in 1908. Blasius was a student of Prandtl, and his flat-plate solution using Eq. (17.48) was the first practical application of Prandtl's boundary-layer hypothesis since its announcement in 1904. Examine Eq. (17.48) closely. Amazingly enough it is an *ordinary differential equation.* Look what has happened! Starting with the partial differential equations for a flat-plate boundary layer given by Eqs. (17.34) to (17.36), and transforming *both* the independent and dependent variables through Eqs. (17.37) and (17.44), we obtain an ordinary differential equation for $f(\eta)$. In the same breath, we can say that Eq. (17.48) is also an equation for the velocity u because $u = V_\infty f'(\eta)$. Because Eq. (17.48) is a single ordinary differential equation, it is simpler to solve than the original boundary-layer equations. However, it is still a nonlinear equation and must be solved numerically, subject to the transformed boundary conditions,

At $\eta = 0$: $\qquad\qquad\qquad\qquad f = 0, f' = 0$

At $\eta \to \infty$: $\qquad\qquad\qquad\qquad f' = 1$

[Note that at the wall where $\eta = 0$, $f' = 0$ because $u = 0$, and therefore $f = 0$ from Eq. (17.46) evaluated at the wall.]

Equation (17.48) is a third-order, nonlinear, ordinary differential equation; it can be solved numerically by means of standard techniques, such as the Runge-Kutta method (such as that described in Ref. 52). The integration begins at the wall and is carried out in small increments Δy in the direction of increasing y away from the wall. However, since Eq. (17.48) is third order, three boundary conditions must be known at $\eta = 0$; from the above, only two are specified. A third boundary condition, namely, some value for $f''(0)$, must be *assumed*; Eq. (17.48) is then integrated across the boundary layer to a large value of η. The value of f' at large eta is then examined. Does it match the boundary condition at the edge of the boundary layer, namely, is $f' = 1$ satisfied at the edge of the boundary layer? If not, assume a different value of $f''(0)$ and integrate again. Repeat this process until convergence is obtained. This numerical approach is

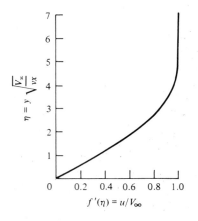

FIGURE 17.8

Incompressible velocity profile for a flat plate; solution of the Blasius equation.

called the "shooting technique"; it is a classical approach, and its basic philosophy and details are discussed at great length in Sec. 16.4. Its application to Eq. (17.48) is more straightforward than the discussion in Sec. 16.4, because here we are dealing with an incompressible flow and only one equation, namely, the momentum equation as embodied in Eq. (17.48).

The solution of Eq. (17.48) is plotted in Fig. 17.8 in the form of $f'(\eta) = u/V_\infty$ as a function of η. Note that this curve is the *velocity profile* and that it is a function of η only. Think about this for a moment. Consider two different x stations along the plate, as shown in Fig. 17.9. In general, $u = u(x, y)$, and the velocity profiles in terms of $u = u(y)$ at given x stations will be *different*. Clearly, the variation of u normal to the wall will change as the flow progresses downstream. However, when plotted versus η, we see that the profile, $u = u(\eta)$, is the same for *all x* stations, as illustrated in Fig. 17.9. This result is an example of a *self-similar solution*—solutions where the boundary-layer profiles, when plotted

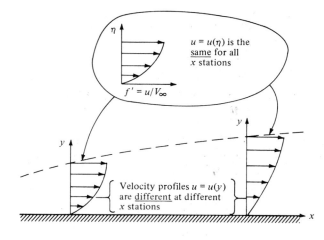

FIGURE 17.9

Velocity profiles in physical and transformed space, demonstrating the meaning of self-similar solutions.

versus a similarity variable, η, are the same for all x stations. For such self-similar solutions, the governing boundary-layer equations reduce to one or more ordinary differential equations in terms of a transformed independent variable. Self-similar solutions occur only for certain special types of flows—the flow over a flat plate is one such example. In general, for the flow over an arbitrary body, the boundary-layer solutions are nonsimilar; the governing partial differential equations cannot be reduced to ordinary differential equations.

Numerical values of f, f', and f'' tabulated versus η can be found in Ref. 42. Of particular interest is the value of f'' at the wall; $f''(0) = 0.332$. Consider the local skin friction coefficient defined as $c_f = \tau_w / \frac{1}{2}\rho_\infty V_\infty^2$. From Eq. (15.1), the shear stress at the wall is given by

$$\tau_w = \mu \left(\frac{\partial u}{\partial y}\right)_{y=0} \tag{17.49}$$

However, from Eqs. (17.42) and (17.44),

$$\frac{\partial u}{\partial y} = V_\infty \frac{\partial f'}{\partial y} = V_\infty \sqrt{\frac{V_\infty}{\nu x}} \frac{\partial f'}{\partial \eta} = V_\infty \sqrt{\frac{V_\infty}{\nu x}} f'' \tag{17.50}$$

Evaluating Eq. (17.50) at the wall, where $y = \eta = 0$, we obtain

$$\left(\frac{\partial u}{\partial y}\right)_{y=0} = V_\infty \sqrt{\frac{V_\infty}{\nu x}} f''(0) \tag{17.51}$$

Combining Eqs. (17.49) and (17.51), we have

$$c_f = \frac{\tau_w}{\frac{1}{2}\rho_\infty V_\infty^2} = \frac{2\mu}{\rho_\infty V_\infty^2} V_\infty \sqrt{\frac{V_\infty}{\nu x}} f''(0)$$

$$= 2\sqrt{\frac{\mu}{\rho_\infty V_\infty x}} f''(0) = \frac{2f''(0)}{\sqrt{\mathrm{Re}_x}} \tag{17.52}$$

where Re_x is the local Reynolds number. Since $f''(0) = 0.332$ from the numerical solution of Eq. (17.48), then Eq. (17.52) yields

$$\boxed{c_f = \frac{0.664}{\sqrt{\mathrm{Re}_x}}} \tag{17.53}$$

which is a classic expression for the local skin friction coefficient for the incompressible laminar flow over a flat plate—a result that stems directly from boundary-layer theory. Its validity has been amply verified by experiment. Note that $c_f \propto \mathrm{Re}_x^{-1/2} \propto x^{-1/2}$; i.e., c_f decreases inversely proportional to the square root of distance from the leading edge. Examining the flat plate sketched in Fig. 17.7, the total drag on the top surface of the entire plate is the integrated contribution of $\tau_w(x)$ from $x = 0$ to $x = c$. Letting C_f denote the skin friction drag coefficient, we obtain from Eq. (1.16)

$$C_f = \frac{1}{c} \int_0^c c_f \, dx \tag{17.54}$$

Substituting Eq. (17.53) into (17.54), we obtain

$$C_f = \frac{1}{c}(0.664)\sqrt{\frac{\mu}{\rho_\infty V_\infty}}\int_0^c x^{-1/2}\,dx = \frac{1.328}{c}\sqrt{\frac{\mu c}{\rho_\infty V_\infty}}$$

or

$$C_f = \frac{1.328}{\sqrt{\mathrm{Re}_c}} \tag{17.55}$$

where Re_c is the Reynolds number based on the total plate length c.

An examination of Fig. 17.8 shows that $f' = 0.99$ at approximately $\eta = 5.0$. Hence, the boundary-layer thickness, which was defined earlier as that distance above the surface where $u = 0.99u_e$, is

$$\eta = y\sqrt{\frac{V_\infty}{\nu x}} = \delta\sqrt{\frac{V_\infty}{\nu x}} = 5.0$$

or

$$\delta = \frac{5.0x}{\sqrt{\mathrm{Re}_x}} \tag{17.56}$$

Note that the boundary-layer thickness is inversely proportional to the square root of the Reynolds number (based on the local distance x). Also, $\delta \propto x^{1/2}$; the laminar boundary layer over a flat plate grows parabolically with distance from the leading edge.

The displacement thickness δ^*, defined by Eq. (17.3), becomes for an incompressible flow

$$\delta^* = \int_0^{y_1}\left(1 - \frac{u}{u_e}\right)dy \tag{17.57}$$

In terms of the transformed variables f' and η given by Eqs. (17.37) and (17.45), the integral in Eq. (17.57) can be written as

$$\delta^* = \sqrt{\frac{\nu x}{V_\infty}}\int_0^{\eta_1}[1 - f'(\eta)]\,d\eta = \sqrt{\frac{\nu x}{V_\infty}}[\eta_1 - f(\eta_1)] \tag{17.58}$$

where η_1 is an arbitrary point above the boundary layer. The numerical solution for $f(\eta)$ obtained from Eq. (17.48) shows that, amazingly enough, $\eta_1 - f(\eta) = 1.72$ for all values of η above 5.0. Therefore, from Eq. (17.58), we have

$$\delta^* = 1.72\sqrt{\frac{\nu x}{V_\infty}}$$

or

$$\delta^* = \frac{1.72x}{\sqrt{\mathrm{Re}_x}} \tag{17.59}$$

Note that, as in the case of the boundary-layer thickness itself, δ^* varies inversely with the square root of the Reynolds number, and $\delta^* \propto x^{1/2}$. Also, comparing Eqs. (17.56) and (17.59), we see that $\delta^* = 0.34\delta$; the displacement thickness is smaller than the boundary-layer thickness, confirming our earlier statement in Sec. 17.2.

The momentum thickness for an incompressible flow is, from Eq. (17.10),

$$\theta = \int_0^{y_1} \frac{u}{u_e} \left(1 - \frac{u}{u_e} \right) dy$$

or in terms of our transformed variables,

$$\theta = \sqrt{\frac{\nu x}{V_\infty}} \int_0^{\eta_1} f'(1-f') \, d\eta \tag{17.60}$$

Equation (17.60) can be integrated numerically from $\eta = 0$ to any arbitrary point $\eta_1 > 5.0$. The result gives

$$\theta = \sqrt{\frac{\eta x}{V_\infty}} (0.664)$$

or

$$\boxed{\theta = \frac{0.664x}{\sqrt{\mathrm{Re}_x}}} \tag{17.61}$$

Note that, as in the case of our previous thicknesses, θ varies inversely with the square root of the Reynolds number and that $\theta \propto x^{1/2}$. Also, $\theta = 0.39\delta^*$, and $\theta = 0.13\delta$. Another property of momentum thickness can be demonstrated by evaluating θ at the trailing edge of the flat plate sketched in Fig. 17.7. In this case, $x = c$, and from Eq. (17.61), we obtain

$$\theta_{x=c} = \frac{0.664c}{\sqrt{\mathrm{Re}_c}} \tag{17.62}$$

Comparing Eqs. (17.55) and (17.62), we have

$$C_f = \frac{2\theta_{x=c}}{c} \tag{17.63}$$

Equation (17.63) demonstrates that the integrated skin friction coefficient for the flat plate is directly proportional to the value of θ evaluated at the trailing edge.

17.5 COMPRESSIBLE FLOW OVER A FLAT PLATE

The properties of the incompressible, laminar, flat-plate boundary layer were developed in Sec. 17.4. These results hold at low Mach numbers where the density is essentially constant through the boundary layer. However, what happens to these properties at high Mach numbers where the density becomes a variable; i.e., what are the compressibility effects? The purpose of the present section is to outline briefly the effects of compressibility on both the derivations and the final results for laminar flow over a flat plate. We do not intend to present much detail; rather, we examine some of the salient aspects which distinguish compressible from incompressible boundary layers.

The compressible boundary-layer equations were derived in Sec. 17.3, and were presented as Eqs. (17.28) to (17.31). For flow over a flat plate, where $dp_e/dx = 0$, these equations become

$$\frac{\partial(\rho u)}{\partial x} + \frac{\partial(\rho v)}{\partial y} = 0 \tag{17.64}$$

$$\rho u \frac{\partial u}{\partial x} + \rho v \frac{\partial u}{\partial y} = \frac{\partial}{\partial y}\left(\mu \frac{\partial u}{\partial y}\right) \tag{17.65}$$

$$\frac{\partial p}{\partial y} = 0 \tag{17.66}$$

$$\rho u \frac{\partial h}{\partial x} + \rho v \frac{\partial h}{\partial y} = \frac{\partial}{\partial y}\left(k \frac{\partial T}{\partial y}\right) + \mu\left(\frac{\partial u}{\partial y}\right)^2 \tag{17.67}$$

Compare these equations with those for the incompressible case given by Eqs. (17.34) to (17.36). Note that, for a compressible boundary layer, (1) the energy equation must be included, (2) the density is treated as a variable, and (3) in general, μ and k are functions of temperature and hence also must be treated as variables. As a result, the system of equations for the compressible case, Eqs. (17.64) to (17.67), is more complex than for the incompressible case, Eqs. (17.34) to (17.36).

It is sometimes convenient to deal with total enthalpy, $h_0 = h + V^2/2$, as the dependent variable in the energy equation, rather than the static enthalpy as given in Eq. (17.67). Note that, consistent with the boundary-layer approximation, where v is small, $h_0 = h + V^2/2 = h + (u+v^2)/2 \approx h + u^2/2$. To obtain the energy equation in terms of h_0, multiply Eq. (17.65) by u, and add to Eq. (17.67), as follows. From Eq. (17.65) multiplied by u,

$$\rho u \frac{\partial(u^2/2)}{\partial x} + \rho v \frac{\partial(u^2/2)}{\partial y} = u \frac{\partial}{\partial y}\left(\mu \frac{\partial u}{\partial y}\right) \tag{17.68}$$

Adding Eq. (17.68) to (17.67), we obtain

$$\rho u \frac{\partial(h+u^2/2)}{\partial x} + \rho v \frac{\partial(h+u^2/2)}{\partial y} = \frac{\partial}{\partial y}\left(k \frac{\partial T}{\partial y}\right) + \mu\left(\frac{\partial u}{\partial y}\right)^2 + u \frac{\partial}{\partial y}\left(\mu \frac{\partial u}{\partial y}\right) \tag{17.69}$$

Recall that for a calorically perfect gas, $dh = c_p\, dT$; hence,

$$\frac{\partial T}{\partial y} = \frac{1}{c_p}\frac{\partial h}{\partial y} = \frac{1}{c_p}\frac{\partial}{\partial y}\left(h_0 - \frac{u^2}{2}\right) \tag{17.70}$$

Substituting Eq. (17.70) into (17.69), we obtain

$$\rho u \frac{\partial h_0}{\partial x} + \rho v \frac{\partial h_0}{\partial y} = \frac{\partial}{\partial y}\left[\frac{k}{c_p}\frac{\partial}{\partial y}\left(h_0 - \frac{u^2}{2}\right)\right] + \mu\left(\frac{\partial u^2}{\partial y}\right)^2 + u \frac{\partial}{\partial y}\left(\mu \frac{\partial u}{\partial y}\right) \tag{17.71}$$

Note that

$$\frac{k}{c_p}\frac{\partial}{\partial y}\left(h_0 - \frac{u^2}{2}\right) = \frac{\mu k}{\mu c_p}\frac{\partial}{\partial y}\left(h_0 - \frac{u^2}{2}\right) = \frac{\mu}{\text{Pr}}\left(\frac{\partial h_0}{\partial y} - u\frac{\partial u}{\partial y}\right) \tag{17.72}$$

and

$$\mu\left(\frac{\partial u}{\partial y}\right)^2 + u\frac{\partial}{\partial y}\left(\mu\frac{\partial u}{\partial y}\right) = \frac{\partial}{\partial y}\left(\mu u\frac{\partial u}{\partial y}\right) \tag{17.73}$$

Substituting Eqs. (17.72) and (17.73) into (17.71), we obtain

$$\rho u\frac{\partial h_0}{\partial x} + \rho v\frac{\partial h_0}{\partial y} = \frac{\partial}{\partial y}\left[\frac{\mu}{\text{Pr}}\frac{\partial h_0}{\partial y} + \left(1 - \frac{1}{\text{Pr}}\right)\mu u\frac{\partial u}{\partial y}\right] \tag{17.74}$$

which is an alternate form of the boundary-layer energy equation. In this equation, Pr is the local Prandtl number, which, in general, is a function of T and hence varies throughout the boundary layer.

For the laminar, compressible flow over a flat plate, the system of governing equations can now be considered to be Eqs. (17.64) to (17.66) and (17.74). These are nonlinear partial differential equations. As in the incompressible case, let us seek a self-similar solution; however, the transformed independent variables must be defined differently:

$$\xi = \rho_e\mu_e u_e x \qquad \xi = \xi(x)$$

$$\eta = \frac{u_e}{\sqrt{2\xi}}\int_0^y \rho\, dy \qquad \eta = \eta(x, y)$$

The dependent variables are transformed as follows:

$$f' = \frac{u}{u_e} \qquad \text{(which is consistent with defining stream function } \psi = \sqrt{2\xi}\, f\text{)}$$

$$g = \frac{h_0}{(h_0)_e}$$

The mechanics of the transformation using the chain rule are similar to that described in Sec. 17.4. Hence, without detailing the precise steps (which are left for your entertainment), Eqs. (17.65) and (17.74) transform into

$$\left(\frac{\rho\mu}{\rho_e\mu_e}f''\right)' + ff' = 0 \tag{17.75}$$

and

$$\left(\frac{\rho\mu}{\rho_e\mu_e}\frac{1}{\text{Pr}}g'\right)' + fg' + \frac{u_e^2}{(h_0)_e}\left[\left(1 - \frac{1}{\text{Pr}}\right)\frac{\rho\mu}{\rho_e\mu_e}f'f''\right]' = 0 \tag{17.76}$$

Examine Eqs. (17.75) and (17.76) closely. They are ordinary differential equations—recall that the primes denote differentiation with respect to η. Therefore, the compressible, laminar flow over a flat plate does lend itself to a self-similar solution, where $f' = f'(\eta)$ and $g = g(\eta)$. That is, the velocity and total enthalpy profiles plotted versus η are the same at any station. Furthermore, the product

$\rho\mu$ is a variable and depends in part on temperature. Hence, Eq. (17.75) is coupled to the energy equation, Eq. (17.76), via $\rho\mu$. Of course, the energy equation is strongly coupled to Eq. (17.75) via the appearance of f, f', and f'' in Eq. (17.76). Hence, we are dealing with a *system* of coupled ordinary differential equations which must be solved simultaneously. The boundary conditions for these equations are

At $\eta = 0$: $\qquad\qquad\qquad f = f' = 0 \qquad g = g_w$

At $\eta \to \infty$: $\qquad\qquad\qquad\quad f' = 1 \qquad g = 1$

Note that the coefficient $u_e^2/(h_0)_e$ appearing in Eq. (17.76) is simply a function of the Mach number:

$$\frac{u_e^2}{(h_0)_e} = \frac{u_e^2}{h_e + u_e^2/2} = \frac{1}{h_e/u_e^2 + \frac{1}{2}} = \frac{1}{c_p h_e/u_e^2 + \frac{1}{2}} = \frac{1}{RT_e/(\gamma-1)u_e^2 + \frac{1}{2}}$$

$$= \frac{1}{1/(\gamma-1)M_e^2 + \frac{1}{2}} = \frac{2(\gamma-1)M_e^2}{2 + (\gamma-1)M_e^2}$$

Therefore, Eq. (17.76) involves as a parameter the Mach number of the flow at the outer edge of the boundary layer, i.e., for the flat-plate case, the freestream Mach number. Hence, we can explicitly see that the compressible boundary-layer solutions will depend on the Mach number. Moreover, because of the appearance of the local Pr in Eq. (17.76), the solutions will also depend on the freestream Prandtl number as a parameter. Finally, note from the boundary conditions that the value of g at the wall, g_w, is a given quantity. Note that at the wall where $u = 0$, $g_w = h_w/(h_0)_e = c_p T_w/(h_0)_e$. Hence, instead of referring to a given enthalpy at the wall, g_w, we usually deal with a given wall temperature, T_w. An alternative to a given value of T_w is the assumption of an *adiabatic wall*, i.e., a case where there is no heat transfer to the wall. If $\dot{q}_w = k(\partial T/\partial y)_w = 0$, then $(\partial T/\partial y)_w = 0$. Hence, for an adiabatic wall, the boundary condition at the wall becomes simply $(\partial T/\partial y)_w = 0$.

In short, we see from the above discussion that a numerical self-similar solution can be obtained for the compressible, laminar flow over a flat plate. However, this solution depends on the Mach number, the Prandtl number, and the condition of the wall (whether it is adiabatic or a constant temperature wall with T_w given). Such numerical solutions have been carried out; see Ref. 43 for details. A classic solution to Eqs. (17.75) and (17.76) is the shooting technique described in Sec. 16.4. The approach here is directly analogous to that used for the solution of compressible Couette flow discussed in Sec. 16.4. Since Eq. (17.75) is third order, we need three boundary conditions at $\eta = 0$. We have only two, namely, $f = f' = 0$. Therefore, *assume* a value for $f''(0)$, and iterate until the boundary condition at the boundary-layer edge, $f' = 1$, is matched. Similarly, Eq. (17.76) is a second-order equation. It requires two boundary conditions at the wall in order to integrate numerically across the boundary layer; we have only one, namely, $g(0) = g_w$. Thus, *assume* $g'(0)$, and integrate Eq. (17.76). Iterate until the outer boundary condition is satisfied; i.e., $g = 1$. Since Eq. (17.75) is coupled to Eq. (17.76), i.e., since $\rho\mu$ in Eq. (17.75) requires a knowledge of the enthalpy (or temperature) profile across the boundary layer, the entire process must be repeated again. This is directly analogous to the two minor iterations nested within the major iteration that was described in the discussion of the shooting method in Sec. 16.4. The approach here is virtually the same philosophy as described in Sec. 16.4, which should be reviewed at this stage. Therefore, no further details will be given here.

Return to Eq. (17.55) for the friction drag coefficient for incompressible flow. The analogous compressible result can be written as

$$C_f = \frac{1.328}{\sqrt{\mathrm{Re}_c}} F\left(M_e, \mathrm{Pr}, \frac{T_w}{T_e}\right) \tag{17.77}$$

In Eq. (17.77), the function F is determined from the numerical solution. Sample results are given in Fig. 17.10, which shows that the product $C_f\sqrt{\mathrm{Re}_c}$ decreases as M_e increases. Moreover, the adiabatic wall is warmer than the wall in the case of $T_w/T_e = 1.0$. Hence, Fig. 17.10 demonstrates that a hot wall also reduces $C_f\sqrt{\mathrm{Re}_c}$.

Return to Eq. (17.56) for the thickness of the incompressible flat-plate boundary layer. The analogous result for compressible flow is

$$\delta = \frac{5.0x}{\sqrt{\mathrm{Re}_x}} G\left(M_e, \mathrm{Pr}, \frac{T_w}{T_e}\right) \tag{17.78}$$

In Eq. (17.78), the function G is obtained from the numerical solution. Sample results are given in Fig. 17.11, which shows that the product $(\delta\sqrt{\mathrm{Re}_x}/x)$ increases as M_e increases. Everything else being equal, boundary layers are thicker at higher Mach numbers. This fact was stated earlier, in Chap. 14 dealing with hypersonic flow. Note also from Fig. 17.11 that a hot wall thickens the boundary layer.

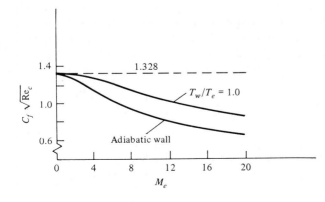

FIGURE 17.10
Friction drag coefficient for laminar, compressible flow over a flat plate, illustrating the effect of Mach number and wall temperature. $\mathrm{Pr} = 0.75$. (*Calculations by E. R. van Driest, NACA Tech. Note 2597.*)

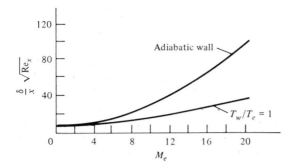

Recall our discussion of Couette flow in Chap. 16. There, we introduced the concept of the recovery factor, r, where

$$h_{aw} = h_e + r\frac{u_e^2}{2} \tag{17.79}$$

This is a general concept, and can be applied to the boundary-layer solutions here. If we assume a constant Prandtl number for the compressible flat-plate flow, the numerical solution shows that

$$r = \sqrt{Pr} \tag{17.80}$$

for the flat plate. Note that Eq. (17.80) is analogous to the result given for Couette flow in that the recovery factor is a function of the Prandtl number only. However, for the flat plate, $r = \sqrt{Pr}$, whereas for Couette flow, $r = Pr$.

Aerodynamic heating for the flat plate can be treated via Reynolds analogy. The Stanton number and skin friction coefficients are defined respectively as

$$C_H = \frac{\dot{q}_w}{\rho_e u_e (h_{aw} - h_w)} \tag{17.81}$$

and

$$c_f = \frac{\tau_w}{\frac{1}{2}\rho_e u_e^2} \tag{17.82}$$

(See our discussion of these coefficients in Chap. 16.) Our results for Couette flow proved that a relation existed between C_H and c_f—namely, Reynolds analogy, given by Eq. (16.59) for Couette flow. Moreover, in this relation, the ratio C_H/c_f was a function of the Prandtl number only. A directly analogous result holds for the compressible flat-plate flow. If we assume that the Prandtl number is constant, then for a flat plate, Reynolds analogy is, from the numerical solution,

$$\boxed{\frac{C_H}{c_f} = \frac{1}{2}Pr^{-2/3}} \tag{17.83}$$

In Eq. (17.83), the local skin friction coefficient, c_f, which is given by Eq. (17.53) for the incompressible flat-plate case, becomes the following form for the compressible flat-plate flow:

$$c_f = \frac{0.664}{\sqrt{Re}} F\left(M_e, Pr, \frac{T_w}{T_e}\right) \qquad (17.84)$$

In Eq. (17.84), F is the same function as appears in Eq. (17.74), and its variation with M_e and T_w/T_e is the same as shown in Fig. 17.10.

17.6 RESULTS FOR TURBULENT BOUNDARY LAYERS

The basic nature of turbulent flow and the transition of laminar to turbulent flow were discussed in Chap. 15. The analysis of turbulent boundary layers is a constantly evolving and heavily worked subject; it is not our purpose to even introduce such analyses here. Rather, in this section, we discuss a few results for the turbulent boundary layer on a flat plate, both incompressible and compressible, simply to provide a basis of comparison with the laminar results described in the previous section. For considerably more detail on the subject of turbulent boundary layers, consult Refs. 42 to 44.

For incompressible flow over a flat plate, the boundary-layer thickness is given approximately by

$$\delta = \frac{0.37x}{Re_x^{1/5}} \qquad (17.85)$$

Note from Eq. (17.85) that the turbulent boundary-layer thickness varies approximately as $Re_x^{-1/5}$ in contrast to $Re_x^{-1/2}$ for a laminar boundary layer. Also, turbulent values of δ grow more rapidly with distance along the surface; $\delta \propto x^{4/5}$ for a turbulent flow in contrast to $\delta \propto x^{1/2}$ for a laminar flow. With regard to skin friction drag, for incompressible turbulent flow over a flat plate, we have

$$C_f = \frac{0.074}{Re_c^{1/5}} \qquad (17.86)$$

Note that for turbulent flow, C_f varies as $Re_c^{-1/5}$ in comparison with the $Re_c^{-1/2}$ variation for laminar flow. Hence, Eq. (17.86) yields larger friction drag coefficients for turbulent flow in comparison with Eq. (17.55) for laminar flow.

The effects of compressibility on Eq. (17.86) are shown in Fig. 17.12, where C_f is plotted versus Re_∞ with M_∞ as a parameter. The turbulent flow results are shown toward the right of Fig. 17.12, at the higher values of Reynolds numbers where turbulent conditions are expected to occur, and laminar flow results are shown toward the left of the figure, at lower values of Reynolds numbers. This type of figure—friction drag coefficient for both laminar and turbulent flow as a

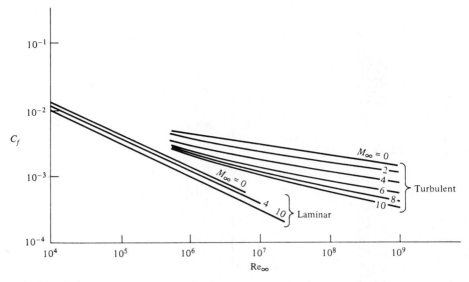

FIGURE 17.12
Turbulent friction drag coefficient for a flat plate as a function of Reynolds and Mach numbers. Adiabatic wall. Pr = 0.75. For contrast, some laminar results are shown. (*Data are from the calculations of van Driest, Ref. 47.*)

function of Re on a log-log plot—is a classic picture, and it allows a ready contrast of the two types of flow. From this figure, we can see that, for the same Re_∞, turbulent skin friction is higher than laminar; also, the slopes of the turbulent curves are smaller than the slopes of the laminar curves—a graphic comparison of the $Re^{-1/5}$ variation in contrast to the laminar $Re^{-1/2}$ variation. Note that the effect of increasing M_∞ is to reduce C_f at constant Re and that this effect is stronger on the turbulent flow results. Indeed, C_f for the turbulent results decreases by almost an order of magnitude (at the higher values of Re_∞) when M_∞ is increased from 0 to 10. For the laminar flow, the decrease in C_f as M_∞ is increased through the same Mach number range is far less pronounced.

17.7 FINAL COMMENTS

This chapter has dealt with boundary layers, especially those on a flat plate. We end with the presentation of a photograph in Fig. 17.13 showing the development of velocity profiles in the boundary layer over a flat plate. The fluid is water, which flows from left to right. The profiles are made visible by the hydrogen bubble technique, the same used for Fig. 16.13. The Reynolds number is low (the freestream velocity is only 0.6 m/s); hence, the boundary-layer thickness is large. However, the thickness of the plate is only 0.5 mm, which means that the boundary layer shown here is on the order of 1 mm thick—still small on an absolute scale. In any event, if you need any further proof of the existence of boundary layers, Fig. 17.13 is it.

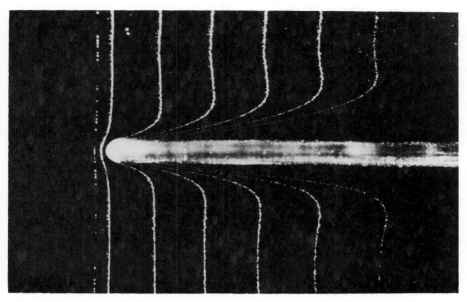

FIGURE 17.13
Photograph of velocity profiles for the laminar flow over a flat plate. Flow is from left to right. (*Courtesy of Yasuki Nakayama, Tokai University, Japan.*)

17.8 SUMMARY

Return to the road map given in Fig. 17.2, and make certain that you feel at home with the material represented by each box. The highlights of our discussion of boundary layers are summarized as follows:

> The basic quantities of interest from boundary-layer theory are the velocity and thermal boundary-layer thicknesses, δ and δ_T, respectively, the shear stress at the wall, τ_w, and heat transfer to the surface, $\dot{q}_w$. In the process, we can define two additional thicknesses: the displacement thickness
>
> $$\delta^* \equiv \int_0^{y_1} \left(1 - \frac{\rho u}{\rho_e u_e}\right) dy \qquad \delta \le y_1 \to \infty \qquad (17.3)$$
>
> and the momentum thickness
>
> $$\theta \equiv \int_0^{y_1} \frac{\rho u}{\rho_e u_e} \left(1 - \frac{u}{u_e}\right) dy \qquad \delta \le y_1 \to \infty \qquad (17.10)$$
>
> Both δ^* and θ are related to decrements in the flow due to the presence of the boundary layer; δ^* is proportional to the decrement in mass flow, and θ is proportional to the decrement in momentum flow. Moreover, δ^* is the distance away from the body surface through which the outer inviscid flow is displaced due to the boundary layer. The body shape plus δ^* defines a new effective body seen by the inviscid flow.

By an order-of-magnitude analysis, the complete Navier-Stokes equations for two-dimensional flow reduce to the following boundary-layer equations:

Continuity:
$$\frac{\partial(\rho u)}{\partial x} + \frac{\partial(\rho v)}{\partial y} = 0 \tag{17.28}$$

x momentum:
$$\rho u \frac{\partial u}{\partial x} + \rho v \frac{\partial u}{\partial y} = -\frac{dp_e}{dx} + \frac{\partial}{\partial y}\left(\mu \frac{\partial u}{\partial y}\right) \tag{17.29}$$

y momentum:
$$\frac{\partial p}{\partial y} = 0 \tag{17.30}$$

Energy:
$$\rho u \frac{\partial h}{\partial x} + \rho v \frac{\partial h}{\partial y} = \frac{\partial}{\partial y}\left(k \frac{\partial T}{\partial y}\right) + u \frac{dp_e}{dx} + \mu \left(\frac{\partial u}{\partial y}\right)^2 \tag{17.31}$$

These equations are subject to the boundary conditions:

At the wall: $y = 0$ $u = 0$ $v = 0$ $h = h_w$

At the boundary-layer edge: $y \to \infty$ $u \to u_e$ $h \to h_e$

Inherent in the above boundary-layer equations are the assumptions that $\delta \ll c$, Re is large, and M_∞ is not inordinately large.

For incompressible laminar flow over a flat plate, the boundary-layer equations reduce to the Blasius equation

$$2f''' + ff'' = 0 \tag{17.48}$$

where $f' = u/u_e$. This produces a self-similar solution where $f' = f'(\eta)$, independent of any particular x station along the surface. A numerical solution of Eq. (17.48) yields numbers which lead to the following results.

Local skin friction coefficient:
$$c_f = \frac{\tau_w}{\frac{1}{2}\rho_\infty V_\infty^2} = \frac{0.664}{\sqrt{Re_x}} \tag{17.53}$$

Integrated friction drag coefficient: $C_f = \dfrac{1.328}{\sqrt{Re_c}}$ (17.55)

Boundary-layer thickness:
$$\delta = \frac{5.0x}{\sqrt{Re_x}} \tag{17.56}$$

Displacement thickness:
$$\delta^* = \frac{1.72x}{\sqrt{Re_x}} \tag{17.59}$$

Momentum thickness:
$$\theta = \frac{0.664x}{\sqrt{Re_x}} \tag{17.61}$$

Compressibility effects are such as to make boundary-layer solutions a function of Mach number, Prandtl number, and wall-to-freestream temperature ratio. Typical compressibility effects are shown in Figs. 17.10 to 17.12. Generally, compressibility reduces C_f and increases δ.

Approximations for turbulent, incompressible flow over a flat plate are

$$\delta = \frac{0.37x}{\mathrm{Re}_x^{1/5}} \qquad (17.79)$$

and

$$C_f = \frac{0.074}{\mathrm{Re}_c^{1/5}} \qquad (17.80)$$

PROBLEMS

Note: The standard sea level value of viscosity coefficient for air is $\mu = 1.7894 \times 10^{-5}$ kg/(m · s) $= 3.7373 \times 10^{-7}$ slug/(ft · s).

17.1. The wing on a Piper Cherokee general aviation aircraft is rectangular, with a span of 9.75 m and a chord of 1.6 m. The aircraft is flying at cruising speed (141 mi/h) at sea level. Assume that the skin friction drag on the wing can be approximated by the drag on a flat plate of the same dimensions. Calculate the skin friction drag:
(*a*) If the flow were completely laminar (which is not the case in real life)
(*b*) If the flow were completely turbulent (which is more realistic)
Compare the two results.

17.2. For the case in Prob. 17.1, calculate the boundary-layer thickness at the trailing edge for
(*a*) Completely laminar flow
(*b*) Completely turbulent flow

17.3. For the case in Prob. 17.1, calculate the skin friction drag accounting for transition. Assume the transition Reynolds number $= 5 \times 10^5$.

17.4. Consider Mach 4 flow at standard sea level conditions over a flat plate of chord 5 in. Assuming all laminar flow and adiabatic wall conditions, calculate the skin friction drag on the plate per unit span.

17.5. Repeat Prob. 17.4 for the case of all turbulent flow.

17.6. Consider a compressible, laminar boundary layer over a flat plate. Assuming Pr = 1 and a calorically perfect gas, show that the profile of total temperature through the boundary layer is a function of the velocity profile via

$$T_0 = T_w + (T_{0,e} - T_w)\frac{u}{u_e}$$

where T_w = wall temperature and $T_{0,e}$ and u_e are the total temperature and velocity, respectively, at the outer edge of the boundary layer. [*Hint*: Compare Eqs. (17.65) and (17.74).]

NAVIER-STOKES SOLUTIONS: SOME EXAMPLES

A numerical simulation of the flow over an airfoil using the Reynolds averaged Navier-Stokes equations can be conducted on today's supercomputers in less than a half hour for less than $1000 cost in computer time. If just one such simulation had been attempted 20 years ago on computers of that time (e.g., the IBM 704 class) and with algorithms then known, the cost in computer time would have amounted to roughly $10 million, and the results for that single flow would not be available until 10 years from now, since the computation would have taken about 30 years to complete.

Dean R. Chapman, NASA, 1977

18.1 INTRODUCTION

This chapter is short. Its purpose is to discuss the third option for the solution of viscous flows as discussed in Sec. 15.7, namely, the exact numerical solution of the complete Navier-Stokes equations. This option is the purview of modern computational fluid dynamics—it is a state-of-the-art research activity which is currently in a rapid state of development. This subject now occupies volumes of modern literature; for a basic treatment, see the definitive text on computational fluid dynamics listed as Ref. 54. We will only list a few sample calculations here.

18.2 THE APPROACH

Return to the complete Navier-Stokes equations, as derived in Chap. 15, and repeated below for convenience:

Continuity:

$$\frac{\partial \rho}{\partial t} = -\left[\frac{\partial(\rho u)}{\partial x} + \frac{\partial(\rho v)}{\partial y} + \frac{\partial(\rho w)}{\partial z}\right] \tag{18.1}$$

x momentum:

$$\frac{\partial u}{\partial t} = -u\frac{\partial u}{\partial x} - v\frac{\partial u}{\partial y} - w\frac{\partial u}{\partial z} + \frac{1}{\rho}\left[-\frac{\partial p}{\partial x} + \frac{\partial \tau_{xx}}{\partial x} + \frac{\partial \tau_{yx}}{\partial y} + \frac{\partial \tau_{zx}}{\partial z} \right] \tag{18.2}$$

y momentum:

$$\frac{\partial v}{\partial t} = -u\frac{\partial v}{\partial x} - v\frac{\partial v}{\partial y} - w\frac{\partial v}{\partial z} + \frac{1}{\rho}\left[-\frac{\partial p}{\partial y} + \frac{\partial \tau_{xy}}{\partial x} + \frac{\partial \tau_{yy}}{\partial y} + \frac{\partial \tau_{zy}}{\partial z} \right] \tag{18.3}$$

z momentum:

$$\frac{\partial w}{\partial t} = -u\frac{\partial w}{\partial x} - v\frac{\partial w}{\partial y} - w\frac{\partial w}{\partial z} + \frac{1}{\rho}\left[-\frac{\partial p}{\partial z} + \frac{\partial \tau_{xz}}{\partial x} + \frac{\partial \tau_{yz}}{\partial y} + \frac{\partial \tau_{zz}}{\partial z} \right] \tag{18.4}$$

Energy:

$$\frac{\partial(e + V^2/2)}{\partial t} = -u\frac{\partial(e + V^2/2)}{\partial x} - v\frac{\partial(e + V^2/2)}{\partial y} - w\frac{\partial(e + V^2/2)}{\partial z} + \dot{q}$$

$$+ \frac{1}{\rho}\left[\frac{\partial}{\partial x}\left(k\frac{\partial T}{\partial x} \right) + \frac{\partial}{\partial y}\left(k\frac{\partial T}{\partial y} \right) + \frac{\partial}{\partial z}\left(k\frac{\partial T}{\partial z} \right) \right.$$

$$- \frac{\partial(pu)}{\partial x} - \frac{\partial(pv)}{\partial y} - \frac{\partial(pw)}{\partial z} + \frac{\partial(u\tau_{xx})}{\partial x} + \frac{\partial(u\tau_{yx})}{\partial y} + \frac{\partial(u\tau_{zx})}{\partial z}$$

$$\left. + \frac{\partial(v\tau_{xy})}{\partial x} + \frac{\partial(v\tau_{yy})}{\partial y} + \frac{\partial(v\tau_{zy})}{\partial z} + \frac{\partial(w\tau_{xz})}{\partial x} + \frac{\partial(w\tau_{yz})}{\partial y} + \frac{\partial(w\tau_{zz})}{\partial z} \right] \tag{18.5}$$

These equations have been written with the time derivatives on the left-hand side and all spatial derivatives on the right-hand side. This is the form suitable to a time-dependent solution of the equations, as discussed in Chaps. 13 and 16. Indeed, Eqs. (18.1) to (18.5) are partial differential equations which have a mathematically "elliptic" behavior; i.e., on a physical basis they treat flow-field information and flow disturbances that can travel throughout the flow field, in both the upstream and downstream directions. The time-dependent technique is particularly suited to such a problem.

The time-dependent solution of Eqs. (18.1) to (18.5) can be carried out in direct parallel to the discussion in Sec. 16.4. It is important for you to return to that section and review our discussion of the time-dependent solution of compressible Couette flow using MacCormack's technique. We suggest doing this before reading further. The approach to the solution of Eqs. (18.1) to (18.5) for other problems is exactly the same. Therefore, we will not elaborate further here.

18.3 EXAMPLES OF SOME SOLUTIONS

In this section, we present samples of a few numerical solutions of the complete Navier-Stokes equations. All these solutions have the following in common:

1. They were obtained by means of a time-dependent solution using MacCormack's technique as described in Sec. 16.4.

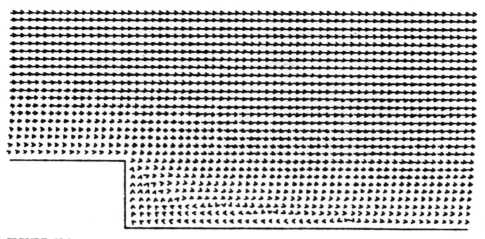

FIGURE 18.1
Velocity vector diagram for the flow over a rearward-facing step. $M = 2.19$, $T = 1005$ K, Re $= 70,000$ (based on step height) (*Ref. 46*). Note the recirculating flow region downstream of the step.

2. They utilize a turbulence model called the Baldwin-Lomax model (e.g., see Ref. 55 for a discussion of this model). Hence, turbulent flow is modeled in these calculations.

3. They require anywhere from thousands to close to a million grid points for their solution. Therefore, these are problems that must be solved on large-scale digital computers.

18.3.1 Flow over a Rearward-Facing Step

The supersonic viscous flow over a rearward-facing step was examined in Ref. 46. Some results are shown in Figs. 18.1 and 18.2. The flow is moving from left to right. In the velocity vector diagram in Fig. 18.1, note the separated, recirculating flow region just downstream of the step. The calculation of such separated flows is the forte of solutions of the complete Navier-Stokes equations. In contrast, the boundary-layer equations discussed in Chap. 17 are not suited for the analysis of separated flows; boundary-layer calculations usually "blow up" in regions of separated flow. Figure 18.2 shows the temperature contours (lines of constant temperature) for the same flow in Fig. 18.1.

18.3.2 Flow over an Airfoil

The viscous compressible flow over an airfoil was studied in Ref. 56. For the treatment of this problem, a nonrectangular finite-difference grid is wrapped around the airfoil, as shown in Fig. 18.3. Equations (18.1) to (18.5) have to be transformed into the new curvilinear coordinate system in Fig. 18.3. The details are beyond the scope of this book; see Ref. 56 for a complete discussion. Some

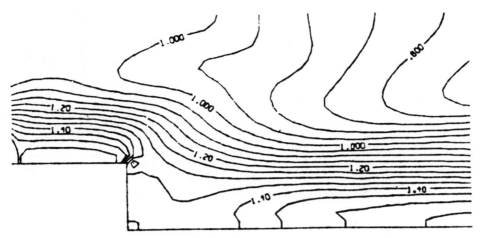

FIGURE 18.2
Temperature contours for the flow shown in Fig. 18.1. The separated region just downstream of the step is a reasonably constant pressure, constant temperature region.

results for the streamline patterns are shown in Fig. 18.4a and b. Here, the flow over a Wortmann airfoil at zero angle of attack is shown. The freestream Mach number is 0.5, and the Reynolds number based on chord is relatively low, Re = 100,000. The completely laminar flow over this airfoil is shown in Fig. 18.4a. Because of the peculiar aerodynamic properties of some low Reynolds number flows over airfoils (see Refs. 51 and 56), we note that the laminar flow separated

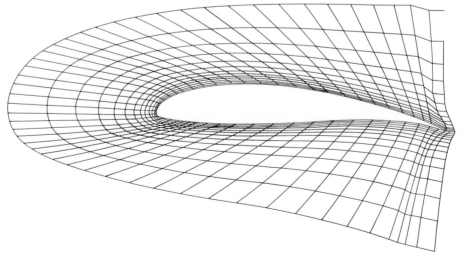

FIGURE 18.3
Curvilinear, boundary-fitted finite-difference grid for the solution of the flow over an airfoil. (*From Kothari and Anderson, Ref. 56.*)

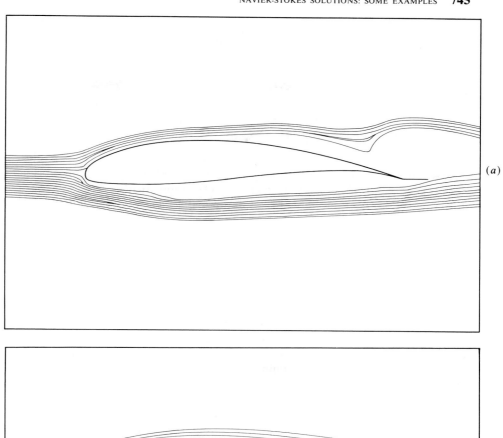

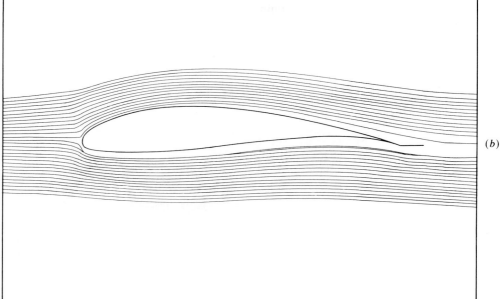

FIGURE 18.4
Streamlines for the low Reynolds flow over a Wortmann airfoil. Re = 100,000. (*a*) Laminar flow.
(*b*) Turbulent flow.

over both the top and bottom surfaces of the airfoil. However, in Fig. 18.4*b*, the turbulence model is turned on for the calculation; note that the flow is now completely attached. The differences in Fig. 18.4*a* and *b* vividly demonstrate the basic trend that turbulent flow resists flow separation much more strongly than laminar flow.

18.3.3 Flow over a Complete Airplane

We end this section by noting a history-making calculation. In Ref. 57, a solution of the complete Navier-Stokes equations for the flow field over an entire airplane was reported—the first such calculation ever made. In this work, Shang and Scherr carried out a time-dependent solution using MacCormack's method—just

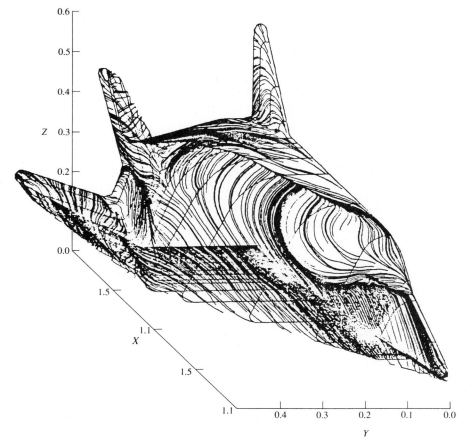

FIGURE 18.5
Surface shear stress lines on the X-24C. (*From Shang and Scherr, Ref. 57.*)

as we have discussed it in Sec. 16.4. See Ref. 57 for the details. Also, a lengthy description of this work can be found in Chap. 8 of Ref. 55. Shang and Scherr applied their calculation to the hypersonic viscous flow over the X-24C hypersonic test vehicle. To illustrate the results, the surface streamline pattern is shown in Fig. 18.5, as calculated in Ref. 57. In reality, since the flow velocity is zero at the surface in a viscous flow (the no-slip condition), the lines shown in Fig. 18.5 are the surface shear stress directions.

18.4 SUMMARY

With this, we end our discussion of viscous flow. The purpose of all of Part IV has been to introduce you to the basic aspects of viscous flow. The subject is so vast that it demands a book in itself—many of which have been written (see, e.g., Refs. 41 through 45). Here, we have presented only enough material to give you a flavor for some of the basic ideas and results. This is a subject of great importance in aerodynamics, and if you wish to expand your knowledge and expertise of aerodynamics in general, we encourage you to read further on the subject.

We are also out of our allotted space for this book. Therefore, we hope that you have enjoyed and benefited from our presentation of the fundamentals of aerodynamics. However, before closing the cover, it might be useful to return once again to Fig. 1.31, which is the block diagram categorizing the different general types of aerodynamic flows. Recall the curious, uninitiated thoughts you might have had when you first examined this figure during your study of Chap. 1, and compare these with the informed and mature thoughts that you now have—honed by the aerodynamic knowledge packed into the intervening pages. Hopefully, each block in Fig. 1.31 has substantially more meaning for you now than when we first started. If this is true, then my efforts as an author have not gone in vain.

#

ISENTROPIC FLOW PROPERTIES

M	$\dfrac{p_o}{p}$	$\dfrac{\rho_o}{\rho}$	$\dfrac{T_o}{T}$	$\dfrac{A}{A^*}$
0.2000 − 01	0.1000 + 01	0.1000 + 01	0.1000 + 01	0.2894 + 02
0.4000 − 01	0.1001 + 01	0.1001 + 01	0.1000 + 01	0.1448 + 02
0.6000 − 01	0.1003 + 01	0.1002 + 01	0.1001 + 01	0.9666 + 01
0.8000 − 01	0.1004 + 01	0.1003 + 01	0.1001 + 01	0.7262 + 01
0.1000 + 00	0.1007 + 01	0.1005 + 01	0.1002 + 01	0.5822 + 01
0.1200 + 00	0.1010 + 01	0.1007 + 01	0.1003 + 01	0.4864 + 01
0.1400 + 00	0.1014 + 01	0.1010 + 01	0.1004 + 01	0.4182 + 01
0.1600 + 00	0.1018 + 01	0.1013 + 01	0.1005 + 01	0.3673 + 01
0.1800 + 00	0.1023 + 01	0.1016 + 01	0.1006 + 01	0.3278 + 01
0.2000 + 00	0.1028 + 01	0.1020 + 01	0.1008 + 01	0.2964 + 01
0.2200 + 00	0.1034 + 01	0.1024 + 01	0.1010 + 01	0.2708 + 01
0.2400 + 00	0.1041 + 01	0.1029 + 01	0.1012 + 01	0.2496 + 01
0.2600 + 00	0.1048 + 01	0.1034 + 01	0.1014 + 01	0.2317 + 01
0.2800 + 00	0.1056 + 01	0.1040 + 01	0.1016 + 01	0.2166 + 01
0.3000 + 00	0.1064 + 01	0.1046 + 01	0.1018 + 01	0.2035 + 01
0.3200 + 00	0.1074 + 01	0.1052 + 01	0.1020 + 01	0.1922 + 01
0.3400 + 00	0.1083 + 01	0.1059 + 01	0.1023 + 01	0.1823 + 01
0.3600 + 00	0.1094 + 01	0.1066 + 01	0.1026 + 01	0.1736 + 01
0.3800 + 00	0.1105 + 01	0.1074 + 01	0.1029 + 01	0.1659 + 01
0.4000 + 00	0.1117 + 01	0.1082 + 01	0.1032 + 01	0.1590 + 01
0.4200 + 00	0.1129 + 01	0.1091 + 01	0.1035 + 01	0.1529 + 01
0.4400 + 00	0.1142 + 01	0.1100 + 01	0.1039 + 01	0.1474 + 01
0.4600 + 00	0.1156 + 01	0.1109 + 01	0.1042 + 01	0.1425 + 01
0.4800 + 00	0.1171 + 01	0.1119 + 01	0.1046 + 01	0.1380 + 01
0.5000 + 00	0.1186 + 01	0.1130 + 01	0.1050 + 01	0.1340 + 01
0.5200 + 00	0.1202 + 01	0.1141 + 01	0.1054 + 01	0.1303 + 01
0.5400 + 00	0.1219 + 01	0.1152 + 01	0.1058 + 01	0.1270 + 01
0.5600 + 00	0.1237 + 01	0.1164 + 01	0.1063 + 01	0.1240 + 01
0.5800 + 00	0.1256 + 01	0.1177 + 01	0.1067 + 01	0.1213 + 01
0.6000 + 00	0.1276 + 01	0.1190 + 01	0.1072 + 01	0.1188 + 01

M	$\dfrac{p_o}{p}$	$\dfrac{\rho_o}{\rho}$	$\dfrac{T_o}{T}$	$\dfrac{A}{A^*}$
0.6200 + 00	0.1296 + 01	0.1203 + 01	0.1077 + 01	0.1166 + 01
0.6400 + 00	0.1317 + 01	0.1218 + 01	0.1082 + 01	0.1145 + 01
0.6600 + 00	0.1340 + 01	0.1232 + 01	0.1087 + 01	0.1127 + 01
0.6800 + 00	0.1363 + 01	0.1247 + 01	0.1092 + 01	0.1110 + 01
0.7000 + 00	0.1387 + 01	0.1263 + 01	0.1098 + 01	0.1094 + 01
0.7200 + 00	0.1412 + 01	0.1280 + 01	0.1104 + 01	0.1081 + 01
0.7400 + 00	0.1439 + 01	0.1297 + 01	0.1110 + 01	0.1068 + 01
0.7600 + 00	0.1466 + 01	0.1314 + 01	0.1116 + 01	0.1057 + 01
0.7800 + 00	0.1495 + 01	0.1333 + 01	0.1122 + 01	0.1047 + 01
0.8000 + 00	0.1524 + 01	0.1351 + 01	0.1128 + 01	0.1038 + 01
0.8200 + 00	0.1555 + 01	0.1371 + 01	0.1134 + 01	0.1030 + 01
0.8400 + 00	0.1587 + 01	0.1391 + 01	0.1141 + 01	0.1024 + 01
0.8600 + 00	0.1621 + 01	0.1412 + 01	0.1148 + 01	0.1018 + 01
0.8800 + 00	0.1655 + 01	0.1433 + 01	0.1155 + 01	0.1013 + 01
0.9000 + 00	0.1691 + 01	0.1456 + 01	0.1162 + 01	0.1009 + 01
0.9200 + 00	0.1729 + 01	0.1478 + 01	0.1169 + 01	0.1006 + 01
0.9400 + 00	0.1767 + 01	0.1502 + 01	0.1177 + 01	0.1003 + 01
0.9600 + 00	0.1808 + 01	0.1526 + 01	0.1184 + 01	0.1001 + 01
0.9800 + 00	0.1850 + 01	0.1552 + 01	0.1192 + 01	0.1000 + 01
0.1000 + 01	0.1893 + 01	0.1577 + 01	0.1200 + 01	0.1000 + 01
0.1020 + 01	0.1938 + 01	0.1604 + 01	0.1208 + 01	0.1000 + 01
0.1040 + 01	0.1985 + 01	0.1632 + 01	0.1216 + 01	0.1001 + 01
0.1060 + 01	0.2033 + 01	0.1660 + 01	0.1225 + 01	0.1003 + 01
0.1080 + 01	0.2083 + 01	0.1689 + 01	0.1233 + 01	0.1005 + 01
0.1100 + 01	0.2135 + 01	0.1719 + 01	0.1242 + 01	0.1008 + 01
0.1120 + 01	0.2189 + 01	0.1750 + 01	0.1251 + 01	0.1011 + 01
0.1140 + 01	0.2245 + 01	0.1782 + 01	0.1260 + 01	0.1015 + 01
0.1160 + 01	0.2303 + 01	0.1814 + 01	0.1269 + 01	0.1020 + 01
0.1180 + 01	0.2363 + 01	0.1848 + 01	0.1278 + 01	0.1025 + 01
0.1200 + 01	0.2425 + 01	0.1883 + 01	0.1288 + 01	0.1030 + 01
0.1220 + 01	0.2489 + 01	0.1918 + 01	0.1298 + 01	0.1037 + 01
0.1240 + 01	0.2556 + 01	0.1955 + 01	0.1308 + 01	0.1043 + 01
0.1260 + 01	0.2625 + 01	0.1992 + 01	0.1318 + 01	0.1050 + 01
0.1280 + 01	0.2697 + 01	0.2031 + 01	0.1328 + 01	0.1058 + 01
0.1300 + 01	0.2771 + 01	0.2071 + 01	0.1338 + 01	0.1066 + 01
0.1320 + 01	0.2847 + 01	0.2112 + 01	0.1348 + 01	0.1075 + 01
0.1340 + 01	0.2927 + 01	0.2153 + 01	0.1359 + 01	0.1084 + 01
0.1360 + 01	0.3009 + 01	0.2197 + 01	0.1370 + 01	0.1094 + 01
0.1380 + 01	0.3094 + 01	0.2241 + 01	0.1381 + 01	0.1104 + 01
0.1400 + 01	0.3182 + 01	0.2286 + 01	0.1392 + 01	0.1115 + 01
0.1420 + 01	0.3273 + 01	0.2333 + 01	0.1403 + 01	0.1126 + 01
0.1440 + 01	0.3368 + 01	0.2381 + 01	0.1415 + 01	0.1138 + 01
0.1460 + 01	0.3465 + 01	0.2430 + 01	0.1426 + 01	0.1150 + 01
0.1480 + 01	0.3566 + 01	0.2480 + 01	0.1438 + 01	0.1163 + 01
0.1500 + 01	0.3671 + 01	0.2532 + 01	0.1450 + 01	0.1176 + 01
0.1520 + 01	0.3779 + 01	0.2585 + 01	0.1462 + 01	0.1190 + 01
0.1540 + 01	0.3891 + 01	0.2639 + 01	0.1474 + 01	0.1204 + 01
0.1560 + 01	0.4007 + 01	0.2695 + 01	0.1487 + 01	0.1219 + 01
0.1580 + 01	0.4127 + 01	0.2752 + 01	0.1499 + 01	0.1234 + 01
0.1600 + 01	0.4250 + 01	0.2811 + 01	0.1512 + 01	0.1250 + 01

M	$\dfrac{p_o}{p}$	$\dfrac{\rho_o}{\rho}$	$\dfrac{T_o}{T}$	$\dfrac{A}{A^*}$
0.1620 + 01	0.4378 + 01	0.2871 + 01	0.1525 + 01	0.1267 + 01
0.1640 + 01	0.4511 + 01	0.2933 + 01	0.1538 + 01	0.1284 + 01
0.1660 + 01	0.4648 + 01	0.2996 + 01	0.1551 + 01	0.1301 + 01
0.1680 + 01	0.4790 + 01	0.3061 + 01	0.1564 + 01	0.1319 + 01
0.1700 + 01	0.4936 + 01	0.3128 + 01	0.1578 + 01	0.1338 + 01
0.1720 + 01	0.5087 + 01	0.3196 + 01	0.1592 + 01	0.1357 + 01
0.1740 + 01	0.5244 + 01	0.3266 + 01	0.1606 + 01	0.1376 + 01
0.1760 + 01	0.5406 + 01	0.3338 + 01	0.1620 + 01	0.1397 + 01
0.1780 + 01	0.5573 + 01	0.3411 + 01	0.1634 + 01	0.1418 + 01
0.1800 + 01	0.5746 + 01	0.3487 + 01	0.1648 + 01	0.1439 + 01
0.1820 + 01	0.5924 + 01	0.3564 + 01	0.1662 + 01	0.1461 + 01
0.1840 + 01	0.6109 + 01	0.3643 + 01	0.1677 + 01	0.1484 + 01
0.1860 + 01	0.6300 + 01	0.3723 + 01	0.1692 + 01	0.1507 + 01
0.1880 + 01	0.6497 + 01	0.3806 + 01	0.1707 + 01	0.1531 + 01
0.1900 + 01	0.6701 + 01	0.3891 + 01	0.1722 + 01	0.1555 + 01
0.1920 + 01	0.6911 + 01	0.3978 + 01	0.1737 + 01	0.1580 + 01
0.1940 + 01	0.7128 + 01	0.4067 + 01	0.1753 + 01	0.1606 + 01
0.1960 + 01	0.7353 + 01	0.4158 + 01	0.1768 + 01	0.1633 + 01
0.1980 + 01	0.7585 + 01	0.4251 + 01	0.1784 + 01	0.1660 + 01
0.2000 + 01	0.7824 + 01	0.4347 + 01	0.1800 + 01	0.1687 + 01
0.2050 + 01	0.8458 + 01	0.4596 + 01	0.1840 + 01	0.1760 + 01
0.2100 + 01	0.9145 + 01	0.4859 + 01	0.1882 + 01	0.1837 + 01
0.2150 + 01	0.9888 + 01	0.5138 + 01	0.1924 + 01	0.1919 + 01
0.2200 + 01	0.1069 + 02	0.5433 + 01	0.1968 + 01	0.2005 + 01
0.2250 + 01	0.1156 + 02	0.5746 + 01	0.2012 + 01	0.2096 + 01
0.2300 + 01	0.1250 + 02	0.6076 + 01	0.2058 + 01	0.2193 + 01
0.2350 + 01	0.1352 + 02	0.6425 + 01	0.2104 + 01	0.2295 + 01
0.2400 + 01	0.1462 + 02	0.6794 + 01	0.2152 + 01	0.2403 + 01
0.2450 + 01	0.1581 + 02	0.7183 + 01	0.2200 + 01	0.2517 + 01
0.2500 + 01	0.1709 + 02	0.7594 + 01	0.2250 + 01	0.2637 + 01
0.2550 + 01	0.1847 + 02	0.8027 + 01	0.2300 + 01	0.2763 + 01
0.2600 + 01	0.1995 + 02	0.8484 + 01	0.2352 + 01	0.2896 + 01
0.2650 + 01	0.2156 + 02	0.8965 + 01	0.2404 + 01	0.3036 + 01
0.2700 + 01	0.2328 + 02	0.9472 + 01	0.2458 + 01	0.3183 + 01
0.2750 + 01	0.2514 + 02	0.1001 + 02	0.2512 + 01	0.3338 + 01
0.2800 + 01	0.2714 + 02	0.1057 + 02	0.2568 + 01	0.3500 + 01
0.2850 + 01	0.2929 + 02	0.1116 + 02	0.2624 + 01	0.3671 + 01
0.2900 + 01	0.3159 + 02	0.1178 + 02	0.2682 + 01	0.3850 + 01
0.2950 + 01	0.3407 + 02	0.1243 + 02	0.2740 + 01	0.4038 + 01
0.3000 + 01	0.3673 + 02	0.1312 + 02	0.2800 + 01	0.4235 + 01
0.3050 + 01	0.3959 + 02	0.1384 + 02	0.2860 + 01	0.4441 + 01
0.3100 + 01	0.4265 + 02	0.1459 + 02	0.2922 + 01	0.4657 + 01
0.3150 + 01	0.4593 + 02	0.1539 + 02	0.2984 + 01	0.4884 + 01
0.3200 + 01	0.4944 + 02	0.1622 + 02	0.3048 + 01	0.5121 + 01
0.3250 + 01	0.5320 + 02	0.1709 + 02	0.3112 + 01	0.5369 + 01
0.3300 + 01	0.5722 + 02	0.1800 + 02	0.3178 + 01	0.5629 + 01
0.3350 + 01	0.6152 + 02	0.1896 + 02	0.3244 + 01	0.5900 + 01
0.3400 + 01	0.6612 + 02	0.1996 + 02	0.3312 + 01	0.6184 + 01
0.3450 + 01	0.7103 + 02	0.2101 + 02	0.3380 + 01	0.6480 + 01
0.3500 + 01	0.7627 + 02	0.2211 + 02	0.3450 + 01	0.6790 + 01

M	$\dfrac{p_o}{p}$	$\dfrac{\rho_o}{\rho}$	$\dfrac{T_o}{T}$	$\dfrac{A}{A^*}$
0.3550 + 01	0.8187 + 02	0.2325 + 02	0.3520 + 01	0.7113 + 01
0.3600 + 01	0.8784 + 02	0.2445 + 02	0.3592 + 01	0.7450 + 01
0.3650 + 01	0.9420 + 02	0.2571 + 02	0.3664 + 01	0.7802 + 01
0.3700 + 01	0.1010 + 03	0.2701 + 02	0.3738 + 01	0.8169 + 01
0.3750 + 01	0.1082 + 03	0.2838 + 02	0.3812 + 01	0.8552 + 01
0.3800 + 01	0.1159 + 03	0.2981 + 02	0.3888 + 01	0.8951 + 01
0.3850 + 01	0.1241 + 03	0.3129 + 02	0.3964 + 01	0.9366 + 01
0.3900 + 01	0.1328 + 03	0.3285 + 02	0.4042 + 01	0.9799 + 01
0.3950 + 01	0.1420 + 03	0.3446 + 02	0.4120 + 01	0.1025 + 02
0.4000 + 01	0.1518 + 03	0.3615 + 02	0.4200 + 01	0.1072 + 02
0.4050 + 01	0.1623 + 03	0.3791 + 02	0.4280 + 01	0.1121 + 02
0.4100 + 01	0.1733 + 03	0.3974 + 02	0.4362 + 01	0.1171 + 02
0.4150 + 01	0.1851 + 03	0.4164 + 02	0.4444 + 01	0.1224 + 02
0.4200 + 01	0.1975 + 03	0.4363 + 02	0.4528 + 01	0.1279 + 02
0.4250 + 01	0.2108 + 03	0.4569 + 02	0.4612 + 01	0.1336 + 02
0.4300 + 01	0.2247 + 03	0.4784 + 02	0.4698 + 01	0.1395 + 02
0.4350 + 01	0.2396 + 03	0.5007 + 02	0.4784 + 01	0.1457 + 02
0.4400 + 01	0.2553 + 03	0.5239 + 02	0.4872 + 01	0.1521 + 02
0.4450 + 01	0.2719 + 03	0.5480 + 02	0.4960 + 01	0.1587 + 02
0.4500 + 01	0.2894 + 03	0.5731 + 02	0.5050 + 01	0.1656 + 02
0.4550 + 01	0.3080 + 03	0.5991 + 02	0.5140 + 01	0.1728 + 02
0.4600 + 01	0.3276 + 03	0.6261 + 02	0.5232 + 01	0.1802 + 02
0.4650 + 01	0.3483 + 03	0.6542 + 02	0.5324 + 01	0.1879 + 02
0.4700 + 01	0.3702 + 03	0.6833 + 02	0.5418 + 01	0.1958 + 02
0.4750 + 01	0.3933 + 03	0.7135 + 02	0.5512 + 01	0.2041 + 02
0.4800 + 01	0.4177 + 03	0.7448 + 02	0.5608 + 01	0.2126 + 02
0.4850 + 01	0.4434 + 03	0.7772 + 02	0.5704 + 01	0.2215 + 02
0.4900 + 01	0.4705 + 03	0.8109 + 02	0.5802 + 01	0.2307 + 02
0.4950 + 01	0.4990 + 03	0.8457 + 02	0.5900 + 01	0.2402 + 02
0.5000 + 01	0.5291 + 03	0.8818 + 02	0.6000 + 01	0.2500 + 02
0.5100 + 01	0.5941 + 03	0.9579 + 02	0.6202 + 01	0.2707 + 02
0.5200 + 01	0.6661 + 03	0.1039 + 03	0.6408 + 01	0.2928 + 02
0.5300 + 01	0.7457 + 03	0.1127 + 03	0.6618 + 01	0.3165 + 02
0.5400 + 01	0.8335 + 03	0.1220 + 03	0.6832 + 01	0.3417 + 02
0.5500 + 01	0.9304 + 03	0.1320 + 03	0.7050 + 01	0.3687 + 02
0.5600 + 01	0.1037 + 04	0.1426 + 03	0.7272 + 01	0.3974 + 02
0.5700 + 01	0.1154 + 04	0.1539 + 03	0.7498 + 01	0.4280 + 02
0.5800 + 01	0.1283 + 04	0.1660 + 03	0.7728 + 01	0.4605 + 02
0.5900 + 01	0.1424 + 04	0.1789 + 03	0.7962 + 01	0.4951 + 02
0.6000 + 01	0.1579 + 04	0.1925 + 03	0.8200 + 01	0.5318 + 02
0.6100 + 01	0.1748 + 04	0.2071 + 03	0.8442 + 01	0.5708 + 02
0.6200 + 01	0.1933 + 04	0.2225 + 03	0.8688 + 01	0.6121 + 02
0.6300 + 01	0.2135 + 04	0.2388 + 03	0.8938 + 01	0.6559 + 02
0.6400 + 01	0.2355 + 04	0.2562 + 03	0.9192 + 01	0.7023 + 02
0.6500 + 01	0.2594 + 04	0.2745 + 03	0.9450 + 01	0.7513 + 02
0.6600 + 01	0.2855 + 04	0.2939 + 03	0.9712 + 01	0.8032 + 02
0.6700 + 01	0.3138 + 04	0.3145 + 03	0.9978 + 01	0.8580 + 02
0.6800 + 01	0.3445 + 04	0.3362 + 03	0.1025 + 02	0.9159 + 02
0.6900 + 01	0.3779 + 04	0.3591 + 03	0.1052 + 02	0.9770 + 02
0.7000 + 01	0.4140 + 04	0.3833 + 03	0.1080 + 02	0.1041 + 03

M	$\dfrac{p_o}{p}$	$\dfrac{\rho_o}{\rho}$	$\dfrac{T_o}{T}$	$\dfrac{A}{A^*}$
$0.7100+01$	$0.4531+04$	$0.4088+03$	$0.1108+02$	$0.1109+03$
$0.7200+01$	$0.4953+04$	$0.4357+03$	$0.1137+02$	$0.1181+03$
$0.7300+01$	$0.5410+04$	$0.4640+03$	$0.1166+02$	$0.1256+03$
$0.7400+01$	$0.5903+04$	$0.4939+03$	$0.1195+02$	$0.1335+03$
$0.7500+01$	$0.6434+04$	$0.5252+03$	$0.1225+02$	$0.1418+03$
$0.7600+01$	$0.7006+04$	$0.5582+03$	$0.1255+02$	$0.1506+03$
$0.7700+01$	$0.7623+04$	$0.5928+03$	$0.1286+02$	$0.1598+03$
$0.7800+01$	$0.8285+04$	$0.6292+03$	$0.1317+02$	$0.1694+03$
$0.7900+01$	$0.8998+04$	$0.6674+03$	$0.1348+02$	$0.1795+03$
$0.8000+01$	$0.9763+04$	$0.7075+03$	$0.1380+02$	$0.1901+03$
$0.9000+01$	$0.2110+05$	$0.1227+04$	$0.1720+02$	$0.3272+03$
$0.1000+02$	$0.4244+05$	$0.2021+04$	$0.2100+02$	$0.5359+03$
$0.1100+02$	$0.8033+05$	$0.3188+04$	$0.2520+02$	$0.8419+03$
$0.1200+02$	$0.1445+06$	$0.4848+04$	$0.2980+02$	$0.1276+04$
$0.1300+02$	$0.2486+06$	$0.7144+04$	$0.3480+02$	$0.1876+04$
$0.1400+02$	$0.4119+06$	$0.1025+05$	$0.4020+02$	$0.2685+04$
$0.1500+02$	$0.6602+06$	$0.1435+05$	$0.4600+02$	$0.3755+04$
$0.1600+02$	$0.1028+07$	$0.1969+05$	$0.5220+02$	$0.5145+04$
$0.1700+02$	$0.1559+07$	$0.2651+05$	$0.5880+02$	$0.6921+04$
$0.1800+02$	$0.2311+07$	$0.3512+05$	$0.6580+02$	$0.9159+04$
$0.1900+02$	$0.3356+07$	$0.4584+05$	$0.7320+02$	$0.1195+05$
$0.2000+02$	$0.4783+07$	$0.5905+05$	$0.8100+02$	$0.1538+05$
$0.2200+02$	$0.9251+07$	$0.9459+05$	$0.9780+02$	$0.2461+05$
$0.2400+02$	$0.1691+08$	$0.1456+06$	$0.1162+03$	$0.3783+05$
$0.2600+02$	$0.2949+08$	$0.2165+06$	$0.1362+03$	$0.5624+05$
$0.2800+02$	$0.4936+08$	$0.3128+06$	$0.1578+03$	$0.8121+05$
$0.3000+02$	$0.7978+08$	$0.4408+06$	$0.1810+03$	$0.1144+06$
$0.3200+02$	$0.1250+09$	$0.6076+06$	$0.2058+03$	$0.1576+06$
$0.3400+02$	$0.1908+09$	$0.8216+06$	$0.2322+03$	$0.2131+06$
$0.3600+02$	$0.2842+09$	$0.1092+07$	$0.2602+03$	$0.2832+06$
$0.3800+02$	$0.4143+09$	$0.1430+07$	$0.2898+03$	$0.3707+06$
$0.4000+02$	$0.5926+09$	$0.1846+07$	$0.3210+03$	$0.4785+06$
$0.4200+02$	$0.8330+09$	$0.2354+07$	$0.3538+03$	$0.6102+06$
$0.4400+02$	$0.1153+10$	$0.2969+07$	$0.3882+03$	$0.7694+06$
$0.4600+02$	$0.1572+10$	$0.3706+07$	$0.4242+03$	$0.9603+06$
$0.4800+02$	$0.2116+10$	$0.4583+07$	$0.4618+03$	$0.1187+07$
$0.5000+02$	$0.2815+10$	$0.5618+07$	$0.5010+03$	$0.1455+07$

APPENDIX
B

NORMAL SHOCK PROPERTIES

M	$\dfrac{p_2}{p_1}$	$\dfrac{\rho_2}{\rho_1}$	$\dfrac{T_2}{T_1}$	$\dfrac{p_{o_2}}{p_{o_1}}$	$\dfrac{p_{o_2}}{p_1}$	M_2
0.1000 + 01	0.1000 + 01	0.1000 + 01	0.1000 + 01	0.1000 + 01	0.1893 + 01	0.1000 + 01
0.1020 + 01	0.1047 + 01	0.1033 + 01	0.1013 + 01	0.1000 + 01	0.1938 + 01	0.9805 + 00
0.1040 + 01	0.1095 + 01	0.1067 + 01	0.1026 + 01	0.9999 + 00	0.1984 + 01	0.9620 + 00
0.1060 + 01	0.1144 + 01	0.1101 + 01	0.1039 + 01	0.9998 + 00	0.2032 + 01	0.9444 + 00
0.1080 + 01	0.1194 + 01	0.1135 + 01	0.1052 + 01	0.9994 + 00	0.2082 + 01	0.9277 + 00
0.1100 + 01	0.1245 + 01	0.1169 + 01	0.1065 + 01	0.9989 + 00	0.2133 + 01	0.9118 + 00
0.1120 + 01	0.1297 + 01	0.1203 + 01	0.1078 + 01	0.9982 + 00	0.2185 + 01	0.8966 + 00
0.1140 + 01	0.1350 + 01	0.1238 + 01	0.1090 + 01	0.9973 + 00	0.2239 + 01	0.8820 + 00
0.1160 + 01	0.1403 + 01	0.1272 + 01	0.1103 + 01	0.9961 + 00	0.2294 + 01	0.8682 + 00
0.1180 + 01	0.1458 + 01	0.1307 + 01	0.1115 + 01	0.9946 + 00	0.2350 + 01	0.8549 + 00
0.1200 + 01	0.1513 + 01	0.1342 + 01	0.1128 + 01	0.9928 + 00	0.2408 + 01	0.8422 + 00
0.1220 + 01	0.1570 + 01	0.1376 + 01	0.1141 + 01	0.9907 + 00	0.2466 + 01	0.8300 + 00
0.1240 + 01	0.1627 + 01	0.1411 + 01	0.1153 + 01	0.9884 + 00	0.2526 + 01	0.8183 + 00
0.1260 + 01	0.1686 + 01	0.1446 + 01	0.1166 + 01	0.9857 + 00	0.2588 + 01	0.8071 + 00
0.1280 + 01	0.1745 + 01	0.1481 + 01	0.1178 + 01	0.9827 + 00	0.2650 + 01	0.7963 + 00
0.1300 + 01	0.1805 + 01	0.1516 + 01	0.1191 + 01	0.9794 + 00	0.2714 + 01	0.7860 + 00
0.1320 + 01	0.1866 + 01	0.1551 + 01	0.1204 + 01	0.9758 + 00	0.2778 + 01	0.7760 + 00
0.1340 + 01	0.1928 + 01	0.1585 + 01	0.1216 + 01	0.9718 + 00	0.2844 + 01	0.7664 + 00
0.1360 + 01	0.1991 + 01	0.1620 + 01	0.1229 + 01	0.9676 + 00	0.2912 + 01	0.7572 + 00
0.1380 + 01	0.2055 + 01	0.1655 + 01	0.1242 + 01	0.9630 + 00	0.2980 + 01	0.7483 + 00
0.1400 + 01	0.2120 + 01	0.1690 + 01	0.1255 + 01	0.9582 + 00	0.3049 + 01	0.7397 + 00
0.1420 + 01	0.2186 + 01	0.1724 + 01	0.1268 + 01	0.9531 + 00	0.3120 + 01	0.7314 + 00
0.1440 + 01	0.2253 + 01	0.1759 + 01	0.1281 + 01	0.9476 + 00	0.3191 + 01	0.7235 + 00
0.1460 + 01	0.2320 + 01	0.1793 + 01	0.1294 + 01	0.9420 + 00	0.3264 + 01	0.7157 + 00
0.1480 + 01	0.2389 + 01	0.1828 + 01	0.1307 + 01	0.9360 + 00	0.3338 + 01	0.7083 + 00
0.1500 + 01	0.2458 + 01	0.1862 + 01	0.1320 + 01	0.9298 + 00	0.3413 + 01	0.7011 + 00
0.1520 + 01	0.2529 + 01	0.1896 + 01	0.1334 + 01	0.9233 + 00	0.3489 + 01	0.6941 + 00
0.1540 + 01	0.2600 + 01	0.1930 + 01	0.1347 + 01	0.9166 + 00	0.3567 + 01	0.6874 + 00
0.1560 + 01	0.2673 + 01	0.1964 + 01	0.1361 + 01	0.9097 + 00	0.3645 + 01	0.6809 + 00
0.1580 + 01	0.2746 + 01	0.1998 + 01	0.1374 + 01	0.9026 + 00	0.3724 + 01	0.6746 + 00

M	$\dfrac{p_2}{p_1}$	$\dfrac{\rho_2}{\rho_1}$	$\dfrac{T_2}{T_1}$	$\dfrac{p_{o_2}}{p_{o_1}}$	$\dfrac{p_{o_2}}{p_1}$	M_2
0.1600 + 01	0.2820 + 01	0.2032 + 01	0.1388 + 01	0.8952 + 00	0.3805 + 01	0.6684 + 00
0.1620 + 01	0.2895 + 01	0.2065 + 01	0.1402 + 01	0.8877 + 00	0.3887 + 01	0.6625 + 00
0.1640 + 01	0.2971 + 01	0.2099 + 01	0.1416 + 01	0.8799 + 00	0.3969 + 01	0.6568 + 00
0.1660 + 01	0.3048 + 01	0.2132 + 01	0.1430 + 01	0.8720 + 00	0.4053 + 01	0.6512 + 00
0.1680 + 01	0.3126 + 01	0.2165 + 01	0.1444 + 01	0.8639 + 00	0.4138 + 01	0.6458 + 00
0.1700 + 01	0.3205 + 01	0.2198 + 01	0.1458 + 01	0.8557 + 00	0.4224 + 01	0.6405 + 00
0.1720 + 01	0.3285 + 01	0.2230 + 01	0.1473 + 01	0.8474 + 00	0.4311 + 01	0.6355 + 00
0.1740 + 01	0.3366 + 01	0.2263 + 01	0.1487 + 01	0.8389 + 00	0.4399 + 01	0.6305 + 00
0.1760 + 01	0.3447 + 01	0.2295 + 01	0.1502 + 01	0.8302 + 00	0.4488 + 01	0.6257 + 00
0.1780 + 01	0.3530 + 01	0.2327 + 01	0.1517 + 01	0.8215 + 00	0.4578 + 01	0.6210 + 00
0.1800 + 01	0.3613 + 01	0.2359 + 01	0.1532 + 01	0.8127 + 00	0.4670 + 01	0.6165 + 00
0.1820 + 01	0.3698 + 01	0.2391 + 01	0.1547 + 01	0.8038 + 00	0.4762 + 01	0.6121 + 00
0.1840 + 01	0.3783 + 01	0.2422 + 01	0.1562 + 01	0.7948 + 00	0.4855 + 01	0.6078 + 00
0.1860 + 01	0.3870 + 01	0.2454 + 01	0.1577 + 01	0.7857 + 00	0.4950 + 01	0.6036 + 00
0.1880 + 01	0.3957 + 01	0.2485 + 01	0.1592 + 01	0.7765 + 00	0.5045 + 01	0.5996 + 00
0.1900 + 01	0.4045 + 01	0.2516 + 01	0.1608 + 01	0.7674 + 00	0.5142 + 01	0.5956 + 00
0.1920 + 01	0.4134 + 01	0.2546 + 01	0.1624 + 01	0.7581 + 00	0.5239 + 01	0.5918 + 00
0.1940 + 01	0.4224 + 01	0.2577 + 01	0.1639 + 01	0.7488 + 00	0.5338 + 01	0.5880 + 00
0.1960 + 01	0.4315 + 01	0.2607 + 01	0.1655 + 01	0.7395 + 00	0.5438 + 01	0.5844 + 00
0.1980 + 01	0.4407 + 01	0.2637 + 01	0.1671 + 01	0.7302 + 00	0.5539 + 01	0.5808 + 00
0.2000 + 01	0.4500 + 01	0.2667 + 01	0.1687 + 01	0.7209 + 00	0.5640 + 01	0.5774 + 00
0.2050 + 01	0.4736 + 01	0.2740 + 01	0.1729 + 01	0.6975 + 00	0.5900 + 01	0.5691 + 00
0.2100 + 01	0.4978 + 01	0.2812 + 01	0.1770 + 01	0.6742 + 00	0.6165 + 01	0.5613 + 00
0.2150 + 01	0.5226 + 01	0.2882 + 01	0.1813 + 01	0.6511 + 00	0.6438 + 01	0.5540 + 00
0.2200 + 01	0.5480 + 01	0.2951 + 01	0.1857 + 01	0.6281 + 00	0.6716 + 01	0.5471 + 00
0.2250 + 01	0.5740 + 01	0.3019 + 01	0.1901 + 01	0.6055 + 00	0.7002 + 01	0.5406 + 00
0.2300 + 01	0.6005 + 01	0.3085 + 01	0.1947 + 01	0.5833 + 00	0.7294 + 01	0.5344 + 00
0.2350 + 01	0.6276 + 01	0.3149 + 01	0.1993 + 01	0.5615 + 00	0.7592 + 01	0.5286 + 00
0.2400 + 01	0.6553 + 01	0.3212 + 01	0.2040 + 01	0.5401 + 00	0.7897 + 01	0.5231 + 00
0.2450 + 01	0.6836 + 01	0.3273 + 01	0.2088 + 01	0.5193 + 00	0.8208 + 01	0.5179 + 00
0.2500 + 01	0.7125 + 01	0.3333 + 01	0.2137 + 01	0.4990 + 00	0.8526 + 01	0.5130 + 00
0.2550 + 01	0.7420 + 01	0.3392 + 01	0.2187 + 01	0.4793 + 00	0.8850 + 01	0.5083 + 00
0.2600 + 01	0.7720 + 01	0.3449 + 01	0.2238 + 01	0.4601 + 00	0.9181 + 01	0.5039 + 00
0.2650 + 01	0.8026 + 01	0.3505 + 01	0.2290 + 01	0.4416 + 00	0.9519 + 01	0.4996 + 00
0.2700 + 01	0.8338 + 01	0.3559 + 01	0.2343 + 01	0.4236 + 00	0.9862 + 01	0.4956 + 00
0.2750 + 01	0.8656 + 01	0.3612 + 01	0.2397 + 01	0.4062 + 00	0.1021 + 02	0.4918 + 00
0.2800 + 01	0.8980 + 01	0.3664 + 01	0.2451 + 01	0.3895 + 00	0.1057 + 02	0.4882 + 00
0.2850 + 01	0.9310 + 01	0.3714 + 01	0.2507 + 01	0.3733 + 00	0.1093 + 02	0.4847 + 00
0.2900 + 01	0.9645 + 01	0.3763 + 01	0.2563 + 01	0.3577 + 00	0.1130 + 02	0.4814 + 00
0.2950 + 01	0.9986 + 01	0.3811 + 01	0.2621 + 01	0.3428 + 00	0.1168 + 02	0.4782 + 00
0.3000 + 01	0.1033 + 02	0.3857 + 01	0.2679 + 01	0.3283 + 00	0.1206 + 02	0.4752 + 00
0.3050 + 01	0.1069 + 02	0.3902 + 01	0.2738 + 01	0.3145 + 00	0.1245 + 02	0.4723 + 00
0.3100 + 01	0.1104 + 02	0.3947 + 01	0.2799 + 01	0.3012 + 00	0.1285 + 02	0.4695 + 00
0.3150 + 01	0.1141 + 02	0.3990 + 01	0.2860 + 01	0.2885 + 00	0.1325 + 02	0.4669 + 00
0.3200 + 01	0.1178 + 02	0.4031 + 01	0.2922 + 01	0.2762 + 00	0.1366 + 02	0.4643 + 00
0.3250 + 01	0.1216 + 02	0.4072 + 01	0.2985 + 01	0.2645 + 00	0.1407 + 02	0.4619 + 00
0.3300 + 01	0.1254 + 02	0.4112 + 01	0.3049 + 01	0.2533 + 00	0.1449 + 02	0.4596 + 00
0.3350 + 01	0.1293 + 02	0.4151 + 01	0.3114 + 01	0.2425 + 00	0.1492 + 02	0.4573 + 00
0.3400 + 01	0.1332 + 02	0.4188 + 01	0.3180 + 01	0.2322 + 00	0.1535 + 02	0.4552 + 00
0.3450 + 01	0.1372 + 02	0.4225 + 01	0.3247 + 01	0.2224 + 00	0.1579 + 02	0.4531 + 00

M	$\dfrac{p_2}{p_1}$	$\dfrac{\rho_2}{\rho_1}$	$\dfrac{T_2}{T_1}$	$\dfrac{p_{o_2}}{p_{o_1}}$	$\dfrac{p_{o_2}}{p_1}$	M_2
0.3500 + 01	0.1412 + 02	0.4261 + 01	0.3315 + 01	0.2129 + 00	0.1624 + 02	0.4512 + 00
0.3550 + 01	0.1454 + 02	0.4296 + 01	0.3384 + 01	0.2039 + 00	0.1670 + 02	0.4492 + 00
0.3600 + 01	0.1495 + 02	0.4330 + 01	0.3454 + 01	0.1953 + 00	0.1716 + 02	0.4474 + 00
0.3650 + 01	0.1538 + 02	0.4363 + 01	0.3525 + 01	0.1871 + 00	0.1762 + 02	0.4456 + 00
0.3700 + 01	0.1580 + 02	0.4395 + 01	0.3596 + 01	0.1792 + 00	0.1810 + 02	0.4439 + 00
0.3750 + 01	0.1624 + 02	0.4426 + 01	0.3669 + 01	0.1717 + 00	0.1857 + 02	0.4423 + 00
0.3800 + 01	0.1668 + 02	0.4457 + 01	0.3743 + 01	0.1645 + 00	0.1906 + 02	0.4407 + 00
0.3850 + 01	0.1713 + 02	0.4487 + 01	0.3817 + 01	0.1576 + 00	0.1955 + 02	0.4392 + 00
0.3900 + 01	0.1758 + 02	0.4516 + 01	0.3893 + 01	0.1510 + 00	0.2005 + 02	0.4377 + 00
0.3950 + 01	0.1804 + 02	0.4544 + 01	0.3969 + 01	0.1448 + 00	0.2056 + 02	0.4363 + 00
0.4000 + 01	0.1850 + 02	0.4571 + 01	0.4047 + 01	0.1388 + 00	0.2107 + 02	0.4350 + 00
0.4050 + 01	0.1897 + 02	0.4598 + 01	0.4125 + 01	0.1330 + 00	0.2159 + 02	0.4336 + 00
0.4100 + 01	0.1944 + 02	0.4624 + 01	0.4205 + 01	0.1276 + 00	0.2211 + 02	0.4324 + 00
0.4150 + 01	0.1993 + 02	0.4650 + 01	0.4285 + 01	0.1223 + 00	0.2264 + 02	0.4311 + 00
0.4200 + 01	0.2041 + 02	0.4675 + 01	0.4367 + 01	0.1173 + 00	0.2318 + 02	0.4299 + 00
0.4250 + 01	0.2091 + 02	0.4699 + 01	0.4449 + 01	0.1126 + 00	0.2372 + 02	0.4288 + 00
0.4300 + 01	0.2140 + 02	0.4723 + 01	0.4532 + 01	0.1080 + 00	0.2427 + 02	0.4277 + 00
0.4350 + 01	0.2191 + 02	0.4746 + 01	0.4616 + 01	0.1036 + 00	0.2483 + 02	0.4266 + 00
0.4400 + 01	0.2242 + 02	0.4768 + 01	0.4702 + 01	0.9948 − 01	0.2539 + 02	0.4255 + 00
0.4450 + 01	0.2294 + 02	0.4790 + 01	0.4788 + 01	0.9550 − 01	0.2596 + 02	0.4245 + 00
0.4500 + 01	0.2346 + 02	0.4812 + 01	0.4875 + 01	0.9170 − 01	0.2654 + 02	0.4236 + 00
0.4550 + 01	0.2399 + 02	0.4833 + 01	0.4963 + 01	0.8806 − 01	0.2712 + 02	0.4226 + 00
0.4600 + 01	0.2452 + 02	0.4853 + 01	0.5052 + 01	0.8459 − 01	0.2771 + 02	0.4217 + 00
0.4650 + 01	0.2506 + 02	0.4873 + 01	0.5142 + 01	0.8126 − 01	0.2831 + 02	0.4208 + 00
0.4700 + 01	0.2560 + 02	0.4893 + 01	0.5233 + 01	0.7809 − 01	0.2891 + 02	0.4199 + 00
0.4750 + 01	0.2616 + 02	0.4912 + 01	0.5325 + 01	0.7505 − 01	0.2952 + 02	0.4191 + 00
0.4800 + 01	0.2671 + 02	0.4930 + 01	0.5418 + 01	0.7214 − 01	0.3013 + 02	0.4183 + 00
0.4850 + 01	0.2728 + 02	0.4948 + 01	0.5512 + 01	0.6936 − 01	0.3075 + 02	0.4175 + 00
0.4900 + 01	0.2784 + 02	0.4966 + 01	0.5607 + 01	0.6670 − 01	0.3138 + 02	0.4167 + 00
0.4950 + 01	0.2842 + 02	0.4983 + 01	0.5703 + 01	0.6415 − 01	0.3201 + 02	0.4160 + 00
0.5000 + 01	0.2900 + 02	0.5000 + 01	0.5800 + 01	0.6172 − 01	0.3265 + 02	0.4152 + 00
0.5100 + 01	0.3018 + 02	0.5033 + 01	0.5997 + 01	0.5715 − 01	0.3395 + 02	0.4138 + 00
0.5200 + 01	0.3138 + 02	0.5064 + 01	0.6197 + 01	0.5297 − 01	0.3528 + 02	0.4125 + 00
0.5300 + 01	0.3260 + 02	0.5093 + 01	0.6401 + 01	0.4913 − 01	0.3663 + 02	0.4113 + 00
0.5400 + 01	0.3385 + 02	0.5122 + 01	0.6610 + 01	0.4560 − 01	0.3801 + 02	0.4101 + 00
0.5500 + 01	0.3512 + 02	0.5149 + 01	0.6822 + 01	0.4236 − 01	0.3941 + 02	0.4090 + 00
0.5600 + 01	0.3642 + 02	0.5175 + 01	0.7038 + 01	0.3938 − 01	0.4084 + 02	0.4079 + 00
0.5700 + 01	0.3774 + 02	0.5200 + 01	0.7258 + 01	0.3664 − 01	0.4230 + 02	0.4069 + 00
0.5800 + 01	0.3908 + 02	0.5224 + 01	0.7481 + 01	0.3412 − 01	0.4378 + 02	0.4059 + 00
0.5900 + 01	0.4044 + 02	0.5246 + 01	0.7709 + 01	0.3180 − 01	0.4528 + 02	0.4050 + 00
0.6000 + 01	0.4183 + 02	0.5268 + 01	0.7941 + 01	0.2965 − 01	0.4682 + 02	0.4042 + 00
0.6100 + 01	0.4324 + 02	0.5289 + 01	0.8176 + 01	0.2767 − 01	0.4837 + 02	0.4033 + 00
0.6200 + 01	0.4468 + 02	0.5309 + 01	0.8415 + 01	0.2584 − 01	0.4996 + 02	0.4025 + 00
0.6300 + 01	0.4614 + 02	0.5329 + 01	0.8658 + 01	0.2416 − 01	0.5157 + 02	0.4018 + 00
0.6400 + 01	0.4762 + 02	0.5347 + 01	0.8905 + 01	0.2259 − 01	0.5320 + 02	0.4011 + 00
0.6500 + 01	0.4912 + 02	0.5365 + 01	0.9156 + 01	0.2115 − 01	0.5486 + 02	0.4004 + 00
0.6600 + 01	0.5065 + 02	0.5382 + 01	0.9411 + 01	0.1981 − 01	0.5655 + 02	0.3997 + 00
0.6700 + 01	0.5220 + 02	0.5399 + 01	0.9670 + 01	0.1857 − 01	0.5826 + 02	0.3991 + 00
0.6800 + 01	0.5378 + 02	0.5415 + 01	0.9933 + 01	0.1741 − 01	0.6000 + 02	0.3985 + 00
0.6900 + 01	0.5538 + 02	0.5430 + 01	0.1020 + 02	0.1635 − 01	0.6176 + 02	0.3979 + 00

M	$\dfrac{p_2}{p_1}$	$\dfrac{\rho_2}{\rho_1}$	$\dfrac{T_2}{T_1}$	$\dfrac{p_{o_2}}{p_{o_1}}$	$\dfrac{p_{o_2}}{p_1}$	M_2
0.7000 + 01	0.5700 + 02	0.5444 + 01	0.1047 + 02	0.1535 − 01	0.6355 + 02	0.3974 + 00
0.7100 + 01	0.5864 + 02	0.5459 + 01	0.1074 + 02	0.1443 − 01	0.6537 + 02	0.3968 + 00
0.7200 + 01	0.6031 + 02	0.5472 + 01	0.1102 + 02	0.1357 − 01	0.6721 + 02	0.3963 + 00
0.7300 + 01	0.6200 + 02	0.5485 + 01	0.1130 + 02	0.1277 − 01	0.6908 + 02	0.3958 + 00
0.7400 + 01	0.6372 + 02	0.5498 + 01	0.1159 + 02	0.1202 − 01	0.7097 + 02	0.3954 + 00
0.7500 + 01	0.6546 + 02	0.5510 + 01	0.1188 + 02	0.1133 − 01	0.7289 + 02	0.3949 + 00
0.7600 + 01	0.6722 + 02	0.5522 + 01	0.1217 + 02	0.1068 − 01	0.7483 + 02	0.3945 + 00
0.7700 + 01	0.6900 + 02	0.5533 + 01	0.1247 + 02	0.1008 − 01	0.7680 + 02	0.3941 + 00
0.7800 + 01	0.7081 + 02	0.5544 + 01	0.1277 + 02	0.9510 − 02	0.7880 + 02	0.3937 + 00
0.7900 + 01	0.7264 + 02	0.5555 + 01	0.1308 + 02	0.8982 − 02	0.8082 + 02	0.3933 + 00
0.8000 + 01	0.7450 + 02	0.5565 + 01	0.1339 + 02	0.8488 − 02	0.8287 + 02	0.3929 + 00
0.9000 + 01	0.9433 + 02	0.5651 + 01	0.1669 + 02	0.4964 − 02	0.1048 + 03	0.3898 + 00
0.1000 + 02	0.1165 + 03	0.5714 + 01	0.2039 + 02	0.3045 − 02	0.1292 + 03	0.3876 + 00
0.1100 + 02	0.1410 + 03	0.5762 + 01	0.2447 + 02	0.1945 − 02	0.1563 + 03	0.3859 + 00
0.1200 + 02	0.1678 + 03	0.5799 + 01	0.2894 + 02	0.1287 − 02	0.1859 + 03	0.3847 + 00
0.1300 + 02	0.1970 + 03	0.5828 + 01	0.3380 + 02	0.8771 − 03	0.2181 + 03	0.3837 + 00
0.1400 + 02	0.2285 + 03	0.5851 + 01	0.3905 + 02	0.6138 − 03	0.2528 + 03	0.3829 + 00
0.1500 + 02	0.2623 + 03	0.5870 + 01	0.4469 + 02	0.4395 − 03	0.2902 + 03	0.3823 + 00
0.1600 + 02	0.2985 + 03	0.5885 + 01	0.5072 + 02	0.3212 − 03	0.3301 + 03	0.3817 + 00
0.1700 + 02	0.3370 + 03	0.5898 + 01	0.5714 + 02	0.2390 − 03	0.3726 + 03	0.3813 + 00
0.1800 + 02	0.3778 + 03	0.5909 + 01	0.6394 + 02	0.1807 − 03	0.4176 + 03	0.3810 + 00
0.1900 + 02	0.4210 + 03	0.5918 + 01	0.7114 + 02	0.1386 − 03	0.4653 + 03	0.3806 + 00
0.2000 + 02	0.4665 + 03	0.5926 + 01	0.7872 + 02	0.1078 − 03	0.5155 + 03	0.3804 + 00
0.2200 + 02	0.5645 + 03	0.5939 + 01	0.9506 + 02	0.6741 − 04	0.6236 + 03	0.3800 + 00
0.2400 + 02	0.6718 + 03	0.5948 + 01	0.1129 + 03	0.4388 − 04	0.7421 + 03	0.3796 + 00
0.2600 + 02	0.7885 + 03	0.5956 + 01	0.1324 + 03	0.2953 − 04	0.8709 + 03	0.3794 + 00
0.2800 + 02	0.9145 + 03	0.5962 + 01	0.1534 + 03	0.2046 − 04	0.1010 + 04	0.3792 + 00
0.3000 + 02	0.1050 + 04	0.5967 + 01	0.1759 + 03	0.1453 − 04	0.1159 + 04	0.3790 + 00
0.3200 + 02	0.1194 + 04	0.5971 + 01	0.2001 + 03	0.1055 − 04	0.1319 + 04	0.3789 + 00
0.3400 + 02	0.1348 + 04	0.5974 + 01	0.2257 + 03	0.7804 − 05	0.1489 + 04	0.3788 + 00
0.3600 + 02	0.1512 + 04	0.5977 + 01	0.2529 + 03	0.5874 − 05	0.1669 + 04	0.3787 + 00
0.3800 + 02	0.1684 + 04	0.5979 + 01	0.2817 + 03	0.4488 − 05	0.1860 + 04	0.3786 + 00
0.4000 + 02	0.1866 + 04	0.5981 + 01	0.3121 + 03	0.3477 − 05	0.2061 + 04	0.3786 + 00
0.4200 + 02	0.2058 + 04	0.5983 + 01	0.3439 + 03	0.2727 − 05	0.2272 + 04	0.3785 + 00
0.4400 + 02	0.2258 + 04	0.5985 + 01	0.3774 + 03	0.2163 − 05	0.2493 + 04	0.3785 + 00
0.4600 + 02	0.2468 + 04	0.5986 + 01	0.4124 + 03	0.1733 − 05	0.2725 + 04	0.3784 + 00
0.4800 + 02	0.2688 + 04	0.5987 + 01	0.4489 + 03	0.1402 − 05	0.2967 + 04	0.3784 + 00
0.5000 + 02	0.2916 + 04	0.5988 + 01	0.4871 + 03	0.1144 − 05	0.3219 + 04	0.3784 + 00

APPENDIX
C

PRANDTL-MEYER FUNCTION AND MACH ANGLE

M	ν	μ	M	ν	μ
0.1000 + 01	0.0000	0.9000 + 02	0.1640 + 01	0.1604 + 02	0.3757 + 02
0.1020 + 01	0.1257 + 00	0.7864 + 02	0.1660 + 01	0.1663 + 02	0.3704 + 02
0.1040 + 01	0.3510 + 00	0.7406 + 02	0.1680 + 01	0.1722 + 02	0.3653 + 02
0.1060 + 01	0.6367 + 00	0.7063 + 02	0.1700 + 01	0.1781 + 02	0.3603 + 02
0.1080 + 01	0.9680 + 00	0.6781 + 02	0.1720 + 01	0.1840 + 02	0.3555 + 02
0.1100 + 01	0.1336 + 01	0.6538 + 02	0.1740 + 01	0.1898 + 02	0.3508 + 02
0.1120 + 01	0.1735 + 01	0.6323 + 02	0.1760 + 01	0.1956 + 02	0.3462 + 02
0.1140 + 01	0.2160 + 01	0.6131 + 02	0.1780 + 01	0.2015 + 02	0.3418 + 02
0.1160 + 01	0.2607 + 01	0.5955 + 02			
0.1180 + 01	0.3074 + 01	0.5794 + 02	0.1800 + 01	0.2073 + 02	0.3375 + 02
			0.1820 + 01	0.2130 + 02	0.3333 + 02
0.1200 + 01	0.3558 + 01	0.5644 + 02	0.1840 + 01	0.2188 + 02	0.3292 + 02
0.1220 + 01	0.4057 + 01	0.5505 + 02	0.1860 + 01	0.2245 + 02	0.3252 + 02
0.1240 + 01	0.4569 + 01	0.5375 + 02	0.1880 + 01	0.2302 + 02	0.3213 + 02
0.1260 + 01	0.5093 + 01	0.5253 + 02	0.1900 + 01	0.2359 + 02	0.3176 + 02
0.1280 + 01	0.5627 + 01	0.5138 + 02	0.1920 + 01	0.2415 + 02	0.3139 + 02
0.1300 + 01	0.6170 + 01	0.5028 + 02	0.1940 + 01	0.2471 + 02	0.3103 + 02
0.1320 + 01	0.6721 + 01	0.4925 + 02	0.1960 + 01	0.2527 + 02	0.3068 + 02
0.1340 + 01	0.7279 + 01	0.4827 + 02	0.1980 + 01	0.2583 + 02	0.3033 + 02
0.1360 + 01	0.7844 + 01	0.4733 + 02			
0.1380 + 01	0.8413 + 01	0.4644 + 02	0.2000 + 01	0.2638 + 02	0.3000 + 02
			0.2050 + 01	0.2775 + 02	0.2920 + 02
0.1400 + 01	0.8987 + 01	0.4558 + 02	0.2100 + 01	0.2910 + 02	0.2844 + 02
0.1420 + 01	0.9565 + 01	0.4477 + 02	0.2150 + 01	0.3043 + 02	0.2772 + 02
0.1440 + 01	0.1015 + 02	0.4398 + 02	0.2200 + 01	0.3173 + 02	0.2704 + 02
0.1460 + 01	0.1073 + 02	0.4323 + 02	0.2250 + 01	0.3302 + 02	0.2639 + 02
0.1480 + 01	0.1132 + 02	0.4251 + 02	0.2300 + 01	0.3428 + 02	0.2577 + 02
0.1500 + 01	0.1191 + 02	0.4181 + 02	0.2350 + 01	0.3553 + 02	0.2518 + 02
0.1520 + 01	0.1249 + 02	0.4114 + 02	0.2400 + 01	0.3675 + 02	0.2462 + 02
0.1540 + 01	0.1309 + 02	0.4049 + 02	0.2450 + 01	0.3795 + 02	0.2409 + 02
0.1560 + 01	0.1368 + 02	0.3987 + 02			
0.1580 + 01	0.1427 + 02	0.3927 + 02	0.2500 + 01	0.3912 + 02	0.2358 + 02
			0.2550 + 01	0.4028 + 02	0.2309 + 02
0.1600 + 01	0.1486 + 02	0.3868 + 02	0.2600 + 01	0.4141 + 02	0.2262 + 02
0.1620 + 01	0.1545 + 02	0.3812 + 02	0.2650 + 01	0.4253 + 02	0.2217 + 02

M	v	μ	M	v	μ
0.2700 + 01	0.4362 + 02	0.2174 + 02	0.5600 + 01	0.8203 + 02	0.1029 + 02
0.2750 + 01	0.4469 + 02	0.2132 + 02	0.5700 + 01	0.8280 + 02	0.1010 + 02
0.2800 + 01	0.4575 + 02	0.2092 + 02	0.5800 + 01	0.8354 + 02	0.9928 + 01
0.2850 + 01	0.4678 + 02	0.2054 + 02	0.5900 + 01	0.8426 + 02	0.9758 + 01
0.2900 + 01	0.4779 + 02	0.2017 + 02	0.6000 + 01	0.8496 + 02	0.9594 + 01
0.2950 + 01	0.4878 + 02	0.1981 + 02	0.6100 + 01	0.8563 + 02	0.9435 + 01
0.3000 + 01	0.4976 + 02	0.1947 + 02	0.6200 + 01	0.8629 + 02	0.9282 + 01
0.3050 + 01	0.5071 + 02	0.1914 + 02	0.6300 + 01	0.8694 + 02	0.9133 + 01
0.3100 + 01	0.5165 + 02	0.1882 + 02	0.6400 + 01	0.8756 + 02	0.8989 + 01
0.3150 + 01	0.5257 + 02	0.1851 + 02	0.6500 + 01	0.8817 + 02	0.8850 + 01
0.3200 + 01	0.5347 + 02	0.1821 + 02	0.6600 + 01	0.8876 + 02	0.8715 + 01
0.3250 + 01	0.5435 + 02	0.1792 + 02	0.6700 + 01	0.8933 + 02	0.8584 + 01
0.3300 + 01	0.5522 + 02	0.1764 + 02	0.6800 + 01	0.8989 + 02	0.8457 + 01
0.3350 + 01	0.5607 + 02	0.1737 + 02	0.6900 + 01	0.9044 + 02	0.8333 + 01
0.3400 + 01	0.5691 + 02	0.1710 + 02	0.7000 + 01	0.9097 + 02	0.8213 + 01
0.3450 + 01	0.5773 + 02	0.1685 + 02	0.7100 + 01	0.9149 + 02	0.8097 + 01
0.3500 + 01	0.5853 + 02	0.1660 + 02	0.7200 + 01	0.9200 + 02	0.7984 + 01
0.3550 + 01	0.5932 + 02	0.1636 + 02	0.7300 + 01	0.9249 + 02	0.7873 + 01
0.3600 + 01	0.6009 + 02	0.1613 + 02	0.7400 + 01	0.9297 + 02	0.7766 + 01
0.3650 + 01	0.6085 + 02	0.1590 + 02	0.7500 + 01	0.9344 + 02	0.7662 + 01
0.3700 + 01	0.6160 + 02	0.1568 + 02	0.7600 + 01	0.9390 + 02	0.7561 + 01
0.3750 + 01	0.6233 + 02	0.1547 + 02	0.7700 + 01	0.9434 + 02	0.7462 + 01
0.3800 + 01	0.6304 + 02	0.1526 + 02	0.7800 + 01	0.9478 + 02	0.7366 + 01
0.3850 + 01	0.6375 + 02	0.1505 + 02	0.7900 + 01	0.9521 + 02	0.7272 + 01
0.3900 + 01	0.6444 + 02	0.1486 + 02	0.8000 + 01	0.9562 + 02	0.7181 + 01
0.3950 + 01	0.6512 + 02	0.1466 + 02	0.9000 + 01	0.9932 + 02	0.6379 + 01
0.4000 + 01	0.6578 + 02	0.1448 + 02	0.1000 + 02	0.1023 + 03	0.5739 + 01
0.4050 + 01	0.6644 + 02	0.1429 + 02	0.1100 + 02	0.1048 + 03	0.5216 + 01
0.4100 + 01	0.6708 + 02	0.1412 + 02	0.1200 + 02	0.1069 + 03	0.4780 + 01
0.4150 + 01	0.6771 + 02	0.1394 + 02	0.1300 + 02	0.1087 + 03	0.4412 + 01
0.4200 + 01	0.6833 + 02	0.1377 + 02	0.1400 + 02	0.1102 + 03	0.4096 + 01
0.4250 + 01	0.6894 + 02	0.1361 + 02	0.1500 + 02	0.1115 + 03	0.3823 + 01
0.4300 + 01	0.6954 + 02	0.1345 + 02	0.1600 + 02	0.1127 + 03	0.3583 + 01
0.4350 + 01	0.7013 + 02	0.1329 + 02	0.1700 + 02	0.1137 + 03	0.3372 + 01
0.4400 + 01	0.7071 + 02	0.1314 + 02	0.1800 + 02	0.1146 + 03	0.3185 + 01
0.4450 + 01	0.7127 + 02	0.1299 + 02	0.1900 + 02	0.1155 + 03	0.3017 + 01
0.4500 + 01	0.7183 + 02	0.1284 + 02	0.2000 + 02	0.1162 + 03	0.2866 + 01
0.4550 + 01	0.7238 + 02	0.1270 + 02	0.2200 + 02	0.1175 + 03	0.2605 + 01
0.4600 + 01	0.7292 + 02	0.1256 + 02	0.2400 + 02	0.1186 + 03	0.2388 + 01
0.4650 + 01	0.7345 + 02	0.1242 + 02	0.2600 + 02	0.1195 + 03	0.2204 + 01
0.4700 + 01	0.7397 + 02	0.1228 + 02	0.2800 + 02	0.1202 + 03	0.2047 + 01
0.4750 + 01	0.7448 + 02	0.1215 + 02	0.3000 + 02	0.1209 + 03	0.1910 + 01
0.4800 + 01	0.7499 + 02	0.1202 + 02	0.3200 + 02	0.1215 + 03	0.1791 + 01
0.4850 + 01	0.7548 + 02	0.1190 + 02	0.3400 + 02	0.1220 + 03	0.1685 + 01
0.4900 + 01	0.7597 + 02	0.1178 + 02	0.3600 + 02	0.1225 + 03	0.1592 + 01
0.4950 + 01	0.7645 + 02	0.1166 + 02	0.3800 + 02	0.1229 + 03	0.1508 + 01
0.5000 + 01	0.7692 + 02	0.1154 + 02	0.4000 + 02	0.1233 + 03	0.1433 + 01
0.5100 + 01	0.7784 + 02	0.1131 + 02	0.4200 + 02	0.1236 + 03	0.1364 + 01
0.5200 + 01	0.7873 + 02	0.1109 + 02	0.4400 + 02	0.1239 + 03	0.1302 + 01
0.5300 + 01	0.7960 + 02	0.1088 + 02	0.4600 + 02	0.1242 + 03	0.1246 + 01
0.5400 + 01	0.8043 + 02	0.1067 + 02	0.4800 + 02	0.1245 + 03	0.1194 + 01
0.5500 + 01	0.8124 + 02	0.1048 + 02	0.5000 + 02	0.1247 + 03	0.1146 + 01

BIBLIOGRAPHY

1. Anderson, John D., Jr.: *Gasdynamic Lasers: An Introduction*, Academic Press, New York, 1976.
2. Anderson, John D., Jr.: *Introduction to Flight*, 3d ed., McGraw-Hill Book Company, New York, 1989.
3. Durand, W. F. (ed): *Aerodynamic Theory*, vol. 1, Springer, Berlin, 1934.
4. Wylie, C. R.: *Advanced Engineering Mathematics*, 4th ed., McGraw-Hill Book Company, New York, 1975.
5. Kreyszig, E.: *Advanced Engineering Mathematics*, John Wiley & Sons, Inc., New York, 1962.
6. Hildebrand, F. B.: *Advanced Calculus for Applications*, 2d ed., Prentice-Hall, Inc., Englewood Cliffs, N.J., 1976.
7. Anderson, D. A., J. C. Tannehill, and R. H. Pletcher: *Computational Fluid Mechanics and Heat Transfer*, McGraw-Hill Book Company, New York, 1984.
8. Prandtl, L., and O. G. Tietjens: *Applied Hydro- and Aeromechanics*, United Engineering Trustees, Inc., 1934; also, Dover Publications, Inc., New York, 1957.
9. Karamcheti, K.: *Principles of Ideal Fluid Aerodynamics*, John Wiley & Sons, Inc., New York, 1966.
10. Pierpont, P. K.: "Bringing Wings of Change," *Astronaut. Aeronaut.*, vol. 13, no. 10, pp. 20–27, October 1975.
11. Abbott, I. H., and A. E. von Doenhoff: *Theory of Wing Sections*, McGraw-Hill Book Company, New York, 1949; also, Dover Publications, Inc., New York, 1959.
12. Munk, Max M.: *General Theory of Thin Wing Sections*, NACA report no. 142, 1922.
13. Bertin, John J., and M. L. Smith: *Aerodynamics for Engineers*, Prentice-Hall, Inc., Englewood Cliffs, N.J., 1979.
14. Hess, J. L., and A. M. O. Smith: "Calculation of potential flow about arbitrary bodies," in *Progress in Aeronautical Sciences*, vol. 8, D. Kucheman (ed.), Pergamon Press, New York, 1967, pp. 1–138.
15. Chow, C. Y.: *An Introduction to Computational Fluid Dynamics*, John Wiley & Sons, Inc., New York, 1979.
16. McGhee, R. J., and W. D. Beasley: *Low-Speed Aerodynamic Characteristics of a 17-Percent-Thick Airfoil Section Designed for General Aviation Applications*, NASA TN D-7428, December 1973.
17. McGhee, R. J., W. D. Beasley, and R. T. Whitcomb: "NASA low- and medium-speed airfoil development," in *Advanced Technology Airfoil Research*, vol. II, NASA CP 2046, March 1980.
18. Glauert, H.: *The Elements of Aerofoil and Airscrew Theory*, Cambridge University Press, London, 1926.

19. Winkelmann, A. E., J. B. Barlow, S. Agrawal, J. K. Saini, J. D. Anderson, Jr., and E. Jones: "The Effects of Leading Edge Modifications on the Post-Stall Characteristics of Wings," AIAA paper no. 80-0199, American Institute of Aeronautics and Astronautics, New York, 1980.

20. Anderson, John D., Jr., Stephen Corda, and David M. Van Wie: "Numerical Lifting Line Theory Applied to Drooped Leading-Edge Wings Below and Above Stall," *J. Aircraft*, vol. 17, no. 12, December 1980, pp. 898–904.

21. Anderson, John D., Jr.: *Modern Compressible Flow: With Historical Perspective*, 2d ed., McGraw-Hill Book Company, New York, 1990.

22. Sears, F. W.: *An Introduction to Thermodynamics. The Kinetic Theory of Gases, and Statistical Mechanics*, 2d ed., Addison-Wesley Publishing Company, Inc., Reading, Mass., 1959.

23. Van Wylen, G. J., and R. E. Sonntag: *Fundamentals of Classical Thermodynamics*, 2d ed., John Wiley & Sons, Inc., New York, 1973.

24. Reynolds, W. C., and H. C. Perkins: *Engineering Thermodynamics*, 2d ed., McGraw-Hill Book Company, New York, 1977.

25. Shapiro, A. H.: *The Dynamics and Thermodynamics of Compressible Fluid Flow*, vols. 1 and 2, The Ronald Press Company, New York, 1953.

26. Leipmann, H. W., and A. Roshko: *Elements of Gasdynamics*, John Wiley & Sons, Inc., New York, 1957.

27. Tsien, H. S.: "Two-Dimensional Subsonic Flow of Compressible Fluids," *J. Aeronaut. Sci.*, vol. 6, no. 10, October 1939, p. 399.

28. von Karman, T. H.: "Compressibility Effects in Aerodynamics," *J. Aeronaut. Sci.*, vol. 8, no. 9, September 1941, p. 337.

29. Laitone, E. V.: "New Compressibility Correction for Two-Dimensional Subsonic Flow," *J. Aeronaut. Sci.*, vol. 18, no. 5, May 1951, p. 350.

30. Stack, John, W. F. Lindsey, and R. E. Littell: *The Compressibility Burble and the Effect of Compressibility on Pressures and Forces Acting on an Airfoil*, NACA report no. 646, 1938.

31. Whitcomb, R. T.: *A Study of the Zero-Lift Drag-Rise Characteristics of Wing-Body Combinations Near the Speed of Sound*, NACA report no. 1273, 1956.

32. Whitcomb, R. T., and L. R. Clark: *An Airfoil Shape for Efficient Flight at Supercritical Mach Numbers*, NASA TMX-1109, July 1965.

33. Anderson, J. D., Jr.: "Computational fluid dynamics—an engineering tool?," in *Numerical/Laboratory Computer Methods in Fluid Dynamics*, A. A. Pouring (ed.), ASME, New York, 1976, pp. 1–12.

34. Owczarek, Jerzy A.: *Fundamentals of Gas Dynamics*, International Textbook Company, Scranton, Pa., 1964.

35. Anderson, J. D., Jr., L. M. Albacete, and A. E. Winkelmann: *On Hypersonic Blunt Body Flow Fields Obtained with a Time-Dependent Technique*, Naval Ordnance Laboratory NOLTR 68-129, 1968.

36. Anderson, J. D., Jr.: "An Engineering Survey of Radiating Shock Layers," *AIAA J.*, vol. 7, no. 9, September 1969, pp. 1665–1675.

37. Cherni, G. G.: *Introduction to Hypersonic Flow*, Academic Press, New York, 1961.

38. Truitt, R. W.: *Hypersonic Aerodynamics*, The Ronald Press Company, New York, 1959.

39. Dorrance, H. W.: *Viscous Hypersonic Flow*, McGraw-Hill Book Company, New York, 1962.

40. Hayes, W. D., and R. F. Probstein: *Hypersonic Flow Theory*, 2d ed., Academic Press, New York, 1966.

41. Vinh, N. X., A. Busemann, and R. D. Culp: *Hypersonic and Planetary Entry Flight Mechanics*, University of Michigan Press, Ann Arbor, 1980.

42. Schlichting, H.: *Boundary Layer Theory*, 7th ed., McGraw-Hill Book Company, New York, 1979.

43. White, F. M.: *Viscous Fluid Flow*, McGraw-Hill Book Company, New York, 1974.

44. Cebeci, T., and A. M. O. Smith: *Analysis of Turbulent Boundary Layers*, Academic Press, New York, 1974.

45. Bradshaw, P., T. Cebeci, and J. Whitelaw: *Engineering Calculation Methods for Turbulent Flow*, Academic Press, New York, 1981.

46. Berman, H. A., J. D. Anderson, Jr., and J. P. Drummond: "Supersonic Flow over a Rearward Facing Step with Transverse Nonreacting Hydrogen Injection," *AIAA J.*, vol. 21, no. 12, December 1983, pp. 1707–1713.

47. Van Driest, E. R.: "Turbulent Boundary Layer in Compressible Fluids," *J. Aeronaut. Sci.*, vol. 18, no. 3, March 1951, p. 145.

48. Loftin, Lawrence K., Jr.: *Quest for Performance: The Evolution of Modern Aircraft*, NASA SP-468, 1985.

49. von Karman, T., and Lee Edson: *The Wind and Beyond: Theodore von Karman, Pioneer in Aviation and Pathfinder in Space*, Little, Brown and Co., Boston, 1967.

50. Nakayama, Y. (ed): *Visualized Flow*, compiled by the Japan Society of Mechanical Engineers, Pergamon Press, New York, 1988.

51. Meuller, Thomas J.: *Low Reynolds Number Vehicles*, AGARDograph no. 288, Advisory Group for Advanced Research and Development, NATO, 1985.

52. Carnahan, B., H. A. Luther, and J. O. Wilkes: *Applied Numerical Methods*, John Wiley & Sons, New York, 1969.

53. Schetz, Joseph A.: *Foundations of Boundary Layer Theory for Momentum, Heat, and Mass Transfer*, Prentice-Hall, Inc., Englewood Cliffs, N.J., 1984.

54. Anderson, Dale, John C. Tannehill, and Richard H. Pletcher: *Computational Fluid Mechanics and Heat Transfer*, Hemisphere Publishing Corp., New York, 1984.

55. Anderson, J. D.: *Hypersonic and High Temperature Gas Dynamics*, McGraw-Hill Book Company, New York, 1989.

56. Kothari, A. P., and J. D. Anderson: "Flows over Low Reynolds Number Airfoils—Compressible Navier-Stokes Solutions," AIAA paper no. 85-0107, January 1985.

57. Shang, J. S., and S. J. Scherr: "Navier-Stokes Solution for a Complete Re-Entry Configuration," *J. Aircraft*, vol. 23, no. 12, December 1986, pp. 881–888.

58. Stevens, V. P.: "Hypersonic Research Facilities at the Ames Aeronautical Laboratory," *J. Appl. Phys.*, vol. 21, 1955, pp. 1150–1155.

59. Hodges, A. J.: "The Drag Coefficient of Very High Velocity Spheres," *J. Aeronaut. Sci.*, vol. 24, 1957, pp. 755–758.

60. Charters, A. C., and R. N. Thomas: "The Aerodynamic Performance of Small Spheres from Subsonic to High Supersonic Velocities," *J. Aeronaut. Sci.*, vol. 12, 1945, pp. 468–476.

61. Cox, R. N., and L. F. Crabtree: *Elements of Hypersonic Aerodynamics*, Academic Press, New York, 1965.

INDEX